IL	interleukin
IMP	inosine-5'-monophosphate
IP_3	inositol-1,4,5-triphosphate
K_m	Michaelis constant
kb	kilobases
kD	kilodalton
LDL	low-density lipoprotein
LHC	light harvesting complex
Man	mannose
NAA	nonessential amino acids
NAD^+	nicotinamide adenine dinucleotide (oxidized form)
NADH	nicotinamide adenine dinucleotide (reduced form)
$NADP^+$	nicotinamide adenine dinucleotide phosphate (oxidized form)
NADPH	nicotinamide adenine dinucleotide phosphate (reduced form)
NDP	nucleoside-5'-diphosphate
NMR	nuclear magnetic resonance
NO	nitric oxide
NTP	nucleoside-5'-triphosphate
P_i	orthophosphate (inorganic phosphate)
PAPS	3'-phosphoadenosine-5'-phosphosulfate
PC	plastocyanin
PDGF	platelet-derived growth factor
PEP	phosphoenolpyruvate
PFK	phosphofructokinase
PIP_2	phosphatidylinositol-4,5-bisphosphate
PP_i	pyrophosphate
PRPP	phosphoribosylpyrophosphate
PS	photosystem
PQ(Q)	plastoquinone (oxidized)
PQH_2 (QH_2)	plastoquinone (reduced)
RER	rough endoplasmic reticulum
RF	releasing factor
RFLP	restriction-frament length polymorphism
RNA	ribonucleic acid
dsRNA	double-stranded RNA
hnRNA	heterogenous nuclear RNA
mRNA	messenger RNA
rRNA	ribosomal RNA
snRNA	small nuclear RNA
ssRNA	single-stranded RNA
tRNA	transfer RNA
snRNP	small ribonucleoprotein particles
RNase	ribonuclease
S	Svedberg unit
SAH	S-adenosylhomocysteine
SAM	S-adenosylmethionine
SDS	sodium dodecyl sulfate
SER	smooth endoplasmic reticulum
SRP	signal recognition particle
T	thymine
THF	tetrahydrofolate
TPP	thiamine pyrophosphate
U	uracil
UDP	uridine-5'-diphosphate
UMP	uridine-5'-monophosphate
UTP	uridine-5'-triphosphate
UQ	ubiquinone (coenzyme Q)(oxidized form)
UQH_2	ubiquinone (reduced form)
VLDL	very low density lipoprotein
XMP	xanthosine-5' monophosphate

BIOCHEMISTRY

An Introduction

Trudy McKee
Thomas Jefferson University

James R. McKee
University of the Sciences in Philadelphia

WCB
McGraw-Hill

Boston   Burr Ridge, IL   Dubuque, IA   Madison, WI   New York   San Francisco   St. Louis
Bangkok   Bogotá   Caracas   Lisbon   London   Madrid
Mexico City   Milan   New Delhi   Seoul   Singapore   Sydney   Taipei   Toronto

WCB/McGraw-Hill

A Division of The McGraw·Hill Companies

BIOCHEMISTRY: AN INTRODUCTION, SECOND EDITION

 This book is printed on recycled, acid-free paper containing 10% postconsumer waste.

34567890 QPH/QPH 06543210

ISBN 0–07–290499–2

Vice president and editorial director: *Kevin T. Kane*
Publisher: *James M. Smith*
Sponsoring editor: *Kent A. Peterson*
Developmental editor: *Margaret B. Horn*
Marketing manager: *Martin J. Lange*
Project manager: *Donna Nemmers*
Senior production supervisor: *Sandra Hahn*
Design manager: *Stuart D. Paterson*
Photo research coordinator: *John C. Leland*
Supplement coordinator: *Stacy A. Patch*
Compositor: *York Graphic Services*
Typeface: *10.5/12 Times Roman*
Printer: *Quebecor Printing Book Group/Hawkins, TN*

Cover design: *Christopher Reese*
Interior design: *Maureen McCutcheon*
Cover image: *BioGrafx*

The credits section for this book begins on page 618 and is considered an extension of the copyright page.

Library of Congress Cataloging-in-Publication Data

McKee, Trudy.
    Biochemistry: an introduction / Trudy McKee, James R. McKee. — 2nd ed.
      p. cm.
    Includes bibliographical references and index.
    ISBN 0-07-290499-2
    1. Biochemistry. I. McKee, James R. (James Robert), 1946– .
  II. Title.
QP514.2.M435   1999
572—dc21   98–4536
                                                 CIP

www.mhhe.com

This book is dedicated to our son
James Adrian McKee

Brief Contents

Extended Contents

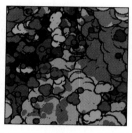

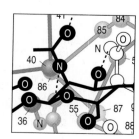

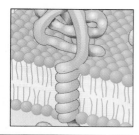

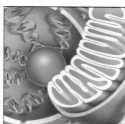

Chapter Thirteen

Nitrogen Metabolism I: Synthesis   359

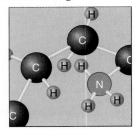

Chapter Fourteen

Nitrogen Metabolism II: Degradation   409

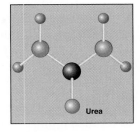

Urea

Chapter Fifteen

Integration of Metabolism   433

Chapter Sixteen

Nucleic Acids   459

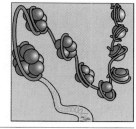

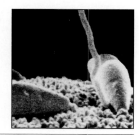

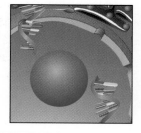

Preface

Life is a mystery with the power to enchant or terrify. This is no less true for biochemists than it is for poets and artists. While many humans appreciate the visual beauty and the majesty of the natural world, biochemists seek to discover the underlying mechanisms of living processes. As life's secrets have been probed, researchers have often been both awed and humbled by the intricacy, sophistication, and resilience of living organisms. In this textbook we have attempted to convey to students the excitement and pleasure we have experienced in our search for knowledge about life.

ORGANIZATION AND APPROACH

This textbook is designed for use by chemistry majors and students in the life sciences. Few assumptions have been made about the chemistry and biology backgrounds the students have. To ensure that all students are sufficiently prepared for acquiring a meaningful understanding of biochemistry, the first four chapters review the principles of relevant topics, such as organic functional groups, noncovalent bonding, thermodynamics, and cell structure. In our experience, students retain new information more efficiently if they can readily see its applications. Therefore, in the remaining chapters of the book, discussions of biomolecules are usually followed immediately by descriptions of their roles in metabolism. For example, chapters devoted to carbohydrate and lipid structure are followed by chapters describing their metabolic functions. A notable theme that begins in the early chapters and continues throughout the book is the relationship between biochemical processes and biological structure. We believe that this approach affords students an interesting holistic view of a subject that so often has been treated as if biochemical reactions occur in some dimension isolated from biology.

During the past 50 years, scientific investigation has increasingly deepened and expanded human understanding of life. In the waning years of the twentieth century, the pace of discovery has become explosive. The continuing avalanche of information, made possible by technological innovations in biochemistry and molecular biology, now allows previously unimaginable insights into the inner workings of living organisms and the causes of disease. The challenge to life sciences educators is how to prepare students for a new century when revolutionary changes created by scientific discovery will affect both their personal and professional lives. The most important tool for students is a clear understanding of biochemistry. The purpose of the second edition of *Biochemistry: An Introduction* remains unchanged from that of the first edition, that is, to present a logical and accessible description of essential biochemical principles.

WHAT IS NEW IN THIS EDITION

As a result of our teaching experience, the suggestions of many students and biochemistry instructors who have used the text, and numerous advances in biochemical research, many changes and improvements have been made, including the following:

The design of the book has been completely revised to make it easier to read and comprehend.

Each chapter has been revised and updated. Examples of topics with improved coverage include protein structure, nucleotide chemistry, DNA synthesis, and DNA repair mechanisms. The art program has also been reevaluated. Some figures have been altered and others have been rerendered to improve clarity.

The sequence of coverage of some topics has been modified to improve comprehension. For example, the chapters devoted to discussions of lipid and membrane structure and metabolism now precede the discussion of energy generation in the aerobic metabolism chapter.

Each paragraph, table, worked problem, and illustration has been reviewed for accuracy.

SUPPLEMENTARY AIDS

1. **Instructors Manual/Test Item File:** Written by the authors, this manual is designed to assist instructors plan and prepare for classes using *Biochemistry: An Introduction.* For each chapter in the text, this manual provides a chapter outline, key words, an extended lecture outline, and enrichment ideas. The test item file contains approximately thirty-five multiple choice, true/false, critical thinking, and mathematical problems per chapter. Suggested answers for the problems appear at the end of each part.

2. **Student Study Guide/Solutions Manual:** This guide accompanies the text and was written by Bruce Morimoto of Purdue University. For each text chapter, a corresponding study guide chapter offers comprehensive reviews, study tips, and additional questions for biochemistry students.

3. **Transparencies:** Accompanying this text, 140 transparencies of key illustrations in the text help the instructor coordinate the lecture to the text.

4. **Microtest:** This computerized classroom management system/service includes a database of test questions, reproducible student self-quizzes, and a grade-recording program. Disks are available for IBM and Macintosh computers and require no program.

5. **Laboratory Manual in Biochemistry (0-697-16735-6):** Written by Dr. Henry Zeiden and Dr. William Dashek, this laboratory manual stresses the theory behind the biochemical and molecular biology techniques and uses an investigative laboratory approach. This manual contains 12 modules, each containing several exercises. Traditional topics such as enzymology, cutting-edge topics, such as molecular biology, and applied topics, such as analysis of carbohydrates, are covered in this laboratory manual.

6. **McGraw-Hill 3D Library of Biomolecules:** Developed by McGraw-Hill, this browser-based CD-Rom takes advantage of virtual reality technology. Instructors and students can view and manipulate three-dimensional animations of more than 100 of the most commonly studied molecules in biochemistry. The library can be accessed via the Netscape Navigator or Internet Explorer browsers on both Windows and Macintosh computers.

7. **How to Study Science:** Written by Fred Drewes of Suffolk County Community College, this excellent workbook offers students helpful suggestions for meeting the considerable challenges of a science course. It offers tips on how to overcome science anxiety. This book's unique design helps to stir students' critical thinking skills and help them develop careful note taking.

ACKNOWLEDGMENTS

The authors wish to express their appreciation for the efforts of the individuals who provided detailed criticism and analysis for the development of both the first and second editions of this text:

Richard Saylor
Shelton State Community College

Craig R. Johnson
Carlow College

Larry D. Martin
Morningside College

Arnulfo Mar
University of Texas at Brownsville

Terry Helser
SUNY College at Oneonta

Edward G. Senkbeil
Salisbury State University

Martha McBride
Norwich University

Ralph Shaw
Southeastern Louisiana University

Clarence Fouche
Virginia Intermont College

Jerome Maas
Oakton Community College

Justine Walhout
Rockford College

William Voige
James Madison University

Carol Leslie
Union University

Harvey Nikkel
Grand Valley State University

Brenda Braaten
Framingham State College

Duane LeTourneau
University of Idaho

William Sweeney
Hunter College

Charles Hosler
University of Wisconsin

Mark Armstrong
Blackburn College

Treva Pamer
Jersey City State College

Bruce Banks
University of North Carolina

David Speckhard
Loras College

Joyce Miller
University of Wisconsin-Platteville

Beulah Woodfin
University of New Mexico

Robley J. Light
Florida State University

Anthony P. Toste
Southwest Missouri State University

Les Wynston
California State University-Long Beach

Alfred Winer
University of Kentucky

Larry L. Jackson
Montana State University

Ivan Kaiser
University of Wyoming

Allen T. Phillips
Pennsylvania State University

Bruce Morimoto
Purdue University

John R. Jefferson
Luther College

Ram P. Singhal
Wichita State University

Craig Tuerk
Morehead State University

Alan Myers
Iowa State University

Allan Bieber
Arizona State University

Scott Pattison
Ball State University

P. Shing Ho
Oregon State University

Charles Englund
Bethany College

Lawrence K. Duffy
University of Alaska-Fairbanks

Paul Kline
Middle Tennessee State University

Christine Tachibana
Pennsylvania State University

The publication of a textbook requires the efforts of many people besides the authors. We are very grateful to our colleagues at WCB/McGraw-Hill, especially Kent Peterson, our sponsoring editor and Margaret Horn, our developmental editor, whose guidance and support have been invaluable. We also express our appreciation for the skilled assistance of Donna Nemmers, project manager. We give a very special thank you to Scott Pattison (Ball State University), Joseph Rabinowitz (Professor Emeritus, University of Pennsylvania), and Ann Randolph (Rosemont College), whose consistent diligence on this project have ensured the accuracy of the text. In addition to their efforts and those of the reviewers, all of the manuscript's narrative and artwork have been reviewed by professional proofreaders. Every word, example, and figure have been independently checked by many individuals.

Finally, we wish to extend our deep appreciation to those individuals who have helped and encouraged us and made this project possible: Nicholas Rosa, Ira Cantor, Joseph and Josephine Rabinowitz, and William and Barbara Morris. To James R. McKee, Sr., and Margaret McKee and our son James Adrian McKee we extend our appreciation for their unfailing patience and encouragement.

Trudy McKee
James R. McKee

Guided Tour Through the Biochemistry Learning System

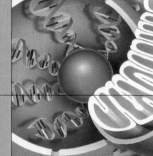

AEROBIC METABOLISM
Chapter Eleven

OUTLINE

In aerobic cells most energy is generated within the mitochondrion. Dioxygen (O_2) is used to oxidize nutrient molecules.

OBJECTIVES

After you have studied this chapter, you should be able to answer these questions:

1. How is energy obtained from the breakdown of acetyl-CoA by the reactions of the citric acid cycle?
2. How is the energy released during the electron transport pathway captured and used to drive biosynthetic processes?
3. How are the pathways of energy metabolism interrelated?
4. How are the pathways of energy metabolism regulated so that each cell's energy requirements are consistently met.
5. Why do calorie-restricted diets promote obesity?
6. What are the toxic products of oxygen metabolism and how do they damage cells?
7. How are toxic oxygen metabolites formed and destroyed?

CHAPTER OUTLINES AND OBJECTIVES

Each chapter begins with an outline that introduces students to the topics to be presented. This outline also provides the instructor with a quick topic summary and help to organize lecture material. A list of objectives, based on major concepts covered in the chapter, enables students to preview the material and become aware of the topics they are expected to master.

DRAMATIC VISUAL PROGRAM

Colorful and informative photographs, illustrations, and tables enhance the learning program. Each chapter begins with an attractive opening photograph or illustration that visually introduces the topics to be discussed.

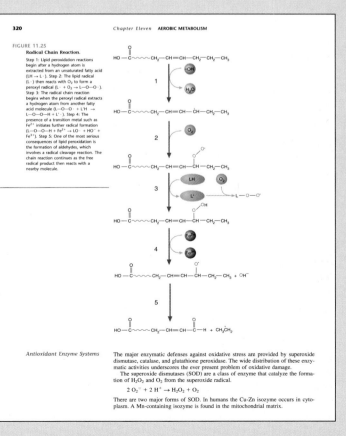

320   *Chapter Eleven* **AEROBIC METABOLISM**

FIGURE 11.25

Radical Chain Reaction.

Step 1: Lipid peroxidation reactions begin after a hydrogen atom is extracted from an unsaturated fatty acid (LH → L·). Step 2: The lipid radical (L·) then reacts with O_2 to form a peroxyl radical (L· + O_2 → L—O—O·). Step 3: The radical chain reaction begins when the peroxyl radical extracts a hydrogen atom from another fatty acid molecule (L—O—O· + L'H → L—O—O—H + L'·). Step 4: The presence of a transition metal such as Fe^{2+} initiates further radical formation (L—O—O—H + Fe^{2+} → LO· + HO^- + Fe^{3+}). Step 5: One of the most serious consequences of lipid peroxidation is the formation of aldehydes, which involves a radical cleavage reaction. The chain reaction continues as the free radical product then reacts with a nearby molecule.

Antioxidant Enzyme Systems

The major enzymatic defenses against oxidative stress are provided by superoxide dismutase, catalase, and glutathione peroxidase. The wide distribution of these enzymatic activities underscores the ever present problem of oxidative damage.

The superoxide dismutases (SOD) are a class of enzyme that catalyze the formation of H_2O_2 and O_2 from the superoxide radical.

$$2 O_2^- + 2 H^+ \rightarrow H_2O_2 + O_2$$

There are two major forms of SOD. In humans the Cu-Zn isozyme occurs in cytoplasm. A Mn-containing isozyme is found in the mitochondrial matrix.

THE EICOSANOIDS

The **eicosanoids** are a diverse group of extremely powerful hormonelike molecules produced in most mammalian tissues. (Because they are generally active within the organ in which they are produced, the eicosanoids are called **autocrine** regulators instead of hormones.) Most eicosanoids are derived from arachidonic acid ($20{:}4^{\Delta5,8,11,14}$), which is also called 5,8,11,14-eicosatetraenoic acid. (Arachidonic acid is synthesized from linoleic acid by adding a two-carbon unit and inserting two additional double bonds.) Production of eicosanoids begins after arachidonic acid is released from membrane phospholipid mol-

ecules by the enzyme phospholipase A_2. The eicosanoids, which include the prostaglandins, thromboxanes, and leukotrienes (Figure 9A), are extremely difficult to study because they are active for short periods (often measured in seconds or minutes). In addition, they are produced only in small amounts.

Prostaglandins are arachidonic acid derivatives that contain a cyclopentane ring with hydroxy groups at C-11 and C-15. Molecules belonging to the E series of prostaglandins have a carbonyl group at C-9, while the F series molecules have an OH group at the same position. The subscript number in a

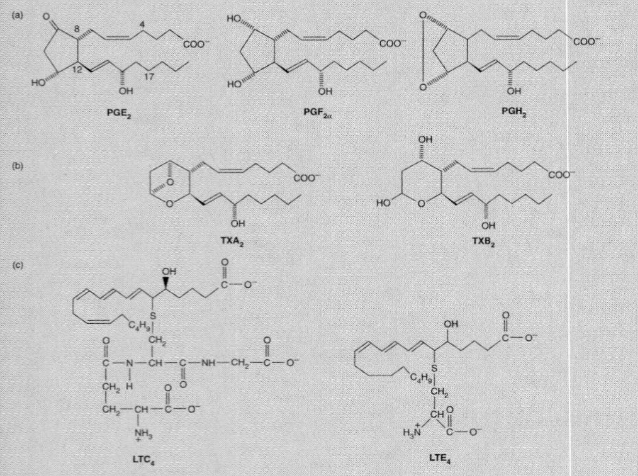

FIGURE 9A
Eicosanoids.
(a) Prostaglandins E_2, $F_{2\alpha}$, and H_2. (b) Thromboxanes A_2 and B_2. (c) Leukotrienes C_4 and E_4.

220

BOLDFACED KEY WORDS

Key words appear in boldface when they are introduced within the text and are immediately defined by the context. All key words are also defined in the glossary.

SPECIAL INTEREST BOXES

These essays which appear throughout the text, help students connect biochemical principles and everyday applications.

KEY CONCEPTS

At the end of sections is a brief summary to help students understand the essential ideas in the section.

IN-CHAPTER PROBLEMS, SOLUTIONS, AND QUESTIONS

Because problem solving is most easily learned by studying examples and practicing, problems with solutions are provided wherever appropriate. In chapter questions help students integrate newly learned material with timely and interesting related information.

Oxidative Stress   323

the nonoxidative phase, ribulose-5-phosphate is primarily converted to either ribose-5-phosphate or xylulose-5-phosphate.

The oxidative phase of the pentose phosphate pathway consists of three reactions (Figure 11.28a). In the first reaction, glucose-6-phosphate dehydrogenase (G-6-PD) catalyzes the oxidation of glucose-6-phosphate. 6-Phosphogluconeolactone and NADPH are products in this reaction. 6-Phosphogluconeolactone is then hydrolyzed to produce 6-phosphogluconate. A second molecule of NADPH is produced during the oxidative decarboxylation of 6-phosphogluconate, a reaction that yields ribulose-5-phosphate.

The nonoxidative phase of the pathway begins with the conversion of ribulose-5-phosphate to ribose-5-phosphate by ribulose-5-phosphate isomerase or to xylulose-5-phosphate by ribulose-5-phosphate epimerase. During the remaining reactions of the pathway (Figure 11.28b), transketolase and transaldolase catalyze the interconversions of trioses, pentoses, and hexoses. Transketolase is a TPP-requiring enzyme that transfers two-carbon units from a ketose to an aldose. Two reactions are catalyzed by transketolase. In the first reaction, the enzyme transfers, a two-carbon unit from xylulose-5-phosphate to ribose-5-phosphate, yielding glyceraldehyde-3-phosphate and sedoheptulose-7-phosphate. In the second transketolase-catalyzed reaction, a two-carbon unit from another xylulose-5-phosphate molecule is transferred to erythrose-4-phosphate to form a second molecule of glyceraldehyde-3-phosphate and fructose-6-phosphate. (Erythrose-4-phosphate is used by some organisms to synthesize aromatic amino acids.) Transaldolase transfers three-carbon units from a ketose to an aldose. In the reaction catalyzed by transaldolase, a three-carbon unit is transferred from sedoheptulose-7-phosphate to glyceraldehyde-3-phosphate. The products formed are fructose-6-phosphate and erythrose-4-phosphate. The result of the nonoxidative phase of the pathway is the synthesis of ribose-5-phosphate and the glycolytic intermediates glyceraldehyde-3-phosphate and fructose-6-phosphate.

The pentose phosphate pathway is regulated to meet the cell's moment-by-moment requirements for NADPH and ribose-5-phosphate. The oxidative phase is very active in cells such as red blood cells or hepatocytes in which demand for NADPH is high. In contrast, the oxidative phase is virtually absent in cells (e.g., muscle cells) that synthesize little or no lipid. G-6-PD catalyzes the rate-limiting step in the pentose phosphate pathway. Its activity is inhibited by NADPH and stimulated by GSSG and glucose-6-phosphate. In addition, diets high in carbohydrate increase the synthesis of both G-6-PD and phosphogluconate dehydrogenase.

KEY *Concepts*
The major enzymatic defenses against oxidative stress are superoxide dismutase, catalase, and glutathione peroxidase. The pentose phosphate pathway produces NADPH and ribose-5-phosphate and several glycolytic intermediates.

QUESTION 11.7

In some regions where malaria is endemic (e.g., the Middle East), fava beans are a staple food. Fava beans are now known to contain two β-glycosides called vicine and convicine.

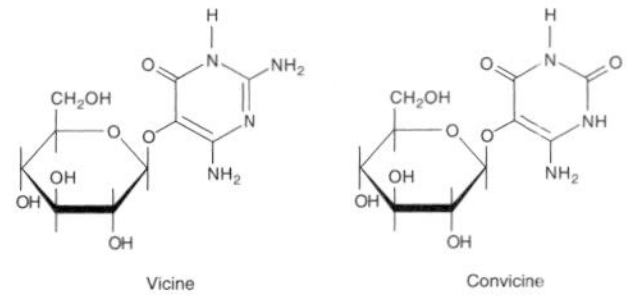

Vicine   Convicine

It is believed that the aglycone components of these substances, called divicine and isouramil, respectively, can oxidize GSH. Individuals who eat fresh fava beans are protected to a certain extent from malaria. A condition known as *favism* results when some G-6-PD-deficient individuals develop a severe hemolytic anemia after eating the beans. Explain why.

SUMMARY

Metabolism includes two major types of biochemical pathways: anabolic and catabolic. In anabolic pathways, large complex molecules are synthesized from smaller precursors. In catabolic pathways, large complex molecules are degraded into smaller, simpler products. Some catabolic reactions release free energy. A fraction of this energy is used to drive certain anabolic reactions. Multicellular organisms maintain an appropriate balance between anabolic and catabolic processes by using intercellular chemical signals. In animals, hormone molecules and neurotransmitters are intercellular signals.

The metabolism of carbohydrates is dominated by glucose because this sugar is an important fuel molecule in most organisms. If cellular energy reserves are low, glucose is degraded by the glycolytic pathway. Glucose molecules that are not required for immediate energy production are stored as either glycogen (in animals) or starch (in plants).

The substrate for glycogen synthesis is UDP-glucose, an activated form of the sugar. UDP-glucose pyrophosphorylase catalyzes the formation of UDP-glucose from glucose-1-phosphate and UTP. Glucose-6-phosphate is converted to glucose-1-phosphate by phosphoglucomutase. To form glycogen requires two enzymes: glycogen synthase and branching enzyme. Glycogen degradation requires glycogen phosphorylase and debranching enzyme. The balance between glycogenesis (glycogen synthesis) and glycogenolysis (glyco-

gen breakdown) is carefully regulated by several hormones (insulin, glucagon, and epinephrine).

During glycolysis, glucose is phosphorylated and cleaved to form two molecules of glyceraldehyde-3-phosphate. Each glyceraldehyde-3-phosphate is then converted to a molecule of pyruvate. A small amount of energy is captured in two molecules of ATP and NADH. In anaerobic organisms, pyruvate is converted to waste products. During this process, NAD^+ is regenerated so that glycolysis can continue. In the presence of O_2, aerobic organisms convert pyruvate to acetyl CoA and then to CO_2 and H_2O. Glycolysis is controlled primarily by allosteric regulation of three enzymes-hexokinase, PFK-1, and pyruvate kinase—and by the hormones glucagon and insulin.

During gluconeogenesis, molecules of glucose are synthesized from noncarbohydrate precursors (lactate, pyruvate, glycerol, and certain amino acids). The reaction sequence in gluconeogenesis is largely the reverse of glycolysis. The three irreversible glycolytic reactions (the synthesis of pyruvate, the conversion of fructose-1,6-bisphosphate to fructose-6-phosphate, and the formation of glucose from glucose-6-phosphate) are bypassed by alternate energetically favorable reactions.

Several sugars other than glucose are important in vertebrate carbohydrate metabolism. These include fructose, galactose, and mannose.

SUGGESTED READINGS

Hallfrisch, J., Metabolic Effects of Dietary Fructose, *FASEB J.,* 4:2652–2660, 1990.

Newsholme, E.A., Challiss, R.A.J., and Crabtree, B., Substrate Cycles: Their Role in Improving Sensitivity in Metabolic Control, *Trends Biochem. Sci.,* 9:277–280, 1984.

Pilkus, S.J., Mahgrabi, M.R., and Claus, T.A., Hormonal Regulation

of Hepatic Gluconeogenesis and Glycolysis, *Ann. Rev. Biochem.,* 57:755–783, 1988.

Shulman, G.I., and Landau, B.R., Pathways of Glycogen Repletion, *Physiol. Rev.,* 72(4):1019–1035, 1992.

VanSchaftingen, E., Fructose-2,6-Bisphosphate, *Adv. Enzymol.,* 59:315–395, 1987.

QUESTIONS

1. Metabolism consists of two major processes. What are they? What function does each one perform? Give two examples of each process.
2. Catabolism consists of three steps. Describe what is accomplished in each step.
3. What is the most important end product of stage 2 metabolism? What is the fate of this molecule in stage 3?
4. What two important components of anabolic processes are produced by catabolic reactions?
5. Briefly define each of the following terms:
   a. amphibolic
   b. steady state
   c. target cell
   d. second messenger
   e. limit dextrin
6. Describe how an enzyme cascade magnifies an initial hormonal signal.
7. Upon entering a cell, glucose is phosphorylated. Give two reasons why this reaction is required.

8. Describe the functions of the following molecules:
   a. insulin
   b. glucagon
   c. fructose-2,6-bisphosphate
9. An individual has a genetic deficiency that prevents the production of hexokinase D. Following a carbohydrate meal, would you expect blood glucose levels to be higher, lower, or about normal? What organ would accumulate glycogen under these circumstances?
10. Glycogen synthesis requires a short primer chain. Explain how new glycogen molecules are synthesized, given this limitation.
11. Describe how epinephrine promotes the conversion of glycogen to glucose.
12. Glycolysis occurs in two stages. Describe what is accomplished in each stage.
13. Why is fructose metabolized more rapidly than glucose?
14. What is the difference between an enol-phosphate ester and a normal phosphate ester that gives PEP such a high phosphate group transfer potential?

CHAPTER SUMMARIES

At the end of each chapter is a summary designed to help students more easily identify important concepts and help them review for quizzes and tests.

SUGGESTED READINGS

At the end of each chapter are suggested references for further study of topics in the text or timely related topics.

END-OF-CHAPTER QUESTIONS

A variety of questions and problems that range in level of difficulty help students measure their mastery of the chapter.

Glossary

acetal the family of organic compounds with the general formula $RCH(OR')_2$; formed from the reaction of two molecules of alcohol with an aldehyde.
acid a molecule that can donate hydrogen ions.
acidosis a condition in which the pH of the blood is below 7.35 for a prolonged time.
activation energy the threshold energy required to produce a chemical reaction.
active site the cleft in the surface of an enzyme

alkyl group a simple hydrocarbon group formed when one hydrogen from the original hydrocarbon (e.g., methyl, $CH_3—$) is removed.
allosteric interaction a regulatory mechanism in which a small molecule, called an effector or modulator, noncovalently binds to a protein and alters its activity.
α-amino acid a molecule in which the amino group is attached to the carbon atom (the

apoenzyme the protein portion of an enzyme that requires a cofactor to function in catalysis.
apoprotein a protein without its prosthetic group.
apoptosis programmed cell death.
aromatic hydrocarbon a molecule that contains a benzene ring or has properties similar to those exhibited by benzene.
atherosclerosis deposition of excess plasma

Techniques in Biochemistry Supplement

Appendix B

These are exciting times for biochemists! During the past fifty years, there has been a continuously accelerating revolution in our understanding of the functioning of living organisms. Much of this knowledge has been possible because of technological innovations. For example, the development of the electron microscope as a biological instrument by Keith Porter and his colleagues in the 1940s led to the discovery of organelles such as mitochondria and lysosomes. Other examples include X-ray crystallography (protein and nucleic acid structure determinations) and radioisotopes (metabolic pathway investigations). In the 1990s biochemists are seeking increasingly more rapid methods for determining DNA base sequences. These and other biochemical techniques exploit the physical and chemical properties of biomolecules.

Scientific research, however, is not a collection of techniques. At the heart of science is the passion and curiosity of the scientist who seeks to understand the natural world. A scientist tests his or her perception of a natural process, sometimes referred to as a paradigm, by designing and performing experiments. Success in scientific investigations depends on three principle factors:

1. The design of experiments that ask clear and well-thought out questions about the living system under investigation.
2. The effective use of currently available technologies.
3. The capacity of the scientist to interpret experimental data, and (if necessary) modify or discard paradigms if they are not supported by this data.

These features of the scientific method are interactive. The availability of a new technology allows scientists to ask new questions

ery of nerve growth factor; recipient of the 1987 Nobel Prize in Physiology and Medicine). Both scientists and artists seek to discern truth. Scientists differ from artists in one respect: they must submit their conceptions of reality (objective reality is the ultimate measure of scientific work) to skeptical colleagues who must be convinced by verifiable experimental results.

The technologies described in this appendix have been chosen because of their seminal importance in modern biochemistry. Because of the intimate relationship between technology and biochemical knowledge, students will find that an understanding of these methods will improve their comprehension of the subject.

Suggested Readings

Bronowski, J., *Science and Human Values,* Harper and Row, New York, 1965.
Burke, J., *The Day the Universe Changed,* Little, Brown, Boston, 1985.
Fischer, E.P. and Lipson, C., *Thinking About Science: Max Delbruck and the Origins of Molecular Biology,* W.W. Norton, New York, 1988.
Hoagland, M., *Toward the Habit of Truth: A Life in the Sciences,* W.W. Norton, New York, 1990.
Keller, E.F. *A Feeling for the Organism: The Life and Work of Barbara McClintock,* W.H. Freeman, 1983.
Kuhn, T.S., *The Structure of Scientific Revolutions,* University of Chicago Press, Chicago, 1970.
Medawar, P.B., *Advice to a Young Scientist,* Harper and Row, New York, 1979.

END-OF-BOOK GLOSSARY OF KEY WORDS

All key words in boldface in the text are defined in the glossary at the end of the textbook.

TECHNIQUES IN BIOCHEMISTRY SUPPLEMENT

From this review of the principle research techniques used to investigate living processes, students can appreciate the relationship between technology and scientific knowledge. Information in this supplement (Appendix B) helps students answer some in-chapter and end-of-chapter questions.

CONCEPT AND APPLICATION ICONS

Throughout the text, students find graphic devices that easily mark the following concepts and applications.

 Plant Biochemistry

 Biomedical Application

 Metabolic Regulation Mechanism

BIOCHEMISTRY: An Introduction

Chapter One

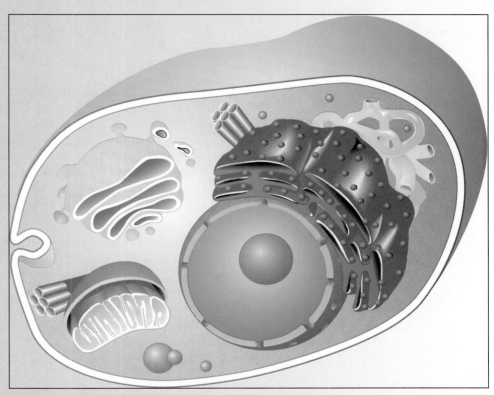

The Living Cell. *Living organisms consist of one or more cells. The capacity of living cells to perform functions such as energy generation, growth, and reproduction is made possible by their complex structures.*

OBJECTIVES

After you have studied this chapter, you should be able to answer these questions:

1. *What principles are central to our understanding of living organisms?*
2. *What characteristics distinguish prokaryotes from eukaryotes?*
3. *What are the four major types of small biomolecules found in cells?*
4. *What are the primary functions of metabolism in living organisms?*
5. *What are the most common types of chemical reactions in living organisms?*
6. *How do cells maintain a high degree of internal order?*

Life has proven to be far more complex than the human imagination could have conceived. The structure of an individual cell is a case in point. Cells are not the bags of protoplasm that scientists envisioned over a century ago; rather, they are structurally complex and dynamic. The story of how our current knowledge about living processes has been acquired, with its myriad plot twists, rivals any piece of detective fiction. The scientists who work to understand the physical reality of the natural world are often amazed at how sophisticated even the simplest organisms are. This chapter and the ones that follow focus on the basic life-sustaining mechanisms that have been discovered thus far.

Biochemistry may be defined as the study of the molecular basis of life. Scientists have long struggled to provide a coherent and accurate view of living processes. In the late nineteenth century, significant progress was made as investigators in the newly emerging science of biochemistry used the concepts of biology, chemistry, physics, and mathematics. After a century of careful observations, biochemists have established several principles that are central to our understanding of living organisms:

1. Cells, the basic structural units of all living organisms, are highly organized. A constant source of energy is required to maintain a cell's ordered state.
2. Living processes consist of thousands of chemical reactions. Precise regulation and integration of these reactions are required to maintain life.
3. Certain fundamental reaction pathways, such as the energy-generating conversion of glucose to pyruvic acid, known as glycolysis, are found in almost all organisms.
4. All organisms use the same types of molecules. Examples of such biomolecules include carbohydrates, lipids, proteins, and nucleic acids.
5. The instructions for growth, development, and reproduction are encoded in each organism's nucleic acids.

The success of the biochemical approach can be measured by observing its influence on such diverse disciplines as medicine, nutrition, pharmacology, environmental studies, and agriculture. In fact, students who are now embarking on careers in any of the biological and biomedical sciences must be well acquainted with biochemical principles. Recent advances in biotechnology have even farther reaching implications, since economists and stockbrokers are also well served by a familiarity with biochemistry.

The objective of this introductory chapter is to provide an overview of biochemical principles. After a brief discussion of diversity in the living world, our examination of biochemistry begins with a discussion of biomolecules. This is followed by an introduction to the fundamental biochemical reactions and metabolic processes.

1.1 THE LIVING WORLD

At least four million species currently live on planet Earth. One currently used classification scheme has divided them into five kingdoms: (1) Monera, (2) Protista, (3) Fungi, (4) Plants, and (5) Animals (Figure 1.1). Although this diversity is breathtaking, all living things are composed of either prokaryotic cells or eukaryotic cells. **Prokaryotic cells** lack a nucleus (*pro* = before, *karyon* = nucleus or kernel). The nucleus found in **eukaryotic cells** (*eu* = true) is a complex membrane-bound structure that contains genetic information in the form of chromosomes, whose principal constituent is DNA. (Deoxyribonucleic acid, or DNA, will be briefly discussed on p. 10.) All species in Kingdom Monera are prokaryotes; all other organisms are eukaryotes.

Prokaryotes

There are two major groups of prokaryotes: eubacteria (usually referred to as bacteria) and the archaea (known until recently as the archaeabacteria). Despite their relatively simple cellular organization, prokaryotes are remarkably abundant and diverse

FIGURE 1.1

Examples from the Five Kingdoms of Living Organisms: Monera, Protista, Fungi, Animals, and Plants.

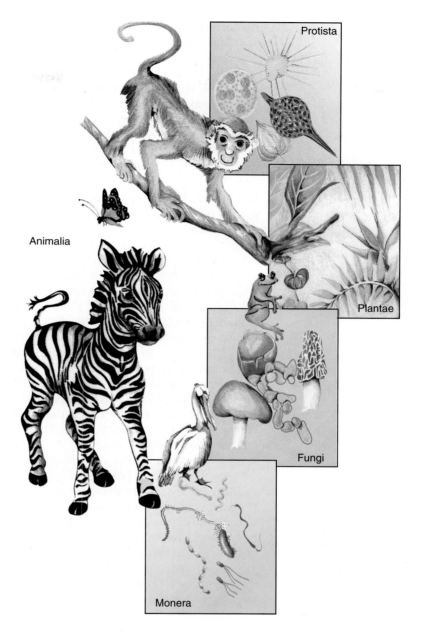

Protista

Animalia

Plantae

Fungi

Monera

organisms. Their biochemical diversity has allowed various species to occupy not only all temperate environments, but also harsh and seemingly lifeless ones. Examples of bacteria include several *Lactobacillus* species, which are beneficial inhabitants of the human intestine, *Staphylococcus aureus,* a toxin-producing organism that causes toxic shock syndrome (among other diseases), and *Oscillatoria,* a member of the cyanobacteria (formerly known as the blue-green algae) commonly found in fresh water. Prominent examples of the archaea include *Halobacterium salinarium,* an inhabitant of salt lakes, and *Thermoplasma acidophilus,* a hydrogen gas-producing organism that thrives near deep sea vents where water temperatures exceed 100°C.

Many prokaryotic species are of great practical interest to humans. In agriculture, bacteria that convert nitrogen gas (N_2) to ammonia (NH_3) in a process called nitrogen fixation contribute greatly to soil fertility. The production of some foods requires certain bacterial species, for example, cheese, yogurt, and sauerkraut. Numerous prokaryotes are very useful in environmental management. For example, several bacterial species that can convert petroleum molecules into harmless products are now routinely used to clean up crude oil spills. In addition, various archeal methanogen (methane-producing) species in association with numerous other microorganisms are now routinely used to convert sewage to methane, a valuable energy source. Prokaryotes have been valuable in biochemical research during most of the twentieth century because of their rapid growth rates and the relative ease of culturing certain bacterial species (especially *Escherichia coli*). The information acquired in research studies of pathogenic microorganisms has been used in medicine to both alleviate and prevent much human suffering. More recently, genetic engineering techniques have made prokaryotes, which are now used to produce hormones, vaccines and other products, more valuable to humans.

Eukaryotes

The second type of living organisms, the eukaryotes, comprises all the remaining species on earth. Although the presence or absence of a true nucleus is the most notable difference between prokaryotes and eukaryotes, there are other significant distinctions between the two groups. Prominent among these are cell volume and complexity.

Eukaryotic cells are substantially larger than prokaryotic cells. The diameter of animal cells, for example, varies between 10 and 30 μm. Such values are approximately ten times higher than those for prokaryotes. Size disparity between the two cell types is more obvious, however, when volume is considered. For example, the volume of a typical eukaryotic cell such as a liver cell (hepatocyte) is between 6,000 and 10,000 μm^3. The volume of *E. coli* cells is several hundred times smaller.

The structural complexity of eukaryotes is remarkable. In addition to a well-formed nucleus, a number of other subcellular structures called **organelles** are present. Each organelle is specialized to perform specific tasks. The compartmentalization afforded by organelles permits the concentration of reactant and product molecules at sites where they can be efficiently used. This plus other factors makes intricate regulatory mechanisms possible. Consequently, the cells of multicellular eukaryotes are able to respond quickly and effectively to the intercellular communications that are required for growth and development.

Most eukaryotes are multicellular (the exception is the unicellular protists). Multicellular organisms have several advantages over unicellular ones. For example, the internal environment in a multicellular body provides a relatively stable, controlled environment for cells. Additionally, a multicellular organism's cells have a division of labor that allows greater complexity in both form and function. These cells, which are often highly specialized, are said to be differentiated. For example, the human body contains approximately 200 different cell types, each of which is uniquely suited to perform a specific function. Multicellular organisms are also able to exploit environmental resources more effectively than single-celled organisms. Trees are a dramatic example. Their massive root systems can absorb large quantities of nutrients and water, and their leaves are relatively efficient collectors of solar energy, which is then converted to the chemical energy required to maintain a substantial biomass. (The term *biomass* is used to describe the total weight of a living organism. Biomass can also refer to the total weight of organisms in an ecosystem.)

Multicellular organisms are not just collections of cells; they are highly ordered living systems. Figure 1.2 illustrates the hierarchical organization of one complex multicellular organism, the human. Anatomical studies have revealed that the human body is composed of organs that together form organ systems. The digestive system, for example, is composed of a number of different organs (i.e., esophagus, stomach, small and large intestines, liver, and pancreas), each of which performs a specific set of activities that contributes to the overall process. Each organ, in turn, contains several types of tissue (e.g., muscle, epithelial, connective, and nervous tissues). A tissue is composed of cells that perform similar functions. For example, glandular tissue is composed of modified epithelial cells that are specialized for secretion. These specialized cells in the glandular tissues of the liver are called hepatocytes.

In the same way that cells are the building blocks of tissues, molecules are the building blocks of cells. Although each cell is a dynamic and interactive set of mol-

KEY *Concepts*

Living organisms may be divided into five kingdoms: Monera, Protista, Fungi, Plants, and Animals. All organisms are composed of either prokaryotic cells or eukaryotic cells. The prokaryotes, which lack a true nucleus, are a biochemically diverse group. Eukaryotes are structurally complex. In addition to a well-formed nucleus, a number of other subcellular structures called organelles are present. Most eukaryotes are multicellular (the exception is the unicellular protists).

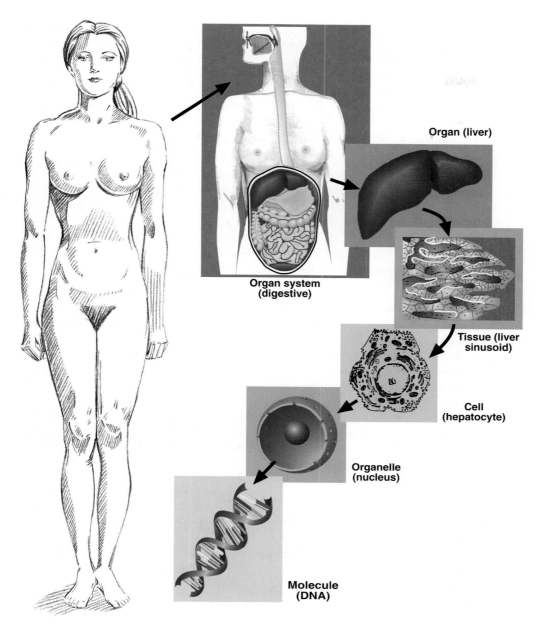

FIGURE 1.2

Hierarchical Organization of a Multicellular Organism: The Human Being.

Multicellular organisms have several levels of organization: Organ systems, organs, tissues, cells, organelles, and molecules. The digestive system and one of its component organs (the liver) are shown. The liver is a multifunctional organ that has several digestive functions. For example, it produces bile, which facilitates fat digestion, and it processes and distributes the food molecules absorbed in the small intestine to other parts of the body. DNA, one molecule found in cells, contains the genetic information that controls cell function.

ecules, the whole is greater than the sum of the parts. Since biochemistry is the study of the chemistry of living organisms, biomolecules and biochemical reactions are meaningful only when viewed in the context of biological structure. For this reason a brief description of cell structure is given in Chapter 2.

1.2 BIOMOLECULES

Animal and plant cells contain approximately 10,000 kinds of molecules (referred to as **biomolecules**). One of these, water, may constitute 50–95% of a cell's content by weight, while ions such as sodium (Na^+), potassium (K^+), magnesium (Mg^{2+}), and calcium (Ca^{2+}) may account for another 1%. Almost all the other kinds of molecules

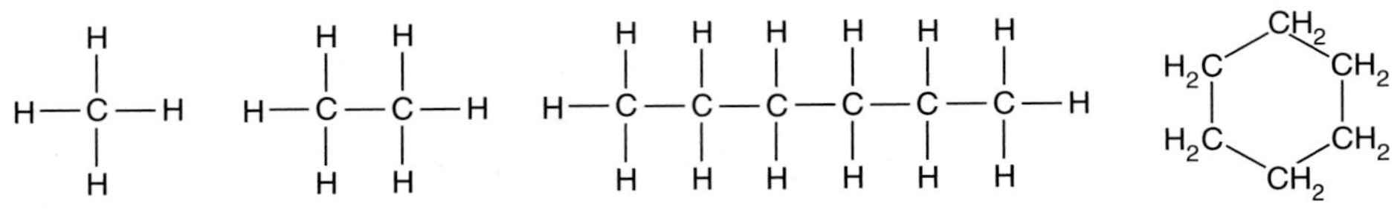

FIGURE 1.3

Structural Formulas of Several Hydrocarbons.

| Methane | Ethane | Hexane | Cyclohexane |

in living organisms are organic. Organic molecules are principally composed of six elements: carbon, hydrogen, oxygen, nitrogen, phosphorus, and sulfur. However, the properties of one element, carbon, are responsible for the almost infinite variety of organic molecules. Carbon atoms can form four strong covalent bonds, either to other carbon atoms or to atoms of other elements.

Of central importance in organic chemistry are the hydrogen- and carbon-containing compounds called the **hydrocarbons** (Figure 1.3). Although hydrocarbons are relatively unimportant in living organisms, several aspects of hydrocarbon chemistry are relevant to biochemistry:

1. Most biomolecules can be considered to be derived from hydrocarbons.
2. Hydrocarbons are nonpolar and therefore not water-soluble. (In **nonpolar** molecules, bonding electrons are shared equally between atoms.) Because hydrocarbons do not interact well with water, they are said to be "hydrophobic." Interactions between biomolecules containing hydrocarbon groups are responsible for the properties of certain cellular components (e.g., membranes).

Functional Groups of Organic Biomolecules

The chemical properties of organic molecules are determined by specific arrangements of atoms called **functional groups.** Different families of organic compounds result when hydrogen atoms on organic molecules are replaced by different functional

TABLE 1.1

Important Functional Groups in Biomolecules

Family Name	Group Structure	Group Name	Significance
Alcohol	R—OH	Hydroxyl	Polar (and therefore water-soluble), forms hydrogen bonds
Aldehyde	R—C(=O)—H	Carbonyl	Polar, found in some sugars
Ketone	R—C(=O)—R'	Carbonyl	Polar, found in some sugars
Acids	R—C(=O)—OH	Carboxyl	Weakly acidic, bears a negative charge when it donates a proton
Amines	R—NH$_2$	Amino	Weakly basic, bears a positive charge when it accepts a proton
Amides	R—C(=O)—NH$_2$	Amido	Polar but does not bear a charge
Thiols	R—SH	Thiol	Does not form hydrogen bonds; therefore less soluble in water than alcohols
Esters	R—C(=O)—O—R	Ester	Found in certain lipid molecules
Double bond	RCH=CHR	Alkene	Important structural component of many biomolecules; e.g., found in lipid molecules

groups (Table 1.1). For example, alcohols result when hydrogen atoms are replaced by hydroxyl groups (—OH). Thus methane (CH_4), a component of natural gas, can be converted into methanol (CH_3OH), a toxic liquid that is used as a solvent in many industrial processes.

Most biomolecules contain more than one functional group. For example, many simple sugar molecules have several hydroxyl groups and an aldehyde group. Amino acids, the building block molecules of proteins, have both an amino group and a carboxyl group. The distinct chemical properties of each functional group contribute to the behavior of any molecule that contains it.

Major Classes of Small Biomolecules

Many of the organic compounds found in cells are relatively small, with molecular weights of less than 1000. Cells contain four families of small molecules: amino acids, sugars, fatty acids, and nucleotides. Members of each group serve several functions. First, they are used in the synthesis of larger molecules, many of which are polymers (e.g., proteins and nucleic acids). Second, some molecules have special biological functions. For example, the nucleotide adenosine triphosphate (ATP) serves as a cellular reservoir of chemical energy. Finally, many small organic molecules are involved in complex reaction pathways. Examples of each class are described below.

Amino Acids   There are 20 commonly occurring α-**amino acids.** As was mentioned previously, each amino acid molecule contains an amino group and a carboxyl group. In α-amino acids the amino group is attached to the carbon atom (the α-carbon) immediately adjacent to the carboxyl group. Also attached to the α-carbon is another group, referred to as the side chain. The chemical properties of each amino acid are determined largely by the identity of its side chain (also called an R group). The general formula for amino acids is

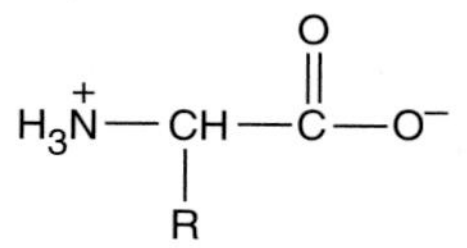

$$H_3\overset{+}{N}-CH-\underset{\underset{R}{|}}{\overset{\overset{O}{\|}}{C}}-O^-$$

The structures of several representative amino acids are shown in Figure 1.4.

FIGURE 1.4

Structural Formulas for Several Amino Acids.

R groups are highlighted. An R group in amino acid structures can be a hydrogen atom (e.g., in glycine), a hydrocarbon group (e.g., the isopropyl group in valine), or a hydrocarbon derivative (e.g., the hydroxymethyl group in serine).

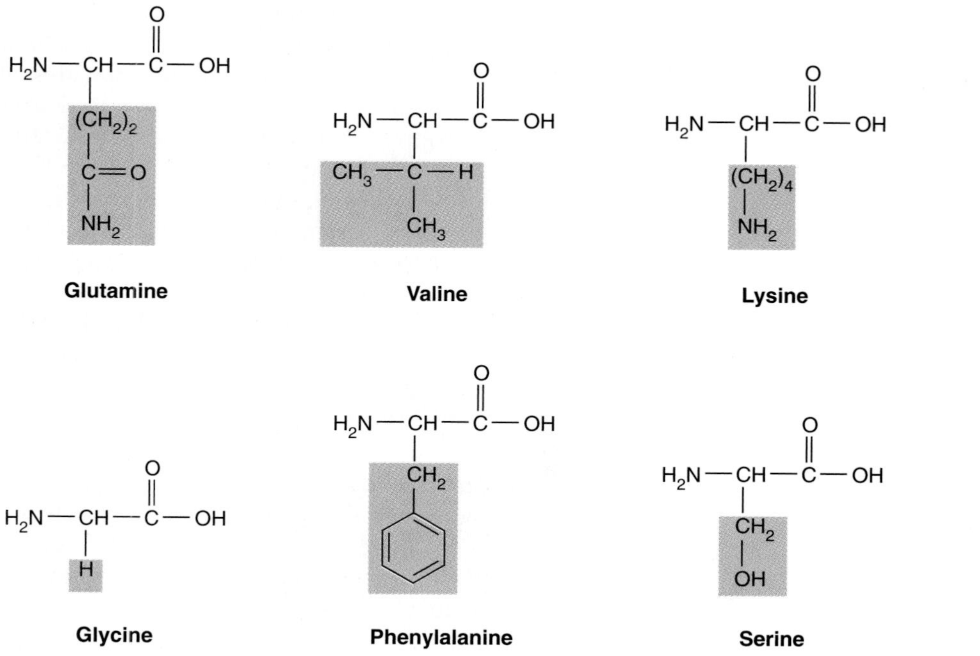

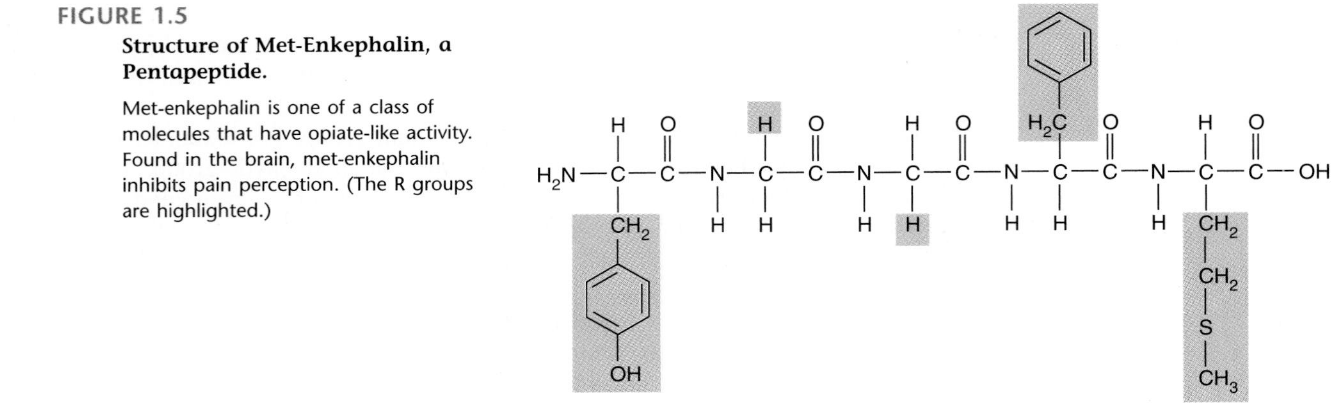

FIGURE 1.5

Structure of Met-Enkephalin, a Pentapeptide.

Met-enkephalin is one of a class of molecules that have opiate-like activity. Found in the brain, met-enkephalin inhibits pain perception. (The R groups are highlighted.)

Amino acid molecules are used primarily in the synthesis of long, complex polymers known as **polypeptides.** Short polypeptides, up to a length of about 50 amino acids, are called **peptides.** Longer polypeptides are often referred to as **proteins.** Polypeptides play a variety of roles in living organisms. Examples of molecules composed of polypeptides include transport proteins, structural proteins, and the enzymes (molecules that accelerate reaction rates). The individual amino acids are connected in polypeptide molecules by an amide bond (a covalent bond formed between the carboxylic acid group of one amino acid and the amino group of the next) (Figure 1.5), often referred to as a peptide bond.

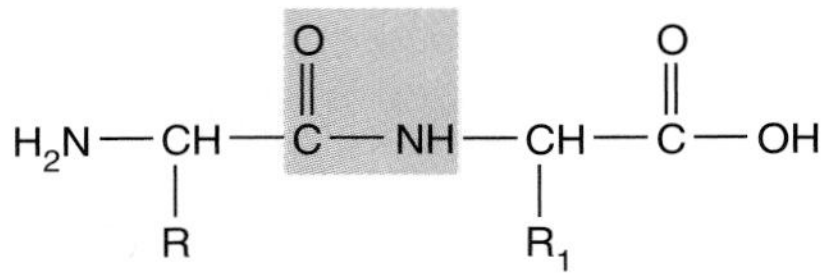

Peptide bond

The structure of the R groups of the amino acids in a protein determines its final three-dimensional structure and therefore its biological function.

Sugars   Carbohydrates, the most abundant organic molecules found in nature, have a wide range of functions, including energy sources and structural components. Carbohydrates also play a role in intercellular communications. The basic units of the carbohydrates are the **monosaccharides,** also known as the simple **sugars.**

Monosaccharides may be described as polyhydroxy aldehydes or ketones and their derivatives that contain at least three carbon atoms. Sugars containing an aldehyde group are called aldoses, while those with a ketone functional group are called ketoses. Typically, sugars are described in terms of both carbon number and functional group. For example, the six-carbon sugar glucose (an important energy source) is an aldohexose. The structures of glucose and several other important monosaccharides are illustrated in Figure 1.6.

Monosaccharides may react with each other to form larger molecules. Polysaccharides are polymers containing a large number of monosaccharides. Examples include glycogen and plant starch (energy storage molecules) as well as cellulose (a plant structural material).

Fatty Acids   **Fatty acids,** an important group of molecules in living systems, are monocarboxylic acids that usually contain an even number of carbon atoms (Figure 1.7). There are two types of fatty acids: **saturated** fatty acids, which contain no carbon-carbon double bonds and **unsaturated** fatty acids, which have one or more double bonds.

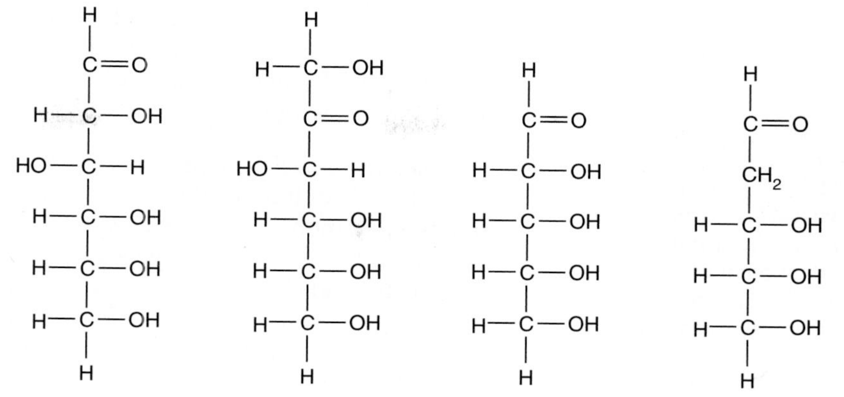

FIGURE 1.6

Several Biologically Important Monosaccharides.

Glucose
(an aldohexose)

Fructose
(a ketohexose)

Ribose
(an aldopentose)

2-Deoxyribose
(an aldopentose)

Fatty acids are components of several lipid molecules. (**Lipids** are a diverse group of substances that are soluble in organic solvents, such as chloroform or acetone, but not soluble in water.) For example, **triacylglycerols** (fats and oils) are esters containing glycerol (a three-carbon alcohol with three hydroxyl groups) and three fatty acids. Certain lipid molecules that resemble triacylglycerols, called **phosphoglycerides,** contain two fatty acids. In these molecules the third hydroxyl group of glycerol is coupled with phosphate, which is in turn attached to small polar compounds such as choline. Phosphoglycerides are an important structural component of cell membranes.

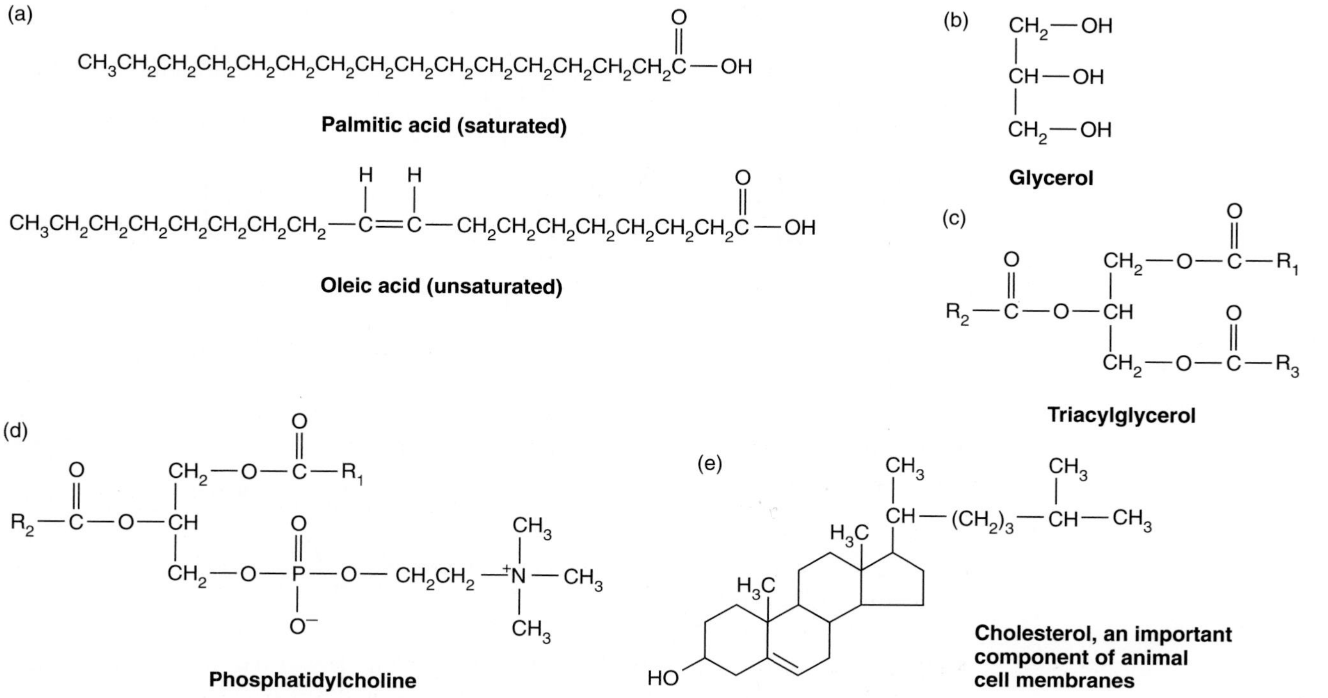

(a)

Palmitic acid (saturated)

Oleic acid (unsaturated)

(b)

Glycerol

(c)

Triacylglycerol

(d)

Phosphatidylcholine

(e)

Cholesterol, an important component of animal cell membranes

FIGURE 1.7

Examples of Lipids.

(a) Fatty acids, (b) glycerol, (c) triacylglycerol, (d) a phospholipid, (e) a steroid.

Examples of other important lipid molecules include the steroids (a group that includes cholesterol, the sex hormones, and cortisol), the fat-soluble vitamins (e.g., vitamins A and E), and the carotenoids (a group of plant pigment molecules that play a role in photosynthesis).

Nucleotides   **Nucleic acids,** the largest molecules in living organisms, are composed of units called **nucleotides.** Each nucleotide contains three components: a five-carbon sugar (either ribose or deoxyribose), a phosphate group, and a nitrogenous base (Figure 1.8). There are two types of nucleic acid: deoxyribonucleic acid (DNA) and ribonucleic acid (RNA). DNA is the repository of an organism's genetic information. RNA is involved in the expression of that information, primarily in protein synthesis.

Nucleotides play many other roles. They participate in a wide variety of biosynthetic and energy-generating reactions. For example, a substantial proportion of the energy obtained from food molecules is used to form the high-energy phosphate bonds of adenosine triphosphate (ATP).

All of the characteristics of living organisms—their complex organization and their capacity to grow and reproduce—belie the fact that they are composed of lifeless molecules. When these molecules are examined individually, they are observed to con-

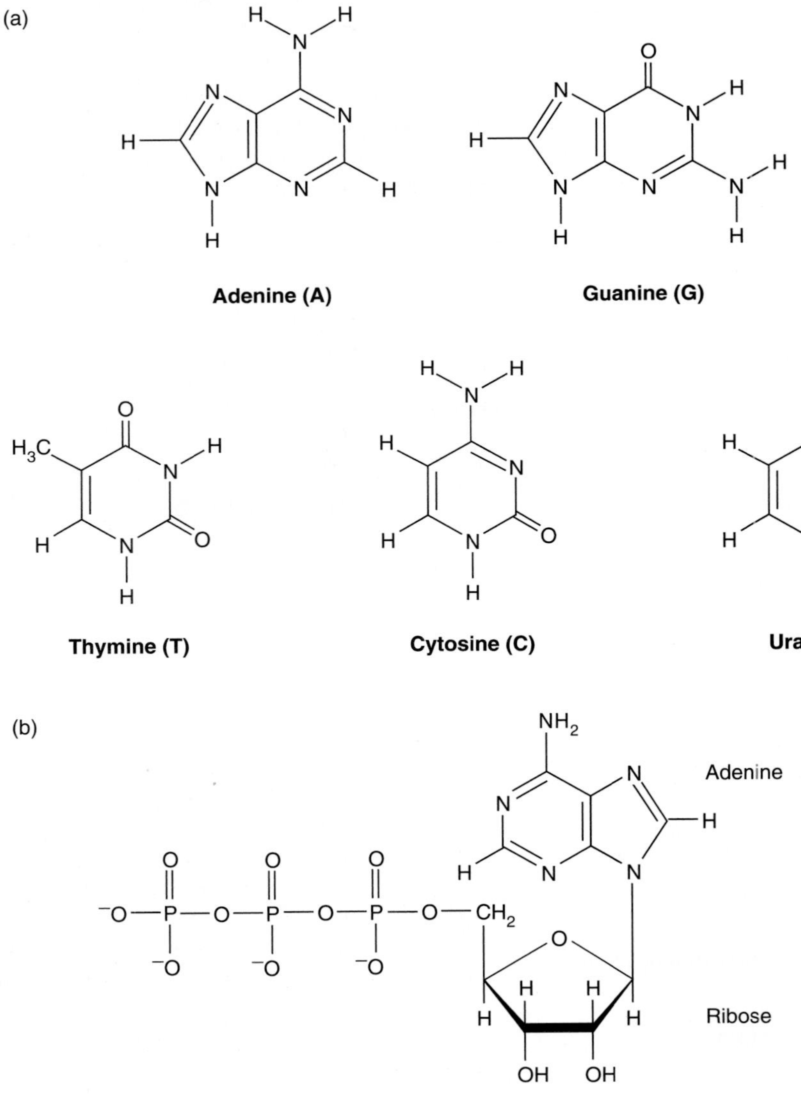

KEY *Concepts*

Most molecules in living organisms are organic. The chemical properties of organic molecules are determined by specific arrangements of atoms called functional groups. Cells contain four families of small molecules: amino acids, sugars, fatty acids, and nucleotides.

1.3 BIOCHEMICAL PROCESSES

FIGURE 1.8

Nucleotides.

(a) The nitrogenous bases, (b) adenosine triphosphate, a ribonucleotide.

form to the physical and chemical laws that govern the universe. Yet at first glance, life appears to be an exception to one such rule known as the *second law of thermodynamics*. This principle, which stipulates that in any system the degree of disorder can only increase, explains our common experience that things left to themselves tend to become disordered. For example, bridges and houses eventually fall apart. Living organisms, however, can maintain highly ordered states for substantial periods of time. They accomplish this feat by consuming inanimate materials acquired from a relatively disorganized environment.

The science of biochemistry seeks to explain how living organisms acquire and maintain their structure and function in the midst of a chaotic universe. Several biochemical principles that have been confirmed by generations of biochemists provide insight into living processes. These themes, which are described at length throughout the remainder of the book, are briefly outlined below.

Biochemical Reactions

All life processes consist of chemical reactions catalyzed by enzymes. The reactions in a living organism, which are known collectively as **metabolism,** result in highly coordinated and purposeful activity. The primary functions of metabolism are: (1) acquisition and utilization of energy, (2) synthesis of molecules needed for cell structure and functioning (i.e., proteins, nucleic acids, lipids, and carbohydrates), and (3) removal of waste products.

At first glance the thousands of reactions that occur in cells appear overwhelmingly complex. However, several characteristics of metabolism allow us to vastly simplify this picture:

1. While the number of reactions is very large, the number of reaction types is relatively small.
2. The mechanisms used in biochemical reactions (i.e., the means by which chemical changes occur) are relatively simple.
3. Reactions of central importance in biochemistry (i.e., those used in energy production and the synthesis and degradation of major cell components) are relatively few in number.

Among the most frequent reactions encountered in biochemical processes are the following: (1) nucleophilic substitution, (2) elimination, (3) addition, (4) isomerization, (5) oxidation-reduction, and (6) hydrolysis. Each will be briefly described.

Nucleophilic Substitution Reactions In **nucleophilic substitution** reactions, as the name suggests, one atom or group is substituted for another:

In the general reaction shown above, the attacking species (A) is called a **nucleophile** ("nucleus-lover"). Nucleophiles are most commonly anions (negatively charged atoms or groups). However, neutral species with unshared electrons can also act as nucleophiles. **Electrophiles** ("electron-lover") are the atoms or groups that are transferred from one nucleophile to another. In the example, nucleophile A is attracted to electrophile B. As the new bond forms between A and B, the old one between B and X breaks. The outgoing nucleophile (in this case, X) is called a **leaving group.**

The reaction of glucose with ATP provides an important example of nucleophilic substitution (Figure 1.9). In this reaction, which is the first step in the utilization of glucose as an energy source, the hydroxyl oxygen on carbon 6 of the sugar molecule is the nucleophile and phosphorus is the electrophile. Adenosine diphosphate is the leaving group.

FIGURE 1.9
A Nucleophilic Substitution.

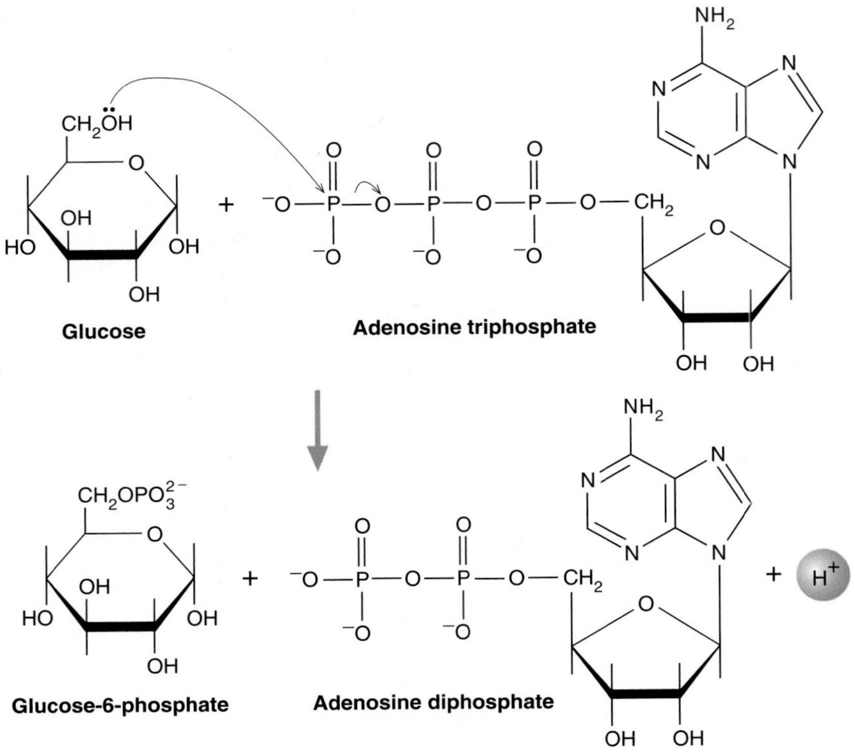

Glucose   Adenosine triphosphate

Glucose-6-phosphate   Adenosine diphosphate

FIGURE 1.10
Dehydration of
2-Phosphoglycerate.

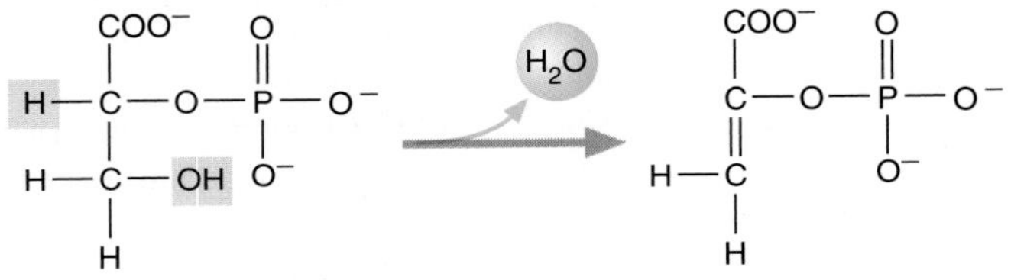

2-Phosphoglycerate   Phosphoenolpyruvate

Elimination Reactions   In **elimination** reactions a double bond is formed when atoms in a molecule are removed:

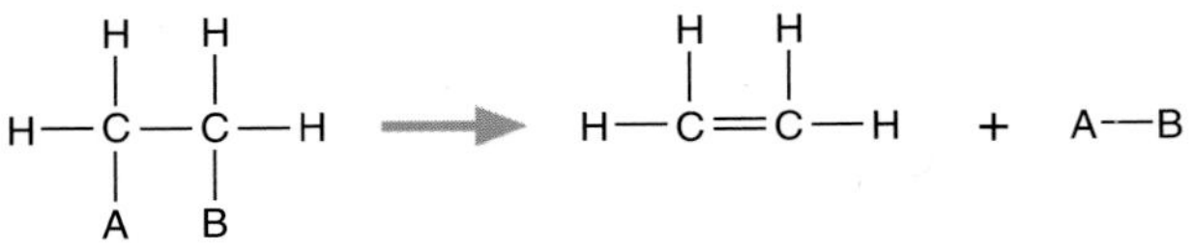

The removal of H_2O from biomolecules containing alcohol functional groups is a commonly encountered elimination reaction. A prominent example of this reaction is the dehydration of 2-phosphoglycerate, an important step in carbohydrate metabolism (Figure 1.10). Other products of elimination reactions include ammonia (NH_3), amines (RNH_2), and alcohols (ROH).

Addition Reactions   In **addition reactions** two molecules combine to form a single product.

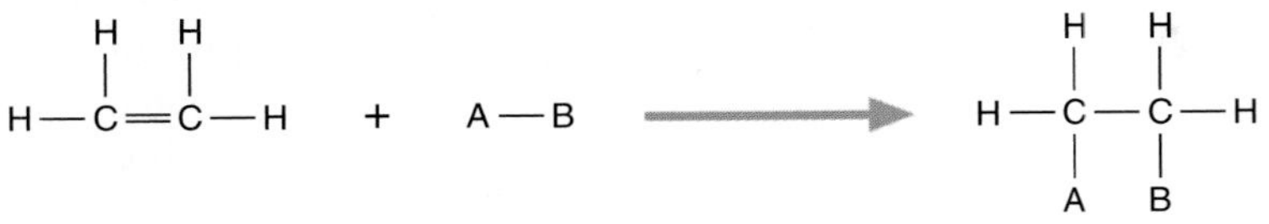

Hydration is one of the most common addition reactions. When water is added to an alkene an alcohol results. The hydration of the metabolic intermediate fumarate to form malate is a typical example (Figure 1.11).

Isomerization Reactions **Isomerization** reactions involve the intramolecular shift of atoms or groups. One of the most common biochemical isomerizations is the interconversion between aldose and ketose sugars (Figure 1.12).

Oxidation-Reduction Reactions **Oxidation-reduction** reactions (also called redox reactions) occur when there is a transfer of electrons from a donor (called the **reducing agent**) to an electron acceptor (called the **oxidizing agent**). When reducing agents donate their electrons, they become **oxidized.** As oxidizing agents accept electrons, they become **reduced.** The two processes always occur simultaneously.

It is not always easy to determine whether biomolecules have gained or lost electrons. However, there are two simple rules that may be used to ascertain whether a molecule has been oxidized or reduced:

1. Oxidation has occurred if a molecule gains oxygen or loses hydrogen:

$$CH_3CH_2-OH \longrightarrow CH_3\overset{\textstyle O}{\underset{\textstyle \|}{C}}-OH$$

Ethyl alcohol   **Acetic acid**

2. Reduction has occurred if a molecule loses oxygen or gains hydrogen:

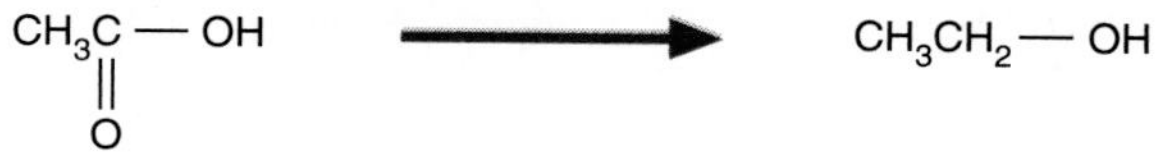

$$CH_3\overset{\textstyle O}{\underset{\textstyle \|}{C}}-OH \longrightarrow CH_3CH_2-OH$$

Acetic acid   **Ethyl alcohol**

FIGURE 1.11
Hydration of Fumarate.

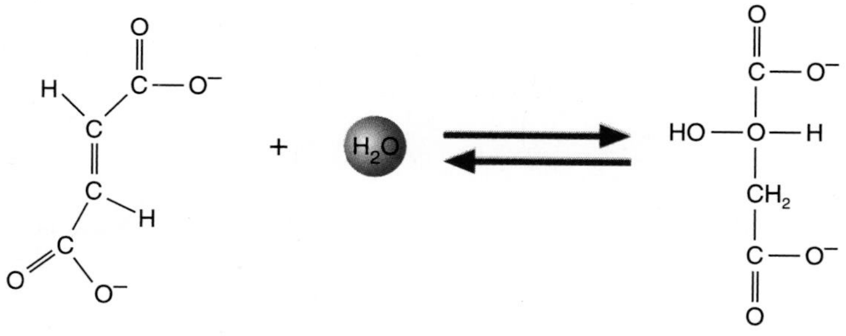

Fumarate   **Malate**

FIGURE 1.12
Isomerization of Sugar Molecules.

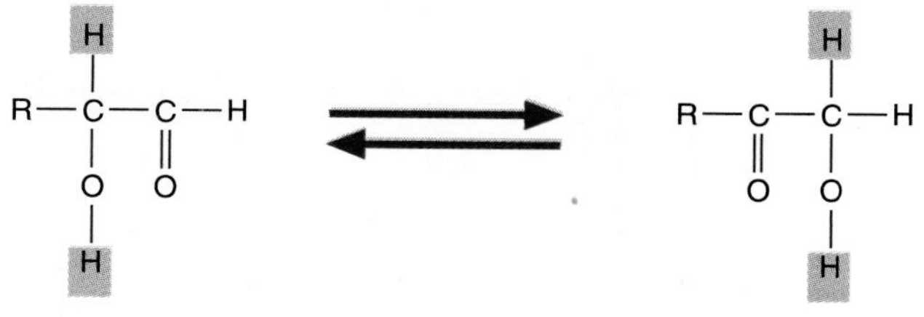

Aldose   **Ketose**

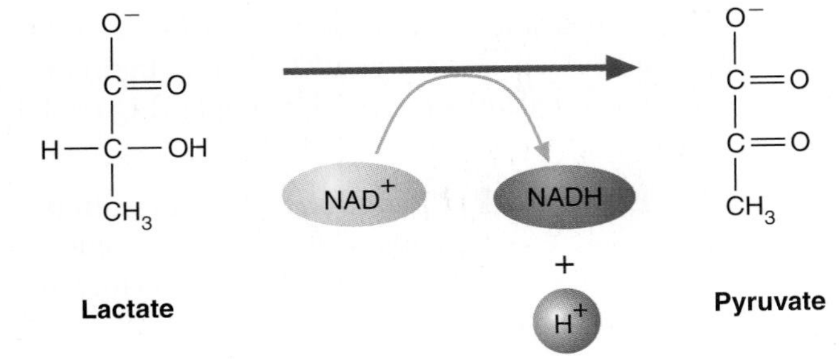

FIGURE 1.13

Oxidation of a Biomolecule.

As lactate is oxidized to form pyruvate, the hydrogen atoms must be transferred to another molecule, in this case a nucleotide known as NAD^+.

In biological redox reactions, electrons are transferred to electron acceptors such as the nucleotide NAD^+ (nicotinamide adenine dinucleotide) (Figure 1.13).

Hydrolysis Reactions   **Hydrolysis** is the cleavage of a covalent bond by water:

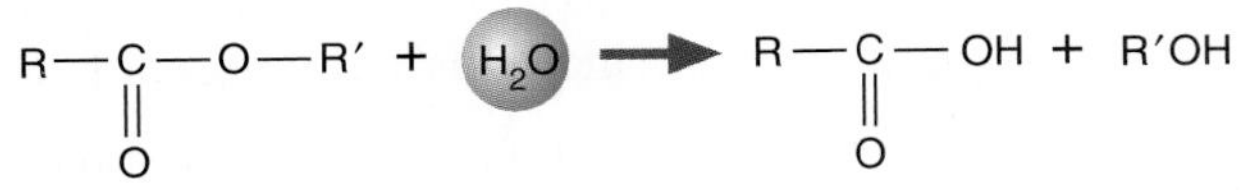

Hydrolytic reactions, which usually involve nucleophilic substitution either at a saturated carbon or a carbonyl carbon, may be catalyzed by acid or base. The digestion of many food molecules involves hydrolysis. For example, proteins are degraded in the stomach in an acid-catalyzed reaction. Another important example is breaking the phosphate bonds of ATP (Figure 1.14). The energy obtained during this reaction is used to drive many cellular processes.

Energy

Living cells are inherently unstable. Only a constant flow of energy prevents them from becoming disorganized (nonliving). **Energy** is defined as the capacity to do work. Although cells have been compared to human-made machines because of their

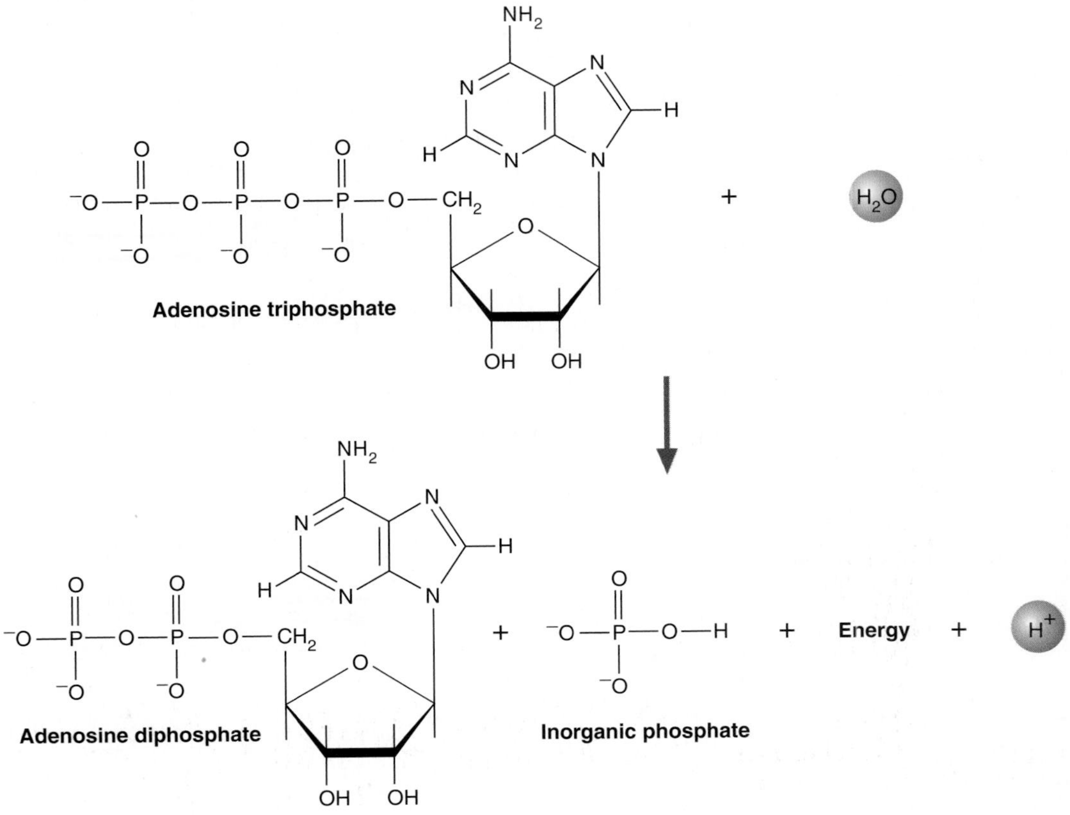

FIGURE 1.14

Hydrolysis of ATP.

The complex structure of cells requires high internal order. This is accomplished by four primary means: (1) synthesis of biomolecules, (2) transport of ions and molecules across cell membranes, (3) production of force and movement, and (4) removal of metabolic waste products and other toxic substances. Each will be discussed briefly.

Synthesis of Biomolecules

Cellular components are synthesized in a vast array of chemical reactions. Many of these reactions are integrated into carefully regulated pathways that involve numerous steps. For example, the nucleotide adenosine monophosphate is synthesized in a 12-step pathway. It should be noted that a large number of biosynthetic reactions require energy, which is supplied directly or indirectly by the simultaneous breaking of the phosphoanhydride bonds of ATP molecules.

The molecules formed in biosynthetic reactions perform several functions. They can be assembled into supramolecular structures (e.g., the proteins and lipids that constitute membranes), serve as informational molecules (e.g., DNA and RNA), or catalyze chemical reactions (i.e., the enzymes).

Transport Across Membranes

Cell membranes regulate the passage of ions and molecules from one compartment to another. For example, the plasma membrane (the cell's outer membrane) is a selective barrier. It is responsible for the transport of certain substances such as nutrients from a relatively disorganized environment into the precisely ordered cellular interior. Similarly, ions and molecules are transported into and out of organelles during biochemical processes. For example, fatty acids are transported into an organelle known as the mitochondrion so that they may be broken down to generate energy.

Much of the cell's transport work is accomplished by membrane-bound protein molecules. When substances are transported against a gradient (i.e., from an area of low concentration to an area of high concentration), energy is required. This process is referred to as **active transport.** For example, the Na^+-K^+ pump, a protein that occurs in the plasma membranes of animal cells, is involved in regulating cell volume. It utilizes at least one-third of available energy to pump Na^+ out of and K^+ into the cell. For every ATP molecule utilized, three Na^+ ions are pumped out and two K^+ ions are pumped in. Because more Na^+ ions are pumped out of the cell, there is a net loss of water. (The concept of osmosis, the passage of water in relation to the concentration of ions or molecules, is discussed in Chapter 3.) If this process is impeded (e.g., if cells are treated with ouabain, a substance that inhibits the Na^+-K^+ pump), the affected cells will swell and burst.

Cell Movement

Organized movement is one of the most obvious characteristics of living organisms. The intricate and coordinated activities that sustain life require the movement of cell components. Examples include cell division and organelle movement. Both of these processes depend to a large extent on the structure and function of a complex network of protein filaments known as the *cytoskeleton.*

The forms of cellular motion profoundly influence the ability of all organisms to grow, reproduce, and compete for limited resources. As examples, consider the movement of protists as they search for food in a pond or the migration of human white blood cells as they pursue foreign cells during an infection. More subtle examples include the movement of specific enzymes along a DNA molecule during the chromosome replication that precedes cell division and the secretion of insulin by certain pancreatic cells.

Waste Removal

All living cells produce waste products. For example, animal cells ultimately convert food molecules, such as sugars and amino acids, into CO_2, H_2O, and NH_3. These molecules, if not disposed of properly, can be toxic. Some substances are readily removed. In animals, for example, CO_2 diffuses out of cells and (after a brief and reversible conversion to bicarbonate in the blood) is quickly exhaled through the respiratory system. Excess H_2O is excreted through the kidneys. Other molecules, however, are sufficiently toxic that elaborate mechanisms have been evolved to provide for their disposal. The urea cycle (described in Chapter 14), used in many animals to dispose of NH_3, converts this extremely harmful substance into urea, a less toxic molecule.

Living cells also contain a wide variety of complex organic molecules that must be disposed of. Plant cells solve this problem by transporting such molecules into a vacuole, where they are either broken down or stored. Animals, however, must use disposal mechanisms that depend on water solubility (e.g., the formation of urine by the kidney). Hydrophobic (water-insoluble) substances such as the steroid hormones, which cannot be broken down into simpler molecules, are converted during a series of reactions into water-soluble derivatives. This mechanism is also used to solubilize exogenous organic molecules such as drugs and environmental contaminants.

capacity to convert fuel into chemical, mechanical, and electrical work, there are significant differences between the two. Among the most prominent of these is the inability of cells, which are composed of relatively fragile organic molecules, to withstand conditions that are typically associated with machines (e.g., electric currents and high temperatures and pressures). Because of these and other constraints, living organisms have evolved more subtle mechanisms for generating and using energy.

The ultimate source of the energy used by all life on earth is the sun. Photosynthetic organisms capture light energy and use it to transform carbon dioxide (CO_2)

into sugar and other biomolecules. These substances are, in turn, consumed by organisms such as animals that use them as energy sources and structural materials. At each step, as molecular bonds are rearranged to generate energy, some of the energy is lost as heat. Eventually, all captured light energy is transformed into heat, but before this happens, cells use the energy to maintain their complex structures and activities.

One of the principal mechanisms by which cells obtain energy from chemical bonds is by the oxidation of biomolecules. In energy-transforming reactions, electrons are transferred from one molecule to another, and in this process, electrons lose energy. A portion of this energy is captured and used to maintain highly organized cellular structures and functions. The methods by which cellular order is accomplished are discussed in Special Interest Box 1.1.

KEY WORDS

active transport, 15
addition reaction, 12
amino acid, 7
biomolecule, 5
electrophile, 11
elimination, 12
energy, 14
eukaryotic cell, 2
fatty acid, 8
functional group, 6

hydration, 13
hydrocarbon, 6
hydrolysis, 14
isomerization, 13
leaving group, 11
lipid, 9
metabolism, 11
monosaccharide, 8
nonpolar, 6
nucleic acid, 10

nucleophile, 11
nucleophilic substitution, 11
nucleotide, 10
organelle, 4
oxidation-reduction (redox), 13
oxidize, 13
oxidizing agent, 13
peptide, 8
phosphoglyceride, 9
polypeptide, 8

prokaryotic cell, 2
protein, 8
reduce, 13
reducing agent, 13
saturated, 8
sugar, 8
triacylglycerol, 9
unsaturated, 8

SUMMARY

Biochemistry may be defined as the study of the molecular basis of life. Biochemists have established several principles that are central to our understanding of living organisms: (1) cells, the basic structural units of all living things, are highly organized; (2) living processes consist of hundreds of chemical reactions; (3) certain fundamental reaction pathways are found in all living organisms; (4) all organisms use the same types of molecules; and (5) the instructions for growth, development, and reproduction are encoded in an organism's nucleic acids.

One currently used classification scheme has divided the estimated four million species on planet Earth into five kingdoms: (1) Monera, (2) Protista, (3) Fungi, (4) Plants, and (5) Animals. All living things are composed of either prokaryotic cells or eukaryotic cells. Prokaryotes, which include eubacteria and the archaea, lack a membrane-bound cellular organelle called a nucleus. The eukaryotes consist of all the remaining species. These cells contain a nucleus as well as complex structures that are not observed in prokaryotes.

Most eukaryotes are multicellular. Multicellular organisms have several advantages over unicellular ones. These include (1) the provision of a relatively stable environment for most of the organism's cells, (2) the capacity for greater complexity in an organism's form and function, and (3) the ability to exploit environmental resources more effectively than single-celled organisms can.

Animal and plant cells contain approximately 10,000 kinds of molecules. Water constitutes 50–95% of a cell's content by weight, while ions such as Na^+, K^+, and Ca^{2+} may account for another 1%. Almost all the other kinds of biomolecules are organic. The proper-

ties of the element carbon are responsible for the almost infinite variety of organic molecules. The chemical properties of organic molecules are determined by specific arrangements of atoms called functional groups. Different families of organic molecules result when hydrogen atoms on hydrocarbon molecules are replaced by functional groups. Most biomolecules contain more than one functional group.

Many of the biomolecules found in cells are relatively small, with molecular weights of less than 1000. Cells contain four families of small molecules: amino acids, sugars, fatty acids, and nucleotides. Members of each group serve several functions: (1) they are used in the synthesis of larger molecules, (2) some small molecules have special biological functions, and (3) many small molecules are components in complex reaction pathways.

All life processes consist of chemical reactions catalyzed by enzymes. The reactions of a living cell, which are known collectively as metabolism, result in highly coordinated and purposeful activity. Among the most frequent reactions encountered in biochemical processes are (1) nucleophilic substitution, (2) elimination, (3) isomerization, (4) oxidation-reduction, and (5) hydrolysis.

Living cells are inherently unstable. Only a constant flow of energy prevents them from becoming disorganized. One of the means by which cells obtain energy is oxidation of biomolecules.

The complex structure of cells requires a high degree of internal order. This is accomplished by four primary means: (1) synthesis of biomolecules, (2) transport of ions and molecules across cell membranes, (3) production of movement, and (4) removal of metabolic waste products and other toxic substances.

SUGGESTED READINGS

Atkens, P.W., Molecules, *Sci. Amer.*, New York, 1987.

Curtis, H., and Bannes, N.S., *Biology,* 5th ed., Worth, New York, 1989.

Lewis, R., *Life,* 3rd ed., WCB/McGraw-Hill, Dubuque, Iowa, 1998.

Mader, S., *Biology,* 6th ed., WCB/McGraw-Hill, Dubuque, Iowa, 1998.

Margulis, L.M., and Schwartz, K.V., *Five Kingdoms: An Illustrated Guide to the Phyla of Life on Earth,* 2nd ed., W.H. Freeman, New York, 1988.

QUESTIONS

1. List five principles that are central to our current understanding of living organisms.

2. Describe the major differences between prokaryotic and eukaryotic cells.

3. Identify the functional groups in the following molecules:

a. $CH_3\overset{O}{\underset{\|}{C}}-H$

b. $HO-\overset{O}{\underset{\|}{C}}-CH_2CH_2\underset{\underset{NH_2}{|}}{CH}-\overset{O}{\underset{\|}{C}}-OH$

c. $CH_3CH_2\underset{\underset{SH}{|}}{CH}CH_3$

d. $CH_3\overset{O}{\underset{\|}{C}}-O-CH_3$

e. $\underset{H}{\overset{CH_3}{}}C=C\underset{H}{\overset{CH_2CH_3}{}}$

f. $CH_3\overset{O}{\underset{\|}{C}}-\underset{\underset{H}{|}}{N}-CH_2CH_3$

g. $CH_3\overset{O}{\underset{\|}{C}}CH_3$

h. $\underset{CH_2OH}{\overset{CH_2OH}{\underset{|}{\overset{|}{CH-OH}}}}$

4. Name four classes of small biomolecules. In what larger biomolecules are they found?

5. Define the following terms:
   a. biochemistry
   b. oxidation
   c. reduction
   d. active transport
   e. leaving group
   f. elimination
   g. isomerization
   h. nucleophilic substitution
   i. reducing agent
   j. oxidizing agent

6. List two functions for each of the following biomolecules:
   a. fatty acids
   b. sugars
   c. nucleotides

7. What are the roles of DNA and RNA?

8. How do cells obtain energy from chemical bonds?

9. How do plants dispose of waste products?

10. List the five kingdoms of living organisms. How many examples from each kingdom can you name? Which organisms are prokaryotic? What are eukaryotic? In which kingdoms are multicellular organisms found?

11. What is the difference between an unsaturated and a saturated hydrocarbon?

12. It is often assumed that biochemical processes in prokaryotes and eukaryotes are basically similar. Is this a safe assumption?

13. What advantages do multicellular organisms have over unicellular organisms?

14. Assign each of the following compounds to one of the major classes of biomolecule:

a.

b.

c. $CH_3-(CH_2)_9-CH_2-\overset{O}{\underset{\|}{C}}-OH$

d.

e.

15. Define the following terms:
    a. metabolism
    b. nucleophile
    c. cytoskeleton
    d. electrophile
    e. energy
16. What are organelles? In general, what advantages do they provide to eukaryotes?
17. Much of what is known about biochemical processes is a direct result of research using prokaryotic organisms. Most organisms, however, are eukaryotic. Can you suggest any reasons why so many research efforts have used prokaryotes? Why not use eukaryotes directly?

18. What are the primary functions of metabolism?
19. Give an example of each of the following reaction processes:
    a. nucleophilic substitution
    b. elimination
    c. oxidation-reduction
    d. hydrolysis
20. Life on earth arose from nonliving molecules. Life is highly organized, and nonliving materials are disorganized. Is this seemingly spontaneous increase in organization a violation of the second law of thermodynamics? Explain.
21. List several important ions that are found in living organisms.
22. What are the common types of chemical reactions found in living cells?
23. Biochemical reactions have been viewed as exotic versions of organic reactions. Can you suggest any problems with this assumption?
24. Describe several functions of polypeptides.
25. Carbohydrates are widely recognized as sources of metabolic energy. What are two other critical roles that carbohydrates play in living organisms?
26. What are the largest biomolecules? What functions do they serve in living organisms?
27. Nucleotides have roles in addition to being components of DNA and RNA. Give an example.
28. How is order maintained within living cells?
29. Name several waste products that animal cells produce.

LIVING CELLS

Chapter Two

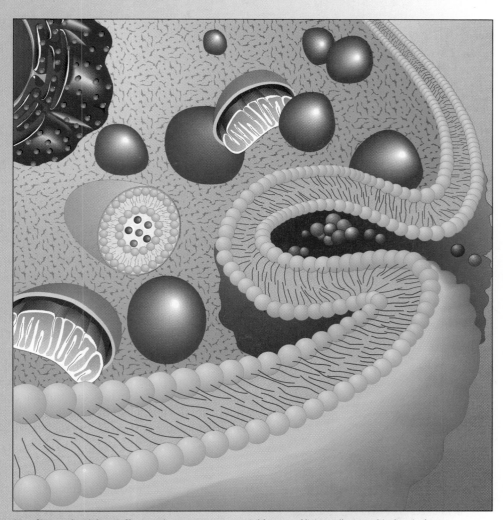

Membranes in Living Cells. *Membranes are an essential feature of living cells. Most biochemical processes occur in or near these dynamic and complex supermolecular structures.*

OBJECTIVES

After you have studied this chapter, you should be able to answer these questions:

1. *How are prokaryotic and eukaryotic cells alike and how are they different?*

2. *Which organelles carry out the processes required to maintain the living state in eukaryotic cells?*

3. *How might eukaryotic cells have evolved from prokaryotic cells?*

4. *What technologies have aided scientists in discovering how cells function?*

Cells are the structural units of all living organisms. One remarkable feature of cells is their diversity. For example, the human body contains about 200 types of cells. This great variation reflects the variety of functions that cells can perform. However, no matter what their shape, size, or species, cells are also amazingly similar. They are all self-contained units surrounded by a membrane that separates them from their environment. They are all composed of the same types of molecules.

Cells are the fundamental units of life. They are self-reproducing living entities that are bounded by a selectively permeable plasma membrane. The two types of cells in existence (prokaryotic and eukaryotic) are believed to have evolved from primordial cells that existed over three billion years ago. Evidence of this relationship lies in the many "family resemblances" among modern cells. These include a similar chemical composition and the use of DNA as genetic material.

Recall that one of the most obvious differences between prokaryotic and eukaryotic cells is that of structural complexity. Presumably because of their small sizes and relatively simple structures, prokaryotes multiply quickly. Rapid reproduction coupled with a diverse set of genetic mechanisms (Chapter 17) has contributed to the capacity of various prokaryotic species to occupy virtually every ecological niche in the biosphere. In fact, prokaryotes have survived and flourished for eons. In contrast, the most notable feature of eukaryotes is their extraordinarily complex internal structure. Because eukaryotes carry out their various metabolic functions in a variety of membrane-bound organelles, they are capable of a more sophisticated intracellular metabolism. The diverse metabolic regulatory mechanisms made possible by this complexity promote two important lifestyle features required by multicellular organisms: cell specialization and intercellular cooperation. Consequently, it is not surprising that the majority of eukaryotes are multicellular organisms composed of numerous types of specialized cells.

The objective of this chapter is to provide an overview of cell structure. This is a valuable exercise because biochemical reactions do not occur in isolation. It is becoming increasingly obvious that our understanding of living processes is incomplete without some knowledge of their cellular context. After a brief discussion of prokaryotic structure, the eukaryotic organelles will be described in relationship to their roles in living cells.

2.1 STRUCTURE OF PROKARYOTIC CELLS

Bacteria, the most thoroughly studied prokaryotes, have three shapes: rodlike (bacilli), spheroidal (cocci), and helically coiled (spirilla) (Figure 2.1). A rigid *cell wall* maintains the organism's shape and protects it from mechanical injury. The strength of this structure is due in large part to the presence of complex peptide- and carbohydrate-containing polymers called peptidoglycans. In addition to the peptidoglycan layer, some bacteria possess an outer lipid bilayer with embedded proteins and attached polysaccharides. Cells that lack this outer layer have a thick peptidoglycan layer and can retain a specific dye. These cells are designated Gram-positive. Cells that cannot retain the dye, referred to as Gram-negative, possess a thin peptidoglycan layer and an outer lipid bilayer. The Gram stain is an important tool used to identify bacteria.

Directly underneath the cell wall is the **plasma membrane,** a dynamic structure that controls the flow of substances into and out of the cell. Numerous crucial reactions are catalyzed here. In addition, special receptor molecules facing the outer surface allow the organism to respond to the presence of food molecules and other chemicals in the environment. Although prokaryotes lack the complex membranous internal structures of eukaryotes, an invagination of the plasma membrane, known as the *mesosome,* is found in many species (Figure 2.2). The functions of mesosomes remain unresolved.

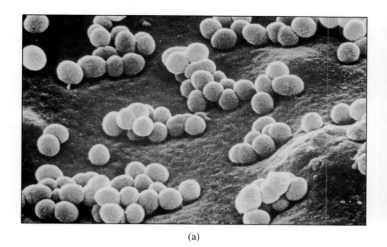

(a)

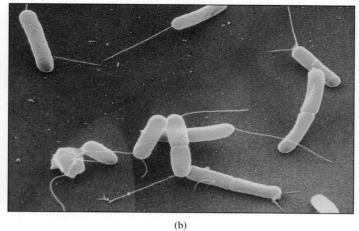

(b)

FIGURE 2.1

Bacteria.

The three most common bacterial shapes are (a) cocci (sphere-like), (b) bacilli (rod-shaped), and (c) spirilla (spiral).

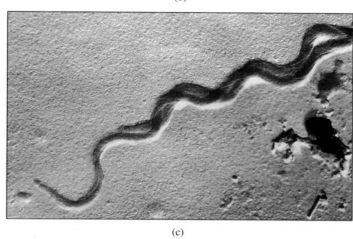

(c)

FIGURE 2.2

A Dividing Bacterial Cell.

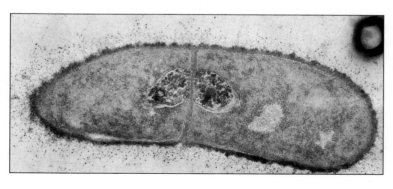

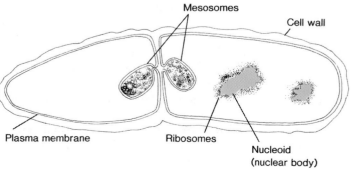

Mesosomes

Cell wall

Plasma membrane

Ribosomes

Nucleoid (nuclear body)

FIGURE 2.3

Bacterial Pili.

In this electron micrograph, a sex pilus connects two conjugating *E. coli* cells. Note the numerous smaller pili covering the surface of one of the cells.

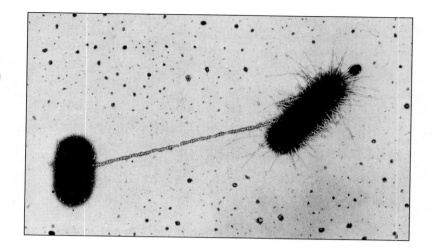

The prokaryotic cell has, instead of a nucleus, a circular DNA molecule called a **chromosome** that is located in an irregularly shaped region called the nucleoid. Many bacteria contain additional small circular DNA molecules called **plasmids** that exist separately from the cell's chromosome. Although they are not required for growth or reproduction, plasmids may provide the cell with an advantage over cells that lack plasmids. For example, segments of DNA coding for antibiotic resistance are often found on plasmids. In the presence of an antibiotic, resistant cells synthesize a protein that inactivates the antibiotic before it can damage the cell. Such cells continue to grow and reproduce, while susceptible cells die.

Prokaryotic cytoplasm appears to be relatively unstructured. Its grainy appearance is due to the large number of ribosomes that it contains. Other factors associated with protein synthesis, such as various types of RNA, are typically present in abundance, as are a variety of enzymes. Granules of organic material such as glycogen (an important energy storage molecule) and inorganic substances such as polyphosphate (a reservoir for phosphate, an essential component of DNA and RNA) may also be a prominent feature of cytoplasm.

Many bacterial cells have external appendages. *Pili* (singular: pilum) are fine, hairlike structures that may allow cells to attach to food sources and host tissues. *Sex pili* are used by some bacteria to transfer genetic information from donor cells to recipients, a process known as *conjugation* (Figure 2.3). Another type of appendage, known as a *flagellum* (plural: flagella), is a flexible corkscrew-shaped protein filament that is used for locomotion. Cells are pushed forward when flagella rotate in a counterclockwise direction, while clockwise rotation results in a stop and tumble motion, allowing the cell to reorient for a subsequent forward run.

KEY *Concepts*

Prokaryotic cells are small and structurally simple. They are bounded by a cell wall and a plasma membrane. They lack a nucleus and other organelles. Their DNA molecules, which are circular, are located in an irregularly shaped region called the nucleoid. Ribosomes are present in an otherwise featureless cytoplasm.

QUESTION 2.1

The most obvious difference between prokaryotic and eukaryotic cells is their sizes. A typical rod-shaped bacterial cell has a diameter of 1 μm and a length of 2 μm. The spherical human hepatocyte (liver cell), a widely studied eukaryotic cell, has a diameter of about 20 μm. Calculate the volume of each cell type. To appreciate the magnitude of the size differences between the two cell types, estimate how many bacterial cells would fit inside the liver cell. (*Hint:* Use the expression $V = \pi r^2 h$ for the volume of a cylinder and $V = 4\pi r^3/3$ for the volume of a sphere.)

2.2 STRUCTURE OF EUKARYOTIC CELLS

The generalized structures of animal and plant cells are illustrated in Figures 2.4 and 2.5. Each cellular component is briefly described in the sections that follow.

Plasma Membrane

The plasma membrane (also known as the cell membrane) separates the eukaryotic cell from its external environment. Despite its deceptively small size (4–5 nm thick), the plasma membrane performs several vital functions for the cell. One is transport,

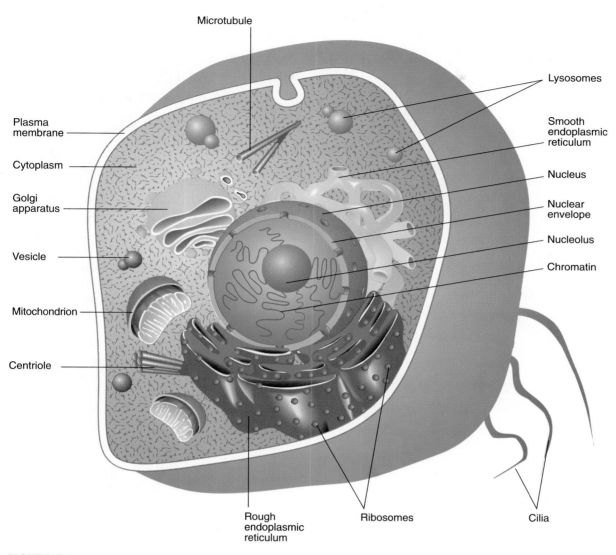

Microtubule

Plasma membrane

Cytoplasm

Golgi apparatus

Vesicle

Mitochondrion

Centriole

Lysosomes

Smooth endoplasmic reticulum

Nucleus

Nuclear envelope

Nucleolus

Chromatin

Rough endoplasmic reticulum

Ribosomes

Cilia

FIGURE 2.4

Animal Cell Structure.

Nucleus

which requires numerous carrier and channel proteins that are embedded in the membrane. Plasma membrane provides some mechanical strength and shape to the cell. It also plays a role in communication between cells, responsiveness to extracellular signals, and specialized functions such as muscle contraction and nerve impulse conduction.

The plasma membrane, like all cell membranes, is composed of lipid and protein molecules (Chapter 9). The basic structure of biological membranes is provided by lipids (Figure 2.6). Roles that plasma membrane proteins perform include transport of ions and nutrients such as glucose, an important energy source. A variety of enzymes are also found in membrane. Other membrane proteins act as **receptors,** structures that bind specific molecules (e.g., hormones) on the external surface of the cell. This type of interaction initiates a programmed response by the cell. For example, binding of the hormone insulin to insulin receptors results in a series of changes in the cell's activity. The most obvious of these is the transport of glucose into the cell.

The importance of the **nucleus** in regulating cell function has been long appreciated. However, its mechanism of control was not understood until the significance of its major component, DNA, was discovered (Chapter 16). The nucleus is now known to

FIGURE 2.5

Plant Cell Structure.

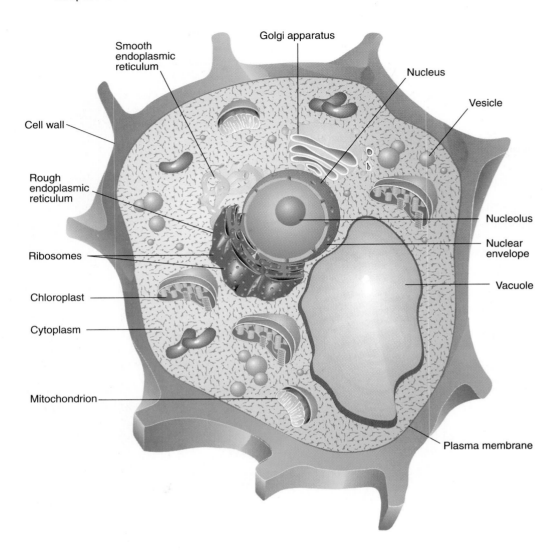

Smooth endoplasmic reticulum

Golgi apparatus

Nucleus

Vesicle

Cell wall

Rough endoplasmic reticulum

Ribosomes

Chloroplast

Cytoplasm

Mitochondrion

Nucleolus

Nuclear envelope

Vacuole

Plasma membrane

FIGURE 2.6

Membrane Structure.

Biological membranes are bilayers of phospholipid molecules in which numerous proteins are suspended. Some proteins extend completely across the membrane.

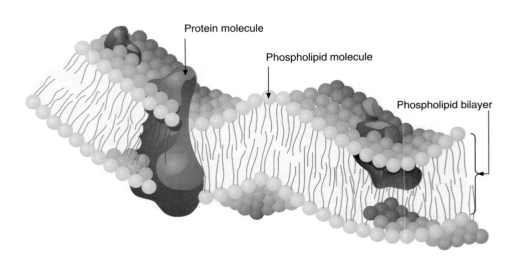

Protein molecule

Phospholipid molecule

Phospholipid bilayer

FIGURE 2.7

The Nucleus.

The nucleus is enclosed by a nuclear envelope, which consists of inner and outer membranes. A large number of nuclear pores perforate the surface of the envelope. The nucleolus, which is not surrounded by a membrane, is essentially a cluster of looped chromosomal segments.

KEY_Concepts_

The nucleus contains the cell's genetic information and the machinery for converting that information into protein molecules. The nucleolus plays an important role in the synthesis of ribosomal RNA. RER is primarily involved in protein synthesis. SER lacks attached ribosomes and is involved in lipid synthesis and biotransformation.

Endoplasmic Reticulum

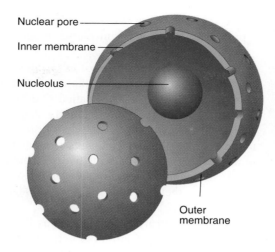

FIGURE 2.7

The Nucleus.

The nucleus is enclosed by a nuclear envelope, which consists of inner and outer membranes. A large number of nuclear pores perforate the surface of the envelope. The nucleolus, which is not surrounded by a membrane, is essentially a cluster of looped chromosomal segments.

perform two critical functions for the cell. First, it contains the cell's blueprints, the cell's hereditary information. Second, the nucleus exerts a profound influence over all cellular metabolic activities. This influence, which is exerted by directing the synthesis of protein cell components, is in turn affected by the passage of molecules back and forth between the cytoplasm and the nucleus.

The nucleus (Figure 2.7) consists of *nucleoplasm* that is bounded by the **nuclear envelope.** A prominent feature of nucleoplasm is a network of **chromatin fibers** composed of DNA and DNA packaging proteins known as histones. The nuclear envelope consists of two chemically distinct membranes perforated at regular intervals by **nuclear pores.** Large molecules such as proteins and RNA pass through the pores. Smaller molecules are often transported through the membrane by specific carriers.

When nuclei are stained with certain dyes, one or more spherical structures called **nucleoli** (singular: nucleolus) become visible. The nucleolus plays a major role in the synthesis of ribosomal RNA (discussed below).

The **endoplasmic reticulum (ER)** is a system of interconnected membranous tubules, vesicles, and large flattened sacs. A hint of its importance in cell function is that it often constitutes more than half of a cell's total membrane. The repeatedly folded, continuous sheets of ER membrane enclose an internal space called the ER *lumen.* This compartment, which is often referred to as the *cisternal space,* is entirely separated from the cytoplasm by ER membrane.

There are two forms of ER. The **rough ER** (RER), which is primarily involved in the synthesis of membrane proteins and protein for export from the cell, is so named because of the numerous ribosomes that stud its cytoplasmic surface (Figure 2.8). The second form lacks attached ribosomes and is called **smooth ER** (SER) (Figure 2.9). Although the SER membranes are continuous with those of RER, their physical appearances may be significantly different. In hepatocytes (the predominant cell type in liver), for example, SER consists of a tubular network that penetrates large regions

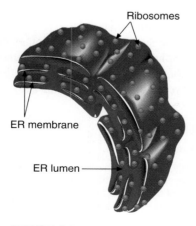

FIGURE 2.8

Rough Endoplasmic Reticulum.

FIGURE 2.9

Smooth Endoplasmic Reticulum.

of the cytoplasm. Functions of SER include lipid synthesis and **biotransformation,** a process in which water-insoluble organic molecules are prepared for excretion.

Ribosomes

The cytoplasmic **ribosomes** of eukaryotes are relatively small organelles (20 nm in diameter), whose function is the biosynthesis of proteins. Composed of a variety of proteins and a type of RNA called ribosomal RNA (rRNA), ribosomes are complex structures containing two irregularly shaped subunits of unequal size (Figure 2.10). They come together to form whole ribosomes when protein synthesis is initiated; when not in use, the ribosomal subunits separate. The number and distribution of ribosomes in any cell depend on the relative metabolic activity and the proteins being synthesized. Although eukaryotic ribosomes are larger and more complex than those of prokaryotes, they are similar in overall shape and function.

Golgi Apparatus

The **Golgi apparatus** (also known as the **Golgi complex**) is named for the Italian cell biologist Count Camillo Golgi, who first described it in 1898. Formed from relatively large, flattened, saclike membranous vesicles that resemble a stack of plates, the Golgi apparatus (called **dictyosomes** in plants) is involved in the packaging and distribution of cell products to internal and external compartments (Figure 2.11).

The Golgi apparatus has two faces. The plate (or *cisterna*) positioned closest to the ER is on the forming (*cis*) face, while the one on the maturing (*trans*) face is typically close to the portion of the cell's plasma membrane that is engaged in secretion. Small membranous vesicles containing newly synthesized protein and lipid bud off from the ER and fuse with the *cis* Golgi membrane. These molecules are transported from one Golgi sac to the next by vesicles, where they are further processed by enzymes. Once the products reach the *trans* face, they are then targeted to other parts of the cell. Secretory products, such as digestive enzymes or hormones, are concentrated within *secretory vesicles* (also known as *secretory granules*) that bud off from the *trans* face. Secretory granules remain in storage in the cytoplasm until the cell is stimulated to secrete them. The secretory process, referred to as **exocytosis,** consists of the fusion of the membrane-bound granules with the plasma membrane (Figure 2.12). The contents of the granules are then released into the extracellular space. In plants, the functions of the Golgi apparatus include transport of substances into the cell wall and expansion of the plasma membrane during cell growth.

KEY *Concepts*

Formed from relatively large, flattened, saclike membranous vesicles, the Golgi apparatus is involved in packaging and secretion of cell products.

Lysosomes

Lysosomes function in intracellular and extracellular digestion. Both the ER and Golgi apparatus are involved in the formation of lysosomes, which are small, membranous, saclike organelles. Lysosomes are capable of degrading most biomolecules. They participate in the life of a cell in three fundamental ways: (1) digestion of food molecules or other substances taken into the cell by **endocytosis** (a process illustrated in Figure 2.13), (2) digestion of worn out or unnecessary cell components, and (3) breakdown of extracellular material.

Although the appearance of lysosomes differs from one cell type to another, they are typically spherical with an average diameter of 500 nm. Bounded by a single membrane, lysosomes contain granules that are aggregates of digestive enzymes. These proteins are often referred to as *acid hydrolases* because they require an acidic envi-

FIGURE 2.10

The Eukaryotic Ribosome.

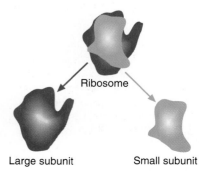

Ribosome

Large subunit   Small subunit

FIGURE 2.11

The Golgi Apparatus.

The Golgi apparatus is usually located near the nucleus. Note the curved structure of the cisternae. There is a continuous flow of substances through the Golgi apparatus. The Golgi apparatus is responsible for sorting and packaging several types of protein, small molecules, and new membrane components.

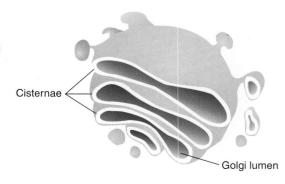

Cisternae

Golgi lumen

FIGURE 2.12

Exocytosis.

Exocytosis is a process whereby membrane-bound vesicles fuse with the cell's plasma membrane, releasing their contents outside the cell. Proteins to be secreted are produced in the ER and transported to the Golgi apparatus, where they are processed further and packaged into vesicles. These vesicles then migrate to the plasma membrane and merge with it, often in response to an extracellular signal.

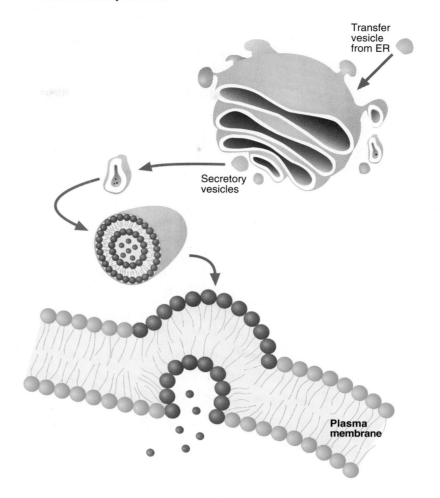

Transfer vesicle from ER

Secretory vesicles

Plasma membrane

KEY *Concepts*

The function of lysosomes is intracellular and extracellular digestion. These spherical, membranous organelles contain a group of enzymes called acid hydrolases, which degrade most biomolecules.

ronment to function properly and they use water molecules to split large molecules into fragments. (Because the plant vacuole contains acid hydrolases, it is considered to function to a certain extent like a lysosome. Plant vacuoles are membranous sacs that store a wide variety of substances.)

Two properties of lysosomal membrane are especially interesting. First, certain membrane proteins acting as proton pumps create the required acidic environment within the lysosomes. Second, under certain circumstances lysosomal enzymes leak into other parts of the cell. Such an occurrence would ordinarily have devastating consequences, because all of the cell contents would eventually be degraded. In several pathological conditions such as rheumatoid arthritis and gout, there is a release of lysosomal enzymes by macrophages (a white blood cell that plays an important role in inflammatory responses). The release of these enzymes into the affected tissue contributes to further inflammation and tissue destruction.

Although lysosomal function has common features in various tissues, its specific role differs. For example, macrophage lysosomes are prominent components in the normal immunological process by which damaged cells and foreign organisms are degraded. Lysosomal enzymes secreted from osteoclast cells are largely responsible for the resorption phase of bone remodeling.

QUESTION 2.2

In many genetic disorders, a lysosomal enzyme required to degrade a specific molecule is missing or defective. One example of these maladies, often referred to as lysosomal storage diseases, is Tay-Sachs disease. Afflicted individuals inherit a defective gene from each parent that codes for an enzyme that degrades a complex lipid molecule. Symptoms include severe mental retardation and death before the age of five years. What is happening in the patient's cells? (*Hint:* Synthesis of the lipid molecule continues at a normal rate.)

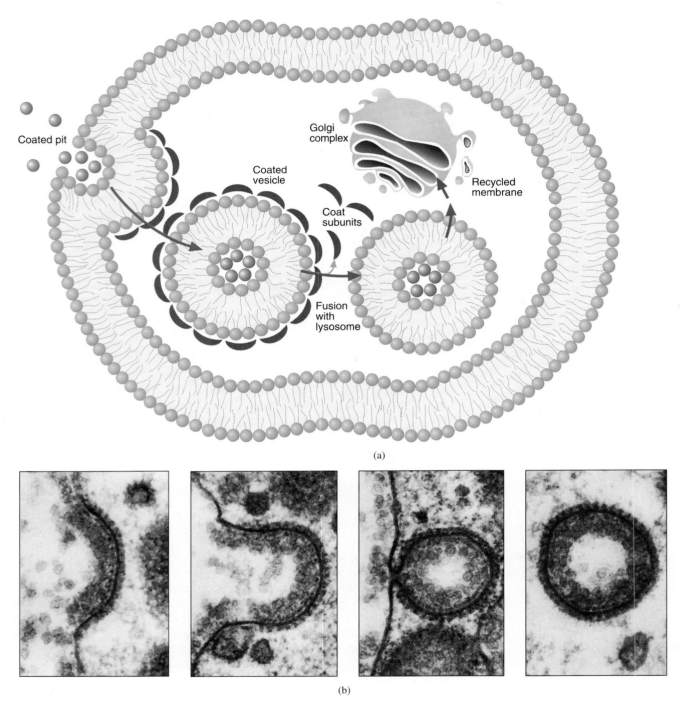

(a)

(b)

FIGURE 2.13

Endocytosis.

(a) Extracellular substances may enter the cell during endocytosis, a process in which specialized regions of plasma membrane called coated pits progressively invaginate to form closed vesicles. After each vesicle fuses with a lysosome, the contents are digested by lysosomal enzymes. The vesicle membrane is then recycled. It is eventually returned to the plasma membrane during exocytosis. (b) Electron micrographs illustrating the initial events in endocytosis.

Peroxisomes

Peroxisomes are small spherical membranous organelles that contain oxidative enzymes (proteins that catalyze the transfer of electrons). These organelles, whose enzymatic composition varies among species and cells within individual organisms, are most noted for their involvement in the generation and breakdown of toxic molecules known as peroxides. For example, hydrogen peroxide (H_2O_2) is generated when mo-

FIGURE 2.14

Peroxisome in a Tobacco Leaf Cell.

The granular substance surrounding the crystal-like core is called the matrix.

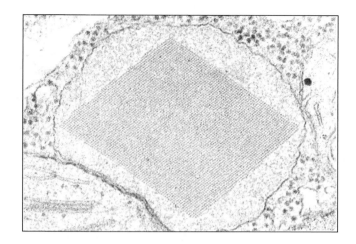

lecular oxygen (O_2) is used to remove hydrogen atoms from specific organic molecules. (Once formed, H_2O_2 must be immediately destroyed before it damages the cell.) This process is especially important in liver and kidney cells, which have an important detoxifying role in animal bodies. For example, peroxisomes are involved in the oxidation of ingested ethanol.

Two types of peroxisomes have been identified in plants. One, found in leaves, is responsible for an oxygen-consuming process known as *photorespiration* in which carbon dioxide (CO_2) is produced. The other type of peroxisome (often called **glyoxysomes**) is found in germinating seed. In these structures, lipid molecules are converted into carbohydrate, which provides energy for growth and development (Figure 2.14).

KEY *Concepts*

Peroxisomes contain oxidative enzymes. They are most noted for their involvement in the generation and breakdown of toxic molecules known as peroxides.

Mitochondria

All eukaryotic cells utilize O_2 to generate the energy for cellular activities. This process, known as **aerobic respiration,** takes place in **mitochondria** (singular: mitochondrion). Mitochondria are often depicted as sausage-shaped structures (Figure 2.15), but their appearance varies considerably among different species and cell types. Their configuration also changes with the physiological status of the cell. For exam-

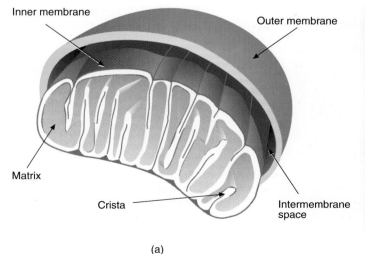

Inner membrane
Outer membrane
Matrix
Crista
Intermembrane space

(a)

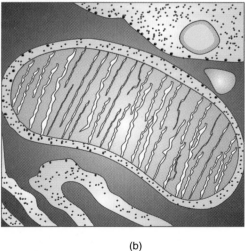

(b)

FIGURE 2.15

The Mitochondrion.

(a) Membranes and crista, (b) cross section.

ple, the internal appearance of liver mitochondria has been observed to change dramatically during active respiration (Figure 2.16). Additionally, the fragmentation or inordinate swelling of mitochondria is a very sensitive indicator of cell injury.

Each mitochondrion is bounded by two membranes. The smooth **outer membrane** is relatively porous, since it is permeable to most molecules with masses less than 10,000 daltons. The **inner membrane,** which is impermeable to ions and a variety of organic molecules, projects inward into folds that are called cristae (singular: crista). Embedded in this membrane are structures called *respiratory assemblies* (described in Chapter 11), which are responsible for the synthesis of adenosine triphosphate (ATP), the cell's energy storage molecule. Also present are a series of proteins that are responsible for the transport of specific molecules and ions.

Together, both membranes create two separate compartments: (1) the *intermembrane space* and (2) the *matrix.* The intermembrane space contains several enzymes involved in nucleotide metabolism, while the gellike matrix consists of high concentrations of enzymes and ions, as well as a myriad of small organic molecules. The matrix also contains several circular DNA molecules and all components for protein synthesis. (See Special Interest Box 2.1.)

Plastids

Plastids, structures that are found only in plants, algae, and some protists, are bounded by a double membrane. Although the inner membrane is not folded as in mitochondria, another separate intricately arranged internal membrane is often present. In plants, all plastids develop from *proplastids,* which are small, nearly colorless structures found in the meristem (a special region in plants made up of undifferentiated cells from which new tissues arise). Proplastids develop according to the requirements of each differentiated cell. Mature plastids are of two types: (1) *leucoplasts,* which store substances such as starch or proteins in storage organs (e.g., roots or tubers), and (2) **chromoplasts,** which accumulate the pigments that are responsible for the colors of leaves, flower petals, and fruits.

Chloroplasts are a type of chromoplast that are specialized for the conversion of light energy into chemical energy. (In this process, called **photosynthesis,** which will be described in Chapter 12, light energy is used to drive the synthesis of carbohydrate from CO_2.) The structure of chloroplasts (Figure 2.17) is similar in several respects to that of mitochondria. For example, the outer membrane is highly permeable, while the relatively impermeable internal membrane contains special carrier proteins that control molecular traffic into and out of the organelle.

An intricately folded internal membrane system, called the **thylakoid membrane,** is responsible for the metabolic function of chloroplasts. For example, chlorophyll molecules, which capture light energy during photosynthesis, are bound to thylakoid

FIGURE 2.16

Rat Liver Mitochondria in the (a) Low-Energy (Orthodox), and (b) High-Energy (Condensed) Conformations.

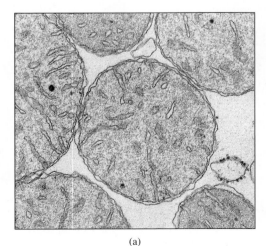

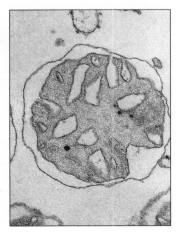

(a)   (b)

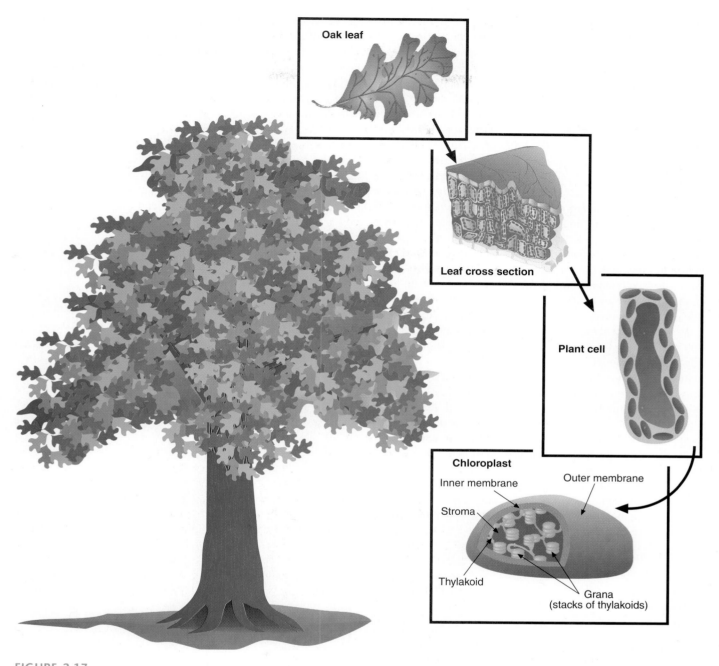

FIGURE 2.17

The Chloroplast.

Chloroplasts are one type of organelle found in multicellular plants.

membrane proteins. Certain portions of thylakoid membrane form tightly stacked structures called **grana** (singular: granum), while the entire membrane encloses a compartment known as the *thylakoid lumen* (or channel). Surrounding the thylakoid membrane is a dense enzyme-filled substance, analogous to the mitochondrial matrix, called the **stroma.** In addition to enzymes, the stroma contains DNA, RNA, and ribosomes. Membrane segments that connect adjacent grana are referred to as stroma lamellae (singular: lamella).

ENDOSYMBIOSIS

Symbiosis, defined as the living together of two dissimilar organisms in an intimate relationship, is a common biological phenomenon. Such associations vary from the parasitic, in which one organism derives benefit at the other's expense, to the mutualistic, in which both organisms benefit. One example of the former involves humans and trypanosomes, the protozoa that cause African sleeping sickness. The relationship between humans and the intestinal bacterium *Lactobacillus acidophilus* is an example of the latter. A variety of *Lactobacillus* species, obtained from the consumption of fermented milk products, exchange protection, warmth, and nutrition for several beneficial effects to humans. These include protection against pathogenic microorganisms such as *Clostridium difficile,* the lowering of serum cholesterol, and perhaps some anticancer effects.

In her book *The Origin of Eukaryotic Cells,* published in 1970, Lynn Margulis uses the concept of symbiosis to explain a major unresolved biological problem. She proposed that the mitochondria and chloroplasts, as well as cilia and flagella of eukaryotic cells, evolved from prokaryotic cells. According to the *endosymbiotic hypothesis,* eukaryotic cells began as large anaerobic organisms. (The term "anaerobic" indicates that oxygen is not used to generate energy.) Mitochondria arose when small aerobic (oxygen-utilizing) bacteria were ingested by the larger cells. In exchange for benefits such as protection and a constant nutrient supply, the smaller cell provided its host with energy generated by a process known as aerobic respiration. As time passed, the bacteria lost their independence because of the transfer of several genes (genetic coding units) to the host cell nucleus. Similarly, chloroplasts are believed to descend from cells that were similar to modern cyanobacteria, while cilia and flagella derive from ancient spiral prokaryotes.

The endosymbiotic hypothesis is supported by a considerable amount of indirect evidence:

1. Mitochondria and chloroplasts are similar in size to many modern prokaryotes.
2. These two organelles reproduce by binary fission, as do bacteria and the archaea (Figure 2A).
3. The genetic information (DNA) and the protein-synthesizing capability of mitochondria and chloroplasts are similar to those of prokaryotes. For example, both mitochondrial and chloroplast DNA are circular and "naked" (i.e., not complexed with proteins as nuclear DNA is). (There is insufficient genetic information on these chromosomes to account for all organelle components. However, the nuclear genes that are responsible for synthesis of mitochondrial components resemble prokaryotic genes.)
4. The ribosomes of chloroplasts and mitochondria are similar in size and function to those of prokaryotes. For example, drugs such as the antibiotic chloramphenicol, which kill certain bacteria by inhibiting the protein-synthesizing activities of ribosomes, also inhibit chloroplast and mitochondrial ribosomal function.
5. Traces of the other nucleic acid, RNA, have been found in the basal bodies of cilia and flagella. This evidence is considered by some researchers to support the idea that such eukaryotic structures arose by a symbiotic union.
6. Many modern organisms contain intracellular symbiotic bacteria, cyanobacteria, or algae, that is, such associations are not difficult to establish. For example, a primitive freshwater animal called *Chlorohydra* owes its green color to endosymbiont algae. The hydra's nutrition is supplemented by the photosynthetic activity of the algae.

The endosymbiotic hypothesis is illustrated in Figure 2B.

FIGURE 2A

Replication of a Mitochondrion by Binary Fission.

QUESTION

2.3   *Cyanophora paradoxa* is a eukaryotic organism that incorporates cyanobacteria, which are aerobic photosynthesizing prokaryotic organisms, into its cells. Describe the benefits that both species derive from this relationship.

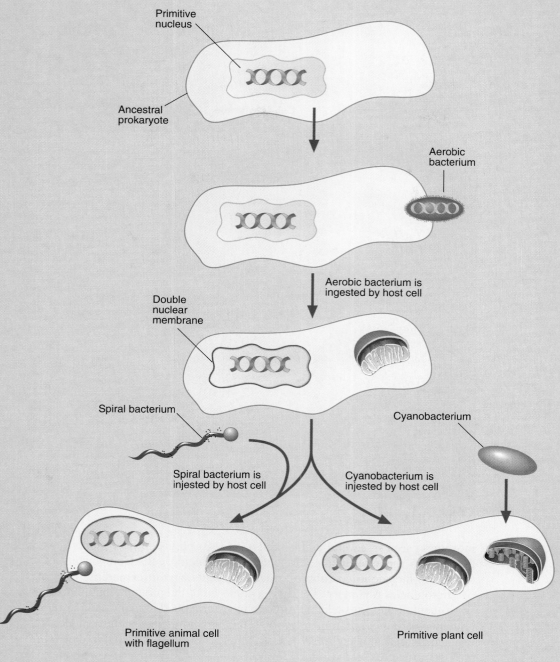

Primitive
nucleus

Ancestral
prokaryote

Aerobic
bacterium

Aerobic bacterium is
ingested by host cell

Double
nuclear
membrane

Spiral bacterium

Cyanobacterium

Spiral bacterium is
injested by host cell

Cyanobacterium is
injested by host cell

Primitive animal cell
with flagellum

Primitive plant cell

FIGURE 2B

The Endosymbiotic Hypothesis.

The first eukaryotes are believed to have evolved at least 1.5 billion years ago. The transition from ancient prokaryotic to eukaryotic cell structure is arguably the most important one in evolution, except for the origin of life itself. The endosymbiotic hypothesis is an interesting and compelling view of this transition.

Cytoskeleton

Cytoplasm was once believed to be a structureless solution in which the nucleus was suspended. Experiments have revealed not only the extensive membrane system and the membranous organelles described above, but also an intricate supportive network of proteinaceous fibers and filaments called the **cytoskeleton** (Figure 2.18). The highly developed framework of the cytoskeleton contributes to the life of cells in several ways: (1) maintenance of overall cell shape, (2) facilitation of coherent cellular movements (e.g., the cytoplasmic streaming observed in plant cells and the ameboid

FIGURE 2.18

Diagrammatic View of the Cytoskeleton.

For the sake of clarity only microtubule components of the cytoskeleton are illustrated.

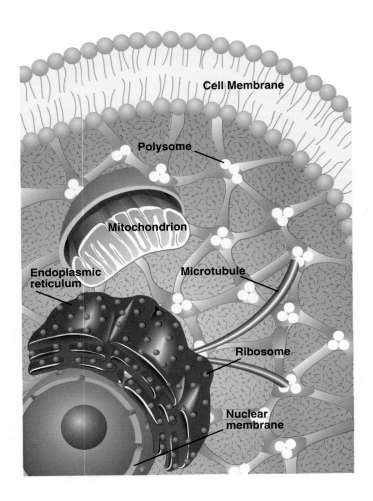

movement seen in several animal cells), and (3) provision of a supporting structure that guides the movement of organelles within the cell. Components of the cytoskeleton include **microtubules, microfilaments,** and **intermediate fibers.**

Microtubules (diameter = 25 nm), composed of the protein tubulin, are the largest constituent of the cytoskeleton. Although they are found in many cellular regions, microtubules are most prominent in long, thin structures that require support (e.g., the extended axons and dendrites of nerve cells). They are also found in the *mitotic spindle* (the structure formed in dividing cells that is responsible for the equal dispersal of chromosomes into daughter cells) and the slender, hairlike organelles of locomotion known as cilia and flagella (Figure 2.19).

FIGURE 2.19

Cilia and Flagella.

The microtubules are arranged in the classic 9 + 2 pattern. Two central microtubules are surrounded by an outer ring of nine pairs of microtubules.

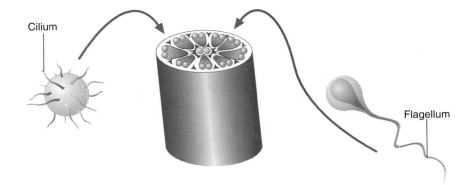

KEY Concepts

The cytoskeleton, a highly structured network of proteinaceous filaments, is responsible for maintenance of overall cell shape and facilitation of cell movements. Components of the cytoskeleton include microtubules, microfilaments, and intermediate fibers.

Microfilaments, which are small (5–7 nm in diameter) fibers composed of the protein *actin*, perform their functions by interacting with certain cross-linking proteins. Important roles of microfilaments include involvement in cytoplasmic streaming (a process that is most easily observed in plant cells in which cytoplasmic currents rapidly displace organelles such as chloroplasts) and ameboid movement (a type of locomotion created by the formation of temporary cytoplasmic protrusions).

Intermediate fibers (10 nm in diameter) are proteinaceous structures with a heterogeneous composition. Despite this variation, their structural organization is similar in many cell types. Involved primarily in the maintenance of cell shape, intermediate fibers are especially prominent in cells that are subjected to mechanical stress. For example, one type, known as keratin filaments, is found in the outermost cell layers of the skin.

QUESTION 2.4

Without referring to this chapter, can you draw diagrams of a prokaryotic and a eukaryotic cell and label each structural component? Describe the function of each structure in one sentence.

CELL TECHNOLOGY

During the past 50 years, our understanding of the functioning of living organisms has undergone a revolution. Much of our current knowledge of biochemical processes is due directly to technological innovations. For example, the development of the electron microscope (EM) as a biological instrument by Keith Porter and his colleagues in the 1940s was responsible for the resolution of the fine structure of organelles. Structures such as the mitochondria and lysosomes were not discovered until this time; under the light microscope they appeared as mere granules. The electron microscope also revealed that the membranes of the Golgi apparatus are often continuous with those of the ER. This discovery is especially significant because of the role both organelles play in protein synthesis. Refer to Appendix B, Supplement 2 for a discussion of the most important cellular techniques used in biochemical research. This information is useful in answering questions in subsequent chapters.

KEY WORDS

aerobic respiration, *29*
biotransformation, *26*
chloroplast, *30*
chromatin fiber, *25*
chromoplast, *30*
chromosome, *22*
cytoskeleton, *33*
dictyosome, *26*
endocytosis, *26*
endoplasmic reticulum (ER), *25*
exocytosis, *26*

glyoxysome, *29*
Golgi apparatus (Golgi complex), *26*
granum, *31*
inner membrane, *30*
intermediate fiber, *34*
lysosome, *26*
microfilament, *34*
microsome (Appendix B, Supplement 2)
microtubule, *34*

mitochondrion, *29*
nuclear envelope, *25*
nuclear pore, *25*
nucleolus, *25*
nucleus, *23*
outer membrane, *30*
peroxisome, *28*
photosynthesis, *30*
plasma membrane, *20*
plasmid, *22*
plastid, *30*

receptor, *23*
ribosome, *26*
rough ER (RER), *25*
smooth ER (SER), *25*
stroma, *31*
symbiosis, *32*
thylakoid membrane, *30*

SUMMARY

Cells are the structural units of all living organisms. There are two types of cells found in all currently existing organisms: prokaryotic and eukaryotic. Prokaryotes are simpler in structure than eukaryotes. They also have a vast biochemical diversity across species lines, since almost any organic molecule can be used as a food source by some species of prokaryote. Unlike the prokaryotes, the eukaryotes carry out their metabolic functions in membrane-bound compartments called organelles.

In most prokaryotes a rigid cell wall maintains the organism's shape and protects it from mechanical injury.

Although the prokaryotic cell lacks a nucleus, a circular DNA molecule called a chromosome is located in an irregularly shaped region called the nucleoid. Many bacteria contain additional small circular DNA molecules called plasmids. Plasmids may carry genes for special function proteins that provide protection, metabolic specialization, or reproductive advantages to the organism.

The plasma membrane of both prokaryotes and eukaryotes performs several vital functions. The most important of these is controlled molecular transport, which is facilitated by carrier and channel proteins.

The nucleus of eukaryotes contains DNA, the cell's genetic information. Ribosomal RNA is synthesized in the nucleolus.

The endoplasmic reticulum (ER) is a system of interconnected membranous tubules, vesicles, and large flattened sacs. There are two forms of ER. The rough ER, which is primarily involved in protein synthesis, is so named because of the numerous ribosomes that stud its cytoplasmic surface. The second form lacks attached ribosomes and is called smooth ER. Functions of the smooth ER include lipid synthesis and biotransformation.

The cytoplasmic ribosomes of eukaryotes are relatively small organelles that synthesize proteins. Ribosomes are complex structures composed of a variety of proteins and a type of RNA called ribosomal RNA.

Formed from relatively large, flattened, saclike membranous vesicles that resemble a stack of plates, the Golgi apparatus is involved in the packaging and secretion of cell products.

Peroxisomes are small spherical membranous organelles that contain a variety of oxidative enzymes. These organelles are most noted for their involvement in the generation and breakdown of peroxides.

Aerobic respiration, a process by which cells use O_2 to generate energy, takes place in mitochondria. Each mitochondrion is bounded by two membranes. The smooth outer membrane is permeable to most molecules with masses less than 10,000 daltons. The inner membrane, which is impermeable to ions and a variety of organic molecules, projects inward into folds that are called cristae. Embedded in this membrane are structures called respiratory assemblies that are responsible for the synthesis of ATP.

Plastids, structures that are found only in plants, algae, and some protists, are bounded by a double membrane. Another separate intricately arranged internal membrane is also often present. Chromoplasts accumulate the pigments that are responsible for the color of leaves, flower petals, and fruits. Chloroplasts are a type of chromoplast that are specialized to convert light energy into chemical energy.

The cytoskeleton, a supportive network of fibers and filaments, is involved in the maintenance of cell shape, facilitation of cellular movement, and the intracellular transport of organelles.

SUGGESTED READINGS

Alberts, B., Bray, D., Lewis, J., Raff, M., Roberts, K., and Watson, J.D., *Molecular Biology of the Cell,* Garland, New York, 1994.

de Duve, C., *A Guided Tour of the Living Cell,* Scientific American, New York, 1984.

de Duve, C., *Blueprint for a Cell,* Carolina Biological Supply Co., Burlington, N.C., 1991.

Goodsell, D.S., *The Machinery of Life,* Springer-Verlag, New York, 1993.

Margulis, L., *Symbiosis in Cell Evolution,* W.H. Freeman, New York, 1981.

QUESTIONS

1. Define the term *cell*.
2. What evidence is there that all cells have a common ancestor?
3. Draw a diagram of a bacterial cell. Label and explain the function of each of the following components:
   a. mesosome
   b. nucleoid
   c. plasmid
   d. cell wall
   e. pili
   f. flagella
4. The outer boundary of most eukaryotic cells is a cell membrane, while the outer boundary of a prokaryotic cell is a cell wall. How do these structures differ in function?
5. Indicate whether the following structures are present in prokaryotic or eukaryotic cells:
   a. nucleus
   b. plasma membrane
   c. endoplasmic reticulum
   d. mesosome
   e. mitochondria
   f. nucleolus
6. Briefly define the following terms:
   a. exocytosis
   b. biotransformation
   c. grana
   d. symbiosis

e. endosymbiosis
f. proplastids
g. thylakoid

7. How do lysosomes participate in the life of a cell?

8. Plastids, structures found only in _____, are of two types. These are _____, which are used to store starch and protein, and _____, which accumulate pigments.

9. List six pieces of evidence that support the endosymbiotic hypothesis.

10. What functions does the cytoskeleton perform in living cells?

11. Several pathogenic bacteria (e.g., *Bacillus anthracis,* the cause of anthrax) produce an outermost mucoid layer called a capsule. Capsules may be composed of polysaccharide or protein. What effect do you think this "coat" would have on a bacterium's interactions with an animal's immune system?

12. Many eukaryotic cells lack a cell wall. Suggest several reasons why this is an advantage.

13. What are the two essential functions of the nucleus?

14. Eukaryotic cells are more highly specialized than prokaryotic cells. Can you suggest some advantages and disadvantages of specialization?

15. Both peroxisomes and mitochondria consume molecular oxygen. How do their functions differ?

16. A particular organelle found in a eukaryote is thought to have arisen from a free-living organism. The finding of what type of molecule in the organelle would strongly support this hypothesis?

17. What roles do plasma membrane proteins play in cells?

18. Name the two forms of endoplasmic reticulum. What functions do they serve in the cell?

19. Describe the functions of the Golgi apparatus.

20. The endosymbiotic hypothesis proposes that mitochondria and chloroplasts are derived from aerobic bacteria. Is there any structural feature of these organelles that precludes their having been developed by eukaryotic cells?

21. In addition to providing support, the cytoskeleton also immobilizes enzymes and organelles in the cytoplasm. What advantage does this immobilization have over allowing the cell contents to freely diffuse in the cytoplasm?

WATER: The Medium of Life

Chapter Three

The Water Planet. *Unique among the planets in the solar system, the earth is an oceanic world. Water's unique properties make life on earth possible.*

OBJECTIVES

After you have studied this chapter, you should be able to answer these questions:

1. How is the molecular structure of water related to its physical and chemical behavior?

2. How does noncovalent bonding affect the chemical and biological properties of water?

3. What is pH and how does it affect living cells?

4. What is a buffer? What role do buffers play in living cells?

5. What are the colligative properties of aqueous solutions?

The most important factor in the evolution of life on earth is the abundant liquid water found on the planet's surface. The unique chemical and physical characteristics of water are so crucial for living systems that life undoubtedly could not have arisen in its absence. Among these characteristics are water's chemical stability and its remarkable solvent properties. Wherever water is found, there are living organisms. Although large bodies of water such as lakes and oceans support diverse and often abundant populations of organisms, certain specially adapted organisms are also found where water is present but in short supply. In deserts, for example, brief and infrequent rainstorms immediately trigger the blooming of certain plants. These plants then race through their entire life cycle before the rainwater has completely evaporated. Even when other conditions are unfavorable (e.g., extremely hot or cold environments), water makes life possible. For example, certain bacterial species thrive in hot springs. Similarly, algae grow on the melting edges of glaciers. Why is water so vital for life? Understanding water's critical role in living processes requires a review of its molecular structure and the unique properties that are the consequences of that structure.

Earth is unique among the planets in our solar system primarily because of its vast oceans of water. Formed over billions of years, water was produced during high-temperature interactions between atmospheric hydrocarbons and the silicate and iron oxides in the earth's mantle. Moisture reached the planet's surface as steam emitted during volcanic eruptions. Oceans formed as the steam condensed and fell back to earth as rain. This first rain may have lasted over 60,000 years.

Over millions of years, water has profoundly affected our planet. For example, it has sculpted the earth's surface. Whether falling as rain or flowing in rivers, water has eroded the hardest rocks and transformed the mountains and continents.

Life is now believed by most scientists to have arisen in the ancient seas. It is not an accident that life arose in association with water, since this substance has several unusual properties that suits it to be the medium of life. Among these are its thermal properties and unusual solvent characteristics. Water's properties are directly related to its molecular structure.

3.1 MOLECULAR STRUCTURE OF WATER

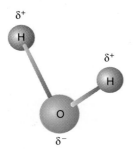

FIGURE 3.1

Charges on a Water Molecule.

The two hydrogen atoms in each molecule carry partial positive charges. The oxygen atom carries a partial negative charge.

The water molecule (H_2O) is composed of two atoms of hydrogen and one of oxygen. Each hydrogen atom is linked to the oxygen atom by a single covalent bond (Figure 3.1). Oxygen is more electronegative than hydrogen (i.e. oxygen has a greater capacity to attract electrons). The oxygen atom bears a slightly negative charge (δ^-) and each of the two hydrogen atoms bears a slightly positive charge (δ^+). Consequently, there is a separation of charge within the molecule. The electron distribution in oxygen-hydrogen bonds is described as **polar** or asymmetrical. If water molecules were linear, then the bond polarities would balance each other and water would be nonpolar. However, water molecules are bent; the bond angle is 104.5° (Figure 3.2). In contrast, carbon dioxide (O—C—O), which also possesses polar covalent bonds, is nonpolar because the molecule is linear.

Molecules such as water, in which charge is separated, are called **dipoles.** Such molecules have opposite charges on two ends. When molecular dipoles are subjected to an electric field, they orient themselves in the direction opposite to that of the field (Figure 3.3).

Because of the large difference in electronegativity of hydrogen and oxygen, the hydrogens of one water molecule are attracted to the unshared pairs of electrons of another water molecule. (Hydrogens attached to nitrogen and fluorine behave the same way.) This noncovalent bond is called a **hydrogen bond** (Figure 3.4). (It should be

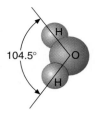

FIGURE 3.2

Space-Filling Model of a Water Molecule.

Because the water has a bent geometry, the distribution of charge within the molecule is asymmetric. Water is therefore polar.

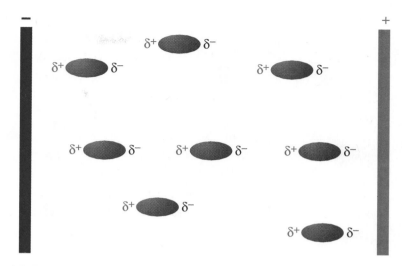

FIGURE 3.3

Molecular Dipoles in an Electric Field.

FIGURE 3.4

The Hydrogen Bond.

A hydrogen bond is a weak attraction between an electronegative atom in one molecule and a hydrogen atom in another molecule. The hydrogen bonds between water molecules are represented by short parallel lines.

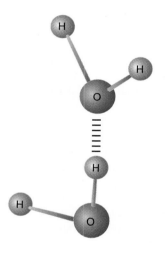

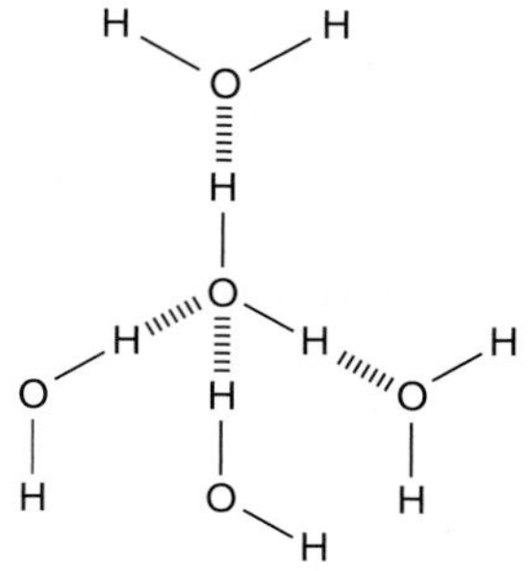

FIGURE 3.5

Hydrogen Bonding Between Water Molecules.

In water, each molecule can form hydrogen bonds with four other water molecules.

noted that the term hydrogen bond describes an electrostatic interaction rather than a covalent bond.) In addition to hydrogen bonds, three other types of noncovalent interactions affect the interactions of water with other molecules. These are **electrostatic interactions, van der Waals forces,** and **hydrophobic interactions.** Because biological reactions take place in a water medium, an understanding of noncovalent bonding is important.

3.2 NONCOVALENT BONDING

The physical and chemical properties of water make it remarkably suitable for its numerous functions in living organisms. Noncovalent interactions play a vital role in determining these properties. For example, hydrogen bonding is largely responsible for water's thermal properties. Noncovalent bonds also have a significant effect on the structure and (therefore the function) of biomolecules. Prominent examples include proteins (Chapter 5) and nucleic acids (Chapter 16).

Hydrogen Bonds

Covalent bonds between hydrogen and oxygen or nitrogen are sufficiently polar so that the hydrogen nucleus is weakly attracted to the lone pair electrons of an oxygen or nitrogen of a neighboring molecule (Figure 3.5). The resulting intermolecular "bond" acts as a bridge between water molecules. Although considerably weaker than ionic and covalent bonds, hydrogen bonds are stronger than most noncovalent bonds.

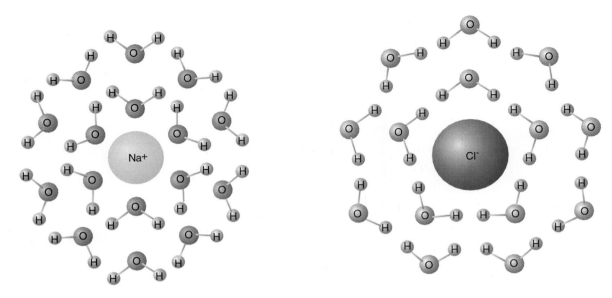

FIGURE 3.6

Solvation Spheres of Water Molecules Around Na$^+$ and Cl$^-$ Ions.

When an ionic compound is dissolved in water, its ions separate because the polar water molecules attract the ions more than the ions attract each other.

Electrostatic Interactions

Electrostatic interactions occur between oppositely charged atoms or groups. Like other noncovalent bonds, they are of great significance in determining the shape and function of biomolecules. For example, the attraction between the —COO$^-$ and —NH$_3^+$ groups is a factor in determining the three-dimensional structure of proteins. An important aspect of all electrostatic interactions in aqueous solution is the hydration of ions. Because water molecules are polar, they are attracted to charged ions. Shells of water molecules, referred to as **solvation spheres,** cluster around both positive and negative ions (Figure 3.6). As ions become hydrated, the attractive force between them is reduced, and the charged species dissolves in the water. Electrostatic interactions between charged groups or polar groups on biomolecules are most significant in water depleted regions since water competes with great success with electrostatic interactions between other groups. The *dielectric constant* of a solvent is a measure of its capacity to reduce the attractive force between ions. Water, sometimes called the *universal solvent* because of the large variety of ionic and polar substances it can dissolve, has a very large dielectric constant.

Van der Waals Forces

Van der Waals forces are relatively weak, transient electrostatic interactions. They occur between permanent and/or induced dipoles. They may be attractive or repulsive, depending on the distance between the atoms or groups involved. The attraction between molecules is greatest at a distance called the van der Waals radius. If molecules approach each other more closely, a repulsive force develops. The magnitude of van der Waals forces depend on how easily an atom is polarized. Electronegative atoms with unshared pairs of electrons are easily polarized.

There are three types of van der Waals forces:

1. **Dipole–dipole interactions.** These forces, which occur between molecules containing electronegative atoms, cause molecules to orient themselves so that the positive end of one molecule is directed toward the negative end of another (Figure 3.7a). Hydrogen bonds (described above) are an especially strong type of dipole–dipole interaction.

2. **Dipole–induced dipole interactions.** A permanent dipole induces a transient dipole in a nearby molecule by distorting its electron distribution (Figure 3.7b).

FIGURE 3.7

Van der Waals Forces.

There are three types of van der Waals forces: (a) dipole–dipole interactions, (b) dipole–induced dipole interactions, and (c) induced dipole–induced dipole interactions. The relative ease with which electrons respond to an electric field determines the magnitude of van der Waals forces. Dipole–dipole interactions are the strongest; induced dipole–induced dipole interactions are the weakest.

(a)   Dipole - dipole interactions

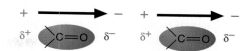

(b)   Dipole - induced dipole interactions

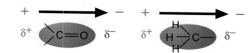

(c)   Induced dipole - induced dipole interactions

FIGURE 3.8

Hydrophobic Interactions Between Water and a Nonpolar Substance.

As soon as nonpolar substances (e.g., hydrocarbons) are mixed with water (a), they coalesce into droplets (b) because water-water interactions are stronger than water-hydrocarbon interactions.

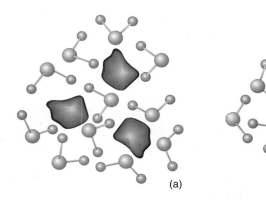

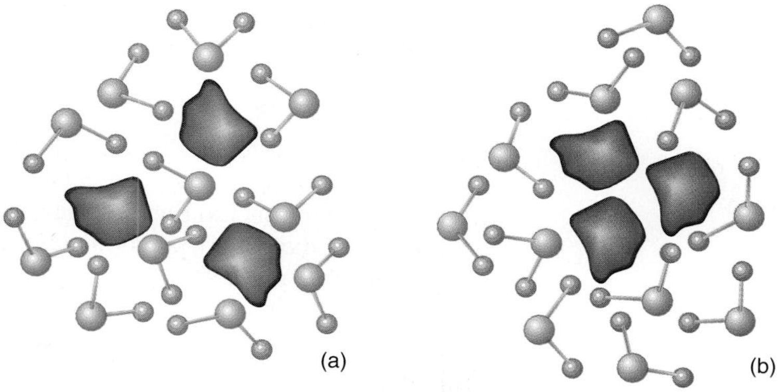

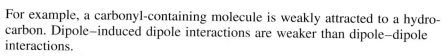

(a)   (b)

For example, a carbonyl-containing molecule is weakly attracted to a hydrocarbon. Dipole–induced dipole interactions are weaker than dipole–dipole interactions.

3. **Induced dipole–induced dipole interactions.** The motion of electrons in nearby nonpolar molecules results in transient charge imbalances in adjacent molecules (Figure 3.7c). A transient dipole in one molecule polarizes the electrons in a neighboring molecule. This attractive interaction, often called **London dispersion forces,** is extremely weak.

Hydrophobic Interactions

KEY *Concepts*

Noncovalent bonds (i.e., hydrogen bonds, electrostatic interactions, van der Waals forces, and hydrophobic interactions) play important roles in determining the physical and chemical properties of water. They also have a significant effect on the structure and function of biomolecules.

When mixed with water, small amounts of nonpolar substances coalesce into droplets. This process results from **hydrophobic interactions.** Hydrophobic ("water-hating") molecules, such as the hydrocarbons, are virtually insoluble in water. Their association into droplets (or, in larger amounts, into a separate layer) results from the solvent properties of water, not from the relatively weak attraction between the associating nonpolar molecules. When nonpolar molecules enter an aqueous environment, hydrogen-bonded water molecules attempt to form a cagelike structure around them (Figure 3.8). Sufficient energy is not available in the surroundings to form this cage-like structure, and nonpolar molecules are expelled. The droplets that form result from the most energetically favorable configuration of the surrounding water molecules.

Hydrophobic interactions have a profound effect on living cells. For example, they are primarily responsible for the structure of membranes and the stability of proteins.

3.1

Proteins are amino acid polymers. Noncovalent bonding plays an important role in determining the three-dimensional structures of proteins. Typical examples of noncovalent interactions (shaded areas) between amino acid side chains are shown below.

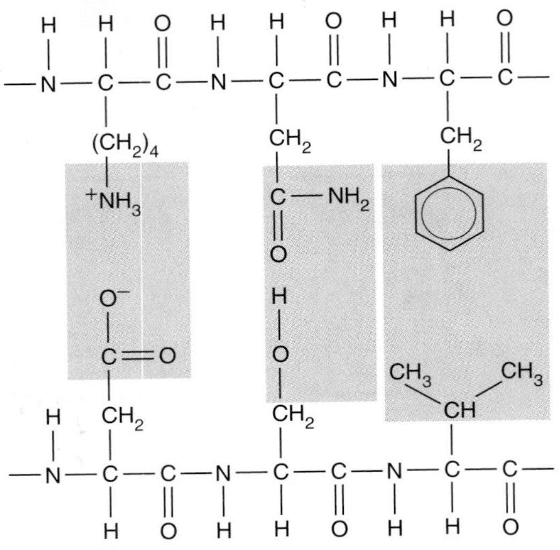

Which noncovalent bond is primarily responsible for the interactions indicated in the figure?

3.3 THERMAL PROPERTIES OF WATER

Perhaps the oddest property of water is that it is a liquid at room temperature. If water is compared with related molecules of similar molecular weight, it becomes apparent that water's melting and boiling points are exceptionally high (Table 3.1). If water followed the pattern of compounds such as hydrogen sulfide, it would melt at $-100°C$ and boil at $-91°C$. Under these conditions, most of the earth's water would be steam, making life unlikely. However, water actually melts at $0°C$ and boils at $+100°C$. Consequently, it is a liquid over most of the wide range of temperatures typically found on the earth's surface. Hydrogen bonding is responsible for this anomalous behavior.

Because of its molecular structure, each water molecule can form hydrogen bonds with four other water molecules. Each of the latter molecules can form hydrogen bonds with other water molecules. The maximum number of hydrogen bonds form when water has frozen into ice (Figure 3.9). Energy is required to break these bonds. When ice is warmed to its melting point, approximately 15% of the hydrogen bonds break. Liquid water consists of icelike clusters of molecules whose hydrogen bonds are continuously breaking and forming. As the temperature rises, the movement and

TABLE 3.1

Melting and Boiling Points of Water and Other Group VI Hydrogen-Containing Compounds

Name	Formula	Molecular Weight (daltons*)	Melting Point (°C)	Boiling Point (°C)
Water	H_2O	18	0	100
Hydrogen sulfide	H_2S	34	−85.5	−60.7
Hydrogen selenide	H_2Se	81	−60.4	−41.5
Hydrogen telluride	H_2Te	129.6	−49	−2

*1 dalton = 1 atomic mass unit

FIGURE 3.9

Hydrogen Bonding Between Water Molecules in Ice.
Hydrogen bonding in ice produces a very open structure. Ice is less dense than water in its liquid state.

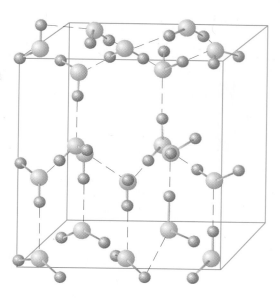

TABLE 3.2

Heat of Fusion of Water and Other Group VI Hydrogen-Containing Compounds

Name	Formula	Molecular Weight (daltons)	Heat of Fusion (cal/g)	Heat of Fusion (J/g)
Water	H_2O	18	80	335
Hydrogen sulfide	H_2S	34	16.7	69.9
Hydrogen selenide	H_2Se	81	7.4	31

The heat of fusion is the amount of heat required to change 1 g of a solid into a liquid at its melting point. 1 cal = 4.187 J.

vibrations of the water molecules accelerate, and additional hydrogen bonds are broken. When the boiling point is reached, the water molecules break free from one another and vaporize. The energy required to raise water's temperature is substantially higher than expected (Table 3.2). In addition to the energy absorbed in increasing molecular agitation, a significant amount of energy is dissipated by the rapid vibration of shared hydrogens back and forth between oxygen atoms.

One consequence of water's high *heat of vaporization* (the energy required to vaporize one mole of a liquid at a pressure of one atmosphere) and high *heat capacity* (the energy that must be added or removed to change the temperature by one degree Celsius) is that water acts as an effective modulator of climatic temperature. Water can absorb and store solar heat and release it slowly. Consider, for example, the relatively moderate temperature transitions when seasons change on land masses near the oceans. In contrast, in the interior of large land masses such as the American Midwest, seasons begin abruptly, and seasonal extremes of temperature may be dramatic. In exceptionally dry areas such as the Sahara Desert, daily changes in temperature may vary by as much as 100°F. Heat absorbed during the day in a desert is quickly lost by reradiation at night.

Not surprisingly, water plays an important role in the thermal regulation of living organisms. Water's high heat capacity, coupled with the high water content found in most organisms (between 50% and 95%, depending on species), helps maintain an organism's internal temperature. The evaporation of water is used as a cooling mechanism, since it permits large losses of heat. For example, an adult human may elim-

KEY *Concepts*

Hydrogen bonding is responsible for water's unusually high freezing and boiling points. Because water has a high heat capacity, it can absorb and release heat slowly. Water plays an important role in regulating heat in living organisms.

inate as much as 1200 g of water daily in expired air, sweat, and urine. The associated heat loss may amount to approximately 20% of the total heat generated by metabolic processes.

QUESTION

3.2   The ammonia molecule is similar to water. It also has a higher boiling point than would be expected from its molecular weight, a high heat of fusion (the energy required for a substance to change between the solid and liquid states), and a high heat capacity. Draw the structure for solid ammonia. Would you expect this "ice" to be more or less dense than liquid ammonia?

3.4 SOLVENT PROPERTIES OF WATER

KEY *Concepts*

Water's dipolar structure and its capacity to form hydrogen bonds enable water to dissolve many ionic and polar substances. Because nonpolar molecules cannot form hydrogen bonds, they cannot dissolve in water. Amphipathic molecules, such as fatty acid salts, spontaneously rearrange themselves in water to form micelles.

Water is a remarkable solvent. Water's dipolar structure and its capacity to form hydrogen bonds enable water to dissolve many ionic and polar substances. Salts such as sodium chloride (NaCl) are held together by ionic (or electrostatic) forces. They dissolve easily in water because dipolar water molecules are attracted to the Na^+ and Cl^- ions.

Organic molecules with ionizable groups and many neutral organic molecules with polar functional groups also dissolve in water, primarily because of the hydrogen bonding capacity of water. Such associations form, for example, between water and the carbonyl groups of aldehydes and ketones and the hydroxyl groups of alcohols.

As mentioned, nonpolar compounds are not soluble in water. Because they lack polar functional groups, such molecules cannot form hydrogen bonds.

A large number of molecules, referred to as **amphipathic**, contain both polar and nonpolar groups. This property significantly affects their behavior in water. For example, fatty acid salts are amphipathic molecules because they contain ionized carboxyl groups and nonpolar hydrocarbon groups. Having both polar and nonpolar groups, fatty acid salts form structures called **micelles** (Figure 3.10) in water. In micelles, the charged species (i.e., the carboxylate groups), called polar heads, orient themselves so that they are in contact with the external aqueous environment. The nonpolar hydrocarbon tails are sequestered in the hydrophobic interior. The tendency of amphipathic molecules to spontaneously rearrange themselves in water is responsible for the structural characteristics of numerous cell components. For example, a

FIGURE 3.10

Formation of Micelles.

The polar heads of amphipathic molecules orient themselves so that they are in contact with water molecules. The nonpolar tails aggregate in the center, away from water.

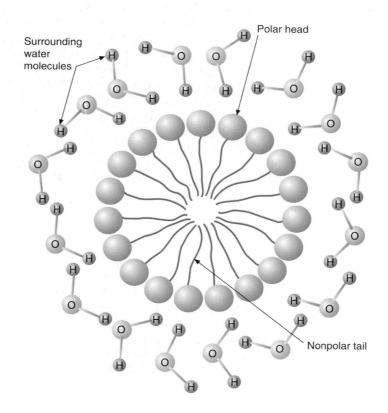

group of bilayer-forming phospholipid molecules is a major constituent of cell membranes.

3.5 IONIZATION OF WATER

Liquid water molecules have a limited capacity to ionize to form a hydrogen ion (H^+) and a hydroxide ion (OH^-). (H^+ does not actually exist in aqueous solution. In water a proton combines with a water molecule to form the hydrated hydrogen ion, H_3O^+, commonly called a *hydronium ion*. For convenience, however, the hydrated proton is usually represented as H^+.) The dissociation of water

$$H_2O \rightleftarrows H^+ + OH^-$$

may be expressed according to the concepts of the law of mass action as follows:

$$K_{eq} = \frac{[H^+][OH^-]}{[H_2O]}$$

where K_{eq} is the equilibrium constant for the reaction. Since the molar concentration of pure water (55.5 M) is considerably larger than that of any solutes, it too is considered a constant. (The concentration of water, $[H_2O]$, is obtained by dividing the number of grams in 1 liter of water, 1000 g, by the molecular weight of water, 18 g/mole.) After this value is substituted into this equation, it may be rewritten as follows:

$$K_{eq} \times 55.5 = [H^+][OH^-]$$

The term $K_{eq} \times 55.5$ is called the *ion product of water* (K_w). Since the equilibrium constant for the reversible ionization of water is equal to 1.8×10^{-16} M (at 25°C), the above relationship yields

$$K_w = (1.8 \times 10^{-16})(55.5) = [H^+][OH^-]$$
$$= 1.0 \times 10^{-14}$$

This means that the product of $[H^+]$ and $[OH^-]$ in any water solution (at 25°C) is always 1×10^{-14}. Since $[H^+]$ is equal to $[OH^-]$ when pure water dissociates,

$$[H^+] = [OH^-] = 1 \times 10^{-7} \text{ M}$$

Thus the hydrogen ion concentration of pure water is equal to 1×10^{-7} M.

When a solution contains equal amounts of H^+ and OH^-, it is said to be *neutral*. When an ionic or polar substance is dissolved in water, it may change the relative numbers of H^+ and OH^-. Solutions with an excess of H^+ are *acidic*, while those with a greater number of OH^- are *basic*. Hydrogen ion concentration varies over a very wide range: commonly between 10^0 and 10^{-14}. Because such numbers are cumbersome, the pH scale has been devised.

Acids, Bases, and pH

Many biomolecules have acidic or basic properties. There are several possible definitions of these important classes of ionic compounds. For our purposes, however, an **acid** may be defined as a hydrogen ion donor, and a **base** as a hydrogen ion acceptor. Strong acids (e.g., HCl) and bases (e.g., NaOH) ionize almost completely in water:

$$HCl \rightarrow H^+ + Cl^-$$
$$NaOH \rightarrow Na^+ + OH^-$$

Many acids and bases, however, do not dissociate completely. Organic acids (compounds with carboxyl groups) do not completely dissociate in water. They are referred to as **weak acids.** Organic bases have a small but measurable capacity to combine with hydrogen ions. Many common **weak bases** contain amino groups.

The dissociation of an organic acid is described by the following reaction:

$$\underset{\text{Weak acid}}{HA} \rightleftarrows H^+ + \underset{\substack{\text{Conjugate Base} \\ \text{of HA}}}{A^-}$$

Note that the unprotonated product of the dissociation reaction is referred to as a **conjugate base.** For example, acetic acid (CH_3COOH) dissociates to form the conjugate base acetate (CH_3COO^-).

The strength of a weak acid (i.e., the number of hydrogen ions released) may be determined by using the following expression:

$$K_a = \frac{[H^+][A^-]}{[HA]}$$

where K_a is the acid dissociation constant. The larger the value of K_a, the stronger the acid is. Because K_a values vary over a wide range, they are expressed by using a logarithmic scale:

$$pK_a = -\log K_a$$

Dissociation constants and pK_a values for several common weak acids are given in Table 3.3.

The hydrogen ion is one of the most important ions in biological systems. The concentration of this ion affects most cellular and organismal processes. For example, the structure and function of proteins and the rates of most biochemical reactions are strongly affected by hydrogen ion concentration. Additionally, hydrogen ions play a major role in processes such as energy generation (Chapter 11) and endocytosis.

The **pH scale** (Figure 3.11) conveniently expresses hydrogen ion concentration. pH is defined as the negative logarithm of the concentration of hydrogen ions:

$$pH = -\log [H^+]$$

On the pH scale, neutrality is defined as pH 7 (i.e., $[H^+]$ is equal to 1×10^{-7} M). Solutions with pH values less than 7 (i.e., $[H^+]$ greater than 1×10^{-7} M) are acidic. Those with pH values greater than 7 are basic or alkaline.

Note that while pH and pK_a use similar mathematical expressions, pH (the negative log of the hydrogen ion concentration of a system) may vary, but pK_a (the negative log of the acid ionization constant) does not vary at constant temperature.

Buffers

Because hydrogen ion concentration affects living processes so profoundly, it is not surprising that regulating pH is a universal and essential activity of living organisms. Hydrogen ion concentration must typically be kept within narrow limits. For example, normal human blood has pH 7.4. It may vary between 7.35 and 7.45, although blood normally contains many acidic or basic waste products dissolved in it. Certain disease processes cause pH changes that, if not corrected, can be disastrous. **Acidosis,** a condition that occurs when human blood pH falls below 7.35, results from an

TABLE 3.3

Dissociation Constants and pK_a Values for Common Weak Acids

Acid	HA	A⁻	K_a	pK_a
Acetic acid	CH_3COOH	CH_3COO^-	1.76×10^{-5}	4.76
Carbonic acid	H_2CO_3	HCO_3^-	4.30×10^{-7}	6.37
Bicarbonate	HCO_3^-	CO_3^{2-}	5.61×10^{-11}	10.25
Lactic acid	$\begin{array}{c} CH_3CHCOOH \\ \mid \\ OH \end{array}$	$\begin{array}{c} CH_3CHCOO^- \\ \mid \\ OH \end{array}$	1.38×10^{-4}	3.86
Phosphoric acid	H_3PO_4	$H_2PO_4^-$	7.25×10^{-3}	2.14
Dihydrogen phosphate	$H_2PO_4^-$	HPO_4^{2-}	6.31×10^{-8}	7.20

Equilibrium constants should be expressed in terms of activities rather than concentrations (activity is the *effective* concentration of a substance in a solution). However, in dilute solutions, concentrations may be substituted for activities with reasonable accuracy.

FIGURE 3.11

The pH Scale and the pH of Common Fluids.

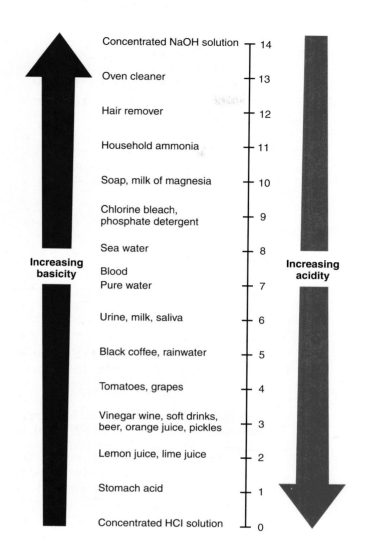

excessive production of acid in the tissues, loss of base from body fluids, or the failure of the kidneys to excrete acidic metabolites. Acidosis occurs in certain diseases (e.g., diabetes mellitus) and during starvation. If blood pH drops below 7, the central nervous system becomes depressed. This results in coma and eventually death. When blood pH rises above 7.45, **alkalosis** results. This condition, brought on by prolonged vomiting or by ingestion of excessive amounts of alkaline drugs, overexcites the central nervous system. Muscles then go into a state of spasm. If uncorrected, convulsions and respiratory arrest develop.

Living systems regulate hydrogen ion concentration by substances that act as buffers. Because of their capacity to combine with H^+ ions or to release H^+ under different conditions, **buffers** help maintain a relatively constant hydrogen ion concentration. The most common buffers consist of weak acids and their conjugate bases. A buffered solution can resist pH changes because an equilibrium between the buffer's components is established. Therefore buffers obey **Le Chatelier's principle,** which states that if a stress is applied to a reaction at equilibrium, the equilibrium will be displaced in the direction that relieves the stress. Consider a solution containing acetate buffer, which consists of acetic acid and sodium acetate (Figure 3.12). The buffer is created by neutralizing acetic acid with the base NaOH:

$$CH_3COOH + OH^- \rightarrow CH_3COO^- + H_2O$$

FIGURE 3.12

Titration of Acetic Acid with NaOH.

The shaded band indicates the pH range over which acetate buffer functions effectively. A buffer is most effective at or near its pK_a value.

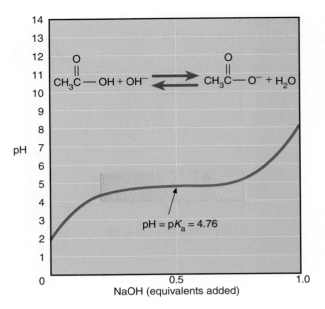

If hydrogen ions are added to a solution containing acetate buffer, they combine with the acetate anion to form acetic acid:

$$H^+ + CH_3COO^- \rightarrow CH_3COOH$$

This reaction removes the added H$^+$ from solution and forces the pH toward its original value. If more OH$^-$ ions are added, more acetic acid dissociates to furnish additional H$^+$ ions to form water.

Again the H$^+$ ion concentration remains unchanged.

Buffering Capacity   The capacity of a buffer to maintain a specific pH depends on two factors: (1) the molar concentration of the acid-conjugate base pair and (2) the ratio of their concentrations. Buffering capacity is directly proportional to the concentration of the buffer components. In other words, the more molecules of buffer present, the more H$^+$ and OH$^-$ ions can be absorbed without changing the pH. The concentration of the buffer is defined as the sum of the concentration of the weak acid and its conjugate base. For example, a 0.2 M acetate buffer may contain 0.1 mole of acetic acid and 0.1 mole of sodium acetate in 1 liter of H$_2$O. Such a buffer may also consist of 0.05 mole of acetic acid and 0.15 mole of sodium acetate in 1 liter of H$_2$O. The most effective buffers are those that contain equal concentrations of both components. However, there are exceptions. Bicarbonate buffer (p. 52), one of the most important physiological buffers, significantly deviates from such a ratio.

Henderson-Hasselbalch Equation   In choosing or making a buffer, the concepts of both pH and pK_a are useful. The relationship between these two quantities is expressed in the Henderson-Hasselbalch equation, which is derived from the equilibrium expression below:

$$K_a = \frac{[H^+][A^-]}{[HA]}$$

Solving for [H$^+$] results in

$$[H^+] = K_a \frac{[HA]}{[A^-]}$$

Taking the negative logarithm of each side, we obtain

$$-\log\,[H^+] = -\log\,K_a - \log\frac{[HA]}{[A^-]}$$

Defining $-\log\,[H^+]$ as pH and $-\log\,K_a$ as pK_a gives

$$pH = pK_a - \log\frac{[HA]}{[A^-]}$$

If the log term is inverted, thereby changing its sign, the following relationship, called the *Henderson-Hasselbalch equation,* is obtained:

$$pH = pK_a + \log\frac{[A^-]}{[HA]}$$

Notice that when [A$^-$] = [HA], the equation becomes

$$pH = pK_a + \log\,1$$
$$= pK_a + 0$$

Under this circumstance, pH is equal to pK_a.

Typical buffer problems along with their solutions are given below.

PROBLEMS

3.1

Calculate the pH of a mixture of 0.25 M acetic acid and 0.1 M sodium acetate. The pK_a of acetic acid is 4.76.

Solution

$$pH = pK_a + \log\frac{[\text{acetate}]}{[\text{acetic acid}]}$$

$$pH = 4.76 + \log\frac{0.1}{0.25} = 4.76 - 0.398 = 4.36$$

3.2

What is the pH in the preceding problem if the mixture consists of 0.1 M acetic acid and 0.25 M sodium acetate?

Solution

$$pH = 4.76 + \log\frac{0.25}{0.1} = 4.76 + 0.398 = 5.16$$

3.3

Calculate the ratio of lactic acid and lactate required in a buffer system of pH 5.00. The pK_a of lactic acid is 3.86.

Solution

The equation

$$pH = pK_a + \log\frac{[\text{lactate}]}{[\text{lactic acid}]}$$

can be rearranged to

$$\log\frac{[\text{lactate}]}{[\text{lactic acid}]} = pH - pK_a$$

$$= 5.00 - 3.86 = 1.14$$

Therefore the required ratio is

$$\frac{[\text{lactate}]}{[\text{lactic acid}]} = \text{antilog } 1.14$$

$$= 13.8$$

For a lactate buffer to have a pH of 5, the lactate and lactic acid components must be present in a ratio of 13.8 to 1.

3.4   During the fermentation of wine, a buffer system consisting of tartaric acid and potassium hydrogen tartrate is produced by a biochemical reaction. Assuming that at some time the concentration of potassium hydrogen tartrate is twice that of tartaric acid, calculate the pH of the wine. The pK_a of tartaric acid is 2.96.

Solution

$$\text{pH} = pK_a + \log \frac{[\text{hydrogen tartrate}]}{[\text{tartaric acid}]}$$

$$= 2.96 + \log 2$$

$$= 2.96 + 0.30$$

$$= 3.26$$

Physiological Buffers

The three most important buffers in the body are the bicarbonate buffer, the phosphate buffer, and the protein buffer. Each is especially useful.

Bicarbonate Buffer   Bicarbonate buffer, one of the more important buffers in blood, has three components. The first of these, carbon dioxide, reacts with water to form carbonic acid:

$$CO_2 + H_2O \rightleftarrows \underset{\text{Carbonic acid}}{H_2CO_3}$$

Carbonic acid then rapidly dissociates to form H^+ and HCO_3^- ions:

$$H_2CO_3 \rightleftarrows H^+ + \underset{\text{Bicarbonate}}{HCO_3^-}$$

Since the concentration of H_2CO_3 is very low in blood, the above equations may be simplified to

$$CO_2 + H_2O \rightleftarrows H^+ + HCO_3^-$$

Recall that buffering capacity is greatest at or near the pK_a of the acid-conjugate base pair. Bicarbonate buffer is clearly unusual in that its pK_a is 6.37. It would appear at first glance that bicarbonate buffer is ill-suited to buffer blood. The ratio of HCO_3^- to CO_2 required to maintain a pH of 7.4 is approximately 11 to 1. In other words, the bicarbonate buffer operates in blood near the limit of its buffering power. In addition, the concentrations of CO_2 and HCO_3^- are not exceptionally high. Despite these liabilities, the bicarbonate buffering system is important because both components can be regulated. Carbon dioxide concentration is adjusted by changes in the rate of respiration.

$$CO_2 + H_2O \rightleftarrows H_2CO_3 \rightleftarrows HCO_3^- + H^+$$

Bicarbonate concentration is regulated by the kidneys. If bicarbonate concentration ($[HCO_3^-]$) decreases, the kidneys remove H^+ from the blood, shifting the equilibrium to the right and increasing $[HCO_3^-]$. Carbonic acid lost in this process is quickly replenished by hydrating CO_2, a waste product of cellular metabolism. When excess amounts of HCO_3^- are produced, they are excreted by the kidney. As acid is added to the body's bicarbonate system, $[HCO_3^-]$ decreases and CO_2 is formed.

FIGURE 3.13

Titration of $H_2PO_4^-$ by Strong Base.

The shaded band indicates the pH range over which the weak acid-conjugate base pair $H_2PO_4^-/HPO_4^{2-}$ functions effectively as a buffer.

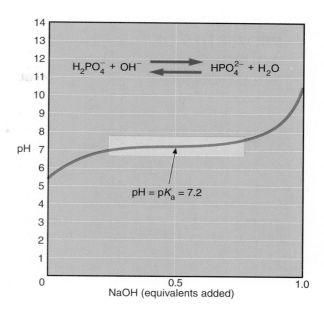

Since the excess CO_2 is exhaled, the ratio of HCO_3^- to CO_2 remains essentially unchanged.

Phosphate Buffer   Phosphate buffer consists of the weak acid-conjugate base pair $H_2PO_4^-/HPO_4^{-2}$ (Figure 3.13):

$$H_2PO_4^- \rightleftarrows H^+ + HPO_4^{2-}$$

Dihydrogen phosphate   Hydrogen phosphate

With pK_a 7.2, it would appear that phosphate buffer is an excellent choice for buffering the blood. Although the blood pH of 7.4 is well within this buffer system's capability, the concentrations of $H_2PO_4^-$ and HPO_4^{2-} in blood are too low to have a major effect. Instead, the phosphate system is an important buffer in intracellular fluids where its concentration is approximately 75 mEq/L. Phosphate concentration in extracellular fluids such as blood is about 4 mEq/L. (An equivalent is defined as the mass of acid or base that can furnish or accept one mole of H^+ ions. The abbreviation mEq indicates a milliequivalent.) Since the normal pH of cell fluids is approximately 7.2 (the range is from 6.9 to 7.4), an equimolar mixture of $H_2PO_4^-$ and HPO_4^{2-} is typically present. Although cells contain other weak acids, these substances are unimportant as buffers. Their concentrations are quite low, and their pK_a values are significantly lower than intracellular pH. For example, lactic acid has a pK_a of 3.86.

Protein Buffer   Proteins are a significant source of buffering capacity. Composed of amino acids linked together by peptide bonds, proteins contain several types of ionizable groups in side chains that can donate or accept protons. Since protein molecules are present in significant concentration in living organisms, they are powerful buffers. For example, the oxygen-carrying protein hemoglobin is the most abundant biomolecule in red blood cells. Because of its structure and high cellular concentration, hemoglobin plays a major role in maintaining blood pH. Also present in high concentrations and buffering the blood are the serum albumins and other proteins.

QUESTION

3.3

Severe diarrhea is one of the most common causes of death in young children. One of the principal effects of diarrhea is the excretion of large quantities of sodium bicarbonate. In which direction does the bicarbonate buffer system shift under this circumstance? What is the resulting condition called?

3.6 COLLIGATIVE PROPERTIES OF AQUEOUS SOLUTIONS

Several physical properties of liquid water can change when solute molecules are dissolved. The **colligative properties** (vapor pressure depression, boiling point elevation, freezing point depression, and osmotic pressure) are grouped together because they depend on the number of dissolved particles (ions or molecules) in a given mass of solvent, not on their identity.

Several colligative properties have an obvious impact on living processes. For example, freshwater fish survive in freezing temperatures because the solute concentration of their blood depresses its freezing point below that of the surrounding water. However, osmotic pressure is the most obviously relevant colligative property for living organisms because it is the main cause of water flow across cellular membranes. Because osmotic pressure is more easily understood in relationship to the other colligative properties, each will be discussed briefly.

Vapor Pressure

If a sample of water is placed in a closed container, the molecules evaporate until they reach an equilibrium. At equilibrium the molecules are condensing into the liquid phase at the same rate as they are evaporating into the gas phase. As the evaporated molecules collide with the side of the container, they exert the **vapor pressure.** Since the vapor pressure of a liquid depends on how readily the molecules escape from its surface, adding nonvolatile solute molecules decreases the pressure. For example, if 5% of the molecules in a solution are nonvolatile (e.g., glucose), then the vapor pressure will be 5% lower than that of pure water. The vapor pressure is lower because the glucose molecules prevent water molecules from escaping into the vapor state (Figure 3.14).

Boiling Point Elevation

Recall that the normal boiling point of water at 1 atm is 100°C. Because adding nonvolatile solutes reduces vapor pressure, a higher temperature is required to raise the vapor pressure sufficiently to boil the solution. The magnitude of the boiling point elevation is proportional to the solute concentration. Each mole of solute dissolved in 1 kg of water at standard pressure (1 atm or 760 mm Hg) elevates the boiling point by 0.51°C.

Freezing Point Depression

The freezing point of a solution can also be considered in terms of vapor pressure. At the freezing point of water, the vapor pressure of ice and liquid water are equal. Because solute molecules lower the vapor pressure of water, the solution must be cooled to below 0°C to freeze it. The freezing point of a 1 m solution is depressed by 1.86°C. In dilute solutions the freezing point depression is directly proportional to molality. The following equation can be used to determine the molality of a solution if the freezing point depression ΔT_f is known:

$$\Delta T_f = -K_f m$$

Freezing point depression can be used to determine the molecular weight of a substance if its concentration is known. However, since the molal freezing point depres-

FIGURE 3.14

Adding Solute Molecules Lowers Vapor Pressure.

Solute molecules (red units) present a barrier to escaping solvent molecules (white units). Vapor pressure decreases.

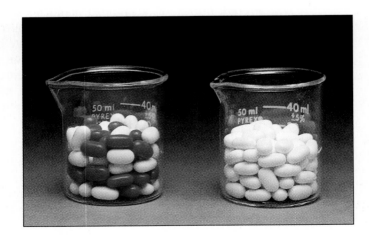

sion constant K_f for water is small (1.855°C kg/mole), solvents such as benzene (K_f = 5.12°C kg/mole) are often used instead.

Osmotic Pressure

Osmosis is the spontaneous process in which solvent molecules pass through a semi-permeable membrane from a solution of lower concentration to a solution of higher concentration (Special Interest Box 3.1). Pores in the membrane are wide enough to allow solvent molecules to pass in both directions but too narrow for the larger solute molecules or ions to pass. Figure 3.15 illustrates the movement of solvent across a membrane. Over a period of time, water moves from side A (more dilute) to side B (more concentrated) because the solute molecules block the water molecules. Therefore, the higher the concentration of water in a solution (i.e., the lower solute concentration), the faster the movement through the membrane. **Osmotic pressure** stops the net flow of water across the membrane. The force generated by osmosis may be considerable. For example, osmosis appears to be largely responsible for the upward flow of sap in trees.

Osmotic pressure depends on solute concentration. A device called an osmometer (Figure 3.16) measures osmotic pressure. Osmotic pressure can also be calculated using the equation

$$\pi = iMRT$$

where

π = osmotic pressure
i = van't Hoff factor
M = molarity
R = gas constant (0.082 L·atm/K·mole)
T = Kelvin temperature

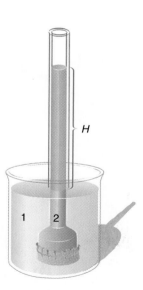

FIGURE 3.16

A Demonstration of Osmosis Using an Osmometer.

Volume 1 contains pure water. Volume 2 contains a solute such as sucrose. The membrane is permeable to water but not to the sucrose. Therefore there will be a net movement of water into the osmometer. The osmotic pressure is proportional to the height H of the solution in the tube.

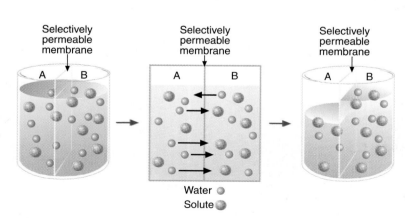

Water ○
Solute ●

FIGURE 3.15

Osmotic Pressure.

Over a period of time water diffuses from side A to side B. Equilibrium between the solutions on both sides of a semipermeable membrane is attained when there is no net movement of water molecules from side A (more dilute) to side B (more concentrated). Osmotic pressure stops the net flow of water across the membrane.

OSMOSIS AND LIVING CELLS

Osmotic pressure creates some critical problems for living organisms. Cells typically contain fairly high concentrations of solutes, that is, small organic molecules and ionic salts, as well as lower concentrations of macromolecules. Consequently, they may gain or lose water because of the concentration of solute in their environment. If cells are placed in an **isotonic solution** (i.e., the concentration of solute and water is the same on both sides of the selectively permeable plasma membrane) there is no net movement of water in either direction across the membrane (Figure 3A). For example, red blood cells are isotonic to a 0.9% NaCl solution. When cells are placed in a solution with a lower solute concentration (i.e., a **hypotonic solution**), water moves into the cells. Red blood cells, for example, swell and rupture in a process called *hemolysis* when they are immersed in pure water. In **hypertonic solutions**, those with higher solute concentrations, cells shrivel because there is a net movement of water out of the cell. The shrinkage of red blood cells in hypertonic solution (e.g., a 3% NaCl solution) is referred to as *crenation*.

Because of their relatively low cellular concentration, macromolecules have little direct effect on cellular osmolarity. However, macromolecules such as the proteins contain a large number of ionizable groups. The large number of ions of opposite charge that are attracted to these groups have a substantial effect on intracellular osmolarity. Unlike most ions, the ionizable groups of proteins cannot penetrate cell membranes. (Cell membranes are not, strictly speaking, osmotic membranes, since they allow various ions, nutrients, and waste products to pass through. The term *dialyzing membrane* gives a more accurate description of their function.) As a result, at equilibrium the concentrations for each ionic species are not the same on both sides of a cell's plasma membrane. Instead, the intracellular concentrations of inorganic ions are higher than outside the cell. The consequences of this phenomenon are called the *Donnan effect*:

1. A constant tendency toward cellular swelling because of water entry due to osmotic pressure
2. The establishment of an electrical gradient called a **membrane potential**

Because of the Donnan effect, cells must constantly regulate their osmolarity. Many cells, for example, animal and bacterial cells, pump out certain inorganic ions such as Na^+. This process, which requires a substantial proportion of cellular energy, controls cell volume. Several species, such as some protozoa and algae, periodically expel water from special contractile vacuoles. Since plant cells have rigid cell walls, plants use the Donnan effect to create an internal hydrostatic pressure, called *turgor pressure* (Figure 3B). This process drives cellular growth and expansion and makes many plant structures rigid.

Artificial semipermeable membranes are routinely used in biochemical laboratories to separate small solutes from larger

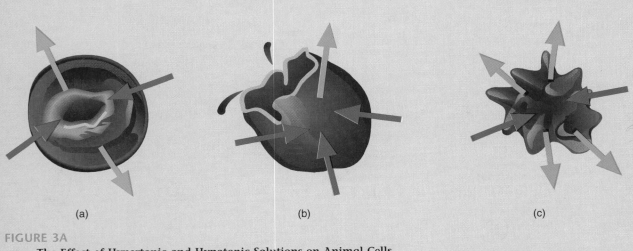

(a) (b) (c)

FIGURE 3A

The Effect of Hypertonic and Hypotonic Solutions on Animal Cells.

(a) Isotonic solutions do not change cell volume; (b) hypotonic solutions cause cell rupture; (c) hypertonic solutions cause cells to shrink (crenation).

The concentration of a solution can be expressed in terms of osmolarity. The unit of *osmolarity* is osmol/liter. From the equation above, the osmolarity is equal to *iM*, where *i* (the *van't Hoff factor*) represents the degree of ionization of the solute species. The degree of ionization of a 1 M NaCl solution is 90% with 10% of the NaCl existing as ions pairs. The value of *i* for this solution, therefore, is 1.9. The value of *i*

solutes. For example, this technique (referred to as **dialysis**) is used as an important early step in protein purification. An impure protein-containing specimen is placed in a cellophane dialysis bag (Figure 3C), which is then suspended in flowing distilled water or in a buffered solution. After a certain time, all the small solutes leave the bag. The protein solution, which may contain many high-molecular-weight impurities, is then ready for further purification.

Hemodialysis is a clinical application of dialysis that removes toxic waste from the blood of patients suffering from temporary or permanent renal failure. All constituents of blood except blood cells and the plasma proteins move freely between blood and the dialyzing fluid. Since dialysis tubing allows passage of nutrients (glucose, amino acids) and essential elec-

trolytes (Na^+, K^+), these and other vital substances are included in the dialyzing fluid. Their inclusion prevents a net loss of these materials from the blood. Because no waste products such as urea and uric acid are in the dialyzing fluid, these substances are lost from the blood in large quantities.

Although hemodialysis is very effective in removing toxic waste from the body, it does not solve all the problems brought on by renal failure. For example, until recently, patients suffering from renal failure often became anemic because they lacked a protein hormone called erythropoietin, which is normally secreted by the kidney. (Erythropoietin stimulates red blood cell synthesis.) Because of DNA technology (Chapter 17), erythropoietin is now readily administered to dialysis patients.

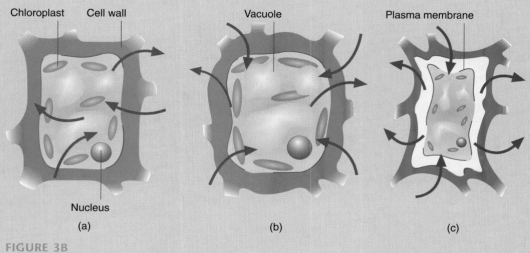

Chloroplast   Cell wall

Vacuole

Plasma membrane

Nucleus

(a)   (b)   (c)

FIGURE 3B

Osmotic Pressure and Plant Cells.

(a) Isotonic solutions do not change cell volume. (b) Plant cells typically exist in a hypotonic environment. When water enters these cells they become swollen. Rigid cell walls prevent the cells from bursting. (c) In a hypertonic environment the cell membrane pulls away from the cell wall because of water loss and the plant wilts.

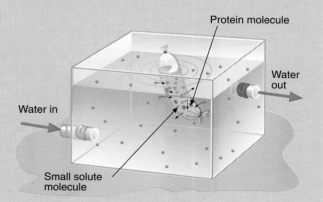

Protein molecule

Water out

Water in

Small solute molecule

FIGURE 3C

Dialysis.

Proteins are routinely separated from low-molecular-weight impurities by dialysis. When a dialysis bag containing a cell extract is suspended in water or a buffered solution, small molecules pass out through the membrane's pores. If the solvent outside the bag is continually renewed, all low-molecular-weight impurities are removed from the inside.

approaches 2 for NaCl solutions as they become increasingly more dilute. The value of *i* for a 1 M solution of a weak acid that undergoes a 10% ionization is *i* of 1.1. The value of *i* for a nonionizable solute is always 1.0. Several osmotic pressure problems appear on page 58.

PROBLEMS

3.5 When 0.1 g of urea (M.W. = 60) is diluted to 100 mL with water, what is the osmotic pressure of the solution? (Assume that the temperature is 25°C.)

Solution

Calculate the osmolarity of the urea solution.

$$\text{Molarity} = 0.1 \text{ g urea} \times \frac{1 \text{ mole}}{60 \text{ g}} \times \frac{1}{0.1 \text{ L}}$$
$$= 1.7 \times 10^{-2} \text{ mole/L}$$

The number of particles produced per mole of solute is 1. The osmotic pressure at 25°C (298 K) is given by the equation

$$\pi = iMRT$$

Urea is a nonelectrolyte, so $i = 1$

$$\pi = (1)\left(1.7 \times 10^{-2} \frac{\text{mole}}{\text{L}}\right)\left(0.0821 \frac{\text{L·atm}}{\text{K·mole}}\right)(298 \text{ K})$$
$$\pi = 0.42 \text{ atm}$$

3.6 Estimate the osmotic pressure of a solution of 0.1 M NaCl at 25°C. Assume 100% ionization of solute.

Solution

A solution of 0.1 M NaCl produces 0.2 mole of particles per liter (0.1 mole of Na^+ and 0.1 mole of Cl^-). The osmotic pressure at 25°C (298 K) is

$$\pi = 2 \times 0.1 \frac{\text{mole}}{\text{L}} \times 0.0821 \frac{\text{L·atm}}{\text{K·mole}} \times 298 \text{ K}$$

QUESTION

3.4 Osmotic pressure may be used to determine the molecular weights of pure substances. It is an especially useful technique because low concentrations of solute produce relatively large osmotic pressures. Although more sophisticated techniques are now available to determine molecular weight, osmotic pressure is still used occasionally because of its simplicity. Determine the molecular weight of myoglobin (an oxygen storage protein that gives muscle its red color). The osmotic pressure of a 1.0-mL water solution containing 1.5×10^{-3} g of the protein was measured as 2.06×10^{-3} atm at 25°C.

KEY WORDS

acid, *47*
acidosis, *48*
alkalosis, *48*
amphipathic molecule, *46*
base, *47*
buffer, *49*
colligative property, *54*
conjugate base, *48*
dialysis, *57*
dipole, *40*
electrostatic interaction, *41*
hydrogen bond, *40*
hydrophobic interaction, *43*
hypertonic solution, *56*
hypotonic solution, *56*
isotonic solution, *56*
le Chatelier's principle, *49*
London dispersion force, *43*
membrane potential, *56*
micelle, *46*
osmosis, *54*
osmotic pressure, *55*
pH scale, *48*
polar, *40*
solvation sphere, *42*
van der Waals force, *41*
vapor pressure, *54*
weak acid, *47*
weak base, *47*

SUMMARY

Water molecules (H_2O) are composed of two atoms of hydrogen and one of oxygen. Each hydrogen atom is linked to the oxygen atom by a single covalent bond. The oxygen-hydrogen bonds are polar and water molecules are dipoles. One consequence of water's polarity is that water molecules are attracted to each other by the electrostatic force between the oxygen of one molecule and the hydrogen of another. This attraction is called a hydrogen bond.

In addition to hydrogen bonds, there are three other noncovalent bonds. These are electrostatic interactions, van der Waals forces, and hydrophobic interactions. Noncovalent interactions also play a vital role in determining the structure (and therefore the function) of biomolecules, especially in proteins and nucleic acids.

Water has an exceptionally high heat capacity. Its boiling and melting points are significantly higher than those of compounds of comparable structure and molecular weight. Hydrogen bonding is responsible for this anomalous behavior.

Water is also a remarkable solvent. Water's dipolar structure and its capacity to form hydrogen bonds enable it to dissolve many ionic and polar substances.

Liquid water molecules have a limited capacity to ionize to form a hydrogen ion (H^+) and a hydroxide ion (OH^-). When a solution contains equal amounts of H^+ and OH^- ions, it is said to be neutral. Solutions with an excess of H^+ are acidic, while those with a greater number of OH^- are basic. Because organic acids do not completely dissociate in water, they are referred to as weak acids. The acid dissociation constant K_a is a measure of the strength of a weak

acid. Because K_a values vary over a wide range, pK_a values ($-\log K_a$) are used instead.

The hydrogen ion is one of the most important ions in biological systems. The pH scale conveniently expresses hydrogen ion concentration. pH is defined as the negative logarithm of the hydrogen ion concentration.

Because hydrogen ion concentration affects living processes so profoundly, it is not surprising that regulating pH is a universal and essential activity of living organisms. Hydrogen ion concentration is typically kept within narrow limits. Because buffers combine with H^+ ions, they help maintain a relatively constant hydrogen ion concentration. The ability of a solution to resist pH changes is called buffering capacity. Most buffers consist of a weak acid and its conjugate base.

Several physical properties of liquid water change when solute molecules are dissolved. The most important of these for living organisms is osmotic pressure, the pressure that prevents the flow of water across cellular membranes.

SUGGESTED READINGS

Montgomery, R., and Swenson, C.A., *Quantitative Problems in Biochemical Sciences,* 2nd ed., W.A. Freeman, New York, 1976.

Pennisi, E., Water, Water Everywhere, *Sci. News* 143:121–125, 1993.

Rand, R.P., Raising Water to New Heights, *Science* 256:618, 1992.

Stewart, P.A., *How to Understand Acid-Base: A Quantitative Acid-Base Primer for Biology and Medicine,* Elsevier, New York, 1981.

Stillinger, F.H., Water Revisited, *Science* 209:451–457, 1980.

Wiggins, P.M., Role of Water in Some Biological Processes, *Microbiol. Rev.* 54:432–449, 1990.

QUESTIONS

1. Which of the following are acid-conjugate base pairs?
   a. H_2CO_3, CO_3^{2-}
   b. $H_2PO_4^-$, PO_4^{3-}
   c. HCO_3^-, CO_3^{2-}
   d. H_2O, OH^-

2. What is the hydrogen ion concentration in a solution at pH 8.3?

3. Consider the following titration curve. Estimate the effective buffer range.

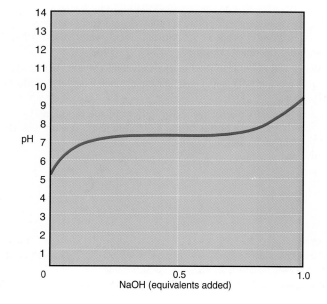

4. Describe how you would prepare a 0.1 M phosphate buffer with a pH of 7.2. What ratio of salt to acid would you use?

5. Which of the following compounds can form hydrogen bonds with like molecules or with water?

   a. $CH_3-CH_2-\overset{\displaystyle O}{\underset{\|}{C}}-O-CH_3$

   b. $CH_3-\overset{\displaystyle \|}{\underset{O}{C}}-N\overset{H}{\underset{CH_3}{\diagup}}$

   c. $CH_3-N\overset{CH_3}{\underset{CH_3}{\diagup}}$

   d. CH_3-O-CH_2-O-H

6. What is the osmolarity of a 1.3 M solution of sodium phosphate (Na_3PO_4)? Assume 85% ionization for this solution.

7. What is the freezing point of a 2.8 M solution of sodium nitrate ($NaNO_3$) and a 3 M solution of glucose? Assume 100% ionization of $NaNO_3$.

8. What direction does water flow when a dialysis bag containing a 3 M solution of the sugar fructose is placed in each of the following solutions?
   a. 1 M sodium lactate
   b. 3 M sodium lactate
   c. 4.5 M sodium lactate

$$CH_3 \!-\! \underset{\underset{OH}{|}}{C} \!-\! \overset{\overset{O}{\|}}{C} \!-\! O^- \quad Na^+$$

(with H above the first C)

9. What interactions occur between the following molecules and ions?
   a. water and ammonia
   b. lactate and ammonium ion
   c. benzene and octane
   d. carbon tetrachloride and chloroform
   e. chloroform and diethylether

10. A solution containing 56 mg of a protein in 30 mL of distilled water exerts an osmotic pressure of 0.01 atm at $T = 25°C$. Determine the molecular weight of the unknown protein.

11. Many fruits can be preserved by candying. The fruit is immersed in a highly concentrated sugar solution, then the sugar is allowed to crystallize. How does the sugar preserve the fruit?

12. Tyrosine is an amino acid.

$$H \!-\! O \!-\! \bigcirc \!-\! CH_2 \!-\! CH \!-\! \overset{\overset{O}{\|}}{C} \!-\! OH$$

(with NH_2 below the CH)

Which atoms in this molecule can form hydrogen bonds?

13. Briefly define the following terms:
   a. hydrogen bond
   b. pH
   c. buffer
   d. osmotic pressure
   e. colligative properties
   f. isotonic
   g. amphipathic
   h. hydrophobic interactions
   i. dipole
   j. induced dipole

14. Which of the following molecules would you expect to have a dipole moment?
   a. CCl_4 d. CH_3OCH_3
   b. $CHCl_3$ e. CH_3CH_3
   c. H_2O f. H_2

15. Explain why water has an unusually high heat capacity. Would you expect ammonia or methane to have high heat capacities?

16. Which of the following molecules would you expect to form micelles?
   a. NaCl   d. $CH_3(CH_2)_{10}COO^-Na^+$
   b. CH_3COOH e. $CH_3(CH_2)_{10}CH_3$
   c. $CH_3COO^-NH_4^+$

17. Which of the following molecules or ions are weak acids? Explain.
   a. HCl   d. HNO_3
   b. $H_2PO_4^-$ e. HSO_4^-
   c. CH_3COOH

18. Which of the following species can form buffer systems?
   a. $NH_4^+Cl^-$
   b. CH_3COOH, HCl
   c. CH_3COOH, $CH_3COO^-Na^+$
   d. H_3PO_4, PO_4^{3-}

19. Describe how you can increase the buffering capacity of a 0.1 M acetate buffer.

20. What effect does hyperventilation have on blood pH?

21. Which of the following are not colligative properties of water?
   a. pH
   b. boiling point elevation
   c. vapor pressure
   d. freezing point depression
   e. osmotic pressure
   f. buffering capacity

22. What is the relationship between osmolarity and molarity?

23. Explain why ice is less dense than water. If ice were not less dense than water, how would the oceans be affected? How would the development of life on earth be affected?

24. Why can't seawater be used to water plants?

25. Explain how the acids produced in metabolism are transported to the liver without greatly affecting the pH of the blood.

26. Explain how water's high heat of vaporization makes steam an effective sterilizing agent.

27. The pH scale is valid only for water. Why is this so?

28. Calculate the boiling point of a 2 M aqueous solution of sodium sulfate (Na_2SO_4). Assume 100% ionization of the electrolyte.

29. Is it possible to prepare a buffer consisting only of carbonic acid and sodium carbonate?

30. Gelatin is a mixture of protein and water that is mostly water. Explain how the water-protein mixture becomes a solid.

31. What is the pH of a solution that is 10^{-8} in HCl?

32. Reverse osmosis is thought to be an effective alternative to distillation for purifying of water. Do you think this process would require more or less energy than distillation? Explain.

33. Many molecules are polar, yet they do not form significant hydrogen bonds. What is so unusual about water that hydrogen bonding becomes possible?

34. Water has been described as the universal solvent. If this statement were strictly true, could life have arisen in a water medium? Explain.

35. Bicarbonate is one of the main buffers of the blood, and phosphate is the main buffer of the cells. Suggest a reason why this observation is true.

ENERGY

Chapter Four

Energy Transformation. *The cheetah, the fastest land animal on earth, transforms the chemical bond energy in food to the energy that sustains its living processes.*

OBJECTIVES

After you have studied this chapter, you should be able to answer these questions:

1. *How do the principles of thermodynamics apply to living organisms?*
2. *How is thermodynamics used to determine whether specific biochemical reactions are spontaneous?*
3. *What is the significance of free energy?*
4. *How do oxidation–reduction reactions generate energy?*
5. *What is the role of ATP in living systems?*

*All living organisms have an unrelenting requirement for energy. The flow of energy that sustains life on earth originates in the sun, where thermonuclear reactions generate radiant energy. A small amount of the solar energy that reaches the earth is captured by organisms such as plants and certain microorganisms. These organisms, called **photoautotrophs**, use solar energy by transforming it into other forms of energy. During photosynthesis, solar energy is converted into the chemical bond energy of sugar molecules. This bond energy is used to produce the vast array of organic molecules found in living organisms and to drive processes such as active transport cell division, endocytosis (discussed in Chapter 9) and muscle contraction. Ultimately, chemical bond energy is converted into heat, which is then dissipated into the environment.*

The study of energy transformations in living organisms is called **bioenergetics.** It is concerned with the initial and final energy states of biomolecules and is useful in determining the direction and extent to which specific biochemical reactions proceed. Chemical reactions are determined by three factors. Two of these, **enthalpy** (total heat content) and **entropy** (disorder), are related to the first and second laws of thermodynamics, respectively. The third factor, called **free energy** (energy available to do chemical work), is derived from a mathematical relationship between enthalpy and entropy.

The chapter begins with a brief description of thermodynamic concepts and their relationship to biochemical reactions. This is followed by a discussion of oxidation–reduction reactions, the cell's primary mechanism for capturing energy. Finally, the role of ATP and other high-energy compounds is described.

4.1 THERMODYNAMICS

Thermodynamics is concerned with heat and energy transformations. Such transformations are considered to take place in a "universe" composed of a system and its surroundings (Figure 4.1). A system is defined according to the interests of the investigator. For example, a system may be defined as an entire organism or a single cell or a reaction occurring in a flask. In an *open system* matter and energy are exchanged between the system and its surroundings. If only energy can be exchanged with the surroundings, then the system is said to be *closed*. Living organisms, which consume nutrients from their surroundings and release waste products into it, are open systems.

Knowledge of thermodynamic functions such as enthalpy, entropy, and free energy enables biochemists to predict whether a process occurs spontaneously. For example, the direction and extent to which a reaction proceeds are determined by the changes in these quantities during the reaction.

Note that several thermodynamic properties are *state functions*, that is, their values do not depend on which pathway is used to make or degrade a specific substance. For example, the energy content of a glucose molecule, which may be a product of photosynthesis or of the breakdown of lactose (milk sugar), does not depend on which pathway is used to make the glucose. Some quantities, such as **work** (the change in energy caused by a force) and heat, are not state functions; rather, they are energy changes occurring during a process and depend on the pathway. For example, living cells capture a portion of the energy in glucose molecules. This energy is used to perform cellular work such as muscle contraction. If glucose molecules are instead ignited in a laboratory dish, the overall reaction is the same but all of the chemical bond energy in the glucose is transformed directly into heat and little or no measurable work is performed. The energy content of the glucose molecules is the same in each process. The work accomplished by each process is different.

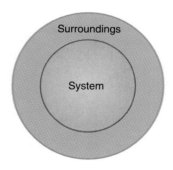

FIGURE 4.1

A Thermodynamic Universe.

A universe consists of a system and its surroundings.

First Law of Thermodynamics

The *first law of thermodynamics* states that energy can be neither created nor destroyed. Although energy may be converted from one form into another, the total energy in the universe must remain constant. The first law of thermodynamics may be written as

$$\Delta E = Q - W \tag{1}$$

where

ΔE = the change in the energy of a system
Q = the heat absorbed by the system
W = the work done by the system

Since biochemical systems operate at relatively constant temperature, pressure, and volume, the first law can be expressed simply as

$$\Delta H = Q - W \tag{2}$$

where ΔH = the change in enthalpy. Since the work done on the system can be defined as

$$W = V\Delta P + P\Delta V \tag{3}$$

if the process takes place at constant pressure and volume, $W = 0$ and ΔH is equal to the flow of heat energy. If ΔH is negative ($\Delta H < 0$), the reaction or process gives off heat and is referred to as **exothermic**. If ΔH is positive ($\Delta H > 0$), heat is absorbed from the surroundings. Such processes are called **endothermic.** In **isothermic** processes ($\Delta H = 0$) heat is not exchanged with the surroundings.

Equation (2) indicates that the total energy change of a biological system is equivalent to the heat evolved or absorbed by the system. Since the enthalpy of a reactant or product is a state function (independent of pathway), then the enthalpy change for any reaction forming that substance can be used to calculate the ΔH of a reaction involving that substance. If the sum of the ΔH values ($\Sigma\Delta H$) for both the reactants and the products is known, then the enthalpy change for the reaction can be calculated by using the following equation:

$$\Sigma\Delta H_{\text{reaction}} = \Sigma\Delta H_{\text{products}} - \Sigma\Delta H_{\text{reactants}} \tag{4}$$

The standard enthalpy of formation per mole, ΔH_f, is commonly used in enthalpy calculations. H_f is the energy evolved or absorbed when one mole of a substance is formed from its most stable elements. Note that equation (4) cannot predict the direction of any chemical reaction. It can determine only the heat flow. An enthalpy calculation for a reaction at constant pressure is given below.

KEY *Concepts*

At constant pressure, a system's enthalpy change ΔH is equal to the flow of heat energy. If ΔH is negative, the reaction or process is exothermic. If ΔH is positive, the reaction or process is endothermic. In isothermic processes no heat is exchanged with the surroundings.

PROBLEM 4.1

Given the following ΔH values, calculate ΔH for the reaction

$$6\ CO_2 + 6\ H_2O \rightarrow C_6H_{12}O_6 + 6\ O_2$$

	ΔH_f*	
	kcal[†]/mole	kJ[‡]/mole
$C_6H_{12}O_6$	−304.7	−1274.9
CO_2	−94.0	−393.5
H_2O	−68.4	−286.2
O_2	0	0

*ΔH_f is the energy evolved to produce a compound from its elements.
[†]1 kcal is the energy required to raise the temperature of 1000 g of water 1°C.
[‡]The joule (J) is a unit of energy that is gradually replacing the calorie (cal) in scientific usage. One calorie is equal to 4.184 joules.

Solution

The total enthalpy for a reaction is equal to the sum of the enthalpy values of the products minus those of the reactants.

$$6 \; CO_2 + 6 \; H_2O \rightarrow \qquad C_6H_{12}O_6 + 6 \; O_2$$

$$\Delta H = \Sigma n\Delta H_{products} - n\Delta H_{reactants}$$

$$6 \; (-94) + 6(-68.4) \qquad -304.7 + 6(0)$$

$$-564 + (-410.4) \qquad -304.7$$

$$\Delta H = -304.7 - (-974.4) = 669.7 = 670 \; kcal/mole$$

The positive ΔH indicates that the reaction is endothermic.

Second Law of Thermodynamics

The *second law of thermodynamics* states that the universe tends to become more disorganized. In other words, any spontaneous process always increases the entropy of the universe. According to the second law, the total entropy change is positive for every process. However, the entropy increase may not take place in the reacting system. Since

$$\Delta S_{univ} = \Delta S_{system} + \Delta S_{surroundings}$$

the increase in total entropy may take place in any part of the universe. Living cells do not increase their internal disorder when they consume and metabolize nutrients. The organism's surroundings increase in entropy instead (Figure 4.2).

Although entropy may be considered to be unusable energy, the formation of entropy is not a useless activity. In spontaneous processes, such as the melting of ice and the burning of wood, entropy always increases. In irreversible processes, entropy is now believed to be a driving force. Entropy directs a system toward equilibrium with its surroundings. (The mechanism by which living organisms avoid reaching equilibrium is discussed below.) To predict whether a process is spontaneous, the sign of ΔS_{univ} must be known. For example, if the value of ΔS_{univ} for a process is positive (i.e., the entropy of the universe increases), then the process is spontaneous. If ΔS_{univ} is negative, the process does not occur, but the reverse process takes place. The opposite process is spontaneous. If ΔS_{univ} is equal to zero, neither process tends to occur. Organisms that are at equilibrium with their surroundings are dead.

Although entropy always increases in a spontaneous process, measuring it is often impractical. A more convenient function for predicting the spontaneity of a process, called **free energy** G, was defined by Josiah Gibbs in the late nineteenth

KEY *Concepts*

The second law of thermodynamics states that the universe tends to become more disorganized. Entropy increases may take place anywhere in the system's universe. For processes in living organisms, the increase in entropy takes place in their surroundings.

Free Energy

FIGURE 4.2

A Living Cell as a Thermodynamic System.

(a) The molecules of the cell and its surroundings are in a relatively disordered state. (b) Heat is released from the cell as a consequence of reactions that create order among the molecules inside the cell. This energy increases the random motion, and therefore the disorder, of the molecules outside the cell. This process causes a net positive entropy change. The cell's decrease in entropy is more than offset by an increase in the entropy of the surroundings.

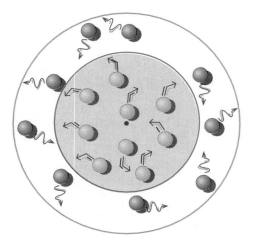

(a)

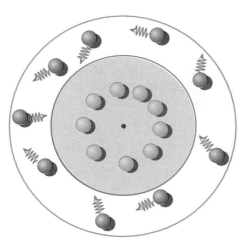

(b)

century. Free energy is a state function that relates the first and second laws of thermodynamics:

$$\Delta G = \Delta H - T\Delta S$$

where

ΔG = Gibbs free energy change
ΔH = enthalpy change
T = Kelvin temperature (°C + 273.15)
ΔS = entropy change

Free energy represents the maximum amount of useful work obtainable from a process. If a process occurs at constant temperature, pressure, and volume, it will be spontaneous in the direction in which free energy decreases ($\Delta G < 0$). Such processes are called **exergonic** (Figure 4.3). If the free energy change is positive ($\Delta G > 0$), the process is called **endergonic.** When the free energy change in a system is zero, the system is at equilibrium. Although the ΔG value is an important indicator of spontaneity, it provides no information about reaction rates. The study of reaction rates is called **kinetics** (Chapter 6).

Standard Free Energy Changes

A convention known as the *standard state* simplifies free energy calculations for biochemical reactions. The standard free energy $\Delta G°$ is a constant for each reaction. It is defined for reactions at 25°C and 1 atm pressure and for which the concentration of all solutes is 1 M.

The change in free energy for a reaction

$$aA + bB \rightleftarrows cC + dD$$

is related to the reaction's equilibrium constant:

$$K_{eq} = \frac{[C]_{eq}^{c}[D]_{eq}^{d}}{[A]_{eq}^{a}[B]_{eq}^{b}}$$

The free energy change ΔG is related to the standard free energy change $\Delta G°$ by the following equation:

$$\Delta G = \Delta G° + 2.303RT \log \frac{[C]^{c}[D]^{d}}{[A]^{a}[B]^{b}}$$

where

R = the gas constant (1.987 cal/mole·K)
T = Kelvin temperature

If the reaction is allowed to go to equilibrium, then $\Delta G = 0$. Because there is no free energy change, the expression reduces to

$$\Delta G° = -2.303RT \log \frac{[C]_{eq}^{c}[D]_{eq}^{d}}{[A]_{eq}^{a}[B]_{eq}^{b}}$$

When the reaction is at equilibrium, the expression for the standard free energy is

$$\Delta G° = -2.303RT \log K_{eq}$$

Because most biochemical reactions take place at or near pH 7, this pH value has been designated the standard pH in bioenergetics. The standard free energy change at pH 7 ($\Delta G°'$) is used in this textbook. However, the criterion for spontaneity of a reaction is ΔG, not $\Delta G°'$. A free energy problem appears on page 66.

FIGURE 4.3

The Gibbs Free Energy Equation.

At constant pressure, enthalpy (H) is essentially equal to the total energy content of the system. A process is spontaneous if it decreases free energy. Free energy changes ΔG are negative if enthalpy decreases or if the entropy term $T\Delta S$ is sufficiently large.

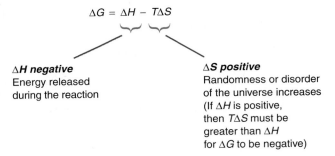

$$\Delta G = \Delta H - T\Delta S$$

ΔH negative
Energy released during the reaction

ΔS positive
Randomness or disorder of the universe increases (If ΔH is positive, then $T\Delta S$ must be greater than ΔH for ΔG to be negative)

4.2

For the reaction $HC_2H_3O_2 \rightleftarrows C_2H_3O_2^- + H^+$, calculate $\Delta G°$ and $\Delta G°'$. Assume that $T = 25°C$. The ionization constant for acetic acid is 1.8×10^{-5}. Is this reaction spontaneous? Recall that

$$K_{eq} = \frac{[C_2H_3O_2^-][H^+]}{[HC_2H_3O_2]}$$

Solution

1. Calculate $\Delta G°$.

$$\begin{aligned}\Delta G° &= -2.303RT \log K_{eq} \\ &= -2.303(1.987\,\text{cal/mole·K})(298\ \text{K})\log (1.8 \times 10^{-5}) \\ &= +6500\ \text{cal/mole}\end{aligned}$$

The $\Delta G°$ indicates that under these conditions the reaction is not spontaneous.

2. Calculate $\Delta G°'$. Use the relation between free energy change and standard free energy change.

$$\Delta G = \Delta G° + 2.303RT \log \frac{[C]^c[D]^d}{[A]^a[B]^b}$$

For this example the expression becomes

$$\Delta G°' = \Delta G° + 2.303RT \log \frac{[C_2H_3O_2^-][H^+]}{[HC_2H_3O_2]}$$

Recall that under standard conditions, [acetate] and [acetic acid] are both 1 molar. Substituting values, we have

$$\begin{aligned}\Delta G°' &= 6500 + 2.303(1.987\ \text{cal/mole·K})(298\ \text{K})(-7) \\ &= 6500 - 9549.66 = -3049.66\ \text{cal/mole} \quad \text{or} \quad -3050\ \text{cal/mole}\end{aligned}$$

Under the conditions specified for $\Delta G°$ (i.e., 1 M concentrations for all reactants including H^+) the ionization of acetic acid is not spontaneous, as indicated by the positive $\Delta G°$. When the pH value is 7, however, the reaction becomes spontaneous. A low $[H^+]$ makes the ionization of a weak acid such as acetic acid a more likely process, as indicated by the negative $\Delta G°'$.

Coupled Reactions

Many chemical reactions within living organisms have positive $\Delta G°'$ values. Fortunately, free energy values are additive in any reaction sequence.

$$A + B \rightleftarrows C + D \qquad\qquad \Delta G°'_{\text{reaction 1}} \qquad\qquad (1)$$
$$C + E \rightleftarrows F + G \qquad\qquad \Delta G°'_{\text{reaction 2}} \qquad\qquad (2)$$

$$A + B + E \rightleftarrows D + F + G \qquad \Delta G°'_{\text{overall}} = \Delta G°'_{\text{reaction 1}} + \Delta G°'_{\text{reaction 2}} \qquad (3)$$

Note that reactions (1) and (2) have a common intermediate. If the net $\Delta G°'$ value ($\Delta G°_{\text{overall}}$) is sufficiently negative, forming the products F and G is an exergonic process.

The conversion of glucose-6-phosphate to fructose-1,6-bisphosphate illustrates the principle of coupled reactions (Figure 4.4). The common intermediate in this reaction sequence is fructose-6-phosphate. Since the formation of fructose-6-phosphate from

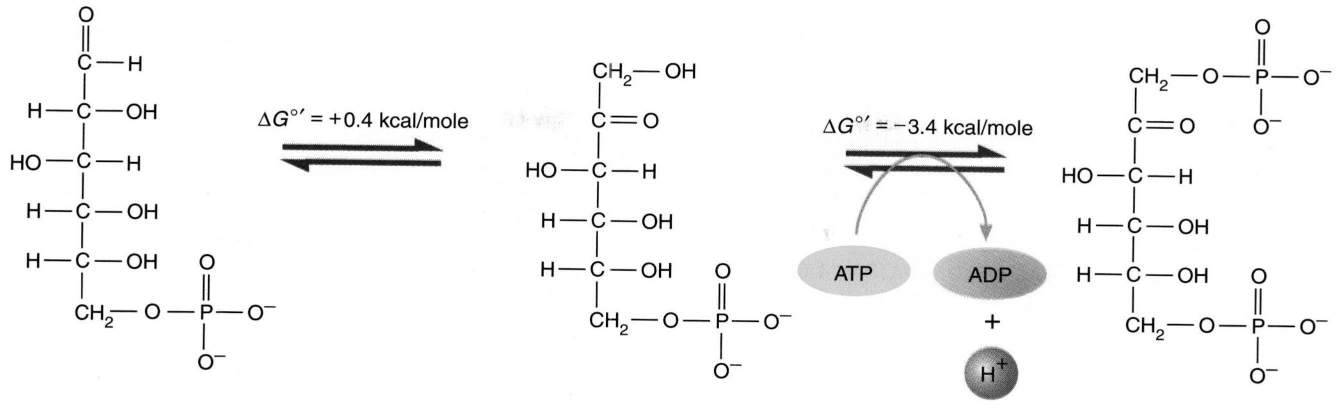

$\Delta G^{\circ\prime} = +0.4$ kcal/mole

$\Delta G^{\circ\prime} = -3.4$ kcal/mole

Glucose-6-phosphate

Fructose-6-phosphate

Fructose-1,6-bisphosphate

FIGURE 4.4

A Coupled Reaction.

The net $\Delta G^{\circ\prime}$ value for the two reactions is -3.0 kcal/mole.

glucose-6-phosphate is endergonic ($\Delta G^{\circ\prime}$ is $+0.4$ kcal/mole), the reaction is not expected to proceed as written (at least under standard conditions). The conversion of fructose-6-phosphate to fructose-1,6-bisphosphate is strongly exergonic because it is coupled to the cleavage of the phosphoanhydride bond of ATP. (The cleavage of ATP's phosphoanhydride bond to form ADP yields approximately -7.3 kcal/mole. ATP in living organisms is discussed in Section 4.3.) Since $\Delta G^{\circ\prime}_{\text{overall}}$ for the coupled reactions is negative, the reactions do proceed in the direction written at standard conditions.

PROBLEM 4.3

Glycogen is synthesized from glucose-1-phosphate. To be incorporated into glycogen, glucose-1-phosphate must be "energized" by being converted to a derivative of the nucleotide uridine diphosphate. The reaction is

Glucose-1-phosphate + UTP $\rightarrow$ UDP-glucose + PP$_i$

If the $\Delta G^{\circ\prime}$ value for this reaction is approximately zero, is this reaction favorable? If PP$_i$ (pyrophosphate) is hydrolyzed,

PP$_i$ + H$_2$O $\rightarrow$ 2 P$_i$

the loss in free energy ($\Delta G^{\circ\prime}$) is -8 kcal. How does this second reaction affect the first one? What is the overall reaction? Determine the $\Delta G^{\circ\prime}_{\text{overall}}$ value.

Solution

The overall reaction is

Glucose-phosphate + UTP $\rightarrow$ UDP-glucose + 2 P$_i$

$\Delta G^{\circ\prime}_{\text{overall}} = \Delta G^{\circ\prime}_{\text{reaction 1}} + \Delta G^{\circ\prime}_{\text{reaction 2}}$

$= 0 + (-8 \text{ kcal})$

$= -8 \text{ kcal}$

The hydrolysis of PP$_i$ drives the formation of UDP-glucose to the right.

QUESTION 4.1

Within living cells the concentrations of ATP and the products of its hydrolysis (ADP and P$_i$) are significantly lower than the standard 1 M concentrations. Therefore the actual free energy of hydrolysis of ATP ($\Delta G'$) differs from the standard free energy ($\Delta G^{\circ\prime}$). Unfortunately, it is difficult to obtain an accurate measure of the concentrations of cellular components. For this reason, only estimates can be made. The following equation includes a correction for nonstandard concentrations:

$$\Delta G' = \Delta G^{\circ\prime} + 2.303RT \log \frac{[ADP][P_i]}{[ATP]}$$

The temperature is 37°C. Assume that the pH is 7. Calculate the free energy of hydrolysis of ATP if the concentrations (mM) within a liver cell are as follows:

$$ATP = 4.0, \quad ADP = 1.35, \quad P_i = 4.65$$
$$\Delta G^{\circ\prime} = -7.3 \text{ kcal/mole}$$

4.2 OXIDATION–REDUCTION REACTIONS

In living organisms, both energy-capturing and energy-releasing processes consist largely of redox reactions. Because of the important role that redox reactions play in living organisms, the principles of redox reactions are reviewed.

Recall that redox reactions occur when electrons are transferred between an electron donor (reducing agent) and an electron acceptor (oxidizing agent). In some redox reactions, only electrons are transferred. For example, in the reaction

$$Cu^+ + Fe^{3+} \rightleftarrows Cu^{2+} + Fe^{2+}$$

Cu^+, the reducing agent, is oxidized to form Cu^{2+}. Meanwhile, Fe^{3+} is reduced to Fe^{2+}. In many reactions, however, both electrons and protons are transferred. In one such reaction, the enzyme lactate dehydrogenase catalyzes the transfer of two protons and two electrons in the reduction of pyruvate by NADH to form lactate (Figure 4.5). (NADH is one example of a class of biomolecules used as intermediates in redox reactions. Their structures and role in metabolism are discussed in Chapters 6 and 11.)

Redox reactions are more easily understood if they are separated into half-reactions. For example, in the reaction between copper and iron the Cu^+ ion lost an electron to become Cu^{2+}.

$$Cu^+ \rightleftarrows Cu^{2+} + e^-$$

This equation indicates that Cu^+ is the electron donor. (Together Cu^+ and Cu^{2+} constitute a **conjugate redox pair.**) As Cu^+ loses an electron, Fe^{3+} gains an electron to form Fe^{2+}:

$$Fe^{3+} + e^- \rightleftarrows Fe^{2+}$$

In this half-reaction, Fe^{3+} is an electron acceptor. The separation of redox reactions emphasizes that electrons are always the common intermediates between half-reactions.

The constituents of half-reactions may be observed in an electrochemical cell (Figure 4.6). Each half-reaction takes place in a separate container or *half-cell*. The movement of electrons generated in the half-cell undergoing oxidation (e.g., $Cu^+ \rightarrow Cu^{2+} + e^-$) generates a voltage (or potential difference) between the two half-cells. The sign of the voltage (measured by a voltmeter) is positive or negative according to the direction of the electron flow. The magnitude of the potential difference is a measure of the energy that drives the reaction.

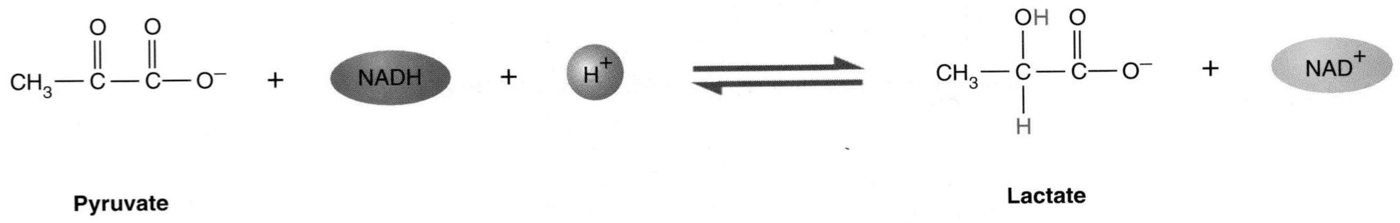

Pyruvate   **Lactate**

FIGURE 4.5

Reduction of Pyruvate by NADH.

In this redox reaction, both protons and electrons are transferred.

FIGURE 4.6

An Electrochemical Cell.

Electrons flow from the copper electrode through the voltmeter to the iron electrode. The salt bridge containing KCl completes the electric circuit. The voltmeter measures the electrical potential as electrons flow from one half-cell to the other.

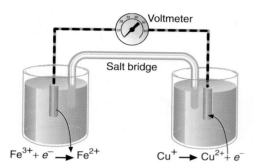

The tendency for a specific substance to lose or gain electrons is called its **redox** or **reduction potential.** The redox potential of a conjugate redox pair is measured in an electrochemical cell against a reference standard, usually a standard hydrogen electrode. The redox potential of the standard hydrogen electrode is 0.0 V at 1 atm, by definition. Substances with a more negative reduction potential will transfer electrons to a substance with a more positive reduction potential and the $\Delta E_0'$ will be positive. In biochemistry the reference half-reaction is

$$2\,H^+ + 2e^- \rightleftarrows H_2$$

when

$$pH = 7$$
$$Temperature = 25°C$$
$$Pressure = 1\ atm$$

Under these conditions the reduction potential of the hydrogen electrode is 0.42 V when measured against the standard hydrogen electrode in which the $[H^+]$ is 1 M. Substances with reduction potentials lower than 0.42 V (i.e., those with more negative values) have a lower affinity for electrons than does H^+. Substances with higher reduction potentials (i.e., those with more positive values) have a greater affinity for electrons (Table 4.1). (The pH in the test electrode is 7.0 for each of the redox half-reactions and the pH of the reference standard electrode is 0.)

Electrons flow spontaneously from a species with a more negative E_0' value to a species with a more positive E_0', so that the $\Delta E_0'$ is positive. The relationship between $\Delta E_0'$ and $\Delta G^{°\prime}$ is

$$\Delta G^{°\prime} = -nF\,\Delta E_0'$$

where

$\Delta G^{°\prime}$ = the standard free energy
n = the number of electrons transferred
F = the Faraday constant (23,062 cal/V·mole)
$\Delta E_0'$ = the difference in reduction potential between the electron donor and the electron acceptor under standard conditions

Most of the living cell's free energy is captured by the mitochondrial electron transport system (Chapter 11). During this process, electrons are transferred from a relatively electronegative redox pair ($NADH/NAD^+$) to more electropositive acceptors. The last component in the system is the $H_2O/\frac{1}{2}\,O_2$ pair.

$$\frac{1}{2}\,O_2 + NADH + H^+ \rightarrow H_2O + NAD^+$$

The free energy released as a pair of electrons passes from NADH to O_2 under standard conditions is calculated as follows:

$$\Delta G^{°\prime} = -2(23{,}062\ cal/V\!\cdot\!mole)[0.815 - (-0.32)]$$
$$= -52.4\ kcal/mole$$

A significant portion of the free energy generated as electrons move from NADH to O_2 in the electron transport system is used to synthesize ATP.

In several metabolic processes, electrons move from redox pairs with more positive reduction potentials to those with more negative reduction potentials. Of course, energy is required. The most prominent example of this phenomenon is photosynthesis (Chapter 12). Photosynthetic organisms use captured light energy to drive electrons from electron donors, such as water, to electron acceptors with more negative reduction potentials (Figure 4.7). The energized electrons eventually flow back to acceptors with more positive reduction potentials, thereby providing energy for ATP synthesis and CO_2 reduction to form carbohydrate.

KEY *Concepts*

In living organisms, both energy-capturing and energy-releasing processes consist primarily of redox reactions. In redox reactions, electrons move between an electron donor and an electron acceptor. In many reactions, both electrons and protons are transferred.

TABLE 4.1

Standard Reduction Potentials*

Redox Half-Reaction	Standard Reduction Potential (E_0') (V)
$2 H^+ + 2 e^- \rightarrow H_2$	-0.42
$NAD^+ + H^+ + 2 e^- \rightarrow NADH$	-0.32
Fumarate $+ 2 H^+ + 2 e^- \rightarrow$ succinate	-0.031
$S + 2 H^+ + 2 e^- \rightarrow H_2S$	-0.23
Acetaldehyde $+ 2 H^+ + 2 e^- \rightarrow$ ethanol	-0.20
Pyruvate $+ 2 H^+ + 2 e^- \rightarrow$ lactate	-0.19
$Cu^+ \rightarrow Cu^{2+} + e^-$	-0.16
Cytochrome b (Fe^{3+}) $+ e^- \rightarrow$ cytochrome b (Fe^{2+})	$+0.075$
Cytochrome c_1 (Fe^{3+}) $+ e^- \rightarrow$ cytochrome c_1 (Fe^{2+})	$+0.22$
Cytochrome c (Fe^{3+}) $+ e^- \rightarrow$ cytochrome c (Fe^{2+})	$+0.235$
Cytochrome a (Fe^{3+}) $+ e^- \rightarrow$ cytochrome a (Fe^{2+})	$+0.29$
$NO_3^- + 2 H^+ + 2 e^- \rightarrow NO_2^- + H_2O$	$+0.42$
$NO_2^- + 8 H^+ + 6 e^- \rightarrow NH_4^+ + 2 H_2O$	$+0.44$
$Fe^{3+} + e^- \rightarrow Fe^{2+}$	$+0.77$
$\frac{1}{2} O_2 + 2 H^+ + 2 e^- \rightarrow H_2O$	$+0.82$

*By convention, redox reactions are written with the reducing agent to the right of the oxidizing agent and the number of electrons transferred. In this table the redox pairs are listed in order of increasing E_0' values. The more negative the E_0' value is for a redox pair, the lower its affinity for electrons. The more positive the E_0' value is, the greater the affinity of the redox pair for electrons. Under appropriate conditions, a redox half-reaction reduces any of the half-reactions below it in the table.

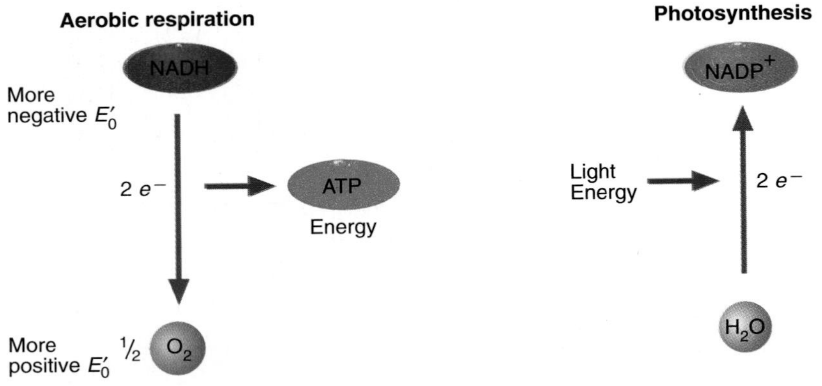

FIGURE 4.7

Electron Flow and Energy.

Electron flow may be used to generate and capture energy in aerobic respiration. Energy may also be used to drive electron flow in photosynthesis. ($NADP^+$ is a phosphorylated version of NAD^+.)

The following is an oxidation–reduction problem.

PROBLEM

4.4

Use the following half-cell potentials to calculate (a) the overall cell potential and (b) $\Delta G^{\circ\prime}$.

$$\text{Succinate} + \frac{1}{2} O_2 \rightarrow \text{fumarate} + H_2O$$

The half-reactions are

$$\text{Fumarate} + 2 H^+ + 2 e^- \rightarrow \text{succinate} \quad (\Delta E_0' = -0.031 \text{ V})$$

$$\frac{1}{2} O_2 + 2 H^+ + 2 e^- \rightarrow H_2O \quad (\Delta E_0' = +0.82 \text{ V})$$

Solution

Write one of the half-reactions as an oxidation (i.e., reverse the equation) and add the two half-reactions.

$$\text{Succinate} \rightarrow \text{fumarate} + 2\ H^+ + 2\ e^- \qquad (\Delta E_0' = +0.031\ \text{V}) \text{ (oxidation)}$$

$$\frac{1}{2}\ O_2 + 2\ H^+ + 2\ e^- \rightarrow H_2O \qquad\qquad (\Delta E_0' = +0.82\ \text{V}) \text{ (reduction)}$$

The net reaction is therefore

$$\text{Succinate} + \frac{1}{2}O_2 \rightarrow \text{fumarate} + H_2O$$

a. The overall potential is the sum of the potentials for each half-cell:

$$+0.82 + 0.031 = +0.85\ \text{V}$$

b. Use the formula to find $\Delta G^{\circ\prime}$.

$$\begin{aligned}
\Delta G^{\circ\prime} &= -nF\Delta E_0' \\
&= (-2)(23.062\ \text{cal/V·mole})(0.85\ \text{V}) \\
&= -39{,}205.4\ \text{cal} \\
&= -39\ \text{kcal/mole}
\end{aligned}$$

4.2

Because redox reactions play an important role in living processes, biochemists need to determine the oxidation state of the atoms in a molecule. In one method, the oxidation state of an atom is determined by assigning numbers to carbon atoms based on the type of groups attached to them. For example, a bond to a hydrogen is assigned the value -1. A bond to another carbon atom is valued at 0, while a bond to an electronegative atom such as oxygen or nitrogen is valued at $+1$. The values of a single carbon atom in a molecule may range from -4 (e.g., CH_4) to $+4$ (CO_2). Note that methane is a high-energy molecule and carbon dioxide is a low-energy molecule. As carbon changes its oxidation state from -4 (methane) to $+4$ (carbon dioxide), a large amount of energy is released. This process is therefore highly exothermic.

Ethanol is degraded in the liver by a series of redox reactions. Identify the oxidation state of the indicated carbon atom in each molecule in the following reaction sequence.

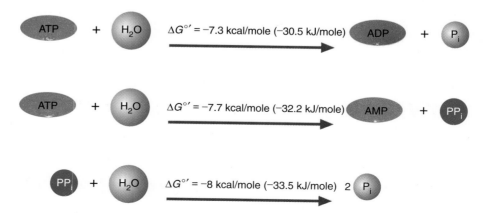

4.3 THE ROLE OF ATP

Adenosine triphosphate (ATP) plays an extraordinarily important role within living cells. The hydrolysis of ATP (Figure 4.8) immediately and directly provides the free energy to drive an immense variety of endergonic biochemical reactions. Pro-

FIGURE 4.8

Hydrolysis of ATP.

ATP may be hydrolyzed to form ADP and P_i (orthophosphate) or AMP (adenosine monophosphate) and PP_i (pyrophosphate). Pyrophosphate may be subsequently hydrolyzed to orthophosphate, releasing additional free energy. The hydrolysis of ATP to form AMP and pyrophosphate is often used to drive reactions with high positive $\Delta G^{\circ\prime}$ values or to ensure that a reaction goes to completion.

REDOX IN THE DEEP

Most autotrophic ("self-feeding") organisms trap light energy during photosynthesis. (Photosynthesizing organisms, referred to as photoautotrophs, include green plants, algae, and cyanobacteria.) A small group of bacterial species use specific inorganic reactions to generate energy to drive their metabolic processes. These organisms are often referred to as **lithotrophs** or **chemolithotrophs** (*lithos* = stone). Found in a diverse set of habitats (e.g., soil and fresh and marine waters), the chemolithotrophs recycle elements in the biosphere. For example, the nitrifying bacteria convert both ammonia and nitrites to nitrates. As an important component of the nitrogen cycle, nitrifying bacteria contribute to soil fertility.

In 1977, scientists discovered a new and unexpected habitat in the Pacific Ocean northeast of the Galápagos Islands. In areas surrounding underwater hot springs, a large number of previously unknown animal species were found living in total darkness. In other marine and terrestrial habitats the photoautotrophs produce the majority of the organic food required to sustain animal life. Yet these hot water springs, 2600 meters below the ocean's surface, are teeming with life. Two especially prominent examples are the giant white clam (*Calyptogena magnifica*) and the giant tube worm (*Riftia pachyptila*).

What conditions could account for these oceanic oases? Researchers made a closer inspection of the hot springs, called hydrothermal vents or calderas, where columns of mineral-laden hot water pour through fissures in the sea floor. (Cold water seeps downward through cracks in the earth's crust and is heated by molten lava. Later, it is forced upward through the chimneylike vents.) The water in the hot springs is rich in hydrogen sulfide (H_2S). (In the extreme heat and pressure deep within the crust, sulfate is reduced to form the high-energy H—S bond.) The water surrounding the vents contains large numbers of sulfur bacteria, which get energy from H_2S. These organisms are the primary organic food for the hydrothermal vent community. As they oxidize H_2S to form H_2SO_4, the sulfur bacteria use the energy of the earth's interior captured by H_2S formation to convert CO_2 into organic nutrients.

To benefit from this process, several vent animals have established endosymbiotic relationships with the sulfur bacteria. (See Special Interest Box 2.1 for a discussion of endosymbiosis.) One of the best-researched examples is *R. pachyptila*. The giant tube worm, which consists primarily of a long, thin sac, has neither a mouth nor a digestive tract. The trophosome, the site of the redox reactions, is the animal's most prominent organ. It is colonized by a large number of sulfur-oxidizing bacteria. In return for a steady supply of H_2S, O_2, and CO_2 provided by the tube worm's circulatory system, the sulfur bacteria provide the organic nutrients required for the worm's growth and development.

---

duced from ADP and P_i using energy released by the breakdown of food molecules and the light reactions of photosynthesis, ATP drives several types of processes (Figure 4.9). These include (1) biosynthesis of biomolecules, (2) active transport of substances across cell membranes, and (3) mechanical work such as muscle contraction.

Because of its structure, ATP is ideally suited to its role as universal energy currency. ATP is a nucleotide composed of adenine, ribose, and a triphosphate unit (Figure 4.10). The two terminal phosphoryl groups ($-PO_3^{2-}$) are linked by phosphoan-

FIGURE 4.9

The Role of ATP.

ATP is an intermediate in the flow of energy from food molecules to the biosynthetic reactions of metabolism.

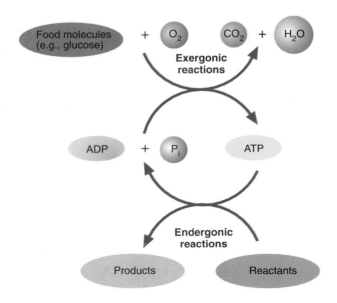

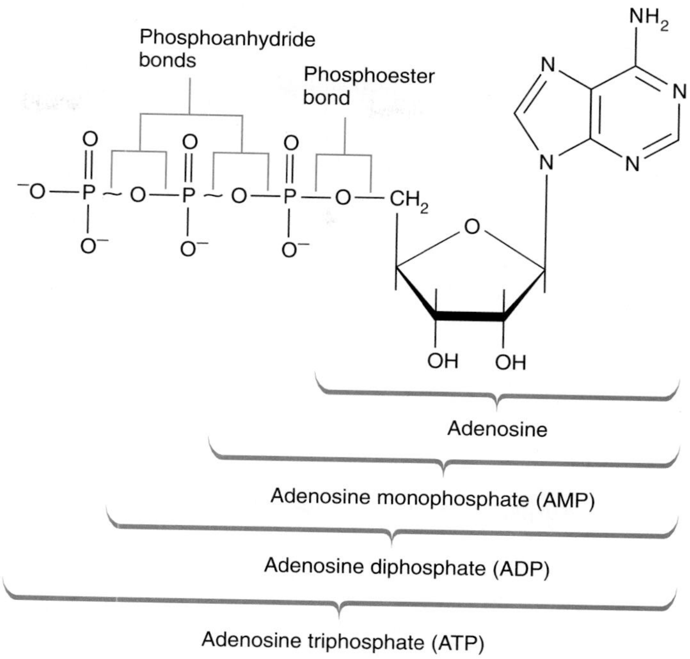

FIGURE 4.10

The Structure of ATP.

The squiggle ($\sim$) in ATP indicates that this nucleotide is a high-energy phosphate compound.

FIGURE 4.10

KEY *Concepts*

The hydrolysis of ATP immediately and directly provides the free energy to drive an immense variety of endergonic biochemical reactions. Because ATP has an intermediate phosphate group transfer potential, it can carry phosphoryl groups from higher-energy compounds to lower-energy compounds. ATP is the energy currency for living systems.

hydride bonds. Although anhydrides are easily hydrolyzed, the phosphoanhydride bonds of ATP are sufficiently stable under mild intracellular conditions. Specific enzymes facilitate ATP hydrolysis.

The tendency of ATP to undergo hydrolysis, also referred to as its **phosphate group transfer potential,** is not unique. A variety of biomolecules can transfer phosphate groups to other compounds. Table 4.2 lists several important examples. Phosphorylated compounds with high $\Delta G^{\circ\prime}$ values of hydrolysis have higher phosphate group transfer potentials than those compounds with lower values. Because ATP has an intermediate phosphate group transfer potential, it can be an intermediate carrier of phosphoryl groups from higher-energy compounds such as phosphoenolpyruvate to low-energy compounds (Figure 4.11). ATP is therefore the "energy currency" for living systems, since cells usually transfer phosphate by coupling reactions to ATP hydrolysis. The two phosphoanhydride bonds of ATP are often referred to as "high-energy." However, the term *high-energy bond* is now considered inappropriate. The

TABLE 4.2

Standard Free Energy of Hydrolysis of Selected Phosphorylated Biomolecules

Molecule	$\Delta G^{\circ\prime}$	
	kcal/mole	kJ/mole
Glucose-6-phosphate	−3.3	−13.8
Fructose-6-phosphate	−3.8	−15.9
Glucose-1-phosphate	−5	−20.9
ATP → AMP + PP$_i$	−7.7	−32.2
ATP → ADP + P$_i$	−7.3	−30.5
PP$_i$	−8.0	−33.5
Phosphocreatine	−10.3	−43.1
Carbamoyl phosphate	−12.3	−51.4
Glycerate-1,3-biphosphate	−11.8	−49.4
Phosphoenolpyruvate	−14.8	−61.9

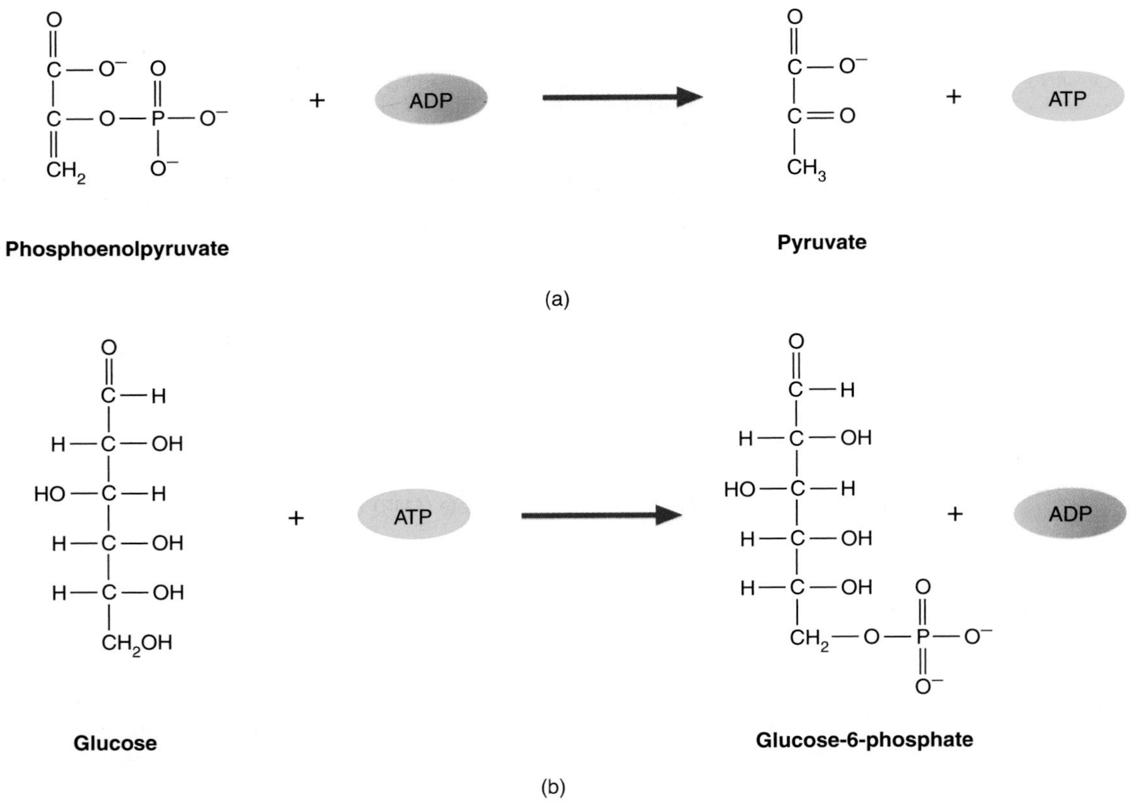

Phosphoenolpyruvate   **Pyruvate**

(a)

Glucose   **Glucose-6-phosphate**

(b)

FIGURE 4.11

Transfer of Phosphoryl Groups

(a) Transfer of a phosphoryl group from phosphoenolpyruvate to ADP. As discussed in Chapter 8, this reaction is one of two steps which form ATP during glycolysis, a reaction pathway that breaks down glucose. (b) Transfer of a phosphoryl group from ATP to glucose. The product of this reaction, glucose-6-phosphate, is the first intermediate formed during glycolysis.

free energy generated by ATP hydrolysis depends on the difference between the free energy of the products and that of the reactants. High-energy bonds are "high" only when compared with something else, that is, the P—O bonds in phosphate. To understand why ATP hydrolysis is so exergonic, several factors must be considered.

1. At typical intracellular pH values, the triphosphate unit of ATP carries three or four negative charges that repel each other. Hydrolysis of ATP reduces electrostatic repulsion.

2. Because of *resonance hybridization,* the products of ATP hydrolysis are more stable than ATP. When a molecule has two or more alternative structures that differ only in the position of electrons, the result is called a **resonance hybrid.** The electrons in a resonance hybrid with several contributing structures possess much less energy than those with fewer contributing structures. The contributing structures of the phosphate resonance hybrid are illustrated in Figure 4.12.

FIGURE 4.12

Contributing Structures of the Resonance Hybrid of Phosphate.

At physiological pH, orthophosphate is HPO_4^{2-}. In this illustration, H^+ is not assigned permanently to any of the four oxygen atoms.

QUESTION 4.3

Walking consumes approximately 100 kcal/mi. For ATP hydrolysis (ATP → ADP + P$_i$), the reaction that drives muscle contraction, $\Delta G^{\circ\prime}$ is -7.3 kcal/mole (-30.5 kJ/mole). Calculate how many grams of ATP must be produced to walk a mile. ATP synthesis is coupled to the oxidation of glucose ($\Delta G^{\circ\prime} = -686$ kcal/mole). How many grams of glucose are actually metabolized to produce this amount of ATP? (Assume that only glucose oxidation is used to generate ATP and that 40% of the energy generated from this process is used to phosphorylate ADP. The gram molecular weight of glucose is 180 g, while that of ATP is 507 g.)

KEY WORDS

bioenergetics, *62*

chemolithotrophs, *72*

conjugate redox pair, *68*

endergonic process, *65*

endothermic reaction, *63*

enthalpy, *62*

entropy, *62*

exergonic process, *65*

exothermic reaction, *63*

free energy, *62*

isothermic reaction, *63*

kinetics, *65*

lithotrophs, *72*

phosphate group transfer potential, *73*

photoautotroph, *62*

redox potential (reduction potential), *69*

resonance hybrid, *74*

work, *62*

SUMMARY

All living organisms unrelentingly require energy. By using bioenergetics, the study of energy transformations, the direction and extent to which biochemical reactions proceed can be determined. Enthalpy (a measure of heat content) and entropy (a measure of disorder) are related to the first and second laws of thermodynamics, respectively. Free energy (the portion of total energy that is available to do work) is related to a mathematical relationship between enthalpy and entropy.

Energy and heat transformations take place in a "universe" composed of a system and its surroundings. In an open system, matter and energy are exchanged between the system and its surroundings. If energy but not matter can be exchanged with the surroundings, then the system is said to be closed. Living organisms are open systems.

Several thermodynamic quantities are state functions, that is, their value does not depend on the pathway used to make or degrade a specific substance. Examples of state functions are total energy, free energy, enthalpy, and entropy. Some quantities, such as work and heat, are not state functions, because they depend on the pathway.

Free energy, a state function that relates the first and second laws of thermodynamics, represents the maximum useful work obtainable from a process. Exergonic processes, that is, processes in which free energy decreases ($\Delta G < 0$), are spontaneous. If the free energy change is positive ($\Delta G > 0$), the process is called endergonic. A system is at equilibrium when the free energy change is zero. The standard free energy (ΔG°) is defined for reactions at 25°C, 1 atm pressure, and 1 M solute concentrations. The standard pH in bioenergetics is 7. The standard free energy change $\Delta G^{\circ\prime}$ at pH 7 is used in this textbook.

Most reactions that capture or release energy are redox reactions. In these reactions, electrons are transferred between an electron donor (reducing agent) and an electron acceptor (oxidizing agent). In some reactions, only electrons are transferred. In other reactions, both electrons and protons are transferred. The tendency for a specific conjugate redox pair to lose an electron is called its redox potential. Electrons flow spontaneously from electronegative redox pairs to those that are more positive. In favorable redox reactions $\Delta E_0{}'$ is positive and $\Delta G^{\circ\prime}$ is negative.

ATP hydrolysis provides most of the free energy required for living processes. ATP is ideally suited to its role as universal energy currency because its phosphoanhydride structure is easily hydrolyzed.

SUGGESTED READINGS

Hanson, R.W., The Role of ATP in Metabolism. *Biochem. Educ.* 17:86–92, 1989.

Harold, F.M., *The Vital Force: A Study of Bioenergetics*, W.H. Freeman, New York, 1986.

Ho, M.W., *The Rainbow and the Worm: The Physics of Organisms*, World Scientific Publishing Co., Singapore, 1993.

Schrödinger, E., *What Is Life?*, Cambridge University Press, Cambridge, England, 1944.

QUESTIONS

1. Define each of the following terms:
   a. thermodynamics
   b. endergonic
   c. enthalpy
   d. free energy
   e. high-energy bond
   f. reduction
   g. redox potential
   h. phosphate group transfer potential

2. Which of the following thermodynamic quantities are state functions? Explain.
   a. work
   b. entropy
   c. enthalpy
   d. free energy

3. Which of the following reactions could be driven by coupling to the hydrolysis of ATP? (The $\Delta G^{\circ\prime}$ value in kcal/mole for each reaction is indicated in parentheses.)

   $ATP + H_2O \rightarrow ADP + P_i$ (-7.3)

   a. glucose-1-phosphate $\rightarrow$ glucose-6-phosphate (-1.7)
   b. glucose $+ P_i \rightarrow$ glucose-6-phosphate $(+3.3)$
   c. acetic acid $\rightarrow$ acetic anhydride $(+23.8)$
   d. glucose $+$ fructose $\rightarrow$ sucrose $(+7)$
   e. glycylglycine $+$ water $\rightarrow$ 2 glycine (-2.2)

4. The K_a for the ionization of formic acid is 1.8×10^{-4}. Calculate ΔG° for this reaction at 25°C.

5. The following reaction is catalyzed by the enzyme glutamine synthetase:

   $ATP + glutamate + NH_3 \rightarrow ADP + P_i + glutamine$

   Use the following equations with $\Delta G^{\circ\prime}$ values given in kcal/mole, to calculate $\Delta G^{\circ\prime}$ for the overall reaction.

   $ATP + H_2O \rightarrow ADP + P_i$ (-7.3)
   Glutamine $+ H_2O \rightarrow$ glutamate $+ NH_3$ (-3.4)

6. Use the following half-reactions and $\Delta E_0{}'$ values given in volts.

   Ethanol $\rightarrow$ acetaldehyde $+ 2 H^+ + 2 e^-$ $(+0.2)$
   NADH $\rightarrow$ NAD$^+ + H^+ + 2 e^-$ $(+0.32)$

   Calculate the standard free energy change of the reaction

   Ethanol $+$ NAD$^+ \rightarrow$ acetaldehyde $+$ NADH $+ H^+$

   Is this reaction exergonic or endergonic?

7. The $\Delta G^{\circ\prime}$ values (kcal/mole) for the following reactions are indicated in parentheses.

   Ethyl acetate $+$ water $\rightarrow$
          ethyl alcohol $+$ acetic acid (-4.7) (i)
   Glucose-6-phosphate $+$ water $\rightarrow$
          glucose $+ P_i$ (-3.3) (ii)

   Which statements are true?
   a. The rate of reaction (i) is greater than the rate of reaction (ii).
   b. The rate of reaction (ii) is greater than the rate of reaction (i).
   c. Neither reaction is spontaneous.
   d. Reaction rates cannot be determined from energy values.

8. Under standard conditions, which statements are true?
   a. $\Delta G = \Delta G^{\circ}$
   b. $\Delta H = \Delta G$
   c. $\Delta G = \Delta G^{\circ} + 2.303RT \log K_{eq}$
   d. $\Delta G^{\circ} = \Delta H - T\Delta S$
   e. $P = 1$ atm
   f. $T = 273$ K
   g. [reactants] = [products] = 1 molar

9. Which statements concerning free energy change are true?
   a. Free energy change is a measure of the rate of a reaction.
   b. Free energy change is a measure of the maximum amount of work available from a reaction.
   c. Free energy change is a constant for a reaction under any conditions.
   d. Free energy change is related to the equilibrium constant for a specific reaction.
   e. Free energy change is equal to zero at equilibrium.
   f. Free energy change of a reaction is related to the redox potential of the reaction.

10. Consider the following reaction:

    Glucose-1-phosphate $\rightarrow$ glucose-6-phosphate
    $\Delta G^{\circ} = -1.7$ kcal/mole

    What is the equilibrium constant for this reaction at 25°C?

11. Pyruvate oxidizes to form carbon dioxide and water and liberates 273 kcal/mole. If electron transport also occurs, approximately 12.5 ATP molecules are produced. The free energy of hydrolysis for ATP is -7.3 kcal/mole. What is the apparent efficiency of ATP production?

12. Which of the following compounds would you expect to liberate the least free energy when hydrolyzed? Explain.
    a. ATP
    b. ADP
    c. AMP
    d. phosphoenolpyruvate
    e. phosphocreatine

13. In the reaction

    $ATP + glucose \rightarrow ADP + glucose-6-phosphate$

    ΔG° is -4 kcal/mole. Assume that the concentration of ATP and ADP are each one molar and $T = 25$°C. What ratio of glucose-6-phosphate to glucose would allow the reverse reaction to begin?

14. Consider the following reduction half-reactions:

    Pyruvate $+ 2 e^- + 2 H^+ \rightarrow$ lactate $(E_0{}' = -0.19$ V)
    Fe$^{3+} + 1 e^- \rightarrow$ Fe^{2+} $(E_0{}' = +0.77$ V)

    Determine the voltage for the following reaction:
    2 Fe$^{2+} +$ pyruvate $+ 2$ H+ $\rightarrow 2$ Fe$^{3+} +$ lactate

15. Which statements are true and which are false?
    a. In a closed system, neither energy nor matter is exchanged with the surroundings.
    b. State functions are independent of the pathway.
    c. A process is isothermic if $\Delta H = 0$.
    d. The sign and magnitude of ΔG give important information about the direction and rate of a reaction.
    e. At equilibrium, $\Delta G = \Delta G°$.
    f. For two reactions to be coupled, they must have a common intermediate.
    g. A reducing agent accepts electrons from an oxidizing agent.

16. Chemolithotropes live in regions of the ocean that receive no sunlight. These organisms convert hydrogen sulfide to sulfate as to obtain energy to incorporate carbon dioxide into organic molecules. Calculate how many moles of hydrogen sulfide are required to produce one mole of glucose. Assume that the efficiency of the reaction is 100%.

$H_2S + 2 O_2 \rightarrow H_2SO_4$ ($\Delta H_f = -189.1$ kcal/mole)

$6 CO_2 + 6 H_2O \rightarrow C_6H_{12}O_6 + 6 O_2$ ($\Delta H_f = +670$ kcal/mole)

Could these organisms survive without photosynthesis producing additional oxygen near the ocean surface?

17. Thermodynamics is based on the behavior of large numbers of molecules. Yet within a cell there may only be a few molecules of a particular type at a time. Do the laws of thermodynamics apply under these circumstances?

18. Frequently, when salts dissolve in water, the solution becomes warm. Such a process is exothermic. When other salts, such as sodium chloride, dissolve in water, the solution becomes cold, indicating an endothermic process. Since endothermic processes are usually not spontaneous, why does the latter process proceed?

19. Of the three thermodynamic quantities ΔH, ΔG, and ΔS, which provides the most useful criterion of spontaneity in a reaction? Explain.

20. What factors make ATP suitable as an "energy currency" for the cell?

21. To determine the $\Delta G°'$ of a reaction within a cell, what information would you need?

PEPTIDES AND PROTEINS

Chapter Five

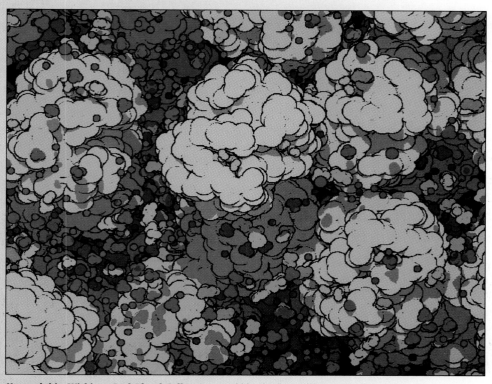

Hemoglobin Within a Red Blood Cell. *Human red blood cells are filled almost to bursting with the oxygen-carrying protein hemoglobin. The large pink structures are hemoglobin molecules. Sugar and amino acids are shown in green. Positive ions are blue. Negative ions are red. The large blue molecule is an enzyme.*

OBJECTIVES

After you have studied this chapter, you should be able to answer these questions:

1. *How does the structure of amino acids affect the proteins that contain them?*
2. *In addition to their role as protein building blocks, what other functions do amino acids have?*
3. *What are peptides and what functions do they have?*
4. *What bonds are important in protein structure?*
5. *What are the four levels of protein structure?*
6. *What functions do proteins have in living organisms?*
7. *What is protein denaturation and how does it occur?*
8. *What unique properties do fibrous proteins typically have?*
9. *What roles do globular proteins play in living organisms?*
10. *What is allosteric regulation?*

Proteins are essential constituents of all organisms. Most tasks performed by living cells require proteins. The variety of functions that they perform is astonishing. In animals, for example, proteins are the primary structural components of muscle, connective tissue, feathers, nails, and hair. In addition to serving as structural materials in all living organisms, proteins are involved in such diverse functions as metabolic regulation, transport, defense, and catalysis. The functional diversity of these biopolymers is directly related to the variety of sizes and shapes they can assume.

Proteins are immensely complex molecules. These polymers may be composed of 20 different amino acids. The molecular weights of amino acid polymers, often referred to as **polypeptides,** range from several thousand to several million daltons. Those with low molecular weights, typically consisting of fewer than 50 amino acids, are called **peptides.** The term **protein** describes molecules with more than 50 amino acids. Each protein consists of one or more polypeptide chains. The distinction between proteins and peptides is often unclear. For example, some biochemists define oligopeptides as polymers consisting of two to ten amino acids and polypeptides as having more than ten residues. Proteins, in this view, have molecular weights greater than 10,000 daltons (D). In addition, the terms "protein" and "polypeptide" are often used interchangeably. Throughout this textbook the terms "peptide" and "protein" will be used as defined above. The term "polypeptide" will be used whenever the topic under discussion applies to both peptides and proteins.

In this chapter the structure of peptides and proteins is examined. After a discussion of the structure and chemical properties of the amino acids, the structures of peptides and proteins are described. Several examples of fibrous and globular proteins are examined. The emphasis throughout the chapter is on the intimate relationship between the structure and function of polypeptides. In Chapter 6 the functioning of the enzymes, an especially important group of proteins, is discussed. Protein synthesis is covered in Chapter 18.

5.1 AMINO ACIDS

Proteins may be broken into their constituent monomer molecules by hydrolysis. The amino acid products of the reaction constitute the *amino acid composition* of the protein. The structures of the 20 amino acids that are commonly found in proteins are shown in Figure 5.1. These amino acids are referred to as *standard* amino acids. (Common abbreviations for the standard amino acids are listed in Table 5.1.) *Nonstandard* amino acids consist of amino acid residues that have been chemically modified after they have been incorporated into a polypeptide and amino acids that occur in living organisms but are not found in proteins. Note that 19 of the standard amino acids have the same general structure (Figure 5.2). These molecules contain a central carbon atom (the α-carbon) to which an amino group, a carboxylate group, a hydrogen atom, and an R (side chain) group are attached. The twentieth amino acid is proline, which is actually an α-imino acid, since its nitrogen is bonded to the α-carbon and the side chain group. Because of its unusual structure, proline gives a polypeptide important structural features.

When the pH is 7, the carboxyl group of an amino acid is in its conjugate base form ($-COO^-$) and the amino group is in its conjugate acid form ($-NH_3^+$). Thus each amino acid can behave as either an acid or a base. The term **amphoteric** is used to describe this property. Neutral molecules that bear an equal number of positive and negative charges simultaneously are called **zwitterions.** The R group, however gives each amino acid its unique properties.

In addition to their primary function as components of protein, amino acids have several other biological roles.

1. Several α-amino acids or their derivatives act as chemical messengers (Figure 5.3). For example, glycine, γ-amino butyric acid (GABA, a derivative of gluta-

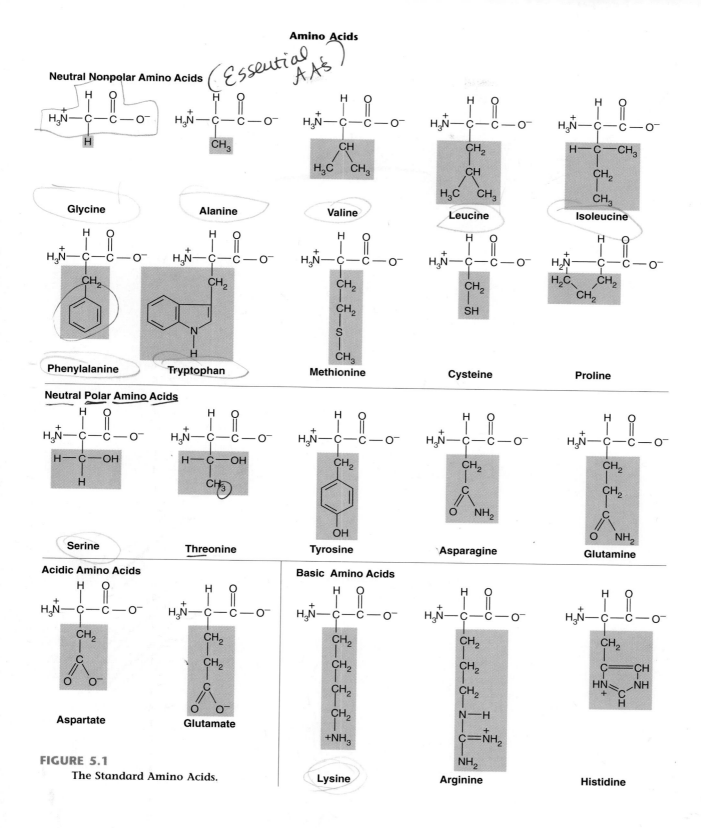

FIGURE 5.1

The Standard Amino Acids.

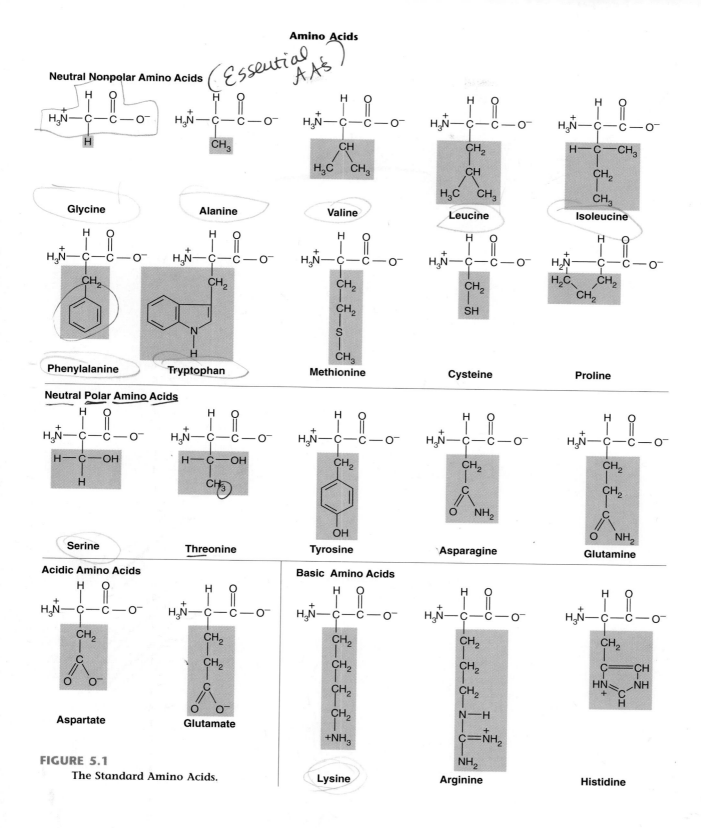

FIGURE 5.2

General Structure of the
α-Amino Acids.

mate), and serotonin and melatonin (derivatives of tryptophan) are **neurotransmitters** (substances released from one nerve cell that influence the function of a second nerve cell or a muscle cell). Thyroxine (a tyrosine derivative produced in the thyroid gland of animals) and indole acetic acid (a tryptophan derivative found in plants) are two examples of hormones. (**Hormones** are chemical signal molecules produced in one cell that regulate the function of other cells.)

TABLE 5.1
Names and Abbreviations of the Standard Amino Acids

Amino Acid	Three-Letter Abbreviations	One-Letter Abbreviations
Alanine	Ala	A
Arginine	Arg	R
Asparagine	Asn	N
Aspartic acid	Asp	D
Cysteine	Cys	C
Glutamic acid	Glu	E
Glutamine	Gln	Q
Glycine	Gly	G
Histidine	His	H
Isoleucine	Ile	I
Leucine	Leu	L
Lysine	Lys	K
Methionine	Met	M
Phenylalanine	Phe	F
Proline	Pro	P
Serine	Ser	S
Threonine	Thr	T
Tryptophan	Trp	W
Tyrosine	Tyr	Y
Valine	Val	V

FIGURE 5.3
Some Derivatives of Amino Acids.

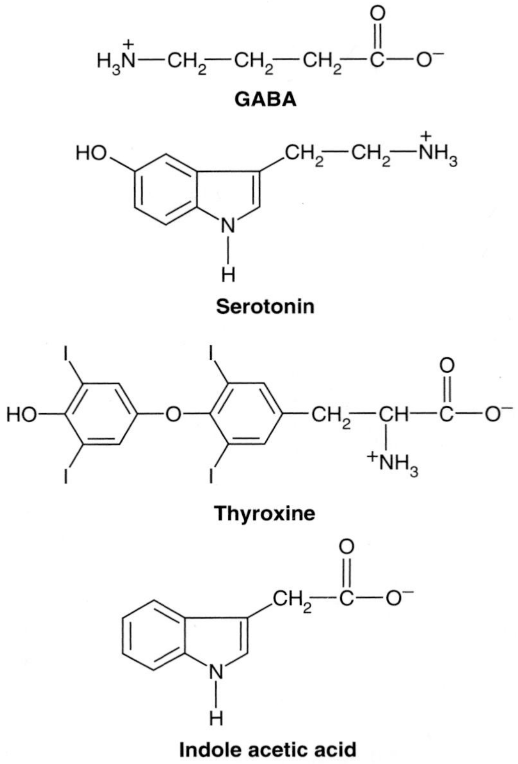

GABA

Serotonin

Thyroxine

Indole acetic acid

KEY *Concepts*

Each protein is composed of building blocks called amino acids. Amino acids are amphoteric molecules, that is, they can act as either an acid or a base. In addition to their primary function as components of proteins, amino acids have several important biological roles.

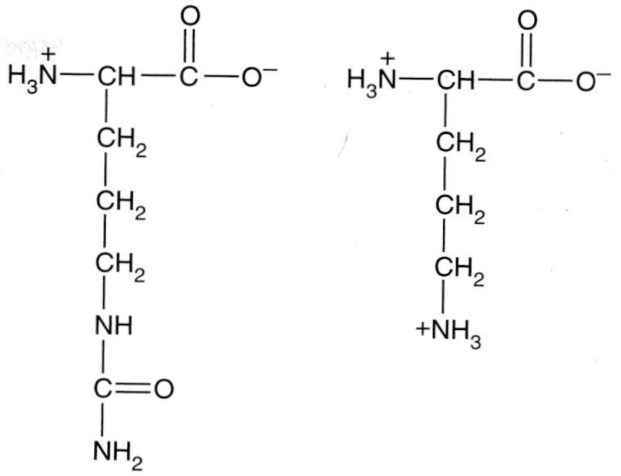

FIGURE 5.4
Citrulline and Ornithine.

Citrulline

Ornithine

2. Amino acids are precursors of a variety of complex nitrogen-containing molecules. Examples include the nitrogenous base components of nucleotides and the nucleic acids, heme (the iron-containing organic group required for the biological activity of several important proteins), and chlorophyll (a pigment of critical importance in photosynthesis).

3. Several standard and nonstandard amino acids act as metabolic intermediates. For example, arginine, citrulline, and ornithine (Figure 5.4) are components of the urea cycle (Chapter 14). The synthesis of urea, a molecule formed in vertebrate livers, is the principal mechanism for the disposal of nitrogenous waste.

Amino Acid Classes

Because the structure of amino acids determines the final three-dimensional configuration of each protein, their structures are examined carefully below. Amino acids are classified according to their capacity to interact with water. By using this criterion, four classes may be distinguished: (1) nonpolar and neutral, (2) polar and neutral, (3) acidic, and (4) basic.

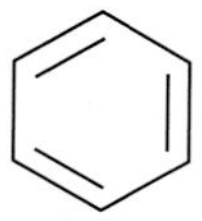

Neutral Nonpolar Amino Acids   The neutral nonpolar amino acids contain hydrocarbon R groups. The term *neutral* is used because these R groups do not bear positive or negative charges. Because they interact poorly with water, nonpolar (i.e., hydrophobic) amino acids play an important role maintaining the three-dimensional structure of proteins. Two types of hydrocarbon side chains are found in this group: aromatic and aliphatic. (**Aromatic** hydrocarbons contain cyclic structures that constitute a class of unsaturated hydrocarbons with unique properties. Benzene is one of the simplest aromatic hydrocarbons (Figure 5.5). The term **aliphatic** refers to nonaromatic hydrocarbons such as methane and cyclohexane.) Phenylalanine and tryptophan contain aromatic ring structures. Glycine, alanine, valine, leucine, isoleucine, and proline have aliphatic R groups. A sulfur atom appears in the aliphatic side chains of methionine and cysteine. In methionine the nonbonding electrons of the sulfur atom can form bonds with electrophiles such as metal ions. Although the sulfhydryl (—SH) group of cysteine is nonpolar, it can form weak hydrogen bonds with oxygen and nitrogen. Sulfhydryl groups, which are highly reactive, are important components of many enzymes. Additionally, the sulfhydryl groups of two cysteine molecules may oxidize spontaneously to form a disulfide compound called cystine. (See p. 91 for a discussion of this reaction.)

Neutral Polar Amino Acids   Because polar amino acids have functional groups capable of hydrogen bonding, they easily interact with water. (Polar amino acids are described as "hydrophilic" or "water-loving.") Serine, threonine, tyrosine, asparagine, and glutamine belong in this category. Serine, threonine, and tyrosine contain a polar hydroxyl group, which enables them to participate in hydrogen bonding, an important factor in protein structure. The hydroxyl groups serve other functions in

FIGURE 5.5
Benzene.

KEY *Concepts*

Amino acids are classified according to their capacity to interact with water. By using this criterion, four classes may be distinguished: nonpolar, polar, acidic, and basic.

proteins. For example, the formation of the phosphate ester of tyrosine is a common regulatory mechanism. Additionally, the —OH groups of serine and threonine are points for attaching carbohydrates. Asparagine and glutamine are amide derivatives of the acidic amino acids aspartate and glutamate. Because the amide functional group is highly polar, the hydrogen-bonding capability of asparagine and glutamine has a significant effect on protein stability.

Acidic Amino Acids   Two standard amino acids have side chains with carboxylate groups. Because the side chains of aspartic acid and glutamic acid are negatively charged at physiological pH, they are often referred to as aspartate and glutamate.

Basic Amino Acids   Basic amino acids bear a positive charge at physiological pH. They can therefore form ionic bonds with acidic amino acids. Lysine, which has a side chain amino group, accepts a proton from water to form an ammonium ion ($-NH_3^+$). When lysine's side chain in proteins such as collagen is oxidized, strong intramolecular and intermolecular cross-linkages are formed. Because the guanidino group of arginine is almost as strong a base as NaOH, it is protonated at physiological pH and, therefore, does not function in acid-base reactions. Histidine, on the other hand, is a weak base, since it is only partially ionized at pH 7. Consequently, histidine residues act as a buffer. They also play an important role in the catalytic activity of numerous enzymes.

Modified Amino Acids in Proteins

Several proteins contain amino acid derivatives that are formed after a polypeptide chain has been synthesized. Among these modified amino acids is γ-carboxyglutamic acid (Figure 5.6), a calcium-binding amino acid residue found in the blood-clotting protein prothrombin. Both 4-hydroxyproline and 5-hydroxylysine are important structural components of collagen, the most abundant protein in connective tissue. Phosphorylation of the hydroxyl-containing amino acids serine, threonine and tyrosine is often used to regulate the activity of proteins. For example, the synthesis of glycogen is significantly curtailed when the enzyme glycogen synthase is phosphorylated.

Amino Acid Stereoisomers

Because the α-carbons of 19 of the 20 standard amino acids are attached to four different groups (i.e., a hydrogen, a carboxyl group, an amino group, and an R group), they are referred to as **asymmetric** or **chiral carbons**. (The twentieth amino acid, glycine, is a symmetrical molecule because its α-carbon is attached to two hydrogens.) There are two possible arrangements for molecules with an asymmetric carbon. Molecules that differ only in the spatial arrangement of their atoms are called **stereoisomers.** There are two types of stereoisomers with asymmetric carbons: D-isomers and L-isomers. Three-dimensional representations of amino acid stereoisomers are illustrated in Figure 5.7. Notice in the figure that the ammonium group is drawn to the left of the chiral carbon in the L-isomer and to the right in the D-isomer. (See Chapter 7 for further discussion of this concept.) The atoms of the two isomers are bonded together in the same pattern except for the position of the ammonium

FIGURE 5.6
    Some Modified Amino Acid Residues Found in Polypeptides.

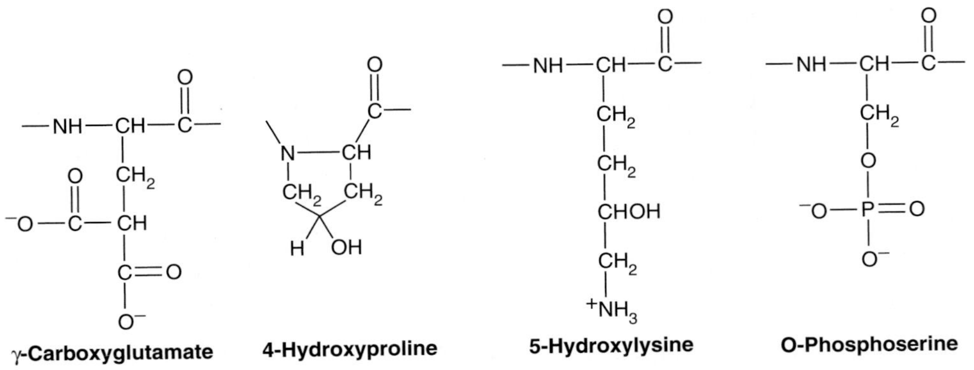

γ-Carboxyglutamate   4-Hydroxyproline   5-Hydroxylysine   O-Phosphoserine

KEY*Concepts*

Molecules that differ only in the spatial arrangement of some of their atoms are called stereoisomers. Stereoisomers with an asymmetric carbon atom have two nonsuperimposable mirror-image forms called enantiomers. Most asymmetric molecules in living organisms have only one stereoisomeric form.

FIGURE 5.7

Two Enantiomers.

L-Alanine and D-alanine are mirror images of each other.

L-Alanine   D-Alanine

group and the hydrogen atom. These two isomers are mirror images of each other. Such molecules, called **enantiomers,** cannot be superimposed on each other. Most asymmetric molecules found in living organisms occur in only one stereoisomeric form, either D or L. For example, with few exceptions, only L-amino acids are found in proteins.

Titration of Amino Acids

Because amino acids contain ionizable groups, the predominant ionic form of these molecules in solution depends on the pH. Titration of an amino acid illustrates the effect of pH on amino acid structure (Figure 5.8a). Titration is also a useful tool in determining the reactivity of amino acid side chains. Consider alanine, a simple amino acid, which has two titratable groups. During titration with a strong base such as NaOH, alanine loses two protons in a stepwise fashion. In a strongly acidic solution (e.g., at pH 0), alanine is present mainly in the form in which the carboxyl group is uncharged. Under this circumstance the molecule's net charge is +1, since the ammonium group is protonated. Lowering of the H^+ concentration results in the carboxyl group losing its proton to become a negatively charged carboxylate group. (In a polyprotic acid, the protons are first lost from the group with the lowest pK_a.) At this point, alanine has no net charge and is electrically neutral. The pH at which this occurs is called the **isoelectric point** (pI). Because there is no net charge at the isoelectric point, amino acids are least soluble at this pH. (Zwitterions crystallize

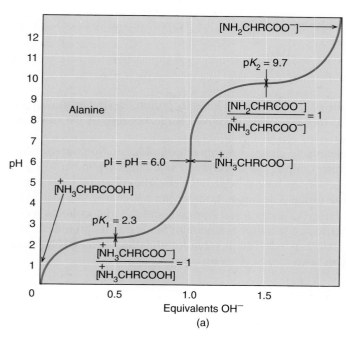

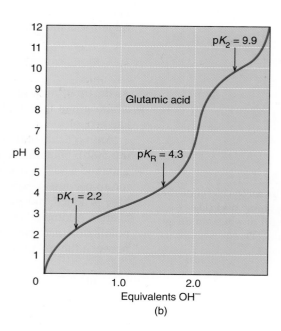

FIGURE 5.8

Titration of (a) Alanine and (b) Glutamic Acid.

TABLE 5.2

pKa Values for the Ionizing Groups of the Amino Acids

Amino Acid	pK_1 — carboxyl group	pK_2 — base amino group	pK_R
Glycine	2.34	9.6	
Alanine	2.34	9.69	
Valine	2.32	9.62	
Leucine	2.36	9.6	
Isoleucine	2.36	9.6	
Serine	2.21	9.15	
Threonine	2.63	10.43	
Methionine	2.28	9.21	
Phenylalanine	1.83	9.13	
Tryptophan	2.83	9.39	
Asparagine	2.02	8.8	
Glutamine	2.17	9.13	
Proline	1.99	10.6	
Cysteine	1.71	10.78	8.33
Histidine	1.82	9.17	6.0
Aspartic acid	2.09	9.82	3.86
Glutamic acid	2.19	9.67	4.25
Tyrosine	2.2	9.11	10.07
Lysine	2.18	8.95	10.79
Arginine	2.17	9.04	12.48

relatively easily.) The isoelectric point for alanine may be calculated as follows:

$$pI = \frac{pK_1 + pK_2}{2}$$

The pK_1 and pK_2 values for alanine are 2.34 and 9.7, respectively (Table 5.2). The pI value for alanine is therefore

$$pI = \frac{2.34 + 9.7}{2} = 6.0$$

As the titration continues, the ammonium group loses its proton, leaving an uncharged amino group. The molecule then has a net negative charge because of the carboxylate group.

Amino acids with ionizable side chains have more complex titration curves (Figure 5.8b). Glutamic acid, for example, has a carboxyl side chain group. At low pH, glutamic acid has net charge +1. As base is added, the α-carboxyl group loses a proton to become a carboxylate group. Glutamate now has no net charge. As still more base is added, the second carboxyl group loses a proton, and the molecule has a −1 charge. Adding additional base results in the ammonium ion losing its proton. At this point, glutamate has a net charge of −2. The pI value for glutamate is the pH halfway between the pK values for the two carboxyl groups:

$$pI = \frac{2.19 + 4.25}{2} = 3.22$$

The isoelectric point for histidine is the pH value halfway between the pK values for the two nitrogen-containing groups. The pK_a and pI values of amino acids in peptides

KEY *Concepts*

Titration is useful to determine the reactivity of amino acid side chains. The pH at which an amino acid has no net charge is called its isoelectric point.

and proteins differ somewhat from those of free amino acids, principally because most of the α-amino and α-carboxyl groups are not ionized but are covalently joined in peptide bonds.

Sample titration problems along with their solutions are given below.

5.1
PROBLEM

Consider the following amino acid and its pK_a values:

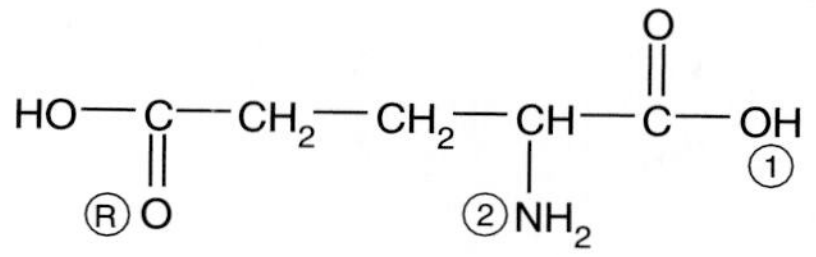

$$HO-\underset{\underset{\textcircled{R}\,O}{\|}}{C}-CH_2-CH_2-\underset{\underset{\textcircled{2}\,NH_2}{|}}{CH}-\underset{\overset{O}{\|}}{C}-OH_{\textcircled{1}}$$

$$pK_{a1} = 2.19, \qquad pK_{aR} = 4.25, \qquad pK_{a2} = 9.67$$

a. Draw the structure of the amino acid as the pH of the solution changes from highly acidic to strongly basic.

Solution

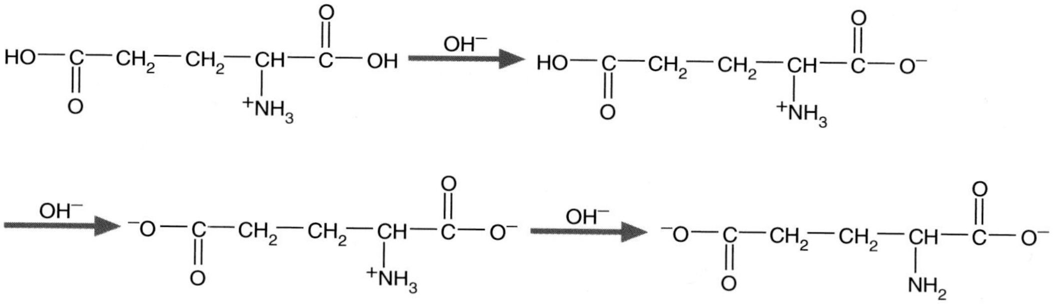

$$HO-\underset{\underset{O}{\|}}{C}-CH_2-CH_2-\underset{\underset{^+NH_3}{|}}{CH}-\underset{\overset{O}{\|}}{C}-OH \xrightarrow{OH^-} HO-\underset{\underset{O}{\|}}{C}-CH_2-CH_2-\underset{\underset{^+NH_3}{|}}{CH}-\underset{\overset{O}{\|}}{C}-O^-$$

$$\xrightarrow{OH^-} {}^-O-\underset{\underset{O}{\|}}{C}-CH_2-CH_2-\underset{\underset{^+NH_3}{|}}{CH}-\underset{\overset{O}{\|}}{C}-O^- \xrightarrow{OH^-} {}^-O-\underset{\underset{O}{\|}}{C}-CH_2-CH_2-\underset{\underset{NH_2}{|}}{CH}-\underset{\overset{O}{\|}}{C}-O^-$$

The ionizable hydrogens are lost in order of acidity, the most acidic ionizing first.

b. Which form of the amino acid is present at the isoelectric point?

Solution

The form present at the isoelectric point is electrically neutral:

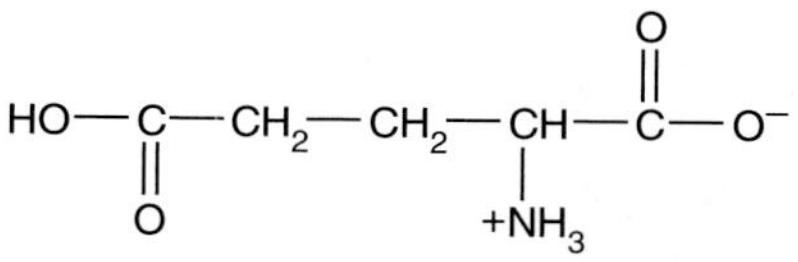

$$HO-\underset{\underset{O}{\|}}{C}-CH_2-CH_2-\underset{\underset{^+NH_3}{|}}{CH}-\underset{\overset{O}{\|}}{C}-O^-$$

c. Calculate the isoelectric point.

Solution

The isoelectric point is the average of the two pK_a's bracketing the isoelectric structure:

$$pI = \frac{pK_{a1} + pK_{a2}}{2} = \frac{4.25 + 2.19}{2} = 3.22$$

d. Sketch the titration curve for the amino acid.

Solution

Plateaus appear at the pK_a's and are centered about 0.5 Eq, 1.5 Eq, and 2.5 Eq of base. There is a sharp rise at 1 Eq, 2 Eq, and 3 Eq. The isoelectric point is midway on the sharp rise between pK_{a1} and pK_{a2}

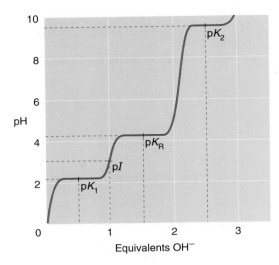

e. In what direction does the amino acid move when placed in an electric field at the following pH values: 1, 3, 5, 7, 9, 12?

Solution

At pH values below the pI, the amino acid is positively charged and moves to the cathode (negative electrode). At pH values above the pI, the amino acid is negatively charged and moves toward the anode (positive electrode). At the isoelectric point the amino acid has no net charge and therefore does not move in the electric field.

5.2   Consider the following tetrapeptide:

Lys—Ser—Asp—Ala

a. Determine the pI for the peptide.

Solution

The structure of the tetrapeptide in its most acidic form is shown below.

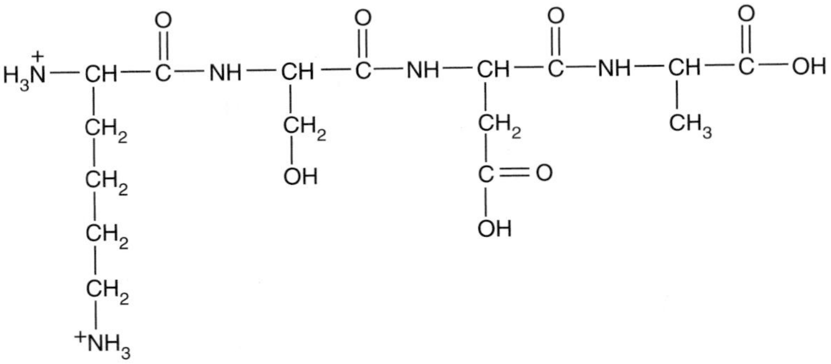

Refer to Table 5.2 for the pK_a values for lysine and aspartic acid, both of which have ionizable side chains in addition to containing terminal α-amino and α-carboxyl groups, respectively. These values are as follows:

Lysine: α-amino = 8.95,   amino side chain = 10.79
Aspartic acid: carboxyl side chain = 3.86
Alanine: α-carboxyl = 2.34

(These values are approximations, since the behavior of amino acids is affected by the presence of other groups.) The electrically neutral peptide is formed after both carboxyl groups have lost their protons but before either ammonium group has lost any protons. The isoelectric point is calculated as follows:

$$pI = \frac{3.86 + 8.95}{2} = \frac{12.81}{2} = 6.4$$

b. In what direction does the peptide move when placed in an electric field at the following pHs: 4 and 9?

Solution

At pH = 4 the peptide is positively charged and moves toward the negative electrode (cathode). At pH = 9 the peptide is negatively charged and will move toward the positive electrode (anode).

Amino Acids Reactions

The functional groups of organic molecules determine which reactions they may undergo. Amino acids with their carboxyl groups, amino groups, and various R groups can undergo numerous chemical reactions. However, two reactions (i.e., peptide bond and disulfide bridge formation) are of special interest because of their effect on protein structure.

Peptide Bond Formation Polypeptides are linear polymers composed of amino acids linked together by peptide bonds. **Peptide bonds** (Figure 5.9) are amide linkages formed when the unshared electron pair of the α-amino nitrogen atom of one

FIGURE 5.9

Formation of a Dipeptide.

(a) A peptide bond forms when the α-carboxyl group of one amino acid reacts with the amino group of another. (b) A water molecule is lost in the reaction.

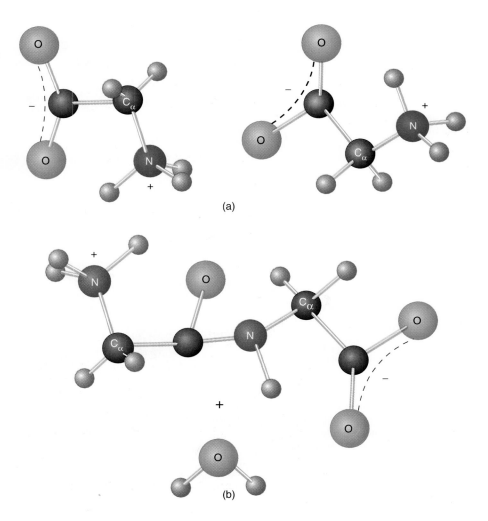

amino acid attacks the α-carboxyl carbon of another in a nucleophilic acyl substitution reaction. Because this reaction is a dehydration, that is, a water molecule is removed, the linked amino acids are referred to as *amino acid residues*. When two amino acid molecules are linked, the product is called a dipeptide. For example, glycine and serine can form the dipeptides glycylserine or serylglycine. As amino acids are added and the chain lengthens, the prefix reflects the number of residues. For example, a tripeptide contains three amino acid residues, a tetrapeptide four, and so on. By convention the amino acid residue with the free amino group is called the *N-terminal* residue and is written to the left. The free carboxyl group on the *C-terminal* residue appears on the right. Peptides are named by using their amino acid sequences, beginning from their N-terminal residue. For example,

$$\text{H}_2\text{N—Tyr—Ala—Cys—Gly—COOH}$$

is a tetrapeptide named tyrosylalanylcysteinylglycine.

Large polypeptides have well defined three-dimensional structures. This structure, referred to as the molecule's native conformation, is a direct consequence of its *amino acid sequence* (the order in which the amino acids are linked together). Since all the linkages connecting the amino acid residues consist of single bonds, it might be expected that each polypeptide undergoes constant conformational changes due to rotation around the single bonds. However, most polypeptides spontaneously fold into a single biologically active form. In the early 1950s, Linus Pauling and his colleagues proposed an explanation. Using X-ray diffraction studies, they determined that the peptide bond is rigid and planar (flat) (Figure 5.10). Having discovered that the C—N

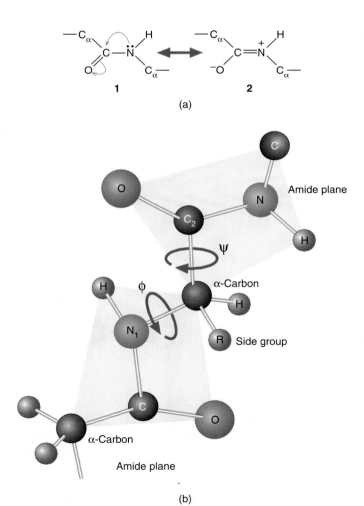

FIGURE 5.10

The Peptide Bond.

(a) Resonance forms of the peptide bond. (b) Dimensions of a dipeptide. Because peptide bonds are rigid, the conformational degrees of freedom of a polypeptide chain are limited to rotations around the C_α—C_2 and C_α—N_1 bonds. The corresponding rotations are represented by ψ and ϕ, respectively.

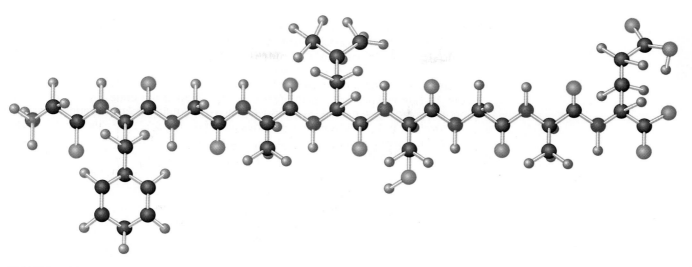

FIGURE 5.11

A Polypeptide Chain.

In polypeptides, successive R groups occur on alternate sides of the peptide bonds. This illustration is a diagrammatic view of an extended polypeptide chain, not a representation of native structure.

FIGURE 5.12

Oxidation of Two Cysteine Molecules to Form Cystine.

The disulfide bond in a polypeptide is called a disulfide bridge.

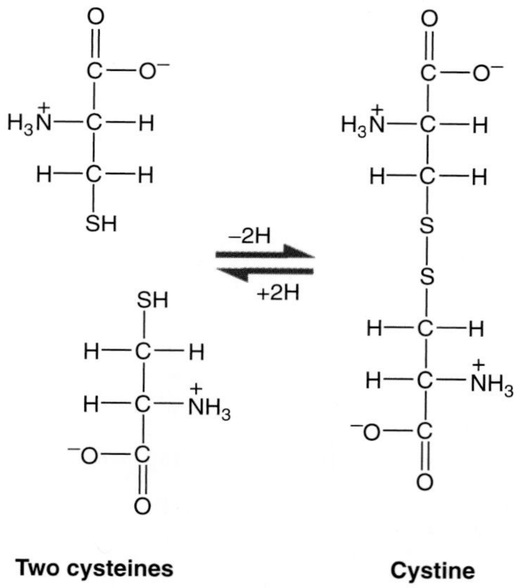

Two cysteines   Cystine

bonds joining each two amino acids are shorter than other types of C—N bonds, Pauling deduced that peptide bonds have a partial double bond character. (This indicates that peptide bonds are resonance hybrids.) The rigidity of the peptide bond has several consequences. Because fully one-third of the bonds in a polypeptide backbone chain cannot rotate freely, there are limits on the number of conformational possibilities. Another consequence is that in extended segments of polypeptide successive R groups appear on opposite sides (Figure 5.11).

Cysteine Oxidation   The sulfhydryl group of cysteine is highly reactive. The most common reaction of this group is a reversible oxidation that forms a disulfide. Oxidation of two molecules of cysteine forms cystine, a molecule that contains a disulfide bond (Figure 5.12). When two cysteine residues form such a bond, it is referred to as a **disulfide bridge.** This bond can occur in a single chain to form a ring or between two separate chains to form an intermolecular bridge. Disulfide bridges help stabilize many polypeptides and proteins.

5.1

In extracellular fluids such as blood, the sulfhydryl groups of cysteine are quickly oxidized to form cystine. Unfortunately, cystine is the least soluble of the amino acids. In a genetic disorder known as *cystinuria,* defective membrane transport of cystine results in excessive excretion of cystine into the urine. Crystallization of the amino acid results in formation of calculi (stones), in the kidney, ureter, or urinary bladder. The stones may cause pain, infection, and blood in the urine. Cystine concentration in the kidney is reduced by massively increasing fluid intake and administering D-penicillamine. It is believed that penicillamine (Figure 5.13) is effective because penicillamine-cysteine disulfide, which is substantially more soluble than cystine, is formed. What is the structure of the penicillamine-cysteine disulfide?

FIGURE 5.13
Structure of Penicillamine.

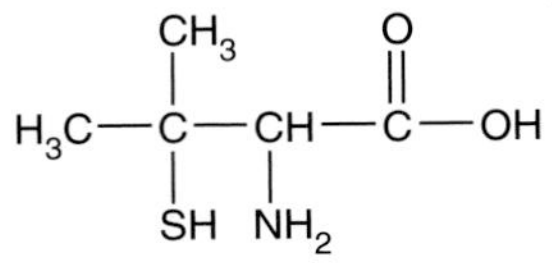

$$H_3C-\underset{\underset{SH}{|}}{\overset{\overset{CH_3}{|}}{C}}-\underset{\underset{NH_2}{|}}{CH}-\overset{\overset{O}{\|}}{C}-OH$$

5.2   PEPTIDES

Although their structures are less complex than the larger protein molecules, peptides have significant biological activities. The structure and function of several interesting examples, presented in Table 5.3, are now discussed.

The tripeptide glutathione (γ-glutamyl-L-cysteinylglycine) contains an unusual γ-amide bond. (Note that the γ-carboxyl group of the glutamic acid residue, not the α-carboxyl group, contributes to the peptide bond.) Found in almost all organisms, glutathione is involved in many important biological processes. Among these are protein and DNA synthesis, drug and environmental toxin metabolism, and amino acid transport. One group of glutathione's functions exploits its effectiveness as a reducing agent. (Because the reducing component of the molecule is the —SH group of the cysteine residue, the abbreviation for glutathione is GSH.) Glutathione protects cells from the destructive effects of oxidation by reacting with substances such as peroxides (R—O—O—R, byproducts of O_2 metabolism). For example, in red blood cells, hydrogen peroxide (H_2O_2) oxidizes the iron of hemoglobin to its ferric form (Fe^{3+}).

TABLE 5.3
Selected Biologically Important Peptides

Name	Amino Acid Sequence
Glutathione	(a) 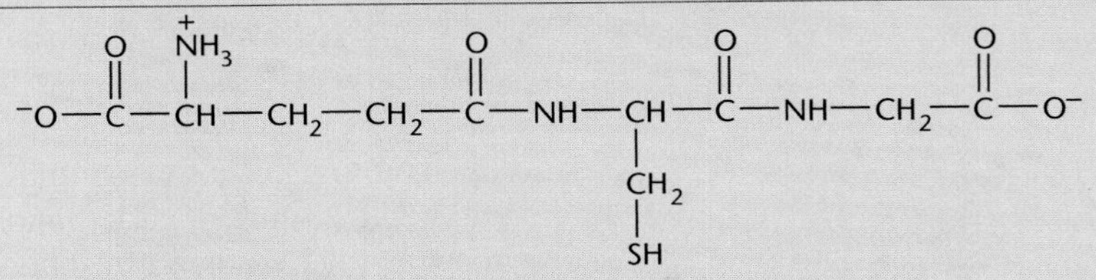 $^-O-\overset{\overset{O}{\|}}{C}-\underset{\underset{NH_3^+}{|}}{CH}-CH_2-CH_2-\overset{\overset{O}{\|}}{C}-NH-\underset{\underset{\underset{SH}{|}}{CH_2}}{CH}-\overset{\overset{O}{\|}}{C}-NH-CH_2-\overset{\overset{O}{\|}}{C}-O^-$
Oxytocin	(b) Cys—Tyr—Ile—Gln—Asn—Cys—Pro—Leu—Gly—NH$_2$, with S—S bridge between the two Cys
Vasopressin	(c) Cys—Tyr—Phe—Gln—Asn—Cys—Pro—Arg—Gly—NH$_2$, with S—S bridge between the two Cys
Met-enkephalin	Tyr—Gly—Gly—Phe—Met
Leu-enkephalin	Tyr—Gly—Gly—Phe—Leu
Atrial natriuretic factor	Ser[1]—Leu—Arg—Arg—Ser—Ser—Cys—Phe—Gly—Gly[10]—Arg—Met—Asp—Arg—Ile—Gly—Ala—Gln—Ser—Gly—Leu—Gly—Cys—Asn—Ser—Phe—Arg—Tyr[28]
Substance P	Arg—Pro—Lys—Pro—Gln—Phe—Phe—Gly—Leu—Met—NH$_2$
Bradykinin	Arg—Pro—Pro—Gly—Phe—Ser—Pro—Phe—Arg

Methemoglobin, the product of this reaction, is incapable of binding O_2. Glutathione protects against the formation of methemoglobin by reducing H_2O_2 in a reaction catalyzed by the enzyme glutathione peroxidase. In the oxidized product GSSG, two tripeptides are linked by a disulfide bond:

$$2 \text{ GSH} + H_2O_2 \rightarrow \text{GSSG} + 2 \text{ } H_2O$$

Because of the high GSH:GSSG ratio normally present in cells, glutathione is an important intracellular reducing agent.

The amino acid sequences of the peptide hormones oxytocin and vasopressin differ only by two residues. Both molecules contain nine amino acids and are produced by the cleavage of polypeptide precursors within different specialized cells in the hypothalamus. (The hypothalamus is a small structure in the vertebrate brain that regulates functions such as water balance, hunger, thirst, and sleep.) After synthesis, oxytocin and vasopressin are transported down nerve tracts into the posterior pituitary gland, where they are stored. Then each peptide is secreted in response to specific signals from the hypothalamus. Oxytocin stimulates contraction of uterine muscles during childbirth and the ejection of milk by mammary glands during lactation. (Recently, oxytocin production by the uterus itself during childbirth has been detected.) In males, oxytocin may have a regulatory role in the synthesis of the sex hormone testosterone. Vasopressin, also known as antidiuretic hormone (ADH), is secreted in response to low blood pressure or a high blood Na^+ concentration. ADH acts by stimulating the kidneys to retain water. Because oxytocin and vasopressin have similar structures, it is not surprising that the functions of the two molecules overlap. Oxytocin has a mild antidiuretic activity, and vasopressin has some oxytocinlike activity.

Met-enkephalin and leu-enkephalin belong to a group of peptides called the **opioid peptides,** found predominantly in nervous tissue cells. Opioid peptides are molecules that relieve pain and produce pleasant sensations. They were discovered after researchers suspected that the physiological effects of opiate drugs such as morphine resulted from their binding to nerve cell receptors for endogenous molecules. Leu-enkephalin and met-enkephalin are pentapeptides that differ only in their C-terminal amino acid residues.

Peptides are one class of messenger molecules that multicellular organisms use to regulate their complex activities. Recall that multicellular organisms, consisting of several hundred cell types, must coordinate a huge number of biochemical processes. A stable internal environment, called *homeostasis,* is maintained by the dynamic interplay between opposing processes. Peptide molecules with opposing functions are now known to affect the regulation of several processes. For example, atrial natriuretic factor, a peptide produced by specialized cells in the heart and the nervous system, stimulates the production of a dilute urine (an effect opposite to that of vasopressin). For another example, substance P and bradykinin stimulate the perception of pain, an effect opposed by the opioid peptides. (Pain is a protective mechanism in animals that warns of tissue damage.)

5.3 PROTEINS

Of all the molecules encountered in living organisms, proteins have the most diverse functions, as the following list suggests:

1. **Catalysis.** *Enzymes* are proteins that direct and accelerate thousands of biochemical reactions in such processes as digestion, energy capture, and biosynthesis. These molecules have remarkable properties. For example, they can increase reaction rates by factors of between 10^6 and 10^{12}. They can perform this feat under the mild conditions of pH and temperature because they can induce or stabilize strained reaction intermediates. Examples include ribulose bisphosphate carboxylase, an important enzyme in photosynthesis, and nitrogenase, a protein complex that is responsible for nitrogen fixation.

2. **Structure.** Some proteins provide protection and support. Structural proteins often have very specialized properties. For example, collagen (the major component of connective tissues) and fibroin (silk protein) have significant

mechanical strength. Elastin, the rubberlike protein found in elastic fibers, is found in several tissues in the body (e.g., blood vessels and skin) that must be elastic to function properly.

3. **Movement.** Proteins are involved in all cell movements. For example, actin, tubulin, and other proteins compose the cytoskeleton. Cytoskeletal proteins are active in cell division, endocytosis, exocytosis, and the ameboid movement of white blood cells.

4. **Defense.** A wide variety of proteins are protective. Examples found in vertebrates include keratin, the protein found in skin cells that aids in protecting the organism against mechanical and chemical injury. The bloodclotting proteins fibrinogen and thrombin prevent blood loss when blood vessels are damaged. The immunoglobulins (or antibodies) are produced by lymphocytes when foreign organisms such as bacteria invade an organism. Binding antibodies to an invading organism is the first step in its destruction.

5. **Regulation.** Binding a hormone molecule to its target cell changes cellular function. Examples of peptide hormones include insulin and glucagon, which both regulate blood glucose levels. Growth hormone stimulates cell growth and division.

6. **Transport.** Many proteins function as carriers of molecules or ions across membranes or between cells. Examples of membrane proteins include the Na^+-K^+ ATPase and the glucose transporter. Other transport proteins include hemoglobin, which carries O_2 to the tissues from the lungs, and the lipoproteins LDL and HDL, which transport lipids from the liver and intestines to other organs.

Because of their diversity, proteins are often classified in two additional ways: (1) shape and (2) composition. Proteins are classified into two major groups based on their shape. As their name suggests, **fibrous proteins** are long, rod-shaped molecules that are insoluble in water and physically tough. Fibrous proteins, such as the keratins found in skin, hair, and nails, have structural and protective functions. **Globular proteins** are compact spherical molecules that are usually water-soluble. Typically, globular proteins have dynamic functions. For example, nearly all enzymes have globular structures. Other examples include the immunoglobulins and the transport proteins hemoglobin and albumin (a carrier of fatty acids in blood).

On the basis of composition, proteins are classified as simple or conjugated. Simple proteins, such as serum albumin and keratin, contain only amino acids. In contrast, each **conjugated protein** consists of a simple protein combined with a nonprotein component. The nonprotein component is called a **prosthetic group.** (A protein without its prosthetic group is called an **apoprotein.** A protein molecule combined with its prosthetic group is referred to as a **holoprotein.**) Prosthetic groups typically play an important, even crucial, role in the function of proteins. Conjugated proteins are classified according to the nature of their prosthetic groups. For example, **glycoproteins** contain a carbohydrate component, **lipoproteins** contain lipid molecules, and **metaloproteins** contain metal ions. Similarly, **phosphoproteins** contain phosphate groups, and **hemoproteins** possess heme groups (p. 109) (see below).

Protein Structure

Proteins are extraordinarily complex molecules. Complete models depicting even the smallest of the polypeptide chains are almost impossible to comprehend. Simpler images that highlight specific features of a molecule are useful. Two methods of conveying structural information about proteins are presented in Figure 5.14. Another structural representation, referred to as a ball and stick model, can be seen on the opening page of Chapter 6 (p. 118).

Biochemists have distinguished several levels of the structural organization of proteins. **Primary structure,** the amino acid sequence, is specified by genetic information. As the polypeptide chain folds, it forms certain localized arrangements of adjacent amino acids that constitute **secondary structure.** The overall three-dimensional shape that a polypeptide assumes is called the **tertiary structure.** Proteins that consist of two or more polypeptide chains (or subunits) are said to have a **quaternary structure.**

FIGURE 5.14

The Enzyme Adenylate Kinase.

(a) A space-filling model illustrates the volume occupied by molecular components and overall shape. (b) In a ribbon model β-pleated segments are represented by flat arrows. The α-helices appear as spiral ribbons.

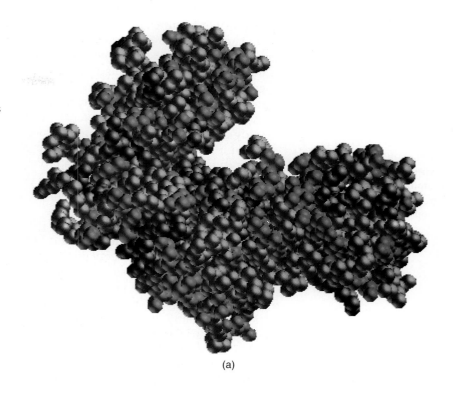

(a)

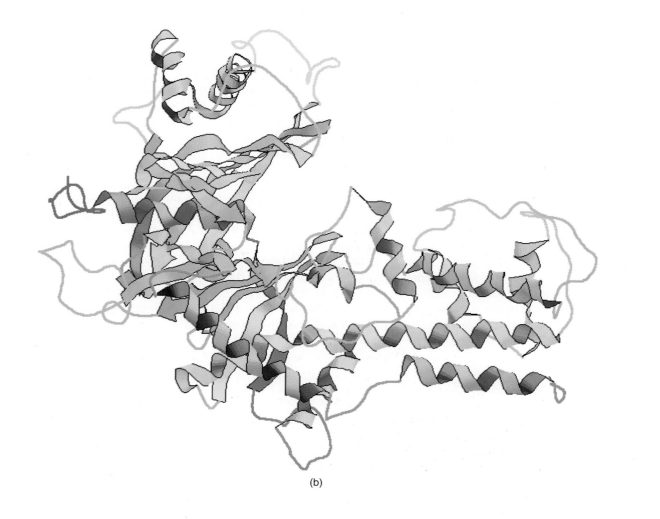

(b)

Primary Structure Every polypeptide has a specific amino acid sequence. The interactions between amino acid residues determine the protein's three-dimensional structure and suggests its functional role and relationship to other proteins. Polypeptides that have similar amino acid sequences and functions are said to be **homologous.** Sequence comparisons among homologous polypeptides have been used to trace the genetic relationships of different species. For example, the sequence homologies of the mitochondrial redox protein cytochrome c have been used extensively in the study of evolution. Sequence comparisons of cytochrome c among numerous species reveal a significant amount of sequence conservation. The amino acid residues that are identical in all homologues of a protein, referred to as *invariant,* are presumed to be essential for the protein's function. (In cytochrome c the invariant residues interact with heme, a prosthetic group, or certain other proteins involved in energy generation.)

KEY Concepts

The primary structure of a polypeptide is its amino acid sequence. The amino acids are connected by peptide bonds. Amino acid residues that are essential for the molecule's function are referred to as invariant. Proteins with similar amino acid sequences and functions are said to be homologous.

Primary Structure, Evolution, and Molecular Diseases Because of the essential role of cytochrome c in energy production, individual organisms with amino acid substitutions at invariant positions are not viable. Mutations (alterations in the DNA sequences that code for a protein's amino acid sequence) are random and spontaneous events. Therefore, primary sequence changes that do not affect a polypeptide's function significantly occur over time. Some of these substitutions are said to be *conservative* because an amino acid with a chemically similar side chain has been substituted. For example, at certain sequence positions, leucine and isoleucine, which both contain hydrophobic side chains, may be substituted for each other without affecting function. Some sequence positions are significantly less stringent. These residues, referred to as *variable,* apparently perform nonspecific roles in the polypeptide's function.

Substitutions at conservative and variable sites have been used to trace evolutionary relationships. These studies assume that the longer the time span since two species diverged from each other, the larger the number of differences in a certain polypeptide's primary structure. For example, humans and chimpanzees are believed to have diverged only recently (perhaps as few as four million years ago). This presumption, based principally on fossil and anatomical evidence, is supported by cytochrome c primary sequence data, since the protein is identical in both species. Animals such as kangaroos, whales, and sheep, whose cytochrome c molecules each differ by ten residues from the human protein, are all believed to have evolved from a common ancestor that lived over 50 million years ago.

Some mutations are deleterious without being immediately lethal. Sickle-cell anemia, which is caused by a mutant hemoglobin, is a classic example of a group of maladies that Linus Pauling and his colleagues referred to as **molecular diseases.** (Dr. Pauling first demonstrated that sickle-cell patients have a mutant hemoglobin through the use of electrophoresis.) Human adult hemoglobin (HbA) is composed of two identical α-chains and two identical β-chains. Sickle-cell anemia results from a single amino acid substitution in the β-chain of HbA. Analysis of the hemoglobin molecules of sickle-cell patients reveals that the only difference between HbA and sickle-cell hemoglobin (HbS) is at amino acid residue 6 in the β-chain (Figure 5.15). Because of the substitution of a hydrophobic valine for a negatively charged glutamic acid, HbS molecules aggregate to form rigid rodlike structures in the oxygen-free state. The

FIGURE 5.15 **Segments of β-Chain in HbA and HbS.**

Individuals possessing the gene for sickle-cell hemoglobin produce β-chains with valine instead of glutamic acid at residue 6.

Hb A	Val—	His—	Leu—	Thr—	Pro—	Glu—	Glu—	Lys—
Hb S	Val—	His—	Leu—	Thr—	Pro—	Val—	Glu—	Lys—
	1	2	3	4	5	6	7	8

patient's red blood cells become sickle-shaped and are susceptible to hemolysis resulting in severe anemia. These red blood cells have a lower than normal oxygen-binding capacity. Intermittent clogging of capillaries by sickled cells also causes tissues to be deprived of oxygen. Sickle-cell anemia is characterized by excruciating pain, eventual organ damage, and earlier death.

Until recently, because of the debilitating nature of sickle-cell disease, affected individuals rarely survived beyond childhood. It might be predicted that the deleterious mutational change that causes this affliction would be rapidly eliminated from human populations. However, the sickle-cell gene is not as rare as would be expected. Sickle-cell disease occurs in individuals who have inherited two copies of the sickle-cell gene. (A **gene** is a DNA sequence that codes for a polypeptide.) Such individuals, who are said to be *homozygous,* inherit one copy of the defective gene from each parent. Each of the parents, referred to as *heterozygous* because they have one normal HbA gene and one defective HbS gene, is said to have the *sickle-cell trait.* Such people lead normal lives even though about 40% of their hemoglobin is HbS. The incidence of sickle-cell trait is especially high in some regions of Africa. In these regions the disease malaria, caused by the *Amopheles* mosquito-borne parasite *Plasmodium* is a serious health problem. Individuals with the sickle-cell trait are less vulnerable to malaria because their red blood cells are a less favorable environment for the growth of the parasite than are normal cells. Because sickle-cell trait carriers are more likely to survive malaria than normal individuals, the incidence of the sickle-cell gene has remained high. (In some areas, as many as 40% of the native populations have the sickle-cell trait.)

QUESTION 5.2

A genetic disease called *glucose-6-phosphate dehydrogenase deficiency* is inherited in a manner similar to that of sickle-cell anemia. The defective enzyme cannot keep erythrocytes supplied with sufficient amounts of the antioxidant molecule NADPH (Chapter 11). NADPH protects cell membranes and other cellular structures from oxidation. Describe in general terms the inheritance pattern of this molecular disease. Why do you think that the antimalarial drug primaquine, which stimulates peroxide formation, results in such devastating cases of hemolytic anemia in carriers of the defective gene? Does it surprise you that this genetic anomaly is commonly found in African and Mediterranean populations?

Secondary Structure   The secondary structure of polypeptides consists of several repeating patterns. The most commonly observed types of secondary structure are the α-helix and the β-pleated sheet. Both α-helix and β-pleated sheet patterns are stabilized by hydrogen bonds between the carbonyl and N—H groups in the polypeptide's backbone. These patterns occur when all the ϕ (phi) angles in a polypeptide segment are equal and all the ψ (psi) bond angles are equal (Figure 5.10b). Because peptide bonds are rigid, the α-carbons are swivel points for the polypeptide chain. Several properties of the R groups (e.g., size and charge, if any) attached to the α-carbon influence the ϕ and ψ angles. Certain amino acids (discussed below) foster or inhibit specific secondary structure patterns. Many fibrous proteins are composed almost entirely of secondary structural patterns.

The *α-helix* is a rigid, rodlike structure that forms when a polypeptide chain twists into a right-handed helical conformation (Figure 5.16). Hydrogen bonds form between the N—H group of each amino acid and the carbonyl group of the amino acid four residues away. There are 3.6 amino acid residues per turn of the helix, and the pitch (the distance between corresponding points per turn) is 54 nm. Amino acid R groups extend outward from the helix. Because of several structural constraints (i.e., the rigidity of peptide bonds and the allowed limits on the values of the ϕ and ψ angles), certain amino acids do not foster α-helix formation. For example, glycine's R group (a hydrogen atom) is so small that the polypeptide chain may be too flexible. Proline, on the other hand, contains a rigid ring that prevents the N—C bond from rotating.

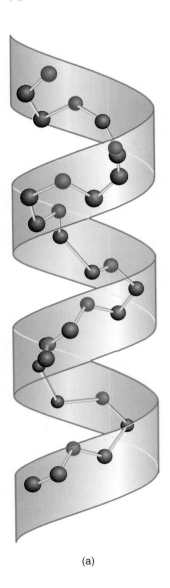

(a)

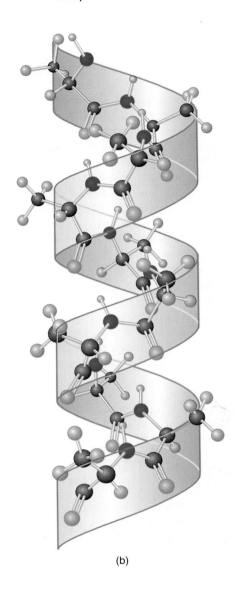

(b)

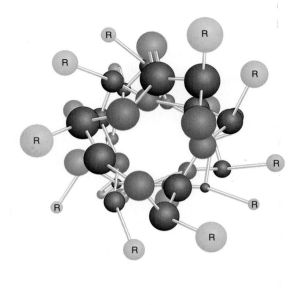

(c)

FIGURE 5.16

The α-Helix.

(a) The helical backbone. (b) A more complete model. Hydrogen bonds form between carbonyl and N—H groups along the long axis of the helix. (c) A top view of the α-helix. The R groups point away from the long axis of the helix.

In addition, proline has no N—H group available to form intrachain hydrogen bonds. Amino acid sequences with large numbers of charged amino acids (e.g., glutamate and aspartate) and bulky R groups (e.g., tryptophan) are also incompatible with α-helix structures.

β-Pleated sheets form when two or more polypeptide chain segments line up side by side (Figure 5.17). Each individual segment is referred to as a *β-strand*. Rather than being coiled, each β-strand is fully extended. β-Pleated sheets are stabilized by hydrogen bonds that form between the polypeptide backbone N—H and carbonyl groups of adjacent chains. There are two β-pleated sheets: parallel and antiparallel. In *parallel* β-pleated sheet structures, the polypeptide chains are arranged in the same direction. *Antiparallel* chains run in opposite directions. Antiparallel β-sheets are more stable than parallel β-sheets because fully colinear hydrogen bonds form. Occasionally, mixed parallel-antiparallel β-sheets are observed.

Many globular proteins contain combinations of α-helix and β-pleated sheet secondary structures (Figure 5.18). These patterns are called **supersecondary structures.** In the βαβ *unit,* two parallel β-pleated sheets are connected by an α-helix segment. In the β-*meander* pattern, two antiparallel β-sheets are connected by polar amino acids and glycines to effect an abrupt change in direction of the polypeptide chain called *reverse or* β-*turns.* In αα-*units,* two successive α-helices separated by a loop

FIGURE 5.17

β-Pleated Sheet.

(a) Two forms of β-pleated sheet: Antiparallel and parallel. Hydrogen bonds are represented by dotted lines. (b) A more detailed view of antiparallel β-pleated sheet.

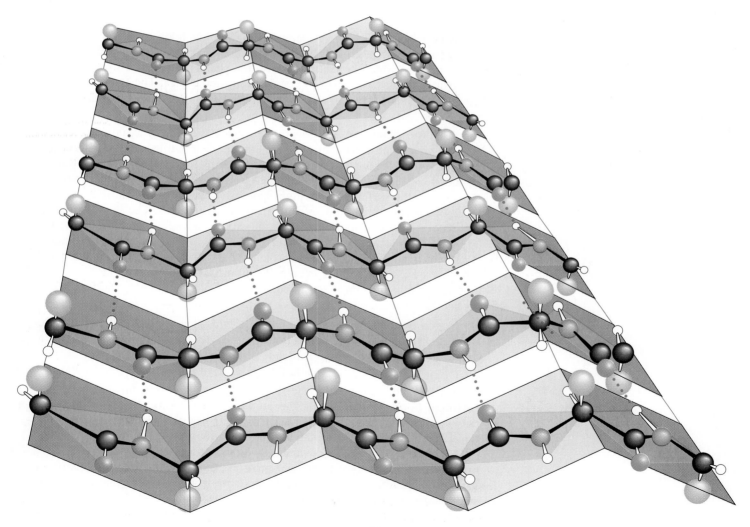

(a)

Antiparallel

Parallel

(b)

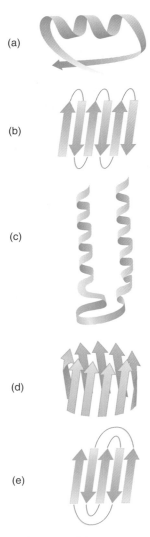

(a)

(b)

(c)

(d)

(e)

FIGURE 5.18

Selected Supersecondary Structures.

(a) $\beta\alpha\beta$ units, (b) β-meander, (c) $\alpha\alpha$-unit, (d) β-barrel, and (e) Greek key.

or nonhelical segment become enmeshed because of compatible side chains. Several β-*barrel* arrangements are formed when various β-sheet configurations fold back on themselves. When an antiparallel β-sheet doubles back on itself in a pattern that resembles a common Greek pottery design, the motif is called the *Greek key*.

Tertiary Structure   Although globular proteins often contain significant numbers of secondary structural elements, several other factors contribute to their structure. The term "tertiary structure" refers to the unique three-dimensional conformations that globular proteins assume as a consequence of the interactions between the side chains in their primary structure. Tertiary structure has several important features:

1. Many polypeptides fold in such a fashion that amino acid residues that are distant from each other in the primary structure come into close proximity.

2. Because of efficient packing as the polypeptide chain folds, globular proteins are compact. During this process, most water molecules are excluded from the protein's interior making interactions between both polar and nonpolar groups possible.

3. Large globular proteins (i.e., those with more than 200 amino acid residues) often contain several compact units called domains. Domains (Figure 5.19) are typically structurally independent segments that have specific functions (e.g., binding an ion or small molecule).

The following types of interactions stabilize tertiary structure (Figure 5.20):

1. **Hydrophobic interactions.** As a polypeptide folds, hydrophobic R groups are brought into close proximity because they are excluded from water. Then the highly ordered water molecules in solvation shells are released from the interior, increasing the disorder (entropy) of the water molecules. The favorable entropy change is a major driving force in protein folding. (Protein folding is described in Chapter 18.)

2. **Electrostatic interactions.** The strongest electrostatic interaction in proteins occurs between ionic groups of opposite charge. Referred to as **salt bridges,** these noncovalent bonds are significant only in regions of the protein where water is excluded because of the energy required to remove water molecules from ionic groups near the surface. Salt bridges have been observed to contribute to the interactions between adjacent subunits in complex proteins. The same is true for the weaker electrostatic interactions (ion-dipole, dipole-dipole, van der Waals). They are significant in the interior of the folded protein and between subunits or in protein-ligand interactions. (In proteins that consist of more than one polypeptide chain, each polypeptide is called a **subunit. Ligands** are molecules that bind to specific sites on larger molecules such as proteins.) Ligand binding pockets are water depleted regions of the protein.

3. **Hydrogen bonds.** A significant number of hydrogen bonds form within a protein's interior and on its surface. In addition to forming hydrogen bonds with one another, the polar amino acid side chains may interact with water or with the polypeptide backbone. Again, the presence of water precludes the formation of hydrogen bonds with other species.

4. **Covalent bonds.** Covalent linkages are created by chemical reactions that alter a polypeptide's structure during or after its synthesis. (Examples of these reactions, referred to as posttranslational modifications, are described in Section 18.2.) The most prominent covalent bonds in tertiary structure are the disulfide bridges found in many extracellular proteins. In extracellular environments these strong linkages partly protect protein structure from adverse changes in pH or salt concentrations. Intracellular proteins do not contain disulfide bridges because of high cytoplasmic concentrations of reducing agents.

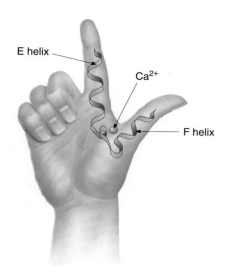

(a) EF hand

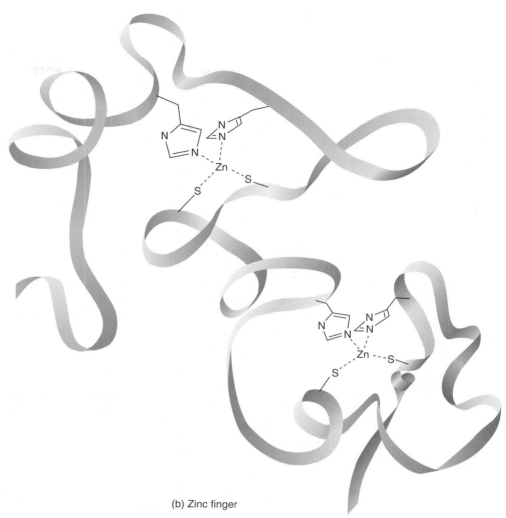

(b) Zinc finger

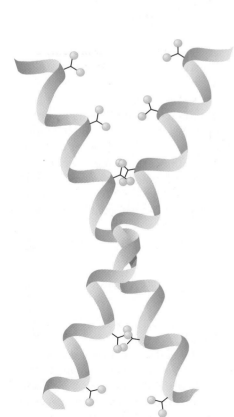

(c) Leucine zipper

FIGURE 5.19

Three Domains Found in Several Proteins.

(a) The EF Hand, which consists of a helix-loop-helix configuration, binds specifically to Ca^{2+}. (The hand motif helps the viewer to comprehend the three-dimensional structure of the calcium-binding domain.) (b) The zinc finger motif is commonly found in DNA-binding proteins. DNA binding proteins often possess several zinc fingers, each of which promotes protein-DNA interactions. (See Chapter 17.) Cysteine residues provide the binding sites for the zinc ions. (c) The leucine zipper is another DNA-binding domain. The knobs on the leucine zipper represent leucine side chains.

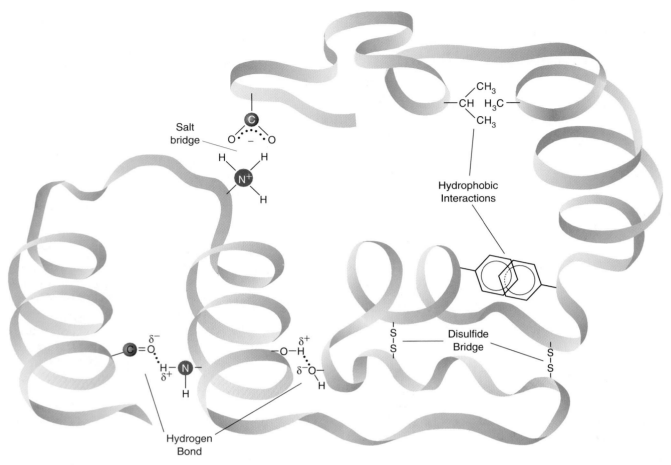

FIGURE 5.20
Interactions That Maintain Tertiary Structure.

Quaternary Structure Many proteins, especially those with high molecular weights, are composed of several polypeptide chains. As mentioned each polypeptide component is called a subunit. Subunits in a protein complex may be identical or quite different. Multisubunit proteins in which some or all subunits are identical are referred to as **oligomers.** Oligomers are composed of **protomers** which may consist of one or more subunits. A large number of oligomeric proteins contain 2 or 4 subunit protomers, referred to as dimers and tetramers, respectively. There appear to be several reasons for the common occurrence of multisubunit proteins:

1. Synthesis of separate subunits may be more efficient than substantially increasing the length of a single polypeptide chain.
2. In supermolecular complexes such as collagen fibers, replacement of smaller, worn-out or damaged components can be managed more effectively.
3. The complex interactions of multiple subunits help regulate a protein's biological function.

Polypeptide subunits assemble and are held together by noncovalent interactions such as hydrophobic interactions, electrostatic interactions, and hydrogen bonds, as well as covalent cross-links. As with protein folding, hydrophobic interactions are clearly the most important because the structures of the complementary interfacing surfaces between subunits are similar to those observed in the interior of globular protein domains. Although they are less numerous, covalent cross links significantly stabilize certain multisubunit proteins. Prominent examples include the disulfide bridges in the

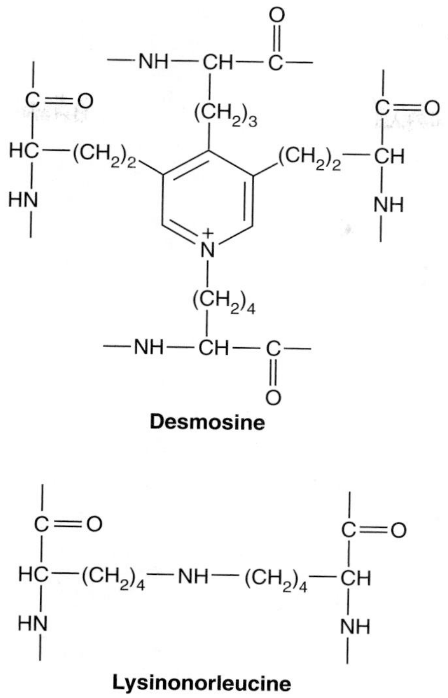

FIGURE 5.21

Desmosine and Lysinonorleucine Linkages.

Desmosine

Lysinonorleucine

KEY *Concepts*

Biochemists distinguish four levels of the structural organization of proteins. In primary structure, the amino acid residues are connected by peptide bonds. The secondary structure of polypeptides is stabilized by hydrogen bonds. Prominent examples of secondary structure are α-helices and β-pleated sheets. Tertiary structure is the unique three-dimensional conformation that a protein assumes because of the interactions between amino acid side chains. Several types of interactions stabilize tertiary structure: hydrophobic interactions, electrostatic interactions, hydrogen bonds, and certain covalent bonds. Proteins that consist of several separate polypeptide subunits exhibit quaternary structure. Both noncovalent and covalent bonds hold the subunits together.

immunoglobulins, and the desmosine and lysinonorleucine linkages in certain connective tissue proteins. *Desmosine* (Figure 5.21) cross-links connect four polypeptide chains in the rubberlike connective tissue protein elastin. They are formed as a result of a series of reactions involving the oxidation of lysine side chains. A similar process results in the formation of *lysinonorleucine*, a cross-linking structure that is found in elastin and collagen.

Quite often the interactions between subunits are affected by the binding of ligands. In **allostery,** the control of protein function through ligand binding, binding a ligand to a specific site in a protein triggers a conformational change that alters its affinity for other ligands. Ligand-induced conformational changes in such proteins are called **allosteric transitions** and the ligands which trigger them are called **effectors** or **modulators.** Allosteric effects can be positive or negative, depending on whether effector binding increases or decreases the protein's affinity for other ligands. One of the best understood examples of allosteric effects, the binding of O_2 and other ligands to hemoglobin is described on p. 111-114. (Because allosteric emzymes play a key role in the control of metabolic processes, allostery is discussed further in Sections 6.3 and 6.5.)

Loss of Protein Structure Protein structure is especially sensitive to environmental factors. Many physical and chemical agents can disrupt a protein's native conformation. The process of structure disruption is called **denaturation.** (Denaturation is not usually considered to include the breaking of peptide bonds.) Depending on the degree of denaturation, the molecule may partially or completely lose its biological activity. Denaturation often results in easily observable changes in the physical properties of proteins. For example, soluble and transparent egg albumin (egg white) becomes insoluble and opaque upon heating. Like many denaturations, cooking eggs is an irreversible process. The following example of a reversible denaturation was demonstrated by Christian Anfinsen in the 1950s. Bovine pancreatic ribonuclease (a digestive enzyme from cattle that degrades RNA) is denatured when treated with β-mercaptoethanol and 8 M urea. During this process, ribonuclease, composed of a single polypeptide with four disulfide bridges, completely unfolds and loses all biological activity. Careful removal of the denaturing agents with dialysis results in a spontaneous and correct refolding of the polypeptide and reformation of the disulfide bonds. The fact that Anfinsen's experiment resulted in a full restoration of the enzyme's catalytic activity, served as early evidence that three-dimensional structure is determined by a polypeptide's amino acid sequence. However, most proteins treated similarly do not renature.

Denaturing conditions include the following:

1. **Strong acids or bases.** Changes in pH result in protonation of some protein side groups which alters hydrogen bonding and salt bridge patterns. As a protein approaches its isoelectric point, it becomes insoluble and precipitates from solution.

2. **Organic solvents.** Water-soluble organic solvents such as ethanol interfere with hydrophobic interactions because they interact with nonpolar R groups and form

PROTEIN POISONS

Some pathogenic organisms damage humans by producing poisonous substances called toxins. *Toxins,* many of which are proteins, exert their effects by several methods:

1. damage to cell membranes,
2. disruption of various intracellular functions, and
3. inhibition of function at nerve cell synapses.

Toxins that act directly on cell membranes, called cytolytic toxins, disturb and ultimately kill the target cells. Produced by many organisms (e.g., bacteria, fungi, plants, fish, and snakes), cytolytic toxins may cause damage in several ways. For example, streptolysin O (67,000 D), produced by the bacterium *Streptococcus pyogenes,* causes pores to form in the target cell membranes. Affected cells are rapidly lysed because the cell membrane is much more permeable to ions such as Na^+. Streptolysin O is believed to cause some of the damage in rheumatic fever.

Cell membrane destruction may also be caused by toxic enzymes. For example, many organisms secrete enzymes, called phospholipases, which cause the hydrolysis of membrane lipid molecules. Phospholipase A2 is found in the venom of several snakes.

Many toxins interfere with intracellular functions. The best-characterized of these are diphtheria toxin and cholera toxin, produced by the bacteria *Corynebacterium diptheriae* and *Vibrio cholerae,* respectively. Both of these toxins contain two subunits, called A and B. The A subunit is responsible for the toxic effect, while the B subunit binds to the target cell. Once diphtheria toxin has entered the target cell, the A and B subunits split apart. The A subunit, which is an enzyme, catalyzes a reaction that prevents protein synthesis. The cell dies because it cannot synthesize proteins. The host organism dies because cardiac, kidney, and nervous tissue are destroyed.

The B subunit of cholera toxin, which is made up of five identical subunits (Figure 5A), attaches to the membranes of intestinal cells. The A subunit is then inserted into these cells, where it activates an enzyme that increases the concentration of a nucleotide called cyclic AMP (cAMP). High sustained concentrations of cAMP, a molecule that opens membrane chloride

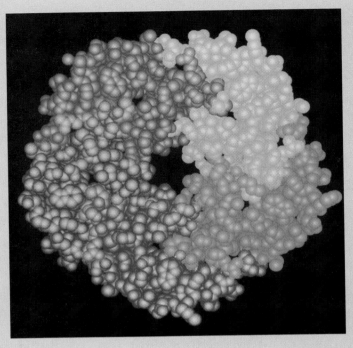

FIGURE 5A

The β-Subunit of the Cholera Toxin.

channels, causes severe diarrhea. (Loss of chloride results in water loss because of osmotic pressure. Several gallons of fluid per day may be lost.) If left untreated, severe dehydration may cause death within 48 hours, by circulatory shock brought on by low blood volume.

Several toxic proteins act as neurotoxins by disrupting the activity of synapses. (A synapse is a junction between two neurons or between a neuron and a muscle cell.) The pain, tremors, and irritability that result from black widow spider bites are caused by α-latrotoxin (125,000 D). This molecule, a single polypeptide, stimulates a massive release of the neurotransmitter acetylcholine (ACh). In contrast, ACh release is inhibited by botulinum toxin, a mixture of several proteins produced by

hydrogen bonds with water and polar protein groups. Nonpolar solvents also disrupt hydrophobic interactions.

3. **Detergents.** These amphipathic molecules disrupt hydrophobic interactions, causing proteins to unfold into extended polypeptide chains. (**Amphipathic molecules** contain both hydrophobic and hydrophilic components.)
4. **Reducing agents.** In the presence of reagents such as urea, reducing agents such as β-mercaptoethanol convert disulfide bridges to sulfhydryl groups. Urea disrupts hydrogen bonds and hydrophobic interactions.
5. **Salt concentration.** The solubility of proteins varies considerably. Fibrous proteins, for example, are virtually insoluble in water. The addition of small amounts of salt, a process called *salting in,* often improves solubility significantly. The binding of salt ions to the protein's ionizable groups decreases interaction between oppositely charged groups on the protein molecules. Water molecules then can form solvation spheres around these groups. When large amounts of salt are added to a protein in solution, a precipitate forms. The large

the bacterium *Clostridium botulinum.* Botulism, a malady most commonly caused by eating contaminated canned food, is characterized by vomiting, dizziness, and sometimes paralysis and death. A related species, *C. tetani,* produces another deadly neurotoxin. Tetanus toxin causes a severe paralysis by blocking neurotransmitter release (primarily glycine and γ-aminobutyric acid) in the central nervous system.

Cholera: A Short History

Toxin-producing organisms, such as *Vibrio cholerae,* do not kill only individual humans. Under certain circumstances they can affect an entire civilization. Cholera has had lasting effects on the Western world. Because of improved transportation, the 1817 cholera epidemic in India reached Europe. Traveling at an average speed of five miles a day, the disease left millions dead. Cholera claimed its first British victim in the port city of Sunderland in October 1831. The 22,000 deaths that followed dur-

ing the next two years were due largely to horrendous living conditions during the Industrial Revolution (Figure 5B). Despite intense but misdirected efforts, cholera epidemics often occurred during the decades that followed. How the disease spread was not discovered until the newly emerging science of statistics revealed that poor sanitation and polluted water were responsible. Public pressure, driven by the seemingly endless deaths caused by cholera, eventually forced the British government to assume some responsibility for public health, a relatively modern concept. In 1859 the British Parliament contracted to build an elaborate sewer system in the city of London, at that time the largest municipal project of its kind ever undertaken. Cholera never returned to that city, and the relationship between sanitation and infectious disease was firmly established. In 1883 the German researcher Robert Koch identified the causative agent.

Royal Society of Medicine

FIGURE 5B

Devastating Effect of Cholera on Britain in the Mid-Nineteenth Century.

(a) A poster attests to the severity of the epidemic. (b) Burning tar to kill infection in the air in Exeter. Other "methods" for preventing the spread of cholera included tobacco fumes, and cleaning with turpentine and vinegar.

number of salt ions can effectively compete with the protein for water molecules, that is, the solvation spheres surrounding the protein's ionized groups are removed. The protein molecules aggregate and then precipitate. This process is referred to as *salting out.* Because salting out is usually reversible, it is often used as an early step in protein purification.

6. **Heavy metal ions.** Heavy metals such as mercury (Hg^{2+}) and lead (Pb^{2+}) affect protein structure in several ways. They may disrupt salt bridges by forming ionic bonds with negatively charged groups. Heavy metals also bond with sulfhydryl groups, a process that may result in significant changes in protein structure and function. For example, Pb^{2+} binds to sulfhydryl groups in two enzymes in the heme synthetic pathway (Chapter 13). The resultant decrease in the hemoglobin synthesis causes severe anemia. (In anemia the number of red blood cells or the hemoglobin concentration is lower than normal.) Anemia is one of the most easily measured symptoms of lead poisoning (Special Interest Box 13.4).

7. **Temperature changes.** As the temperature increases, the rate of molecular vibration increases. Eventually, weak interactions such as hydrogen bonds are

disrupted and the protein unfolds. Some proteins are more resistant to heat denaturation and this fact can be used in purification procedures.

8. **Mechanical stress.** Stirring and grinding actions disrupt the delicate balance of forces that maintain protein structure. For example, the foam formed when egg white is beaten vigorously contains denatured protein.

Fibrous Proteins

Fibrous proteins typically contain high proportions of regular secondary structures, such as α-helices or β-pleated sheets. As a consequence of their rodlike or sheetlike shapes, many fibrous proteins have structural rather than dynamic roles. Examples of fibrous proteins include α-keratin, collagen, and silk fibroin.

α-Keratin   In fibrous proteins, bundles of helical polypeptides are commonly twisted together into larger bundles. The structural unit of the α-keratins, a class of proteins found in hair, wool, skin, horns, and fingernails, is an α-helical polypeptide. Each polypeptide has three domains: an amino terminal "head" domain, a central rod-like α-helical domain, and a carboxyl-terminal "tail." There are two families of α-keratin polypeptides: type I and type II. One polypeptide from each family associates to form a coiled coil dimer (Figure 5.22). Two staggered antiparallel rows of these dimers form a left-handed supercoiled structure called a protofilament. Hydrogen bonds and disulfide bridges are the principal interactions between protofilament subunits. Hundreds of filaments, each containing four protofilaments, are packed together to form a macrofibril. Each hair cell, also called a *fiber,* contains several macrofibrils. A strand of hair therefore consists of numerous dead cells packed with keratin molecules.

Many of the physical properties of the α-keratins are reflected in their amino acid compositions. They have a regular α-helical structure because they lack helix-breaking amino acids such as proline and have helix-promoting residues such as alanine and leucine. Because R groups are on the outside of the α-helices, α-keratin's high hydrophobic amino acid content makes this group of proteins very insoluble in

FIGURE 5.22

Keratin.

The α-helical rodlike domains of a Type I polypeptide and a Type II polypeptide form a coiled coil. Two staggered antiparallel rows of these dimers form a supercoiled protofilament. Hundreds of filaments, each containing four protofilaments, form a macrofibril.

α-Helix

Coiled coil of two α- helices

Protofilament (pair of coiled coils)

Filament (four right-hand twisted protofilaments)

water. Its cysteine residues and the formation of interhelix disulfide bridges make α-keratin relatively resistant to stretching. "Hard" keratins, such as those found in horns and nails, have considerably more disulfide bridges than their softer counterparts found in skin. Humans take advantage of the disulfide bridge content of hair during the permanent waving process. After the hair strands are arranged in the desired shape, disulfide bonds are broken with a reducing agent. New disulfide bonds are then formed by an oxidizing agent, thus creating curled hair.

Collagen   Collagen is the most abundant protein in vertebrates. Synthesized by connective tissue cells, collagen molecules are secreted into extracellular matrices. Collagen includes many closely related proteins that have diverse functions. The genetically distinct collagen molecules in skin, bones, tendons, blood vessels, and corneas impart to these structures many of their special properties (e.g., the tensile strength of tendons and the transparency of corneas).

Collagen is composed of three left-handed polypeptide helices that are twisted around each other to form a right-handed superhelix (Figure 5.23). Type I collagen molecules, found in teeth, bone, skin, and tendons, are about 300 nm long and approximately 1.5 nm wide. (There are twenty major families of collagen molecules. Approximately 90% of the collagen found in humans is type I.) Twisting the polypeptides of the superhelix in opposite directions is principally responsible for collagen's strength.

The amino acid composition of collagen is distinctive. Glycine constitutes approximately one-third of the amino acid residues. Proline and 4-hydroxyproline may account for as much as 30% of a collagen molecule's amino acid composition. Small amounts of 3-hydroxyproline and 5-hydroxylysine also occur. (Specific proline and lysine residues in collagen's primary sequence are hydroxylated within the rough ER after the polypeptides have been synthesized. These reactions, which are discussed in Chapter 18, require the antioxidant ascorbic acid.)

Collagen's amino acid sequence primarily consists of large numbers of repeating triplets with the sequence of Gly—X—Y, where X and Y are often proline and

FIGURE 5.23

Collagen Fibrils.

The bands are formed by staggered collagen molecules. Cross striations are about 680 Å apart. Each collagen molecule is about 3000 Å long.

From Voet and Voet, *Biochemistry,* 2nd edition. Copyright © John Wiley & Sons, Inc., New York, NY.

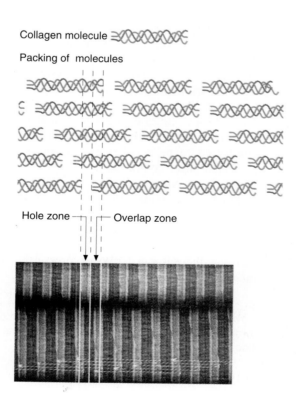

Collagen molecule

Packing of molecules

Hole zone   Overlap zone

FIGURE 5.24

Molecular Model of Silk Fibroin.

Note that the R groups of alanine on one side of each β-pleated sheet interdigitate with similar residues on the adjacent sheet.

© Irving Geis (illustration) from *Biochemistry*, 2nd ed. by Donald Voet and Judith Voet, published by John Wiley & Sons, Inc.

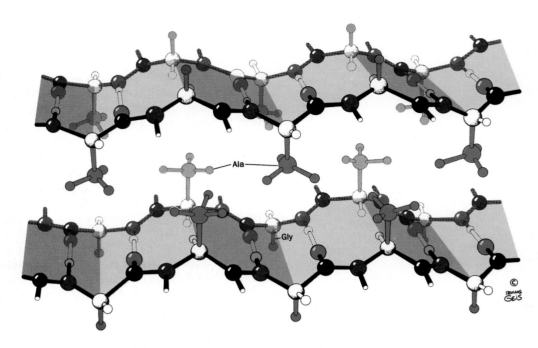

hydroxyproline. Hydroxylysine is also found in the Y position. Simple carbohydrate groups are often attached to the hydroxyl group of hydroxylysine residues. It has been suggested that collagen's carbohydrate components are required for *fibrilogenesis*, the assembly of collagen fibers in their extracellular locations, such as tendons and bone. Cross-linkages between hydroxylysine-linked carbohydrate groups and the ε-amino group of other lysine or hydroxylysine residues on adjacent molecules involves an Amadori rearrangement. (Amadori rearrangements are discussed in Special Interest Box 15.2.)

Glycine is prominent in collagen sequences because the triple helix is formed by interchain hydrogen bonding involving the glycine residues. Therefore, every third residue is in close contact with the other two chains. Glycine is the only amino acid with an R group sufficiently small for the space available. Larger R groups would destabilize the superhelix structure. The triple helix is further strengthened by hydrogen bonding between the polypeptides (due principally to the large number of hydroxyproline residues) as well as lysinonorleucine linkages that stabilize the orderly arrays of triple helices in the final collagen fibril.

Silk Fibroin   Several insects and spiders produce silk, a substance that consists of the fibrous protein fibroin embedded in an amorphous matrix. In fibroin, which is considered to be a β-keratin, the polypeptide chains are arranged in antiparallel β-pleated sheet conformations (Figure 5.24). β-Pleated sheets form because of fibroin's large content of amino acids with relatively small R groups such as glycine and alanine or serine. (Bulky R groups would distort the almost crystalline regularity of silk protein.) Silk is a strong fabric because the chains are fully extended. (Stretching them further would require breaking strong covalent bonds in the polypeptide backbones.) Because the pleated sheets are loosely bonded to each other (primarily with weak van der Waals forces), they slide over each other easily. This arrangement gives silk fibers their flexibility.

QUESTION 5.3   Covalent cross-links contribute to the strength of collagen. The first reaction in cross-link formation is catalyzed by the copper-containing enzyme lysyl oxidase, which converts lysine residues to the aldehyde allysine:

$$
\begin{array}{ccc}
\overset{|}{C}{=}O & & \overset{|}{C}{=}O \qquad O \\
| & \text{Lysyl} & | \qquad\qquad \parallel \\
CH-(CH_2)_4-NH_2 & \xrightarrow{\text{oxidase}} & CH-(CH_2)_3-C-H \\
| & & | \\
NH & & NH \\
| & & |
\end{array}
$$

Lysine residue   Allysine residue

Allysine then reacts with other side chain aldehyde or amino groups to form cross-linkages. For example, two allysine residues react to form an aldol cross-linked product:

$$
\begin{array}{ccc}
\overset{|}{C}{=}O \qquad O & & O \qquad \overset{|}{C}{=}O \\
| \qquad\qquad \parallel & & \parallel \qquad\qquad | \\
CH-(CH_2)_3-C-H & + & H-C-(CH_2)_3-CH \\
| & & | \\
NH & & NH \\
| & & |
\end{array}
$$

Allysine residue   Allysine residue

$$
\begin{array}{ccc}
\overset{|}{C}{=}O & & \overset{|}{C}{=}O \\
| & & | \\
CH-(CH_2)_2-C{=}CH-(CH_2)_3-CH \\
| \qquad\qquad | & & | \\
NH \qquad\qquad C-H & & NH \\
| \qquad\qquad \parallel & & | \\
\qquad\qquad O
\end{array}
$$

Aldol cross-link

(A reaction in which aldehydes form an α, β-unsaturated aldehyde linkage is referred to as an **aldol condensation.** In condensation reactions a small molecule, in this case H_2O, is removed.)

In a disease called *lathyrism,* which occurs in humans and several other animals, a toxin (β-aminopropionitrile) found in sweet peas (*Lathyrus odoratus*) inactivates lysyl oxidase. Consider the abundance of collagen in animal bodies and suggest some likely symptoms of this malady.

Globular Proteins

The biological functions of globular proteins usually involve the precise binding of small ligands or large macromolecules such as nucleic acids or other proteins. Each protein possesses a unique and complex surface that contains cavities or clefts whose structure is complementary to specific ligands. After ligand binding, a conformational change occurs in the protein that is linked to a biochemical event. For example, the binding of ATP to myosin in muscle cells, is a critical event in muscle contraction.

The oxygen-binding proteins myoglobin and hemoglobin are interesting and well-researched examples of globular proteins. They are both members of the hemoproteins, a specialized group of proteins that contain the prosthetic group heme. Although the heme group (Figure 5.25) in both proteins is responsible for the reversible binding of molecular oxygen, the physiological roles of myoglobin and hemoglobin are significantly different. The chemical properties of heme are dependent on the Fe (II) ion in the center of the prosthetic group. Fe (II) forms six coordinate bonds. The iron atom is bound to the four nitrogens in the center of the protoporphyrin ring. Two other coordinate bonds are available, one on each side of the planar heme structure. In myoglobin and hemoglobin, the fifth coordination bond is to the nitrogen atom in a histidine residue, and the sixth coordination bond is available for binding oxygen. In addition to serving as a reservoir for oxygen within muscle cells, myoglobin facilitates the diffusion of oxygen in metabolically active cells. The role of hemoglobin, the

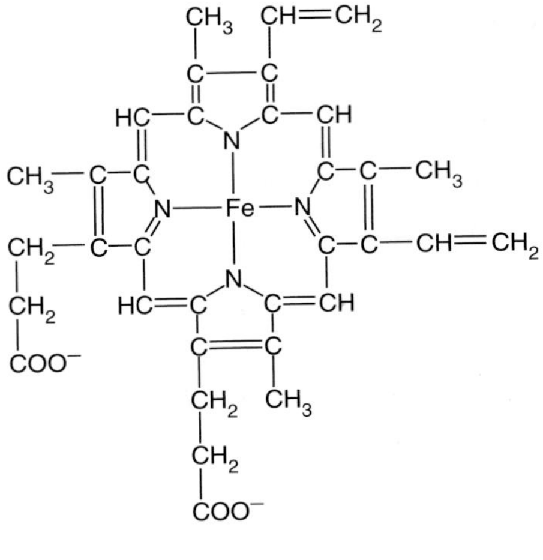

FIGURE 5.25

Heme.

Heme consists of a porphyrin ring (composed of four pyrroles) with Fe II in the center.

primary protein of red blood cells, is to deliver oxygen to cells throughout the body. A comparison of the structures of these two proteins illustrates several important principles of protein structure, function, and regulation.

Myoglobin Myoglobin, found in high concentration in skeletal and cardiac muscle, gives these tissues their characteristic red color. Diving mammals such as whales, which remain submerged for long periods of time, possess high myoglobin concentrations in their muscles. Because of the extremely high concentrations of myoglobin, such muscles are typically brown. The protein component of myoglobin, called globin, is a single polypeptide chain that contains eight sections of α-helix (Figure 5.26). The folded globin chain forms a crevice that almost completely encloses a heme

FIGURE 5.26

Myoglobin.

With the exception of the side chain groups of two histidine residues only the α-carbon atoms of the globin polypeptide are shown. Myoglobin's eight helices are designated A through H. The heme group has an iron atom that binds reversibly with oxygen. To improve clarity one of heme's propionic acid side chains has been displaced.

© Irving Geis

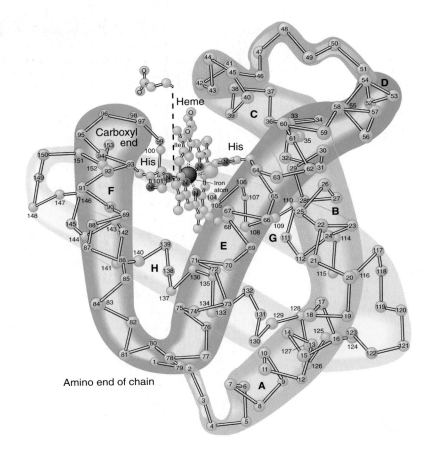

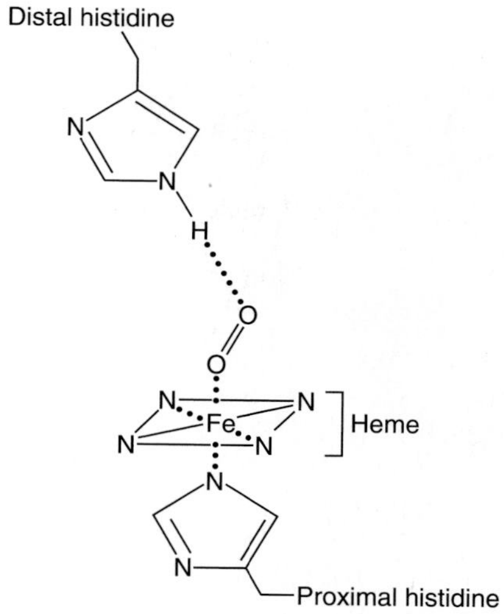

Distal histidine

FIGURE 5.27

The Oxygen-Binding Site of Heme Created by a Folded Globin Chain.

From H.R. Horton, et al., *Principles of Biochemistry*, © 1992, p. 432, fig. 4.27. Reprinted by permission of Prentice-Hall, Inc., Upper Saddle River, NJ.

group. Free heme [Fe(II)] has a high affinity for O_2 and is irreversibly oxidized to form hematin [Fe(III)]. Hematin cannot bind O_2. Noncovalent interactions between amino acid side chains and the nonpolar porphyrin ring within the oxygen-binding crevice decrease heme's affinity for O_2. The decreased affinity protects Fe(II) from oxidation and allows for the reversible binding of O_2. All of the heme-interacting amino acids are nonpolar, with the exception of two histidines, one of which (the proximal histidine) binds directly to the heme iron atom (Figure 5.27). The other (the distal histidine) stabilizes the oxygen-binding site.

Hemoglobin Hemoglobin is a roughly spherical molecule found in red blood cells, where its primary function is to transport oxygen from the lungs to every tissue in the body. Recall that HbA is composed of two α-chains and two β-chains (Figure 5.28). The HbA molecule is commonly designated $\alpha_2\beta_2$. (There is another type of adult hemoglobin. Approximately 2% of human hemoglobin is HbA_2, which contains δ (delta)-chains instead of β-chains.) Before birth, several additional hemoglobin polypeptides are synthesized. The ϵ (epsilon)-chain, which appears in early embryonic life, and the γ-chain found in the fetus closely resemble the β-chain. Because both $\alpha_2\epsilon_2$ and $\alpha_2\gamma_2$ hemoglobins have a greater affinity for oxygen than does HbA, the fetus can preferentially absorb oxygen from the maternal bloodstream.

Although the three-dimensional configurations of myoglobin and the α- and β-chains of hemoglobin are very similar, their amino acid sequences have many differences. Comparison of these molecules from dozens of species has revealed nine invariant amino acid residues. Several invariant residues directly affect the oxygen-binding site, while others stabilize the α-helical peptide segments. The remaining residues may vary considerably. However, most substitutions are conservative. For example, each polypeptide's interior remains nonpolar.

The four chains of hemoglobin are arranged in two identical dimers (designated as $\alpha_1\beta_1$ and $\alpha_2\beta_2$. Each globin polypeptide has a heme-binding unit similar to that described for myoglobin. Although both myoglobin and hemoglobin bind oxygen reversibly, the latter molecule has a complex structure and more complicated binding properties. The numerous noncovalent interactions (mostly hydrophobic) between the subunits in each $\alpha\beta$-dimer remain largely unchanged when hemoglobin interconverts between its oxygenated and deoxygenated forms (Figure 5.29). In contrast, the

FIGURE 5.28

Hemoglobin.

The protein contains four subunits, designated α and β. Each subunit contains a heme group that binds reversibly with oxygen.

© Irving Geis

FIGURE 5.29

Three-Dimensional Structure of
(a) Oxyhemoglobin and
(b) Deoxyhemoglobin.

The β-chains are on top. In the oxy-deoxy transformation, the $\alpha_1\beta_1$ and $\alpha_2\beta_2$ dimers move as units relative to each other. This allows 2,3-bisphosphoglycerate (discussed on p. 114) to bind to the larger central cavity in the deoxyconformation.

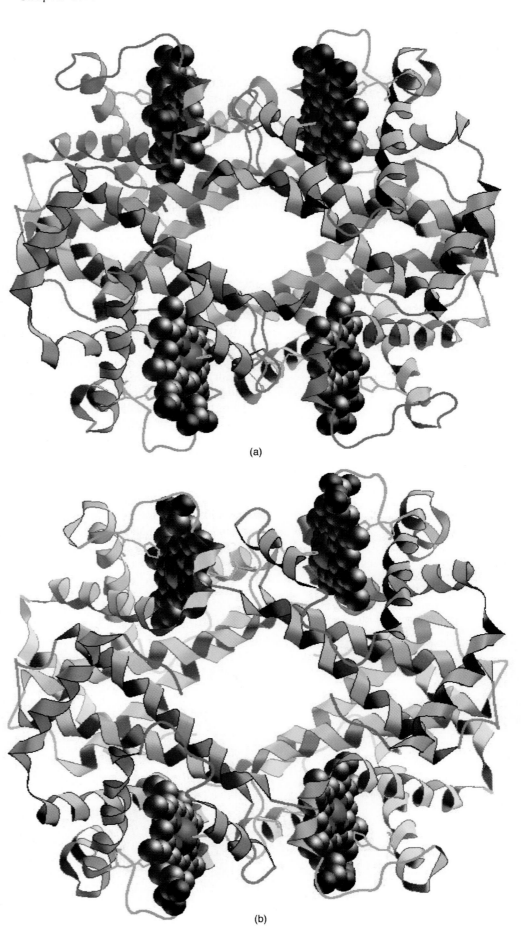

(a)

(b)

FIGURE 5.30

The Hemoglobin Allosteric Transition

When hemoglobin is oxygenated, the $\alpha_1\beta_1$ and $\alpha_2\beta_2$ dimers slide by each other and rotate 15°. (a) Deoxyhemoglobin, (b) oxyhemoglobin.

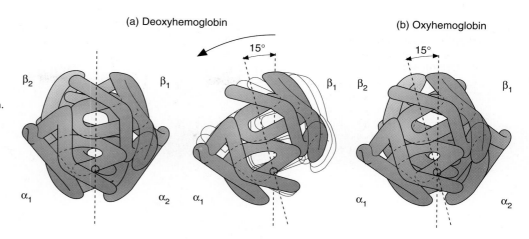

(a) Deoxyhemoglobin   (b) Oxyhemoglobin

relatively small number of interactions between the two dimers change substantially during this transition. When hemoglobin is oxygenated, the salt bridges and certain hydrogen bonds are ruptured as the $\alpha_1\beta_1$ and $\alpha_2\beta_2$ dimers slide by each other and rotate 15° (Figure 5.30). The deoxygenated conformation of hemoglobin (deoxyHb) is often referred to as the T(taut) state and oxygenated hemoglobin (oxyHb) is said to be in the R(relaxed) state. The oxygen induced readjustments in the interdimer contacts is almost simultaneous. In other words, a conformational change in one subunit is rapidly propagated to the other subunits. Consequently, hemoglobin alternates between two stable conformations, the T and R states.

Because of subunit interactions, the oxygen-dissociation curve of hemoglobin has a sigmoidal shape (Figure 5.31). As the first O_2 binds to hemoglobin, the binding of additional O_2 to the same molecule is enhanced. This binding pattern, called **cooperative binding,** results from changes in hemoglobin's three-dimensional structure that are initiated when the first O_2 binds. The binding of the first O_2 facilitates the binding of the remaining three O_2 molecules to the tetrameric hemoglobin molecules. In the lungs, where O_2 tension is high, hemoglobin is quickly saturated (converted to the R state). In tissues depleted of O_2, hemoglobin gives up about half of its oxygen. In contrast to hemoglobin, myoglobin's oxygen-dissociation curve has a hyperbolic shape. This simpler binding pattern, a consequence of myoglobin's simpler structure, reflects several aspects of this protein's role in oxygen storage. Because its dissociation curve is well to the left of the hemoglobin curve, myoglobin gives up oxygen only when the muscle cell's oxygen concentration is very low (i.e., during strenuous exercise). In addition, because myoglobin has a greater affinity for oxygen than does hemoglobin, oxygen moves from blood to muscle.

The binding of ligands other than oxygen affects hemoglobin's oxygen-binding properties. For example, the dissociation of oxygen from hemoglobin is enhanced if pH decreases. By this mechanism called the *Bohr effect,* oxygen is delivered to cells in proportion to their needs. Metabolically active cells, which require large amounts of oxygen for energy generation, also produce large amounts of the waste products H^+ and CO_2. As CO_2 diffuses into blood, it reacts with water to form HCO_3^- and H^+. (The bicarbonate buffer was discussed on p. 52.) The subsequent binding of H^+ to several ionizable groups on hemoglobin molecules enhances the dissociation of O_2 by converting hemoglobin to its T state. (Hydrogen ions bind preferentially to deoxyHb. Any increase in H^+ concentration stabilizes the deoxy conformation of the protein and therefore speeds its formation.) When a small number of CO_2 molecules bind to uncharged amino groups on hemoglobin (forming carbamate or $—NHCOO^-$ groups) the deoxy form (T state) of the protein is additionally stabilized.

FIGURE 5.31

Equilibrium Curves Measure the Affinity of Hemoglobin and Myoglobin for Oxygen.

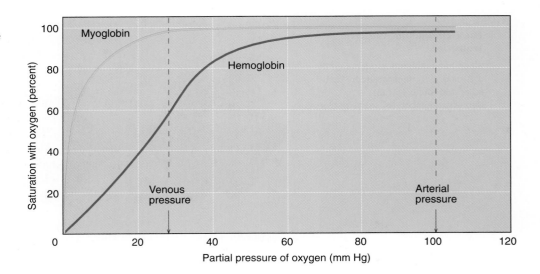

FIGURE 5.32

2,3-Bisphosphoglycerate (BPG) Decreases the Affinity Between Oxygen and Hemoglobin.

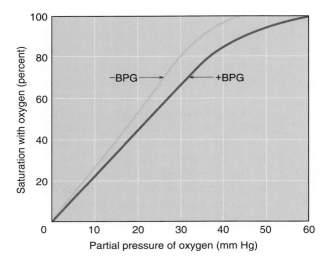

2,3-Bisphosphoglycerate (BPG) (also called glycerate-2,3-bisphosphate) is also an important regulator of hemoglobin function. (Although most cells contain only trace amounts of BPG, red blood cells contain a considerable amount. BPG is derived from glycerate-1,3-bisphosphate, an intermediate in the breakdown of the energy-rich compound glucose.) In the absence of BPG, hemoglobin has a very high affinity for oxygen (Figure 5.32). As with H^+ and CO_2, binding BPG stabilizes deoxyHb. A negatively charged BPG molecule binds in a central cavity within hemoglobin that is lined with positively charged amino acids.

In the lungs the process is reversed. A high oxygen concentration drives the conversion from the deoxyHb configuration to that of oxyHb. The change in the protein's three-dimensional structure initiated by the binding of the first oxygen molecule releases bound CO_2, H^+, and BPG. The H^+ recombine with HCO_3^- to form carbonic acid, which then dissociates to form CO_2 and H_2O. CO_2 subsequently diffuses from the blood into the alveoli.

QUESTION 5.4

Fetal hemoglobin (HbF) binds to BPG to a lesser extent than does HbA. Why do you think HbF has a greater affinity for oxygen than does maternal hemoglobin?

Protein Technology Because most genetic information is expressed through proteins it is not surprising that considerable time, effort, and funding have been devoted to investigating their properties. Since the amino acid sequence of bovine insulin was determined by Frederick Sanger in 1953, the structures of several thousand proteins have been elucidated. In addition to providing insight into the molecular mechanisms of various proteins, information about protein structure has led to a deeper understanding of the evolutionary relationships between species. Since some inherited diseases are now known to be caused by alterations in the amino acid sequence of specific proteins, new diagnostic tests and clinical therapies have been developed. **Some topics in protein technology are described in Appendix B, Supplement 5. Information provided in this discussion is needed to answer several questions at the end of this chapter.**

KEY WORDS

aldol condensation, *109*	disulfide bridge, *91*	isoelectric point, *85*	primary structure, *94*
aliphatic hydrocarbon, *83*	effector, *103*	ligand, *100*	protomer, *102*
allosteric transition, *103*	electrostatic interaction, *100*	lipoprotein, *94*	prosthetic group, *94*
allostery, *103*	enantiomer, *85*	metaloprotein, *94*	protein, *80*
amphipathic molecule, *104*	fibrous protein, *94*	modulator, *103*	quaternary structure, *94*
amphoteric molecule, *80*	gene, *97*	molecular disease, *96*	salt bridges, *100*
apoprotein, *94*	globular protein, *94*	neurotransmitter, *81*	secondary structure, *94*
aromatic hydrocarbon, *83*	glycoprotein, *94*	oligomer, *102*	stereoisomer, *84*
asymmetric carbon, *84*	hemoprotein, *94*	opioid peptide, *93*	subunit, *100*
chiral carbon, *84*	holoprotein, *94*	peptide, *80*	supersecondary structure, *98*
conjugated protein, *94*	homologous polypeptide, *96*	peptide bond, *89*	tertiary structure, *94*
cooperative binding, *113*	hormones, *81*	phosphoprotein, *94*	zwitterions, *80*
denaturation, *103*	hydrophobic interaction, *100*	polypeptide, *80*	

SUMMARY

Proteins have a vast array of functions in living organisms. In addition to serving as structural materials, proteins are involved in metabolic regulation, transport, defense, and catalysis. Polypeptides are amino acid polymers. Proteins may consist of one or more polypeptide chains.

Each amino acid contains a central carbon atom (the α-carbon) to which an amino group, a carboxylate group, a hydrogen atom, and an R group are attached. In addition to composing protein, amino acids have several other biological roles. According to their capacity to interact with water, amino acids may be separated into four classes: (1) nonpolar and neutral, (2) polar and neutral, (3) acidic, and (4) basic.

Titration of amino acids and peptides illustrates the effect of pH on their structures. The pH at which a molecule has no net charge is called its isoelectric point.

Amino acids undergo several chemical reactions. Two reactions are especially important: peptide bond formation and cysteine oxidation.

The functions of protein include catalysis, structural components, movement, defense, regulation, and transport. Proteins are also clas-

sified according to their shape and composition. Fibrous proteins (e.g., collagen) are long, rod-shaped molecules that are insoluble in water and physically tough. Globular proteins (e.g., hemoglobin) are compact, spherical molecules that are usually water soluble.

Biochemists have distinguished four levels of protein structure. Primary structure, the amino acid sequence, is specified by genetic information. As the polypeptide chain folds, local folding patterns constitute the protein's secondary structure. The overall three-dimensional shape that a polypeptide assumes is called the tertiary structure. Proteins that consists of two or more polypeptides have quaternary structure. Many physical and chemical conditions disrupt protein structure. Denaturing agents include strong acids or bases, reducing agents, organic solvents, detergents, high salt concentrations, heavy metals, temperature changes, and mechanical stress.

The biological activity of complex multisubunit proteins is often regulated by allosteric interactions in which small ligands bind to the protein. Any change in the protein's activity is due to changes in the interactions among the protein's subunits. Effectors can increase or decrease the function of a protein.

SUGGESTED READINGS

Chathea, C., Principles that Determine the Structure of Proteins, *Ann. Rev. Biochem.,* 53:537–572, 1984.

Doolittle, R. F., Proteins, *Sci. Amer.,* 253(10)88–96, 1985.

Eyre, D. R., Collagen: Molecular Diversity in the Body's Protein Scaffold, *Science,* 207:1315–1322, 1980.

Karplus, M., and McCannon, J. A., The Dynamics of Proteins, *Sci. Amer.,* 254(4):42–51, 1986.

Kosaka, H. and Seiyama, A., Physiological Role of Nitric Oxide as an Enhancer of Oxygen Transfer from Erythrocytes to Tissues, *Biochem. Biophys. Res. Commun.* 218:749–752, 1996.

Pauling L., and Corey, R. B., Configurations of Polypeptide Chains with Favored Orientations Around Single Bonds: Two New Pleated Sheets, *Proc. Nat. Acad. Sci. USA,* 37:729–740, 1953.

Petruzzelli, R., Aureli, G., Lania, A., Galtieri, A., Desideri, A. and Giardina, B., Diving Behavior and Haemoglobin Function: the Primary Structure of the α- and β-Chains of the Sea Turtle (*Caretta caretta*) and its Functional Inplications, *Biochem. J.* 316:959–965, 1996.

Thorne, J.L., Goldman, N. and Jones, D.T., Combining Protein Evolution and Secondary Structure, *Mol. Biol. Evol.* 13(5): 666–673, 1996.

QUESTIONS

1. Distinguish between proteins, peptides, and polypeptides.
2. Indicate which of the following amino acids are polar, nonpolar, acidic, or basic:
   a. glycine
   b. tyrosine
   c. glutamic acid
   d. histidine
   e. proline
   f. lysine
   g. cysteine
   h. asparagine
   i. valine
   j. leucine
3. Arginine has the following pK_a values:

   $pK_1 = 2.17$, $pK_2 = 9.04$,   and $pK_R = 12.48$

   What is the structure and net charge of arginine at the following pH values? 1, 4, 7, 10, 12
4. Shown below is the titration curve for histidine:

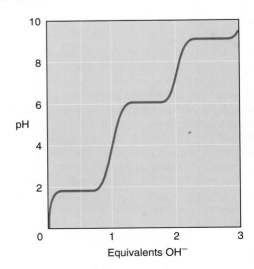

   a. What species are present at each plateau?
   b. Using the titration curve, determine the pK_a of each ionization of histidine.
   c. What is the isoelectric point of histidine?

5. Consider the following molecule:

   a. Name it.
   b. Using the three-letter symbols for the amino acids, how would this molecule be represented?
6. Rotation about the peptide bond in glycylglycine is hindered. Draw the resonance forms of the peptide bond and explain why.
7. List six functions of proteins in the body.
8. Differentiate the terms in each pair below:
   a. globular and fibrous proteins
   b. simple and conjugated proteins
   c. apoprotein and holoprotein
9. Define the following terms:
   a. α-carbon
   b. isoelectric point
   c. peptide bond
   d. hydrophobic amino acid
10. Indicate the level(s) of protein structure to which each of the following contributes:
    a. amino acid sequence
    b. β-pleated sheet
    c. hydrogen bond
    d. disulfide bond
11. What type of secondary structure would the following amino acid sequences be *most* likely to have?
    a. polyproline
    b. polyglycine
    c. Ala—Val—Ala—Val—Ala—Val—
    d. Gly—Ser—Gly—Ala—Gly—Ala
12. List three factors that do not foster α-helix formation.
13. Denaturation is the loss of protein function from structural change or chemical reaction. At what level of protein structure

or through what chemical reaction does each of the following denaturation agents act?

a. heat
b. strong acid
c. saturated salt solution
d. organic solvents (e.g., alcohol or chloroform)

14. Residues such as valine, leucine, isoleucine, methionine, and phenylalanine are often found in the interior of proteins, while arginine, lysine, aspartic acid, and glutamic acid are often found on the surface of proteins. Suggest a reason for this observation. Where would you expect to find glutamine, glycine, and alanine?

15. A polypeptide has a high pI value. Suggest which amino acids might compose it.

16. Proteins that are synthesized by living organisms adopt a biologically active conformation. Yet when such molecules are prepared in the laboratory, they usually fail to spontaneously adopt their active conformations. Can you suggest why?

17. The active site of an enzyme contains sequences that are conserved because they participate in the protein's catalytic activity. The bulk of an enzyme, however, is not part of the active site. Because a substantial amount of energy is required to assemble enzymes, why are they usually so large?

18. Structural protein may incorporate large amounts of immobilized water as part of its structure. Can you suggest how protein molecules "freeze" the water in place and make it part of the protein structure?

19. The peptide bond is a stronger bond than that of esters. What structural feature of the peptide bond gives it additional bond strength?

20. Because of their tendency to avoid water, nonpolar amino acids play an important role in forming and maintaining the three-dimensional structure of proteins. Can you suggest how these molecules accomplish this feat?

21. Chymotrypsin is an enzyme that cleaves other enzymes during sequencing. Why don't chymotrypsin molecules attack each other?

22. Outline the steps to isolate typical protein. What is achieved at each step?

23. Outline the steps to purify a protein. What criteria are used to evaluate purity?

24. List the types of chromatography used to purify proteins. Describe how each separation method works.

25. In sequencing a protein using carboxypeptidase, the protein is first broken down into smaller fragments, which are then separated from one another. Each fragment is then individually sequenced. If this initial fragmentation were not carried out, amino acid residues would build up in the reaction medium. How would these residues inhibit sequencing?

26. In an amino acid analysis, a large protein is broken down into overlapping fragments by using specific enzymes. Why must the sequences be overlapping?

27. Hydrolysis of β-endorphin (a peptide containing 31 amino acid residues) produces the following amino acids:

Tyr (1), Gly (3), Phe (2), Met, Thr (3), Ser (2), Lys (5), Gln (2), Pro, Leu (2), Val (2), Asn (2), Ala (2), Ile, His, and Glu

Treatment with carboxypeptidase liberates Gln. Treatment with DNFB liberates DNP-Tyr. Treatment with trypsin produces the following peptides:

Lys, Gly—Gln, Asn—Ala—Ile—Val—Lys,
Tyr—Gly—Gly—Phe—Met—Thr—Ser—Glu—Lys,
Asn—Ala—His—Lys, Ser—Gln—Thr—Pro—Leu—
Val—Thr—Leu—Phe—Lys

Treatment with chymotrypsin produces the following peptides:

Lys—Asn—Ala—Ile—Val—Lys—Asn—Ala—
His—Lys—Lys—Gly—Gln
Tyr—Gly—Gly—Phe
Met—Thr—Ser—Glu—Lys—Ser—Gln—Thr—Pro—
Leu—Val—Thr—Leu—Phe

What is the primary sequence of β-endorphin?

28. Consider the following tripeptide:

Gly—Ala—Val

a. What is the approximate isoelectric point?
b. In which direction will the tripeptide move when placed in an electric field at the following pH values? 1, 5, 10, 12

29. The following is the amino acid sequence of bradykinin, a peptide released by certain organisms in response to wasp stings:

Arg—Pro—Pro—Gly—Phe—Ser—Pro—Phe—Arg

What amino acids or peptides are produced when bradykinin is treated with each of the following reagents?
a. carboxypeptidase
b. chymotrypsin
c. trypsin
d. DNFB

30. Most amino acids appear blue when treated with ninhydrin reagent. Proline and hydroxyproline appear yellow. Suggest a reason for the difference.

ENZYMES

Chapter Six

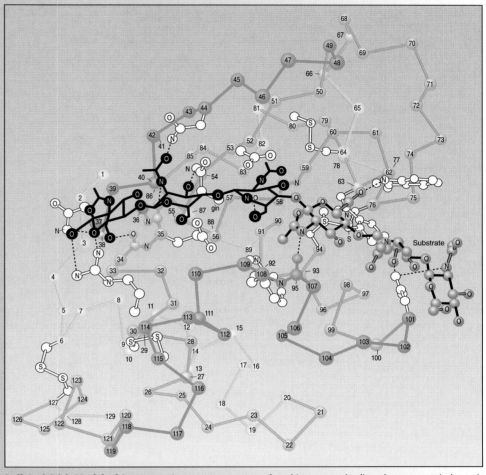

Ball-And-Stick Model of Lysozyme. *Lysozyme, an enzyme found in tears and saliva, destroys certain bacteria by hydrolyzing cell wall polysaccharide. In this model the polypeptide's backbone is shown with a bound segment of polysaccharide (dark blue and black).*

© Irving Geis

OBJECTIVES

After you have studied this chapter, you should be able to answer these questions:

1. *How do enzymes work?*
2. *How do enzymes differ from inorganic catalysts?*
3. *How does protein structure affect enzyme activity?*
4. *How are enzymes named and classified?*
5. *What is enzyme kinetics and how does it provide information about the behavior of enzymes?*
6. *What is enzyme inhibition?*
7. *What catalytic mechanisms do enzymes use to increase reaction rates?*
8. *What role do metals and small organic groups play in biochemical reactions?*
9. *What are allosteric enzymes?*
10. *How and why are enzyme activities regulated in the living cell?*

Life is inconceivable without enzymes. Most of the thousands of biochemical reactions that sustain living processes would occur at imperceptible rates without enzymes. Enzymes are enormously powerful catalysts exhibiting high specificity. Their catalytic activities can be precisely regulated.

One of the most important functions of proteins is their role as catalysts. (Until recently, all enzymes were considered to be proteins. Several examples of catalytic RNA molecules have now been verified. See Chapter 17.) Recall that living processes consist almost entirely of biochemical reactions. Without catalysts these reactions would not occur fast enough to sustain life.

To proceed at a viable rate, most chemical reactions require an initial input of energy. In the laboratory this energy is usually supplied as heat. At temperatures above absolute zero ($-273.1°C$), all molecules possess vibrational energy, which increases as the molecules are heated. Consider the following reaction:

$$A + B \rightarrow C$$

As the temperature rises, vibrating molecules (A and B) are more likely to collide. A chemical reaction occurs when the colliding molecules possess a minimum amount of energy called the **activation energy.** Not all collisions result in chemical reactions because only a fraction of the molecules have sufficient energy or the correct orientation to react (i.e., to break bonds or rearrange atoms into the product molecules). Another way of increasing the likelihood of collisions, thereby increasing the formation of product, is to increase the concentration of the reactants.

In living systems, however, elevated temperatures may harm delicate biological structures and reactant concentrations are usually quite low. Living organisms circumvent these problems by using enzymes.

Enzymes have several remarkable properties. First, the rates of enzymatically catalyzed reactions are often phenomenally high. (Rate increases by factors of 10^6 or greater are common.) Second, in marked contrast to inorganic catalysts, the enzymes are highly specific to the reactions they catalyze. Side products are rarely formed. Finally, because of their complex structures, enzymes can be regulated. This is an especially important consideration in living organisms, which must conserve energy and raw materials.

Because enzymes are involved in so many aspects of living processes, any understanding of biochemistry depends on an appreciation of these remarkable catalysts. This chapter examines their structure and function.

6.1 PROPERTIES OF ENZYMES

How do enzymes work? The answer to this question requires a review of the role of catalysts. By definition a catalyst is a substance that enhances the rate of a chemical reaction but is not permanently altered by the reaction. Catalysts perform this feat because they decrease the activation energy required for a chemical reaction. In other words, catalysts provide an alternative reaction pathway that requires less energy (Figure 6.1). A **transition state** occurs at the apex of both reaction pathways in Figure 6.1. During any chemical reaction reactants with sufficient energy will attain transition state configuration (a strained intermediate form). For biochemical systems, this occurs when the substrate binds to the enzyme. For example, the transition state in the reaction in the oxidation of ethanol to acetaldehyde,

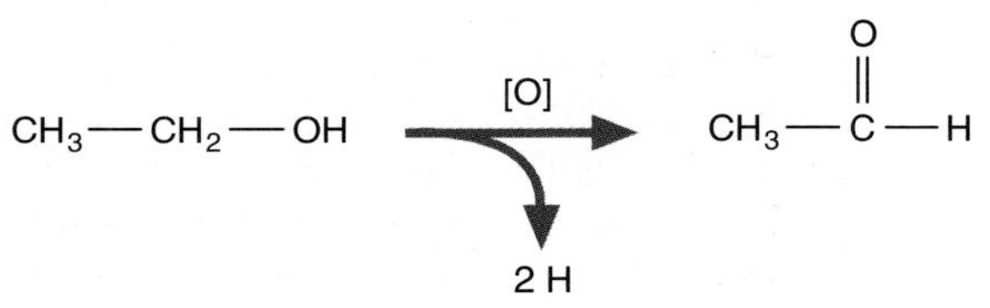

$$CH_3 - CH_2 - OH \xrightarrow[\quad 2\,H \quad]{[O]} CH_3 - \overset{\displaystyle O}{\underset{\displaystyle \|}{C}} - H$$

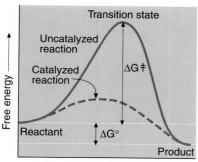

FIGURE 6.1

A Catalyst Reduces the Activation Energy of a Reaction.

A catalyst alters the free energy of activation $\Delta G\ddagger$ and not the standard free energy $\Delta G°$ of the reaction.

might look like

$$CH_3-\underset{\underset{H}{|}}{\overset{\overset{H}{|}}{C}}-O^-$$

Protein catalysts provide a specialized surface to which reactants bind. Some enzymes modify their conformation to force the transition state configuration on the substrate, while others bind only those substrate molecules already in a near-transition state configuration. The enzyme stabilizes the transition state so that a reaction can occur. Enzymes, like all catalysts, cannot alter the equilibrium of the reaction, but they can increase the rate toward equilibrium. Consider the following reversible reaction:

$$A \rightleftarrows B$$

Without a catalyst, the reactant A is converted into the product B at a certain rate. Because this is a reversible reaction, B is also converted into A. The rate expression for the forward reaction is $k_F[A]^n$, and that for the reverse reaction is to $k_F[B]^m$. The superscripts n and m represent the order of a reaction. Reaction order reflects the mechanism by which A is converted to B and vice versa. A reaction order of 2 for the conversion of A to B indicates that it is a bimolecular process and the molecules of A must collide for the reaction to occur. (See Section 6.3.) At equilibrium, the rates for the forward and reverse reactions must be equal:

$$k_F[A]^n = k_R[B]^m \tag{1}$$

which rearranges to

$$\frac{k_F}{k_R} = \frac{[B]^m}{[A]^n} \tag{2}$$

The ratio of the forward and reverse rate constants is the equilibrium constant.

$$K_{eq} = \frac{[B]^m}{[A]^n} \tag{3}$$

In equation (3), if $m = n = 1$ and $k_F = 1 \times 10^{-3}\ s^{-1}$ and $k_R = 1 \times 10^{-6}\ s^{-1}$, then

$$K_{eq} = \frac{10^{-3}}{10^{-6}} = 10^3$$

At equilibrium, therefore, the ratio of products to reactants is 1000 to 1. In the presence of the catalyst, equilibrium is attained in seconds or minutes instead of hours or days.

Even with an inorganic catalyst, most laboratory reactions require an input of energy usually in the form of heat. In addition, most of these catalysts are nonspecific, that is, they accelerate a wide variety of reactions. Enzymes perform their work at moderate temperatures and are quite specific in the reactions that each one catalyzes. The difference between inorganic catalysts and enzymes is directly related to their structures. In contrast to inorganic catalysts, each type of enzyme molecule contains a unique, intricately shaped binding surface called an **active site.** Reactant molecules, called **substrates,** bind to the enzyme's active site, which is typically a small cleft or crevice on a large protein molecule. The active site is not just a binding site, however. Many of the amino acid side chains that line the active site actively participate in the catalytic process.

The *lock-and-key model* of enzyme action, introduced by Emil Fischer in 1890, partially accounts for enzyme specificity. Each enzyme binds to a single type of substrate because the active site and the substrate have complementary structures. The substrate's overall shape and charge distribution allow it to enter and interact with the enzyme's active site. In a modern variation by Daniel Koshland of the lock-and-key

The Induced Fit Model.

Substrate binding causes a conformational change in the enzyme. The conformations of the enzyme hexokinase (a single polypeptide with two domains) (a) before glucose binding and (b) after glucose binding (glucose molecule is not shown). The domains move relative to each other to close around the glucose molecule.

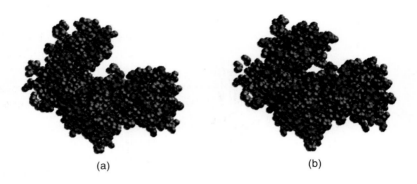

(a)   (b)

KEY *Concepts*

Enzymes are catalysts. Catalysts modify the rate of a reaction because they provide an alternative reaction pathway that requires less energy than the uncatalyzed reaction. Most enzymes are proteins.

model, called the *induced-fit model,* the flexible structure of proteins is taken into account (Figure 6.2). In this model, substrate does not fit precisely into a rigid active site. Instead, noncovalent interactions between the enzyme and substrate change the three-dimensional structure of the active site, conforming the shape of the active site to the shape of the substrate.

Although the catalytic activity of some enzymes depends only on interactions between active site amino acids and the substrate, other enzymes require nonprotein components for their activities. Enzyme **cofactors** may be ions, such as Mg^{2+} or Zn^{2+}, or complex organic molecules, referred to as **coenzymes.** The protein component of an enzyme that lacks an essential cofactor is called an **apoenzyme.** Intact enzymes with their bound cofactors are referred to as **holoenzymes.**

The activities of some enzymes can be regulated to maintain a stable intracellular environment. For example, adjustments in the rates of enzyme-catalyzed reactions allow cells to respond effectively to changes in the concentrations of nutrients. Organisms may control enzyme activities directly, principally through the binding of activators or inhibitors, or indirectly, by regulating enzyme synthesis. (Control of enzyme synthesis requires gene regulation, a topic covered in Chapters 17 and 18.)

6.2 CLASSIFICATION OF ENZYMES

In the early days of biochemistry, enzymes were named at the whim of their discoverers. Often, enzyme names provided no clue to their function (e.g., trypsin), or several names were used for the same enzyme. Enzymes were often named by adding the suffix "-ase" to the name of the substrate. For example, urease catalyzes the hydrolysis of urea. To eliminate confusion, the International Union of Biochemistry (IUB) instituted a systematic naming scheme for enzymes. Each enzyme is now classified and named according to the type of chemical reaction it catalyzes. In this scheme an enzyme is assigned a four-number classification and a two-part name called a *systematic name.* In addition, a shorter version of the systematic name, called the *recommended name,* is suggested by the IUB for everyday use. For example, alcohol:NAD^+ oxidoreductase (E.C. 1.1.1.1) is usually referred to as alcohol dehydrogenase. (The letters E.C. are an abbreviation for enzyme commission.) Because many enzymes were discovered before the institution of the systematic nomenclature, many of the old well-known names have been retained.

The following are the six major enzyme categories:

1. **Oxidoreductases. Oxidoreductases** catalyze oxidation-reduction reactions. Subclasses of this group include the dehydrogenases, oxidases, oxygenases, reductases, peroxidases, and hydroxylases.

2. **Transferases. Transferases** catalyze reactions that involve the transfer of groups from one molecule to another. Examples of such groups include amino, carboxyl, carbonyl, methyl, phosphoryl, and acyl ($RC=O$). Common trivial names for the transferases often include the prefix "trans." Examples include the transcarboxylases, transmethylases, and transaminases.

3. **Hydrolases. Hydrolases** catalyze reactions in which the cleavage of bonds is accomplished by adding water. The hydrolases include the esterases, phosphatases, and peptidases.

4. **Lyases. Lyases** catalyze reactions in which groups (e.g., H_2O, CO_2, and NH_3) are removed to form a double bond or are added to a double bond. Decarboxylases, hydratases, dehydratases, deaminases, and synthases are examples of lyases.

5. **Isomerases.** This is a heterogeneous group of enzymes. **Isomerases** catalyze several types of intramolecular rearrangements. The epimerases catalyze the inversion of asymmetric carbon atoms. Mutases catalyze the intramolecular transfer of functional groups.

6. **Ligases. Ligases** catalyze bond formation between two substrate molecules. The energy for these reactions is always supplied by ATP hydrolysis. The names of many ligases include the term synthetase. Several other ligases are called carboxylases.

6.3 ENZYME KINETICS

Recall from Chapter 4 that the principles of thermodynamics can predict whether a reaction is spontaneous. However, thermodynamic quantities do not provide any information regarding reaction rates. To be useful to an organism, biochemical reactions must occur at reasonable rates. The rate or **velocity** of a biochemical reaction is defined as the change in the concentration of a reactant or product per unit time. The initial velocity v of the reaction $A \rightarrow P$ is

$$v = \frac{-\Delta[A]}{\Delta t} = \frac{\Delta[P]}{\Delta t} \tag{4}$$

where

$[A]$ = concentration of substrate
$[P]$ = concentration of product
t = time

Initial velocity, the rate of the reaction immediately after mixing the enzyme and substrate, is measured because it can be assumed that the reverse reaction, if possible, (i.e., conversion of product into substrate) has not yet occurred to any appreciable extent.

The quantitative study of enzyme catalysis, referred to as **enzyme kinetics,** provides information about reaction rates. Kinetic studies also measure the affinity of enzymes for substrates and inhibitors and provide insight into reaction mechanisms. Enzyme kinetics has several practical applications. These include a greater comprehension of the forces that regulate metabolic pathways and the design of improved therapies.

The rate of the above reaction is proportional to the frequency with which the reacting molecules form product. The reaction rate is

$$\text{Rate} = k[A]^x \tag{5}$$

where k is a rate constant, which depends on the reaction conditions (e.g., temperature, pH, and ionic strength).

Combining equations (4) and (5), we have

$$\frac{\Delta[A]}{\Delta t} = -k[A]^x \tag{6}$$

Another term that is useful in describing a reaction is the reaction's *order*. Order is determined empirically, that is, by experimentation (Figure 6.3). Order is defined as the sum of the exponents on the concentration terms in the rate expression. Determining the order of a reaction allows an experimenter to draw certain conclusions regarding the reaction's mechanism. A reaction is said to follow *first-order kinetics* if the rate depends on the first power of the concentration of a single reactant and suggests that the rate limiting step is a unimolecular reaction (i.e. no molecular collisions are required). In the reaction $A \rightarrow P$ the experimental rate equation becomes

$$\text{Rate} = k[A]^1 \tag{7}$$

KEY *Concepts*

Enzyme kinetics is the quantitative study of enzyme catalysis. Kinetic studies measure reaction rates and the affinity of enzymes for substrates and inhibitors. Kinetics also provides insight into reaction mechanisms.

FIGURE 6.3

Enzyme Kinetic Studies.

(a) Plot of initial velocity *v* versus substrate concentration [S]. The rate of the reaction is directly proportional to substrate concentration only when [S] is low. When [S] becomes sufficiently high that the enzyme is saturated, the rate of the reaction is zero-order with respect to substrate. At intermediate substrate concentrations, the reaction has a mixed order (i.e., the effect of substrate on reaction velocity is in transition). (b) Conversion of substrate to product per unit time. The slope of the curve at $t = 0$ equals the initial rate of the reaction.

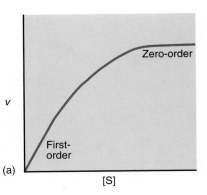

(a)

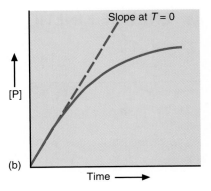

(b)

If [A] is doubled, the rate is observed to double. Reducing [A] by half results in halving the observed reaction rate. In first-order reactions the concentration of the reactant is a function of time, so k is expressed in units of s^{-1}. In any reaction the time required for one-half of the reactant molecules to be consumed is called a *half-life* ($t_{1/2}$).

In the reaction A + B → P, if the order of A and B is 1 each, then the reaction is said to be *second-order* and A and B must collide for product to form (a bimolecular reaction):

$$\text{Rate} = k[\text{A}]^1[\text{B}]^1 \tag{8}$$

In this circumstance the reaction rate depends on the concentrations of the two reactants. In other words, both A and B take part in the reaction's rate-determining step. Second-order rate constants are measured in units of $\text{M}^{-1}\,\text{s}^{-1}$. Sometimes, second-order reactions involve reactants such as water that are present in great excess:

$$\text{A} + \text{H}_2\text{O} \rightarrow \text{P}$$

The second-order rate expression is

$$\text{Rate} = k[\text{A}]^1[\text{H}_2\text{O}]^1 \tag{9}$$

Because water is present in excess, however, the reaction appears to be first-order. Such reactions are said to be *pseudo-first-order*. Another possibility is that only one of two reactants is involved in the rate-determining step, that is, rate = $k[\text{A}]^2$. The other reactant still participates in the mechanism, but not in the rate-determining step.

When the addition of a reactant does not alter a reaction rate, the reaction is said to be *zero-order*. For the reaction A → P the experimentally determined rate expression is

$$\text{Rate} = k[\text{A}]^0 = k \tag{10}$$

The rate is constant because reactant concentration is high enough to saturate all the catalytic sites on the enzyme molecules.

An example of an order determination is given below.

6.1

Consider the following reaction:

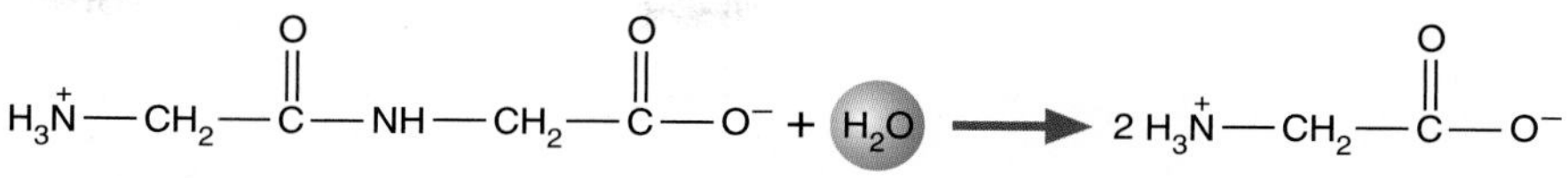

$$H_3\overset{+}{N}-CH_2-\underset{\overset{\|}{O}}{C}-NH-CH_2-\underset{\overset{\|}{O}}{C}-O^- + H_2O \longrightarrow 2\ H_3\overset{+}{N}-CH_2-\underset{\overset{\|}{O}}{C}-O^-$$

Given the following rate data, determine the order in each reactant and the overall order of the reaction. The concentrations are in moles per liter. Enzyme rates are measured in millimoles per second.

[Glycylglycine]	[H₂O]	Rate
0.1	0.1	1×10^2
0.2	0.1	2×10^2
0.1	0.2	2×10^2
0.2	0.2	4×10^2

Solution

The overall rate expression is

$$\text{Rate} = k[\text{Glycylglycine}]^x[\text{H}_2\text{O}]^y$$

To evaluate x and y, determine the effect on the rate of the reaction by increasing the concentration of one reactant while keeping the concentration of the other constant.

For this experiment, doubling the concentration of gly—gly doubles the rate of the reaction; therefore x is equal to 1. Doubling the concentration of water doubles the rate of the reaction. So y is also equal to 1. The rate expression then is

$$\text{Rate} = k[\text{Glycylglycine}]^1[\text{H}_2\text{O}]^1$$

The reaction is first-order in both reactants and second-order overall.

Michaelis-Menten Kinetics

One of the most useful models in the systematic investigation of enzyme rates was proposed by Leonor Michaelis and Maud Menten in 1913. The concept of the enzyme-substrate complex, first enunciated by Victor Henri in 1903, is central to Michaelis-Menten kinetics. When the substrate S binds in the active site of an enzyme E, an intermediate complex (ES) is formed. During the transition state, the substrate is converted into product. After a brief time, the product dissociates from the enzyme. This process can be summarized as follows:

$$E + S \underset{k_2}{\overset{k_1}{\leftrightarrows}} ES \overset{k_3}{\to} E + P \tag{11}$$

where

k_1 = rate constant for ES formation
k_2 = rate constant for ES dissociation
k_3 = rate constant for product formation and release from the active site

(This equation ignores the reversibility of the step in which the ES complex is converted into enzyme and product. This simplifying assumption is allowed if the reaction rate is measured while [P] is still very low. Recall that initial velocities are measured in most kinetic studies.)

According to the Michaelis-Menten model, as currently conceived, it is assumed that (1) k_2 is negligible when compared with k_1 and (2) the rate of formation of ES is equal to the rate of its degradation over most of the course of the reaction. (The latter premise is referred to as the *steady-state assumption*.)

$$\text{Rate} = \frac{\Delta P}{\Delta t} = k_3[ES] \tag{12}$$

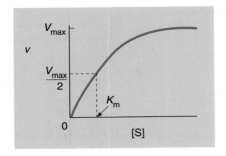

FIGURE 6.4

Reaction Velocity v and Substrate Concentration [S] for a Typical Enzyme-Catalyzed Reaction.

K_m is the substrate concentration at which the enzyme has half-maximal velocity.

KEY *Concepts*

The Michaelis-Menten kinetic model explains several aspects of the behavior of many enzymes. Each enzyme has a K_m value characteristic of that enzyme under specified conditions.

To be useful, a reaction rate must be defined in terms of [S] and [E]. The rate of formation of ES is equal to $k_1[E][S]$, while the rate of ES dissociation is equal to $(k_2 + k_3)[ES]$. The steady-state assumption equates these two rates.

$$k_1[E][S] = (k_2 + k_3)[ES] \tag{13}$$

$$[ES] = \frac{[E][S]}{(k_2 + k_3)/k_1} \tag{14}$$

Michaelis and Menten introduced a new constant, K_m (now referred to as the *Michaelis constant*):

$$K_m = \frac{k_2 + k_3}{k_1} \tag{15}$$

They also derived the equation

$$v = \frac{V_{max}[S]}{[S] + K_m} \tag{16}$$

where V_{max} = maximum velocity that the reaction can attain. This equation, now referred to as the *Michaelis-Menten equation,* has proven to be very useful in defining certain aspects of enzyme behavior. For example, when [S] is equal to K_m, the denominator in equation (16) is equal to 2[S], and v is equal to $V_{max}/2$ (Figure 6.4). The experimentally determined value K_m is considered a constant that is characteristic of the enzyme and the substrate under specified conditions. It may reflect the affinity of the enzyme for its substrate. (If k_3 is much smaller than k_2, that is, $k_3 << k_2$, then the K_m value approximates k_2/k_1. In this circumstance, K_m is the dissociation constant for the ES complex.) The lower the value of K_m, the greater the affinity of the enzyme for ES complex formation.

An enzyme's kinetic properties can also be used to determine its catalytic efficiency. The **turnover number** (k_{cat}) of an enzyme is

$$k_{cat} = \frac{V_{max}}{[E_{total}]} \tag{17}$$

This quantity is the number of moles of substrate converted to product each second per mole of enzyme. Several examples of turnover numbers are given in Table 6.1. Enzyme activity is measured in *international units* (I.U.). One I.U. is defined as the amount of enzyme that produces 1 μmol of product per minute. An enzyme's **specific activity,** a quantity that is used to monitor enzyme purification, is defined as the number of international units per milligram of protein. (A new unit for measuring enzyme activity called the *katal* has recently been introduced. One katal (kat) indicates the transformation of 1 mole of substrate per second.)

TABLE 6.1

Turnover Numbers of Selected Enzymes

Enzyme	k_{cat} (s^{-1})
Catalase	10,000,000
Carbonic anhydrase	1,000,000
Acetylcholinesterase	14,000
Urease	10,000
Lactate dehydrogenase	1,000
Chymotrypsin	100
Phosphoglucomutase	21
Lysozyme	0.5

An example of a kinetic determination is given below.

PROBLEM

6.2

Consider the Michaelis-Menten plot illustrated in Figure 6.5. Identify the following points on the curve

a. V_{max}

b. K_m

FIGURE 6.5

A Michaelis-Menten Plot.

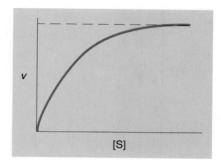

Solution

a. V_{max} is the maximum rate the enzyme can attain. Further increases in substrate concentration further increase the rate.

b. $K_m = [S]$ at $V_{max}/2$

Lineweaver-Burk Plots

K_m and V_{max} values for an enzyme are determined by measuring initial reaction velocities at various substrate concentrations. Approximate values of K_m and V_{max} can be obtained by constructing a graph, as shown in Figure 6.4. A more accurate determination of these values results from an algebraic transformation of the data. The Michaelis-Menten equation, whose graph is a hyperbola,

$$v = \frac{V_{max}[S]}{[S] + K_m}$$

can be rearranged by taking its reciprocal.

$$\frac{1}{v} = \frac{K_m}{V_{max}} \frac{1}{[S]} + \frac{1}{V_{max}} \tag{18}$$

The reciprocals of the initial velocities are plotted as functions of the reciprocals of substrate concentrations. In such a graph, referred to as a *Lineweaver-Burk double-reciprocal plot,* the straight line that is generated has the form $y = mx + b$, where y and x are variables ($1/v$ and $1/[S]$, respectively) and m and b are constants (K_m/V_{max} and $1/V_{max}$, respectively). The slope of the straight line is K_m/V_{max} (Figure 6.6). As

FIGURE 6.6

Lineweaver-Burk Plot.

If an enzyme obeys Michaelis-Menten kinetics, a plot of the reciprocal of the reaction velocity $1/v$ as a function of the reciprocal of the substrate concentration $1/[S]$ will fit a straight line. The slope of the line is K_m/V_{max}. The intercept on the vertical axis is $1/V_{max}$. The intercept on the horizontal axis is $-1/K_m$.

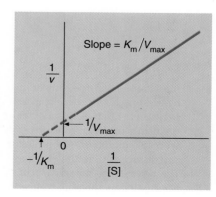

indicated in Figure 6.6, the intercept on the vertical axis is $1/V_{max}$. The intercept on the horizontal axis is $-1/K_m$.

An example of a kinetics problem using the Lineweaver-Burk plot is the following.

6.3

Consider the Lineweaver-Burk plot illustrated in Figure 6.7. Identify:

a. $-1/K_m$ b. $1/V_{max}$ c. K_m/V_{max}

FIGURE 6.7
    A Lineweaver-Burk Plot.

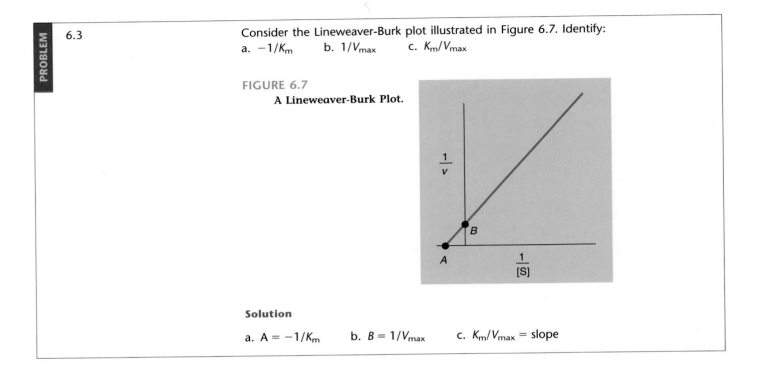

Solution

a. $A = -1/K_m$ b. $B = 1/V_{max}$ c. K_m/V_{max} = slope

Enzyme Inhibition

The activity of enzymes can be inhibited. Study of the mechanisms by which enzymes are inhibited has practical applications. For example, many clinical therapies and biochemical research tools are based on enzyme inhibition.

Many substances can reduce or eliminate the catalytic activity of specific enzymes. Inhibition may be irreversible or reversible. Irreversible inhibitors usually bond covalently to the enzyme, often to a side chain group in the active site. For example, enzymes containing free sulfhydryl groups can react with alkylating agents such as iodoacetate.

$$Enzyme-CH_2-SH \;+\; I-CH_2-\overset{\overset{\displaystyle O}{\|}}{C}-O^- \longrightarrow Enzyme-CH_2-S-CH_2-\overset{\overset{\displaystyle O}{\|}}{C}-O^- \;+\; HI$$

Glyceraldehyde-3-phosphate dehydrogenase, an enzyme in the glycolytic pathway (Chapter 8), is inactivated by alkylation with iodoacetate. Enzymes that use sulfhydryl groups to form covalent bonds with metal cofactors are often irreversibly inhibited by heavy metals (e.g., mercury and lead). The anemia in lead poisoning is caused in part because of lead binding to a sulfhydryl group of ferrochelatase. Ferrochelatase catalyzes the insertion of Fe^{2+} into heme.

In reversible inhibition the inhibitor can dissociate from the enzyme because it binds through noncovalent bonds. The most common forms of reversible inhibition are competitive and noncompetitive (Figure 6.8).

KEY *Concepts*

Inhibition of enzyme activity may be irreversible or reversible. Irreversible inhibitors usually bond covalently to the enzyme. In reversible inhibition, the inhibitor can dissociate from the enzyme. The two most common types of reversible inhibition are competitive and noncompetitive inhibition.

Competitive Inhibition   The structure of a competitive inhibitor closely resembles that of the enzyme's normal substrate. Because of its structure, a competitive inhibitor

FIGURE 6.8

Enzyme Inhibition.

(a) A competitive inhibitor prevents the substrate from binding to the active site. (b) A noncompetitive inhibitor binds to the enzyme (at a site other than the active site), altering the shape of the active site and blocking the conversion of substrate to product.

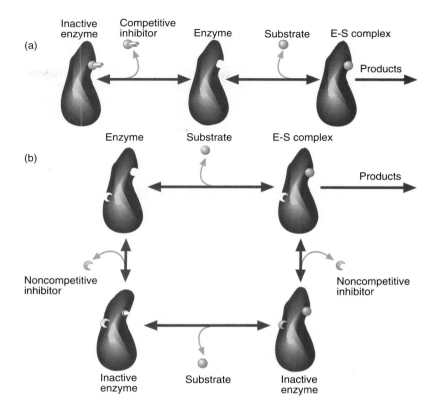

FIGURE 6.9

Michaelis-Menten Plot of Uninhibited Enzyme Activity Versus Competitive Inhibition.

Initial velocity v is plotted against substrate concentration [S]. With competitive inhibition, V_{max} stays constant and K_m increases.

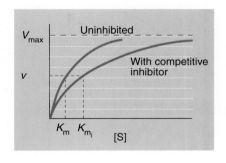

binds reversibly to the enzyme's active site. The inhibitor forms an enzyme-inhibitor complex (EI) that is equivalent to the ES complex.

$$E + I \underset{k_{I2}}{\overset{k_{I1}}{\rightleftharpoons}} EI$$

The concentration of EI complex depends on the concentration of free inhibitor and on the dissociation constant K_I:

$$K_I = \frac{[E][I]}{[EI]} = \frac{k_{I2}}{k_{I1}}$$

Because the EI complex readily dissociates the empty active site is then available for substrate binding. Because no productive reaction occurs during the limited time that the EI complex exists, the enzyme's activity declines (Figure 6.9). The effect of a competitive inhibitor on activity is reversed by increasing the concentration of substrate. At high [S], all the active sites are filled with substrate, and reaction velocity reaches the value observed without an inhibitor.

Succinate dehydrogenase, an enzyme in the Krebs citric acid cycle (Chapter 11), catalyzes the following redox reaction.

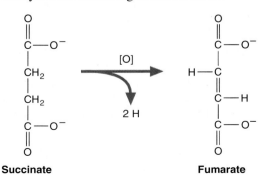

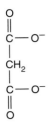

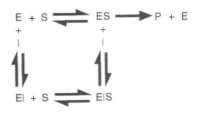

Malonate

FIGURE 6.10

    Malonate.

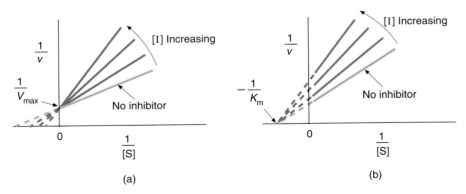

FIGURE 6.11

    Noncompetitive Inhibition.

Noncompetitive inhibitors do not
compete directly with the substrate for
binding to the enzyme.

This reaction is inhibited by malonate (Figure 6.10). Malonate binds to the enzyme's
active site but cannot be converted to product.

Noncompetitive Inhibition   In noncompetitive inhibition, the inhibitor binds to the
enzyme at a site other than the active site. Both EI and EIS complexes form (Figure
6.11). Inhibitor binding alters the enzyme's three-dimensional configuration and
blocks the reaction. For example, AMP is a noncompetitive inhibitor of fructose bisphosphate phosphatase, the enzyme that catalyzes the conversion of fructose-1,6-
bisphosphate to fructose-6-phosphate (Chapter 8). Noncompetitive inhibition is not
reversed by increasing the concentration of substrate (Figure 6.12).

Kinetic Analysis of Enzyme Inhibition   Competitive and noncompetitive inhibition
can be easily distinguished with double-reciprocal plots (Figure 6.13). In two sets of
rate determinations, enzyme concentration is held constant. In the first experiment, the
velocity of the uninhibited enzyme is established. In the second experiment, a constant
amount of inhibitor is included in each enzyme assay. Figure 6.13 illustrates that competitive and noncompetitive inhibitors have different effects on enzymes. Competitive
inhibition changes the K_m of the enzyme but not the V_{max}. (This is shown in the
double-reciprocal plot as a shift in the horizontal intercept.) In noncompetitive inhibition, V_{max} is lowered (i.e., the vertical intercept is shifted), and the K_m is unaffected.

FIGURE 6.12

    **Michaelis-Menten Plot of Uninhibited
Enzyme Activity Versus Noncompetitive
Inhibition.**

Initial velocity *v* is plotted against substrate
concentration [S]. With noncompetitive inhibition
V_{max} decreases and K_m stays constant.

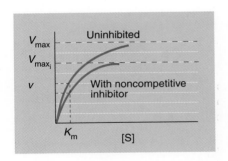

      (a)   (b)

FIGURE 6.13

    Kinetic Analysis of Enzyme Inhibition.

(a) Competitive inhibition. Plots of $1/v$ versus $1/[S]$ in the presence of several concentrations of the
inhibitor intersect at the same point on the **vertical** axis, $1/V_{max}$. (b) Noncompetitive inhibition.
Plots of $1/v$ versus $1/[S]$ in the presence of several concentrations of the inhibitor intersect at the
same point on the **horizontal** axis, $-1/K_m$. In noncompetitive inhibition the dissociation
constants for ES and EIS are assumed to stay the same.

An example of a problem concerned with enzyme inhibition is the following.

PROBLEM 6.4

Consider the Lineweaver-Burk plot illustrated in Figure 6.14.

    Line *A* = Normal enzyme-catalyzed reaction
    Line *B* = Compound B added
    Line *C* = Compound C added

Identify the type of inhibitory action shown by compounds B and C.

FIGURE 6.14

  A Lineweaver-Burk Plot.

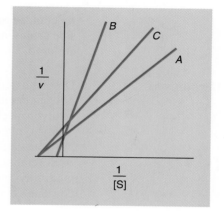

Solution

Compound B is a competitive inhibitor. Compound C is a noncompetitive inhibitor.

 Allosteric Enzymes   Although the Michaelis-Menten model is an invaluable tool, it does not explain the kinetic properties of many enzymes. For example, plots of reaction velocity versus substrate concentration for many enzymes with multiple subunits are sigmoidal rather than hyperbolic, as predicted by the Michaelis-Menten model (Figure 6.15). Such effects are seen in an important group of enzymes called the **allosteric enzymes.** The substrate-binding curve in Figure 6.15 resembles the oxygen-binding curve of hemoglobin. There are several other similarities between allosteric enzymes and hemoglobin. Most allosteric enzymes are multisubunit proteins. The binding of substrate to one protomer in an allosteric enzyme affects the binding properties of adjacent protomers. In addition, the activity of allosteric enzymes is affected by effector molecules that bind to additional sites called *allosteric* or *regulatory sites.* Allosteric enzymes generally catalyze key regulatory steps in biochemical pathways. (Regulation of allosteric enzymes is discussed on pp. 143.)

FIGURE 6.15

  The Kinetic Profile of an Allosteric Enzyme.

  The sigmoidal binding curve displayed by many allosteric enzymes resembles the curve for the cooperative binding of O_2 to hemoglobin.

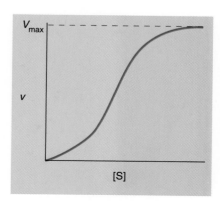

QUESTION 6.1

> Drinking methanol can cause blindness in humans, as well as a severe acidosis that may be life-threatening. Methanol is toxic because it is converted in the liver to formaldehyde and formic acid by the enzymes alcohol dehydrogenase and aldehyde dehydrogenase. Methanol poisoning is treated with dialysis and infusions of bicarbonate and ethanol. Explain why each treatment is used.

6.4 CATALYSIS

However valuable kinetic studies are, they reveal little about how enzymes catalyze biochemical reactions. Biochemists use other techniques to investigate the catalytic mechanisms of enzymes. (A *mechanism* is a set of steps in a chemical reaction.) Enzyme mechanism investigations seek to relate enzyme activity to the structure and function of the active site. X-ray crystallography, chemical inactivation of active site side chains, and studies using simple model compounds as substrates and as inhibitors are used.

Catalytic Mechanisms

Despite extensive research, the mechanisms of only a few enzymes are known in significant detail. However, it has become increasingly clear that enzymes use the same catalytic mechanisms as nonenzymatic catalysts. Enzymes achieve significantly higher catalytic rates because their active sites possess structures that are uniquely suited to promote catalysis.

Several factors contribute to enzyme catalysis. The most important of these are (1) proximity and strain effects, (2) electrostatic effects, (3) acid-base catalysis, and (4) covalent catalysis. Each factor will be described briefly.

Proximity and Strain Effects   For a biochemical reaction to occur, the substrate must come into close proximity to catalytic functional groups (side chain groups involved in a catalytic mechanism) within the active site. In addition, the substrate must be precisely oriented to the catalytic groups. Once the substrate is correctly positioned, a change in the enzyme's conformation may result in a strained enzyme-substrate complex. This strain helps to bring the enzyme-substrate complex into the transition state. In general, the more tightly the active site can bind the substrate while it is in its transition state, the greater the rate of the reaction.

Electrostatic Effects   Recall that the strength of electrostatic interactions is related to the capacity of surrounding solvent molecules to reduce the attractive forces between chemical groups (Chapter 3). Because water is largely excluded from the active site as the substrate binds, the local dielectric constant is often low. The charge distribution in the relatively anhydrous active site may influence the chemical reactivity of the substrate. In addition, weak electrostatic interactions, such as those between permanent and induced dipoles in both the active site and the substrate, are believed to contribute to catalysis. A more efficient binding of substrate lowers the free energy of the transition state, which accelerates the reaction.

Acid-Base Catalysis   Chemical groups can often be made more reactive by adding or removing a proton. Enzyme active sites contain side chain groups that act as proton donors or acceptors. These groups are referred to as general acids or general bases. (*General acids* and *general bases* are substances that can release a proton or accept a proton, respectively. Enzymes almost always use general acids or general bases in preference to protons or hydroxide groups. For the sake of simplicity, however, the symbols H^+ and OH^- are often used in illustrations of reaction mechanisms.) For example, the side chain of histidine (referred to as an imidazole group) often participates in catalytic mechanisms because its pK_a is approximately 6. Therefore the histidine side chain ionizes within the physiological pH range. The protonated form of histidine is a general acid. Once it loses its proton (and becomes a conjugate base), histidine is a general base.

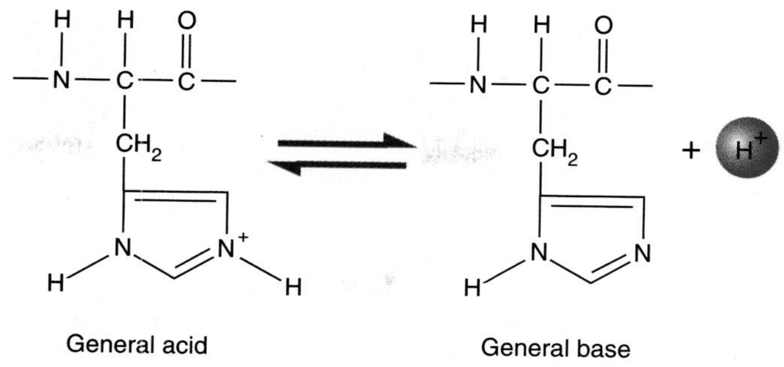

General acid   General base

Histidine

Transfers of protons are a common feature of chemical reactions. For example, consider the hydrolysis of an ester.

Because water is a weak nucleophile, ester hydrolysis is relatively slow in neutral solution. Ester hydrolysis takes place much more rapidly if the pH is raised. As hydroxide ion attacks the polarized carbon atom of the carbonyl group (Figure 6.16a), a tetrahedral intermediate is formed. As the intermediate breaks down, a proton is transferred from a nearby water molecule. The reaction is complete when the alcohol

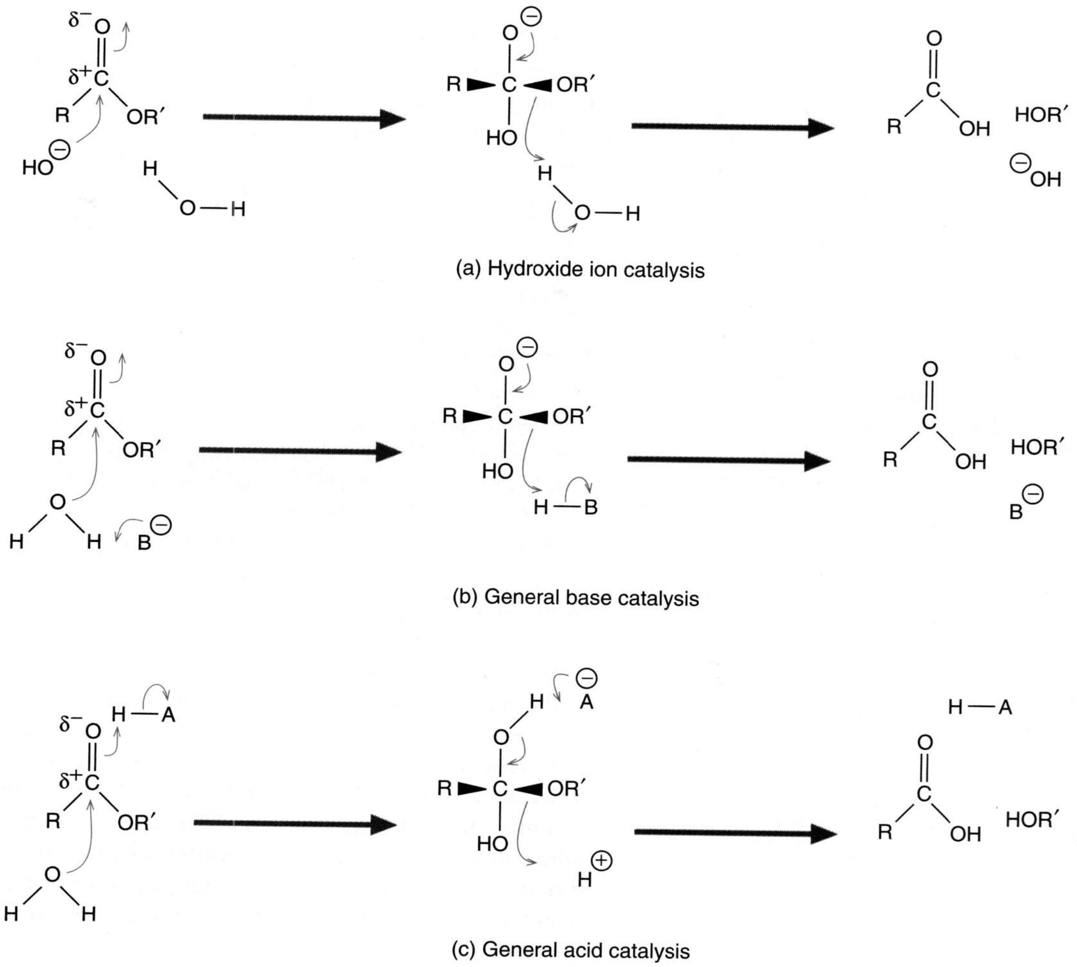

FIGURE 6.16

Ester Hydrolysis.

Esters can be hydrolyzed in several ways: (a) catalysis by free hydroxide ion, (b) general base catalysis, and (c) general acid catalysis. A colored arrow represents the movement of an electron pair during each mechanism.

(a) Hydroxide ion catalysis

(b) General base catalysis

(c) General acid catalysis

is released. However, hydroxide ion catalysis is not practical in living systems. Enzymes use several functional groups that behave as general bases to transfer protons efficiently. Such groups can be precisely positioned in relation to the substrate (Figure 6.16b). Ester hydrolysis can also be catalyzed by a general acid (Figure 6.16c). As the oxygen of the ester's carbonyl group binds to the proton, the carbon atom becomes more positive. The ester then becomes more susceptible to the nucleophilic attack of a water molecule.

Covalent Catalysis   In some enzymes a nucleophilic side chain group forms an unstable covalent bond with the substrate. The enzyme-substrate complex then forms product. A class of enzymes called the **serine proteases** use the —CH_2—OH group of serine as a nucleophile to hydrolyze peptide bonds. (Examples of serine proteases include the digestive enzymes trypsin and chymotrypsin and the blood-clotting enzyme thrombin.) During the first step, the nucleophile attacks the carbonyl group. As the ester bond is formed, the peptide bond is broken. The resulting acyl-enzyme intermediate is hydrolyzed in a second reaction by water.

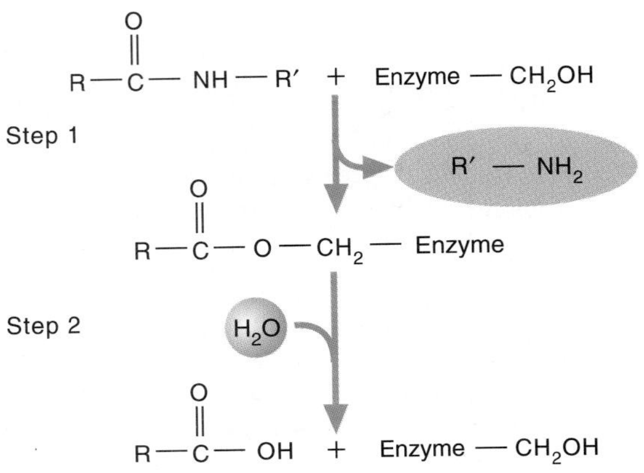

Several other amino acid side chains may act as nucleophiles. The sulfhydryl group of cysteine, the carboxylate groups of aspartate and glutamate, and the imidazole group of histidine can play this role.

Several metal cations and coenzymes also aid catalysis. Their role in facilitating enzyme function is described next.

The Role of Cofactors in Enzyme Catalysis

Active site amino acid side chains are primarily responsible for catalyzing proton transfers and nucleophilic substitutions. To catalyze other reactions, enzymes require nonprotein cofactors, that is, metal cations and the coenzymes. The structural properties and chemical reactivities of each group are briefly discussed.

Metals   The important metals in living organisms fall into two classes: the transition metals (e.g., Fe^{2+} and Cu^{2+}) and the alkali and alkaline earth metals (e.g., Na^+, K^+, Mg^{2+} and Ca^{2+}). Because of their electronic structures, the transition metals are most often involved in catalysis. Although they have important functions in living organisms, the alkali and alkaline earth metals are only rarely found in tight complexes with proteins. This discussion is therefore concerned with the properties of the transition metals.

Several properties of transition metals make them useful in catalysis. Metal ions provide a high concentration of positive charge that is especially useful in binding small molecules. Because transition metals act as *Lewis acids* (electron pair acceptors), they are effective electrophiles. (Amino acid side chains are poor electrophiles because they cannot accept unshared pairs of electrons.) Because their directed valences allow them to interact with two or more ligands, metal ions help orient the substrate within the active site. As a consequence, the substrate–metal ion complex polarizes the substrate and promotes catalysis. For example, when the Zn^{2+} cofactor of carbonic anhydrase polarizes a water molecule, a Zn^{2+}-bound OH group forms. (Strictly speaking, zinc is not a transition metal. However, its properties are similar to those of the transition metals.) The OH group (acting as a nucleophile) attacks CO_2, converting it into HCO_3^-.

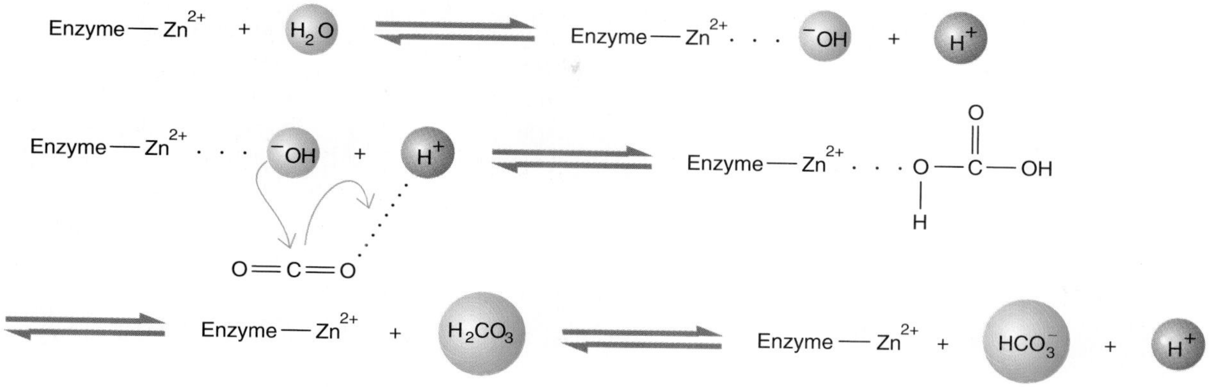

Finally, because transition metals have two or more valence states, they can mediate oxidation-reduction reactions. For example, the reversible oxidation of Fe^{2+} to form Fe^{3+} is important in the function of cytochrome P_{450}. Cytochrome P_{450} is a microsomal enzyme found in animals that processes toxic substances (Chapter 19).

QUESTION 6.2

Copper is a cofactor in several enzymes, including lysyl oxidase and superoxide dismutase. Ceruloplasmin, a deep-blue glycoprotein, is the principal copper-containing protein in blood. It is used to transport Cu^{2+} and maintain appropriate levels of Cu^{2+} in the body's tissues. Ceruloplasmin also catalyzes the oxidation of Fe^{2+} to Fe^{3+}, an important reaction in iron metabolism. Because the metal is widely found in foods, copper deficiency is rare in humans. Deficiency symptoms include anemia, leukopenia (reduction in blood levels of white blood cells), bone defects, and weakened arterial walls. The body is partially protected from exposure to excessive copper (and several other metals) by metalothionein, a small metal-binding protein that possesses a large proportion of cysteine residues. Certain metals (most notably zinc and cadmium) induce the synthesis of metalothionein in the intestine and liver.

In *Menkes' syndrome* intestinal absorption of copper is defective. How could affected infants be treated to avoid the symptoms of the disorder, which include seizures, retarded growth, and brittle hair?

In another rare inherited disorder, called *Wilson's disease,* excessive amounts of copper accumulate in liver and brain tissue. A prominent symptom of the disease is the deposition of copper in greenish-brown layers surrounding the cornea, called Kayser-Fleischer rings. Wilson's disease is now known to be caused by a defective ATP-dependent protein that transports copper across cell membranes. Apparently, the copper transport protein is required to incorporate copper into ceruloplasmin and to excrete excess copper. In addition to a low copper diet, Wilson's disease is treated with zinc sulfate and the chelating agent penicillamine (p. 92). Describe how these treatments work. (*Hint:* Metalothionein has a greater affinity for copper than for zinc.)

Coenzymes Most coenzymes are derived from vitamins. **Vitamins** (organic nutrients required in small amounts in the human diet) are divided into two classes: water-soluble and lipid-soluble. In addition there are certain vitaminlike nutrients (e.g., lipoic acid, carnitine, and p-aminobenzoic acid) that can be synthesized in small amounts and that facilitate enzyme-catalyzed processes. Table 6.2 lists the vitamins, their coenzyme forms, and the reactions they promote. The structure and function of the coenzyme forms of nicotinic acid (niacin) and riboflavin are described in this chapter. The other coenzymes and vitaminlike nutrients are discussed in later chapters.

TABLE 6.2

Vitamins and Their Coenzyme Forms

Vitamin	Coenzyme Form	Reaction or Process Promoted
WATER SOLUBLE VITAMINS		
Thiamine (B$_1$)	Thiamine pyrophosphate	Decarboxylation, aldehyde group transfer
Riboflavin (B$_2$)	FAD and FMN	Redox
Pyridoxine (B$_6$)	Pyridoxal phosphate	Amino group transfer
Nicotinic acid (niacin)	NAD and NADP	Redox
Pantothenic acid	Coenzyme A	Acyl transfer
Biotin	Biocytin	Carboxylation
Folic acid	Tetrahydrofolic acid	One-carbon group transfer
Vitamin B$_{12}$	Deoxyadenosylcobalamin, methylcobalamin	Intramolecular rearrangements
Ascorbic acid (vitamin C)	Unknown	Hydroxylation
LIPID-SOLUBLE VITAMINS		
Vitamin A	Retinal	Vision, growth, and reproduction
Vitamin D	1,25-Dihydroxycholecalciferol	Calcium and phosphate metabolism
Vitamin E	Unknown	Lipid antioxidant
Vitamin K	Unknown	Blood clotting

KEY *Concepts*

The amino acid side chains in the active site of enzymes catalyze proton transfers and nucleophilic substitutions. Other reactions require a group of nonprotein cofactors, that is, metal cations and the coenzymes. Metal ions are effective electrophiles, and they help orient the substrate within the active site. In addition, certain metal cations mediate redox reactions. Coenzymes are organic molecules that have a variety of functions in enzyme catalysis.

There are two coenzyme forms of nicotinic acid: nicotinamide adenine dinucleotide (NAD) and nicotinamide adenine dinucleotide phosphate (NADP). These coenzymes occur in oxidized forms (NAD$^+$ and NADP$^+$) and reduced forms (NADH and NADPH). The structures of NAD$^+$ and NADP$^+$ both contain adenosine and the N-ribosyl derivative of nicotinamide, which are linked together through a pyrophosphate group (Figure 6.17a). NADP has an additional phosphate attached to the 2'—OH group of adenosine. (The ring atoms of the sugar in a nucleotide are designated with a prime to distinguish them from atoms in the base.) Both NAD$^+$ and NADP$^+$ carry electrons for several enzymes in a group known as the dehydrogenases. (Dehydrogenases catalyze oxidation-reduction and hydrogen removal reactions. Many dehydrogenases that catalyze reactions involved in energy generation use the coenzyme NADH. Those enzymes that require NADPH usually catalyze biosynthetic reactions. A small number of dehydrogenases can use either NADH or NADPH.)

Alcohol dehydrogenase catalyzes the reversible oxidation of ethanol to form acetaldehyde:

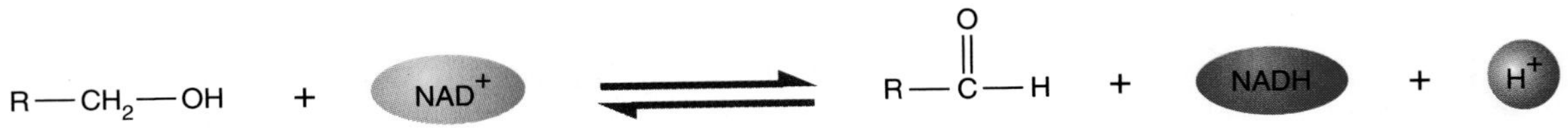

$$\text{R}—\text{CH}_2—\text{OH} + \boxed{\text{NAD}^+} \rightleftharpoons \text{R}—\overset{\displaystyle \overset{\text{O}}{\|}}{\text{C}}—\text{H} + \boxed{\text{NADH}} + \boxed{\text{H}^+}$$

During this reaction, NAD$^+$ accepts a hydride ion (a proton with two electrons) from ethanol, the substrate molecule undergoing oxidation. Note that the equivalent of two hydrogen atoms are removed from the substrate molecules, so an H$^+$ is produced in addition to the hydride ion. The reversible reduction of NAD$^+$ is illustrated in Figure 6.17b.

In most reactions catalyzed by dehydrogenases, the NAD$^+$ (or NADP$^+$) is bound only transiently to the enzyme. After the reduced version of the coenzyme is released from the enzyme, it donates the hydride ion to another molecule, called an *electron acceptor*. The high-energy bond between the hydrogen and the nicotinamide ring provides the energy for the transfer of the hydride ion.

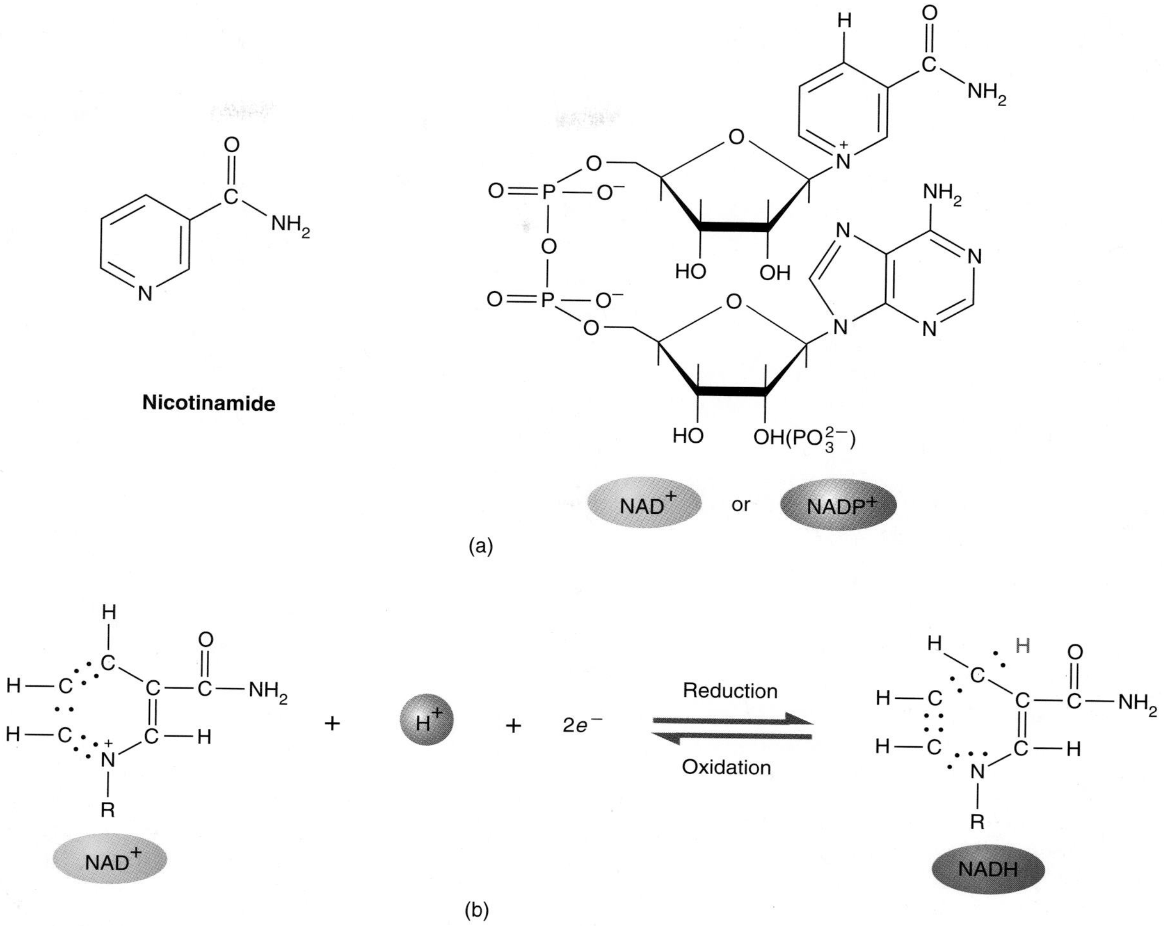

(a)

(b)

FIGURE 6.17

Nicotinamide Adenine Dinucleotide (NAD).

(a) Nicotinamide and NAD(P)$^+$. (b) Reversible reduction of NAD$^+$ to NADH. To simplify the equation, only the nicotinamide ring is shown. The rest of the molecule is designated R.

Riboflavin (Vitamin B$_2$) is a component of two coenzymes: flavin mononucleotide (FMN) and flavin adenine dinucleotide (FAD) (Figure 6.18). FMN and FAD function as tightly bound prosthetic groups in a class of enzymes known as the **flavoproteins.** Flavoproteins are a diverse group of catalysts; they function as dehydrogenases, oxidases, and hydroxylases. These enzymes, which catalyze oxidation-reduction reactions, use the isoalloxazine group of FAD or FMN as a donor or acceptor of two hydrogen atoms. Succinate dehydrogenase is a prominent example of the flavoproteins. It catalyzes the oxidation of succinate to form fumarate, an important reaction in energy production.

Effects of Temperature and pH on Enzyme-Catalyzed Reactions

Any environmental factor that disturbs protein structure may change enzymatic activity. Enzymes are especially sensitive to changes in temperature and pH.

Temperature All chemical reactions are affected by temperature. The higher the temperature, the higher the reaction rate. The reaction velocity increases because more molecules have sufficient energy to enter into the transition state. The rates of enzyme-catalyzed reactions also increase with increasing temperature. However, enzymes are proteins that become denatured at high temperatures. Each enzyme has an

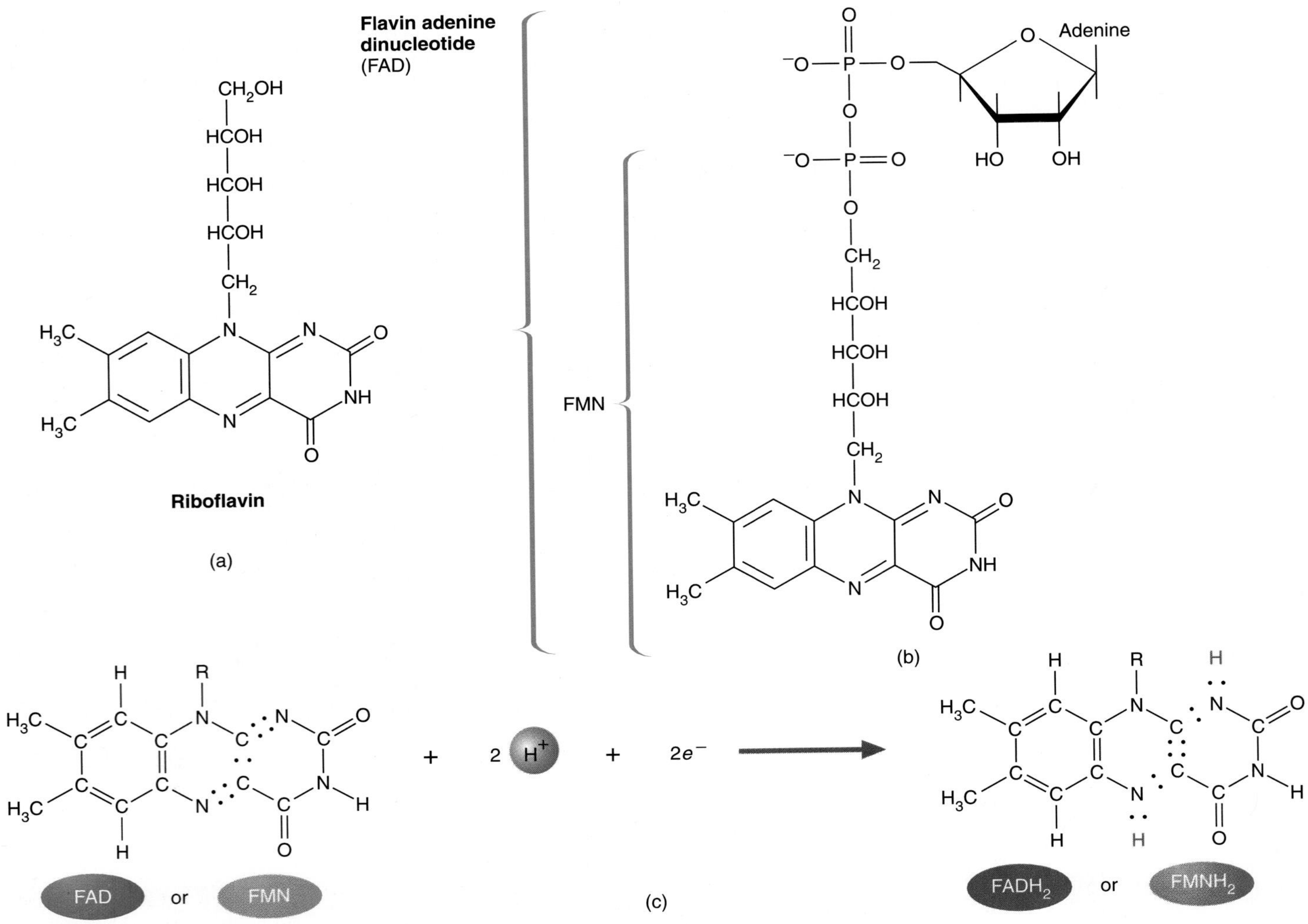

FIGURE 6.18

Flavin Coenzymes.

(a) The vitamin riboflavin. (b) FAD and FMN. (c) Reversible reduction of flavin coenzymes. To simplify the equation, only the isoalloxazine ring system is shown. The rest of the coenzyme is designated R.

optimum temperature at which it operates at maximal efficiency (Figure 6.19). (Because enzymes are proteins, optimum temperature values depend on pH and ionic strength.) If the temperature is raised beyond the optimal temperature, the activity of many enzymes declines abruptly. An enzyme's optimum temperature is usually close to the normal temperature of the organism it comes from. For example, the temperature optima of most human enzymes are close to 37°C.

pH Hydrogen ion concentration affects enzymes in several ways. First, catalytic activity is related to the ionic state of the active site. Changes in hydrogen ion concentration can affect the ionization of active site groups. For example, the catalytic activity of a certain enzyme requires the protonated form of a side chain amino group. If the pH becomes sufficiently alkaline that the group loses its proton, the enzyme's activity may be depressed. In addition, substrates may also be affected. If a substrate contains an ionizable group, a change in pH may alter its capacity to bind to the active site. Second, changes in ionizable groups may change the tertiary structure of the enzyme. Drastic changes in pH often lead to denaturation.

FIGURE 6.19

Effect of Temperature on Enzyme Activity.

Modest increases in temperature increase the rate of enzyme-catalyzed reactions because of an increase in the number of collisions between enzyme and substrate. Eventually, increasing the temperature decreases the reaction velocity. Catalytic activity is lost because heat denatures the enzyme.

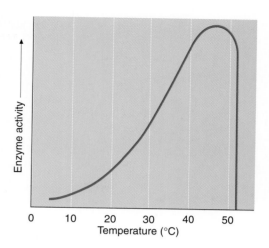

FIGURE 6.20

Effect of pH on Two Enzymes.

Each enzyme has a certain pH at which it is most active. A change in pH can alter the ionizable groups within the active site or affect the enzyme's conformation.

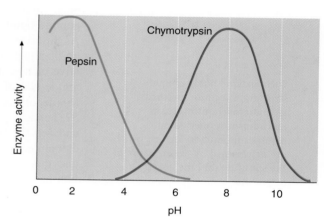

Although a few enzymes tolerate large changes in pH, most enzymes are active only within a narrow pH range. For this reason, living organisms employ buffers to closely regulate pH. The pH value at which an enzyme's activity is maximal is called the **pH optimum** (Figure 6.20). The pH optima of enzymes vary considerably. For example, the optimum pH of pepsin, a proteolytic enzyme produced in the stomach, is approximately 2. For chymotrypsin, which digests protein in the small intestine, the optimum pH is approximately 8.

Detailed Mechanisms of Enzyme Catalysis

More than 2000 enzymes have been studied, each of which has a unique structure, substrate specificity, and reaction mechanism. Each reaction mechanism is affected by the catalysis-promoting factors described in the preceding section. The mechanisms of a variety of enzymes have been investigated intensively over the past several decades. The catalytic mechanisms of two well-characterized enzymes follows.

Chymotrypsin   Chymotrypsin is a 27,000 D protein that belongs to the serine proteases. The active sites of all serine proteases contain a characteristic set of amino acid residues. In the chymotrypsin numbering system these are His 57, Asp 102, and Ser 195. Studies of crystallized enzyme bound to substrate analogs reveal that these residues are close to each other in the active site. The active site serine residue plays an especially important role in the catalytic mechanisms of this group of enzymes. Serine proteases are irreversibly inhibited by diisopropylfluorophosphate (DFP). In DFP-inhibited enzymes, the inhibitor is covalently bound only to Ser 195 and not to any of the other 29 serines. The special reactivity of Ser 195 is attributed to the proximity of His 57 and Asp 102. The imidazole ring of His 57 lies between the carboxyl group of Asp 102 and the —CH$_2$OH group of Ser 195. The carboxyl group of Asp 102 polarizes His 57, thus allowing it to act as a general base (i.e., the abstraction of a proton by the imidazole group is promoted).

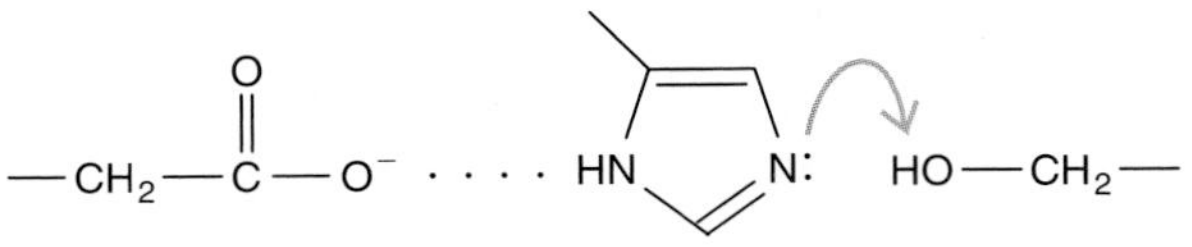

$$-CH_2-\overset{\overset{\displaystyle O}{\|}}{C}-O^- \cdots \cdots HN \diagup N: \quad HO-CH_2-$$

Removing the proton from the serine OH group converts it into a more effective nucleophile.

Chymotrypsin catalyzes the hydrolysis of peptide bonds adjacent to aromatic amino acids. The probable mechanism for this reaction is illustrated in Figure 6.21. Step (a) of the figure shows the initial enzyme-substrate complex. Asp 102, His 57, and Ser 195 are aligned as described above. In addition, the aromatic ring of the substrate's phenylalanine residue is seated in a hydrophobic binding pocket, and the substrate's peptide bond is hydrogen-bonded to the amide NH groups of Ser 195 and Gly 193. As described above, the nucleophilic hydroxyl oxygen of Ser 195 launches a nucleophilic attack on the carbonyl carbon of the substrate. The **oxyanion** that forms as negative charge moves to the carbonyl oxygen is stabilized by hydrogen bonds to the amide NH of Ser 195 and Gly 193. The tetrahedral intermediate, illustrated in Step (b), decomposes to form the covalently bound acyl-enzyme intermediate (Step c). His 57, acting as a general acid, is believed to facilitate this decomposition. In Steps (d) and (e), the two previous steps are reversed. With water acting as an attacking nucleophile, a tetrahedral (oxyanion) intermediate is formed. By Step (f) (the final enzyme-substrate complex) the bond between the serine oxygen and the carbonyl carbon has been broken. Serine is again hydrogen-bonded to His 57.

Alcohol Dehydrogenase   Recall that alcohol dehydrogenase, an enzyme found in many eukaryotic cells (e.g., animal liver, plant leaves, and yeast), catalyzes the reversible oxidation of an alcohol to form an aldehyde:

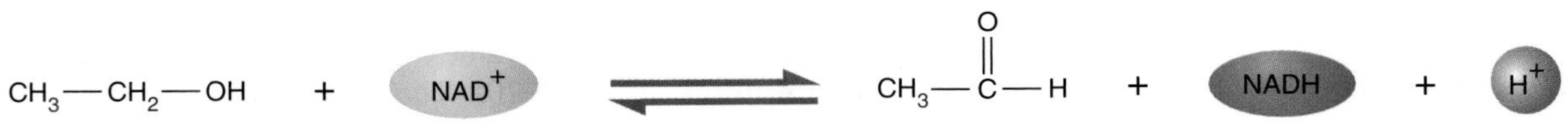

$$CH_3-CH_2-OH \; + \; NAD^+ \; \rightleftharpoons \; CH_3-\overset{\overset{\displaystyle O}{\|}}{C}-H \; + \; NADH \; + \; H^+$$

In this reaction, which involves the removal of two electrons and two protons, the coenzyme NAD acts as the electron acceptor.

The active site of alcohol dehydrogenase contains two cysteine residues (Cys 48 and Cys 174) and a histidine residue (His 67), all of which are coordinated to a zinc ion (Figure 6.22a). After NAD$^+$ binds to the active site, the substrate ethanol enters and binds to the Zn^{2+} as the alcoholate anion (Figure 6.22b). The electrostatic effect of Zn^{2+} stabilizes the transition state. As the intermediate decomposes, the hydride ion is transferred from the substrate to the nicotinamide ring of NAD$^+$ (Figure 6.22c). After the aldehyde product is released from the active site, NADH also dissociates.

6.5 ENZYME REGULATION

The thousands of enzyme-catalyzed chemical reactions in living cells are organized into a series of *biochemical* (or *metabolic*) *pathways*. Each pathway consists of a sequence of catalytic steps. The product of the first reaction becomes the substrate of the next and so on. The number of reactions varies from one pathway to another. For example, animals form glutamine from α-ketoglutarate in a pathway that has two sequential steps, while the synthesis of tryptophan by *Escherichia coli* requires 13 steps. Frequently, biochemical pathways have branch points. For example, chlorismate, a metabolic intermediate in tryptophan biosynthesis, is also a precursor of phenylalanine and tyrosine.

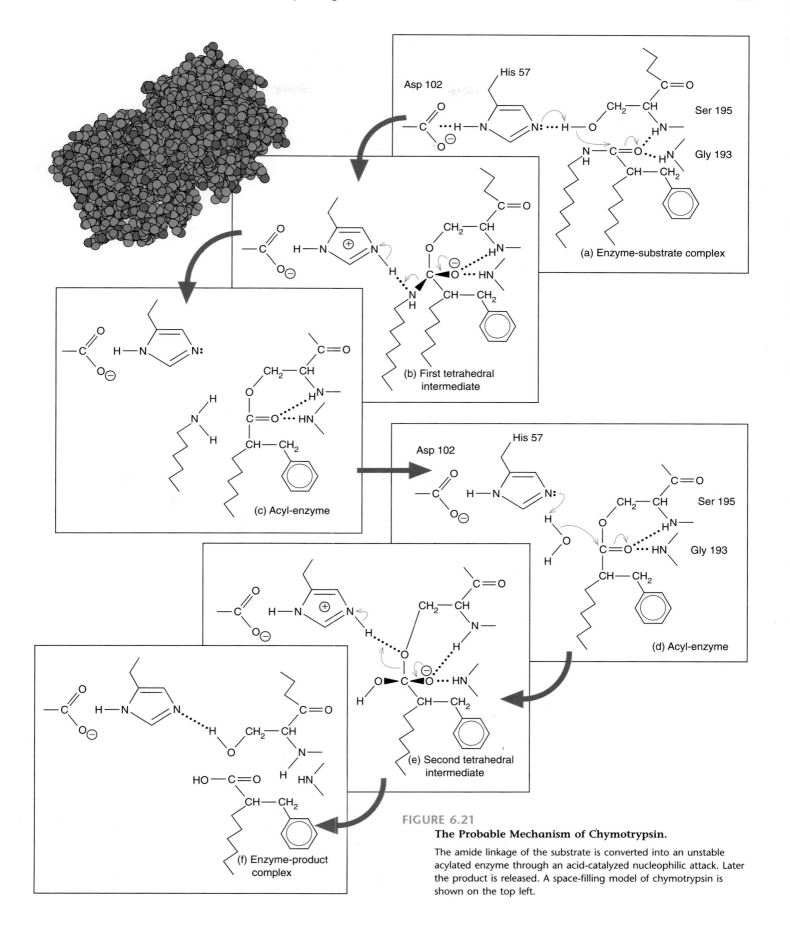

(a) Enzyme-substrate complex

(b) First tetrahedral intermediate

(c) Acyl-enzyme

(d) Acyl-enzyme

(e) Second tetrahedral intermediate

(f) Enzyme-product complex

Asp 102   His 57   Ser 195   Gly 193

FIGURE 6.21

The Probable Mechanism of Chymotrypsin.

The amide linkage of the substrate is converted into an unstable acylated enzyme through an acid-catalyzed nucleophilic attack. Later the product is released. A space-filling model of chymotrypsin is shown on the top left.

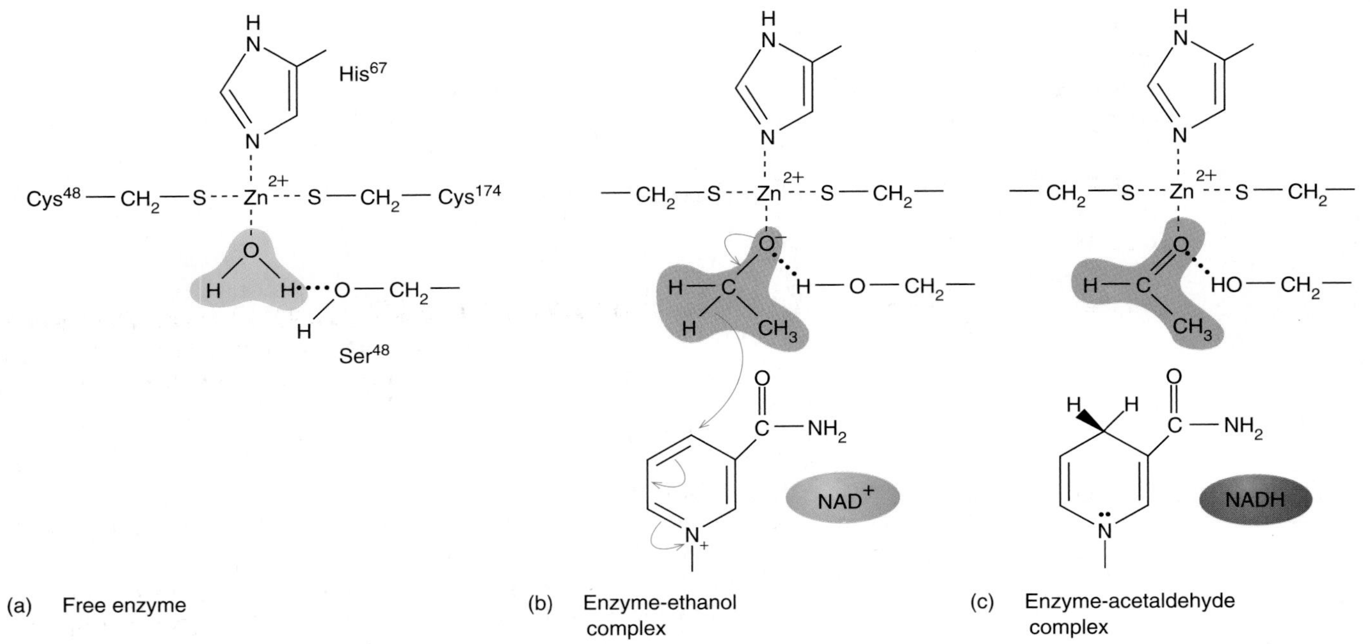

(a)   Free enzyme

(b)   Enzyme-ethanol
        complex

(c)   Enzyme-acetaldehyde
        complex

FIGURE 6.22

Functional Groups of the Active Site of Alcohol Dehydrogenase.

(a) Without a substrate, a molecule of water is one of the ligands of the Zn^{2+} ion. (b) The substrate ethanol probably binds to the Zn^{2+} as the alcoholate anion, displacing the water molecules. (c) NAD^+ accepts a hydride ion from the substrate and the aldehyde product is formed.

Living organisms have evolved sophisticated mechanisms for regulating biochemical pathways. Regulation is essential for several reasons:

1. **Maintenance of an ordered state.** Regulation of each pathway results in the production of the substances required to maintain cell structure and function in a timely fashion and without wasting resources.
2. **Conservation of energy.** Cells constantly control energy-generating reactions so that they consume just enough nutrients to meet their energy requirements.
3. **Responsiveness to environmental changes.** Cells can make relatively rapid adjustments to changes in temperature, pH, ionic strength, and nutrient concentrations because they can increase or decrease the rates of specific reactions.

The regulation of biochemical pathways is complex. It is achieved primarily by adjusting the concentrations and activities of certain enzymes. Control is accomplished by (1) genetic control, (2) covalent modification, (3) allosteric regulation, and (4) compartmentation.

Genetic Control

The synthesis of enzymes in response to changing metabolic needs, a process referred to as **enzyme induction,** allows cells to respond efficiently to changes in their environment. For example, *E. coli* cells grown without the sugar lactose initially cannot metabolize this nutrient when it is introduced into the bacterium's growth medium. After its introduction in the absence of glucose, the enzymes needed to utilize lactose as an energy source are produced. After all of the lactose is consumed, synthesis of these enzymes is terminated.

The synthesis of certain enzymes may also be specifically inhibited. In a process called *repression,* the end product of a biochemical pathway may inhibit the synthesis of a key enzyme in the pathway. For example, in *E. coli* the products of some amino acid synthetic pathways regulate the synthesis of key enzymes. The mechanism of this form of metabolic control is discussed in Chapter 17.

Covalent Modification

Some enzymes are regulated by the reversible interconversion between their active and inactive forms. Several covalent modifications of enzyme structure cause these changes in function. Many such enzymes have specific residues that may be phosphorylated and dephosphorylated. For example, glycogen phosphorylase (Chapter 8) catalyzes the first reaction in the degradation of glycogen, a carbohydrate energy storage molecule. In a process controlled by hormones, the inactive form of the enzyme (glycogen phosphorylase b) is converted to the active form (glycogen phosphorylase a) by adding a phosphate group to a specific serine residue. Other types of reversible covalent modification include methylation, acetylation, and adenylation (the covalent addition of the nucleotide adenosine monophosphate).

Several enzymes are produced and stored as inactive precursors called **proenzymes** or **zymogens.** Zymogens are converted into active enzymes by the irreversible cleavage of one or more peptide bonds. For example, chymotrypsinogen is produced in the pancreas. After chymotrypsinogen is secreted into the small intestine, it is converted to its active form in several steps (Figure 6.23). Initially, trypsin (another proteolytic enzyme) cleaves the peptide bond between Arg 15 and Ile 16. Later, chymotrypsin cleaves other peptide bonds, creating the catalytically active enzyme that assists in the digestion of dietary protein. Other enzymes activated by partial proteolysis include pepsin, trypsin, elastase, collagenase, and the blood-clotting enzyme thrombin.

Allosteric Regulation

In each biochemical pathway at least one enzyme acts as a pacemaker, that is, it sets the rate for the entire pathway. **Pacemaker enzymes** (or **regulatory enzymes**) usually catalyze the first unique (or "committed") step in a pathway. Another typical control point is the first step of a branch in a pathway that leads to an alternate product. There are two major strategies for regulating pacemaker enzymes: covalent modification (discussed above) and allosteric regulation.

Cells use allosteric regulation to respond effectively to changes in intracellular conditions. Recall that allosteric enzymes are usually composed of several protomers whose properties are affected by effector molecules. The binding of an effector to an allosteric enzyme can affect the binding of substrate to that enzyme. If the ligands are identical (e.g., the binding of a substrate influences the binding of additional substrate), then the allosteric effects are referred to as *homotropic.* The term *heterotropic effects* indicates that the binding of a ligand affects the binding of a different ligand (e.g., an inhibitor decreases an enzyme's affinity for its substrate). Homotropic allostery is also referred to as *cooperativity.*

FIGURE 6.23

The Activation of Chymotrypsinogen.

The inactive zymogen chymotrypsinogen is activated in several steps. After its secretion into the small intestine, chymotrypsinogen is converted into π-chymotrypsin when trypsin, another proteolytic enzyme, cleaves the peptide bond between Arg-15 and Ile-16. Later chymotrypsin cleaves several other peptide bonds and conformational changes cause the formation of α-chymotrypsin.

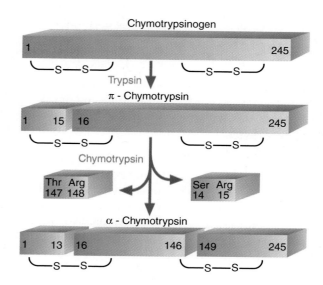

FIGURE 6.24

The Rate of an Enzyme-Catalyzed Reaction as a Function of Substrate Concentration.

The activity of allosteric enzymes is affected by positive effectors (activators) and negative effectors (inhibitors).

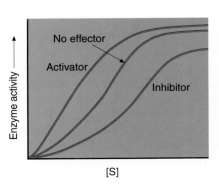

Allosteric effects may be positive or negative. Recall that the binding curves for allosteric enzymes are sigmoidal. The binding of an effector shifts the curve to a higher or lower activity, depending on whether it is an activator or inhibitor (Figure 6.24). For example, aspartate transcarbamoylase (ATCase) in *E. coli* is an allosteric enzyme that catalyzes the first step in a reaction pathway that leads to the synthesis of the pyrimidine nucleotide cytidine triphosphate (CTP). CTP acts as an inhibitor of ATCase activity. This is an example of **negative feedback inhibition,** a process in which the product of a pathway inhibits the activity of the pacemaker enzyme (Figure 6.25). The purine nucleotide ATP acts as an activator. ATP activation of ATCase makes sense because nucleic acid biosynthesis requires relatively equal amounts of purine and pyrimidine nucleotides. When ATP concentration is higher than that of CTP, ATCase is activated. When ATP concentration is lower than that of CTP, the net effect on ATCase is inhibitory. The inhibitor CTP shifts the curve to the right, indicating an increase in the apparent K_m of ATCase. The activator ATP shifts the curve to the left, indicating a lower K_m.

Two theoretical models that attempt to explain the behavior of allosteric enzymes are the concerted model and the sequential model. In the *concerted* (or *symmetry*) model, it is assumed that the enzyme exists in only two states: T(aut) and R(elaxed). Substrates and activators bind more easily to the R conformation, while inhibitors favor the T conformation. The term "concerted" is applied to this model because the conformations of all the protein's protomers are believed to change simultaneously when the first effector binds. (This rapid concerted change in conformation maintains the protein's overall symmetry.) The binding of an activator shifts the equilibrium in favor of the R form. An inhibitor shifts the equilibrium toward the T conformation.

The behavior of the enzyme phosphofructokinase appears to be consistent with the concerted model. Phosphofructokinase catalyzes the transfer of a phosphate group from ATP to the OH group on C-1 of fructose-6-phosphate.

$$\text{Fructose-6-phosphate} + \text{ATP} \rightarrow \text{fructose-1,6-bisphosphate} + \text{ADP}$$

This reaction is the most important regulatory control point in glycolysis, an important energy-generating biochemical pathway. Phosphofructokinase contains four identical subunits, each of which has an active site and an allosteric site. The enzyme is stimulated by ADP, AMP, and other metabolites. It is inhibited by phosphoenolpyruvate (an intermediate in glycolysis), citrate (an intermediate in the citric acid cycle, a related biochemical pathway) and ATP (ATP is both a substrate and an inhibitor, that is, it can bind to both the active site and the allosteric site.) The binding of the allosteric effectors alters the rate of glycolysis in response to changes in the cell's energy needs and the availability of other fuels. Kinetic data suggest that phosphofructokinase has two conformations: T and R. When fructose-6-phosphate binds, all four subunits convert from the T conformation to the R conformation.

Although the concerted model explains several aspects of allosteric enzymes, it has certain limitations. First of all, the concerted model is too simple to account for the complex behavior of many enzymes. For example, it cannot account for **negative cooperativity,** a phenomenon observed in a few enzymes in which the binding of the first ligand reduces the affinity of the enzyme for similar ligands. The concerted model accounts only for **positive cooperativity,** in which the first ligand increases subsequent ligand binding. In addition, the concerted model makes no allowances for hybrid conformations.

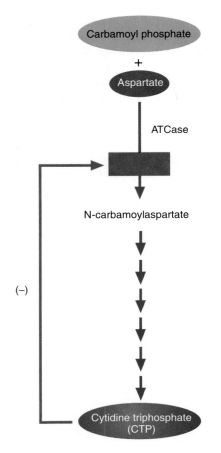

FIGURE 6.25

Feedback Inhibition.

ATCase (aspartate transcarbamoylase) catalyzes the committed step in the synthesis of CTP. CTP, the product of the pathway, inhibits ATCase.

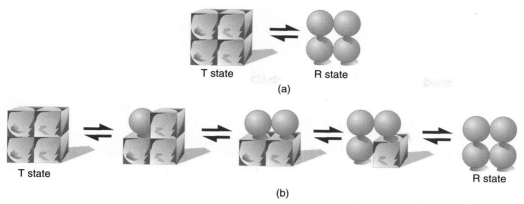

FIGURE 6.26

Allosteric Interaction Models.

(a) In the concerted model, the enzyme exists in only two conformations. Substrates and activators have a greater affinity for the R state. Inhibitors favor the T state. (b) In the sequential model, one protomer assumes an R conformation as it binds to substrate. As the first protomer changes its conformation, the affinity of nearby protomers for ligand is affected.

In the sequential model, it is assumed that proteins are flexible. The binding of a ligand to one protomer in a multisubunit protein prompts a conformational change that is sequentially transmitted to adjacent protomers. The more sophisticated sequential model (Figure 6.26) allows for the intermediate conformations believed to be a closer approximation than the conformations in the simpler concerted model. Negative cooperativity has also been observed. A ligand binding to one protomer can induce conformation changes in adjacent protomers that might make ligand binding less likely.

It should be emphasized that the concerted and sequential models are theoretical models, i.e. the behavior of many allosteric proteins appears to be more complex than can be accounted for by either model. For example, the cooperative binding of O_2 by hemoglobin (the most thoroughly researched allosteric protein) appears to exhibit features of both models. The binding of the first O_2 initiates a concerted $T \rightarrow R$ transition that involves small changes in the conformation of each subunit (a feature of the sequential model). In addition, hemoglobin species with only one or two bound O_2 have been observed.

Compartmentation

In eukaryotic cells, biochemical pathways are segregated into different organelles. One purpose for this physical separation is that opposing processes are easier to control if they occur in different compartments. For example, fatty acid biosynthesis occurs in the cytoplasm, while the energy-generating reactions of fatty acid oxidation occur within the mitochondria. By using hormones and other mechanisms, metabolic control over fatty acid metabolism is exerted by regulating the transport of specific molecules across mitochondrial membrane. Close coordination between the two processes prevents excess waste of energy due to significant overlap of synthesis and oxidation of fatty acids. Net synthesis occurs when the cell's energy is high and net oxidation occurs when the cell's energy is low.

Another factor related to cellular compartmentation is that each organelle can concentrate specific substances such as substrates and coenzymes. In addition, special microenvironments are often created within organelles. For example, lysosomes contain hydrolytic enzymes that require a relatively high concentration of hydrogen ions for optimum activity. (Optimal lysosomal enzyme activity occurs at pH 5. The cell's cytoplasmic pH is approximately 7.2.) Lysosomes can concentrate H^+ because the lysosomal membrane, which is itself impervious to H^+, contains an energy-driven H^+ pump.

KEY *Concepts*

All biochemical pathways are regulated to maintain the ordered state of living cells. Regulation is accomplished by genetic control, covalent modification of enzymes, allosteric regulation, and cell compartmentation.

ENZYME TECHNOLOGY: MEDICAL APPLICATIONS

As knowledge concerning enzymes has grown, they have increasingly been used to solve problems. The earliest uses of enzymes were in food processing. For example, rennin, a proteolytic enzyme obtained from calf stomachs, has been used for thousands of years to produce cheese. More recently, certain enzymes have become invaluable tools in medicine. A new era in medicine began in 1954 when researchers discovered that aspartate aminotransferase (ASAT, also known as serum glutamate-oxaloacetate transaminase, or SGOT) is elevated in the blood serum of patients with myocardial infarction (heart attack). Shortly thereafter, it was discovered that the serum levels of both ASAT and alanine aminotransferase (ALAT, also known as serum glutamate-pyruvate transaminase, or SGPT) become elevated after the liver is damaged. In the ensuing years, medical scientists have investigated dozens of enzymes. Enzyme technology currently plays a role in two aspects of medical practice: diagnosis and therapy. Examples of each are briefly discussed below.

Diagnostic Uses of Enzymes

Enzymes are useful in modern medical practice for several reasons. Enzyme assays provide important information concerning the presence and severity of disease. In addition, enzymes often provide a means of monitoring a patient's response to therapy. Genetic predispositions to certain diseases may also be determined by measuring specific enzyme activities.

In the clinical laboratory, enzymes are used in two ways. First, the activity of certain enzymes may be measured directly. For example, the measurement of blood levels of acid phosphatase activity is used to diagnose prostatic carcinoma (a urinary tract tumor that occurs in males). Second, several enzymes are used as reagents. Because purified enzymes are available, detecting certain metabolites is more accurate and cost-effective. For example, the enzyme urate oxidase is used to measure blood levels of uric acid, a metabolite whose concentration is usually high in patients suffering from gout.

For an enzyme to be used in diagnosis, several conditions must be met.

1. **Ease of measurement.** An enzyme assay should be both accurate and convenient. The development of such an assay involves determining the saturating levels of substrate and cofactors, as well as good temperature and pH control. (Recall that under zero-order conditions, velocity is proportional to the enzyme's concentration.)

2. **A convenient method for obtaining clinically useful specimens.** Specimens of blood, urine, and (to a lesser extent) cerebrospinal fluid are readily available. However, blood, which contains a variety of enzymes and metabolites, is most often used in clinical diagnoses. Enzymes are measured by using blood *plasma* (the liquid remaining after the blood cells have been removed) or blood *serum* (the straw-colored liquid that results when blood has been allowed to clot).

Blood plasma contains two types of enzymes. The enzymes that are in the highest concentrations are specific to plasma. They include the enzymes involved in the blood-clotting process (e.g., thrombin and plasmin) and lipoprotein metabolism (Chapter 10). The nonspecific plasma enzymes have no physiological role in plasma and are normally present in low concentrations. In the normal turnover of cells, intracellular enzymes are released. An organ damaged by disease or trauma may elevate nonspecific plasma enzymes. For useful assays of these enzymes, measured enzyme activity should be commensurate with the damage. Because few enzymes are specific to one organ, the activities of several enzymes must often be measured. The procedure to confirm a diagnosis of myocardial infarction illustrates the use of enzymes in diagnosis.

Myocardial Infarction

Interruption of the heart's blood supply leads to the death of cardiac muscle cells. The symptoms of myocardial infarction include pain in the left chest that may radiate to the neck, left shoulder, and arm, and irregular breathing. The initial diagnosis is based on these and other symptoms. Therapy is instituted immediately. Physicians then use several enzyme assays to confirm the diagnosis and to monitor the course of treatment. The enzymes most commonly assayed are creatine kinase (CK) and lactate dehydrogenase (LDH). Each enzyme's activity shows a characteristic time profile in terms of its release from damaged cardiac muscle cells and rate of clearance from blood (Figure 6A).

The blood levels of both enzymes must be monitored. Neither enzyme concentration gives sufficient information for the duration of treatment. For example, CK is the first enzyme detected during a myocardial infarction. CK's serum concentration rises and falls so rapidly that it is of little clinical use after several days. LDH, whose serum concentration rises later, is used to monitor the later stages of heart damage. Monitoring the activity of several enzymes also prevents misdiagnosis. Re-

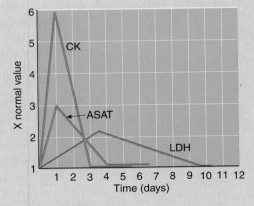

FIGURE 6A

Characteristic Pattern of Serum Cardiac Enzyme Concentrations Following a Myocardial Infarction.

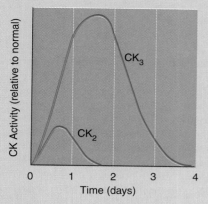

FIGURE 6B

Characteristic Pattern of Serum Creatine Kinase Concentrations Following a Myocardial Infarction.

call that few enzymes are specific to a particular organ. Until recently, serum levels of ASAT and ALAT were used to distinguish between heart and liver damage. (The ratios of the activities of the two enzymes is different in the two organs.) However, a technique employing variants of CK and LDH is now considered to be more reliable.

Both CK and LDH occur in multiple forms called isozymes. **Isozymes** are active forms of an enzyme with slightly different amino acid sequences. Isozymes can be distinguished from each other because they migrate differently during electrophoresis. CK, which occurs as a dimer, has two types of protomer: muscle type (M) and brain type (B). Heart muscle contains CK_2 (MB) and CK_3 (MM). Only the MB isozyme is found exclusively in heart muscle. Its concentration in blood reaches a maximum within a day following the infarction. CK_3 (MM), which is also found in other tissues, peaks a day after CK_2 (Figure 6B). LDH is a tetramer composed of two protomers: heart type (H) and muscle type (M). There are five different LDH isozymes. LDH 1 (H_4) and LDH 2 (H_3M) are found only in heart muscle and red blood cells. LDH 5 (M_4) occurs in both liver and skeletal muscle. LDH_3 and LDH_4 are found in other organs. The difference in the migration patterns of LDH isozymes of a normal individual and a patient suffering from myocardial infarction are illustrated in Figure 6C. Information generated from measuring the blood levels and migration patterns of both CK

and LDH virtually guarantees that a correct diagnosis will be made.

Therapeutic Uses of Enzymes

The use of enzymes in medical therapy has been limited. When administered to patients, enzymes are often rapidly inactivated or degraded. The large amounts of enzyme that are often required to sustain a therapy may provoke allergic reactions. There are, however, several examples of successful enzyme therapies.

Streptokinase is a proteolytic enzyme produced by *Streptococcus pyogenes* (a bacterium that causes throat and skin infections). Streptokinase promotes the growth of the bacterium in tissue because it digests blood clots. It is currently used with significant success in the treatment of myocardial infarction, which results from the occlusion of the coronary arteries. If administered soon after the beginning of a heart attack, streptokinase can often prevent or significantly reduce further damage to the heart. Streptokinase catalyzes the conversion of plasminogen to plasmin, the trypsinlike enzyme that digests fibrin (the primary component of blood clots). Human tissue plasminogen activator (tPA), a product of recombinant DNA technology (Chapter 17 and Appendix B, Supplement 17) that acts in a similar fashion, is also used to treat myocardial infarction.

The enzyme asparaginase is used to treat several types of cancer. It catalyzes the following reaction:

$$\text{L-Asparagine} + H_2O \rightarrow \text{L-aspartate} + NH_3$$

Asparaginase occurs in plants, vertebrates, and bacteria but not in human blood. Unlike most normal cells, the cells in certain kinds of tumor, such as in several forms of adult leukemia, cannot synthesize asparagine. Infusing asparaginase reduces the blood's concentration of asparagine and often causes tumor regression. (The regression of these tumors is not completely understood. Presumably, lack of asparagine inhibits protein synthesis, causing cell death.) Asparaginase therapy has several serious side effects, such as allergic reactions and liver damage. Unfortunately, after several asparaginase treatments some patients may develop resistance. (*Resistance* is a condition in which tumor cells grow despite the presence of a toxic substance that previously caused tumor regression). Apparently, some tumor cells can induce the synthesis of asparagine synthetase, an enzyme that converts aspartate to asparagine.

FIGURE 6C

Electrophoresis Pattern of Lactate Dehydrogenase Isozymes.

(a) LDH isozymes from a normal individual. (b) LDH isozymes from a myocardial infarction patient.

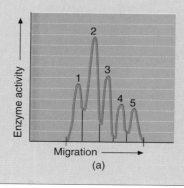
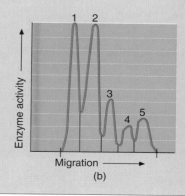

QUESTION 6.3

Drugs are chemicals that alter or enhance physiological processes. For example, aspirin suppresses pain, and antibiotics kill infectious organisms. Once a drug is consumed, it is absorbed and distributed to the tissues, where it performs its function. Eventually, drug molecules are processed (primarily in the liver) and excreted. The dosage of each drug that physicians prescribe is based on the amount required to achieve a therapeutic effect and the drug's average rate of excretion from the body. Various reactions prepare drug molecules for excretion. (See Chapter 19.) Examples include oxidation, reduction, and conjugation reactions. (In conjugation reactions, small polar or ionizable groups are attached to a drug molecule to improve its solubility.) Not surprisingly, enzymes play an important role in drug metabolism. The amount of certain enzymes directly affect a patient's ability to metabolize a specific drug. For example, isoniazid is an antituberculosis agent. (Tuberculosis is a highly infectious, chronic, debilitating disease caused by *Mycobacterium tuberculosis*.) It is metabolized by N-acetylation. (In N-acetylation an amide bond forms between the substrate and an acetyl group.) The rate at which isoniazid is acetylated determines its clinical effectiveness.

Two tuberculosis patients with similar body weights and symptoms are given the same dose of isoniazid. Although both take the drug as prescribed, one patient fails to show a significant clinical improvement. The other patient is cured. Because genetic factors appear to be responsible for the differences in drug metabolism, can you suggest a reason why these patients reacted so differently to isoniazid? How can physicians improve the percentage of patients who are cured?

KEY WORDS

activation energy, *120*
active site, *121*
allosteric enzyme, *131*
apoenzyme, *122*
coenzyme, *122*
cofactor, *122*
enzyme induction, *142*
enzyme kinetics, *123*
flavoprotein, *137*

holoenzyme, *122*
hydrolase, *122*
isomerase, *123*
isozyme, *147*
ligase, *123*
lyase, *123*
negative cooperativity, *144*
negative feedback, *144*
oxidoreductase, *122*

oxyanion, *140*
pacemaker enzyme, *143*
pH optimum, *139*
positive cooperativity, *144*
proenzyme, *143*
regulatory enzyme, *143*
serine protease, *134*
specific activity, *126*
steady state, *125*

substrate, *121*
transferase, *122*
transition state, *120*
turnover number, *126*
velocity, *123*
vitamin, *135*
zymogen, *143*

SUMMARY

Enzymes are biological catalysts. They enhance reaction rates because they provide an alternative reaction pathway that requires less energy than an uncatalyzed reaction. In contrast to some inorganic catalysts, most enzymes catalyze reactions at mild temperatures. In addition, enzymes are specific to the types of reactions they catalyze. Each type of enzyme has a unique, intricately shaped binding surface called an active site. Substrate binds to the enzyme's active site, which is a small cleft or crevice in a large protein molecule. In the lock-and-key model of enzyme action, the structures of the enzyme's active site and the substrate are complementary. In the induced-fit model, the protein molecule is assumed to be flexible.

Each enzyme is currently classified and named according to the type of reaction it catalyzes. There are six major enzyme categories: oxidoreductases, transferases, hydrolases, lyases, isomerases, and ligases.

Enzyme kinetics is the quantitative study of enzyme catalysis. According to the Michaelis-Menten model, when the substrate S binds in the active site of an enzyme E, an ES transition state complex is formed. During the transition state, the substrate is converted into product. After a time the product dissociates from the enzyme. In the Michaelis-Menten equation,

$$v = \frac{V_{max}[S]}{[S] + K_m}$$

V_{max} is the maximal velocity for the reaction, and K_m is a rate constant. Experimental determinations of K_m and V_{max} are made with Lineweaver-Burk double-reciprocal plots.

Enzyme inhibition may be reversible or irreversible. Irreversible inhibitors usually bind covalently to enzymes. In reversible inhibition, the inhibitor can dissociate from the enzyme. The most common types of reversible inhibition are competitive and noncompetitive.

The kinetic properties of allosteric enzymes are not explained by the Michaelis-Menten model. Most allosteric enzymes are multisubunit proteins. The binding of substrate or effector to one subunit affects the binding properties of other protomers.

Enzymes use the same catalytic mechanisms as nonenzymatic catalysts. Several factors contribute to enzyme catalysis: proximity and strain effects, electrostatic effects, acid-base catalysis, and covalent catalysis. Combinations of these factors affect enzyme mechanisms.

Active site amino acid side chains are primarily responsible for catalyzing proton transfers and nucleophilic substitutions. Nonprotein cofactors (metals and coenzymes) are used by enzymes to catalyze other types of reactions.

Enzymes are sensitive to environmental factors such as temperature and pH. Each enzyme has an optimum temperature and an optimum pH.

The chemical reactions in living cells are organized into a series of biochemical pathways. The pathways are controlled primarily by adjusting the concentrations and activities of enzymes through genetic control, covalent modification, allosteric regulation, and compartmentation.

SUGGESTED READINGS

Dische, Z., The Discovery of Feedback Inhibition, *Trends Biochem. Sci.,* 1:269–270, 1976.

Kraut, J., How Do Enzymes Work? *Science,* 242:533–540, 1988.

Miller, J.A., Women in Chemistry, in Kass-Simon, G., and Fannes, P. (Eds.), *Women in Science: Righting the Record,* pp. 300–334, Indiana University Press, Bloomington, 1990.

Monod, J., Changeux, J.P., and Jacob F., Allosteric Proteins and Cellular Control Systems, *J. Mol. Biol.,* 6:306–329, 1963.

Perutz, M.F., Mechanisms of Cooperativity and Allosteric Regulation in Proteins, *Quart. Rev. Biophys.,* 22:139–151, 1989.

Schultz, P.G., The Interplay Between Chemistry and Biology in the Design of Enzymatic Catalysts, *Science,* 240:426–433, 1988.

QUESTIONS

1. Clearly define the following terms:
   a. activation energy
   b. catalyst
   c. active site
   d. coenzyme
   e. velocity of a chemical reaction
   f. half-life
   g. turnover number
   h. katal
   i. noncompetitive inhibitor
   j. repression

2. What are four important properties of enzymes?

3. Living things must regulate the rate of catalytic processes. Explain how the cell regulates enzymatic reactions.

4. Determine the class of enzyme that is most likely to catalyze each of the following reactions:

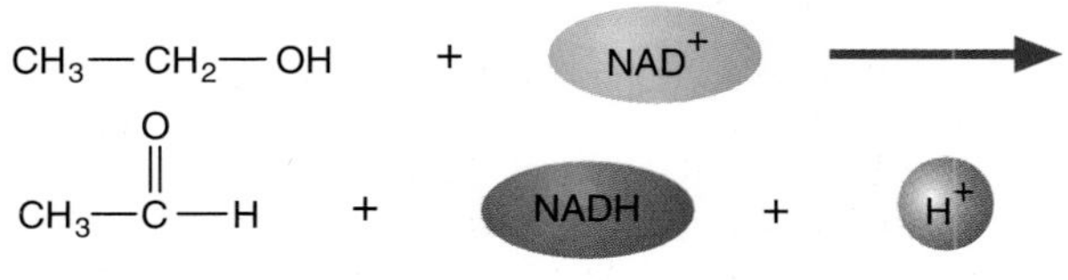

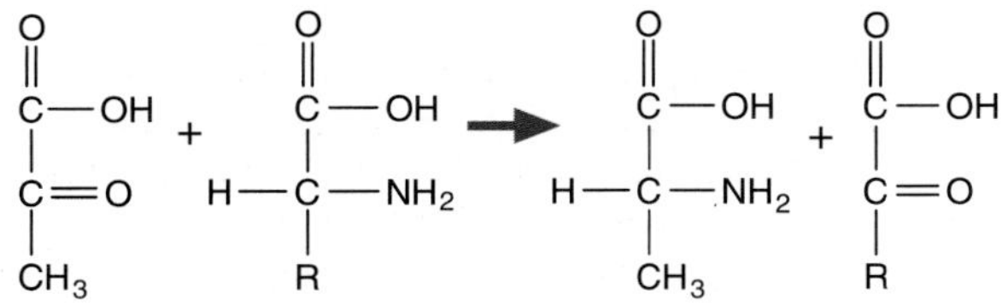

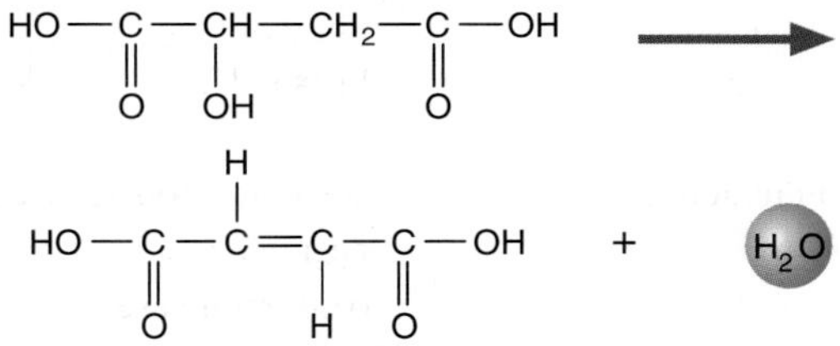

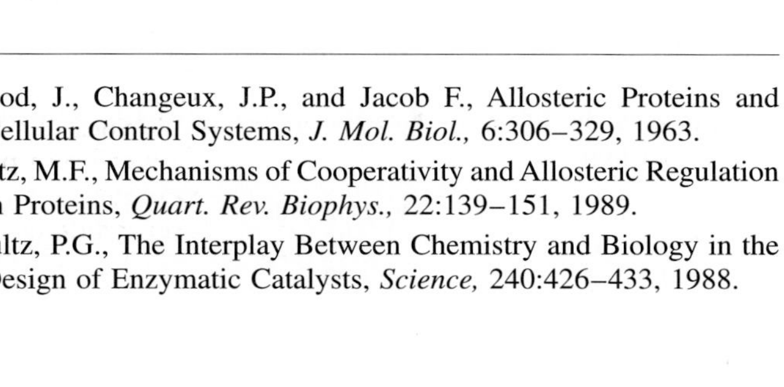

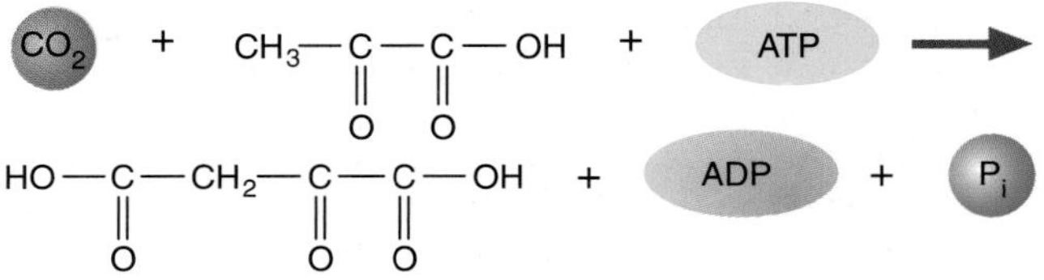

5. Consider the following reaction:

Pyruvate

Using the following data, determine the order of the reaction for each substrate and the overall order of the reaction.

Experiment	[Pyruvate]*	[ADP]	[P_i]	Rate
1	0.1	0.1	0.1	8×10^{-4}
2	0.2	0.1	0.1	1.6×10^{-3}
3	0.2	0.2	0.1	3.2×10^{-3}
4	0.1	0.1	0.2	3.2×10^{-3}

*Concentrations are in moles per liter. Enzyme rates are measured in moles per liter per second.

6. Consider the following data for an enzyme-catalyzed hydrolysis reaction:

[Substrate] (M)	V (μmol/min)	V_I (μmol/min)
6×10^{-6}	20.8	4.2
1×10^{-5}	29	5.8
2×10^{-5}	45	9
6×10^{-5}	67.6	13.6
1.8×10^{-4}	87	16.2

Using a Michaelis-Menten plot, determine K_m for the uninhibited reaction.

7. Using the data in Question 6:
   a. Generate a Lineweaver-Burk plot for the data.
   b. Explain the significance of the (i) horizontal intercept, (ii) vertical intercept, (iii) slope.
   c. What type of inhibition is being measured?

8. What are the two types of enzyme inhibitors? Give an example of each.

9. Two experiments were performed with the enzyme ribonuclease. In experiment 1 the effect of increasing substrate concentration on reaction velocity was measured. In experiment 2 the reaction mixtures were identical to those in experiment 1 except for the addition of 0.1 mg of an unknown compound to each tube. Plot the data according to the Lineweaver-Burk method. Determine the effect of the unknown compound on the enzyme's activity. (Substrate concentration is measured in millimoles per liter. Velocity is measured in the change in optical density per hour.)

Experiment 1		Experiment 2	
[S]	V	[S]	V
0.5	0.81	0.5	0.42
0.67	0.95	0.67	0.53
1	1.25	1	0.71
2	1.61	2	1.08

10. Several factors contribute to enzyme catalysis. What are they? Briefly explain the effect of each.

11. List three reasons why the regulation of biochemical processes is important.

12. Describe negative feedback inhibition.

13. Describe the two models that explain the binding of allosteric enzymes. Use either model to explain the binding of oxygen to hemoglobin.

14. What are the major coenzymes? Briefly describe the function of each.

15. What properties of transition metals make them useful as enzyme cofactors?

16. Describe the mechanism for chymotrypsin. Show how the amino acid residues of the active site participate in the reaction.

17. The ΔH for the following reaction is +41.6 cal:

$$CH_3CH_2OH + O_2 \rightarrow CH_3CHO + H_2O$$
Ethanol   Acetaldehyde

Explain why ethyl alcohol is stable in an oxygen atmosphere for appreciable periods of time.

18. Enzymes act by reducing the activation energy of a reaction. Describe several ways in which this is accomplished.

19. In enzyme kinetics, why are measurements made at the start of a reaction?

20. Histidine is frequently used as a general acid or general base in enzyme catalysis. Consider the pK_a's of the side groups of the amino acids listed in Table 5.2 to suggest a reason why this is so.

21. Suggest a reason why enzymes can be partially protected from thermal denaturation by high concentrations of substrate.

22. Enzymes are stereochemically specific, that is, they often convert only one stereoisomeric form of substrate into product. Why is such specificity inherent in their structure?

CARBOHYDRATES

Chapter Seven

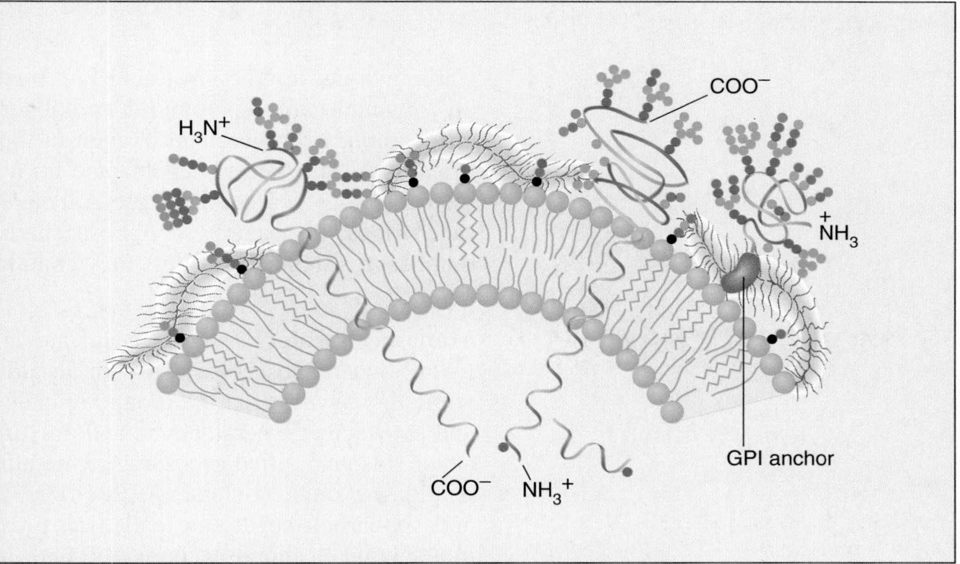

The Cell Surface. *Significant amounts of carbohydrate are attached to membrane protein and lipids on the external surface of cells.*

OBJECTIVES

After you have studied this chapter, you should be able to answer these questions:

1.　*How are carbohydrates classified?*

2.　*What are optical isomers and why are they important in living organisms?*

3.　*What information do Fischer projection, Haworth, conformational, and space-filling models of carbohydrates provide?*

4.　*What are the most common disaccharides found in nature and what purpose do they serve?*

5.　*What are the general structural features of oligosaccharides and polysaccharides?*

6.　*What structural properties of cellulose account for its widespread occurrence in plants?*

7.　*What are the most common homopolysaccharides and heteropolysaccharides and what are their functions?*

8.　*What structural properties distinguish proteoglycans from glycoproteins?*

9.　*What are the functions of proteoglycans and glycoproteins?*

Carbohydrates are not just an important source of rapid energy production for living cells. They are also structural building blocks of cells and components of numerous metabolic pathways. A broad range of cellular phenomena, such as cell recognition and binding (e.g., by other cells, hormones, and viruses), depend on carbohydrates. This chapter describes the structures and chemistry of typical carbohydrate molecules found in living organisms.

Carbohydrates are the most abundant organic molecules in nature. More than half of all "organic" carbon is found in carbohydrates. The majority of these substances contain carbon, hydrogen, and oxygen in the ratio $C_n(H_2O)_n$, hence the name "hydrate of carbon." They have been adapted for a wide variety of biological functions, which include energy sources and structural elements. Carbohydrates are classified as monosaccharides, disaccharides, oligosaccharides, and polysaccharides according to the number of simple sugar units they contain.

7.1 MONOSACCHARIDES

Monosaccharides or simple sugars are defined as polyhydroxy aldehydes or ketones. Recall that monosaccharides with an aldehyde functional group are called **aldoses**, while those with a ketone group are called *ketoses* (Figure 7.1). The simplest aldose and ketose are glyceraldehyde and dihydroxyacetone, respectively (Figure 7.2). Sugars are also classified according to the number of carbon atoms they contain. For example, the smallest sugars, called *trioses*, contain three carbon atoms. Four-, five-, and six-carbon sugars are called *tetroses, pentoses,* and *hexoses,* respectively. The most abundant monosaccharides found in living cells are the pentoses and hexoses. Often, class names such as aldohexoses and ketopentoses, which combine information about carbon number and functional groups, describe monosaccharides. For example, glucose, a six-carbon aldehyde-containing sugar, is referred to as an aldohexose.

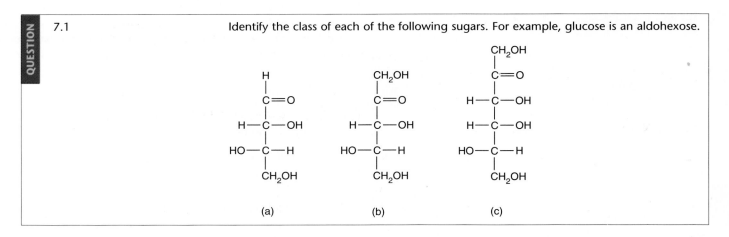

QUESTION 7.1 Identify the class of each of the following sugars. For example, glucose is an aldohexose.

FIGURE 7.1
General Formula for Aldose and Ketose Forms of Monosaccharides.

Monosaccharide Structure

$$H$$
$$|$$
$$C=O$$
$$|$$
$$H—C—OH$$
$$|$$
$$CH_2OH$$

Glyceraldehyde

$$CH_2OH$$
$$|$$
$$C=O$$
$$|$$
$$CH_2OH$$

Dihydroxyacetone

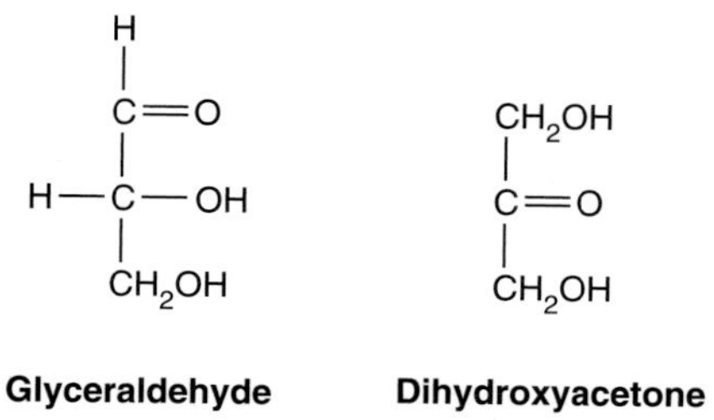

FIGURE 7.2

Glyceraldehyde and Dihydroxyacetone.

Glyceraldehyde, one of the simplest monosaccharides, occurs in two stereoisomeric forms (Figure 7.3). The two types of glyceraldehyde differ in one physical property. They rotate plane-polarized light in opposite directions. Molecules that possess this property are called **optical isomers.** (In plane-polarized light, produced by passing unpolarized light through a special filter, the light waves vibrate in only one plane.) For example, one glyceraldehyde isomer rotates the light beam in a clockwise direction and is said to be dextrorotary (designated by +). The other glyceraldehyde isomer, referred to as levorotary (designated by −), rotates the beam in the opposite direction to an equal degree. Optical isomers are often designated as D or L, for example, D-glucose and L-alanine. The D or L indicates the similarity of the arrangement of atoms around a molecule's asymmetric carbon to the asymmetric carbon in either of the glyceraldehyde isomers. (Glyceraldehyde is the reference compound for optical isomers.) Because most carbohydrates have two or more chiral centers, the letters D and L refer only to a molecule's structural relationship to either of the glyceraldehyde isomers, not to the direction in which it rotates plane-polarized light. (Longer-chain sugars can be thought of as having additional H—C—OH groups between the carbonyl carbon and the chiral carbon of glyceraldehyde.) For example, because D-fructose rotates light to the left, it can be designated as D(−)-fructose. The D family of aldoses, which contains several biologically important sugars, is illustrated in Figure 7.4.

QUESTION 7.2

When viewed in two dimensions (as on the printed page), the structural differences between optical isomers may appear trivial. However, many biomolecules are optically active, and the capacity of enzymes to distinguish between D and L substrate molecules is an important feature of the chemistry of living cells. For example, most of the enzymes that break down and use dietary carbohydrates can bind to D-sugars but not to their L-isomers. Can you convince yourself that the D- and L-isomers of an optically active molecule are indeed different in three-dimensional space? Make models of D- and L-glyceraldehyde with an organic chemistry model kit or with colored styrofoam balls and toothpicks.

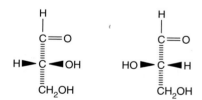

D-Glyceraldehyde   **L-Glyceraldehyde**

FIGURE 7.3

D- and L-Glyceraldehyde.

These molecules are mirror images of each other.

As the number of chiral carbon atoms increases in optically active compounds, the number of possible isomers also increases. The total number of possible isomers can be determined by using Van't Hoff's rule: A compound with n asymmetric carbon atoms has a maximum of 2^n possible stereoisomers.

Stereoisomers that are not enantiomers (mirror-image isomers) are called **diastereomers.** For example, the aldopentoses D-ribose and L-ribose are enantiomers, as are D-arabinose and L-arabinose (Figure 7.5). The sugars D-ribose and D-arabinose are diastereomers because they are isomers but not mirror images.

Diastereomers that differ in the configuration at a single asymmetric carbon atom (but not the reference carbon) are called **epimers.** (In optical isomers the reference carbon is the asymmetric carbon that is most remote from the carbonyl carbon. Its configuration is similar to that of the asymmetric carbon in either D- or L-glyceraldehyde.) For example, D-glucose and D-galactose are epimers because their structures differ only in the configuration of the OH group at carbon 4 (Figure 7.4). D-Mannose and D-galactose are not epimers because their configurations differ at more than one carbon.

The sugar structures shown in Figures 7.4 and 7.5 are known as Fischer projections (in honor of the great Nobel prize–winning German chemist Emil Fischer). In these structures the carbohydrate backbone is drawn vertically with the most highly oxidized carbon usually shown at the top. The horizontal lines are understood to project toward the viewer, and the vertical lines recede from the viewer. The family of the sugar, whether D or L, is indicated by the direction of the hydroxyl group attached to the reference carbon. For example, in the D family the OH attached to the most remote asymmetric carbon (e.g., carbon 5 in a six-carbon sugar) points to the right.

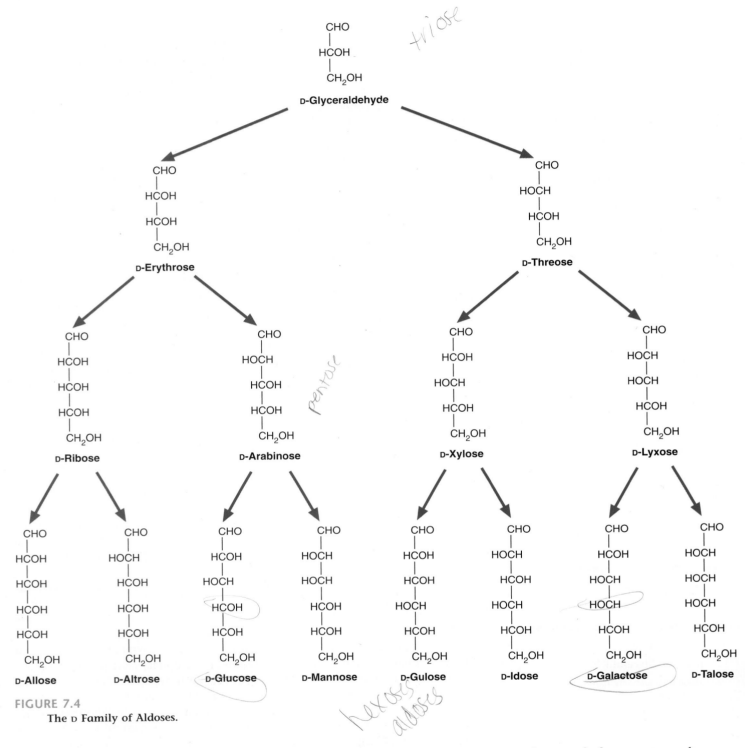

FIGURE 7.4
The D Family of Aldoses.

Cyclic Structure of Monosaccharides   Sugars that contain four or more carbons exist primarily in cyclic forms. Ring formation occurs in aqueous solution because aldehydes and ketones react reversibly with hydroxyl groups present in the sugar to form cyclic **hemiacetals** (Figure 7.6). Ordinary hemiacetals (molecules with hydroxyl and ether groups on the same carbon) are unstable and easily revert to the aldehyde or ketone forms. Note that as cyclization occurs, the carbonyl carbon becomes a new chiral center. This carbon is called the *anomeric carbon atom*. The two possible diastereomers that may form during the cyclization reaction are called **anomers.**

FIGURE 7.5

The Optical Isomers D- and L-Ribose and D- and L-Arabinose.

D-Ribose and D-arabinose are diastereomers; that is, they are not mirror images.

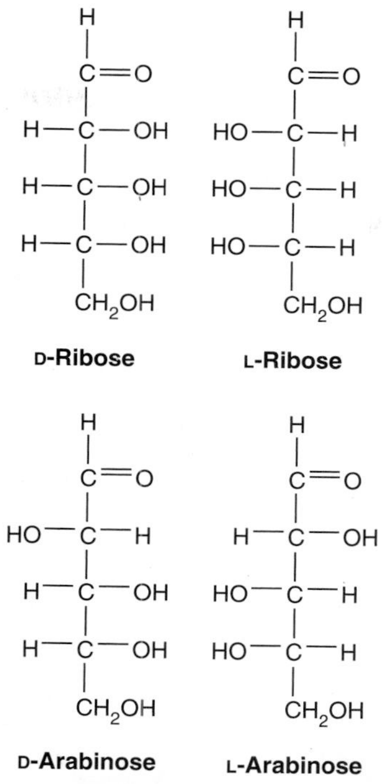

D-Ribose   L-Ribose

D-Arabinose   L-Arabinose

FIGURE 7.6

Formation of Hemiacetals.

(a) From an aldehyde. (b) From a ketone.

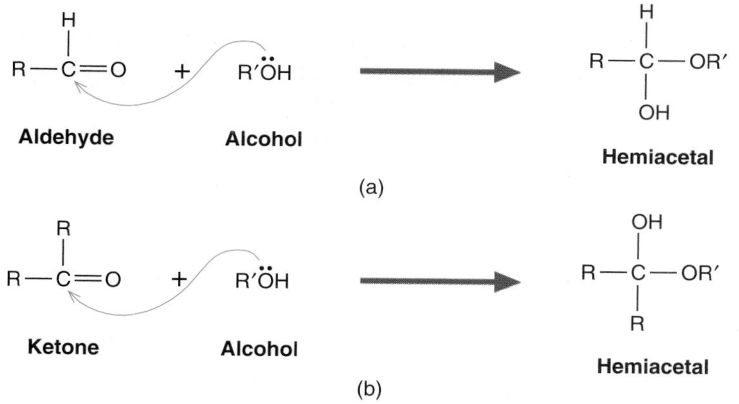

Aldehyde   Alcohol   Hemiacetal

(a)

Ketone   Alcohol   Hemiacetal

(b)

In aldose sugars the hydroxyl group of the newly formed hemiacetal occurs on carbon 1 and may occur either above the ring (in the "up" position) or below the ring (in the "down" position). When the hydroxyl is down, the structure is in the α-anomeric form. If the hydroxyl is up, the structure is in the β-anomeric form. The anomers are defined relative to the D and L classification of sugars. The above rules apply only to D-sugars, the most common ones found in nature. In the L-sugars the α-anomeric OH group is above the ring.

Haworth Structures   Fischer representations of cyclic sugar molecules use a long bond to indicate ring structure (Figure 7.7a). A more accurate picture of carbohydrate structure was developed by the English chemist W.N. Haworth (Figure 7.7b).

Haworth structures more closely depict proper bond angles and lengths than do Fischer representations. To convert from the traditional Fischer formula of a D-pentose or D-hexose to a Haworth formula, the following steps should be followed:

FIGURE 7.7

Monosaccharide Structure.

(a) Formation of the hemiacetal structure of glucose. (b) Haworth structure of glucose.

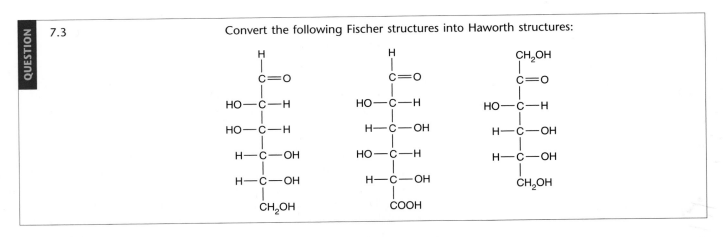

1. Draw a five- or six-membered ring with the oxygen placed as shown below:

Five-membered ring   **Six-membered ring**

2. Starting with the anomeric carbon to the right of the ring oxygen, place hydroxyl groups either above or below the plane of the ring. Groups that are pointing to the left in the Fischer projection formula should go above the plane of the ring, and those that are pointing to the right in the Fischer projection formula should go below the ring.

QUESTION 7.3   Convert the following Fischer structures into Haworth structures:

Five-membered hemiacetal rings are called *furanoses* because of their structural similarity to furan (Figure 7.8). For example, the cyclic form of fructose depicted in Figure 7.9 is called fructofuranose. Six-membered rings are called *pyranoses* because of their similarity to pyran. Glucose, in the pyranose form, is called glucopyranose.

Conformational Structures   Although Haworth projection formulas often represent carbohydrate structure, they are oversimplifications. Bond angle analysis and X-ray analysis demonstrate that *conformational formulas* are more accurate representations

KEY *Concepts*

Monosaccharides, defined as polyhydroxy aldehydes or ketones, are either aldoses or ketoses. Sugars that contain four or more carbons primarily have cyclic forms. Cyclic aldoses or ketoses are hemiacetals.

FIGURE 7.8
Furan and Pyran.

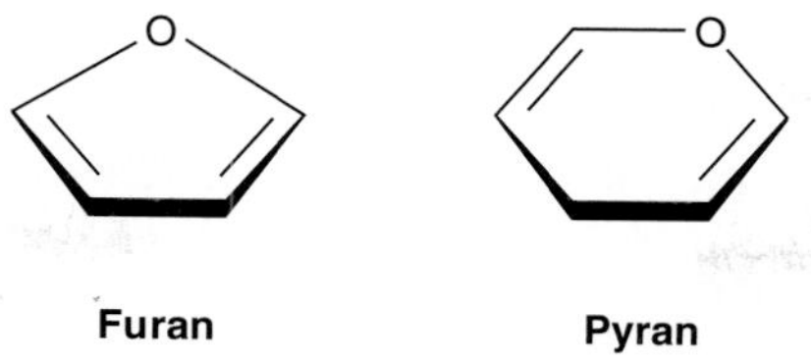

Furan   Pyran

FIGURE 7.9
Fischer and Haworth Forms of D-Fructose.

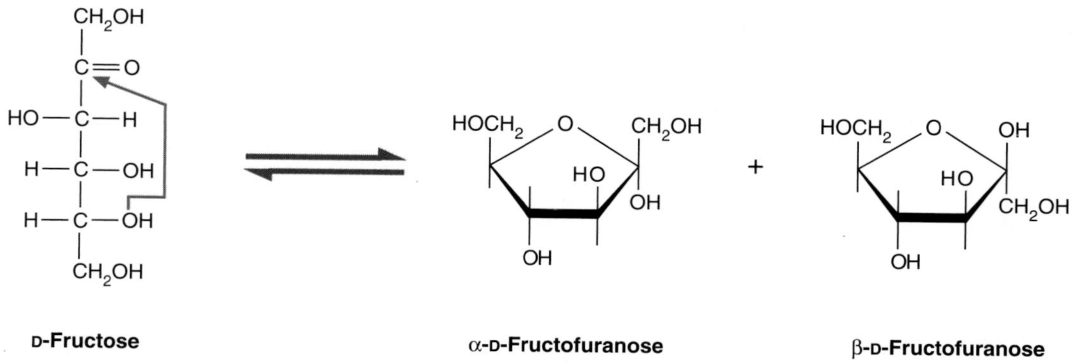

D-Fructose   α-D-Fructofuranose   β-D-Fructofuranose

FIGURE 7.10
α- and β-Glucose.

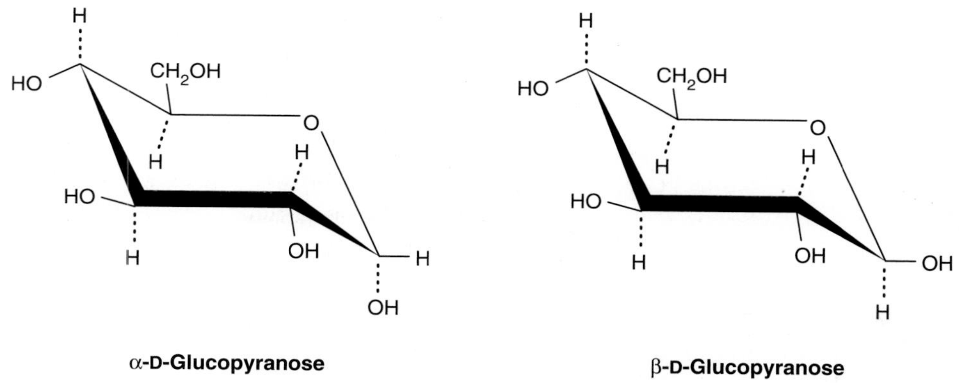

α-D-Glucopyranose   β-D-Glucopyranose

of monosaccharide structure (Figure 7.10). Conformational structures are more accurate because they illustrate the puckered nature of sugar rings.

Space-filling models, whose dimensions are proportional to the radius of the atoms, also give useful structural information. (See Figures 7.19, 7.20, and 7.21.)

Reactions of Monosaccharides

Mutarotation The α and β forms of monosaccharides are readily interconverted when dissolved in water. This spontaneous process, called *mutarotation,* produces an equilibrium mixture of α- and β-forms in both furanose and pyranose ring structures. The proportion of each form differs with each sugar type. Glucose, for example, exists primarily as a mixture of α- and β-pyranoses (Figure 7.11). Fructose is predominantly found in the α- and β-furanose forms. The open chain formed during mutarotation can participate in oxidation-reduction reactions, as discussed below.

Oxidation-Reduction Reactions In the presence of oxidizing agents, metal ions, such as Cu^{2+}, and certain enzymes, monosaccharides readily undergo several oxidation reactions. Oxidation of an aldehyde group yields an **aldonic acid,** while oxidation of a terminal CH_2OH group (but not the aldehyde group) gives an **uronic acid.** Oxidation of both the aldehyde and CH_2OH gives an **aldaric acid** (Figure 7.12).

FIGURE 7.11

Equilibrium Mixture of D-Glucose.

When glucose is dissolved in water at 25°C, the anomeric forms of the sugar undergo very rapid interconversions. When equilibrium is reached (i.e., there is no net change in the occurrence of each form), the glucose solution contains the percentages shown.

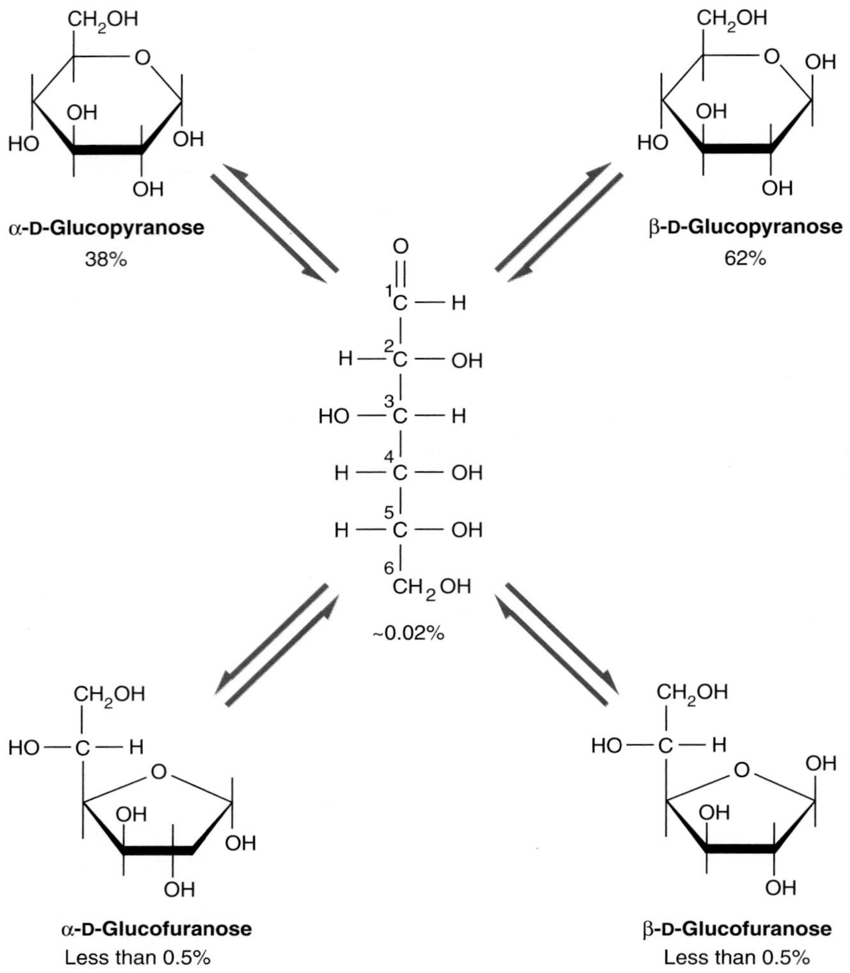
α-D-Glucopyranose
38%
β-D-Glucopyranose
62%
~0.02%
α-D-Glucofuranose
Less than 0.5%
β-D-Glucofuranose
Less than 0.5%

FIGURE 7.12

Oxidation Products of Glucose.

The newly oxidized groups are highlighted.

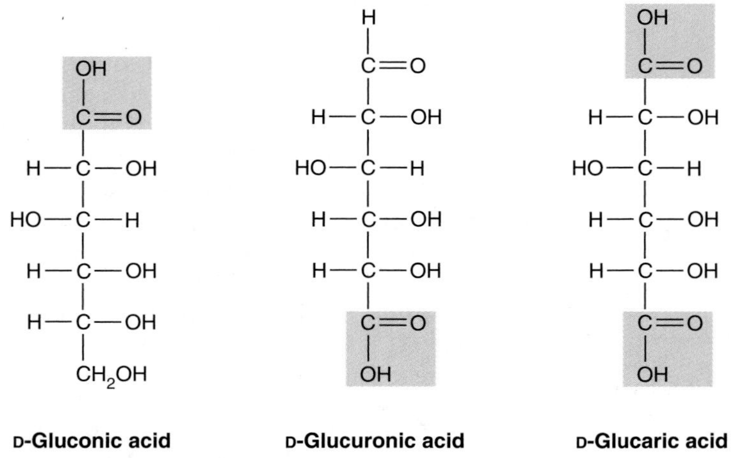
D-Gluconic acid
D-Glucuronic acid
D-Glucaric acid

The carbonyl groups in both aldonic and uronic acids can react with an OH group in the same molecule to form a cyclic ester known as a **lactone:**

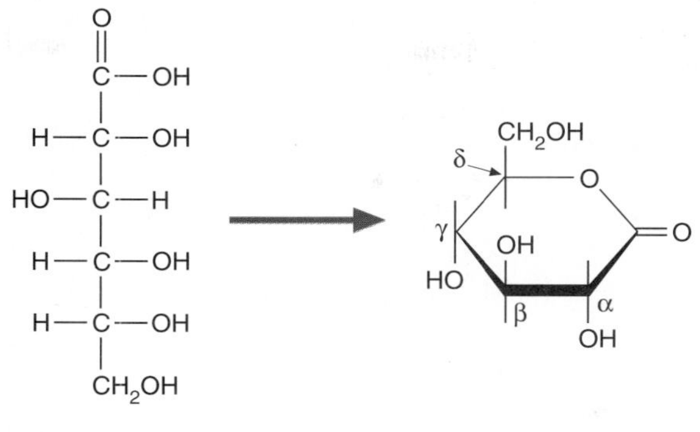

D-Gluconic acid
(An aldonic acid)

D-Glucono-δ-lactone

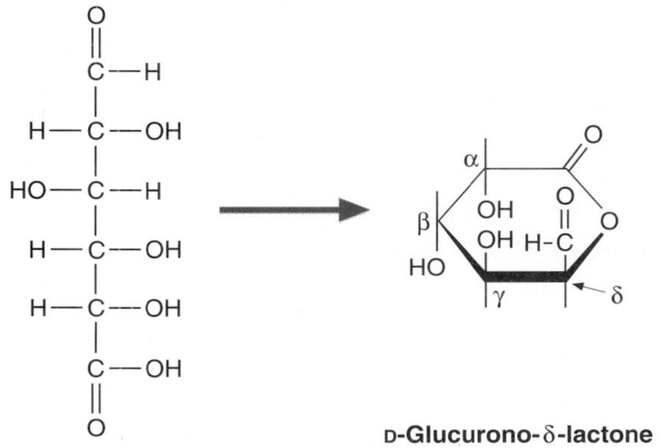

D-Glucuronic acid
(A uronic acid)

D-Glucurono-δ-lactone

Lactones are commonly found in nature. For example, L-ascorbic acid (vitamin C) is a lactone derivative of D-glucuronic acid. (See Special Interest Box 7.1.)

Sugars that can be oxidized by weak oxidizing agents such as Benedict's reagent are called **reducing sugars** (Figure 7.13). Since the reaction occurs only with sugars that can revert to the open chain form, all monosaccharides are reducing sugars.

Reduction Reduction of the aldehyde and ketone groups of monosaccharides yields the sugar alcohols **(alditols).** Reduction of D-glucose, for example, yields D-glucitol, also known as D-sorbitol (Figure 7.14). Sugar alcohols are used commercially in processing foods and pharmaceuticals. Sorbitol, for example, improves the shelf life of

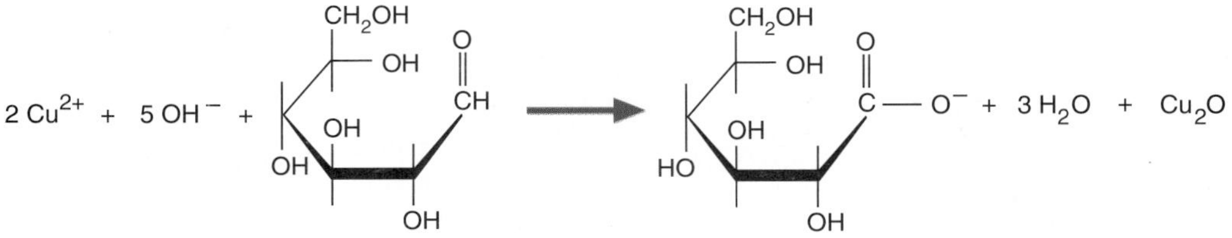

FIGURE 7.13

Reaction of Glucose with Benedict's Reagent.

Benedict's reagent, copper (II) sulfate in a solution of sodium carbonate and sodium citrate, is reduced by the monosaccharide glucose. Glucose is oxidized to form the salt of gluconic acid. The reaction also forms the reddish-brown precipitate Cu_2O and other oxidation products.

$$2\,Cu^{2+} + 5\,OH^- + \cdots \longrightarrow \cdots + 3\,H_2O + Cu_2O$$

ASCORBIC ACID

Ascorbic acid, a lactone with a molecular weight of 176.1, is a powerful reducing agent. It is synthesized by all mammals except for guinea pigs, apes, fruit-eating bats, and, of course, humans. These species must obtain ascorbic acid (also called *vitamin* C) in their diet. The three enzymes necessary to synthesize ascorbic acid have been isolated from the microsomal fractions of liver tissue from species that can synthesize the molecule. In humans and guinea pigs, however, one of the enzymes (gulonolactone oxidase) has not been detected. (Antibodies that bind specifically to gulonolactone oxidase do not bind to human or guinea pig hepatic microsomes.) Presumably, lack of this enzyme prevents these species from synthesizing ascorbic acid.

Scurvy (Figure 7A), the disease caused by a significant depletion of the body's stores of ascorbic acid, was common in Europe before the introduction of root crops at the end of the Middle Ages. However, as Europeans took increasingly longer voyages to search for a trade route to the Far East, outbreaks of the malady were dramatic (and often documented). For example, a ship's log written during Vasco da Gama's expedition (1498) reveals an outbreak of scurvy on the return journey across the Arabian Sea. The sailors "again suffered from their gums, their legs also swelled and other parts of the body and these swellings spread until the sufferer died, without exhibiting symptoms of any other disease." During the expedition the sailors became convinced that oranges were a curative for scurvy, whose symptoms did not appear until ten weeks after setting sail.

When James Lind, a surgeon in the British Royal Navy, performed his famous experiment (perhaps the first controlled clinical trial) in 1746, he compared the curative effects of several treatments on 12 sailors who had scurvy. Two men were assigned to each of six daily treatments for two weeks. Briefly, the treatments were (1) one quart of hard cider, (2) 25 mL of elixir of vitriol three times a day, (3) 18 mL of vinegar three times a

FIGURE 7A

Scurvy Grass Promoted as a Cure for the Disease.

The title page of the English version of Moellenbrok's seventeenth-century book and his illustration.

day, (4) 0.3 L of seawater, (5) two oranges and one lemon, and (6) 4 mL of a medicinal paste containing garlic and mustard seed, among other herbs. Contrary to popular opinion, Lind's work did not prove the efficacy of citrus fruits in the treatment of scurvy; these had already been an accepted treatment for several hundred years. Actually, citrus fruits were Lind's positive controls. He was apparently interested in testing the effectiveness of the official treatment of the Royal Navy: vinegar (treatment 3). The results of the trial: Those who received the oranges and lemons were fit for duty after six days. The other treatments varied widely in outcome but were all significantly less effective than citrus fruits.

candy because it helps prevent moisture loss. Adding sorbitol syrup to artificially sweetened canned fruit reduces the unpleasant aftertaste of the artificial sweetener saccharin. Once consumed, sorbitol is converted into fructose in the liver.

Isomerization Monosaccharides undergo several types of isomerization. For example, after several hours an alkaline solution of D-glucose also contains D-mannose and D-fructose. Both isomerizations involve an intramolecular shift of a hydrogen atom and a relocation of a double bond (Figure 7.15). The intermediate formed is called an **enediol.** The reversible transformation of glucose to fructose is an example of an aldose-ketose interconversion. Because the configuration at a single asymmetric carbon changes, the conversion of glucose to mannose is referred to as an **epimerization.** Several enzyme-catalyzed reactions involving enediols occur in carbohydrate metabolism (Chapter 8).

Esterification Like all free OH groups, those of carbohydrates can be converted to esters by reactions with acids. Esterification often dramatically changes a sugar's chemical and physical properties. Phosphate and sulfate esters of carbohydrate molecules are among the most common ones found in nature.

FIGURE 7.14

Laboratory Reduction of Glucose to Form D-Glucitol (Sorbitol).

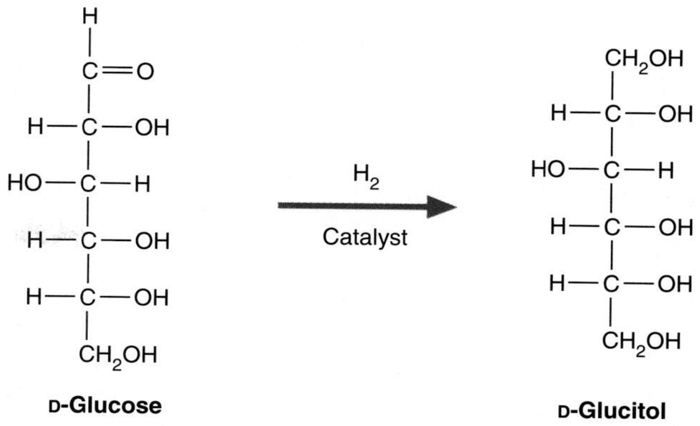

D-Glucose H_2 / Catalyst   D-Glucitol

FIGURE 7.15

Isomerization of D-Glucose to Form D-Mannose and D-Fructose.

An enediol intermediate is formed in this process.

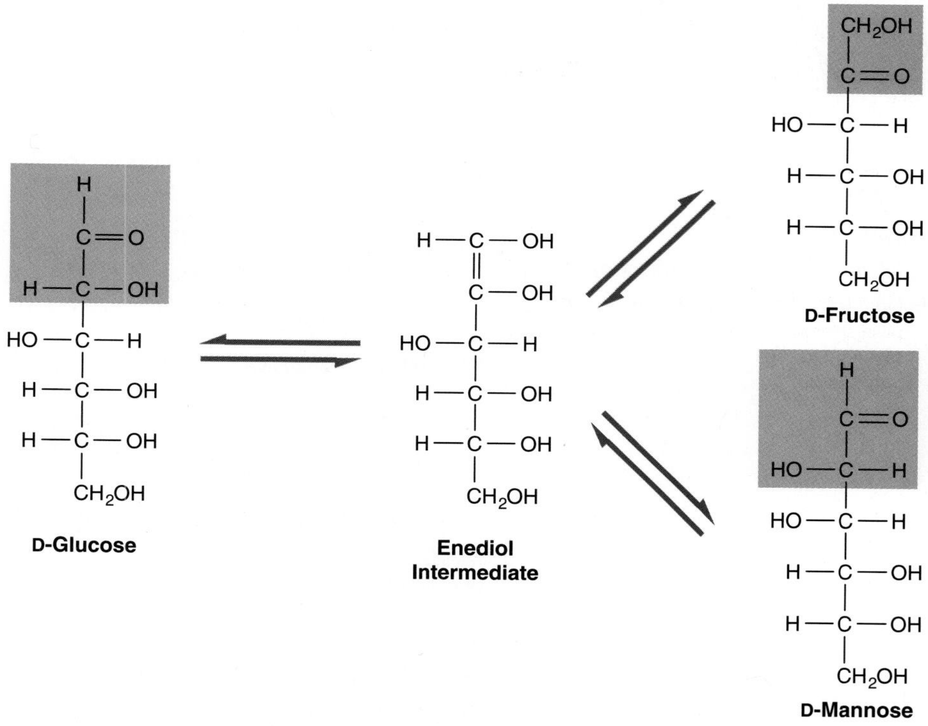

D-Glucose   Enediol Intermediate   D-Fructose   D-Mannose

Phosphorylated derivatives of certain monosaccharides are important metabolic components of living cells. They are frequently formed during reactions with ATP. They are important because many biochemical transformations use nucleophilic substitution reactions. Such reactions require a leaving group. In a carbohydrate molecule this group would most likely be an OH group. However, because OH groups are poor leaving groups, any substitution reaction would be unlikely. The problem is solved by converting an appropriate OH group to a phosphate ester, which can then be displaced by an incoming nucleophile. As a consequence, a slow reaction now occurs much more rapidly.

Sulfate esters of carbohydrate molecules are found predominantly in the proteoglycan components of connective tissue (Special Interest Box 7.3). Because sulfate esters are charged, they bind large amounts of water and small ions. They also participate in forming salt bridges between carbohydrate chains.

FIGURE 7.16
Formation of Acetals.

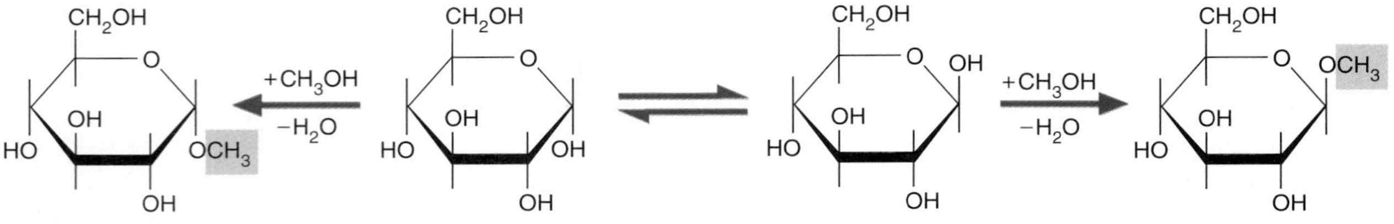

α-Methyl glucoside

α-Glucose

β-Glucose

β-Methyl glucoside

FIGURE 7.17

Methyl Glucoside Formation.

Noncarbohydrate components of glycosides are called aglycones. The highlighted methyl groups are aglycones.

Glycoside Formation   Hemiacetals react with alcohols to form **acetals** (Figure 7.16). The new linkage is called a **glycosidic linkage,** and the compound is called a **glyco-side.** The name of the glycoside specifies the sugar component. For example, acetals of glucose and fructose are called *glucosides,* and *fructosides,* respectively. Additionally, glycosides derived from sugars with five-membered rings are called *furanosides;* those from six-membered rings are called *pyranosides.* A relatively simple example shown in Figure 7.17 illustrates the reaction of glucose with methanol to form two types of methyl glucosides. Because glycosides are acetals, they are stable in basic solutions. Carbohydrate molecules that contain only acetal groups do not test positive with Benedict's reagent. (Acetal formation "locks" a ring so it cannot undergo oxidation or mutarotation.) Only hemiacetals act as reducing agents.

If an acetal linkage is formed between the hemiacetal hydroxyl group of one monosaccharide and a hydroxyl group of another monosaccharide, the resulting glycoside is called a **disaccharide.** A molecule containing a large number of monosaccharides linked by glycosidic linkages is called a **polysaccharide.**

QUESTION 7.4

Glycosides are often found in nature. One example is salicin (Figure 7.18), a compound found in willow tree bark. Conversion of salicin to salicylic acid is one step in the production of the modern analgesic aspirin (acetylsalicylic acid). Can you identify the carbohydrate and aglycone (noncarbohydrate) components of salicin?

FIGURE 7.18
Salicin.

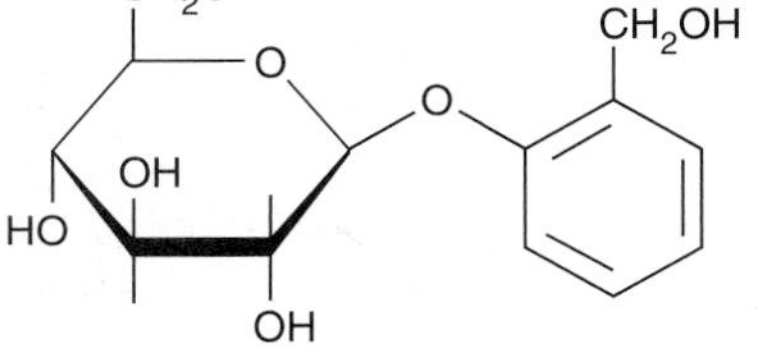

FIGURE 7.19

α-D-**Glucopyranose.**

Compare the information provided by the (a) space-filling model and (b) the Haworth structure. Carbon atoms are green, oxygen atoms are red, and hydrogen atoms are white.

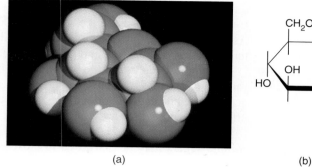

(a)   (b)

Important Monosaccharides

Glucose   D-Glucose, also known as dextrose, is found in large quantities throughout the living world (Figure 7.19). It is the primary fuel for living cells. In animals, glucose is the preferred energy source of brain cells and cells that have few or no mitochondria, such as erythrocytes. Cells that have a limited oxygen supply, such as those in the eyeball, also use large amounts of glucose to generate energy. Dietary sources include plant starch and the disaccharides lactose, maltose, and sucrose.

Fructose   D-Fructose, or levulose, is often referred to as fruit sugar because of its high content in fruit. It is also found in some vegetables as well as in honey (Figure 7.20). This molecule is an important member of the ketose family of sugars. On a per gram basis, fructose is twice as sweet as sucrose. It can therefore be used in smaller amounts. For this reason, fructose is often used as a sweetening agent in processed food products. Large amounts of fructose are used in the male reproductive tract. It is synthesized in the seminal vesicles and then incorporated into semen. Sperm use the sugar as an energy source.

Galactose   Galactose is necessary to synthesize a variety of biomolecules (Figure 7.21). These include lactose (in lactating mammary glands), glycolipids, certain phospholipids, proteoglycans, and glycoproteins. Synthesis of these substances is not

KEY *Concepts*

Glucose, fructose, and galactose are among the most important monosaccharides in living organisms.

FIGURE 7.20

β-D-**Fructofuranose.**

(a) Space-filling model and (b) Haworth structure.

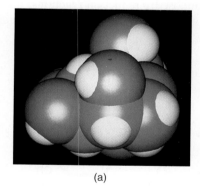

(a)   (b)

FIGURE 7.21

α-D-**Galactopyranose.**

(a) Space-filling model and (b) Haworth structure.

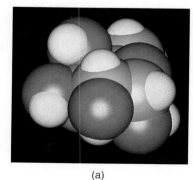

(a)   (b)

diminished by diets that lack galactose or the disaccharide lactose (the principal dietary source of galactose), since the sugar is readily synthesized from glucose-1-phosphate. As was mentioned previously, galactose and glucose are epimers at carbon 4. The interconversion of galactose and glucose is catalyzed by an enzyme called an epimerase.

In *galactosemia,* a genetic disorder, an enzyme required to metabolize galactose is missing. Galactose, galactose-1-phosphate, and galactitol (a sugar alcohol derivative) accumulate and cause liver damage, cataracts, and severe mental retardation. The only effective treatment is early diagnosis and a diet free of galactose.

Monosaccharide Derivatives

Simple sugars may be converted to closely related chemical compounds. Several of these are important metabolic and structural components of living organisms.

Uronic Acids   Recall that uronic acids are formed when the terminal CH_2OH group of a monosaccharide is oxidized. Two uronic acids are important in animals: D-glucuronic acid and its epimer, L-iduronic acid (Figure 7.22). In liver cells, glucuronic acid is combined with molecules such as steroids, certain drugs, and bilirubin (a degradation product of the oxygen-carrying protein hemoglobin) to improve water solubility. This process helps remove them from the body. Both D-glucuronic acid and L-iduronic acid are abundant in connective tissue carbohydrate components.

Amino Sugars   In amino sugars a hydroxyl group (most commonly on carbon 2) is replaced by an amino group (Figure 7.23). These compounds are common constituents of the complex carbohydrate molecules found attached to cellular proteins and lipids. The most common amino sugars of animal cells are D-glucosamine and D-galactosamine. Amino sugars are often acetylated. One such molecule is N-acetylglucosamine. N-acetyl-neuraminic acid (the most common form of sialic acid) is a condensation product of D-mannosamine and pyruvic acid, a 2-ketocarboxylic acid. Sialic acids are ketoses containing nine carbon atoms that may be amidated with acetic or glycolic acid (hydroxyacetic acid).

Deoxysugars   Monosaccharides in which an —H has replaced an —OH group are known as *deoxysugars.* Two important deoxysugars found in cells are L-fucose (formed from D-mannose by reduction reactions) and 2-deoxy-D-ribose (Figure 7.24). Fucose is often found among the carbohydrate components of glycoproteins, such as those

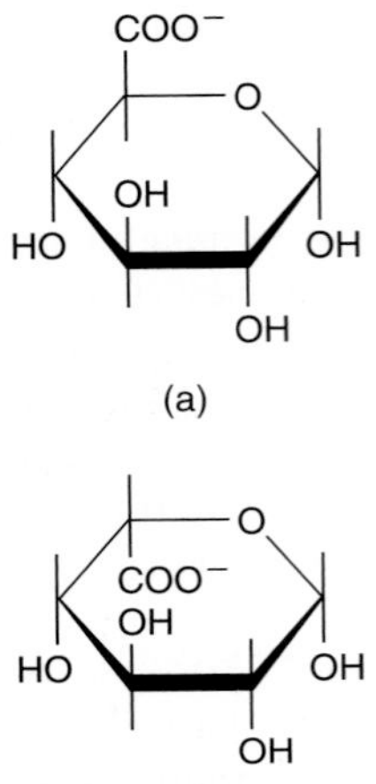

FIGURE 7.22

α-D-Glucuronate (a) and β-L-Iduronate (b).

FIGURE 7.23

Amino Sugars.

(a) α-D-Glucosamine, (b) α-D-galactosamine, (c) N-acetyl-α-D-glucosamine, and (d) N-acetylneuraminic acid (sialic acid).

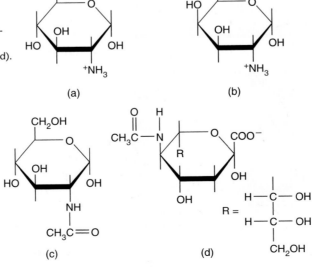

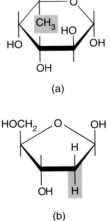

FIGURE 7.24

Deoxy Sugars.

(a) β-L-Fucose (6-deoxymannose) and (b) 2-deoxy-β-D-ribose. The carbon atoms that have —OH groups replaced by —H are highlighted

of the ABO blood group determinates on the surface of red blood cells. As was mentioned in Section 1.2, 2-deoxyribose is the pentose sugar component of DNA.

7.2 DISACCHARIDES

Disaccharides are glycosides composed of two monosaccharide units. Found in abundance in nature, disaccharides provide a significant source of calories in many human diets. Examples of important disaccharides include lactose and sucrose.

Digestion of disaccharides is mediated by enzymes synthesized by cells lining the small intestine. Deficiency of any of these enzymes causes unpleasant symptoms when the undigestible disaccharide sugar is ingested. Because carbohydrates are absorbed principally as monosaccharides, any undigested disaccharide molecules pass into the large intestine, where osmotic pressure draws water from the surrounding tissues (diarrhea). Colonic bacteria digest the disaccharides (fermentation), thus producing gas (bloating and cramps). The most commonly known deficiency is *lactose intolerance,* which may occur in most human adults except those with ancestors from northern Europe and/or certain African groups. Caused by lack of the enzyme lactase, lactose intolerance is treated by eliminating the sugar from the diet or (in some cases) by treating food with the enzyme lactase.

Lactose Lactose (milk sugar) is a disaccharide found in milk. It is composed of one molecule of galactose linked through the hydroxyl group on carbon 1 in a β-glycosidic linkage to the hydroxyl group of carbon 4 of a molecule of glucose (Figure 7.25). Because the anomeric carbon of galactose is in the β-configuration, the linkage between the two monosaccharides is designated as $\beta(1,4)$. The free anomeric carbon of the glucose residue can exist in either the α- or β-configuration. Because the glucose component contains a hemiacetal group, lactose is a reducing sugar.

Maltose Maltose, also known as malt sugar, is an intermediate product of starch hydrolysis and does not appear to exist freely in nature. Maltose is a disaccharide with an $\alpha(1,4)$ glycosidic linkage between two D-glucose molecules. In solution the free anomeric carbon undergoes mutarotation, which results in an equilibrium mixture of α- and β-maltoses (Figure 7.26).

FIGURE 7.25
α- and β-Lactose.

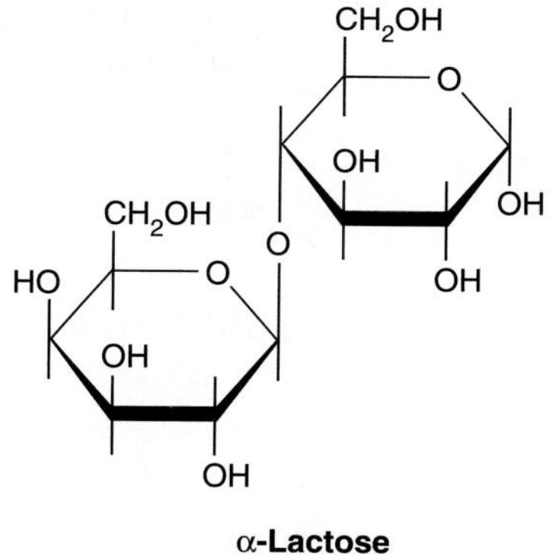

α-Lactose

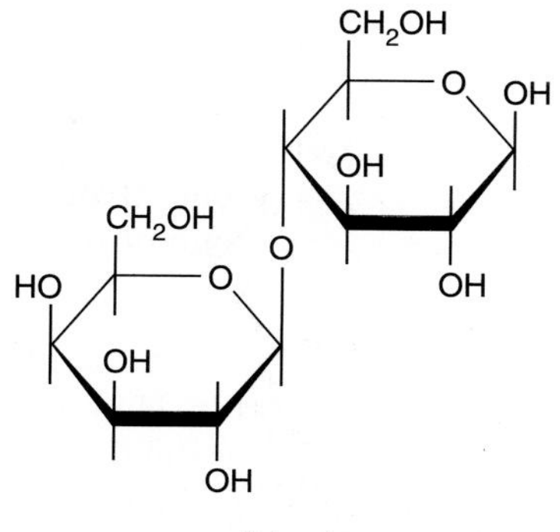

β-Lactose

FIGURE 7.26
α- and *β-*Maltose.

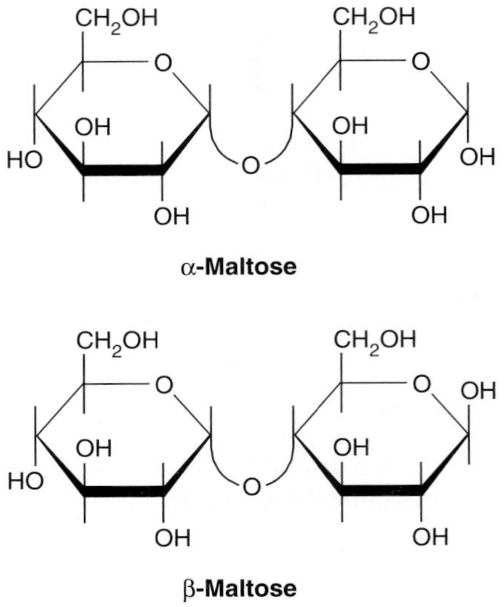

α-Maltose

β-Maltose

FIGURE 7.27
*β-*Cellobiose.

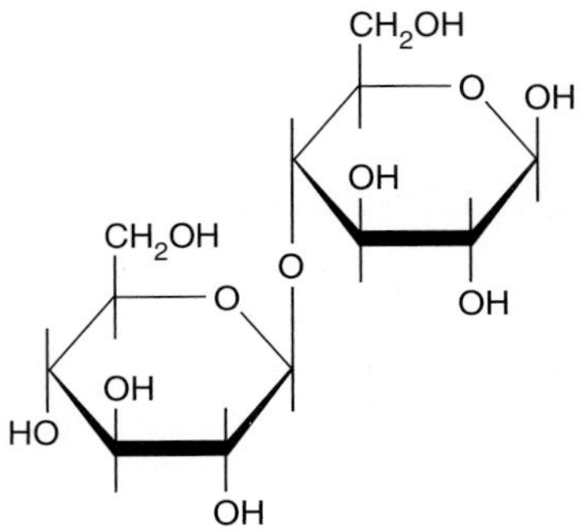

FIGURE 7.28
Sucrose.

The glucose and fructose residues are linked by an $\alpha,\beta(1,2)$ glycosidic bond.

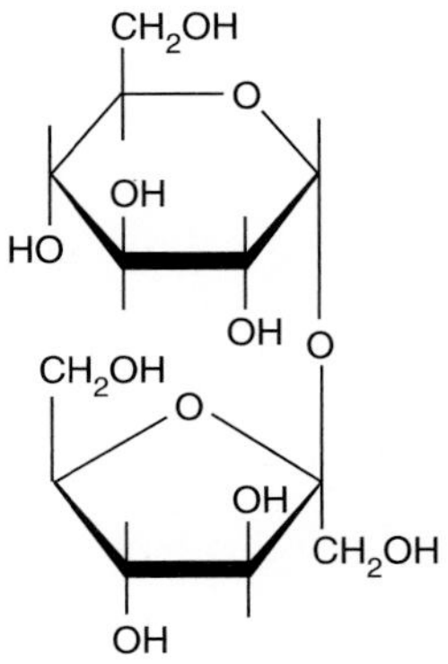

Cellobiose   Cellobiose, a degradation product of cellulose, contains two molecules of glucose linked by a $\beta(1,4)$ glycosidic bond (Figure 7.27). Like maltose, whose structure is identical except for the direction of the glycosidic bond, cellobiose does not freely exist in nature.

Sucrose   Sucrose (common table sugar: cane sugar or beet sugar) is produced in the leaves and stems of plants. It is a transportable energy source throughout the entire plant. Containing both α-glucose and β-fructose residues, sucrose differs from the previously described disaccharides in that the monosaccharides are linked through a glycosidic bond between both anomeric carbons (Figure 7.28). Because neither monosaccharide ring can revert to the open chain form, sucrose is a nonreducing sugar.

KEY *Concepts*

Disaccharides are glycosides composed of two monosaccharide units. Examples of disaccharides are maltose, lactose, cellobiose, and sucrose.

7.3 OLIGOSACCHARIDES

The term **oligosaccharide** is often used for carbohydrates that consist of two to ten monosaccharide units. These small polymers are most often found attached to polypeptides in glycoproteins (pp. 174–175) and some glycolipids (Chapter 9). Among the best-characterized oligosaccharide groups are those attached to membrane and secretory proteins found in the endoplasmic reticulum and Golgi complex of various cells. There are two broad classes of oligosaccharides: N-linked and O-linked. N-linked oligosaccharides are attached to polypeptides by an N-glycosidic bond with the side chain amide group of the amino acid asparagine. There are three major types of asparagine-linked oligosaccharides: high-mannose, hybrid, and complex (Figure 7.29). O-linked oligosaccharides are attached to polypeptides by the side chain hydroxyl group of the amino acid serine or threonine in polypeptide chains or the hydroxyl group of membrane lipids.

7.4 POLYSACCHARIDES

Polysaccharide molecules are used as storage forms of energy or as structural materials. They are composed of large numbers of monosaccharide units connected by glycosidic linkages. Most common polysaccharides are large molecules containing from hundreds to thousands of sugar units. These molecules may have a linear structure

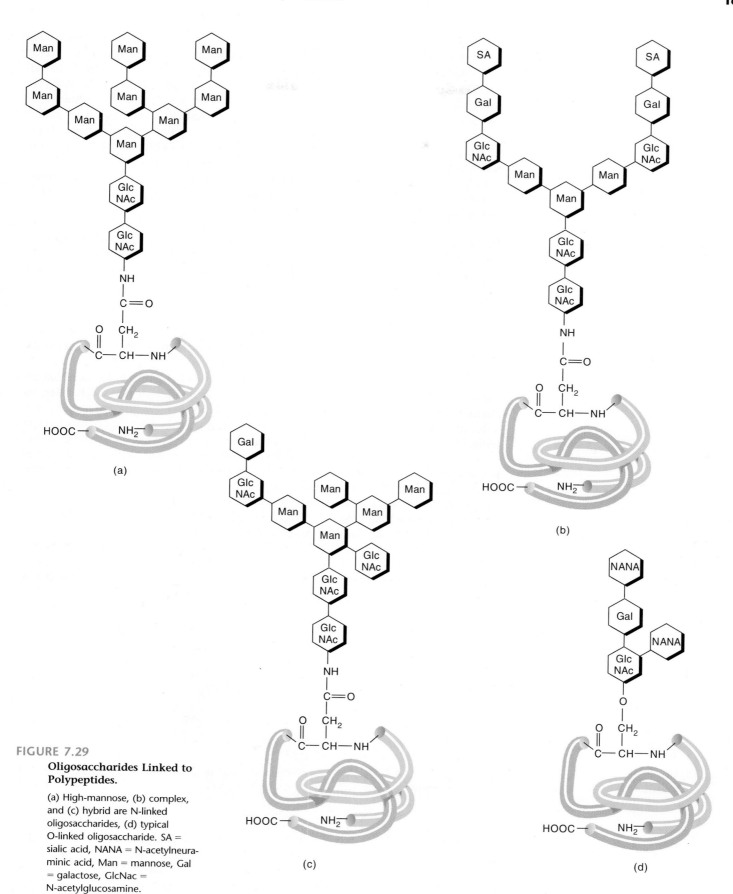

FIGURE 7.29

Oligosaccharides Linked to Polypeptides.

(a) High-mannose, (b) complex, and (c) hybrid are N-linked oligosaccharides, (d) typical O-linked oligosaccharide. SA = sialic acid, NANA = N-acetylneuraminic acid, Man = mannose, Gal = galactose, GlcNac = N-acetylglucosamine.

(a)

(b)

(c)

(d)

like that of cellulose or amylose, or they may have branched shapes like those found in glycogen and amylopectin. Unlike nucleic acids and proteins, which have specific molecular weights, the molecular weights of many polysaccharides have no fixed values. The size of such molecules is a reflection of the metabolic state of the cell producing them. For example, when blood sugar levels are high (e.g., after a meal), the liver synthesizes glycogen. Glycogen molecules in a well-fed animal may have molecular weights as high as 2×10^7. When blood sugar levels fall, the liver enzymes begin breaking down the glycogen molecules, releasing glucose into blood. If the animal continues to fast, the process continues until glycogen reserves are almost used up.

Polysaccharides may be divided into two classes: *homopolysaccharides,* which are composed of one type of monosaccharide, and *heteropolysaccharides,* which contain two or more types of monosaccharides.

Homopolysaccharides

The homopolysaccharides found in abundance in nature are starch, glycogen, and cellulose. All of these substances yield D-glucose when they are hydrolyzed. Starch and glycogen are glucose storage molecules in plants and animals, respectively. Cellulose is the primary structural component of plant cells.

Starch   Starch, the energy reservoir of plant cells, is a significant source of carbohydrate in the human diet. Much of the nutritional value of the world's major foodstuffs (e.g., potatoes, rice, corn, and wheat) comes from starch. Two polysaccharides occur together in starch: amylose and amylopectin.

Amylose is composed of long, unbranched chains of D-glucose residues that are linked with $\alpha(1,4)$ glycosidic bonds (Figure 7.30). A number of polysaccharides, including both types of starch, have one *reducing end* in which the ring can open to form a free aldehyde group with reducing properties. The internal anomeric carbons in these molecules are involved in acetal linkages and are not free to act as reducing agents.

Amylose molecules, which typically contain several thousand glucose residues, vary in molecular weight from 150,000 to 600,000. Because the linear amylose molecule forms long, tight helices, its compact shape is ideal for its storage function. The common iodine test for starch works because molecular iodine inserts itself into these helices. (The intense blue color of a positive test comes from electronic interactions between iodine molecules and the helically arranged glucose residues of the amylose.)

The other form of starch, **amylopectin,** is a branched polymer containing both $\alpha(1,4)$ and $\alpha(1,6)$ glycosidic linkages. The $\alpha(1,6)$ branch points may occur every 20–25 glucose residues and prevent helix formation (Figure 7.31a). The number of glucose units in amylopectin may vary from a few thousand to a million.

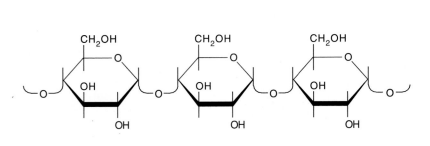

(a)

(b)

FIGURE 7.30

Amylose.

(a) The D-glucose residues of amylose are linked through $\alpha(1,4)$ glycosidic bonds. (b) The amylose polymer forms a left-handed helix.

Starch digestion begins in the mouth, where the salivary enzyme α-amylase initiates hydrolysis of the glycosidic linkages. Digestion continues in the small intestine, where pancreatic α-amylase randomly hydrolyses all the $\alpha(1,4)$ glycosidic bonds except those next to the branch points. The products of α-amylase are maltose, the trisaccharide maltotriose, and the α-limit dextrins (oligosaccharides that typically contain eight glucose units with one or more $\alpha(1,6)$ branch points). Several enzymes secreted by cells that line the small intestine convert these intermediate products into glucose. Glucose molecules are then absorbed into intestinal cells. After passage into the bloodstream, they are transported to the liver and then to the rest of the body.

Glycogen   **Glycogen** is the carbohydrate storage molecule in vertebrates. It is found in greatest abundance in liver and muscle cells. (Glycogen may make up as much as 8–10% of the wet weight of liver cells and 2–3% of that of muscle cells.) Glycogen (Figure 7.31b) is similar in structure to amylopectin except that it has more branch points, possibly at every fourth glucose residue in the core of the molecule. In the outer regions of glycogen molecules, branch points are not so close together (approximately every 8–12 residues). Because the molecule is more compact than other

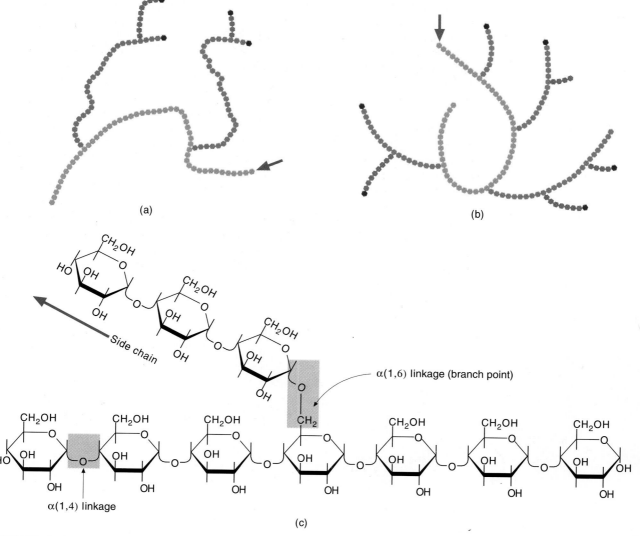

(a)

(b)

$\alpha(1,6)$ linkage (branch point)

Side chain

$\alpha(1,4)$ linkage

(c)

FIGURE 7.31

Amylopectin (a) and Glycogen (b).

Each hexagon represents a glucose molecule. Notice that each molecule has only one reducing end (arrow). (c) Detail from (a) or (b).

polysaccharides, it takes up little space, an important consideration in mobile animal bodies. The many nonreducing ends in glycogen molecules allow for stored glucose to be rapidly mobilized when the animal demands much energy.

Cellulose   Cellulose is a polymer composed of D-glucopyranose residues linked by $\beta(1,4)$ glycosidic bonds (Figure 7.32). It is the most important structural polysaccharide of plants. Because cellulose comprises about one-third of plant biomass, it is the most abundant organic substance on earth. Approximately 100 trillion kg of cellulose is produced each year.

Pairs of unbranched cellulose molecules, which may contain as many as 12,000 glucose units each, are held together by hydrogen bonding to form sheetlike strips called *microfibrils* (Figure 7.33). Each bundle of microfibrils contains approximately 40 of these pairs. These structures are found in both plant primary and secondary cell walls, where they provide a structural framework that both protects and supports cells.

The ability to digest cellulose is found only in microorganisms that possess the enzyme cellulase. Certain animal species (e.g., termites and cows) use such organisms in their digestive tracts to digest cellulose. The breakdown of the cellulose makes glucose available to both the microorganisms and their host. Although many animals cannot digest cellulose-containing plant materials, these substances play a vital role in nutrition. Cellulose is one of several plant products that make up the dietary fiber that is now believed to be important for good health.

Because of its structural properties, cellulose has enormous economic importance. Products such as wood, paper, and textiles (e.g., cotton, linen, and ramie) owe many of their unique characteristics to their cellulose content.

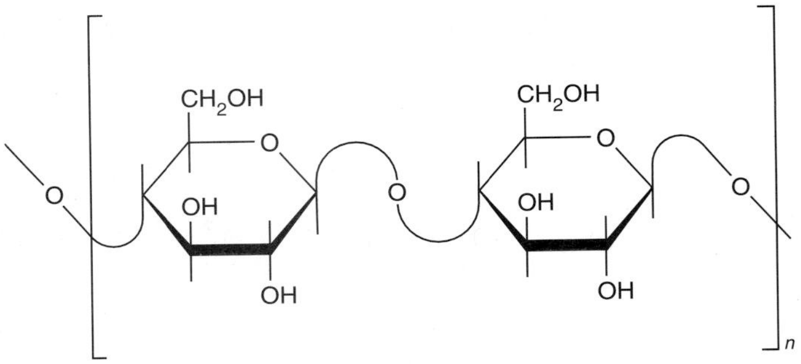

FIGURE 7.32
Cellulose.

FIGURE 7.33

Cellulose Microfibrils.

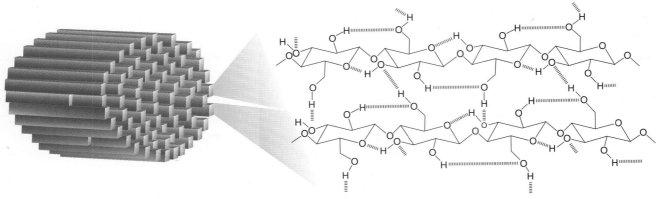

Intermolecular hydrogen bonds between adjacent cellulose molecules are largely responsible for the great strength of cellulose.

Weaving plant fibers is a relatively recent development in human history, apparently for technological reasons. Before our ancestors learned to spin and weave, they discovered which plants contained useful fibers and how these fibers could be extracted. According to archeological evidence, one of the earliest plants used for fiber was flax (*Linum usitatissimum*). It was woven into linen at least 8000 years ago. The cultivation and uses of flax were beautifully illustrated on the walls of Egyptian tombs (Figure 7B).

The chemical properties of cellulose contribute to the quali-ties that make linen such an attractive fabric (i.e., its smoothness, strength, and water absorbency). Bast fibers, found in the phloem (a component of the plant's vascular system) are used to make linen. They contain thicker cell walls than most of the other plant tissues. In a chemical process called *retting*, bast fibers are harvested after the rest of the plant is decomposed by bacteria. Because of the large amount of cellulose within their cell walls, bast fibers can withstand the numerous corrosive chemical reactions of decomposition.

FIGURE 7B

Harvesting Flax.

From a wall painting in the tomb of Sennedjem, Deir-el-Medina, 20th dynasty, about 1150 B.C.

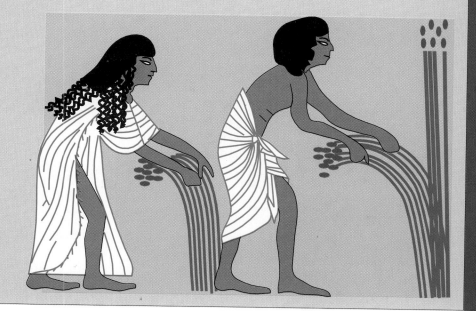

QUESTION 7.5

It has been estimated that two high-energy phosphate bonds must be expended to incorporate one glucose molecule into glycogen. Why is glucose stored in muscle and liver in the form of glycogen, not as individual glucose molecules? In other words, why is it advantageous for a cell to expend metabolic energy to polymerize glucose molecules? (Hint: Besides the reasons given above, refer to Chapter 3 for another problem that glucose polymerization solves.)

Heteropolysaccharides

Heteropolysaccharides are high-molecular-weight carbohydrate polymers that contain more than one kind of monosaccharide. Because many of the sugar residues are amino derivatives, these substances are often referred to as **glycosaminoglycans** (GAGs). GAGs, the principal components of the proteoglycans (See Special Interest Box 7.3), are classified according to their sugar residues, the linkages between these residues, and the presence and location of sulfate groups. Five classes are distinguished: *hyaluronic acid, chondroitin sulfate, dermatan sulfate, heparin* and *heparan sulfate*, and *keratan sulfate*.

Glycosaminoglycans are linear polymers with disaccharide repeating units (Table 7.1). These units contain a hexuronic acid (a uronic acid containing six carbon atoms), except for keratan sulfate, which contains galactose. Usually an N-acetylhexosamine sulfate is also present, except in hyaluronic acid, which contains N-acetylglucosamine. Many disaccharide units contain both carboxyl and sulfate functional groups. All GAGs therefore have many negative charges at physiological pH. The charge repulsion keeps GAGs separated from each other. Additionally, the relatively inflexible polysaccharide chains are strongly hydrophilic. GAGs occupy a huge volume relative to their mass because they attract large volumes of water. For example, hydrated hyaluronic acid may occupy a volume 1000 times greater than its dry state.

TABLE 7.1

Disaccharide Repeating Units in Selected Glycosaminoglycans

Name	Repeating Unit	Molecular Weight	Comments
Chondroitin sulfate		5000–50,000	May also have sulfate on carbon 6. Important component of cartilage.
	(1,4)-O-β-D-Glucopyranosyluronic acid-(1,3)-2-acetamido-2-deoxy-6-O-sulfo-β-D-galactopyranose		
Dermatan sulfate		15,000–40,000	Varying amounts of D-glucuronic acid may be present. Concentration increases during aging process.
	(1,4)-O-α-L-Idopyranosyluronic acid-(1,3)-2-acetamido-2-deoxy-4-O-sulfo-β-D-galactopyranose		
Heparin		6000–25,000	Anticoagulant activity. Found in mast cells. Also contains D-glucuronic acid.
	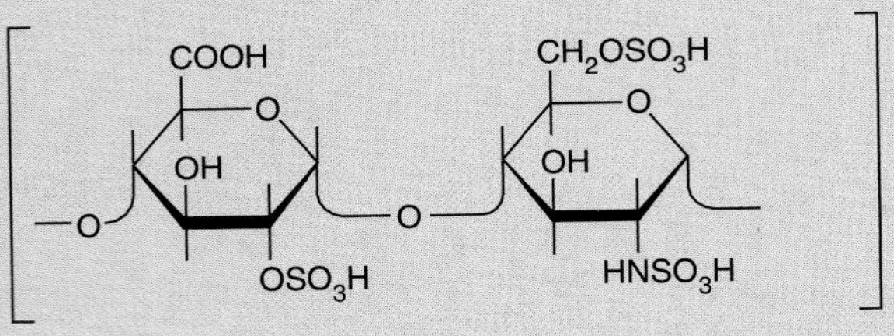 (1,4)-O-α-D-Glucopyranosyluronic acid-2-sulfo-(1,4)-2-sulfamido-2-deoxy-6-O-sulfo-α-D-glucopyranose		
Keratan sulfate		4000–19,000	Minor constituent of proteoglycans. Found in cornea, cartilage, and intervertebral disks.
	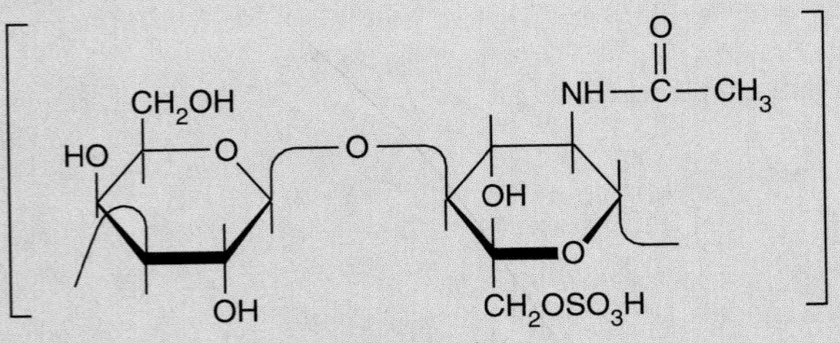 (1,3)-O-β-D-Galactopyranose-(1,4)-2-acetamido-2-deoxy-6-O-sulfo-β-D-glucopyranose		
Hyaluronic acid	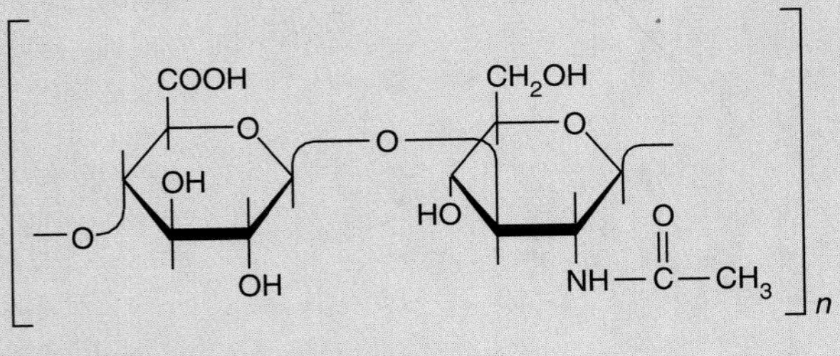	4000	Most abundant GAG in the vitreous humor of the eye and the synovial fluid of joints.
	(1,4)-O-β-D-Glucopyranosyluronic acid-(1,3)-2-acetamido-2-deoxy-β-D-glucopyranose		

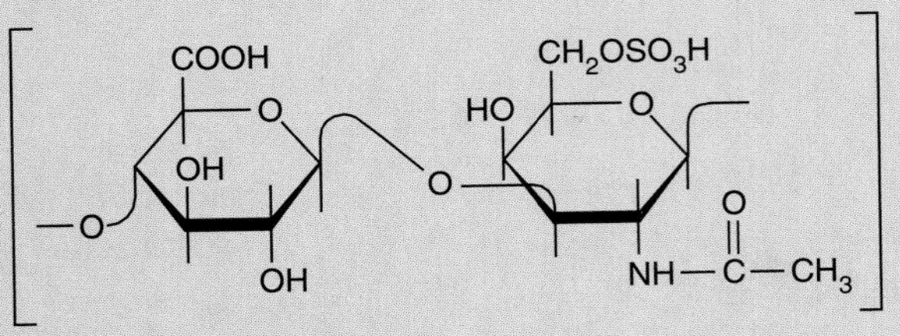

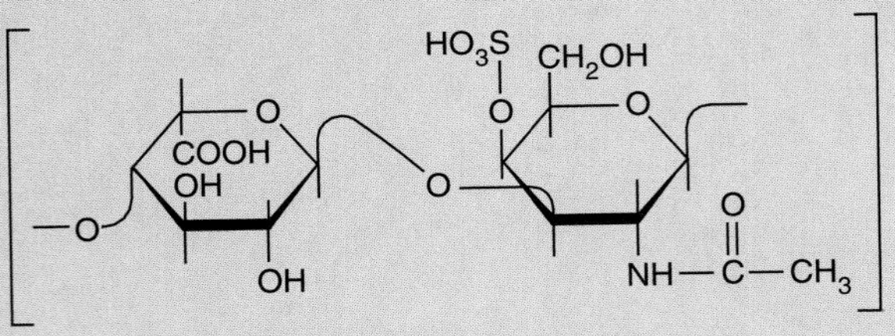

The compounds that result from the covalent linkages of carbohydrate molecules to both proteins and lipids are collectively known as the **glycoconjugates.** These substances have profound effects on the function of individual cells, as well as the cell-cell interactions of multicellular organisms. There are two classes of carbohydrate-protein conjugate: proteoglycans and glycoproteins. Although both molecular types contain carbohydrate and protein, their structures and functions appear, in general, to be substantially different. The *glycolipids,* which are oligosaccharide-containing lipid molecules, are found predominantly on the outer surface of plasma membranes. A discussion of their structure is deferred until Chapter 9.

Proteoglycans

Proteoglycans are distinguished from the more common glycoproteins by their extremely high carbohydrate content, which may constitute as much as 95% of the dry weight of such molecules. These molecules are found predominantly in the extracellular matrix (intercellular material) of tissues. All proteoglycans contain GAG chains. The GAG chains are linked to protein molecules (known as *core proteins*) by N- and O-glycosidic linkages. The diversity of proteoglycans results from both the number of different core proteins and the large variety of classes and lengths of the carbohydrate chains (Figure 7C).

Because proteoglycans contain large numbers of GAGs, which are polyanions, large volumes of water and cations are trapped within their structure. Proteoglycan molecules occupy space that is thousands of times as large as a densely packed molecule of the same mass. Proteoglycans contribute support and elasticity to tissues. Consider, for example, the strength, flexibility, and resilience of cartilage. The structural diversity of proteoglycans allows them to play a variety of structural and

FIGURE 7C

Proteoglycan.

Proteoglycan aggregates are typically found in the extracellular matrix of connective tissue. The noncovalent attachment of each proteoglycan to hyaluronic acid is mediated by two linker proteins (not shown).

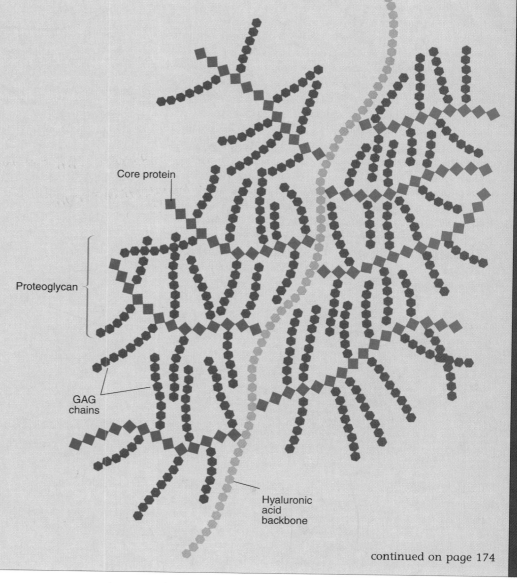

Core protein

Proteoglycan

GAG chains

Hyaluronic acid backbone

continued on page 174

functional roles in living organisms. Together with matrix proteins such as collagen, fibronectin, and laminin, they form an organized meshwork that provides strength and support to multicellular tissues. Proteoglycans are also present at the surface of cells, where they are directly bound to the plasma membrane. Although the function of these latter molecules is not yet clear, they may play an important role in membrane structure and cell-cell interactions.

A number of genetic diseases associated with proteoglycan metabolism, known as *mucopolysaccharidoses,* have been identified. Because proteoglycans are constantly being synthesized and degraded, their excessive accumulation (due to missing or defective lysosomal enzymes) has very serious consequences. For example, in *Hurler's syndrome,* an autosomal recessive disorder (a disease in which one copy of the defective gene is inherited from each parent), deficiency of a specific enzyme causes dermatan sulfate to accumulate. Symptoms include mental retardation, skeletal deformity, and death in early childhood.

Glycoproteins

Glycoproteins are commonly defined as proteins that are covalently linked to carbohydrate through N- or O-linkages. The carbohydrate composition of glycoprotein varies from 1% to more than 85% of total weight. The carbohydrates found include monosaccharides and disaccharides such as those attached to the structural protein collagen and the branched oligosaccharides on plasma glycoproteins. Although the glycoproteins are sometimes considered to include the proteoglycans, structural reasons seem to be sufficient to examine them separately, such as the relative absence in glycoproteins of uronic acids, sulfate groups, and the disaccharide repeating units that are typical of proteoglycans.

The carbohydrate groups of glycoproteins are linked to the polypeptide by either (1) an N-glycosidic linkage between N-acetylglucosamine (GlcNAc) and the amino acid asparagine (Asn) or (2) an O-glycosidic linkage between N-acetylgalactosamine (GalNAc) and the hydroxyl group of the amino acids serine (Ser) or threonine (Thr). The former glycoprotein class is sometimes referred to as *asparagine-linked;* the latter is often called *mucin-type.*

Asparagine-Linked Carbohydrate

As mentioned, three structural forms of asparagine-linked oligosaccharide occur in glycoproteins: high-mannose, complex, and hybrid. *High-mannose type* is composed of GlcNAc and mannose. *Complex-type* may contain fucose, galactose, and sialic acid in addition to GlcNAc and mannose. *Hybrid-type* oligosaccharides contain features of both complex and high-mannose-type species (Figure 7.29). Despite these differences, the core of all N-linked oligosaccharides is the same. This core, which is constructed on a membrane-bound lipid molecule, is covalently linked to asparagine during ongoing protein synthesis (Chapter 18). Several additional reactions, in the lumen of the endoplasmic reticulum and the Golgi complex, form the final N-linked oligosaccharide structures.

Mucin-Type Carbohydrate

While all N-linked oligosaccharides are bound to protein via GlcNAc-Asn, the linking groups of O-glycosidic oligosaccharides are of several types. The most common of these is GalNAc-Ser (or GalNAc-Thr). Mucin-type carbohydrate units vary considerably in size and structure, from disaccharides such as Gal-1, 3-GalNAc, found in the antifreeze glycoprotein of antarctic fish (Figure 7D), to the complex oligosaccharides of blood groups such as those of the ABO system.

Glycoprotein Functions

Glycoproteins are a diverse group of molecules that are ubiquitous constituents of most living organisms (Table 7.2). They occur in cells, in both soluble and membrane-bound form, as well as in extracellular fluids. Vertebrate animals are particularly rich in glycoproteins. Examples of such substances include the metal-transport proteins transferrin and ceruloplasmin, the blood-clotting factors, and many of the components of complement (proteins involved in cell destruction during immune reactions). A number of hormones (chemicals produced by certain cells that are transported by blood to other cells, where they exert regulatory effects) are glycoproteins. Consider, for example, follicle-stimulating hormone (FSH), produced by the anterior pituitary gland. FSH stimulates the development of both eggs and sperm. Additionally, many enzymes are glycoproteins. Ribonuclease (RNase), the enzyme that degrades ribonucleic acid, is a well-researched example. Other glycoproteins are in-

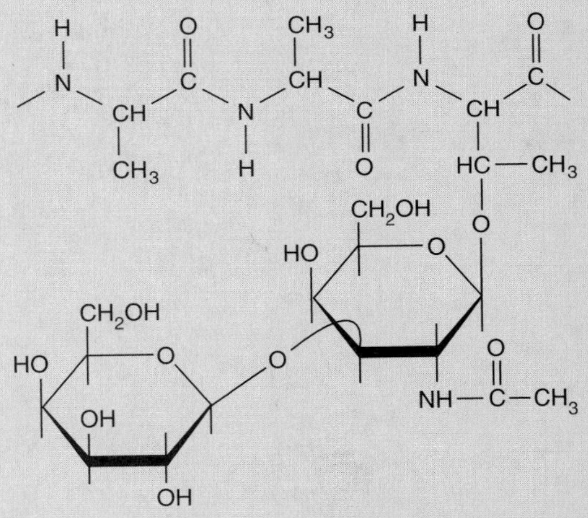

FIGURE 7D

Antifreeze Glycoprotein.

This segment is a recurring tripeptide unit.

TABLE 7.2

Glycoproteins

Type	Example	Source	Molecular Weight	Percent Carbohydrate
Enzyme	Ribonuclease B	Bovine	14,700	8
Immunoglobulin	IgA	Human	160,000	7
	IgM	Human	950,000	10
Hormones	Chorionic gonadotropin	Human placenta	38,000	31
	FSH	Human	34,000	20
Membrane protein	Glycophorin	Human RBC	31,000	60
Lectins (carbohydrate-binding proteins)	Lima bean lectin	Lima bean	124,000	0
	Potato lectin	Potato	50,000	50

tegral membrane proteins (see Chapter 9). Of these, Na$^+$-K$^+$-ATPase (an ion pump found in the plasma membrane of animal cells) and the major histocompatibility antigens (cell surface markers used to cross-match organ donors and recipients) are especially interesting examples.

Although many glycoproteins have been studied, the role of carbohydrate is still not clearly understood. Despite challenging technical problems, some progress has been made in discerning how the carbohydrate component contributes to biological activity. Recent research has focused on how carbohydrate stabilizes protein molecules and functions in recognition processes in multicellular organisms.

The presence of carbohydrate on protein molecules protects them from denaturation. For example, bovine RNase A is more susceptible to heat denaturation than its glycosylated counterpart RNase B. Several other studies have shown that sugar-rich glycoproteins are relatively resistant to proteolysis (splitting of polypeptides by enzyme-catalyzed hydrolytic reactions). Because the carbohydrate is on the molecule's surface, it may shield the polypeptide chain from proteolytic enzymes.

The carbohydrates in glycoproteins seem to affect biological function. In some glycoproteins this contribution is more easily discerned than in others. For example, a large content of sialic acid residues is responsible for the high viscosity of salivary mucins (the lubricating glycoproteins of saliva). Another interesting example is the antifreeze glycoproteins of antarctic fish. Apparently, their disaccharide residues form hydrogen bonds with water molecules. This process retards the growth of ice crystals.

Glycoproteins are now known to be important in complex recognition phenomena such as cell-molecule, cell-virus, and cell-cell interactions. Prime examples of glycoprotein involvement in cell-molecule interactions include the insulin receptor, whose binding to insulin facilitates the transport of glucose into numerous cell types. It does so, in part, by recruiting glucose transporters to the plasma membrane. In addition, the glucose transporter that is directly responsible for transporting the sugar into cells is also a glycoprotein. The interaction between gp120, the target cell binding glycoprotein of HIV (the AIDS virus), and host cells is a fascinating example of cell-virus interaction. The attachment of gp120 to the CD4 receptor found on the surface of several human cell types is now considered to be an early step in the infective process. Removal of carbohydrate from purified gp120 significantly reduces the binding of the viral protein to the CD4 receptor. Paradoxically, the oligosaccharides attached to the glycoprotein on the surface of vesicular stomatitis virus are not required for viral infectivity.

Cell surface glycoproteins, components of the *glycocalyx* (also known as the *cell coat*), are now recognized as players in cellular adhesion. This process is a critical event in the cell-cell interactions of growth and differentiation (Figure 7E). The best-characterized of these substances are called cell adhesion molecules (CAMs). CAMs are now believed to be involved in the embryonic development of the mouse nervous system. Sialic residues in the N-linked oligosaccharides of several CAMs have been shown to be important in this phenomenon.

Recent improvements in technology have led to an increased appreciation of the importance of carbohydrate in glycoproteins. Consequently, there is currently a heightened interest in studying cellular glycosylation patterns. Carbohydrate structure is now used to investigate normal processes such as nerve development (mentioned above) and certain disease processes. For example, changes in the galactose content of the antibody class IgG have recently been shown to be directly related to the severity (i.e., the degree of inflammation) of juvenile arthritis. Additionally, recent evidence indicates that changes in glycosylation patterns accompany changes in the behavior of cancer cells. This knowledge is currently making tumor detection and the metastatic process (the spread of cancerous cells from a tumor to other body parts), more accessible to investigation.

continued on page 176

FIGURE 7E

The Glycocalyx.

(a) Electron micrograph of the surface of a lymphocyte stained to reveal the glycocalyx (cell coat). (b) The glycocalyx. (A GPI (glycosylphosphatidylinositol) anchor is a specialized structure that attaches several diverse types of oligosaccharide to the plasma membrane of some eukaryotic cells.)

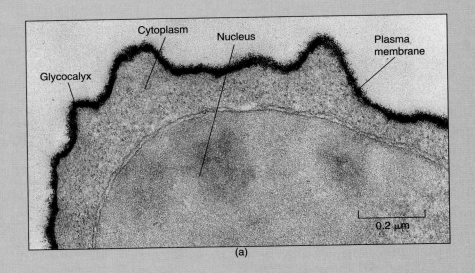

(a)

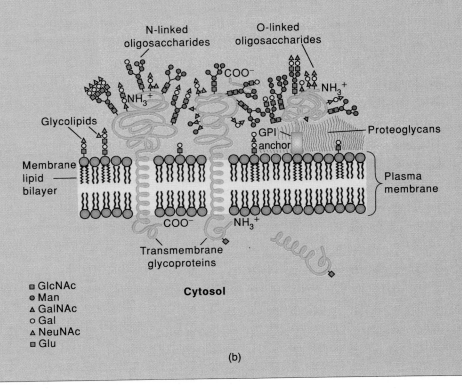

- ▫ GlcNAc
- ● Man
- ▲ GalNAc
- ○ Gal
- ▲ NeuNAc
- ▫ Glu

(b)

KEY WORDS

acetal, *162*

aldaric acid, *157*

alditol, *159*

aldonic acid, *157*

aldose, *152*

amylopectin, *168*

amylose, *168*

anomer, *154*

diastereomers, *153*

disaccharide, *162*

enediol, *160*

epimerization, *160*

epimer, *153*

glycoconjugate, *173*

glycogen, *169*

glycosaminoglycan, *171*

glycoside, *162*

glycosidic linkage, *162*

hemiacetal, *154*

lactone, *159*

monosaccharide, *152*

oligosaccharide, *166*

optical isomer, *153*

polysaccharide, *162*

proteoglycan, *173*

reducing sugars, *159*

uronic acid, *157*

SUMMARY

Carbohydrates, the most abundant organic molecules in nature, are classified as monosaccharides, disaccharides, oligosaccharides, and polysaccharides according to the number of simple sugar units they contain. Monosaccharides with an aldehyde functional group are called aldoses; those with a ketone group are known as ketoses. Simple sugars belong to either the D or L family, according to whether the configuration of the asymmetric carbon farthest from the aldehyde or ketone group resembles the D- or L-isomer of glyceraldehyde. The D family contains most biologically important sugars.

Sugars containing five or six carbons exist in cyclic forms that result from reactions between hydroxyl groups with either aldehyde or ketone groups. In both five-membered rings (furanoses) and six-membered rings (pyranoses), the hydroxyl group attached to the anomeric carbon lies either below (α) or above (β) the plane of the ring. The spontaneous interconversion between α- and β-forms is called mutarotation.

Simple sugars undergo a variety of chemical reactions. Derivatives of these molecules, such as uronic acids, amino sugars, deoxy sugars, and phosphorylated sugars, have important roles in cellular metabolism.

Glycosidic (acetal) bonds form between the hydroxyl group of one monosaccharide and the anomeric carbon of another. Disaccharides are carbohydrates composed of two monosaccharides. Oligosaccharides, carbohydrates that typically contain as many as ten monosaccharide units, are often attached to proteins and lipids. Poly-saccharide molecules, which are composed of large numbers of monosaccharide units, may have a linear structure like cellulose and amylose or a branched structure like glycogen and amylopectin. Polysaccharides may consist of only one sugar type (homopolysaccharides) or multiple types (heteropolysaccharides).

The three most common homopolysaccharides found in nature (starch, glycogen, and cellulose) all yield D-glucose when hydrolyzed. Cellulose is a plant structural material; starch and glycogen are storage forms of glucose in plant and animal cells, respectively.

Glycoconjugates form from the covalent linkages of carbohydrate molecules to either protein or lipids. There are two classes of carbohydrate-protein conjugate: proteoglycans and glycoproteins. The enormous heterogeneity of proteoglycans, which are found predominantly in the extracellular matrix of tissues, allows them to play diverse, but as yet poorly understood, roles in living organisms. Glycoproteins occur in cells, in both soluble and membrane-bound form, as well as in extracellular fluids. Glycoproteins are now known to be important in multicellular organisms.

SUGGESTED READINGS

Drickamer, K., and Carrier, J., Carbohydrates and Glycoconjugates: Upwardly Mobile Sugars Gain Status as Information-Bearing Molecules, *Curr. Opin. Struc. Biol.*, 2(5):653–654, 1992.

Dwek, R.A., Glycobiology: More Functions for Oligosaccharides, *Science*, 269:1234–1235, 1995.

Rademacher, T.W., Parekh, R.B., and Dwek, R.A., Glycobiology, *Ann. Rev. Biochem.*, 57:785–838, 1988.

Ruoslahti, E. Structure and Biology of Proteoglycans, *Ann. Rev. Cell Biol.*, 4:229–255, 1988.

Sharon, N., Carbohydrates, *Sci. Amer.*, 243(5):90–116, 1980.

Sharon, N., and Halina, L., Carbohydrates in Cell Recognition, *Sci. Amer.*, 268(1):82–89, 1993.

Stoop, J.M.H., Williamson, J.D., and Pharr, D.M., Mannitol Metabolism in Plants: A Method for Coping with Stress, *Trends Plant Sci.*, 1(5):139–144, 1996.

Yeh, O., and Feeney, R.E., Antifreeze Proteins: Structures and Mechanisms of Function, *Chem. Rev.*, 96:601–617, 1996.

QUESTIONS

1. Write Haworth structures for the following compounds:
   a. α-D-glucopyranose and β-D-glucofuranose
   b. sucrose
   c. D-lactose
   d. sialic acid
   e. pyranose form of D-mannose
   f. chondroitin sulfate, repeating unit

2. Give an example of each of the following:
   a. epimer
   b. glycosidic linkage
   c. reducing sugar
   d. monosaccharide
   e. anomer
   f. diastereomer

3. What structural relationship is indicated by the term D-sugar? Why are ($+$) glucose and ($-$) fructose both classified as D-sugars?

4. Name an example of each of the following classes of compounds:
   a. glycoprotein
   b. proteoglycan
   c. disaccharide
   d. glycosaminoglycan

5. What is the difference between a heteropolysaccharide and a homopolysaccharide? Give examples.

6. Convert each of the following Fischer representations to Haworth formulas:

a.
$$\begin{array}{c} O \\ \| \\ CH \\ | \\ H-C-OH \\ | \\ HO-C-H \\ | \\ H-C-OH \\ | \\ CH_2OH \end{array}$$

b.
$$\begin{array}{c} O \\ \| \\ CH \\ | \\ HO-C-H \\ | \\ HO-C-H \\ | \\ H-C-OH \\ | \\ H-C-OH \\ | \\ CH_2OH \end{array}$$

c.
$$\begin{array}{c} O \\ \| \\ CH \\ | \\ H-C-OH \\ | \\ HO-C-H \\ | \\ HO-C-H \\ | \\ H-C-OH \\ | \\ CH_2OH \end{array}$$

7. Which of the following carbohydrates are reducing and which nonreducing?
   a. starch
   b. cellulose
   c. fructose
   d. sucrose
   e. ribose

8. What structural differences characterize starch, cellulose, and glycogen?

9. β-Galactosidase is an enzyme that hydrolyses only β-1,4 linkages of lactose. An unknown trisaccharide is converted by β-galactosidase into maltose and galactose. Draw the structure of the trisaccharide.

10. Steroids are large, complex, lipid-soluble molecules that are very insoluble in water. Reaction with glucuronic acid makes a steroid much more water-soluble and enables transport through the blood. What structural feature of the glucuronic acid increases the solubility?

11. Many bacteria are surrounded by a proteoglycan coat. Use your knowledge of the properties of this substance to suggest a function for such a coat.

12. What shape do carbohydrate chains linked with $\alpha(1,4)$ glycosidic bonds generally have?

13. Determine the number of possible stereoisomers for the following compounds:

Ribulose   Sedoheptulose

14. Raffinose is the most abundant trisaccharide found in nature.

   a. Name the three monosaccharide units of raffinose.
   b. There are two glycosidic linkages. Are they α or β?
   c. Is raffinose a reducing or nonreducing sugar?
   d. Is raffinose capable of mutarotation?

15. Give at least one function of each of the following:
   a. glycogen
   b. glycosaminoglycans
   c. glycoconjugates
   d. proteoglyans
   e. glycoproteins
   f. polysaccharides

16. Carbohydrates are frequently phosphorylated before being attached to another molecule. Suggest a reason for this phenomenon.

17. The polymer chains of glycosaminoglycans are widely spread apart and bind large amounts of water.
   a. What two functional groups of the polymer make this binding of water possible?
   b. What type of bonding is involved?

18. Alginic acid, isolated from seaweed and used as a thickening agent for ice cream and other foods, is a polymer of D-mannuronic acid with $\beta(1,4)$ glycosidic linkages.
   a. Draw the structure of alginic acid.
   b. Why does this substance act as a thickening agent?

D-Mannuronic acid

19. What is the maximum number of stereoisomers for mannuronic acid (Question 18)?

20. In glycoproteins, what are the three amino acids to which the carbohydrate groups are most frequently linked?

21. Chondroitin sulfate chains have been likened to a large fishnet, passing small molecules through their matrix but excluding large ones. Use the structure of chondroitin sulfate and proteoglycans to explain this analogy.

22. The polysaccharide chitin is the chief component of the shells of arthropods (e.g., lobsters and grasshoppers) and of mollusks (e.g., oysters and snails). Chitin can be obtained from these sources by soaking the shells in cold dilute hydrochloric acid to dissolve the calcium carbonate. The threadlike substance formed is composed of linear long-chain molecules. Hydrolysis with boiling acid gives D-glucosamine and acetic acid in equimolar amounts. Milder enzymatic hydrolysis gives N-acetyl-D-glucosamine as the sole product. Chitin's linkages are identical to those of cellulose. What is the structure of chitin?

23. Define the term *reducing sugar*. What structural feature does a reducing sugar have?

24. Compare the structures of proteoglycans and glycoproteins. How are structural differences related to their functions?

25. What role is carbohydrate thought to play in maintaining glycoprotein stability?

26. How does the structure of cellulose differ from starch and glycogen?

27. Determine which of the following sugar pairs are enantiomers, diastereomers, epimers or an aldose-ketose pair.
    a. D-erythrose and D-threose
    b. D-glucose and D-mannose
    c. D-ribose and L-ribose
    d. D-allose and D-galactose
    e. D-glyceraldehyde and dihydroxyacetone

28. Proteoglycan aggregates in tissues form hydrated, viscous gels. Can you think of any obvious mechanical reason why their capacity to form gels is important to cell function? (Hint: Liquid water is virtually incompressible.)

CARBOHYDRATE METABOLISM

Chapter Eight

Products of Fermentation. *Humans use certain microorganisms to metabolize sugar in the absence of oxygen to produce cheese, wine, and bread.*

OBJECTIVES

After you have studied this chapter, you should be able to answer these questions:

1. *What are the two major types of metabolic pathways in living organisms?*

2. *What biochemical reactions are responsible for the breakdown of food molecules in energy production?*

3. *How are metabolic processes regulated so that the energy and biosynthetic requirements of organisms are consistently met?*

4. *How is glycogen alternately synthesized and degraded to provide animals a consistent supply of glucose?*

5. *How do cells extract energy from glucose?*

6. *What is the difference between aerobic respiration and fermentation?*

7. *How do glycolysis and gluconeogenesis differ? How are they similar?*

8. *What regulatory mechanism ensures that glycolysis and gluconeogenesis do not occur simultaneously to any great extent?*

Carbohydrates play several crucial roles in the metabolic processes of living organisms. As was described previously, they serve as energy sources and as structural elements in living cells. This chapter focuses on the role of carbohydrates in energy production. Because the monosaccharide glucose is a prominent energy source in almost all living cells, a major emphasis is placed on its synthesis, degradation, and storage. The use of other sugars is also discussed.

Living cells are in a state of ceaseless activity. As described, to maintain its "life" each cell depends on highly coordinated and complex biochemical reactions. In this chapter, several major reaction pathways of the carbohydrate metabolism of animals are discussed. During **glycolysis,** an ancient pathway found in almost all organisms, glucose is converted to two molecules of pyruvate and a small amount of energy. Glycogen, a storage form of glucose in vertebrates, is synthesized by **glycogenesis** when glucose levels are high and degraded by **glycogenolysis** when glucose is in short supply. Glucose can also be synthesized from noncarbohydrate precursors by reactions referred to as **gluconeogenesis.** In later chapters, other related pathways are discussed. *Photosynthesis,* a process in which light energy is captured to drive carbohydrate synthesis, is described in Chapter 12. In Chapter 11 the glyoxylate cycle and the pentose phosphate pathway are considered. In the *glyoxylate cycle* some organisms (primarily plants) manufacture carbohydrate from fatty acids. The *pentose phosphate pathway* enables cells to produce ribose-5-phosphate (the sugar found in nucleic acids) and NADPH, an important cellular reducing agent.

The reactions of carbohydrate metabolism are obviously of crucial importance to the life of living organisms. These processes must be viewed in a larger context. The chapter begins with an overview of the metabolism of cells and multicellular organisms.

Because glucose is the major fuel of most organisms, its synthesis and use are the focus of any discussion of carbohydrate metabolism. In vertebrates, glucose is transported throughout the body in the blood. If cellular energy reserves are low, glucose

FIGURE 8.1

Major Pathways in Carbohydrate Metabolism.

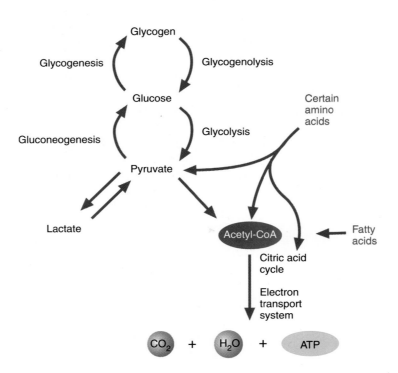

is degraded by the glycolytic pathway. Glucose molecules not required for immediate energy production are stored as glycogen in liver and muscle. Depending on a cell's metabolic requirements, glucose can also be used to synthesize, for example, other monosaccharides, fatty acids, and certain amino acids. Figure 8.1 summarizes the major pathways of carbohydrate metabolism in animals.

8.1 OVERVIEW OF METABOLISM

Metabolism, the sum of all the enzyme-catalyzed reactions in a living organism, is a dynamic, coordinated activity. Many of these reactions are organized into pathways. Each biochemical pathway consists of several reactions that occur sequentially; that is, the product of one reaction is the substrate for the one that follows. There are two major types of biochemical pathways: anabolic and catabolic. In **anabolic** or biosynthetic pathways, large complex molecules are synthesized from smaller precursors. Building block molecules (e.g., amino acids, sugars, and fatty acids) produced or acquired from the diet are incorporated into larger, more complex molecules. Because biosynthesis increases order and complexity, anabolic pathways require free energy. Examples of anabolic processes include the synthesis of polysaccharides and proteins from sugars and amino acids, respectively. During **catabolic pathways** (Figure 8.2),

FIGURE 8.2

The Three Stages of Catabolism.

During stage 1, nutrient molecules are hydrolyzed to their building block units. In stage 2, building blocks are converted to easily oxidized forms (primarily acetyl CoA). In stage 3, acetyl CoA is completely oxidized to form CO_2 and H_2O. Energy is captured when ATP synthesis is linked to the electron transport pathway.

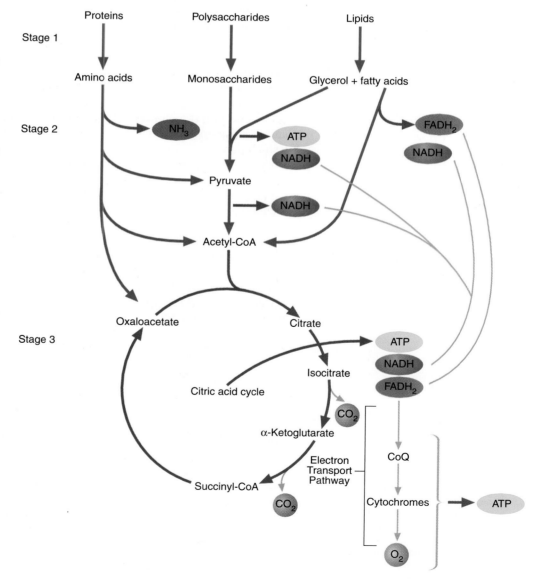

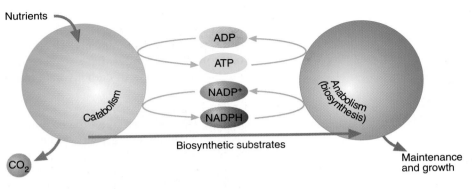

FIGURE 8.3

Anabolism and Catabolism.

In aerobic heterotrophs, catabolic pathways convert nutrients to small-molecule starting materials. Energy (ATP) and reducing power (NADPH) generated during catabolism drive biosynthetic reactions.

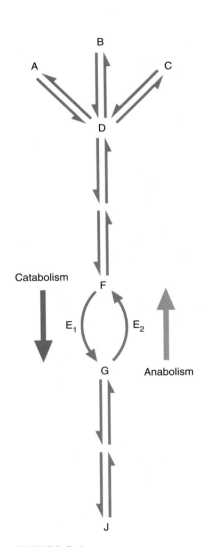

FIGURE 8.4

Dual Function Pathway.

Amphibolic pathways operate in anabolism and catabolism. In each, at least one step is irreversible. Activating or inhibiting the enzymes that catalyze these competing reactions control the amphibolic pathway.

large complex molecules are degraded into smaller, simpler products. Some catabolic pathways release free energy. A fraction of this energy is captured and used to drive anabolic reactions. β-Oxidation, a catabolic process that degrades fatty acids, is a prominent example of an energy-generating pathway.

In aerobic cells, catabolism consists of three stages. In stage 1 the major nutrient molecules (proteins, fats, and polysaccharides) are degraded to building block molecules in a process called *digestion.* During digestion, proteins are converted into amino acids; fats (triacylglycerols) are hydrolyzed to form monoacylglycerols and free fatty acids; and polysaccharides are degraded to simple sugar molecules. In stage 2 the amino acids, monosaccharides, and fatty acids produced in stage 1 are converted to a small number of simpler molecules. The most important product of stage 2 is acetyl-CoA. (In an acetyl-CoA molecule, an acetyl group is attached to coenzyme A, a carrier of acyl groups.) During stage 3 the acetyl group of acetyl-CoA enters the citric acid cycle and is completely oxidized. (The **citric acid cycle** occurs in eukaryotic mitochondria and the cytoplasm of some prokaryotes.) The carbon atoms of the acetyl group are completely oxidized to form CO_2, while the energy-rich hydrogen atoms are transferred to the coenzymes NAD^+ and FAD. A significant amount of energy is released during stage 3, when NADH and $FADH_2$ are oxidized by the electron transport system. (The **electron transport system** is a series of linked oxidation-reduction reactions in the inner membrane of mitochondria and the cell membrane of certain bacteria. Oxygen is the terminal electron acceptor.) A portion of the energy generated during electron transport is used to synthesize ATP.

The relationship between anabolic and catabolic processes is illustrated in Figure 8.3. As nutrient molecules are degraded, free energy and reducing power are conserved in ATP and NADH molecules, respectively. Biosynthetic processes utilize metabolites of catabolism, synthesized ATP and NADPH, to create complex structure and function.

In some dual-function pathways, called **amphibolic pathways,** the reactions are reversible. (Each enzyme catalyzes both the forward and reverse reactions.) For example, the critic acid cycle is both a mechanism for oxidizing acetyl groups and a source of biosynthetic precursors. In some dual-function pathways (e.g., glycolysis and gluconeogenesis in liver, p. 204), one or more irreversible reactions function as highly regulated control points. At each control point different allosteric enzymes catalyze the forward and the reverse reactions (Figure 8.4).

Throughout the life of an organism, a precise balance is struck between anabolic and catabolic processes. As a young animal grows and matures, for example, the rate of anabolic processes is greater than that of catabolic processes. As a healthy adulthood is reached, anabolic processes slow, and growth essentially stops. Throughout the remainder of its life (except during illness or pregnancy) the animal's tissues exist in a metabolic steady state. In a **steady state** the rate of anabolic processes is approximately equal to that of catabolic processes. Consequently, the appearance and functioning of the animal change little from one day to the next. Only over long periods of time do the inevitable signs of aging appear.

How are animals (or other multicellular organisms) able to maintain a balance between anabolic and catabolic processes as they respond and adapt to changes in their environment? The answer to this question is not fully understood. However, various forms of intercellular communication are believed to play an important role. Most intercellular communication occurs by means of chemical signals. Once released into the extracellular environment, each chemical signal is recognized by specific cells (called **target cells**), which then respond in a specific manner. Most chemical signals are modified amino acids, fatty acid derivatives, peptides, proteins, or steroids. However, recent research indicates that other molecules may act as chemical messengers. For example, nitric oxide (NO), a toxic gas, is now believed to be involved in the regulation of several biological processes (Section 13.3).

In animals the nervous and endocrine systems are primarily responsible for coordinating metabolism. (Hormonal regulation in plants is briefly described in Chapter 12.) The nervous system provides a rapid and efficient mechanism for acquiring and processing environmental information. Nerve cells, called neurons, release neurotransmitters (Section 13.3) at the end of long cell extensions called axons into tiny intercellular spaces called synapses. The neurotransmitter molecules bind to nearby cells, evoking specific responses from those cells.

Metabolic regulation by the endocrine system is achieved by secreting chemical signals called hormones directly into the blood. (The endocrine system, described briefly in Chapter 15, is composed of specialized cells, many of which are found in glands.) After hormone molecules are secreted, they travel through the blood until they reach a target cell. Most hormone-induced changes in cell function result from alterations in the activity or concentrations of enzymes. Hormones interact with cells by binding to specific receptor molecules. The receptors for most water-soluble hormones (e.g., polypeptides and epinephrine) are located on the surface of target cells. Binding these hormones to membrane-bound receptors triggers an intracellular response. The actions of many hormones are mediated by a group of molecules referred to as **second messengers.** (The hormone molecule is the first messenger.) Several second messengers have been identified. These include the nucleotides cyclic AMP (cAMP) and cyclic GMP (cGMP), calcium ions, and the inositol phospholipid system. (Second messengers are described in Chapter 15.) Most second messengers act to modulate allosteric enzymes, often by a powerful amplification device called an enzyme cascade. In an *enzyme cascade* (Figure 8.5) enzymes present in their inactive forms are converted to their active forms in a sequentially expanding array leading to a substantial amplification of the original signal. This process is often initiated when a second messenger binds to a specific enzyme. For example, binding cAMP to inactive protein kinase A converts it to active protein kinase A which, in turn, modifies the activity of many target enzymes through phosphorylation. The original signal generates an amplified and diversified response.

Steroid hormones are lipid-soluble molecules that act by a different mechanism. Once a steroid hormone has diffused into a cell, it binds to a specific receptor protein in the cytoplasm. The hormone-receptor complex moves into the nucleus, where it binds to specific sites on DNA. Steroids alter a cell's pattern and rate of protein synthesis. (This topic is discussed in Chapter 17.) The thyroid hormones act similarly.

Research increasingly shows that the distinction between the nervous and endocrine systems is not as clear as was once thought. For example, certain nerve cells, referred to as neurosecretory cells, synthesize and release hormones into the blood. Oxytocin and vasopressin (p. 93) are two prominent examples. In addition, several neurotransmitters act through second messengers. Depending on circumstances, epinephrine, whose mechanism involves cAMP, acts as a neurotransmitter and a hormone.

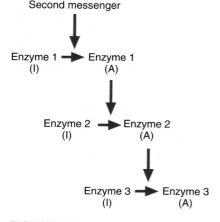

FIGURE 8.5

An Enzyme Cascade.

An enzyme cascade is a powerful mechanism in which a series of enzymes are sequentially activated. (Activation is often initiated by a second messenger molecule.) At each step in the cascade, the number of enzyme molecules increases exponentially. This amplification of a signal initiated by a few hormone molecules allows a rapid cellular response. I = inactive; A = active.

8.2 GLYCOGEN METABOLISM

The synthesis and degradation of glycogen are carefully regulated so that sufficient glucose is available for the body's energy needs. Both glycogenesis and glycogenolysis are controlled primarily by three hormones: insulin, glucagon, and epinephrine.

Glycogenesis Glycogen synthesis occurs after a meal, when blood glucose levels are high. Immediately after entering a cell, glucose and other sugar molecules are phosphorylated. (Phosphorylation prevents transport of glucose out of the cell and increases the nucleophilicity of the oxygen in the resulting phosphoester bond.) The phosphorylation of sugars may also facilitate binding these substrates in enzyme active sites. Several enzymes, called the hexokinases, catalyze the phosphorylation of hexoses in all cells in the body. ATP, a cosubstrate in the reaction, is complexed with Mg^{2+}. (ATP-Mg^{2+} complexes are common in kinase-catalyzed reactions.) Under intracellular conditions the reaction is irreversible.

Glucose   Glucose-6-phosphate

KEY *Concepts*

During glycogenesis, glycogen synthase catalyzes the transfer of the glucosyl group of UDP-glucose to the nonreducing ends of glycogen and glycogen branching enzyme catalyzes the formation of branch points. Glycogenolysis requires glycogen phosphorylase and debranching enzyme. Glycogen metabolism is regulated by the actions of three hormones: glucagon, insulin, and epinephrine.

Hexokinase D (or glucokinase), an isoenzyme found only in liver, has special properties. The other hexokinases have high affinities for glucose relative to its concentration in blood (i.e., they are half-saturated at concentrations of less than 0.1 mM). (Blood glucose levels are approximately 4–5 mM.) In addition, these enzymes are inhibited by glucose-6-phosphate. In contrast, hexokinase D requires much higher glucose concentrations for optimal activity (about 10 mM). Because hexokinase D activity is not inhibited by glucose-6-phosphate, this enzyme significantly helps the liver regulate blood glucose. After a carbohydrate meal, blood glucose levels rise. When glucose enters liver cells, it is immediately converted into glucose-6-phosphate. Because of the kinetic properties of hexokinase D, hepatocytes remove excess glucose from the blood. It has long been recognized that the consumption of a carbohydrate meal is followed promptly by liver glycogenesis. Until recently, it was presumed that blood glucose is the sole direct precursor in this process. However, it now appears that an indirect pathway involving gluconeogenesis may synthesize more liver glycogen than has previously been recognized. This pathway is discussed on pp. 201–208.

In the next step of glycogenesis, glucose-6-phosphate is reversibly converted to glucose-1-phosphate.

Glucose-6-phosphate   Glucose-1,6-bisphosphate   Glucose-1-phosphate

Phosphoglucomutase, which catalyzes this reaction, contains a phosphoryl group attached to a reactive serine residue. The enzyme's phosphoryl group is transferred to glucose-6-phosphate, forming glucose-1,6-bisphosphate. As glucose-1-phosphate forms, the phosphoryl group attached to C-6 is transferred to the enzyme's serine residue.

Because the formation of glycosidic linkages is endergonic, it must be coupled to an energy-releasing reaction. For this reason, many sugar transformations involve sugar-nucleotide derivatives. (Sugar-nucleotide derivatives are common metabolic intermediates. They are valuable in synthetic reactions because the nucleotide group promotes selective binding of the sugar in the active site of only certain enzymes. It also acts as a good "leaving group" when glycosidic bonds form.) Because uridine diphosphate glucose (UDP-glucose) contains two phosphoryl bonds, it is a highly energized molecule. Formation of UDP-glucose, whose $\Delta G^{\circ\prime}$ value is near zero, is a reversible reaction catalyzed by UDP-glucose pyrophosphorylase.

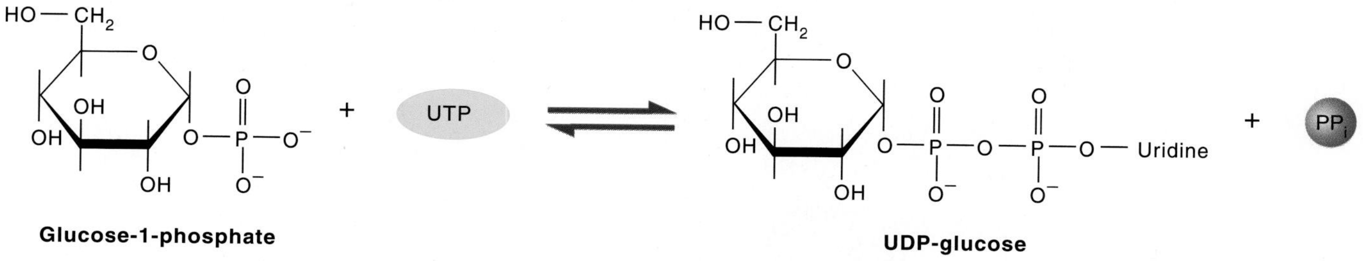

Glucose-1-phosphate   **UDP-glucose**

However, the reaction is driven to completion because pyrophosphate (PP_i) is immediately and irreversibly hydrolyzed by pyrophosphatase with a large loss of free energy ($\Delta G^{\circ\prime} = -8$ kcal/mole).

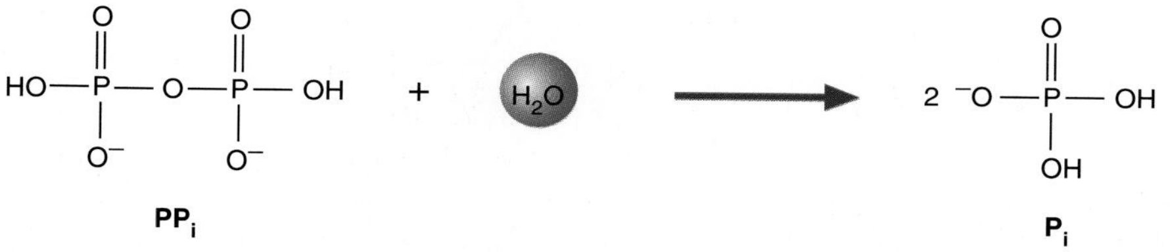

PP_i P_i

(Recall that removing product shifts the reaction equilibrium to the right. This cellular strategy is common.)

The formation of glycogen from UDP-glucose requires two enzymes:

1. glycogen synthase, which catalyzes the transfer of the glucosyl group of UDP-glucose to the nonreducing ends of glycogen (Figure 8.6a), and

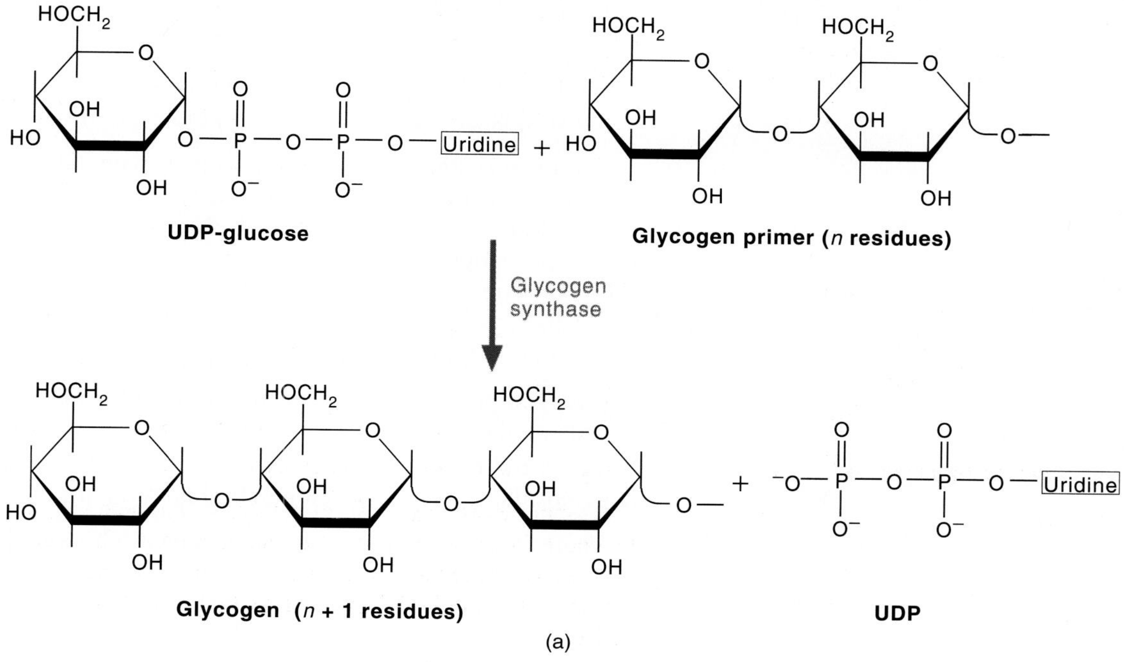

FIGURE 8.6

Glycogen Synthesis.

(a) The enzyme glycogen synthase breaks the phosphodiester linkage of UDP-glucose and forms an $\alpha(1,4)$ glycosidic bond between glucose and the growing glycogen chain.

UDP-glucose   **Glycogen primer (*n* residues)**

Glycogen synthase

Glycogen (*n* + 1 residues)   **UDP**

(a)

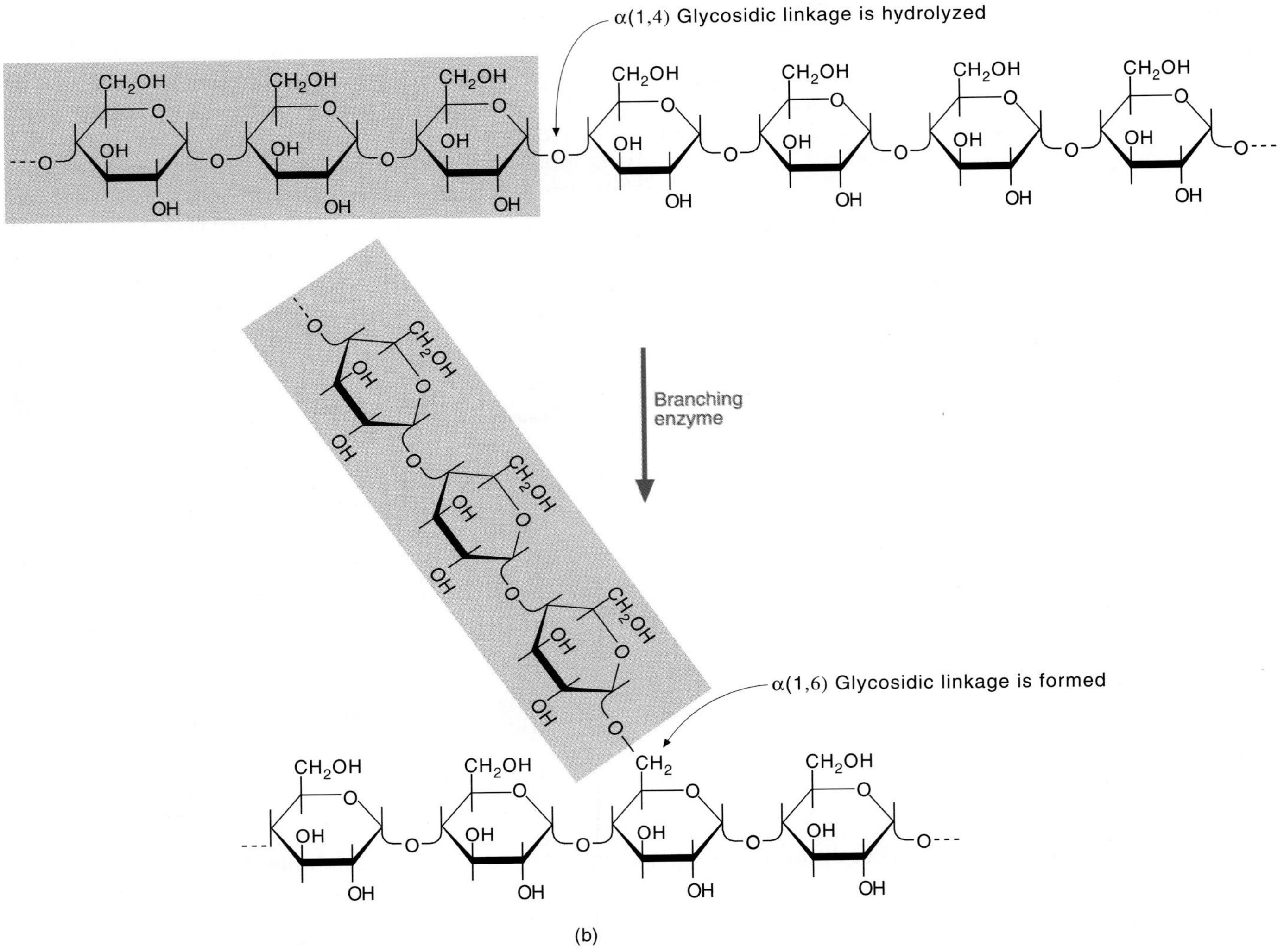

α(1,4) Glycosidic linkage is hydrolyzed

Branching enzyme

α(1,6) Glycosidic linkage is formed

(b)

FIGURE 8.6 (continued)
Glycogen Synthesis.

(b) Branching enzyme is responsible for the synthesis of α(1,6) linkages in glycogen.

2. amylo-α(1,4→1,6)-glucosyl transferase (branching enzyme), which creates the α(1,6) linkages for branches in the molecule (Figure 8.6b).

Glycogen synthesis requires a glycogen chain. Glycogen synthesis is now believed to be initiated by the transfer of glucose from UDP-glucose to a specific tyrosine residue in a "primer" protein called *glycogenin*. Large glycogen granules, each consisting of a single highly branched glycogen molecule, can be observed in the cytoplasm of liver and muscle cells of well-fed animals. The enzymes responsible for glycogen synthesis and degradation coat each granule's surface.

Glycogen degradation requires

1. glycogen phosphorylase, which removes glucose residues from the nonreducing ends on the outer branches of glycogen (Figure 8.7a), and

2. amylo-α(1,6)-glucosidase (debranching enzyme), which hydrolyzes the α(1,6) glycosidic bonds at branch points (Figure 8.7b).

Using inorganic phosphate (P_i), glycogen phosphorylase cleaves the α(1,4) linkages to yield glucose-1-phosphate. Glycogen phosphorylase stops when it comes within four glucose residues of a branch point. (A glycogen molecule that has been

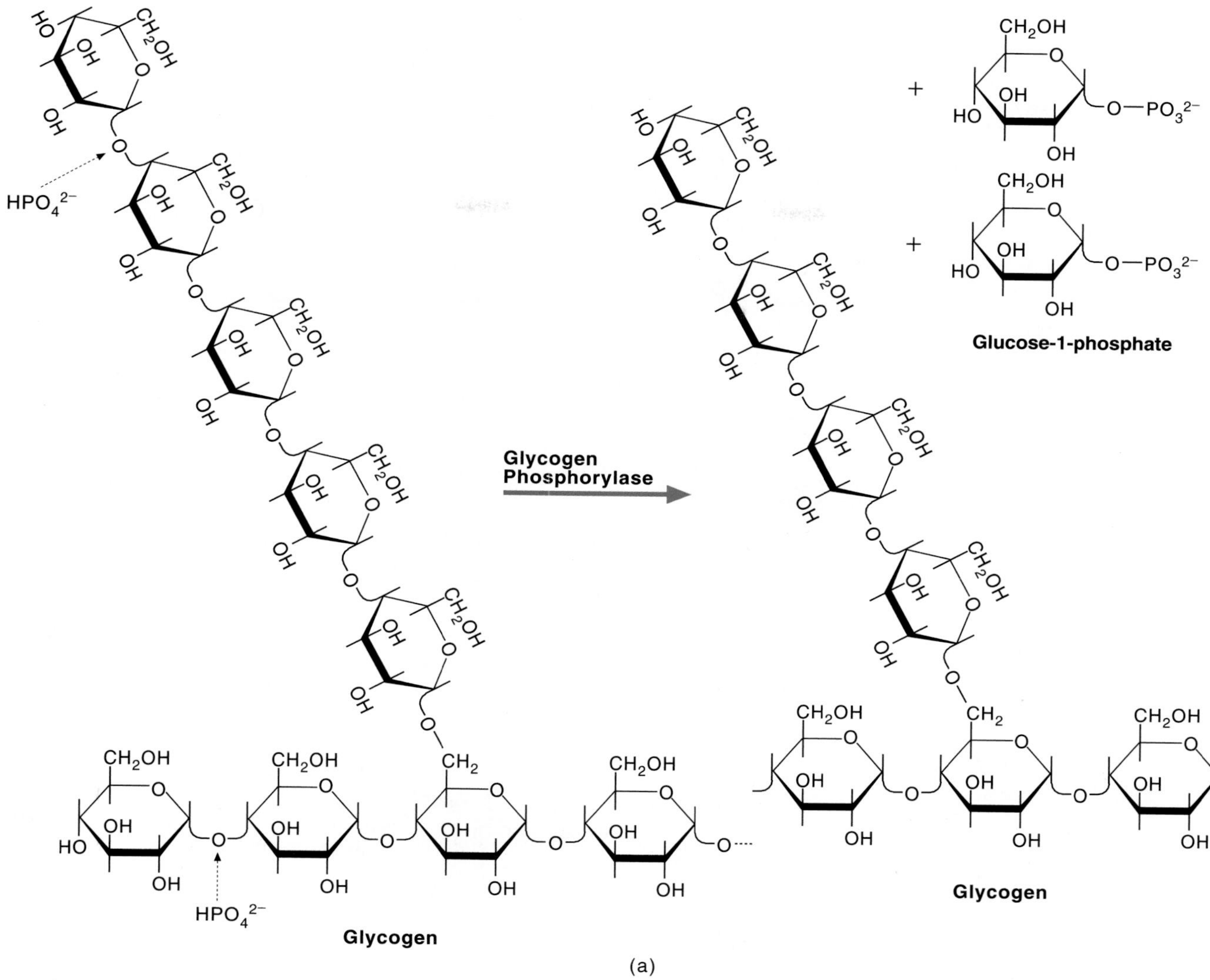

FIGURE 8.7

Glycogen Degradation.

(a) Glycogen phosphorylase catalyzes the removal of a glucose residue from a nonreducing end of a glycogen chain.

degraded to its branch points is called a *limit dextrin*.) Debranching enzyme begins the removal of $\alpha(1,6)$ branch points by transferring the outer three of the four glucose residues attached to the branch point to a nearby nonreducing end. It then removes the single glucose residue attached at each branch point. The product of this latter reaction is free glucose. A summary of glycogenolysis is shown in Figure 8.8.

Regulation of Glycogen Metabolism Glycogen metabolism is carefully regulated to avoid wasting energy. Both synthesis and degradation are controlled through a complex mechanism involving insulin, glucagon, and epinephrine. These hormones initiate processes that control several sets of enzymes. The binding of glucagon to liver cells stimulates glycogenolysis and inhibits glycogenesis. As blood glucose levels drop in the hours after a meal, glucagon ensures that glucose will be released into the bloodstream. After glucagon binds to its receptor, adenylate cyclase (a cell membrane enzyme) is stimulated to convert ATP to the second messenger cAMP. Then cAMP initiates a reaction cascade (described below) that amplifies the original signal. Within seconds a few glucagon molecules have initiated release of thousands of glucose molecules.

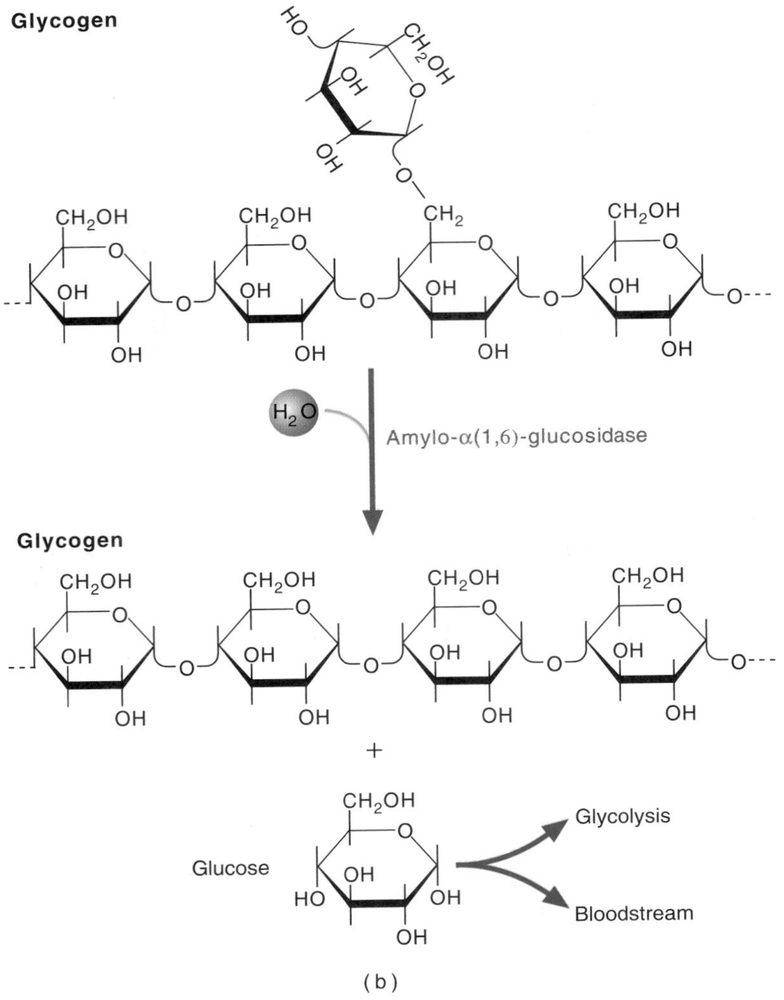

FIGURE 8.7 (continued)

Glycogen Degradation.

(b) Branch points in glycogen are removed by debranching enzyme (amylo-α(1,6) glucosidase).

(b)

When occupied, the insulin receptor becomes an active tyrosine kinase enzyme that causes a phosphorylation cascade which ultimately has the opposite effect of the glucagon/cAMP system: the enzymes of glycogenolysis are inhibited and the enzymes of glycogenesis are activated. Insulin also increases the rate of glucose uptake into several types of target cells, but not liver or brain cells.

Emotional or physical stress releases epinephrine from the adrenal medulla. Epinephrine promotes glycogenolysis and inhibits glycogenesis. In emergency situations, when epinephrine is released in relatively large quantities, massive production of glucose provides the energy required to manage the situation. This effect is referred to as the flight-or-fight response. Epinephrine initiates the process by activating adenylate cyclase in liver and muscle cells. Two other second messengers, calcium ions and inositol trisphosphate (Chapter 15), are also believed to be involved in epinephrine's action.

Glycogen synthase and glycogen phosphorylase have both active and inactive conformations. These forms are interconverted by covalent modification. The active form of glycogen synthase, known as the I (independent) form, is converted to the inactive or D (dependent) form by phosphorylation. In contrast, the inactive form of glycogen phosphorylase (phosphorylase b) is converted to the active form (phosphorylase a) by the phosphorylation of a specific serine residue. The phosphorylating enzyme is called phosphorylase kinase. Phosphorylation of both glycogen synthase and phosphorylase kinase is catalyzed by a protein kinase, which is activated by cAMP. Glycogen synthesis occurs when glycogen synthase and glycogen phosphorylase have been

FIGURE 8.8

Glycogen Degradation.

Glycogen phosphorylase cleaves the
α(1,4) linkages of glycogen to yield
glucose-1-phosphate until it comes
within four glucose residues of a branch
point. Debranching enzyme transfers
three of these residues to a nearby
nonreducing end and releases the fourth
residue as free glucose. The repeated
actions of both enzymes can lead to the
complete degradation of glycogen.

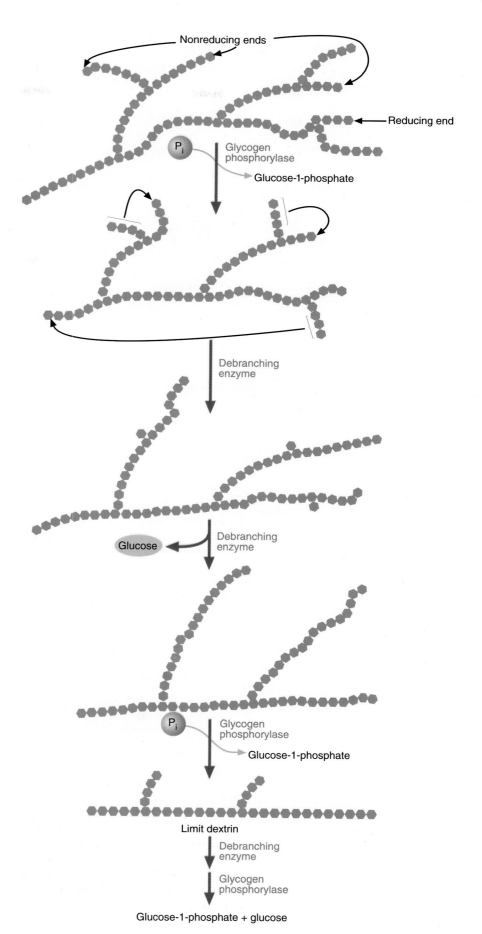

dephosphorylated. This conversion is catalyzed by phosphoprotein phosphatase. Phosphoprotein phosphatase also inactivates phosphorylase kinase. The major factors in this complex process are summarized in Figure 8.9.

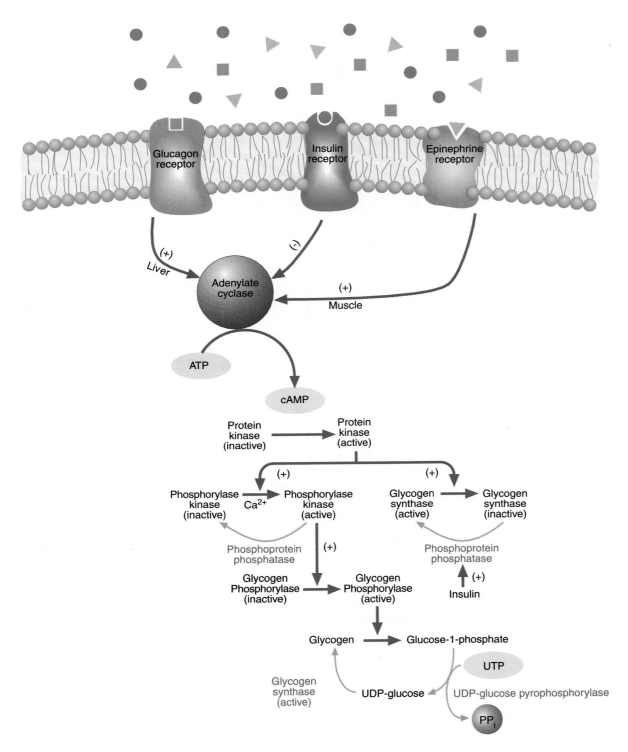

FIGURE 8.9

Major Factors Affecting Glycogen Metabolism.

Glucagon and epinephrine initiate a reaction cascade that converts glycogen to glucose-1-phosphate and inhibits glycogenesis. Insulin inhibits glycogenolysis and stimulates glycogenesis in part by decreasing the synthesis of cAMP and activating phosphoprotein phosphatase.

QUESTION **8.1**   Glycogen storage diseases are caused by inherited defects of one or more enzymes involved in glycogen synthesis or degradation. Patients with *Cori's disease,* caused by a deficiency of debranching enzyme, have enlarged livers (*hepatomegaly*) and low blood sugar concentrations (**hypoglycemia**). Can you suggest what causes these symptoms?

8.3 GLYCOLYSIS

Glycolysis, a series of reactions that occurs in almost every living cell, is believed to be among the oldest of all the biochemical pathways. Several features of glycolysis indicate that it is ancient. First of all, that glycolysis occurs in the cytoplasm of higher cells indicates that this essential pathway arose before eukaryotic organelles. In addition, the energy-generating reactions of glycolysis do not require oxygen, which was conspicuously absent from the earth's early atmosphere.

In glycolysis, also referred to as the *Embden-Meyerhof-Parnas pathway,* each glucose molecule is split and converted to two three-carbon units (pyruvate). During this process several carbon atoms are oxidized. The small amount of energy captured during glycolytic reactions (about 5% of the total available) is stored temporarily in two molecules each of ATP and NADH. The subsequent metabolic fate of pyruvate depends on the organism being considered and its metabolic circumstances. In **anaerobic organisms** (those that do not use oxygen to generate energy), pyruvate may be converted to waste products. Examples include ethanol, lactic acid, acetic acid, and similar molecules. Using oxygen as a terminal electron acceptor, aerobic organisms completely oxidize pyruvate to form CO_2 and H_2O in an elaborate stepwise mechanism known as **aerobic respiration.**

Glycolysis, which consists of ten reactions, occurs in two stages:

1. Glucose is phosphorylated and cleaved to form two molecules of glyceraldehyde-3-phosphate (G-3-P). The two ATP molecules consumed during this stage are like an investment, because the phosphoryl groups in the two G-3-P molecules are later used in ATP synthesis.

2. Glyceraldehyde-3-phosphate is converted to pyruvate. Four ATP molecules and two NADH are produced. Because two ATP were consumed in stage 1, the net production of ATP per glucose molecule is 2.

The glycolytic pathway can be summed up in the following equation:

$$\text{D-Glucose} + 2\ \text{ADP} + 2\ P_i + 2\ \text{NAD}^+ \rightarrow$$
$$2\ \text{pyruvate} + 2\ \text{ATP} + 2\ \text{NADH} + 2\ \text{H}^+ + 2\ H_2O$$

The Reactions of the Glycolytic Pathway

Glycolysis is summarized in Figure 8.10. Glycolysis begins by forming glucose-6-phosphate. As described, glucose-6-phosphate is produced from blood glucose by hexokinase. (Phosphoglucomutase catalyzes the conversion of glucose-1-phosphate, the product of glycogenolysis, to glucose-6-phosphate.) During reaction 2 of glycolysis, the aldose glucose-6-phosphate is converted by phosphoglucose isomerase (PGI) into the ketose fructose-6-phosphate in a readily reversible reaction.

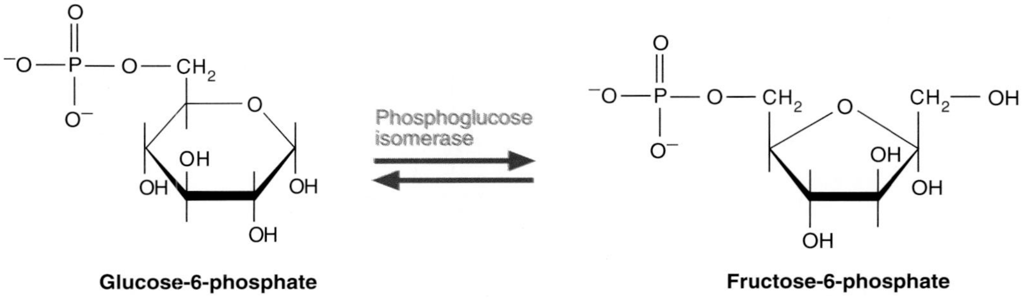

Glucose-6-phosphate   Fructose-6-phosphate

Recall that the isomerization reaction of glucose and fructose involves an enediol intermediate (Figure 7.15). This transformation makes C-1 of the fructose product available for phosphorylation.

FIGURE 8.10

The Glycolytic Pathway.

In glycolysis each glucose molecule is converted into two pyruvate molecules. In addition, two molecules each of ATP and NADH are produced.

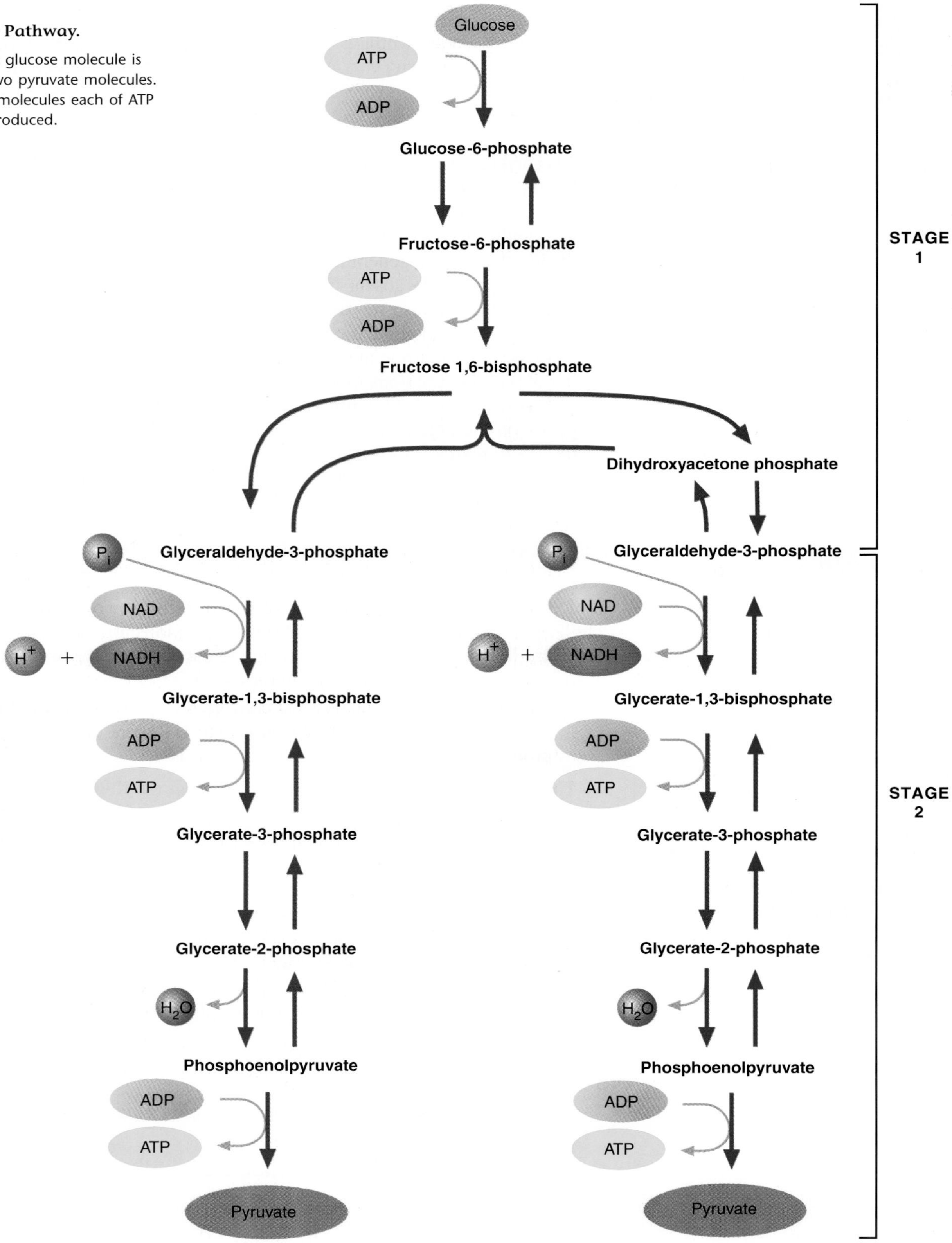

Phosphofructokinase-1 (PFK-1) catalyzes the phosphorylation of fructose-6-phosphate to form fructose-1,6-bisphosphate.

Fructose-6-phosphate   **Fructose-1,6-bisphosphate**

Investing a second molecule of ATP serves several purposes. First of all, because ATP is used as the phosphorylating agent, the reaction proceeds with a large loss of free energy. Under physiological conditions, the reaction is irreversible. After fructose-1,6-bisphosphate has been synthesized, the cell is committed to degrading glucose (rather than converting it to another hexose or storage in a glycogen molecule). Because fructose-1,6-bisphosphate eventually splits into two trioses, another purpose for phosphorylation is to prevent any later product from diffusing out of the cell. (Recall that unphosphorylated sugars leak out of cells.)

PFK-1 is a major regulatory enzyme. Its catalytic activity is controlled by several molecules. In addition to inhibition by the reaction products fructose-1,6-bisphosphate and ADP, PFK-1 is also inhibited by high concentrations of ATP and citrate (a citric acid cycle intermediate). A high concentration of these molecules is a signal that the cell has sufficient energy reserves. In contrast, an elevated AMP concentration, an indicator of an energy deficit, is a PFK-1 activator. In liver, fructose-2,6-bisphosphate is an important activator of PFK-1. Fructose-2,6-bisphosphate is an intracellular indicator that the cell's glucose concentration is high. Formed by the phosphorylation of fructose-6-phosphate by the liver enzyme phosphofructokinase-2 (PFK-2), fructose-2,6-bisphosphate stimulates glycolysis. The reverse reaction is catalyzed by fructose-2,6-bisphosphatase, a catalytic activity on the same molecule as PFK-2.

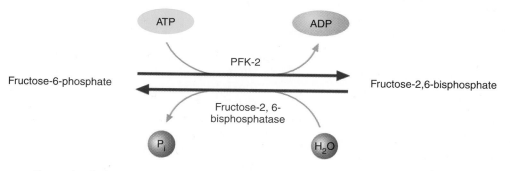

Stage 1 of glycolysis ends by cleaving fructose-1,6-bisphosphate into two three-carbon molecules: glyceraldehyde-3-phosphate (G-3-P) and dihydroxyacetone phosphate (DHAP). This reaction is an **aldol cleavage,** hence the name of the enzyme: aldolase. Aldol cleavages are the reverse of aldol condensations, described in Section 5.3. In aldol cleavages an aldehyde and a ketone are products.

Fructose-1,6-bisphosphate   **Dihydroxyacetone**   **Glyceraldehyde-3-phosphate**
                                                 phosphate

Although the cleavage of fructose-1,6-bisphosphate is energetically unfavorable ($\Delta G^{\circ\prime} = +5.7$ kcal/mole), the reaction proceeds because the products are rapidly removed.

Of the two products of the aldolase reaction, only G-3-P serves as a substrate for the next reaction in glycolysis. To prevent the loss of the other three-carbon unit from the glycolytic pathway, triose phosphate isomerase catalyzes the interconversion of DHAP and G-3-P.

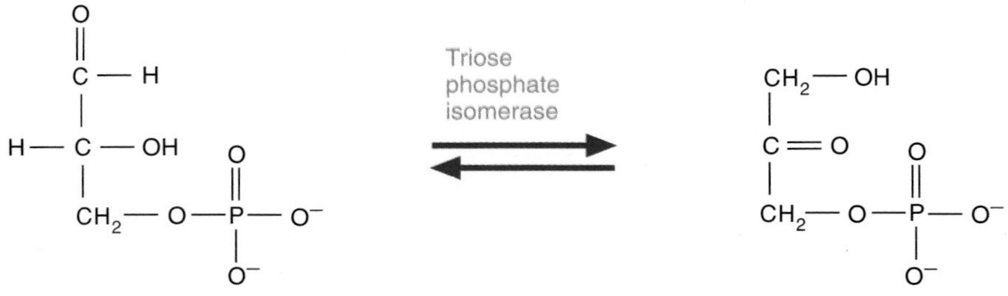

Glyceraldehyde-3-phosphate   **Dihydroxyacetone phosphate**

After this reaction, the original molecule of glucose has been converted to two molecules of G-3-P.

During reaction 6 of glycolysis, G-3-P undergoes oxidation and phosphorylation. The product, glycerate-1,3-bisphosphate, contains a high-energy bond, which may be used in the next reaction to generate ATP.

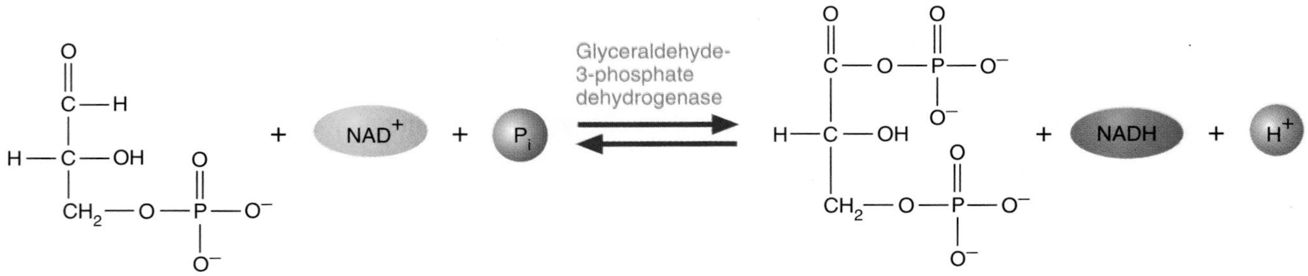

Glyceraldehyde-3-phosphate   **Glycerate-1,3-bisphosphate**

This complex process is catalyzed by glyceraldehyde-3-phosphate dehydrogenase, a tetramer composed of four identical subunits. Each subunit contains one binding site for G-3-P and another for NAD^+. (NAD is permanently bound to this enzyme; most enzymes that bind NAD do so reversibly.) The reaction mechanism has several steps (Figure 8.11). A pair of electrons and one proton are transferred to NAD^+. The reduction of NAD^+ is accompanied by the oxidation of an aldehydic carbon to a carboxylate carbon. After the product dissociates from the enzyme, NADH donates a hydride ion to an unbound NAD^+ in the cytoplasm.

In the next reaction (reaction 7), ATP is synthesized as phosphoglycerate kinase catalyzes the transfer of the high-energy phosphoryl group of glycerate-1,3-bisphosphate to ADP.

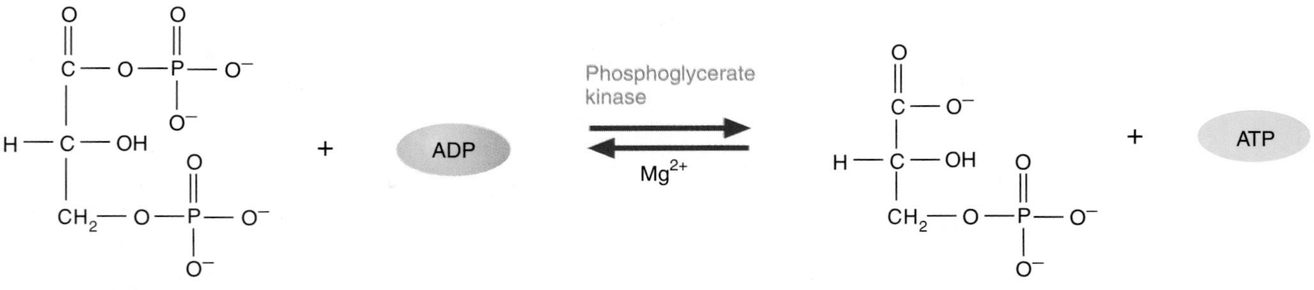

Glycerate-1,3-bisphosphate   **Glycerate-3-phosphate**

FIGURE 8.11

Glyceraldehyde-3-Phosphate Dehydrogenase.

In the first step, the enzyme catalyzes the reaction of the substrate with a sulfhydryl group of a cysteine residue. The substrate is then oxidized (Step 2). Displacement of the enzyme by inorganic phosphate liberates the product, glycerate-1,3-bisphosphate (Step 3). The bound NADH is then reoxidized by the transfer of a hydride ion to a cytoplasmic NAD+ (Step 4).

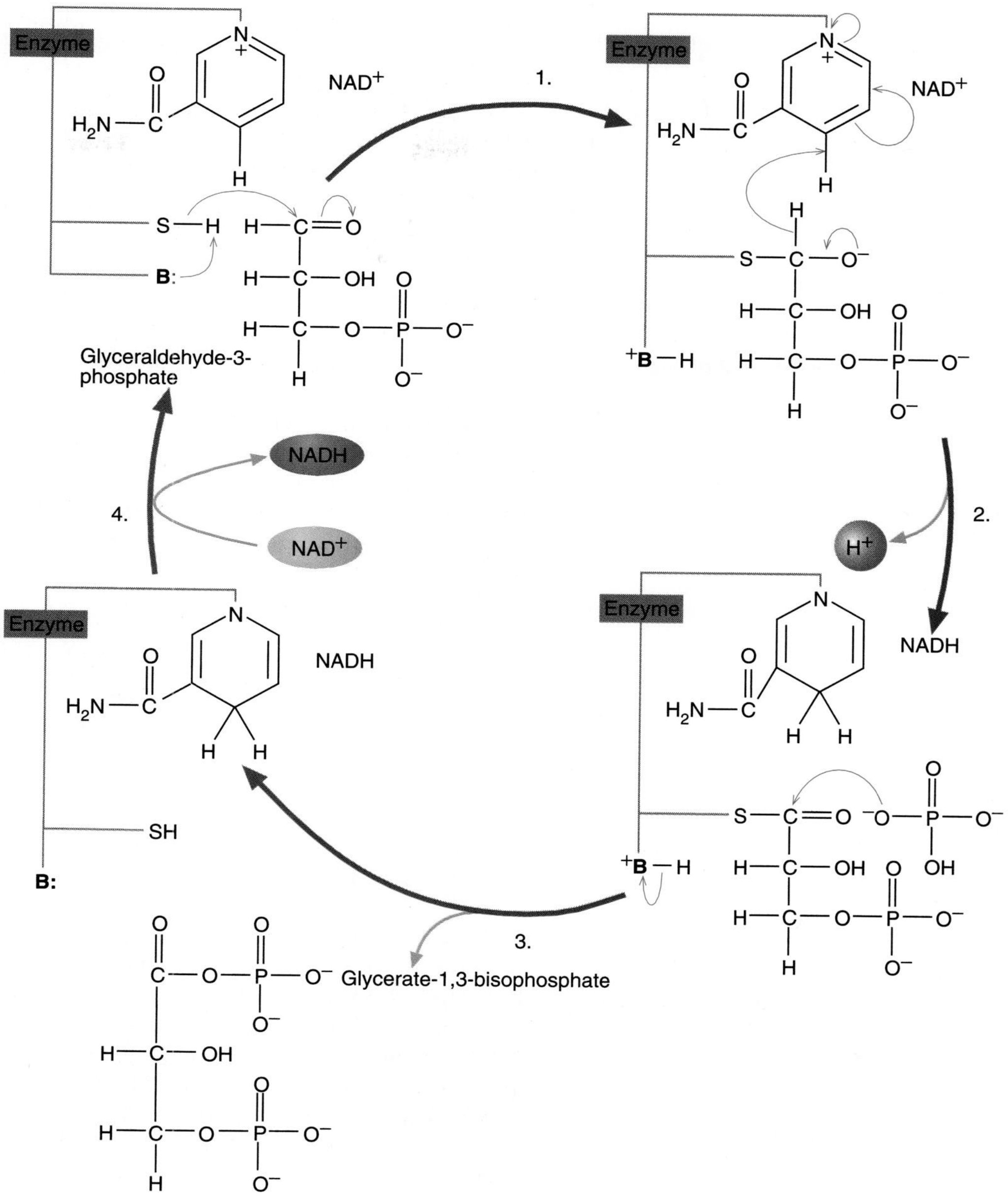

This reaction is an example of a substrate-level phosphorylation. (Because the synthesis of ATP is endergonic, it requires an energy source. In **substrate-level phosphorylations,** ATP is produced because of the transfer of a phosphoryl group from a substrate with a high phosphoryl group transfer potential.) Since two molecules of glycerate-1,3-bisphosphate are formed for every glucose molecule, this reaction produces two ATP molecules and the investment of phosphate bond energy is recovered. Any later ATP synthesis may be considered interest on this investment.

Glycerate-3-phosphate has a low phosphoryl group transfer potential. As such, it is a poor candidate for further ATP synthesis. Cells convert glycerate-3-phosphate with its energy-poor phosphate ester to phosphoenolpyruvate (PEP), which has an exceptionally high phosphoryl group transfer potential. (The standard free energies of hydrolysis of glycerate-3-phosphate and PEP are -3 kcal/mole and -14 kcal/mole, respectively.) In the first step in this conversion (reaction 8), phosphoglycerate mutase catalyzes the transfer of the phosphoryl group from C-3 to C-2.

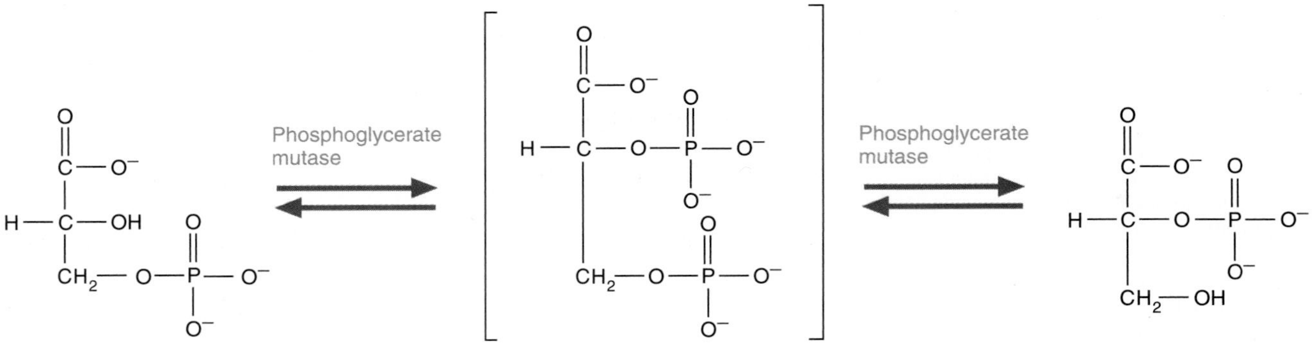

Glycerate-3-phosphate   **Glycerate-2,3-bisphosphate**   **Glycerate-2-phosphate**

The reversible interconversion of glycerate-3-phosphate and glycerate-2-phosphate is similar to the reaction catalyzed by phosphoglucomutase. In both reactions the enzyme is reversibly phosphorylated. In addition, both reactions have an obligatory intermediate containing two phosphoryl groups.

Enolase catalyzes the dehydration of glycerate-2-phosphate to form PEP (reaction 9).

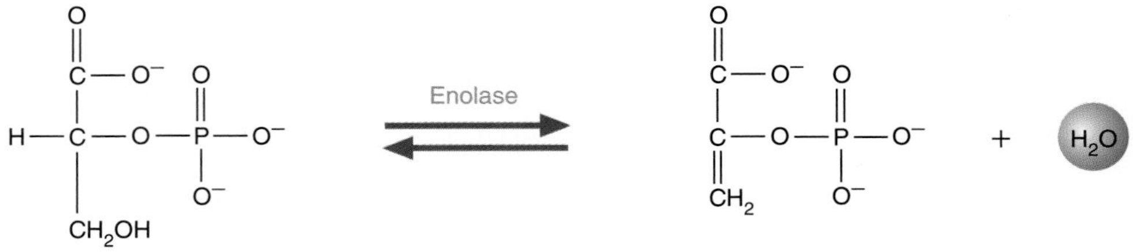

Glycerate-2-phosphate   **Phosphoenolpyruvate (PEP)**

PEP has a higher phosphoryl group transfer potential than does glycerate-2-phosphate because it contains an enol-phosphate group instead of a phosphate ester of an alcohol. The reason for this difference is made apparent in the next reaction. (Aldehydes and ketones have two isomeric forms. The *enol* form contains a carbon-carbon double bond and a hydroxyl group. Enols exist in equilibrium with the more stable carbonyl-containing *keto* form. The interconversion of keto and enol forms, also called **tautomers,** is referred to as **tautomerization.**)

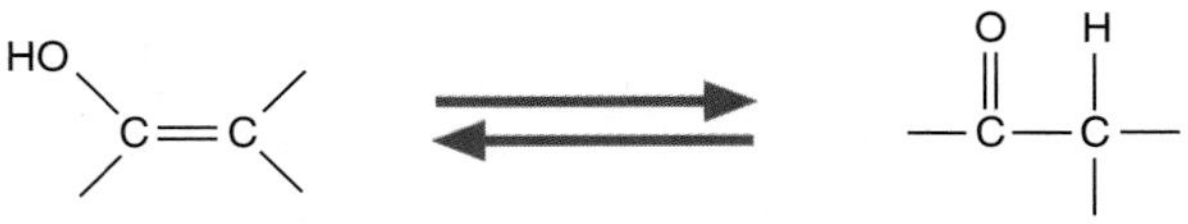

Enol form   **Keto form**

In the final reaction of glycolysis (reaction 10), pyruvate kinase catalyzes the transfer of a phosphoryl group from PEP to ADP. Two molecules of ATP are formed for each molecule of glucose.

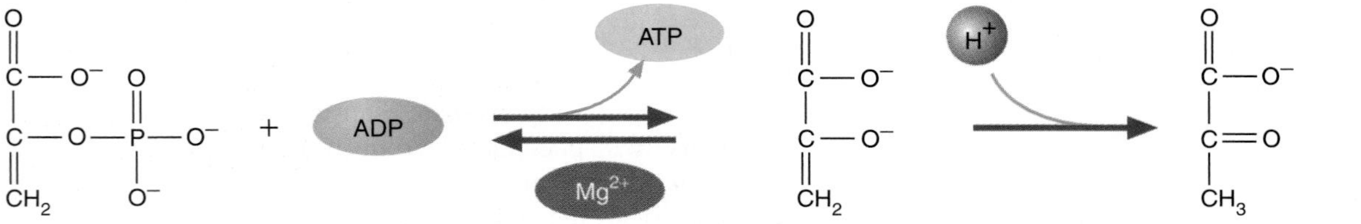

PEP   **Pyruvate (enol form)**   **Pyruvate (keto form)**

Because the free energy of hydrolysis is exceptionally large PEP is irreversibly converted to pyruvate. The free energy loss, which makes the reaction irreversible, is associated with the spontaneous conversion (tautomerization) of the enol form of pyruvate to the more stable keto form.

The Fates of Pyruvate

In terms of energy, the result of glycolysis is the production of two ATPs and two NADHs per molecule of glucose. Pyruvate, the other product of glycolysis, is still an energy-rich molecule, which can produce a substantial amount of ATP. In the presence of O_2, aerobic organisms convert pyruvate to acetyl-CoA. (As described, the acetyl group of acetyl-CoA is then channeled through the citric acid cycle and degraded to CO_2. Concurrently, a series of redox reactions transfer the hydrogen atoms to the electron transport system.) Under anaerobic conditions, additional energy cannot be generated from pyruvate. However, in living organisms several mechanisms use pyruvate to solve the following critical metabolic problem. Each cell has a limited supply of NAD^+. Recall that NAD^+ is reduced to form NADH during Step 5 of glycolysis. To continue glycolysis, cytoplasmic NAD^+ must be constantly regenerated. When O_2 is not present (or the cell cannot use it), NAD^+ is regenerated as pyruvate is converted into another molecule. In this process, referred to as **fermentation,** organisms convert pyruvate into organic molecules. Muscle cells and certain bacterial species (e.g., *Lactobacillus*) produce NAD^+ by transforming pyruvate into lactate.

Pyruvate   **Lactate**

In rapidly contracting muscle cells the demand for energy is high. After the O_2 supply is depleted, *lactic acid fermentation* provides sufficient NAD^+ to allow glycolysis (with its low level of ATP production) to continue for a short time (Figure 8.12).

QUESTION

8.2

The first step in the detoxification of ethanol in the liver is its oxidation to form acetaldehyde. This reaction, catalyzed by alcohol dehydrogenase, produces large amounts of NADH. One common effect of alcohol intoxication is the accumulation of lactate in the blood. Can you explain why this effect occurs?

FIGURE 8.12

Reoxidation of NADH During Anaerobic Glycolysis.

The NADH produced during the conversion of glyceraldehyde-3-phosphate to glycerate-1,3-bisphosphate is oxidized when pyruvate is converted to lactate. This process allows the cell to continue producing ATP for a short time until O_2 is again available.

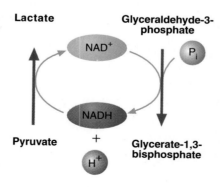

Microorganisms that use lactic acid fermentation to generate energy can be separated into two groups. *Homolactic fermenters* produce only lactate. For example, several species of lactic acid bacteria sour milk. *Heterolactic* or *mixed acid fermenters* produce several organic acids. Certain gases may also be produced. Mixed acid fermentation, for example, occurs in the rumen of cattle. Symbiotic organisms, some of which digest cellulose, synthesize organic acids (e.g., lactic, acetic, propionic, and butyric acids). The organic acids are absorbed from the rumen and used as nutrients. Gases such as methane and carbon dioxide are also produced.

Alcoholic fermentation occurs in yeast and several bacterial species. In yeast, pyruvate is decarboxylated to form acetaldehyde, which is then reduced by NADH to form ethanol. (In a **decarboxylation** reaction, an organic acid loses a carboxyl group as CO_2.)

Pyruvate   **Acetaldehyde**   **Ethanol**

Alcoholic fermentation by certain yeasts is used commercially to produce wine, beer, and bread. (See Special Interest Box 8.1.) Certain bacterial species produce alcohols other than ethanol. For example, *Clostridium acetobutylicum,* an organism related to the causative agents of botulism and tetanus, produces butanol. Until recently, this organism was used commercially to synthesize butanol, an alcohol used to produce detergents and synthetic fibers. A petroleum-based synthetic process has now replaced microbial fermentation.

Regulation of Glycolysis

The regulation of glycolysis is complex because of the crucial role of glucose in energy generation and in the synthesis of numerous metabolites. The rate at which the glycolytic pathway operates is controlled primarily by allosteric regulation of three enzymes: hexokinase, PFK-1, and pyruvate kinase. The reactions catalyzed by these enzymes are irreversible and can be switched on and off by allosteric effectors. In general, allosteric effectors are molecules whose cellular concentrations are sensitive indicators of a cell's metabolic state. In glycolysis there are several types of effectors. As described, several energy-related molecules act as allosteric effectors. For example, a high AMP concentration (an indicator of low energy production) activates PFK-1 and pyruvate kinase. In contrast, a high ATP concentration (an indicator that the cell's energy requirements are being met) inhibits both enzymes. Citrate and acetyl-CoA, which indicate that alternative energy sources are available (e.g., fatty acids), inhibit PFK-1 and pyruvate kinase, respectively. Other allosteric effectors include fructose-6-phosphate, which stimulates PFK-1 activity and glucose-6-phosphate, which inhibits the hexokinases (except hexokinase D). Finally, fructose-2,6-bisphosphate (an indicator of the cell's glucose concentration) stimulates glycolysis by activating PFK-1 and fructose-1,6-biphosphate activates pyruvate kinase. The effects of various allosteric regulators are summarized in Table 8.1.

Control of glycolysis is also affected by hormones. For example, glucagon inhibits glycolysis by repressing the synthesis of fructose-2,6-bisphosphate. Insulin promotes glycolysis by stimulating the synthesis of fructose-2,6-bisphosphate.

KEY *Concepts*

During glycolysis, glucose is converted to two molecules of pyruvate. A small amount of energy is captured in two molecules each of ATP and NADH. In anaerobic organisms, pyruvate is converted to waste products in a process called fermentation. In the presence of oxygen the cells of aerobic organisms convert pyruvate into CO_2 and H_2O.

TABLE 8.1

    Allosteric Regulation of Glycolysis

Enzyme	Activator	Inhibitor
Hexokinase		Glucose-6-phosphate
PFK-1	Fructose-2,6-bisphosphate, Fructose-6-phosphate, AMP	Citrate, ATP
Pyruvate kinase	Fructose-1,6-bisphosphate, AMP	Acetyl CoA, ATP

8.3   Insulin is a hormone secreted by the pancreas when blood sugar increases. Its most easily observable function is to reduce the blood sugar level to normal. The binding of insulin to most body cells promotes the transport of glucose across the plasma membrane. The capacity of an individual to respond to a carbohydrate meal by reducing blood glucose concentration quickly is referred to as *glucose tolerance.* Chromium-deficient animals show a decreased glucose tolerance; that is, they cannot remove glucose from blood quickly enough. The metal is believed to facilitate the binding of insulin to cells. Do you think the chromium is acting as a second messenger, allosteric activator or cofactor?

8.4   Louis Pasteur, the great nineteenth century French chemist and microbiologist, was the first scientist to make the following observation. Cells that can oxidize glucose completely to CO_2 and H_2O use glucose more rapidly in the absence of O_2 than in its presence. O_2 seems to inhibit glucose consumption. Explain in general terms the significance of this finding, now referred to as the **Pasteur effect.**

8.4 GLUCONEOGENESIS

Gluconeogenesis, the formation of new glucose molecules from noncarbohydrate precursors, occurs primarily in the liver. These precursors include lactate, pyruvate, glycerol, and certain α-keto acids (molecules derived from amino acids). In certain situations (i.e., metabolic acidosis or starvation) the kidney can make new glucose. As described, liver glycogen maintains adequate blood glucose levels between meals. When liver glycogen is depleted (e.g., prolonged fasting or vigorous exercise), the gluconeogenesis pathway provides the body with adequate glucose. (Brain and red blood cells rely exclusively on glucose as their energy source. Under exceptional circumstances, brain cells can also use certain fatty acid derivatives to generate energy. Exercising skeletal muscle requires glucose as a principal energy source.)

Gluconeogenesis Reactions

The reaction sequence in gluconeogenesis is largely the reverse of glycolysis. Recall, however, that three glycolytic reactions (the reactions catalyzed by hexokinase, PFK-1, and pyruvate kinase) are irreversible. In gluconeogenesis, alternate energetically favorable reactions catalyzed by different enzymes are used to bypass these obstacles. The reactions unique to gluconeogenesis are summarized below. The entire gluconeogenic pathway and its relationship to glycolysis are illustrated in Figure 8.13. The bypass reactions of gluconeogenesis are as follows:

1. **Synthesis of PEP.** PEP synthesis from pyruvate requires two enzymes: pyruvate carboxylase and PEP carboxykinase. Pyruvate carboxylase, found within mitochondria, converts pyruvate to oxaloacetate (OAA).

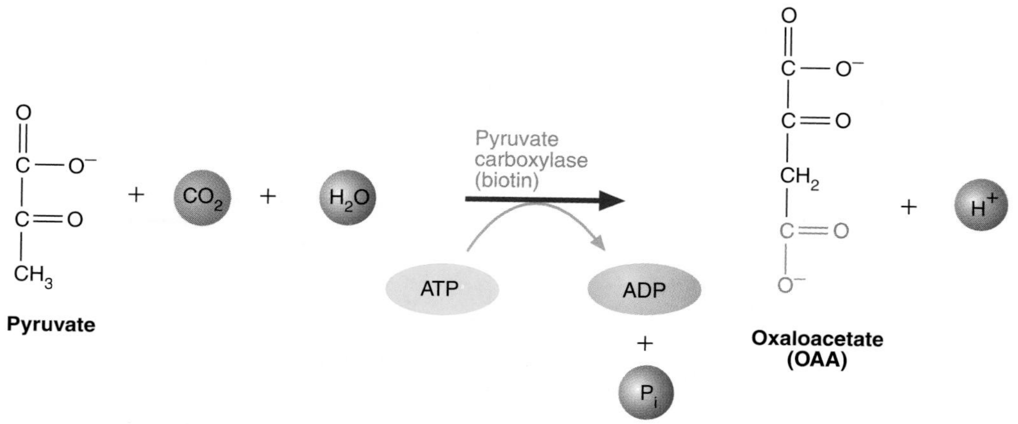

The coenzyme *biotin,* which functions as a CO_2 carrier, is covalently bound to the enzyme through the side chain amino group of a lysine residue. OAA is

FERMENTATION: AN ANCIENT HERITAGE

Alcoholic beverage production has a long and colorful history. Humans probably began making fermented beverages at least 10,000 years ago. However, the archeological evidence is about 5,500 years old. Ancient wine-stained pottery demonstrates that wine making was a flourishing trade in Sumer (now western Iran) by 3500 B.C. By that time, cultivation of the grape vine (*Vitis vinifera*), which originated in central Asia, had spread throughout the Middle East, especially in Mesopotamia (modern Iraq) and Egypt (Figure 8A). Wines were also made from sweet dates and the sap of palm trees.

These ancient peoples also knew how to produce beer by fermenting barley, a starchy grain. (A Sumerian tablet dated approximately 1750 B.C., which contains directions for brewing beer, is probably one of the oldest known recipes.) Beer making was probably a profitable occupation because Sumerian sol-

diers received a portion of their pay as beer. Beer was also popular in ancient Egypt. Numerous references to beer have been found on the walls of ancient tombs. Beer produced in ancient China, Japan, and central Africa was made from millet.

In addition to their intoxicating properties, both wine and beer were valued in the ancient world because of their medicinal properties. Wine was especially esteemed by ancient physicians. For example, Hippocrates (460–370 B.C.), the Greek physician who gave the medical profession its ethical ideals, prescribed wine as a diuretic, as a wound dressing, and (in moderate amounts) as a nourishing beverage.

Although humans have been making alcoholic beverages for thousands of years, fermentation was understood only relatively recently. As their businesses became more competitive in the nineteenth century, commercial producers of wine and

FIGURE 8A
Egyptian Wall Painting Illustrating Wine Production.

decarboxylated and phosphorylated by PEP carboxykinase in a reaction driven by the hydrolysis of guanosine triphosphate (GTP).

PEP carboxykinase is found within the mitochondria of some species and in the cytoplasm of others. In humans this enzymatic activity is found in both compartments. Because mitochondrial membrane is impermeable to OAA, cells that lack

beer in Europe provided substantial financial support for scientific investigations of fermentation. For example, Louis Pasteur was working for the French wine industry when he discovered that wine fermentation is caused by yeast and that wine spoilage (i.e., vinegar formation) is caused by microbial contamination. Pasteur was credited with saving the French wine industry after he discovered that briefly heating wine to 55°C kills unwanted organisms without affecting the wine's taste. This process is now called pasteurization.

Wine Making

Grapes are well suited to the fermentative process because they contain enough sugar to reach a fairly high alcohol content (about 10%). In addition, because the pH of wine is about 3, it is acidic enough to suppress the growth of most other microorganisms.

Wine making begins when ripe grapes are crushed into juice. The crush contains grape skins, seeds, and a liquid referred to as *must*. The must contains sugars (primarily glucose and fructose) in variable amounts (from 12% to 27%), as well as small amounts of several organic acids (e.g., tartaric, malic, and citric acids). White wines are made by using grapes with unpigmented skins or musts from which pigmented grape skins have been removed before fermentation. Red wines result when pigmented grape skins remain in the must throughout fermentation. During fermentation, yeasts not only convert sugar to alcohol but also produce volatile and aromatic molecules that are not present in the original must. Among these are complex esters, long-chain alcohols, various acids, glycerol, and other substances that contribute to the wine's flavor and aroma. (Some of these molecules, referred to as *congeners*, may also contribute to hangovers.)

In commercial wine production, both temperature and oxygen concentration are carefully controlled. At lower temperatures, yeasts produce more of the molecules that enhance flavor and aroma. In addition, other organisms are less likely to flourish during a cool fermentation. A high oxygen concentration at the beginning of a fermentation causes rapid cell division, so more yeast cells are available to ferment sugar. Later, as the oxygen concentration is reduced, the yeast cells excrete larger and larger amounts of alcohol. After fermentation, yeast cells and other particulates are allowed to settle before the wine is carefully decanted. The new wine is then placed in wooden barrels, where it slowly ages. Controlled oxidation results in the complex flavors and aromas typical of fine wines.

Beer Brewing

Because beers are made from starchy grains (primarily barley), the first step in brewing is a process referred to as *malting,* in which starch is broken down into glucose and maltose. During malting, the grain, steeped in water, is allowed to germinate sufficiently that amylase and other enzymes are produced in large quantities. Various enzymes, such as protease, ribonuclease, and phosphatase, make the *wort* (the malt extract that will eventually become beer) a suitable food for yeast. After germination is terminated by drying out the grain, the resulting *malt* is cured at 100°C. (The color and flavor of beer develop, to a significant extent, during curing.) Curing reduces the malt's moisture content to about 2–5% and arrests enzymatic activity. (Amylase is resistant to high temperatures; its optimum temperature is 70°C.)

Brewing begins with *mashing,* in which finely crushed malt is mixed with water and enough enzymes to further degrade any remaining starch and protein. After mashing, the dissolved product (now called the *wort*) is separated from an insoluble residue (referred to as the spent grain) by filtration. (Spent grain is usually sold as cattle fodder.) Afterward, the wort is boiled with hops, the dried cones of the vine *Humulus lupulus,* which give beer its bitter taste. After cooling and removal of the hops, fermentation begins as the wort is inoculated with a pure strain of yeast. (A strain of *Saccharomyces cerevisiae,* sometimes referred to as "brewer's yeast," is often used.) Fermentation is carefully controlled by varying the temperature and other parameters. Fermentation continues until the desired level of alcohol is reached. (In the United States, the amount of alcohol in beer varies between 3.6% and 4.9% by weight.) After the new beer is filtered to remove yeast, it is stored (or "lagered") for several months to permit sedimentation. Beer production ends with filtration and pasteurization.

mitochondrial PEP carboxykinase transfer OAA into the cytoplasm by using, for example, the **malate shuttle.** In this process, OAA is converted into malate by mitochondrial malate dehydrogenase. After the transport of malate across mitochondrial membrane, the reverse reaction is catalyzed by cytoplasmic malate dehydrogenase.

OAA + NADH + H$^+$ ⇌ (Malate dehydrogenase) **Malate** + NAD$^+$

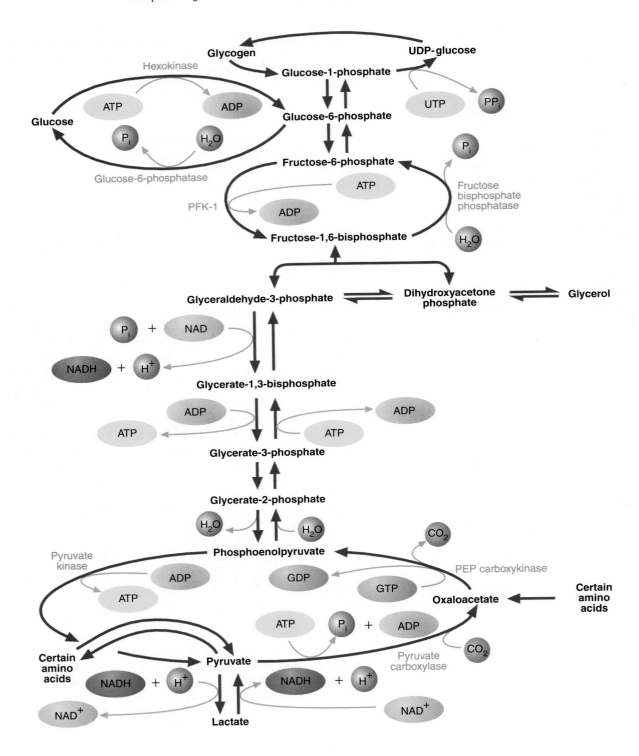

FIGURE 8.13

Glycolysis and Gluconeogenesis.

In gluconeogenesis 7 of the 10 reactions of glycolysis are reversed. Three irreversible glycolytic reactions are bypassed by alternative reactions. The substrates for gluconeogenesis include pyruvate, lactate, certain amino acids, and glycerol.

2. **Conversion of fructose-1,6-bisphosphate to fructose-6-phosphate.** The irreversible PFK-1-catalyzed reaction in glycolysis is bypassed by fructose-1,6-bisphosphatase.

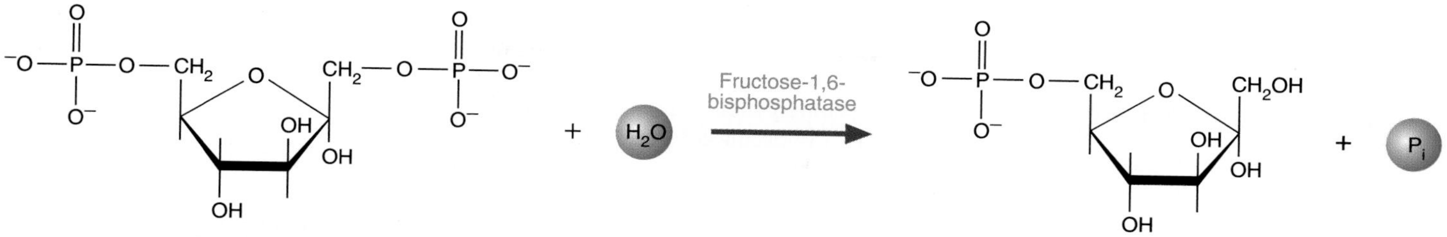

Fructose-1,6-bisphosphate → **Fructose-6-phosphate**

This exergonic reaction ($\Delta G^{\circ\prime} = -4$ kcal/mole) is also irreversible under cellular conditions. ATP is not regenerated. Fructose-1,6-bisphosphatase is an allosteric enzyme. Its activity is stimulated by citrate and inhibited by AMP and fructose-2,6-bisphosphate.

3. **Formation of glucose from glucose-6-phosphate.** Glucose-6-phosphatase, found only in liver and kidney, catalyzes the irreversible hydrolysis of glucose-6-phosphate to form glucose and P_i. Glucose is subsequently released into the blood. (See Special Interest Box 8.2.)

As stated, each of the above reactions is matched by an opposing irreversible reaction in glycolysis. Each set of such paired reactions is referred to as a *substrate cycle*. If both reactions occur simultaneously, energy is wasted. For example, the conversion of glucose to glucose-6-phosphate requires ATP hydrolysis. In the reverse reaction, however, in which glucose-6-phosphate is hydrolyzed, P_i is a product, not ATP. A low level of substrate cycling is now believed to enhance metabolic flexibility. The loss of large amounts of energy, a situation referred to as a **futile cycle,** is usually prevented by metabolic control mechanisms.

Gluconeogenesis is an energy-consuming process. Instead of generating ATP (as in glycolysis), gluconeogenesis requires the hydrolysis of six molecules of ATP.

QUESTIONS

8.5 *Malignant hyperthermia* is a rare, inherited disorder triggered during surgery by certain anesthetics. A dramatic (and dangerous) rise in body temperature (as high as 112°F) is accompanied by muscle rigidity and acidosis. The excessive muscle contraction is initiated by a large release of calcium from the sarcoplasmic reticulum. (The sarcoplasmic reticulum is a calcium-storing organelle in muscle cells.) Acidosis results from excessive lactic acid production. Prompt treatment to reduce body temperature and to counteract the acidosis is essential to save the patient's life. A probable contributing factor to this disorder is futile cycling between glycolysis and gluconeogenesis. Explain why this is a reasonable explanation.

8.6 The reaction summary for gluconeogenesis is shown below. After examining the gluconeogenic pathway, account for each component in the equation. (Hint: The hydrolysis of each nucleotide releases a proton.)

2 $C_3H_8O_3$ + 4 ATP + 2 GTP + 2 NADH + 2 H^+ + 6 H_2O →

(Pyruvic acid)

1 $C_6H_{12}O_6$ + 4 ADP + 2 GDP + 2 NAD^+ + 6 HPO_4^{2-} + 6 H^+

Glucose

QUESTION 8.7

Patients with *Von Gierke's disease* (a glycogen storage disease) lack glucose-6-phosphatase activity. Two prominent symptoms of this disorder are fasting hypoglycemia and lactic acidosis. Can you explain why these symptoms occur?

Gluconeogenesis Substrates

As was previously mentioned, several metabolites are gluconeogenic precursors. Three of the most important substrates are described briefly.

Lactate is released by red blood cells and other cells that lack mitochondria or have low oxygen concentrations. In the **Cori cycle,** lactate is released by skeletal muscle during exercise (Figure 8.14). After lactate is transferred to the liver, it is reconverted to pyruvate by lactate dehydrogenase and then to glucose by gluconeogenesis.

Glycerol, a product of fat metabolism in adipose tissue, is transported to liver in the blood, and then converted to glycerol-3-phosphate by glycerol kinase. (Glycerol kinase is found only in liver.) Oxidation of glycerol-3-phosphate to form DHAP occurs when cytoplasmic NAD^+ concentration is relatively high.

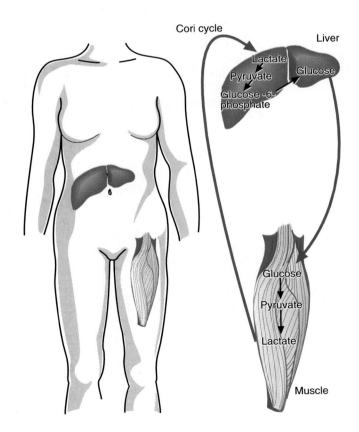

FIGURE 8.14

The Cori Cycle.

During strenuous exercise, lactate is produced in muscle cells under anaerobic conditions. After passing through blood to the liver, lactate is converted to glucose by gluconeogenesis.

Of all the amino acids that can be converted to glycolytic intermediates (molecules referred to as *glucogenic*), alanine is perhaps the most important. (The metabolism of the glucogenic amino acids is described in Chapter 14.) When exercising muscle produces large quantities of pyruvate, some of these molecules are converted to alanine by a transamination reaction involving glutamate.

$$CH_3-\overset{\overset{O}{\|}}{C}-\overset{\overset{O}{\|}}{C}-O^- \quad + \quad ^-O-\overset{\overset{O}{\|}}{C}-CH_2-CH_2-\overset{\underset{+NH_3}{|}}{\underset{|}{C}}-\overset{\overset{O}{\|}}{C}-O^-$$

Pyruvate   **L-Glutamate**

Alanine transaminase

$$CH_3-\overset{\overset{H}{|}}{\underset{\underset{+NH_3}{|}}{C}}-\overset{\overset{O}{\|}}{C}-O^- \quad + \quad ^-O-\overset{\overset{O}{\|}}{C}-CH_2-CH_2-\overset{\overset{O}{\|}}{C}-\overset{\overset{O}{\|}}{C}-O^-$$

L-Alanine   **α-Ketoglutarate**

After it is transported to the liver, alanine is reconverted to pyruvate and then to glucose. The **alanine cycle** (Figure 8.15) serves several purposes. In addition to its role in recycling carbon skeletons between muscle and liver, the alanine cycle is a mechanism for transporting NH_4^+ to the liver. The liver then converts NH_4^+, a very toxic ion, to urea.

According to several recent studies, gluconeogenesis plays a prominent role in liver glycogenesis following a carbohydrate meal. It appears that under physiological conditions, a portion of the new glycogen is formed by a mechanism involving the following sequence: dietary glucose → C_3-molecule → liver glycogen. Lactate and alanine are believed to be the most likely C_3-molecules in this process.

Gluconeogenesis Regulation

As with other metabolic pathways, the rate of gluconeogenesis is affected primarily by substrate availability, allosteric effectors, and hormones. Not surprisingly, gluconeogenesis is stimulated by high concentrations of lactate, glycerol, and amino acids. A high-fat diet, starvation, and prolonged fasting make large quantities of these molecules available.

The four key enzymes in gluconeogenesis (pyruvate carboxylase, PEP carboxykinase, fructose-1,6-bisphosphatase, and glucose-6-phosphatase) are affected to varying degrees by allosteric modulators. For example, fructose-1,6-bisphosphatase is activated by ATP and inhibited by AMP and fructose-2,6-bisphosphate. Acetyl-CoA

FIGURE 8.15

The Glucose-Alanine Cycle.

Alanine is formed from pyruvate in muscle. After it is transported to the liver, alanine is reconverted to pyruvate by alanine transaminase. Eventually pyruvate is used in the synthesis of new glucose. Because muscle cannot synthesize urea from amino nitrogen, the glucose-alanine cycle is used to transfer amino nitrogen to the liver.

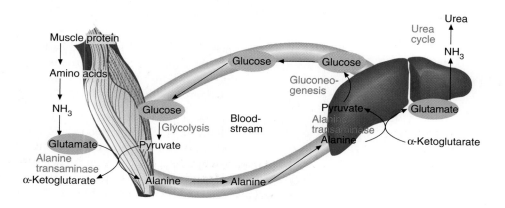

CALCIUM METABOLISM AND BLOOD GLUCOSE

The separation of function allowed by compartmentation within living cells provides sophisticated metabolic control. A recent hypothesis concerned with the relationship between glycogenolysis and calcium metabolism provides an interesting example.

Calcium metabolism is complex. Calcium ions (Ca^{2+}) are involved in an astonishing array of physiological processes. In addition to their role in muscle contraction, endocytosis, blood clotting, and neurotransmitter release (just to name a few), Ca^{2+} modulates the activity of numerous enzymes. Because Ca^{2+} acts as a trigger in these processes, its intracellular concentration is normally kept very low (approximately 10^{-7} M). A steep concentration gradient is achieved by ATP-powered calcium pumps in the plasma membrane and in the membranes of an intracellular calcium-sequestering compartment. (As noted, in muscle cells this compartment is referred to as the sarcoplasmic reticulum. Nonmuscle cells sequester calcium in the *calcisome*, a structure believed to be part of the SER.) Calcium-regulated processes begin after the calcium channels open in both membranes. Calcium is valuable as a metabolic signal because its normally low intracellular concentration can be raised and lowered rapidly. Recent research provides evidence that two processes, calcium signal quenching and blood glucose regulation, are more closely related than was previously thought.

Recall that the product of glycogenolysis is glucose-1-phosphate, which is converted to glucose-6-phosphate by phosphoglucomutase. In liver cells, which maintain blood glucose levels, glucose-6-phosphate is converted to glucose by glucose-6-phosphatase, an enzyme situated on the internal face of SER membrane. (Because glucose-6-phosphate is formed in the cytoplasm, its transport into the SER lumen by a specific transmembrane transport protein has been proposed.) The functional reasons for such an arrangement have been considered a mystery, because all the other reactions of glucose metabolism occur in cytoplasm.

A mechanism has recently been proposed that links the secretion of glucose into blood and calcium signal quenching (Figure 8B). In this model, three systems influence calcium levels in the calcisome: (1) an ATP-dependent calcium uptake pump that promotes calcium influx, (2) calcium channels that allow calcium efflux into the cytoplasm, and (3) a calcium sequestering system that converts glucose-6-phosphate to glucose.

The hormonal signal that stimulates glycogenolysis also opens calcium channels. (See Chapter 15 for a more detailed description of the role of calcium in this process.) As Ca^{2+} levels rise, phosphorylase kinase (the first enzyme in the enzyme cascade that degrades glycogen) is activated. Ultimately, glycogen is converted to glucose-6-phosphate, which is then transported into the calcisome. Membrane-bound glucose-6-phosphatase cleaves off the phosphate group. The uncharged glucose is then free to diffuse out of the cell and into the blood. The phosphate generated from glucose-6-phosphate hydrolysis coprecipitates with Ca^{2+}, which is simultaneously pumped from the cytoplasm by the ATP-calcium pump. The high concentrations of phosphate generated by glucose-6-phosphatase are believed to be largely responsible for the massive translocation of Ca^{2+} out of the cytoplasm that is observed during glucose-6-phosphate hydrolysis. (A calcium-binding protein called *calsequestrin*, found within the calcium-sequestering compartment, also binds to some of the translocated Ca^{2+}.)

KEY *Concepts*

Gluconeogenesis, the synthesis of new glucose molecules from noncarbohydrate precursors, occurs primarily in the liver. The reaction sequence is the reverse of glycolysis except for three reactions that bypass irreversible steps in glycolysis.

activates pyruvate carboxylase. (The concentration of acetyl-CoA, a product of fatty acid degradation, is especially high during starvation.)

As with other biochemical pathways, hormones affect gluconeogenesis by altering the concentrations of allosteric effectors and the rate key enzymes are synthesized. As mentioned previously, glucagon depresses the synthesis of fructose-2,6-bisphosphate, by inhibiting PFK-1 and activating fructose-1,6-bisphosphatase. Another effect of glucagon binding to liver cells is the inactivation of the glycolytic enzyme pyruvate kinase. (Protein kinase, an enzyme activated by cAMP, converts pyruvate kinase to its inactive phosphorylated conformation.) Hormones also influence gluconeogenesis by altering enzyme synthesis. For example, the synthesis of gluconeogenic enzymes is stimulated by cortisol (a steroid hormone produced in the cortex of the adrenal gland). (*Cortisol* facilitates the body's adaptation to stressful situations. Its actions affect carbohydrate, protein, and lipid metabolism.) Finally, insulin suppresses the synthesis of all the key gluconeogenic enzymes

8.5 METABOLISM OF OTHER IMPORTANT SUGARS

Several sugars other than glucose are important in vertebrates. The most notable of these are fructose, galactose, and mannose. Besides glucose, these molecules are the most common sugars found in oligosaccharides and polysaccharides. They are also energy sources. The interconversions among these sugars are illustrated in Figure 8.16. The synthesis of ribose (a structural element of nucleotides and the nucleic acids) is described in Chapter 11.

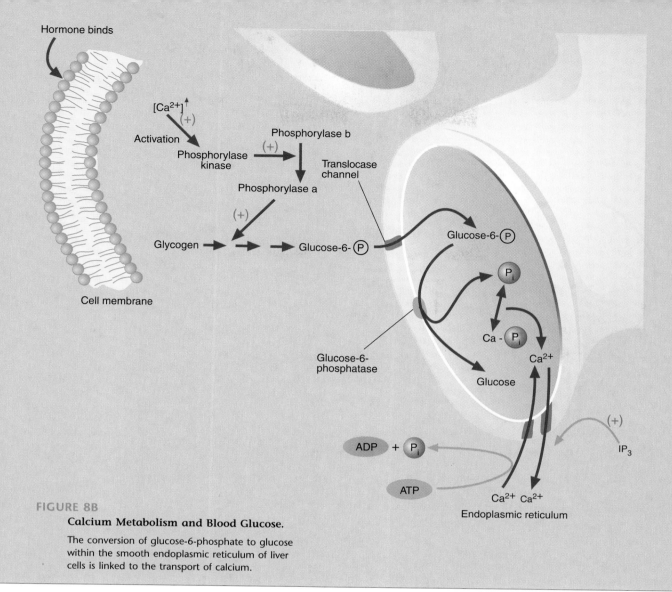

Hormone binds

$[Ca^{2+}]$

$(+)$

Activation

Phosphorylase b

Phosphorylase kinase

$(+)$

Phosphorylase a

Translocase channel

$(+)$

Glycogen → → → Glucose-6-$\boxed{P}$

Glucose-6-$\boxed{P}$

P_i

Ca - P_i

Ca^{2+}

Glucose-6-phosphatase

Glucose

ADP + P_i

$(+)$

IP_3

ATP

Ca^{2+} Ca^{2+}

Endoplasmic reticulum

Cell membrane

FIGURE 8B

Calcium Metabolism and Blood Glucose.

The conversion of glucose-6-phosphate to glucose within the smooth endoplasmic reticulum of liver cells is linked to the transport of calcium.

Fructose Metabolism

As described previously, dietary sources of fructose include fruit, honey, and the disaccharide sucrose. Fructose, a significant source of carbohydrate in the human diet (second only to glucose), can enter the glycolytic pathway by two routes. In the liver, fructose is converted to fructose-1-phosphate by fructokinase.

HO—CH₂   CH₂OH   HO—CH₂   CH₂—O—P—O⁻

OH   OH

OH   OH

Fructokinase

ATP

ADP

OH   OH

Fructose   **Fructose-1-phosphate**

When fructose-1-phosphate enters the glycolytic pathway, it is first split into DHAP and glyceraldehyde by fructose-1-phosphate aldolase. DHAP is then converted to glyceraldehyde-3-phosphate by triose phosphate isomerase. Glyceraldehyde-3-phosphate is generated from glyceraldehyde and ATP by glyceraldehyde kinase.

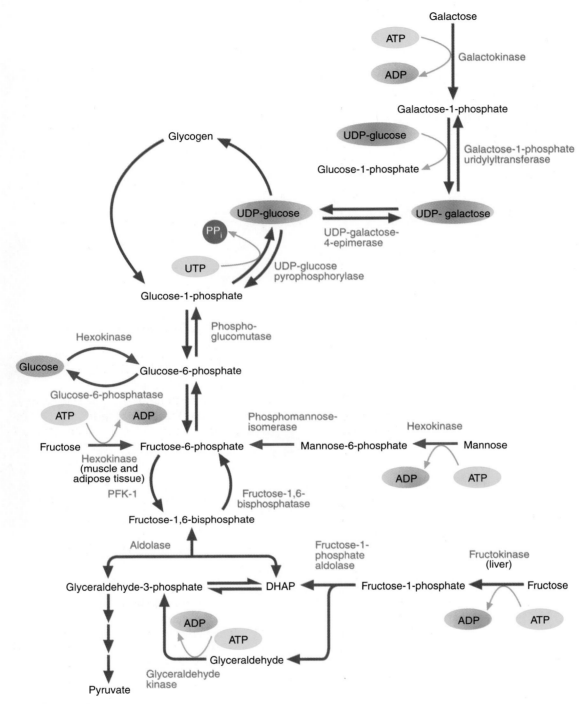

FIGURE 8.16

Carbohydrate Metabolism.

The interconversions of galactose, mannose, and fructose with glucose and its metabolites.

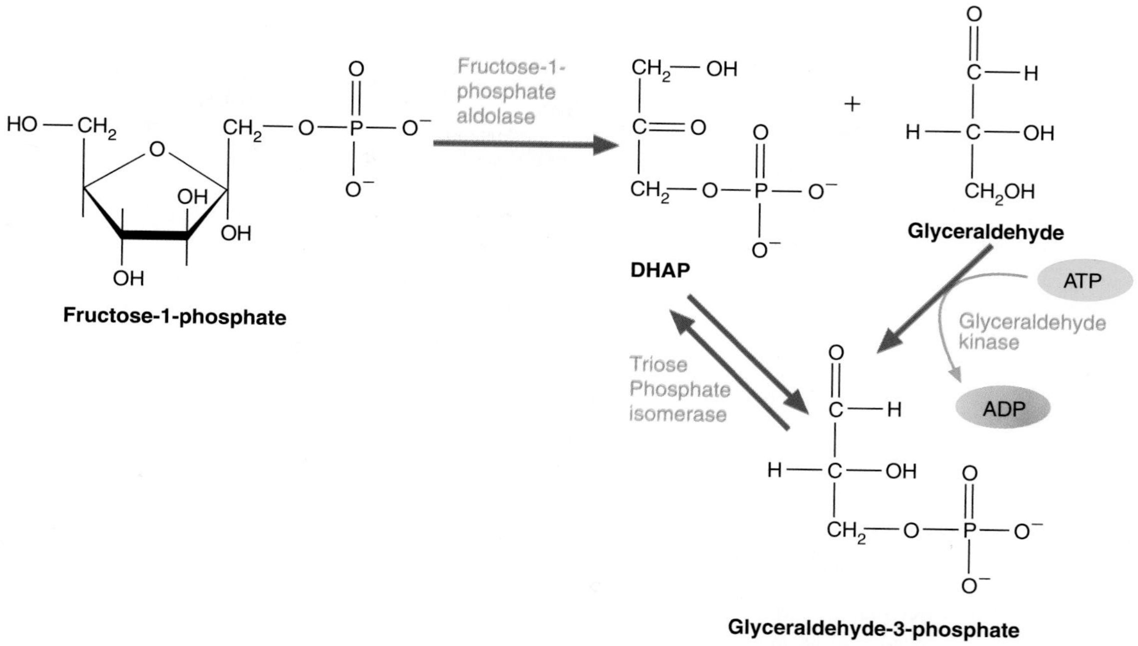

Glyceraldehyde-3-phosphate

The conversion of fructose-1-phosphate into glycolytic intermediates bypasses two regulatory steps (the reactions catalyzed by hexokinase and PFK-1) thus fructose is metabolized more quickly than glucose.

In muscle and adipose tissue, fructose is converted to the glycolytic intermediate fructose-6-phosphate by hexokinase. Because the hexokinases have a low affinity for fructose, this reaction is of minor importance unless fructose consumption is exceptionally high.

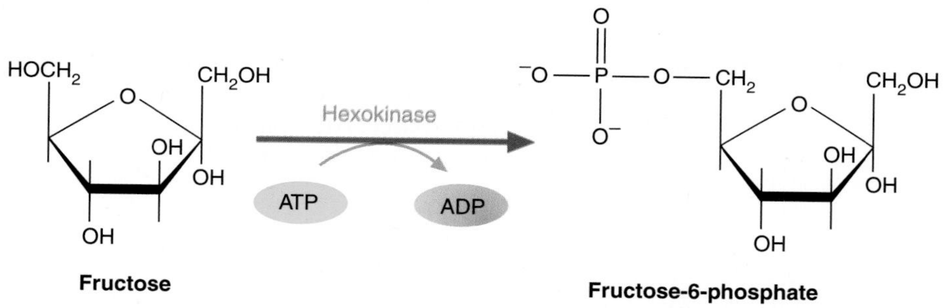

Galactose Metabolism

Although galactose and glucose have similar structures (i.e., they are epimers), several reactions are required for this sugar to enter the glycolytic pathway. Galactose is initially converted to galactose-1-phosphate by galactokinase.

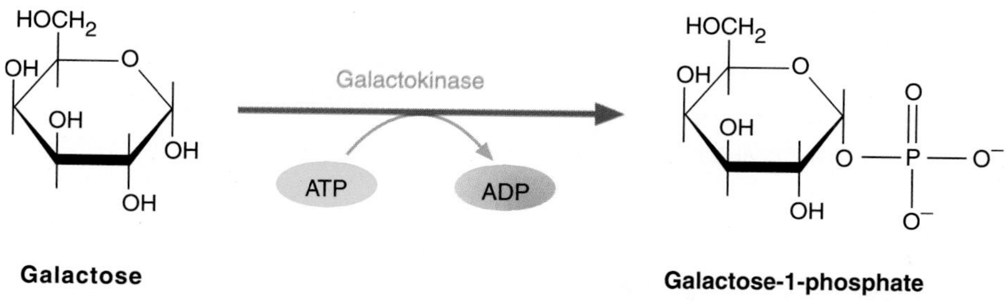

Then galactose-1-phosphate is transformed into the nucleotide derivative UDP-galactose. During fetal development and childhood the first step in this conversion is catalyzed by galactose-1-phosphate uridyltransferase. (The hereditary disorder galactosemia, described on p. 164, is caused by the absence of this enzyme.)

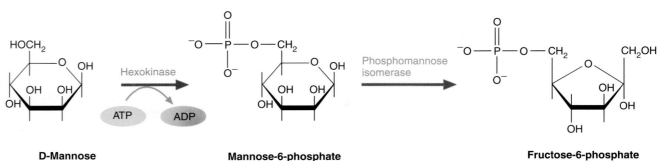

Galactose-1-phosphate → (Galactose-1-phosphate uridyltransferase; UDP-glucose → Glucose-1-phosphate) → **UDP-galactose**

Beginning in adolescence, UDP-galactose is produced in a reaction catalyzed by UDP-galactose pyrophosphorylase.

$$\text{Galactose-1-phosphate} + \text{UTP} \rightleftharpoons \text{UDP-galactose} + \text{PP}_i$$

Then UDP-glucose is formed by the isomerization by galactose catalyzed by UDP-glucose-4-epimerase.

UDP-galactose ⇌ **UDP-glucose**

Depending on the cell's metabolic needs, UDP-glucose is used directly in glycogen synthesis or is converted to glucose-1-phosphate by UDP-glucose pyrophosphorylase. Glucose-1-phosphate enters the glycolytic pathway after its conversion to glucose-6-phosphate by phosphoglucomutase.

Mannose Metabolism

Mannose is an important component of the oligosaccharides that are found in glycoproteins. Because it is a minor component in the diet, mannose is an unimportant energy source. After phosphorylation by hexokinase, mannose enters the glycolytic pathway as fructose-6-phosphate.

D-Mannose → (Hexokinase; ATP → ADP) → **Mannose-6-phosphate** → (Phosphomannose isomerase) → **Fructose-6-phosphate**

KEY WORDS

SUMMARY

Metabolism includes two major types of biochemical pathways: anabolic and catabolic. In anabolic pathways, large complex molecules are synthesized from smaller precursors. In catabolic pathways, large complex molecules are degraded into smaller, simpler products. Some catabolic reactions release free energy. A fraction of this energy is used to drive certain anabolic reactions. Multicellular organisms maintain an appropriate balance between anabolic and catabolic processes by using intercellular chemical signals. In animals, hormone molecules and neurotransmitters are intercellular signals.

The metabolism of carbohydrates is dominated by glucose because this sugar is an important fuel molecule in most organisms. If cellular energy reserves are low, glucose is degraded by the glycolytic pathway. Glucose molecules that are not required for immediate energy production are stored as either glycogen (in animals) or starch (in plants).

The substrate for glycogen synthesis is UDP-glucose, an activated form of the sugar. UDP-glucose pyrophosphorylase catalyzes the formation of UDP-glucose from glucose-1-phosphate and UTP. Glucose-6-phosphate is converted to glucose-1-phosphate by phosphoglucomutase. To form glycogen requires two enzymes: glycogen synthase and branching enzyme. Glycogen degradation requires glycogen phosphorylase and debranching enzyme. The balance between glycogenesis (glycogen synthesis) and glycogenolysis (glycogen breakdown) is carefully regulated by several hormones (insulin, glucagon, and epinephrine).

During glycolysis, glucose is phosphorylated and cleaved to form two molecules of glyceraldehyde-3-phosphate. Each glyceraldehyde-3-phosphate is then converted to a molecule of pyruvate. A small amount of energy is captured in two molecules of ATP and NADH. In anaerobic organisms, pyruvate is converted to waste products. During this process, NAD^+ is regenerated so that glycolysis can continue. In the presence of O_2, aerobic organisms convert pyruvate to acetyl CoA and then to CO_2 and H_2O. Glycolysis is controlled primarily by allosteric regulation of three enzymes-hexokinase, PFK-1, and pyruvate kinase—and by the hormones glucagon and insulin.

During gluconeogenesis, molecules of glucose are synthesized from noncarbohydrate precursors (lactate, pyruvate, glycerol, and certain amino acids). The reaction sequence in gluconeogenesis is largely the reverse of glycolysis. The three irreversible glycolytic reactions (the synthesis of pyruvate, the conversion of fructose-1,6-bisphosphate to fructose-6-phosphate, and the formation of glucose from glucose-6-phosphate) are bypassed by alternate energetically favorable reactions.

Several sugars other than glucose are important in vertebrate carbohydrate metabolism. These include fructose, galactose, and mannose.

SUGGESTED READINGS

Hallfrisch, J., Metabolic Effects of Dietary Fructose, *FASEB J.,* 4:2652–2660, 1990.

Newsholme, E.A., Challiss, R.A.J., and Crabtree, B., Substrate Cycles: Their Role in Improving Sensitivity in Metabolic Control, *Trends Biochem. Sci.,* 9:277–280, 1984.

Pilkus, S.J., Mahgrabi, M.R., and Claus, T.A., Hormonal Regulation of Hepatic Gluconeogenesis and Glycolysis, *Ann. Rev. Biochem.,* 57:755–783, 1988.

Shulman, G.I., and Landau, B.R., Pathways of Glycogen Repletion, *Physiol. Rev.,* 72(4):1019–1035, 1992.

VanSchaftingen, E., Fructose-2,6-Bisphosphate, *Adv. Enzymol.,* 59: 315–395, 1987.

QUESTIONS

1. Metabolism consists of two major processes. What are they? What function does each one perform? Give two examples of each process.

2. Catabolism consists of three steps. Describe what is accomplished in each step.

3. What is the most important end product of stage 2 metabolism? What is the fate of this molecule in stage 3?

4. What two important components of anabolic processes are produced by catabolic reactions?

5. Briefly define each of the following terms:
   a. amphibolic
   b. steady state
   c. target cell
   d. second messenger
   e. limit dextrin

6. Describe how an enzyme cascade magnifies an initial hormonal signal.

7. Upon entering a cell, glucose is phosphorylated. Give two reasons why this reaction is required.

8. Describe the functions of the following molecules:
   a. insulin
   b. glucagon
   c. fructose-2,6-bisphosphate

9. An individual has a genetic deficiency that prevents the production of hexokinase D. Following a carbohydrate meal, would you expect blood glucose levels to be higher, lower, or about normal? What organ would accumulate glycogen under these circumstances?

10. Glycogen synthesis requires a short primer chain. Explain how new glycogen molecules are synthesized given this limitation.

11. Describe how epinephrine promotes the conversion of glycogen to glucose.

12. Glycolysis occurs in two stages. Describe what is accomplished in each stage.

13. Why is fructose metabolized more rapidly than glucose?

14. What is the difference between an enol-phosphate ester and a normal phosphate ester that gives PEP such a high phosphate group transfer potential?

15. In aerobic oxidation, oxygen is the ultimate oxidizing agent (electron acceptor). Name two common oxidizing agents in anaerobic fermentation.

16. Why is it important that gluconeogenesis is not the exact reverse of glycolysis?

17. Where in the body does gluconeogenesis occur? Describe the physiological conditions that activate this important process.

18. The following two reactions constitute a futile cycle:

$$\text{Glucose} + \text{ATP} \rightarrow \text{glucose-6-phosphate}$$
$$\text{Glucose-6-phosphate} + H_2O \rightarrow \text{glucose} + P_i$$

Suggest how such futile cycles are prevented or controlled.

19. What effects do the following molecules have on gluconeogenesis?
    a. lactate
    b. ATP
    c. pyruvate
    d. glycerol
    e. AMP
    f. acetyl-CoA

LIPIDS AND MEMBRANES

Chapter Nine

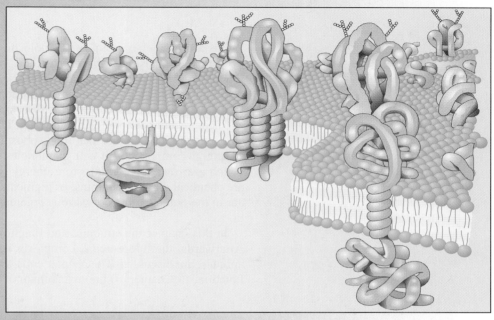

Biological Membrane. Proteins float in a bimolecular lipid layer.

OBJECTIVES

After you have studied this chapter, you should be able to answer these questions:

1. What are the major lipid classes?

2. How do the structural properties of saturated and unsaturated fatty acids affect their biological properties?

3. What are the eicosanoids and why are they important in mammals?

4. How do triacylglycerols store energy efficiently?

5. What role do the wax esters play in living organisms?

6. What structural properties of phospholipids make them the major constituent of biological membranes?

7. How do sphingolipids and glycolipids differ from phospholipids? What role do these molecules play in living organisms?

8. What structural feature do terpenes and steroids have in common? How do these isoprenoid molecules differ from each other? What roles do they play in living organisms?

9. What role do lipids and proteins play in membrane structure and function?

Lipids are naturally occurring substances that dissolve in hydrocarbons but not in water. They perform a stunning array of functions in living organisms. Some lipids are vital energy reserves. Others are the primary structural components of biological membranes. Still other lipid molecules act as hormones, antioxidants, pigments, or vital growth factors and vitamins. This chapter describes the structures and properties of the major lipid classes found in living organisms.

Lipids are a diverse group of biomolecules. Molecules such as fats and oils, phospholipids, steroids, and the carotenoids, which differ widely in both structure and function, are all considered lipids. Because of this diversity, the term **lipid** has an operational rather than a structural definition. Lipids are defined as those substances from living organisms that dissolve in nonpolar solvents such as ether, chloroform, and acetone but not appreciably in water. The functions of lipids are also diverse. Several types of lipid molecules (e.g., phospholipids and sphingolipids) are important structural components in cell membranes. Another type, the fats and oils (both of which are triacylglycerols), store energy efficiently. Other types of lipid molecules are chemical signals, vitamins, or pigments. Finally, some lipid molecules, which occur in the outer coatings of various organisms, have protective or waterproofing functions.

In this chapter the structure and function of each major type of lipid is described. Afterwards, the lipoproteins, complexes of protein and lipid that transport lipids in animals, are discussed. The chapter ends with an overview of membrane structure and function. In Chapter 10 the metabolism of several major lipids is described.

9.1 LIPID CLASSES

Lipids may be classified in many different ways. For this discussion, lipids can be subdivided into the following classes:

1. Fatty acids and their derivatives,
2. Triacylglycerols,
3. Wax esters,
4. Phospholipids (phosphoglycerides and sphingomyelin),
5. Sphingolipids (molecules other than sphingomyelin that contain the amino alcohol sphingosine), and
6. Isoprenoids (molecules that appear to consist of multiple copies of a branched five-carbon hydrocarbon unit called isoprene).

Each class is discussed below.

Fatty Acids and Their Derivatives

As previously described, fatty acids are monocarboxylic acids that contain hydrocarbon chains of variable lengths. (The structures and names of several common fatty acids are illustrated in Table 9.1.) Fatty acids are important components of several types of lipid molecules. They occur primarily in triacylglycerols and several types of membrane-bound lipid molecules.

Certain fatty acids (primarily myristic and palmitic acids) are covalently attached to a wide variety of eukaryotic proteins. Such proteins are referred to as *acylated* proteins. Fatty acid groups (called **acyl groups**) clearly facilitate the interactions between membrane proteins and their hydrophobic environment. However, many aspects of the role of protein acylation are still not understood.

Most naturally occurring fatty acids have an even number of carbon atoms that form an unbranched chain. (Unusual fatty acids with branched or ring-containing chains are found in some species.) Fatty acid chains that contain only carbon-carbon

TABLE 9.1

Examples of Fatty Acids

Common Name	Structure	Abbreviation[*]
Saturated Fatty Acids		
Myristic acid	$CH_3(CH_2)_{12}COOH$	14:0
Palmitic acid	$CH_3(CH_2)_{12}CH_2CH_2COOH$	16:0
Stearic acid	$CH_3(CH_2)_{12}CH_2CH_2CH_2CH_2COOH$	18:0
Arachidic acid	$CH_3(CH_2)_{12}CH_2CH_2CH_2CH_2CH_2CH_2COOH$	20:0
Lignoceric acid	$CH_3(CH_2)_{12}CH_2CH_2CH_2CH_2CH_2CH_2CH_2CH_2CH_2CH_2COOH$	24:0
Cerotic acid	$CH_3(CH_2)_{12}CH_2CH_2CH_2CH_2CH_2CH_2CH_2CH_2CH_2CH_2CH_2CH_2COOH$	26:0
Unsaturated Fatty Acids		
Palmitoleic acid	$CH_3(CH_2)_5\overset{\text{H}}{C}=\overset{\text{H}}{C}(CH_2)_7COOH$	$16:1^{\Delta 9}$
Oleic acid	$CH_3(CH_2)_7\overset{\text{H}}{C}=\overset{\text{H}}{C}(CH_2)_7COOH$	$18:1^{\Delta 9}$
Linoleic acid	$CH_3(CH_2)_4\overset{\text{H}}{C}=\overset{\text{H}}{C}-CH_2-\overset{\text{H}}{C}=\overset{\text{H}}{C}(CH_2)_7COOH$	$18:2^{\Delta 9,12}$
α-Linolenic acid	$CH_3CH_2\overset{\text{H}}{C}=\overset{\text{H}}{C}-CH_2-\overset{\text{H}}{C}=\overset{\text{H}}{C}-CH_2-\overset{\text{H}}{C}=\overset{\text{H}}{C}(CH_2)_7COOH$	$18:3^{\Delta 9,12,15}$
Arachidonic acid	$CH_3(CH_2)_3-\left(CH_2-\overset{\text{H}}{C}=\overset{\text{H}}{C}\right)_4-(CH_2)_3COOH$	$20:4^{\Delta 5,8,11,14}$

[*]In these abbreviations the number to the left of the colon is the number of carbon atoms, and the number to the right is the number of double bonds. A superscript denotes the placement of a double bond. For example, $\Delta 9$ signifies that there are eight carbons between the carboxyl group and the double bond i.e., the double bond occurs between carbons 9 and 10.

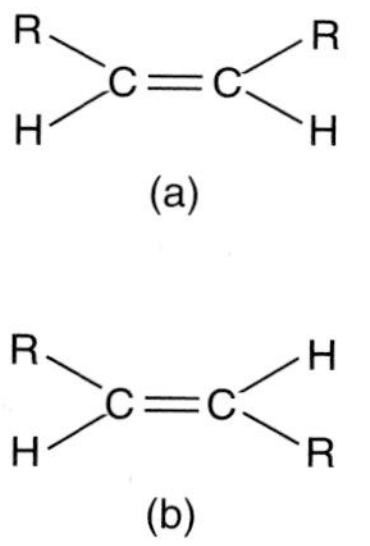

FIGURE 9.1

Isomeric Forms of Unsaturated Molecules.

(a) *cis*-isomer. (b) *trans*-isomer.

single bonds are referred to as *saturated*. Those molecules that contain one or more double bonds are said to *unsaturated*. Because double bonds are rigid structures, molecules that contain them can occur in two isomeric forms: *cis* and *trans*. In **cis-isomers,** for example, similar or identical groups are on the same side of a double bond (Figure 9.1a). When such groups are on opposite sides of a double bond, the molecule is said to be a ***trans*-isomer** (Figure 9.1b).

The double bonds in most naturally occurring fatty acids are in a *cis* configuration. The presence of a *cis* double bond causes an inflexible "kink" in a fatty acid chain (Figure 9.2). Because of this structural feature, unsaturated fatty acids do not pack as closely together as saturated fatty acids. Therefore unsaturated fatty acids have lower melting points and are liquids at room temperature. For example, a sample of palmitic acid (16:0), a saturated fatty acid, melts at 63°C, while palmitoleic acid ($16:1^{\Delta 9}$) melts at 0°C. Interestingly, fatty acids with *trans* double bonds have three-dimensional structures similar to those of saturated fatty acids.

Fatty acids with one double bond are referred to as **monounsaturated** molecules. When two or more double bonds occur in fatty acids, usually separated by methylene groups ($-CH_2-$), they are referred to as **polyunsaturated.** The monounsaturated fatty acid oleic acid ($18:1^{\Delta 9}$) and the polyunsaturated linoleic acid ($18:2^{\Delta 9,12}$) are among the most abundant fatty acids in living organisms.

Organisms such as plants and bacteria can synthesize all the fatty acids they require from acetyl-CoA (Chapter 10). Mammals obtain most of their fatty acids from dietary sources. However, these organisms can synthesize saturated fatty acids and some monounsaturated fatty acids. They can also modify some dietary fatty acids by adding two-carbon units and introducing some double bonds. Fatty acids that can be

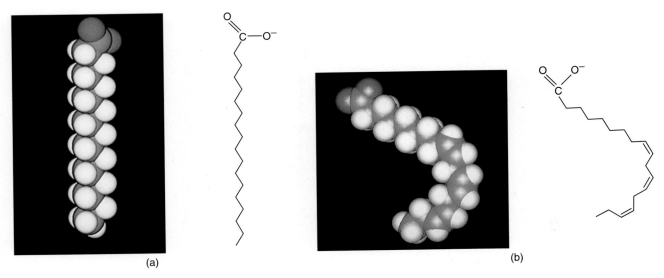

FIGURE 9.2

Space-Filling and Conformational Models.

(a) A saturated fatty acid (stearic acid), and (b) an unsaturated fatty acid (α-linolenic acid).

synthesized are called **nonessential fatty acids.** Because mammals do not possess the enzymes required to synthesize linoleic ($18:2^{\Delta9,12}$) and linolenic ($18:3^{\Delta9,12,15}$) acids, these **essential fatty acids** must be obtained from the diet. (Rich sources of essential fatty acids, which have several critical physiological functions, include some vegetable oils, nuts, and seeds.) In addition to contributing to proper membrane structure, linoleic and linolenic acids are precursors of several important metabolites. The most researched examples of fatty acid derivatives are the eicosanoids (Special Interest Box 9.1).

Fatty acids have several important chemical properties. The reactions that they undergo are typical of short-chain carboxylic acids. For example, fatty acids react with alcohols to form esters.

$$ R-\overset{\overset{\displaystyle O}{\|}}{C}-OH \ +\ R'-OH \ \rightleftharpoons\ R-\overset{\overset{\displaystyle O}{\|}}{C}-O-R' \ +\ H_2O $$

This reaction is reversible; that is, under appropriate conditions a fatty acid ester can react with water to produce a fatty acid and an alcohol. Unsaturated fatty acids with double bonds can undergo hydrogenation reactions to form saturated fatty acids. Finally, unsaturated fatty acids are susceptible to oxidative attack. (This feature of fatty acid chemistry is described in Chapter 11.)

Triacylglycerols

Triacylglycerols are esters of glycerol with three fatty acid molecules (Figure 9.3). (Glycerides with one or two fatty acid groups, called monoacylglycerols and diacylglycerols, respectively, are metabolic intermediates. They are normally present in small amounts.) Because triacylglycerols have no charge (i.e., the carboxyl group of each fatty acid is joined to glycerol through a covalent bond), they are sometimes referred to as **neutral fats.** Most triacylglycerol molecules contain fatty acids of varying lengths, which may be unsaturated, saturated, or a combination (Figure 9.4). Depending on their fatty acid compositions, triacylglycerol mixtures are referred to as fats or oils. *Fats,* which are solid at room temperature, contain a large proportion of saturated fatty acids. *Oils* are liquid at room temperature because of their relatively

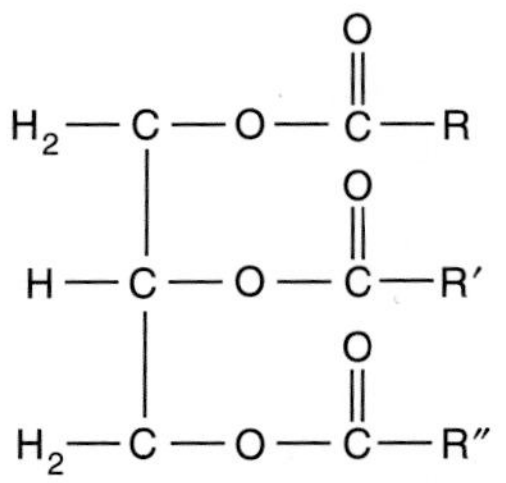

FIGURE 9.3

Triacylgylcerol.

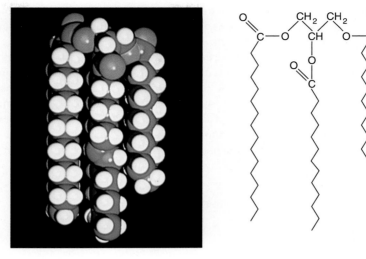

FIGURE 9.4

Space-Filling and Conformational Models of a Triacylglycerol.

Triacylglycerols store energy.

high unsaturated fatty acid content. (Recall that unsaturated fatty acids do not pack together as closely as do saturated fatty acids.)

In animals, triacylglycerols (usually referred to as "fat") have several roles. First, they are the major storage and transport form of fatty acids. Triacylglycerol molecules store energy more efficiently than glycogen for several reasons:

KEY *Concepts*

Triacylglycerols are molecules consisting of glycerol esterified to three fatty acids. In both animals and plants they are a rich energy source.

1. Because triacylglycerols are hydrophobic, they coalesce into compact, anhydrous droplets within cells. A specialized type of cell called the *adipocytes,* found in adipose tissue stores triacylglycerols. Because glycogen (the other major energy storage molecule) binds a substantial amount of water, the anhydrous triacylglycerols store an equivalent amount of energy in about one-eighth of glycogen's volume.

2. Triacylglycerol molecules are less oxidized than carbohydrate molecules. Therefore, triacylglycerols release more energy (9.3 kcal per gram of fat compared with 4.1 kcal per gram of carbohydrate) when they are degraded.

A second important function of fat is to provide insulation in low temperatures. Fat is a poor conductor of heat. Because adipose tissue, with its high triacylglycerol content, is found throughout the body (especially underneath the skin), it prevents heat loss. Finally, in some animals fat molecules secreted by specialized glands make fur or feathers water-repellent.

In plants, triacylglycerols constitute an important energy reserve in fruits and seeds. Because these molecules contain relatively large amounts of unsaturated fatty acids (e.g., oleic and linoleic), they are referred to as plant oils. Seeds rich in oil include peanut, corn, palm, safflower, and soybean. Avocados and olives are fruits with a high oil content.

QUESTION 9.1

Oils can be converted to fats in a commercial nickel-catalyzed process referred to as *partial hydrogenation.* Under relatively mild conditions (180°C and pressures of about 20 lb/in.^{2}) enough double bonds are hydrogenated for liquid oils to solidify. This solid material, oleomargarine, has a consistency like butter. Propose a practical reason why oils are not completely hydrogenated during commercial hydrogenation processes.

THE EICOSANOIDS

The **eicosanoids** are a diverse group of extremely powerful hormonelike molecules produced in most mammalian tissues. (Because they are generally active within the organ in which they are produced, the eicosanoids are called **autocrine** regulators instead of hormones.) Most eicosanoids are derived from arachidonic acid ($20:4^{\Delta 5,8,11,14}$), which is also called 5,8,11,14-eicosatetraenoic acid. (Arachidonic acid is synthesized from linoleic acid by adding a two-carbon unit and inserting two additional double bonds.) Production of eicosanoids begins after arachidonic acid is released from membrane phospholipid mol-

ecules by the enzyme phospholipase A_2. The eicosanoids, which include the prostaglandins, thromboxanes, and leukotrienes (Figure 9A), are extremely difficult to study because they are active for short periods (often measured in seconds or minutes). In addition, they are produced only in small amounts.

Prostaglandins are arachidonic acid derivatives that contain a cyclopentane ring with hydroxy groups at C-11 and C-15. Molecules belonging to the E series of prostaglandins have a carbonyl group at C-9, while the F series molecules have an OH group at the same position. The subscript number in a

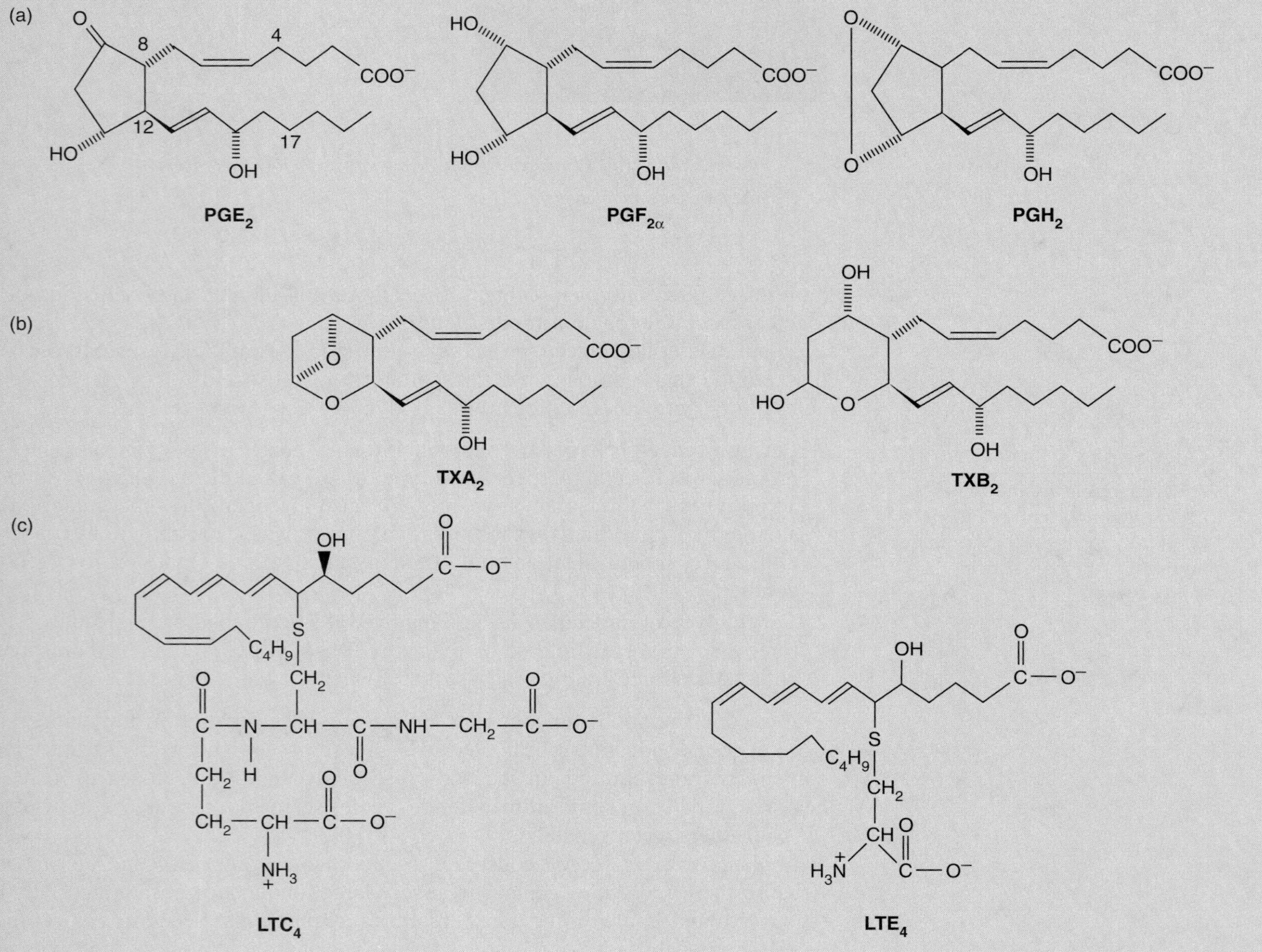

FIGURE 9A

Eicosanoids.

(a) Prostaglandins E_2, $F_{2\alpha}$, and H_2. (b) Thromboxanes A_2 and B_2. (c) Leukotrienes C_4 and E_4.

prostaglandin name indicates the number of double bonds in the molecule. The 2-series, derived from arachidonic acid, appears to be the most important group of prostaglandins in humans. Prostaglandins are involved in a wide range of regulatory functions. For example, prostaglandins promote inflammation, an infection-fighting process that produces pain and fever. They are also involved in reproductive processes (e.g., ovulation and uterine contractions during conception and labor) and digestion (e.g., inhibition of gastric secretion). Additional biological actions of selected prostaglandins are shown in Figure 9B. Prostaglandin metabolism is complex for the following reasons:

1. There are many types of prostaglandins.
2. The types and amounts of prostaglandins are different in each tissue or organ.
3. Certain prostaglandins have opposite effects in different organs. (For example, several E-series prostaglandins cause smooth muscle relaxation in organs such as the intestine and uterus. The same molecules promote contraction of the smooth muscle in the cardiovascular system.)

The **thromboxanes** are also derivatives of arachidonic acid. They differ from other eicosanoids in that their structures have a cyclic ether. Thromboxane A_2 (TxA_2), the most prominent member of this group of eicosanoids, is primarily produced by platelets (blood cells that initiate blood clot formation). Once it is released, TxA_2 promotes platelet aggregation and smooth muscle contraction.

The **leukotrienes** are hydroxy-fatty acid derivatives of arachidonic acid. The name *leukotrienes* stems from their early discovery in white blood cells (leukocytes) and the presence of a *triene* (three conjugated double bonds) in their structures. (The term *conjugated* indicates that carbon-carbon double bonds are separated by one carbon-carbon single bond.) Leukotrienes LTC_4, LTD_4, and LTE_4 have been identified as components of slow-reacting substance of anaphylaxis (SRS-A). (The subscript in a leukotriene name indicates the total number of double bonds in the molecule.) *Anaphylaxis* is an unusually severe allergic reaction that results in respiratory distress, low blood pressure, and shock. During inflammation (a normal response to tissue damage) these molecules increase fluid leakage from blood vessels into affected areas. LTB_4, a potent chemotactic agent, attracts infection-fighting white blood cells to damaged tissue. (Chemotactic agents are also referred to as chemoattractants.)

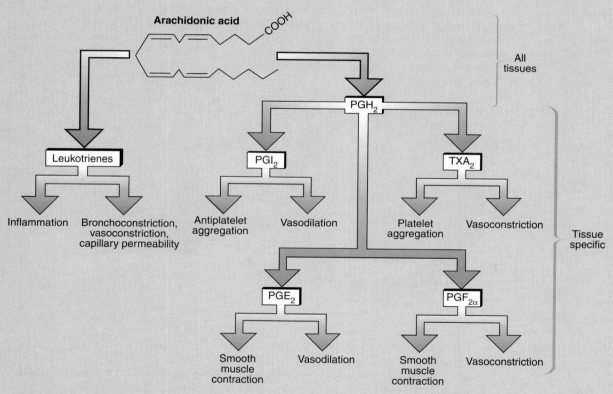

FIGURE 9B

Biological Actions of Selected Eicosanoid Molecules.

The synthesis of these molecules is discussed in Chapter 10.

9.2

Soapmaking is an ancient process. The Phoenicians, a seafaring people who dominated trade in the Mediterranean Sea about 3000 years ago, are believed to have been the first to manufacture soap. Traditionally, soap has been made by heating animal fat with potash. (Potash is a mixture of potassium hydroxide (KOH) and potassium carbonate (K_2CO_3) obtained by mixing wood ash with water.) Currently, soap is made by heating beef tallow or coconut oil with sodium or potassium hydroxide. During this reaction, which is a *saponification,* triacylglycerol molecules are hydrolyzed to give glycerol and the sodium or potassium salts of fatty acids.

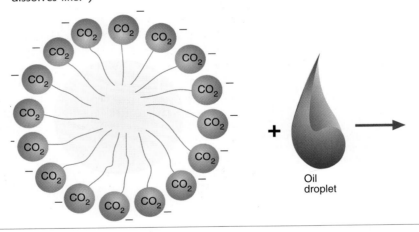

Triacylglycerol   **Soap**   **Glycerol**

Fatty acid salts (soaps) are amphipathic molecules; that is, they spontaneously form into micelles (Figure 3.10). Soap micelles have negatively charged surfaces that repel each other. Because soap can act as an emulsifying agent it is a cleansing agent. (*Emulsifying agents* promote the formation of an emulsion, that is, the dispersal of one substance in another.) When soap and grease are mixed together, an emulsion forms. Complete the following diagram, and explain how this process occurs. (Hint: Recall that "like dissolves like.")

Oil
droplet

Wax Esters

Waxes are complex mixtures of nonpolar lipids. They are protective coatings on leaves, stems, and fruit of plants and the skin and fur of animals. Esters composed of long-chain fatty acids and long-chain alcohols are prominent constituents of most waxes. Well-known examples of waxes include carnauba wax, produced by the leaves of the Brazilian wax palm, and beeswax. The predominant constituent of carnauba wax is the wax ester melissyl cerotate (Figure 9.5). Triacontyl hexadecanoate is one of sev-

FIGURE 9.5

The Wax Ester Melissyl Cerotate.

$$CH_3\!-\!(CH_2)_{24}\!-\!\overset{\overset{\displaystyle O}{\|}}{C}\!-\!O\!-\!(CH_2)_{29}\!-\!CH_3$$

eral important wax esters in beeswax. Waxes also contain hydrocarbons, alcohols, fatty acids, aldehydes, and sterols (steroid alcohols).

Phospholipids

Phospholipids have several roles in living organisms. They are first and foremost structural components of membranes. In addition, several phospholipids are emulsifying agents and surface active agents. (A *surface active agent* is a substance that lowers the surface tension of a liquid, usually water, so that it spreads out over a surface.) Phospholipids are suited to these roles because they are amphipathic molecules. Despite their structural differences, all phospholipids have hydrophobic and hydrophilic domains. The hydrophobic domain is composed largely of the hydrocarbon chains of fatty acids; the hydrophilic domain, called a **polar head group,** contains phosphate and other charged or polar groups.

When phospholipids are suspended in water, they spontaneously rearrange into ordered structures (Figure 9.6). As these structures form, phospholipid hydrophobic groups are buried in the interior to exclude water. Simultaneously, hydrophilic polar head groups are oriented so that they are exposed to water. When phospholipid molecules are present in sufficient concentration, they form bimolecular layers. This property of phospholipids (and other amphipathic lipid molecules) is the basis of membrane structure (pp. 235–239).

There are two types of phospholipids: phosphoglycerides and sphingomyelins. **Phosphoglycerides** are molecules that contain glycerol, fatty acids, phosphate, and an alcohol (e.g., choline). **Sphingomyelins** differ from phosphoglycerides in that they contain sphingosine instead of glycerol. Because sphingomyelins are also classified as sphingolipids, their structures and properties are discussed in the next section.

Phosphoglycerides are the most numerous phospholipid molecules found in cell membranes. The simplest phosphoglyceride, phosphatidic acid, is the precursor for all other phosphoglyceride molecules. Phosphatidic acid is composed of glycerol-3-phosphate that is esterified with two fatty acids. Phosphoglyceride molecules are classified according to which alcohol becomes esterified to the phosphate group. For example, if the alcohol is choline, the molecule is called phosphatidylcholine (PC) (also referred to as lecithin). Other types of phosphoglycerides include phosphatidylethanolamine (PE), phosphatidylserine (PS), diphosphatidylglycerol (dPG), and phosphatidylinositol (PI). (Refer to Table 9.2 for the structures of the common types of phosphoglycerides.) The most common fatty acids in the phosphoglycerides have between 16 and 20 carbons. Saturated fatty acids usually occur at C-1 of glycerol. The fatty acid substituent at C-2 is usually unsaturated. A derivative of phosphatidylinositol, namely, phosphatidyl-4,5-bisphosphate (PIP$_2$), is found in only small amounts in plasma membranes. However, PIP$_2$ is now recognized as an important component of a second messenger system. The *phosphatidylinositol cycle* is initiated when certain hormones bind to membrane receptors.

KEY *Concepts*

Phospholipids are amphipathic molecules that play important roles in living organisms as membrane components, emulsifying agents, and surface active agents. There are two types of phospholipids: phosphoglycerides and sphingomyelins.

FIGURE 9.6

Phospholipid Molecules in Aqueous Solution.

Each molecule is represented as a polar head group attached to one or two fatty acyl chains. (Lysophospholipid molecules possess only one fatty acyl chain.) The monolayer on the surface of the water forms first. As the phospholipid concentration increases, bilayer vesicles begin to form. Because of their smaller size, lysophospholipid molecules form micelles.

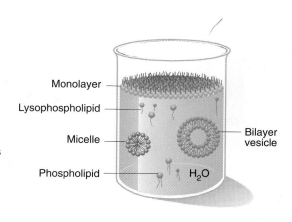

TABLE 9.2

Major Classes of Phosphoglycerides

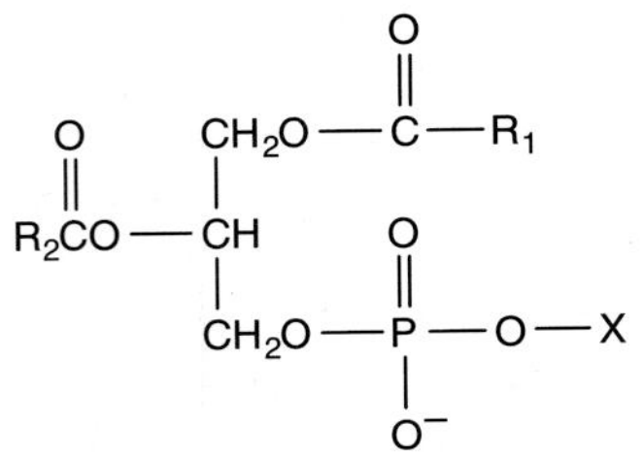

Name of X-OH	Formula of X	Name of Phospholipid
Water	—H	Phosphatidic acid
Choline	$-CH_2CH_2\overset{+}{N}(CH_3)_3$	Phosphatidylcholine (lecithin)
Ethanolamine	$-CH_2CH_2\overset{+}{N}H_3$	Phosphatidylethanolamine (cephalin)
Serine	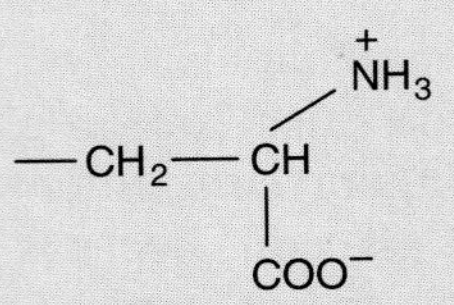 $-CH_2-\overset{\overset{+}{N}H_3}{\underset{COO^-}{CH}}$	Phosphatidylserine
Glycerol	$-CH_2CHCH_2OH$ with OH	Phosphatidylglycerol
Phosphatidylglycerol	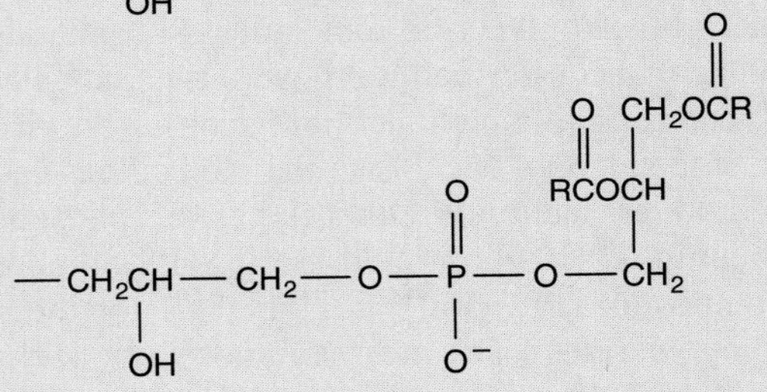 (cardiolipin structure)	Diphosphatidylglycerol (cardiolipin)
Inositol	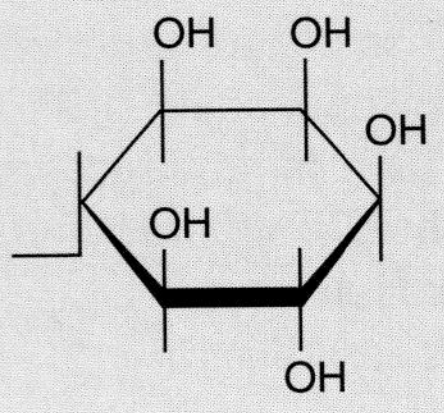 (inositol ring)	Phosphatidylinositol

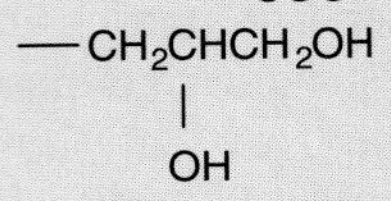

X Substituent

QUESTION 9.3

Dipalmitoylphosphatidylcholine is the major component of *surfactant,* a substance produced by certain cells within the lungs. Surfactant reduces the surface tension of the moist inner surface of the alveoli. (Alveoli, also referred to as alveolar sacs, are the functional units of respiration. Oxygen and carbon dioxide diffuse across the wall of alveolar sacs, which are one cell thick.) The water on alveolar surfaces has a high surface tension because of the attractive forces between the molecules. If the water's surface tension is not reduced, the alveolar sac tends to collapse, making breathing extremely difficult. If premature infants lack sufficient surfactant, they are likely to die of suffocation. This condition is called *respiratory distress syndrome.* Draw the structure of dipalmitoylphosphatidylcholine. Considering the general structural features of phospholipids, propose a reason why surfactant is effective in reducing surface tension.

Sphingolipids

Sphingolipids are important components of animal and plant membranes. All sphingo-lipid molecules contain a long-chain amino alcohol. In animals this alcohol is primarily sphingosine (Figure 9.7). Phytosphingosine is found in plant sphingolipids. The core of each type of sphingolipid is *ceramide*, a fatty acid amide derivative of sphingosine. In *sphingomyelin*, the 1-hydroxyl group of ceramide is esterified to the phosphate group of phosphorylcholine or phosphorylethanolamine (Figure 9.8). Sphingomyelin is found in most animal cell membranes. However, as its name suggests, sphingomyelin is found in greatest abundance in the myelin sheath of nerve cells. (The myelin sheath is formed by successive wrappings of the cell membrane of a specialized myelinating cell around a nerve cell axon. It facilitates the rapid transmission of nerve impulses.)

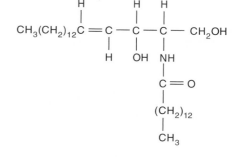

Sphingosine   **Phytosphingosine**   **Ceramide**

FIGURE 9.7

Sphingolipid Components.

FIGURE 9.8

Conformational and Space-Filling Models of Sphingomyelin.

The ceramides are also precursors for the **glycolipids,** sometimes referred to as the *glycosphingolipids* (Figure 9.9). Glycolipids differ from sphingomyelin in that they contain no phosphate. In glycolipids a monosaccharide, disaccharide, or oligosaccharide is attached to a ceramide through an O-glycosidic linkage. The most important glycolipid classes are the cerebrosides, the sulfatides, and the gangliosides. *Cerebrosides* are sphingolipids in which the head group is a monosaccharide. (These molecules, unlike phospholipids, are nonionic.) Galactocerebrosides, the most common example of this class, are almost entirely found in the cell membranes of the brain. If a cerebroside is sulfated, it is referred to as a *sulfatide.* Sulfatides are negatively charged at physiological pH. Sphingolipids that possess oligosaccharide groups

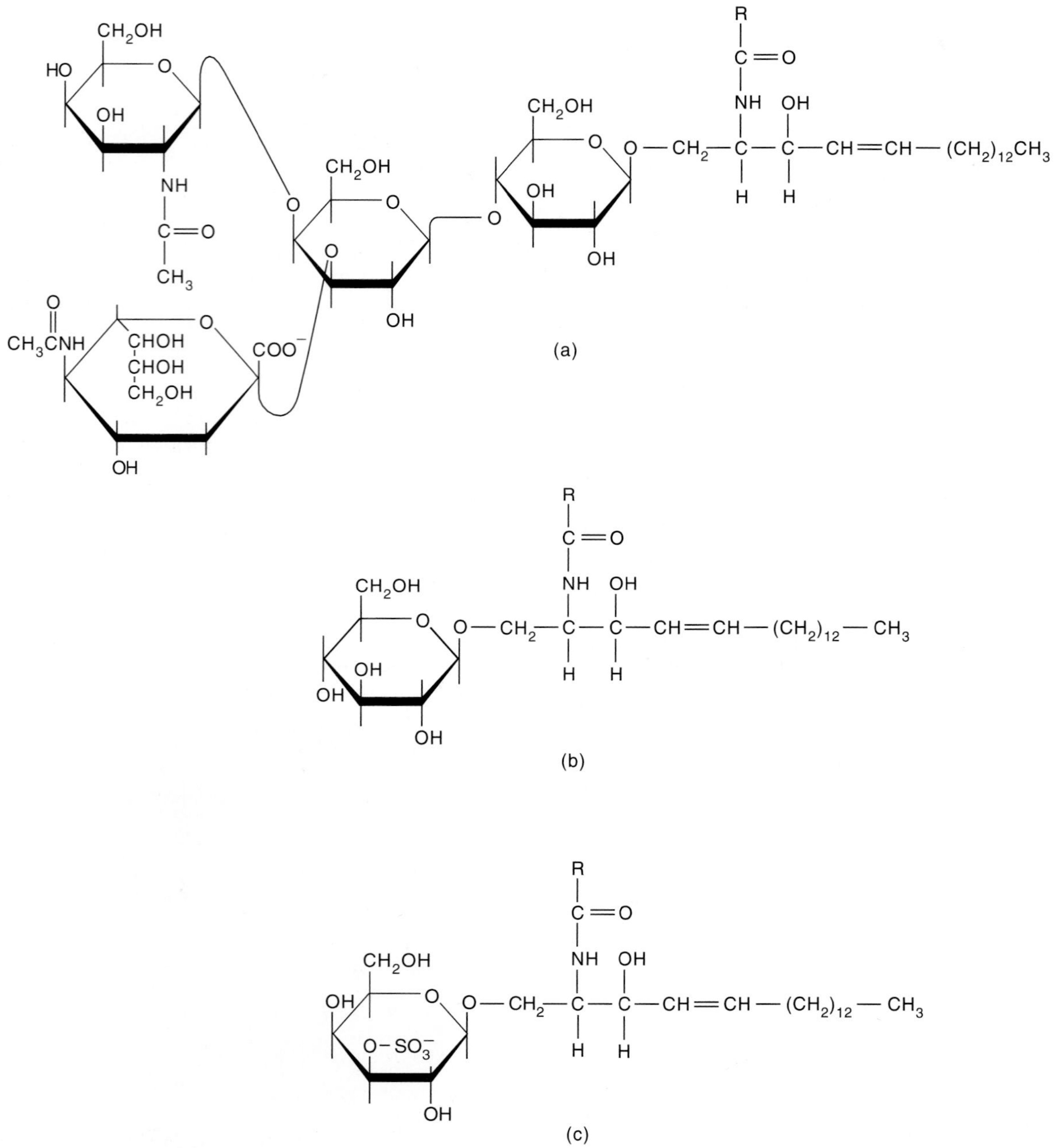

FIGURE 9.9

Selected Glycolipids.

(a) Tay-Sachs ganglioside (GM$_2$), (b) glucocerebroside, and (c) galactocerebroside sulfate (a sulfatide).

Sphingolipids, important membrane components of animals and plants, contain a complex long-chain amino alcohol (either sphingosine or phytosphingosine). The core of each sphingolipid is ceramide, a fatty acid amide derivative of the alcohol molecule. Glycolipids are derivatives of ceramide that possess a carbohydrate component.

Sphingolipid Storage Diseases

Isoprenoids

with one or more sialic acid residues are called *gangliosides.* Although gangliosides were first isolated from nerve tissue, they also occur in most other animal tissues. The names of gangliosides include subscript letter and numbers. The letters M, D, and T indicate whether the molecule contains one, two, or three sialic acid residues, respectively. The numbers designate the sequence of sugars that are attached to ceramide. The Tay-Sachs ganglioside G_{M2} (discussed below) is illustrated in Figure 9.9.

The role of glycolipids is still unclear. Certain glycolipid molecules may bind bacterial toxins, as well as bacterial cells, to animal cell membranes. For example, the toxins that cause cholera, tetanus, and botulism bind to glycolipid cell membrane receptors. Bacteria that have been shown to bind to glycolipid receptors include *Escherichia coli, Streptococcus pneumoniae,* and *Neisseria gonorrhoeae,* the causative agents of urinary tract infections, pneumonia, and gonorrhea, respectively.

Recall that each lysosomal storage disease is caused by a hereditary deficiency of an enzyme required for the degradation of a specific metabolite. Several lysosomal storage diseases are associated with sphingolipid metabolism. Most of these diseases, also referred to as the *sphingolipidoses,* are fatal. The most common sphingolipid storage disease, Tay-Sachs disease, is caused by a deficiency of β-hexosaminidase A, the enzyme that degrades the ganglioside G_{M2}. As cells accumulate this molecule, they swell and eventually die. Tay-Sachs symptoms (i.e., blindness, muscle weakness, seizures, and mental retardation) usually appear several months after birth. Because there is no therapy for Tay-Sachs disease or for any other of the sphingolipidoses, the condition is always fatal (usually by age 3). Examples of the sphingolipidoses are summarized in Table 9.3.

The **isoprenoids** are a vast array of biomolecules that contain repeating five-carbon structural units known as *isoprene units* (Figure 9.10). Isoprenoids are not synthesized from isoprene (methylbutadiene). Instead, their biosynthetic pathways all begin with the formation of isopentenyl pyrophosphate from acetyl-CoA (Chapter 10).

The isoprenoids consist of terpenes and steroids. **Terpenes** are an enormous group of molecules that are found largely in the essential oils of plants. (Essential oils are plant extracts used for thousands of years in perfumes and medicines.) Steroids are derivatives of a complex hydrocarbon ring system.

Terpenes The terpenes are classified according to the number of isoprene residues they contain (Table 9.4). *Monoterpenes* are composed of two isoprene units. Geraniol is a monoterpene found in oil of geranium. Terpenes that contain three isoprenes

TABLE 9.3

Selected Sphingolipid Storage Diseases

Disease	Symptom	Accumulating Sphingolipid	Enzyme Deficiency
Tay-Sachs disease	Blindness, muscle weakness, seizures, mental retardation	Ganglioside G_{M2}	β-Hexosaminidase A
Gaucher's disease	Mental retardation, liver and spleen enlargement, erosion of long bones	Glucocerebroside	β-Glucosidase
Krabbe's disease	Demyelination, mental retardation	Galactocerebroside	β-Galactosidase
Niemann-Pick disease	Mental retardation	Sphingomyelin	Sphingomyelinase

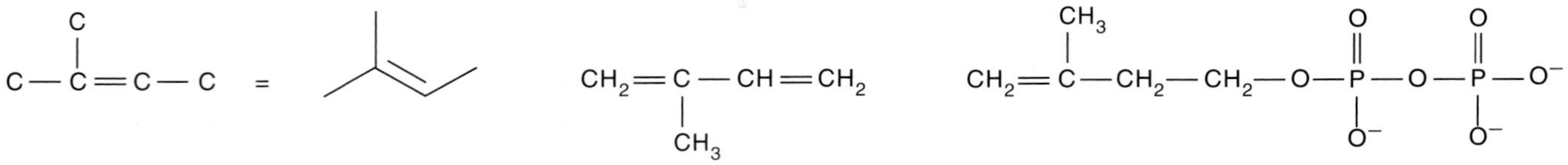

Isoprene unit (a) **Isoprene** (b) **Isopentenyl pyrophosphate** (c)

FIGURE 9.10
Isoprene.
(a) Basic isoprene structure, (b) the organic molecule isoprene, (c) isopentenyl pyrophosphate.

TABLE 9.4
Examples of Terpenes

Type	Number of Isoprene Units	Example: Name	Example: Structure
Monoterpene	2	Geraniol	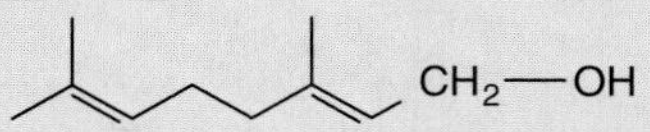
Sesquiterpene	3	Farnesene	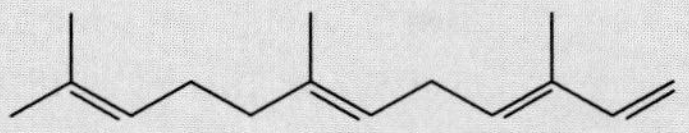
Diterpene	4	Phytol	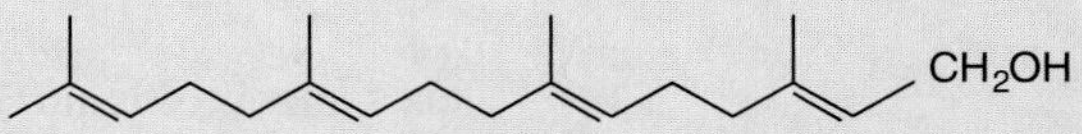
Triterpene	6	Squalene	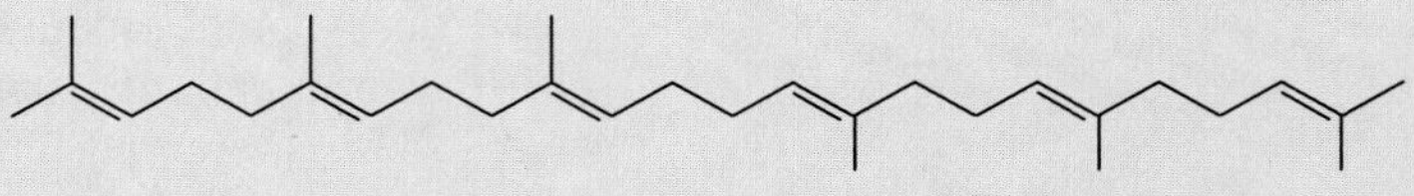
Tetraterpene	8	β-Carotene	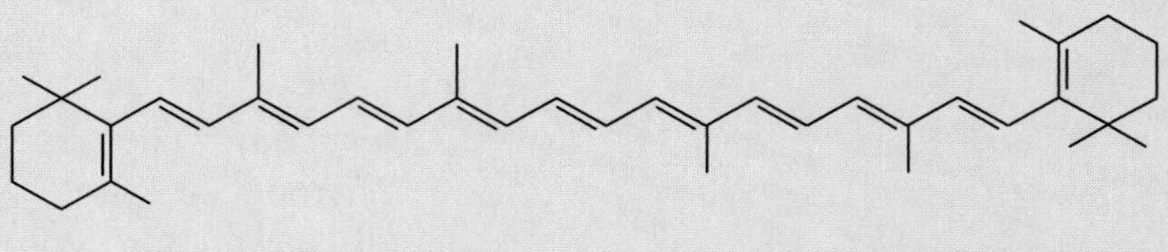
Polyterpene	Thousands	Rubber	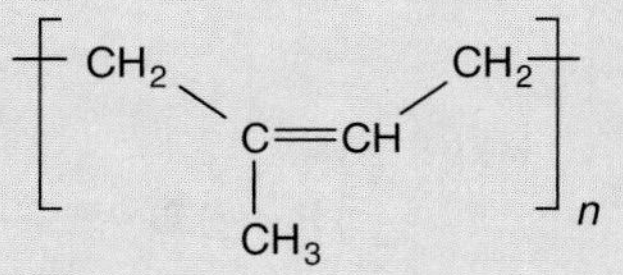

KEY *Concepts*

Isoprenoids are a large group of biomolecules with repeating units derived from isopentenyl-pyrophosphate. There are two types of isoprenoids: terpenes and steroids.

are referred to as *sesquiterpenes.* Farnesene, an important constituent of oil of citronella (a substance used in soap and perfumes), is a sesquiterpene. Phytol, a plant alcohol, is an example of the *diterpenes,* molecules composed of four isoprene units. Squalene, which is found in large quantities in shark liver oil as well as olive oil and yeast, is a prominent example of the *triterpenes.* (Squalene is an intermediate in the synthesis of the steroids.) **Carotenoids,** the orange-colored pigments found in most plants, are the only *tetraterpenes* (molecules composed of eight isoprene units). The *carotenes* are hydrocarbon members of this group. The *xanthophylls* are oxygenated derivatives of the carotenes. *Polyterpenes* are high-molecular-weight molecules composed of hundreds or thousands of isoprene units. Natural rubber is a polyterpene composed of between 3000 and 6000 isoprene units.

Several important biomolecules are composed of nonterpene components attached to isoprenoid groups (often referred to as *prenyl* or *isoprenyl* groups). Examples of these biomolecules, referred to as **mixed terpenoids,** include vitamin E (α-tocopherol) (Figure 11.29a), ubiquinone (Figure 11.14), vitamin K, and some cytokinins (plant hormones) (Figure 9.11).

A variety of proteins in eukaryotic cells are now known to be covalently attached to prenyl groups after their biosynthesis on ribosomes. The prenyl groups most often involved in this process, referred to as **prenylation,** are farnesyl and geranylgeranyl groups (Figure 9.12). The function of protein prenylation is not clear. There is some evidence that it plays a role in the control of cell growth. For example, *Ras proteins,* a group of cell growth regulators, are activated by prenylation reactions.

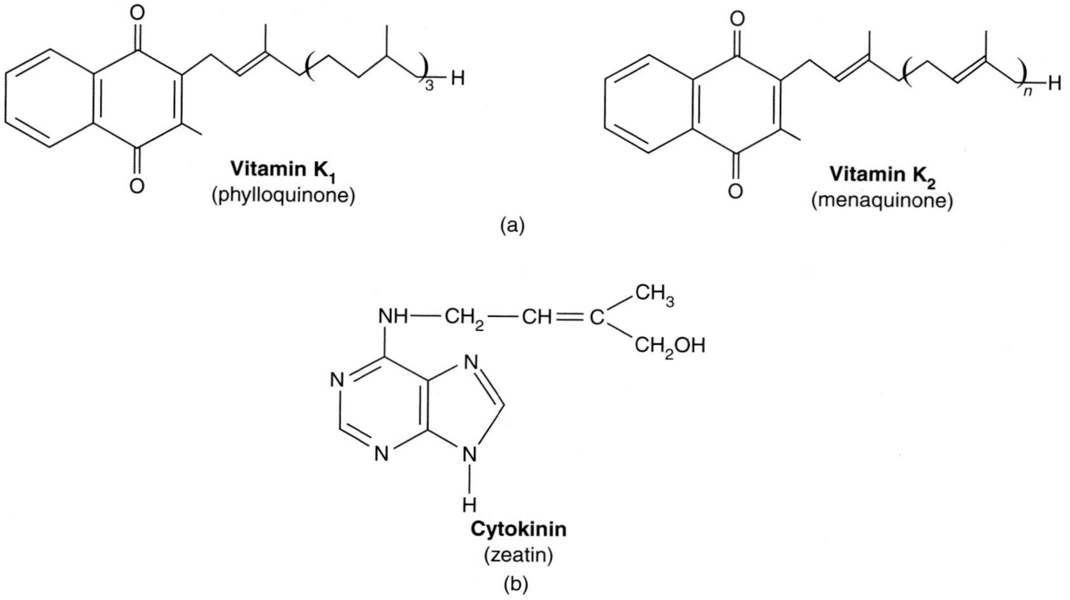

FIGURE 9.11

Selected Mixed Terpenoids.

(a) Vitamin K_1 (phylloquinone) is found in plants, where it acts as an electron carrier in photosynthesis. Vitamin K_2 (menaquinone), synthesized by intestinal bacteria and plays an important role in blood coagulation. (b) Cytokinins are cell-division-promoting substances in plants. Some cytokinins are mixed terpenoids. Zeatin is found in immature maize seeds.

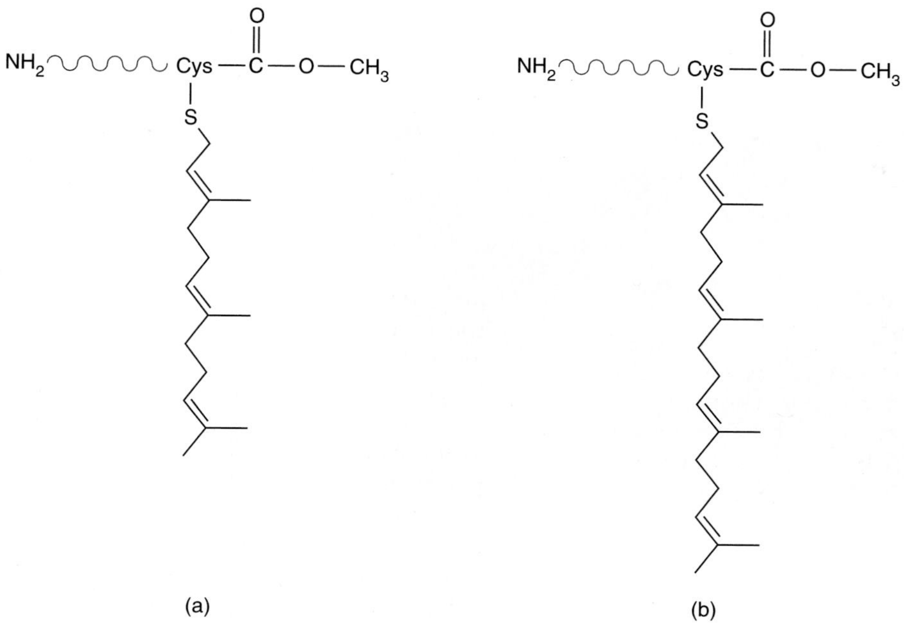

Vitamin K$_1$ (phylloquinone)

Vitamin K$_2$ (menaquinone)

(a)

Cytokinin (zeatin)

(b)

FIGURE 9.12

Prenylated Proteins.

Prenyl groups are covalently attached at the SH group of C terminal cysteine residues. Many prenylated proteins are also methylated at this residue. (a) Farnesylated protein, (b) geranylgeranylated protein.

(a)

(b)

9.4   The majority of terpenes contain one or more ring structures. Consider the following examples. Determine which terpene class they belong to and outline the positions of the isoprene units.

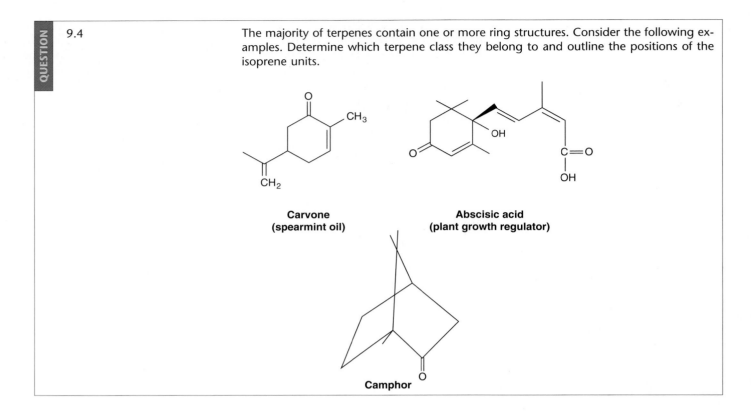

Carvone
(spearmint oil)

Abscisic acid
(plant growth regulator)

Camphor

Steroids   **Steroids** are complex derivatives of triterpenes. They are found in all eukaryotes and a small number of bacteria. Each type of steroid is composed of four fused rings. Steroids are distinguished from each other by the placement of carbon-carbon double bonds and various substituents (e.g., hydroxyl, carbonyl, and alkyl groups).

Cholesterol, an important molecule in animals, is an example of the steroids (Figure 9.13). In addition to being an essential component in animal cell membranes, cholesterol is a precursor in the biosynthesis of all steroid hormones, vitamin D, and bile

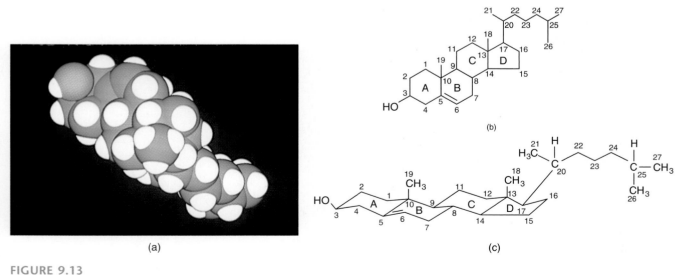

(a)

(b)

(c)

FIGURE 9.13

Structure of Cholesterol.

(a) Space-filling model, (b) conventional view, and (c) conformational model.

salts (Figure 9.14). Cholesterol possesses two methyl substituents (C-18 and C-19), which are attached to C-10 and C-13, respectively, and a Δ5 double bond. A branched hydrocarbon side chain is attached to C-17. Because this molecule has a hydroxyl group (attached to C-3), it is classified as a *sterol*. (Although the term *steroid* is most properly used to designate molecules that contain one or more carbonyl or carboxyl groups, it is often used to describe all derivatives of the steroid ring structure.) Cholesterol is usually stored within cells as a fatty acid ester. The esterification reaction is catalyzed by the enzyme *acyl CoA:cholesterol acyltransferase* (ACAT), located on the cytoplasmic face of the endoplasmic reticulum.

Practically all plant steroid molecules are sterols. The function of plant sterols is still relatively unclear. They undoubtedly play an important role in membrane structure and function. Certain sterol derivatives, such as the cardiac glycosides (see

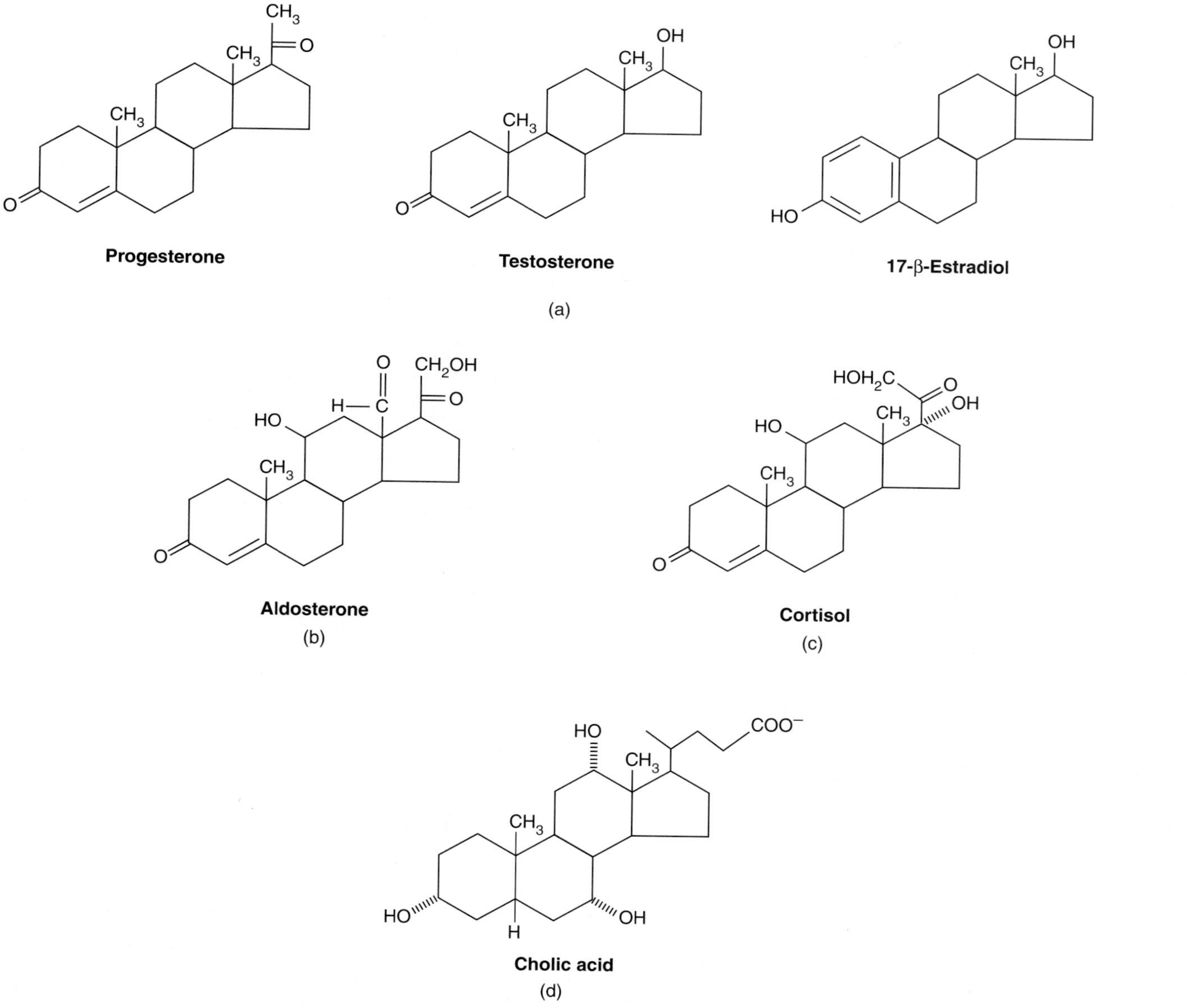

Progesterone

Testosterone

17-β-Estradiol

(a)

Aldosterone

(b)

Cortisol

(c)

Cholic acid

(d)

FIGURE 9.14

Animal Steroids.

(a) Sex hormones (molecules that regulate the development of sexual structures and various reproductive behaviors). (b) A mineralocorticoid (a molecule produced in the adrenal cortex that regulates plasma concentrations of several ions, especially sodium). (c) A glucocorticoid (a molecule that regulates the metabolism of carbohydrates, fats, and proteins). (d) A bile acid (a molecule produced in the liver that helps dietary fats and fat-soluble vitamins in the intestine absorb).

below), are known to protect plants that produce them from predators. Most plant sterols possess a one- or two-carbon substituent attached to C-24. The most abundant sterols in green algae and higher plants are β-sitosterol and stigmasterol (Figure 9.15).

Cardiac glycosides are among the most interesting steroid derivatives. Recall that glycosides are carbohydrate-containing acetals. Although several cardiac glycosides are extremely toxic (e.g., *ouabain*, obtained from the seeds of the plant *Strophanthus gratus*), others have valuable medicinal properties (Figure 9.16). For example, *digi-*

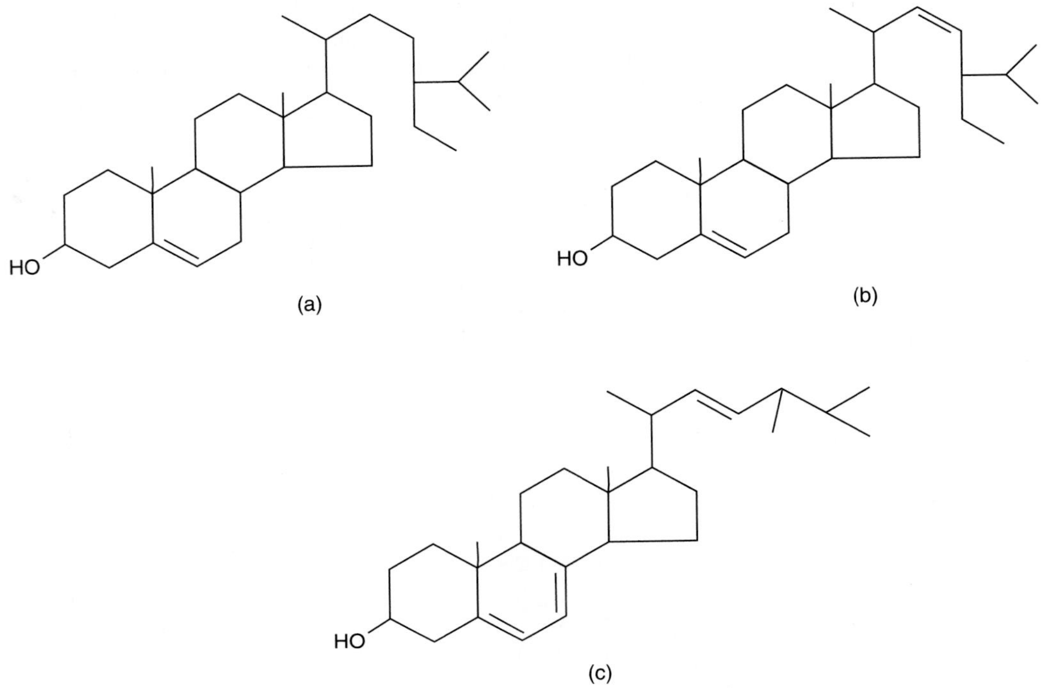

FIGURE 9.15

Plant Steroids.

(a) β-Sitosterol, (b) stigmasterol, and (c) ergosterol (found in fungi). One of the most important roles of plant sterols is the stabilization of cell membranes.

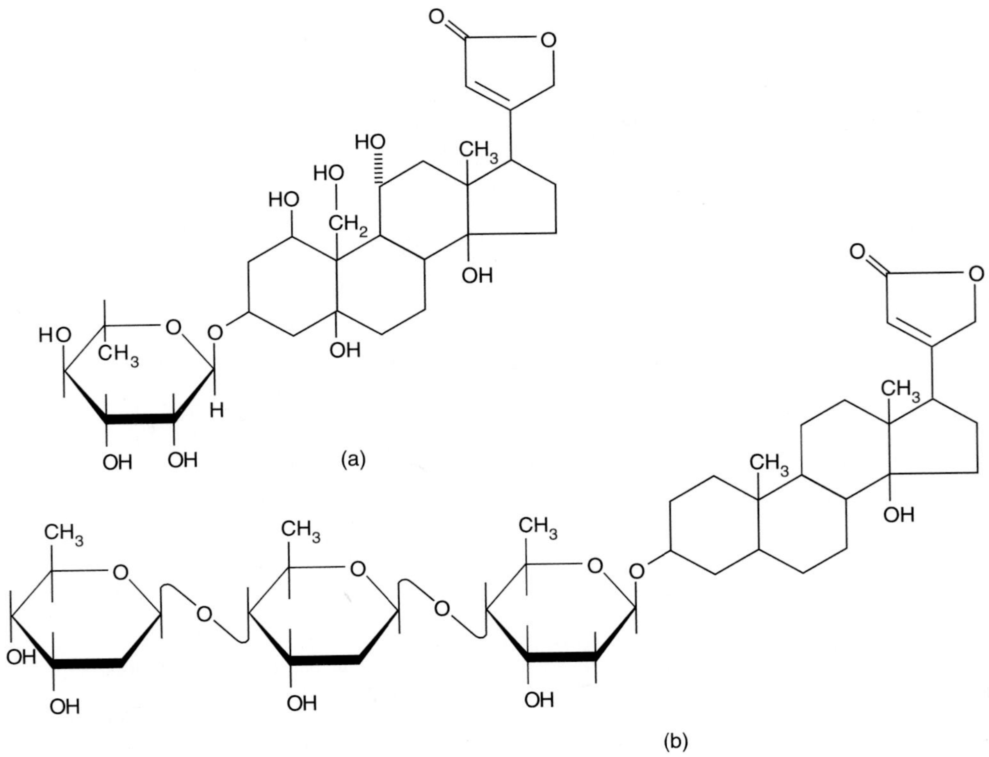

(a)

(b)

(c)

FIGURE 9.16

Cardiac Glycosides.

Each cardiac glycoside possesses a glycone (carbohydrate) and an aglycone component. (a) In ouabain the glycone is one rhamnose residue. The steroid aglycone of ouabain is called ouabagenin. (b) The glycone of digitoxin is composed of three digitoxose residues. The aglycone of digitoxin is called digitoxigenin.

(a)

(b)

talis, an extract of the dried leaves of *Digitalis purpurea* (the foxglove plant), is a time-honored stimulator of cardiac muscle contraction. *Digitoxin,* the major "cardiotonic" glycoside in digitalis, is used to treat congestive heart failure. In higher than therapeutic doses, digitoxin is extremely toxic. Both ouabain and digitoxin inhibit Na^+-K^+ ATPase (Section 9.2).

Lipoproteins

Although the term *lipoprotein* can describe any protein that is covalently linked to lipid groups (e.g., fatty acids or prenyl groups), it is most often used for a group of molecular complexes found in the blood plasma of mammals (especially humans). Plasma lipoproteins transport lipid molecules (triacylglycerols, phospholipids, and cholesterol) through the bloodstream from one organ to another. Lipoproteins also contain several types of lipid-soluble antioxidant molecules (e.g., α-tocopherol and several carotenoids). (The function of *antioxidants,* substances that protect biomolecules from free radicals, will be described in Chapter 11.) The protein components of lipoproteins are called *apolipoproteins* or *apoproteins.* A generalized lipoprotein is shown in Figure 9.17. The relative amounts of lipid and protein components of the major types of lipoprotein are summarized in Figure 9.18.

Lipoproteins are classified according to their density. **Chylomicrons,** which are large lipoproteins of extremely low density, transport dietary triacylglycerols and cholesteryl esters from the intestine to the tissues (especially muscle and adipose tissues). **Very low density lipoproteins** (VLDL) (0.95–1.006 g/cm^3), synthesized in the liver, transport lipids to tissues. As VLDL are transported through the body, they become depleted of triacylglycerols, as well as some apoproteins and phospholipids. Eventually, VLDL are converted to **low-density lipoproteins** (LDL) (1.006–1.063 g/cm^3). LDL carry cholesterol to tissues. In a complex process (Section 9.2) elucidated by Michael Brown and Joseph Goldstein (recipients of the 1985 Nobel Prize for Medicine or Physiology), LDL are engulfed by cells after binding to LDL receptors. The role of **high-density lipoprotein** (HDL) (1.063–1.210 g/cm^3), also produced in the liver, appears to be the scavenging of excessive cholesterol from cell membranes. Cholesteryl esters are formed when the plasma enzyme lecithin:cholesterol acyltransferase (LCAT) transfers a fatty acid residue from lecithin (Figure 9.19). It is now believed that HDL transport these cholesteryl esters to the liver. The liver, the only organ that can dispose of excess cholesterol, converts most of it to bile acids (Chapter 10).

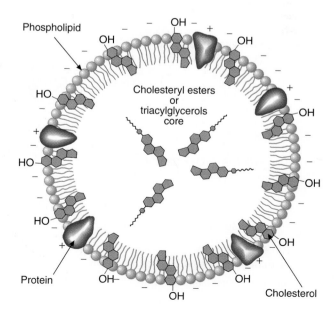

FIGURE 9.17

Plasma Lipoproteins.

Each type of lipoprotein contains a neutral lipid core composed of cholesteryl esters and/or triacylglycerols. This core is surrounded by a layer of phospholipid, cholesterol and protein. Charged and polar residues on the surface of a lipoprotein enable it to dissolve in blood.

FIGURE 9.18

Proportions of Cholesterol, Cholesteryl Ester, Phospholipid, and Protein in Four Major Classes of Plasma Lipoproteins.

Lipoproteins vary in diameter from 10 nm to as much as 1000 nm. Chylomicrons are the largest class of plasma lipoprotein.

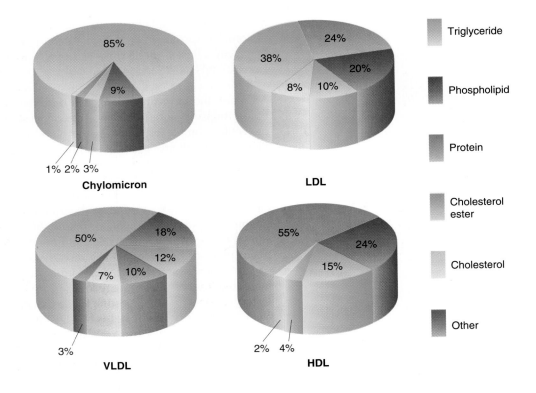

Triglyceride

Phospholipid

Protein

Cholesterol ester

Cholesterol

Other

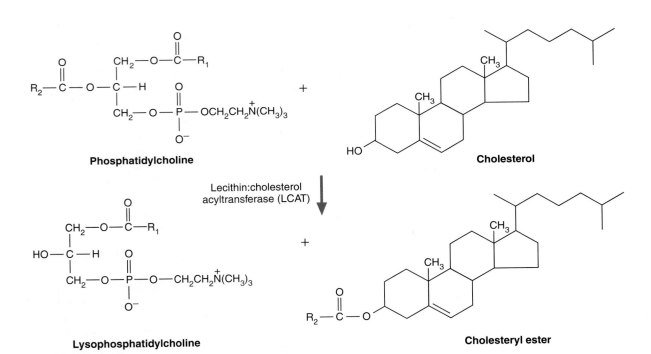

Phosphatidylcholine

Cholesterol

Lecithin:cholesterol acyltransferase (LCAT)

Lysophosphatidylcholine

Cholesteryl ester

FIGURE 9.19

Reaction Catalyzed by Lecithin:Cholesterol Acyltransferase (LCAT).

Cholesteryl ester transfer protein, a protein associated with the LCAT-HDL complex, transfers cholesteryl esters from HDL to VLDL and LDL.

Lipoproteins and
Atherosclerosis

KEY *Concepts*

Plasma lipoproteins transport lipids through the bloodstream. On the basis of density, lipoproteins are classified into four major classes: chylomicrons, VLDL, LDL, and HDL.

Atherosclerosis is a chronic disease in which soft masses, called *atheromas,* accumulate on the inside of arteries. These deposits are also referred to as *plaque.* During plaque formation, which is a progressive process, smooth muscle cells and macrophages as well as various cell debris build up. As macrophages fill with lipid (predominantly cholesterol and cholesteryl esters), they take on a foamlike appearance, hence the name "foam cells." Eventually, atherosclerotic plaque may calcify and protrude sufficiently into arterial lumens that blood flow is impeded. Disruption of vital organ functions, especially those of the brain, heart, and lungs caused by oxygen and nutrient deprivation, usually ensues. In *coronary artery disease,* one of the most common consequences of atherosclerosis, this deprivation damages heart muscle (see Special Interest Box 11.3).

Most of the cholesterol found in plaque is obtained by the ingestion of LDL by foam cells. It is not surprising, therefore, that high plasma LDL levels are directly correlated with high risk for coronary artery disease. (Recall that LDL have a high cholesterol and cholesteryl ester content.) In contrast, a high plasma HDL level is considered to be associated with a low risk for coronary artery disease. Liver cells (hepatocytes) are the only cells that possess HDL receptors. (Other high risk factors include a high-fat diet, smoking, stress, and a sedentary lifestyle.)

LDL play a significant role in atherosclerosis because cells that become foam cells possess LDL receptors. The binding of LDL to LDL receptors initiates endocytosis (Special Interest Box 9.3). Under normal circumstances the cholesterol and other lipids released after LDL enter a cell are used to meet the cell's structural and metabolic needs. Because LDL receptor function is usually highly regulated, the intake of a relatively large number of LDL particles is followed by decreased synthesis of LDL receptors. Macrophages, unlike other cell types, do not exhibit this decrease in LDL receptor synthesis. Although the cause of atherosclerosis is still not understood, recent research has illuminated several aspects. For example, the macrophages found within atherosclerotic plaque possess high levels of LDL receptors with an affinity for oxidized (i.e., damaged) LDL. Clinical trials and animal studies reveal that diets supplemented with ascorbic acid and α-tocopherol, two powerful antioxidants, can retard or arrest plaque formation.

9.2 MEMBRANES

Most of the properties attributed to living organisms (e.g., movement, growth, reproduction, and metabolism) depend, either directly or indirectly, on membranes. All biological membranes have the same general structure. As previously mentioned (Chapter 2), membranes contain lipid and protein molecules. In the currently accepted concept of membranes, referred to as the **fluid mosaic model,** membrane is a bimolecular lipid layer (often referred to as the **lipid bilayer**). The proteins, most of which float within the lipid bilayer, largely determine a membrane's biological functions. Because of the importance of membranes in biochemical processes, the remainder of the chapter is devoted to a discussion of their structure and functions.

Membrane Structure

Because each type of living cell has its own functions, it follows that the structure of its membranes is also unique. Not surprisingly, the proportion of lipid and protein varies considerably among cell types and among organelles within each cell (Table 9.5). The types of lipid and protein found in each membrane also vary.

Membrane Lipids   When amphipathic molecules are suspended in water, they spontaneously rearrange into ordered structures (Figure 9.6). As these structures form, hydrophobic groups become buried in the interior and exclude water. Simultaneously, hydrophilic groups become oriented so that they are exposed to water. As stated previously, phospholipids form into bimolecular layers when sufficiently concentrated. This property of phospholipids (and other amphipathic lipid molecules) is the basis of membrane structure.

AGAINST THE ODDS

Most risk factors for atherosclerosis can be reduced by changes in behavior. Exercise, a low-fat diet, and quitting tobacco use usually decrease the likelihood of coronary artery disease. However, an individual's genetic inheritance can sometimes play a decisive role. For example, some people with poor health habits (e.g., smoking and high-fat diets) do not have high plasma LDL and cholesterol values. Similarly, some individuals with high blood cholesterol never have a heart attack. In contrast, other people, apparently at low risk because of healthy lifestyles, do have heart attacks. Recently, researchers have proposed that a variant of LDL, called lipoprotein (a) (Lp(a)), may be at least partially responsible for these anomalous cases. The amount of Lp(a) in an individual's blood plasma is genetically determined and does not vary markedly throughout life. A high blood plasma Lp(a) concentration is correlated with a high risk of coronary artery disease.

Like LDL, Lp(a) consists of phospholipids, cholesterol, and the apoprotein B-100. An additional apoprotein, a glycoprotein

called apolipoprotein (a), is attached to Lp(a) by a disulfide linkage to apoprotein B-100. The structure of apolipoprotein (a) closely resembles that of plasminogen. (*Plasminogen* is a blood plasma zymogen. When it is activated by a protease, most notably *plasminogen activator,* the fibrin-dissolving protease *plasmin* is produced.)

Because the formation of a blood clot often triggers heart attacks, plasminogen is important in dissolving clots. Some researchers have proposed that under certain conditions, apolipoprotein (a) may promote coronary artery disease by binding to either fibrin or plasminogen activator. Because apolipoprotein (a) does not digest fibrin, blood clot dissolution is undermined.

High plasma levels of Lp(a) have also been shown to cause the proliferation of smooth muscle cells within arterial walls. (Recall that the initial formation of plaque is associated with an increase in the number of such cells in the artery wall.) Lp(a) may interfere with the activity of a growth-retarding substance that normally suppresses cell division in arterial walls.

Membrane lipids are largely responsible for several other important features of biological membranes:

1. Membrane fluidity. The term *fluidity* describes the resistance of membrane components to movement. Rapid lateral movement (Figure 9.20) is apparently responsible for the proper functioning of many membrane proteins. A membrane's fluidity is largely determined by the percentage of unsaturated fatty acids in its phospholipid molecules. (Recall that unsaturated hydrocarbon chains pack less densely than saturated ones.) A high concentration of unsaturated chains results in a more fluid membrane. Cholesterol moderates membrane stability without greatly compromising fluidity because it contains both rigid (ring system) and flexible (hydrocarbon tail) structural elements (Figure 9.21).

2. Selective permeability. Because of their hydrophobic nature, the hydrocarbon chains in lipid bilayers provide a virtually impenetrable barrier to ionic and polar substances. Specific membrane proteins regulate the movement of such substances into and out of cells. To cross a lipid bilayer, a polar substance must shed some or all of its hydration sphere and bind to a carrier protein for membrane translocation or pass through an aqueous protein channel. Both methods shield the hydrophilic molecule from the hydrophobic core of the membrane. Most transmembrane water movement accompanies ion transport. Nonpolar sub-

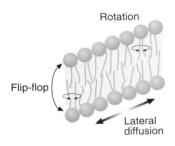

FIGURE 9.20

Lateral Diffusion in Biological Membranes.

Lateral movement of phospholipid molecules is usually relatively rapid. "Flip-flop," the transfer of a lipid molecule from one side of a lipid bilayer to the other, is rare.

TABLE 9.5

Chemical Composition of Some Cell Membranes

Membrane	Protein %	Lipid %	Carbohydrate %
Human erythrocyte plasma membrane	49	43	8
Mouse liver cell plasma membrane	46	54	2–4
Amoeba plasma membrane	54	42	4
Mitochondrial inner membrane	76	24	1–2
Spinach chloroplast lamellar membrane	70	30	6
Halobacterium purple membrane	75	25	0

Source: G. Guidotti, "Membrane Proteins," *Ann. Rev. Biochem. 41*:731, 1972

FIGURE 9.21

Lipid Bilayer.

The flexible hydrocarbon chains in the hydrophobic core (lightly shaded area in the middle) make the membrane fluid. The steroid ring system of cholesterol molecules, positioned in the outer surfaces, stiffen the membrane.

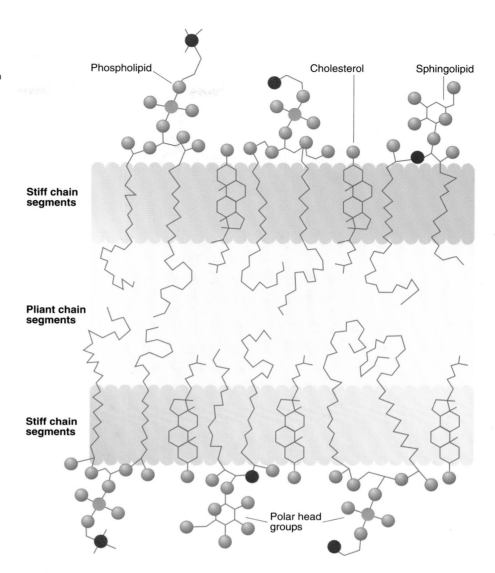

stances simply diffuse through the lipid bilayer down their concentration gradients. Each membrane exhibits its own transport capability or selectivity based on its protein component.

3. Self-sealing capability. When lipid bilayers are disrupted, they immediately and spontaneously reseal (Figure 9.22) because a break in a lipid bilayer exposes the hydrophobic hydrocarbon chains to water. Because breaches in cell membranes can be lethal, this resealing property is critical. (In living cells, certain protein components of membrane and the cytoskeleton, as well as calcium ions, also assist in membrane resealing.)

4. Asymmetry. Biological membranes are asymmetric, that is, the lipid composition of each half of a bilayer is different. For example, the human red blood cell membrane possesses substantially more phosphatidylcholine and sphingomyelin on its outside surface. Most of the membrane's phosphatidylserine and phosphatidylethanolamine are on the inner side. Membrane asymmetry is not unexpected, since each side of a membrane is exposed to a different environment. Asymmetry originates during membrane assembly, since phospholipid biosynthesis occurs on only one side of a membrane. The significance of lipid asymmetry in biological membranes is not yet clear.

FIGURE 9.22

Membrane Self-Sealing.

Disruptions in a lipid bilayer are rapidly resealed. When the hydrophobic tails of lipid molecules are suddenly exposed to polar water molecules, lipids respond by forming hydrophilic edges consisting of polar head groups. As membrane edges draw closer to each other, they fuse and reform the bilayer.

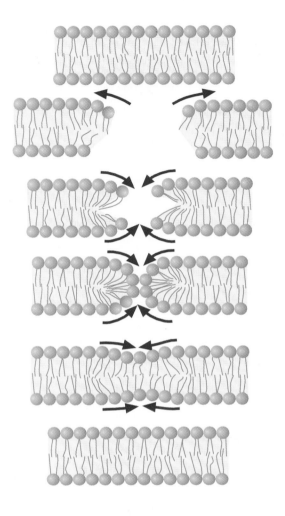

Membrane Proteins As previously mentioned, most of the functions associated with biological membranes require protein molecules. Membrane proteins are often classified by the function they perform. Most of these molecules are structural components, enzymes, hormone receptors, or transport proteins.

Membrane proteins are also classified according to their structural relationship to membrane. Proteins that are embedded in and/or extend through a membrane are referred to as *integral proteins* (Figure 9.23). Such molecules can be extracted only by disrupting the membrane with organic solvents or detergents (Figure 9.24). *Peripheral proteins* are bound to membrane primarily through interactions with integral membrane proteins. Some peripheral proteins interact directly with the lipid bilayer. Typically, peripheral proteins can be released from membrane by relatively gentle methods (e.g., concentrated salt solutions or pH changes). For several technical reasons (see Appendix B, Supplement 9), red blood cell plasma membrane proteins have been widely studied. Examples of red blood cell integral and peripheral proteins are described next.

The two major integral proteins in red blood cell membrane are glycophorin and the anion channel protein (Figure 9.25). *Glycophorin* is a 31-kD glycoprotein with 131 amino acid residues. Approximately 60% of its weight is carbohydrate. (Certain oligosaccharide groups on glycophorin constitute the ABO and MN blood group antigens. However, despite intense research efforts, the function of glycophorin is still unknown.) *Anion channel protein* (also referred to as band 3) is composed of two identical subunits, each consisting of 929 amino acids. This protein channel plays an important role in CO_2 transport in blood. The HCO_3^- ion formed from CO_2 with the aid of carbonic anhydrase diffuses into and out of the red blood cell through the an-

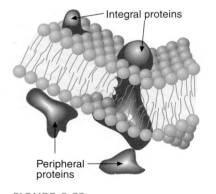

Integral proteins

Peripheral proteins

FIGURE 9.23

Integral and Peripheral Membrane Proteins.

Integral membrane proteins are released only if the membrane is disrupted. Many peripheral proteins can be removed with mild reagents.

FIGURE 9.24

Detergent Solubilization of Membrane Proteins.

As the membrane is mechanically disrupted, detergent forms a complex with the hydrophobic portions of integral membrane proteins and membrane lipids.

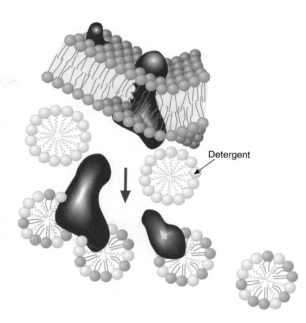

Detergent

FIGURE 9.25

Selected Integral Membrane Proteins.

(a) Glycophorin. Carbohydrate residues (red circles) appear on the N-terminal domain on the outside of the cell. (b) Anion channel protein. In contrast to glycophorin, each subunit of anion channel protein traverses the membrane several times. In addition the N-terminal methionine residue is acetylated.

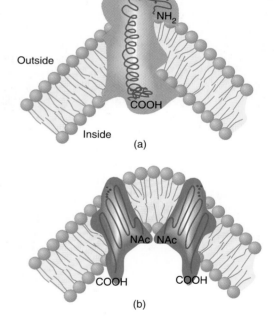

Outside

NH₂

COOH

Inside (a)

NAc NAc

COOH COOH

(b)

All membranes are composed of lipids and proteins. Membrane lipids are mostly amphipathic molecules, which are principally responsible for such properties as membrane fluidity, selective permeability, self-sealing capability, and asymmetry. Membrane proteins are structural components, enzymes, hormone receptors, or transport proteins.

ion channel in exchange for chloride ion (Cl⁻). (The exchange of Cl⁻ for HCO₃⁻, called the *chloride shift*, preserves electrical neutrality within the cell.)

Red blood cell membrane peripheral proteins, composed largely of spectrin, ankyrin, and band 4.1, are primarily involved in preserving the cell's unique biconcave shape. This shape allows rapid diffusion of O₂ throughout the cell. (No hemoglobin molecule is greater than 1 μm away from the cell's surface.) *Spectrin* is a tetramer, composed of two αβ dimers, that binds to ankyrin and band 4.1. *Ankyrin* is a large globular polypeptide (215 kD) that links spectrin to the anion channel protein. (This is a connecting link between the red blood cell's cytoskeleton and its plasma membrane.) *Band 4.1* binds to both spectrin and *actin filaments* (a cytoskeletal component found in many cell types). Because band 4.1 also binds to glycophorin, it too links the cytoskeleton and the membrane.

Membrane Function

Membranes are involved in a bewildering array of functions in living organisms. Among the most important of these are the transport of molecules and ions into and out of cells and organelles and the binding of hormones and other biomolecules. Each of these topics is discussed briefly. A description of receptor-mediated endocytosis follows.

Membrane Transport   Membrane transport mechanisms are vital to living organisms. Ions and molecules constantly move across cell plasma membranes as well as across the membranes of organelles. This flux must be carefully regulated to meet each cell's metabolic needs. For example, a cell's plasma membrane regulates the entrance of nutrient molecules and the exit of waste products. Additionally, it regulates intracellular ion concentrations. Because lipid bilayers are generally impenetrable to ions and polar substances, specific transport components must be inserted into cellular membranes. Several examples of these structures, referred to as transport proteins or permeases, are discussed below.

Biological transport mechanisms are classified according to whether they require energy. Major types of biological transport are illustrated in Figure 9.26. In **passive transport,** there is no direct input of energy. In **simple diffusion,** each solute, propelled by random molecular motion, moves down its concentration gradient (i.e., from an area of high concentration to an area of low concentration). (In this spontaneous process, there is a net movement of solute until an equilibrium is reached. As Figure 9.27 illustrates, a system reaching equilibrium becomes more disordered, that is, entropy increases. Because there is no input of energy, transport occurs with a negative change in free energy.) In general, the higher the concentration gradient, the faster the rate of solute diffusion. The diffusion of gases such as O_2 and CO_2 across membranes is proportional to their concentration gradients. The diffusion of organic molecules also depends on molecular weight and lipid solubility. (In this spontaneous process, there is a net movement of solute until an equilibrium is reached.)

FIGURE 9.26

Transport Across Membranes.

The major transport processes are simple and facilitated diffusion and primary and secondary active transport.

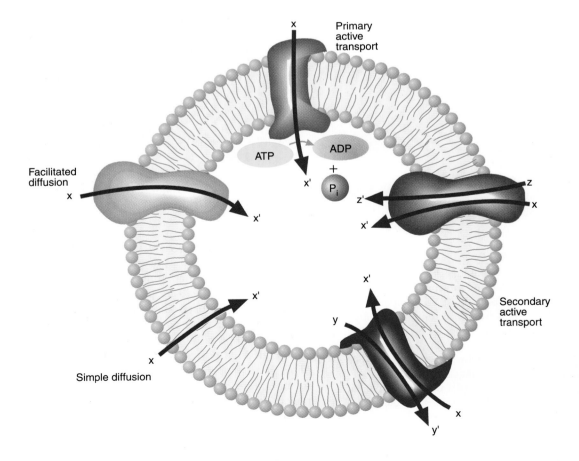

FIGURE 9.27

Simple Diffusion.

By time t, one-fourth of the molecules present on one side of a semipermeable membrane have diffused by random thermal motion of the other side. Eventually, equilibrium is approached. The molecules in this illustration are identical. Their colors indicate their initial locations.

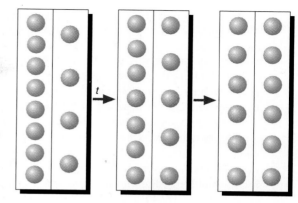

In **facilitated diffusion** the transport of certain large or charged molecules occurs through special channels or carriers. *Channels* are tunnel-like transmembrane proteins. Each type is designed for the transport of a specific solute. Many channels are chemically or voltage-regulated. Chemically regulated channels open or close in response to a specific chemical signal. For example, a *chemically-gated Na^+ channel* in the nicotinic acetylcholine receptor complex (found in muscle cell plasma membranes) opens when acetylcholine binds. Na^+ rushes into the cell and the membrane potential falls. (Recall that membrane potential is an electrical gradient across the membrane. A decrease in membrane potential is membrane *depolarization.*) Local depolarization caused by acetylcholine leads to the opening of nearby Na^+ channels (these are referred to as *voltage-gated Na^+ channels*). *Repolarization,* the reestablishment of the membrane potential, begins with the diffusion of K^+ ions out of the cell through *voltage-gated K^+ channels.* (The diffusion of K^+ ions out of the cell makes the inside less positive, that is, more negative.)

One form of facilitated diffusion involves membrane proteins called *carriers* (sometimes referred to as *passive transporters*). In carrier-mediated transport, a specific solute binds to the carrier on one side of a membrane and causes a conformational change in the carrier. The solute is then translocated across the membrane and released. The red blood cell *glucose transporter* is the best-characterized example of passive transporters. It allows D-glucose to diffuse across the red blood cell membrane for use in glycolysis and the pentose phosphate pathway. Facilitated diffusion increases the rate at which certain solutes move down their concentration gradients. This process cannot cause a net increase in solute concentration on one side of the membrane

Often, cells must accumulate molecules against a concentration gradient. Because this process requires energy, it is referred to as **active transport.** The two forms of active transport are primary and secondary. In *primary active transport,* energy is provided by ATP. Transmembrane ATP-hydrolyzing enzymes use the energy derived from ATP to drive the transport of ions or molecules. The Na^+-K^+ pump (also referred to as the Na^+-K^+ ATPase pump) is a prominent example of a primary transporter. (The Na^+ and K^+ gradients are required for maintaining normal cell volume and membrane potential. Refer to Special Interest Box 3.1.) In *secondary active transport,* concentration gradients generated by primary active transport are harnessed to move substances across membranes. For example, the Na^+ gradient created by the Na^+-K^+ ATPase pump is used in kidney tubule cells and intestinal cells to transport D-glucose (Figure 9.28).

THE LDL RECEPTOR

As mentioned, Brown and Goldstein elucidated the mechanism by which LDL are taken up by cells. As often happens in biomedical research, an inherited disease (in this case, familial hypercholesterolemia) was used to deduce the mechanism of a normal biological process. Patients with *familial hypercholesterolemia* have elevated levels of plasma cholesterol because they have missing or defective LDL receptors. (Recall that LDL transport cholesterol to tissues.) Heterozygous individuals (also referred to as *heterozygotes*) inherit one defective LDL receptor gene. Consequently, they possess half the number of functional LDL receptors. With blood cholesterol values of 300-600 mg/100 mL it is not surprising that heterozygotes have heart attacks as early as the age of 40. They also develop *xanthomas* (cholesterol deposits in the skin) in their 30s. *Homozygotes* (individuals who have inherited a defective LDL receptor gene from both parents) are rare (approximately one in one million). These patients have plasma cholesterol values of 650-1200 mg/100 mL. Both xanthomas and heart attacks occur during childhood or early adolescence. Death usually occurs before the age of 20.

The LDL receptor is a complex glycoprotein (Figure 9C) found on the surface of many cells. When cells need cholesterol for the synthesis of membrane or steroid hormones, they produce LDL receptors and insert them into discrete coated regions of plasma membrane. (Coated membrane regions usually constitute about 2% of a cell's surface. The protein *clathrin*, which has a unique structure referred to as a *triskelion*, is the major protein component of coated regions. It forms a latticelike polymer during the initial stages of endocytosis.) The number of receptors per cell varies from 15,000 to 70,000, depending on cell type and cholesterol requirements.

The process of LDL receptor-mediated endocytosis, presented in Figure 9D, occurs in several steps. It begins within several minutes after LDL have bound to LDL receptors. The coated region surrounding the bound receptor, referred to as a *coated pit*, pinches off and becomes a *coated vesicle*. Subsequently, *uncoated vesicles* are formed as clathrin depolymerizes. Before uncoated vesicles fuse with lysosomes, LDL are uncoupled from LDL receptors as the pH changes from 7 to 5. (This change is created

FIGURE 9C

The LDL Receptor.

The cytoplasmic domain of the LDL receptor plays a critical role in forming of coated pits, an important feature of receptor-mediated endocytosis. Once LDL binds to a receptor, both are rapidly internalized.

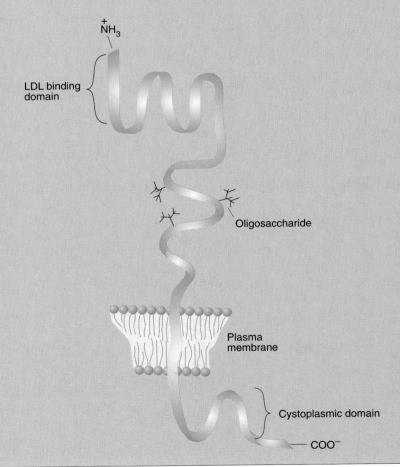

by ATP-driven proton pumps in the vesicle membrane.) LDL receptors are recycled to the plasma membrane, and LDL-containing vesicles fuse with lysosomes. Subsequently, LDL proteins are degraded to amino acids, and cholesteryl esters are hydrolyzed to cholesterol and fatty acids. Under normal circumstances, LDL receptor-mediated endocytosis is a highly regulated process. For example, cholesterol (or a derivative) suppresses the activity of HMG CoA reductase, the enzyme (discussed in Chapter 10) that catalyzes the rate-limiting step in cholesterol synthesis. Additionally, cholesterol stimulates ACAT activity and depresses the synthesis of LDL receptors. The genetic defects that cause familial hypercholesterolemia prevent affected cells from obtaining sufficient cholesterol from LDL. The most common defect is failure to synthesize the receptor. Other defects include ineffective intracellular processing of newly synthesized receptor, defects in the receptor's binding of LDL, and the inability of receptors to cluster in coated pits.

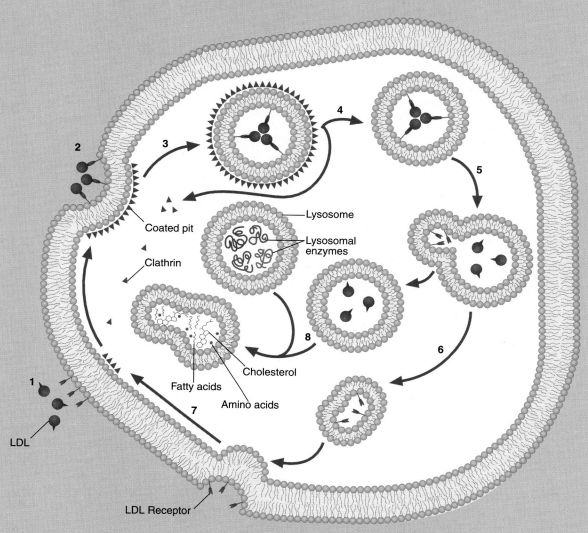

FIGURE 9D

Receptor-Mediated Endocytosis.

LDL receptor-mediated endocytosis begins (Step 1) when LDL binds to an LDL receptor in a coated region of plasma membrane. In Step 2, coated regions bud to form coated pits. Eventually (Step 3), coated vesicles pinch off from the plasma membrane. As clathrin coat protein depolymerizes (Step 4), uncoated vesicles form. A change in the uncoated vesicle's pH causes LDL to separate from its receptor (Step 5). Membrane with LDL receptor is recycled back to the plasma membrane (Steps 6 and 7). Meanwhile, the uncoated vesicle fuses with a lysosome (Step 8) to form a secondary-lysosome. Then enzymes degrade the vesicle's contents.

FIGURE 9.28

The Na⁺-K⁺ ATPase and Glucose Transport.

The Na$^+$-K$^+$ ATPase maintains the Na$^+$ gradient essential to maintain membrane potential. In certain cells, glucose transport depends on the Na$^+$ gradient. Glucose permease transports both Na$^+$ and glucose. Only when both substrates are bound does the protein change its conformation, thus initiating transport.

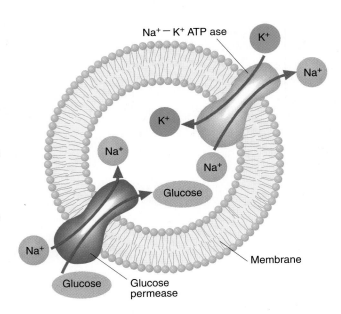

QUESTION 9.5

Transport mechanisms are often categorized according to the number of transported solutes and the direction of solute transport.

1. Uniporters transport one solute.

2. Symporters transport two different solutes simultaneously in the same direction.

3. Antiporters transport two different solutes simultaneously in opposite directions.

After examining the examples of transport discussed in this chapter, determine which of the above categories each one belongs to.

KEY *Concepts*

Membrane transport mechanisms are classified as passive or active according to whether they require energy. In passive transport, solutes moving across membranes move down their concentration gradient. In active transport, energy derived directly or indirectly from ATP hydrolysis or other energy sources is required to move an ion or molecule against its concentration gradient.

Membrane Receptors Membrane receptors play a vital role in the metabolism of all living organisms. They provide mechanisms by which cells monitor and respond to changes in their environment. In multicellular organisms the binding of chemical signals, such as the hormones and neurotransmitters of animals, to membrane receptors is a vital link in intracellular communication. Other receptors are engaged in cell-cell recognition or adhesion. For example, lymphocytes perform a critical role in the immune system function of identifying and then destroying foreign or virus-infected cells when they transiently bind to cell surfaces throughout the body. Similarly, the capacity of cells to recognize and adhere to other appropriate cells in a tissue is of crucial importance in many organismal processes, such as embryonic and fetal development.

The binding of a ligand to a membrane receptor results in a conformational change, which then causes a specific programmed response. Sometimes, receptor responses appear to be relatively straightforward. For example, the binding of acetylcholine to an acetylcholine receptor opens a cation channel. However, most responses are complex. Currently, the most intensively researched example of membrane receptor function is LDL receptor-mediated endocytosis, which is discussed in Special Interest Box 9.3.

Membrane Methods

Recognition of the vast array of functions performed by biological membranes has led to the development of numerous techniques for their investigation. Research accomplished over the past few decades reveals substantial information about membrane structure and function. Several of the most important techniques used in membrane investigations are described in Appendix B, Supplement 9. **Information provided in this discussion is useful in answering several end-of-chapter questions.**

KEY WORDS

active transport, *241*	fluid mosaic model, *235*	neutral fat, *218*	sphingolipid, *225*
acyl group, *216*	glycolipid, *226*	nonessential fatty acid, *218*	sphingomyelin, *223*
atherosclerosis, *235*	high-density lipoprotein, *233*	passive transport, *240*	steroid, *230*
autocrine, *220*	isoprenoid, *227*	phosphoglyceride, *223*	terpene, *227*
carotenoid, *228*	leukotriene, *221*	phospholipid, *223*	thromboxane, *221*
chylomicron, *233*	lipid, *216*	polar head group, *223*	*trans*-isomer, *217*
cis-isomer, *217*	lipid bilayer, *235*	polyunsaturated, *217*	very low density lipoprotein, *233*
eicosanoid, *220*	low-density lipoprotein, *233*	prenylation, *229*	
essential fatty acid, *218*	mixed terpenoid, *229*	prostaglandin, *220*	
facilitated diffusion, *241*	monounsaturated, *217*	simple diffusion, *240*	

SUMMARY

Lipids are a diverse group of biomolecules that dissolve in nonpolar solvents. They can be separated into the following classes: fatty acids and their derivatives, triacylglycerols, wax esters, phospholipids, sphingolipids, and the isoprenoids.

Fatty acids are monocarboxylic acids that occur primarily in triacylglycerols, phospholipids, and sphingolipids. The eicosanoids are a group of powerful hormonelike molecules derived from long chain fatty acids. The eicosanoids include the prostaglandins, thromboxanes, and the leukotrienes.

Triacylglycerols are esters of glycerol with three fatty acid molecules. Triacylglycerols that are solid at room temperature (i.e., they possess mostly saturated fatty acids) are called fats. Those that are liquid at room temperature (i.e., they possess a high unsaturated fatty acid content) are referred to as oils. Triacylglycerols, the major storage and transport form of fatty acids, are an important energy storage form in animals. In plants they store energy in fruits and seeds.

Phospholipids are structural components of membranes. There are two types of phospholipids: phosphoglycerides and sphingomyelins.

Sphingolipids are also important components of animals and plant membranes. They contain a long-chain amino alcohol. In animals this alcohol is sphingosine. Phytosphingosine is found in plant sphingolipids. Glycolipids are sphingolipids that possess carbohydrate groups and no phosphate.

Isoprenoids are molecules that contain repeating five-carbon structural units known as isoprene units. The isoprenoids consist of the terpenes and the steroids.

Plasma lipoproteins transport lipid molecules through the bloodstream from one organ to another. They are classified according to their density. Chylomicrons are large lipoproteins of extremely low density that transport dietary triacylglycerols and cholesteryl esters from the intestine to the tissues. VLDL, which are synthesized in the liver, transport lipids to tissues. As VLDL travel through the bloodstream, they are converted to LDL. LDL are engulfed by cells after binding to LDL receptors on the plasma membrane. HDL, also produced in the liver, scavenge cholesterol from cell membranes. LDL plays an important role in the development of atherosclerosis.

According to the fluid mosaic model, the basic structure of membranes is a lipid bilayer in which proteins float. Membrane lipids (the majority of which are phospholipids) are primarily responsible for the fluidity, selective permeability, and self-sealing properties of membranes. Membrane proteins usually define the biological functions of specific membranes. Depending on their location, membrane proteins can be classified as integral or peripheral. Examples of functions in which membranes are involved include transport and the binding of hormones and other extracellular metabolic signals.

SUGGESTED READINGS

Brown, M.S., and Goldstein, J.L., How LDL Receptors Influence Cholesterol and Atherosclerosis, *Sci. Amer.,* 251(5):58–66, 1984.

Gennis, R.B., *Biomembranes: Molecular Structure and Function,* Springer-Verlag, New York, 1989.

Glomset, J.A., Gelb, M.H., and Farnworth, C.C., Prenyl Proteins in Eukaryotic Cells: A New Type of Membrane Anchor, *Trends Biochem. Sci.,* 15:139–142, 1990.

Gounaris, K., and Barber, J., Monogalactosyldiacylglycerol: The Most Abundant Polar Lipid in Nature, *Trends Biochem. Sci.,* 8:378–381, 1983.

Hakamori, S., Glycosphingolipids, *Sci. Amer.,* 254(5):44–53, 1986.

Lawn, R.M., Lipoprotein (a) in Heart Disease, *Sci. Amer.,* 266(6): 54–60, 1992.

McNeil, P.L., Cell Wounding and Healing, *Amer. Sci.,* 79:222–235, 1991.

Nicholls, D.G., and Ferguson, S.J., *Bioenergetics 2,* Academic Press, New York, 1992.

Scanu, A.M., Lawn, R.M., and Berg, K., Lipoprotein (a) and Atherosclerosis, *Ann. Intern. Med.,* 115(3):209–218, 1991.

Souter, A.K., Familial Hypercholesterolaemia: Mutations in the Gene for the Low Density Lipoprotein Receptor, *Mol. Med. Today* 1(2):90–97, 1995.

Superko, H.R., The Atherogenic Lipoprotein Profile, *Sci. Med.* 4(5):36–45, 1997.

Unwin, N., and Henderson, R., The Structure of Proteins in Biological Membranes, *Sci. Amer.,* 250(2):78–94, 1984.

QUESTIONS

1. Clearly define the following terms:
   a. lipid
   b. autocrine regulators
   c. amphipathic
   d. sesquiterpene
   e. lipid bilayer
   f. prenylation
   g. fluidity
   h. chylomicrons
   i. voltage-gated channel

2. List a major function of each of the following classes of lipid:
   a. phospholipids
   b. sphingolipids
   c. oils
   d. waxes
   e. steroids
   f. carotenoids

3. To which lipid class do each of the following molecules belong?

a.

$$CH_2-O-\overset{\overset{\displaystyle O}{\|}}{C}-(CH_2)_{10}CH_3$$
$$CH-O-\overset{\overset{\displaystyle O}{\|}}{C}-(CH_2)_{10}CH_3$$
$$CH_2-O-\underset{\underset{\displaystyle O}{\|}}{C}-(CH_2)_{10}CH_3$$

b.

c. $CH_3(CH_2)_{10}CH_2-O-\underset{\underset{\displaystyle O}{\|}}{C}-(CH_2)_{10}CH_3$

d. $CH_3(CH_2)_7CH=CH(CH_2)_7COOH$

e.

$$CH_2-O-\overset{\overset{\displaystyle O}{\|}}{C}-(CH_2)_{10}CH_3$$
$$CH-O-\underset{\underset{\displaystyle O}{\|}}{C}-(CH_2)_{10}CH_3$$
$$CH_2-O-\underset{\underset{\displaystyle O}{\diagdown \; O^-}}{\overset{\overset{\displaystyle O}{\|}}{P}}-O-CH_2CH_2\overset{+}{N}(CH_3)_3$$

f.

$$CH_2-O-\underset{\underset{\displaystyle O^-}{|}}{\overset{\overset{\displaystyle O}{\|}}{P}}-O-CH_2CH_2\overset{+}{N}(CH_3)_3$$
$$\underset{\underset{\displaystyle H}{|}}{}$$
$$CH-N-\overset{\overset{\displaystyle}{}}{\underset{\underset{\displaystyle O}{\|}}{C}}-(CH_2)_{10}CH_3$$
$$\underset{\underset{\displaystyle OH}{|}}{HC}-CH=CH-(CH_2)_{12}CH_3$$

4. Mammals in the Arctic (e.g., reindeer) have higher levels of unsaturated fatty acids in their legs and hooves than in the rest of their bodies. Suggest a reason for this phenomenon. Does it have a survival advantage?

5. Explain why entropy increases when a lipid bilayer forms from phospholipid molecules.

6. Sphingomyelins are amphipathic molecules. Draw the structure of a typical sphingomyelin. Determine which regions are hydrophilic and which are hydrophobic.

7. Glycolipids are nonionic lipids that can orient themselves into bilayers as phospholipids do. Suggest a reason why they can accomplish this feat although they lack an ionic group like phospholipids have.

8. What role do plasma lipoproteins play in the human body? Why do plasma lipoproteins require a protein component to accomplish their role?

9. As bacteria are exposed to higher temperatures, their membranes become increasingly fluid. What molecule could a researcher use to restore the organism's original fluidity characteristics? Consider the anatomy of bacteria. Why would this procedure not be feasible?

10. Describe several factors that influence membrane fluidity.

11. The fluid mosaic model of membrane structure has been very useful in explaining membrane behavior. However, the description of membrane as proteins floating in a phospholipid sea is oversimplified. Describe some components of membrane that are restricted in their lateral motion.

12. Which of the following statements or phrases concerning ionophores are true?
    a. forms channels through which ions flow
    b. requires energy
    c. ions may diffuse in either direction
    d. may cause voltage gates
    e. transports all ions with equal ease

13. Explain why spontaneous phospholipid translocation (the movement of a molecule from one side of a bilayer to the other) is so slow.

14. Explain the differences in the ease of lateral movement and *trans* membrane movement of phospholipids.

15. Suggest a reason why elevated LDL levels are a risk factor for coronary artery disease.

16. Explain how potassium moves across a membrane. How are the channels opened? What other ions flow during this process?

17. From what fatty acid are most of the eicosanoids derived? List several medical conditions in which it may appear advantageous to suppress their synthesis.

18. In which of the following processes do the prostaglandins not have a major recognized role?
    a. reproduction
    b. digestion
    c. respiration
    d. inflammation
    e. smooth muscle con- traction and relaxtion

19. Classify each of the following as a monoterpene, diterpene, triterpene, tetraterpene, or sesquiterpene:

a.

b.

(c)

d.

e.

f.

20. Which of the following is not a function of triacylglycerols?
    a. energy storage
    b. insulation
    c. shock absorption
    d. membrane structure

    What molecules perform the roles that are not attributed to triacylglycerols?

21. How does the function of HDL promote the reduction of coronary artery disease risk?

22. For which of the following properties of membranes are lipids not directly responsible? In what features are lipids indirectly involved?
    a. selective permeability
    b. self-sealing capability
    c. fluidity
    d. asymmetry
    e. active transport of ions

23. Describe how glucose is transported across membrane in the kidney. What type of transport is involved?

24. Describe how carrier-mediated transport operates. Give an example.

25. Animal cells are enclosed in a cell membrane. According to the fluid mosaic model, this membrane is held together by hydrophobic interactions. Consider the shear forces involved. Why does this membrane not break every time an animal moves?

26. Species-specific antigens are located on the surface of human and canine cells. If a heterokaryon is formed from red blood cell membrane from both species, what will happen to each set of antigens? What does this suggest about the nature of membranes?

LIPID METABOLISM

Chapter Ten

The energy required for strenuous physical activity is provided primarily by the degradation of fatty acids.

OBJECTIVES

After you have studied this chapter, you should be able to answer these questions:

1. *In what circumstances are triacylglycerols synthesized and degraded?*

2. *How are fatty acids degraded to generate energy?*

3. *What is the difference between the fatty acid catabolic pathways referred to as α-oxidation and β-oxidation? Under what circumstances is each pathway used?*

4. *What are ketone bodies? What purpose do they serve?*

5. *How are the eicosanoids synthesized? Why are specific drugs used to suppress eicosanoid synthesis?*

6. *How and why do cells synthesize fatty acids?*

7. *How are membrane lipids synthesized and degraded? What is membrane remodeling?*

8. *What similarities in isoprenoid synthetic pathways have been observed among living organisms?*

9. *How is cholesterol synthesized and degraded? How is cholesterol metabolism regulated?*

The roles lipids play in living organisms are largely due to their hydrophobic structures. As prominent components of cell membranes, lipids are primarily responsible for the integrity of individual cells, as well as the intracellular compartments that are the hallmark of eukaryotic organisms. The hydrophobic and highly reduced structure of triacylglycerols makes them a compact and rich source of energy for cellular processes. This chapter focuses on the metabolism of the major classes of lipids, that is, how they are synthesized and degraded and how these processes are regulated. A major emphasis is placed on the central metabolite in lipid metabolism: acetyl-Coenzyme A. Because of its prominent role in several human diseases, the metabolism of cholesterol is also discussed.

Acetyl-CoA, an energy-rich molecule composed of coenzyme A (Figure 10.1) and an acetyl group, plays a preeminent role in the metabolism of lipids. In most lipid-related metabolic processes, acetyl-CoA is either a substrate or a product. For example, acetyl-CoA that is not immediately required by a cell in energy production is used in fatty acid synthesis. When fatty acids are degraded to generate energy, acetyl-CoA is produced. Similarly, isopentenyl pyrophosphate, the building block molecule in isoprenoid synthetic reactions, is generated from three acetyl-CoA molecules. Therefore molecules as diverse as the terpenes and steroids found in animals and plants are all synthesized from acetyl-CoA. In this chapter, the metabolism of the major classes of lipid are discussed: fatty acids, triacylglycerols, phospholipids, sphingolipids, and isoprenoids. In addition, the metabolism of several important fatty acid metabolites, that is, the eicosanoids and the ketone bodies, are reviewed. Because of the major role that lipids play in providing energy and structural materials for living cells, several metabolic control mechanisms are discussed throughout the chapter.

10.1 FATTY ACIDS AND TRIACYLGLYCEROLS

As discussed, fatty acids are an important and efficient energy source for many living cells. In animals, most fatty acids are obtained in the diet. For example, in the average U.S. diet, between 30% and 40% of calories ingested are provided by fat. Triacylglycerol molecules are digested within the lumen of the small intestine by

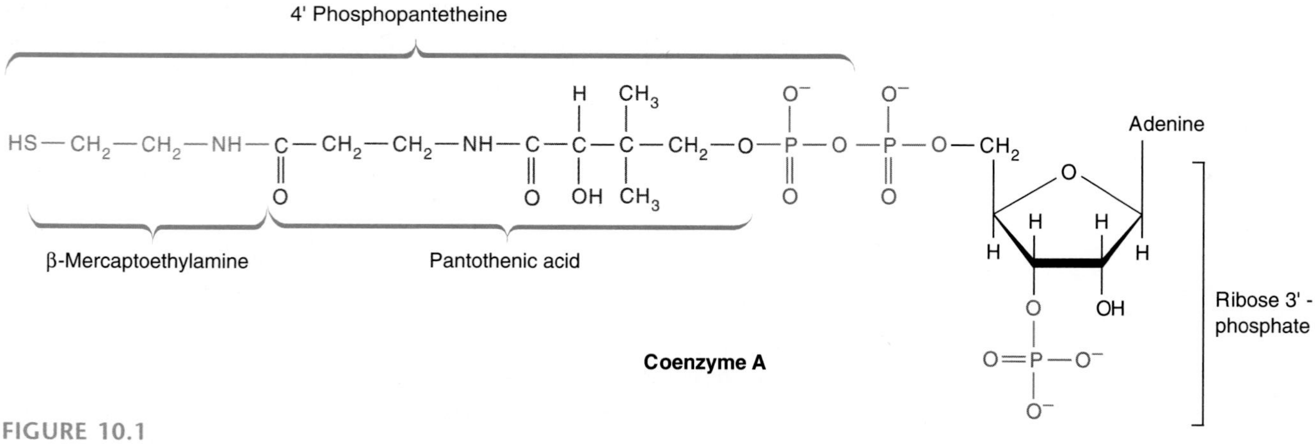

FIGURE 10.1

Coenzyme A.

In coenzyme A a 3'phosphate derivative of ADP is linked to pantothenic acid via a phosphate ester bond. The β-mercaptoethylamine group of coenzyme A is attached to pantothenic acid by an amide bond. Coenzyme A is a carrier of acyl groups which range in size from the acetyl group to long chain fatty acids. Because the reactive SH group forms a thioester bond with acyl groups, coenzyme A is often abbreviated as CoASH. The carbon-sulfur bond of a thioester is more easily cleaved than the carbon-oxygen bond of an ester. Because sulfur is a better leaving group than oxygen, more energy is released when thioesters are hydrolyzed.

pancreatic lipase to form fatty acids and monoacylglycerol. After these latter molecules are transported across the plasma membrane of intestinal wall cells (enterocytes), they are reconverted to triacylglycerols. Enterocytes then combine triacylglycerols with dietary cholesterol and newly synthesized phospholipids and protein to form chylomicrons. After their secretion into the lymph (tissue fluid derived from blood), chylomicrons pass from the lymph into the blood. Most chylomicrons are removed from blood by adipose tissue cells (adipocytes), the body's primary lipid storage depot. Lipoprotein lipase, synthesized by cardiac and skeletal muscle, lactating mammary gland, and adipose tissue, is transferred to the endothelial surface of the capillaries, where it converts the triacylglycerol in chylomicrons into fatty acids and glycerol. (Lipoprotein lipase is activated when it binds to the apoprotein component of chylomicrons. The triacylglycerols in VLDL are also degraded by lipoprotein lipase.) Because adipose tissue cannot use glycerol, this molecule is carried in the blood to the liver, where the enzyme glycerol kinase converts it to glycerol-3-phosphate. (Adipocytes lack glycerol kinase. They derive glycerol-3-phosphate from dihydroxyacetone phosphate, a glycolytic intermediate.) In liver cells, glycerol-3-phosphate can then be used in the synthesis of triacylglycerols, phospholipids, or glucose.

Depending on an animal's current metabolic needs, fatty acids may be converted to triacylglycerols, degraded to generate energy or used in membrane synthesis. For example, serum glucose levels are high immediately after a meal. The hormone insulin promotes triacylglycerol synthesis by facilitating the transport of glucose into adipocytes. (Glucose is a precursor in glycerol-3-phosphate synthesis.) In contrast, adipocytes cannot synthesize triacylglycerol between meals when blood glucose levels are low. Instead, several hormones stimulate the hydrolysis of triacylglycerols within adipose tissue to form glycerol and fatty acids. As discussed, glycerol is a substrate for gluconeogenesis. Fatty acids are degraded by the body's cells to generate energy.

Triacylglycerol synthesis (referred to as **lipogenesis**) is illustrated in Figure 10.2. Glycerol-3-phosphate or dihydroxyacetone phosphate reacts sequentially with three molecules of acyl-CoA (fatty acid esters of CoASH). Acyl-CoA molecules are produced in the following reaction:

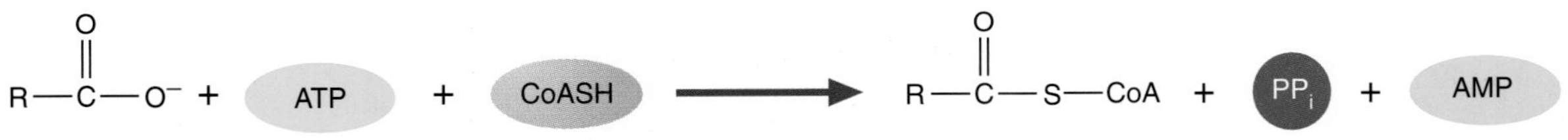

$$R-\overset{\overset{\textstyle O}{\|}}{C}-O^- + \boxed{ATP} + \boxed{CoASH} \longrightarrow R-\overset{\overset{\textstyle O}{\|}}{C}-S-CoA + \boxed{PP_i} + \boxed{AMP}$$

(Note that the reaction is driven to completion by the hydrolysis of pyrophosphate by pyrophosphatase.) In the synthesis of triacylglycerols, phosphatidic acid is formed by two sequential acylations of glycerol-3-phosphate or by a pathway involving the direct acylation of dihydroxyacetone phosphate. In the latter pathway, acyldihydroxyacetone phosphate is later reduced to form lysophosphatidic acid. Phosphatidic acid is then produced when acyldihydroxyacetone phosphate reacts with a second acyl-CoA. Once formed, phosphatidic acid is converted to diacylglycerol by phosphatidic acid phosphatase. A third acylation reaction forms triacylglycerol. Fatty acids derived from both the diet and *de novo* synthesis are incorporated into triacylglycerols. (The term *de novo* is used by biochemists to indicate new synthesis.) *De novo* synthesis of fatty acids is discussed below.

When energy reserves are low, the body's fat stores are mobilized in a process referred to as **lipolysis** (Figure 10.3). Lipolysis occurs during fasting, during vigorous exercise, and in response to stress. Several hormones (e.g., glucagon and epinephrine) bind to specific adipocyte plasma membrane receptors and begin a reaction sequence similar to the activation of glycogen phosphorylase. The enzyme hormone-sensitive lipase is activated by cAMP synthesis. (Triacylglycerols are constantly being synthesized and degraded in adipose tissue. The activation of hormone-sensitive lipase vastly increases the rate of hydrolyzing triacylglycerols.) Both products of lipolysis (i.e., fatty acids and glycerol) are released into the blood. As stated previously, glycerol is transported to the liver,

KEY *Concepts*

When energy reserves are high, triacylglycerols are synthesized in a process called lipogenesis. When energy reserves are low, triacylglycerols are degraded in a process called lipolysis to form fatty acids and glycerol.

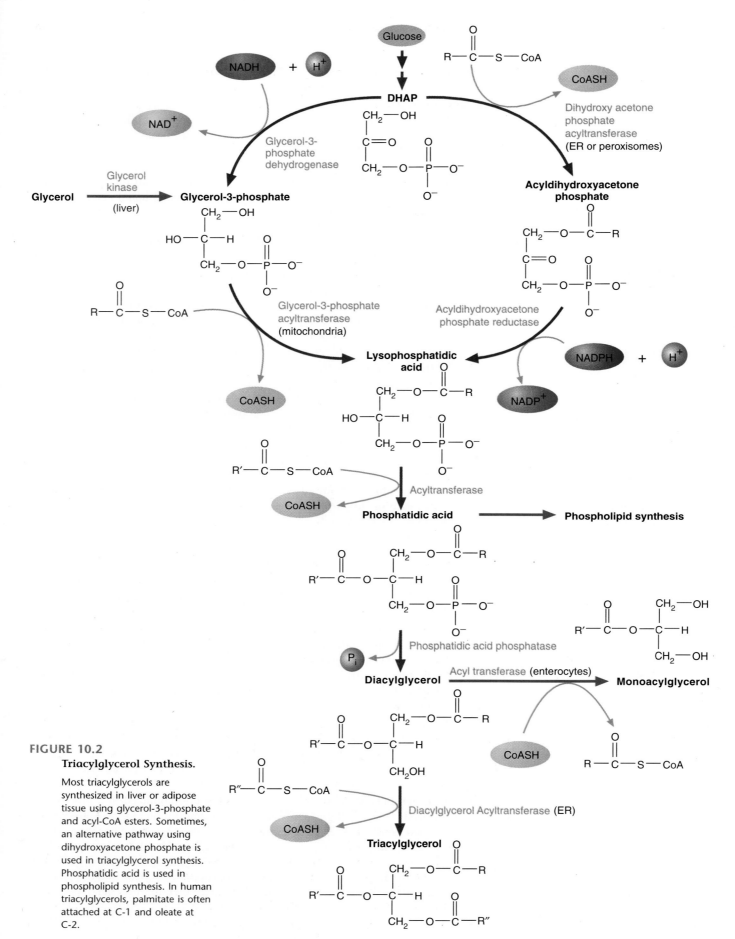

FIGURE 10.2

Triacylglycerol Synthesis.

Most triacylglycerols are synthesized in liver or adipose tissue using glycerol-3-phosphate and acyl-CoA esters. Sometimes, an alternative pathway using dihydroxyacetone phosphate is used in triacylglycerol synthesis. Phosphatidic acid is used in phospholipid synthesis. In human triacylglycerols, palmitate is often attached at C-1 and oleate at C-2.

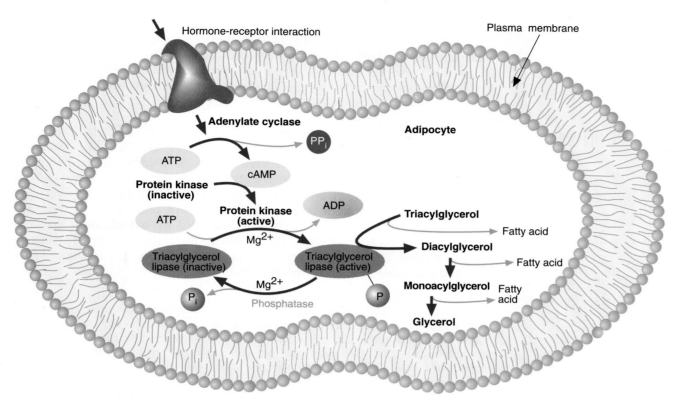

FIGURE 10.3

Lipolysis.

When certain hormones bind to their receptors in adipose tissue, a cascade mechanism releases fatty acids and glycerol from triacylglycerol molecules. Triacylglycerol lipase (sometimes referred to as hormone-sensitive lipase) is activated when it is phosphorylated by protein kinase. Protein kinase is activated by cAMP. After their transport across the plasma membrane, fatty acids are transported in blood to other organs bound to serum albumen.

where it can be used in lipid or glucose synthesis. After their transport across the adipocyte plasma membrane, fatty acids become bound to serum albumin. (Serum albumin is an abundant protein that transports numerous substances in blood. Other examples of hydrophobic albumin ligands include steroids and eicosanoids.) The albumin-bound fatty acids are carried to tissues throughout the body, where they are oxidized to generate energy. Fatty acids are transported into cells by a protein in the plasma membrane. This process is linked to the active transport of sodium. The amount of fatty acid that is transported depends on its concentration in blood and the relative activity of the fatty acid transport mechanism. Cells vary widely in their capacity to transport and use fatty acids. For example, some cells (e.g., brain and red blood cells) cannot use fatty acids as fuel, while others (e.g., cardiac muscle) rely on them for a significant portion of their energy requirements. Once they enter a cell, fatty acids must be transported to their destinations (i.e., mitochondria, ER, and other organelles). Several **fatty acid-binding proteins** may be responsible for this transport.

Most fatty acids are degraded to form acetyl-CoA within mitochondria in a process referred to as β-oxidation. β-Oxidation is also known to occur in peroxisomes. A minor oxidative pathway, referred to as α-oxidation, is also available to degrade certain nonstandard fatty acids (Special Interest Box 10.1).

Fatty acids are synthesized when an organism has met its energy needs and nutrient levels are high. (Glucose and several amino acids are substrates for fatty acid synthesis.) Fatty acids are synthesized from acetyl-CoA in a process that is similar to the reverse of β-oxidation. Although most fatty acids are supplied in the diet, most animal tissues can synthesize some saturated and unsaturated fatty acids. In addition, animals can elongate and desaturate dietary fatty acids. For example, arachidonic acid is produced by adding a two-carbon unit and introducing two double bonds to linoleic acid.

In plants the fatty acids of triacylglycerols are used predominantly as an energy source for germinating seeds. Once synthesized (in a reaction sequence similar to that found in animals), triacylglycerols are stored in vesicles called *oil bodies*. As seeds begin to germinate, the synthesis of lipases causes a massive breakdown of triacylglycerols. Most of the fatty acids released in this process are used to synthesize carbohydrates within glyoxysomes (Section 11.1).

Fatty Acid Degradation

Most fatty acids are degraded by the sequential removal of two-carbon fragments from the carboxyl end of fatty acids. During this process, referred to as β-oxidation, acetyl-CoA is formed as the bond between the α- and β-carbon atoms is broken. (β-Oxidation is so named because the β-carbon of fatty acids, which is two carbons removed from the carboxyl group, is oxidized.) Other mechanisms for degrading fatty acids are known. Most of these degrade unusual fatty acids; for example, odd-chain or branched chain molecules are degraded by α-oxidation. Once it is formed, acetyl-CoA may then be oxidized further to generate energy. β-Oxidation is discussed next. The degradation of odd-chain, branched chain, and unsaturated fatty acids is discussed in Special Interest Box 10.1.

10.1

VLDL secretion by liver cells depends directly on the intracellular concentration of fatty acids. A cytoplasmic fatty acid–binding protein (FABP) may be responsible for the transport of fatty acids to the SER, the site of triacylglycerol synthesis. Because hepatic VLDL secretion has been found to be greater in female rats than in male rats, there may be a connection between an animal's sex hormone status, FABP concentration, and VLDL secretion rate. If this is so, what would you expect to happen if a male rat is injected with estrogen? In general, what mechanisms are involved? (Hint: Review how steroid hormones exert their metabolic effects. Refer to Chapter 8.)

β-Oxidation   β-Oxidation occurs primarily within mitochondria. (β-Oxidation within peroxisomes is discussed below.) Before β-oxidation begins, each fatty acid is activated in a reaction with ATP and CoASH (p. 251). The enzyme that catalyzes this reaction, acyl-CoA ligase, is found in the outer mitochondrial membrane. Because the mitochondrial inner membrane is impermeable to most acyl-CoA molecules, a special carrier called *carnitine* is used to transport acyl groups into the mitochondrion (Figure 10.4). Carnitine-mediated transfer of acyl groups into the mitochondrial matrix is accomplished through the following mechanism (Figure 10.5):

1. Each acyl-CoA molecule is converted to an acylcarnitine derivative.

$$R-\overset{\overset{\displaystyle O}{\|}}{C}-S-CoA \ + \ (CH_3)_3\overset{+}{N}-CH_2-\underset{\underset{\displaystyle OH}{|}}{CH}-CH_2-\overset{\overset{\displaystyle O}{\|}}{C}-O^- \ \rightleftharpoons \ (CH_3)_3\overset{+}{N}-CH_2-\underset{\underset{\underset{\underset{\displaystyle R}{|}}{\underset{\displaystyle C=O}{|}}}{\underset{\displaystyle O}{|}}}{CH}-CH_2-\overset{\overset{\displaystyle O}{\|}}{C}-O^- \ + \ \text{CoASH}$$

Acyl-CoA   **Carnitine**   **Acylcarnitine**

This reaction is catalyzed by carnitine acyltransferase I.

2. A carrier protein within the mitochondrial inner membrane transfers acylcarnitine into the mitochondrial matrix.

3. Acyl-CoA is regenerated by carnitine acyltransferase II.

4. Carnitine is transported back into the intermembrane space by the carrier protein. It then reacts with another acyl CoA.

A summary of the reactions of the β-oxidation of saturated fatty acids is shown in Figure 10.6. The pathway begins with an oxidation-reduction reaction, catalyzed by

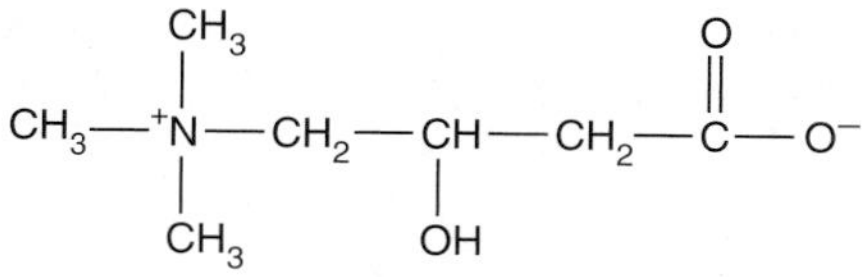

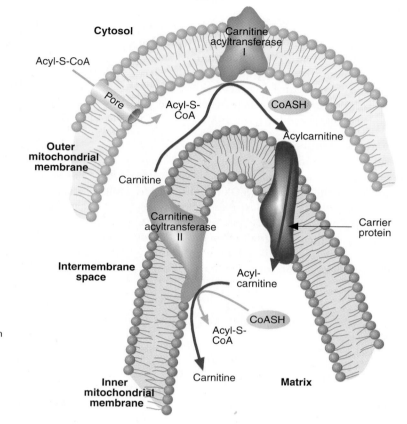

FIGURE 10.4
Carnitine.

FIGURE 10.5

Acyl-CoA Transport into the Mitochondrion.

Acyl-CoA reacts with carnitine to form an acylcarnitine derivative. Carnitine acyltransferase I catalyzes this reaction. Acylcarnitine is then transported across the inner membrane by a carrier protein and is subsequently reconverted to carnitine and acyl-CoA by carnitine acyltransferase II. (Some researchers believe that carnitine acyltransferase I occurs on the outer face of the inner membrane.)

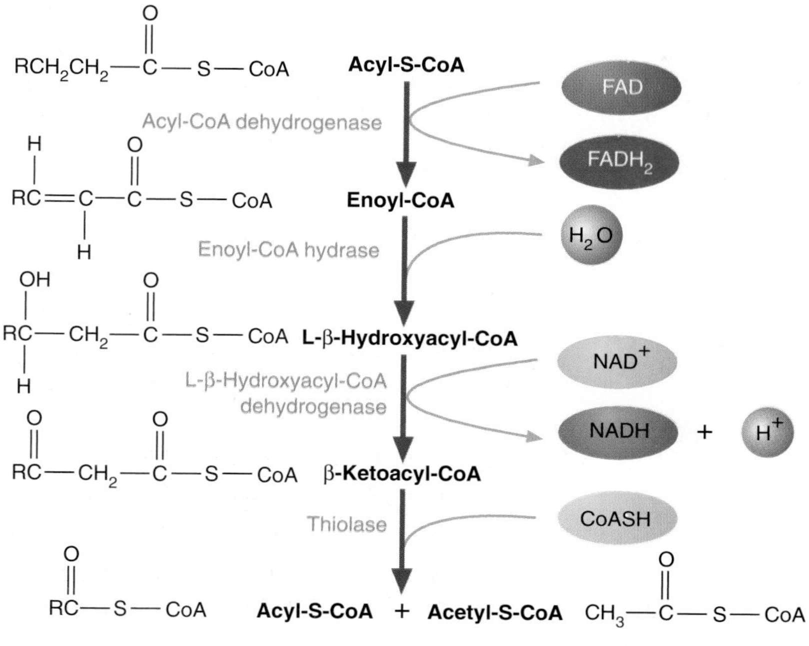

FIGURE 10.6

β-Oxidation of Acyl-CoA.

The β-oxidation of acyl-CoA molecules includes four reactions that occur in the mitochondrial matrix. Each cycle of reactions forms acetyl-CoA and an acyl-CoA that is shorter by two carbons.

FATTY ACID OXIDATION: DOUBLE BONDS AND ODD CHAINS

The β-oxidation pathway degrades saturated fatty acids with an even number of carbon atoms. Other mechanisms are required to degrade unsaturated, odd-chain and branched-chain fatty acids.

Unsaturated Fatty Acid Oxidation

The oxidation of unsaturated fatty acids such as oleic acid requires additional enzymes. They are needed because, unlike the *trans* double bonds introduced during β-oxidation, the double bonds of most naturally occurring unsaturated fatty acids have a *cis* configuration. The enzyme enoyl-CoA isomerase converts the *cis*-β,γ double bond to a *trans*-α,β double bond. The β-oxidation of oleic acid is illustrated in Figure 10A.

Odd-Chain Fatty Acid Oxidation

Although most fatty acids contain an even number of carbon atoms, some organisms (e.g., some plants and microorganisms) produce odd-chain fatty acid molecules. β-Oxidation of such fatty acids proceeds normally until the last β-oxidation cycle, which yields one acetyl-CoA molecule and one propionyl-CoA molecule. Propionyl-CoA is then converted to succinyl-CoA, a citric acid cycle intermediate (Figure 10B). Ruminant animals such as cattle and sheep derive a substantial amount of energy

from the oxidation of odd-chain fatty acids. These molecules are produced by microorganisms in the rumen (stomach).

α-Oxidation

α-Oxidation is a mechanism for degrading branched chain and odd-chain fatty acids. The most important role of α-oxidation in animals is the catabolism of phytanic acid, a branched 20-carbon fatty acid. (Phytanic acid is an oxidation product of phytol, a diterpene alcohol esterified to chlorophyll, the photosynthetic pigment.) Phytol, found in green vegetables, is converted to phytanic acid after it is ingested. Phytanic acid is a component of foods derived from herbivorous (plant-eating) animals (e.g., dairy products). In some plant tissues (e.g., leaves and seeds), α-oxidation is a major mechanism in the degradation of long-chain fatty acids.

β-Oxidation of phytanic acid is blocked by the methyl group substituent on C-3. Consequently, the first step in phytanic acid catabolism is an α-oxidation in which the molecule is converted to a 2-hydroxy fatty acid. (α-Hydroxylating activity has been detected in the ER and in mitochondria.) This reaction is followed by the removal of the carboxyl group (Figure 10C). After activation to a CoA derivative, the product, pristanic acid, can be further degraded by β-oxidation.

FIGURE 10A

β-Oxidation of Oleoyl-CoA.

β-Oxidation progresses until Δ³-*cis*-dodecenoyl-CoA is produced. This molecule is not a suitable substrate for β-oxidation, because it contains a *cis* double bond. After conversion of the β,γ *cis* double bond to a α,β *trans* double bond, β-oxidation recommences.

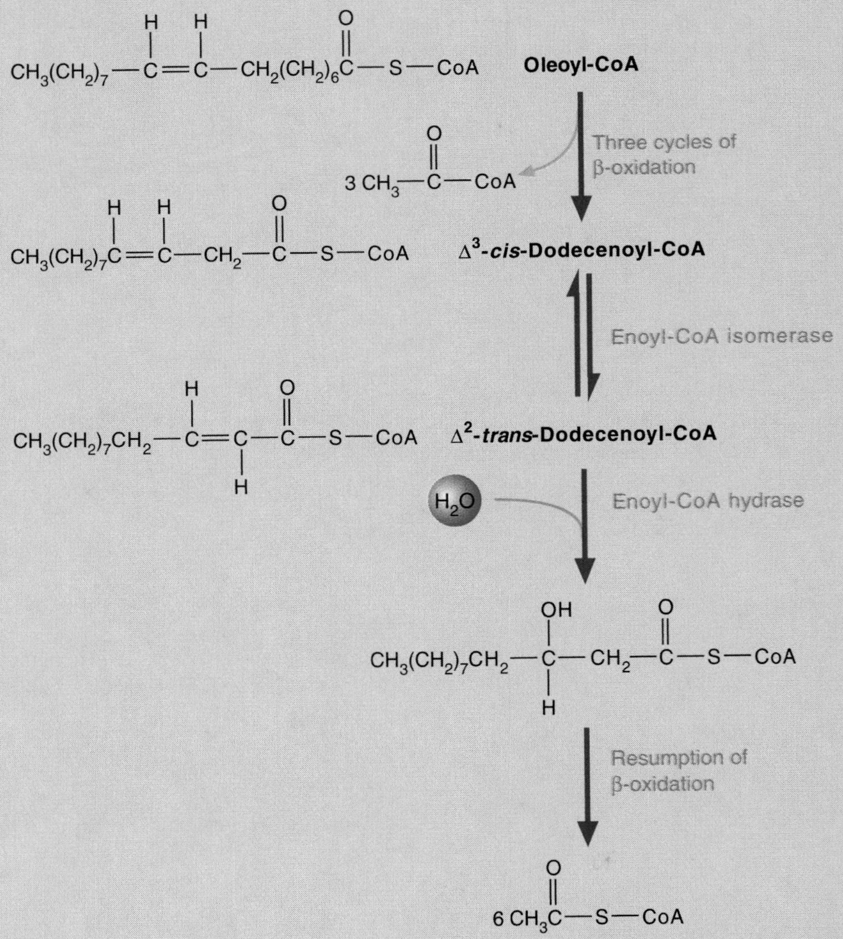

The ability to oxidize phytanic acid is critical because large quantities of this molecule are found in the diet. In *Refsum's disease* (also referred to as *phytanic acid storage syndrome*) a buildup of phytanic acid causes very serious neurological problems. In this rare autosomal recessive condition, nerve damage is caused by a lack of α-hydroxylating activity. The mechanism by which phytanic acid accumulation causes nerve damage is unknown. Eating less phytanic acid-containing foods can significantly reduce nerve damage.

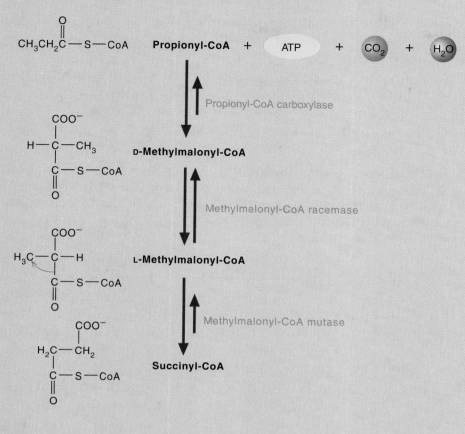

FIGURE 10B

Conversion of Propionyl-CoA to Succinyl-CoA.

In the first step, propionyl-CoA is carboxylated by propionyl-CoA carboxylase, an enzyme with a biotin cofactor. The product, D-methylmalonyl-CoA, is isomerized by methylmalonyl-CoA racemase to form L-methylmalonyl-CoA. In the last step, a hydrogen atom and the carbonyl-CoA group exchange positions. This unusual reaction is catalyzed by methylmalonyl-CoA mutase, an enzyme that requires vitamin B_{12}. (Vitamin B_{12} is 5'-deoxyadenosylcobalamin.)

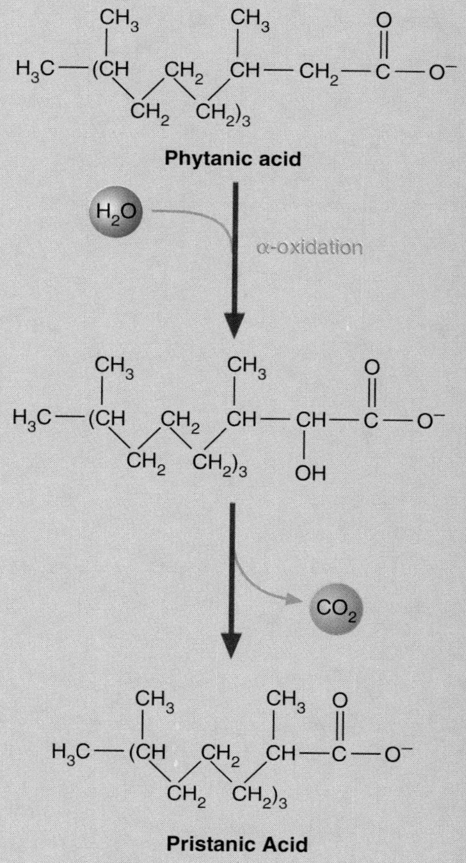

FIGURE 10C

α-Oxidation of Phytanic Acid.

Phytanic acid is converted to pristanic acid by oxidation of the α-carbon and a decarboxylation. Pristanic acid is further degraded to generate a mixture of acetyl-CoA, propionyl-CoA and isobutyl-CoA by the β-oxidation pathway.

Phytanic acid

Pristanic Acid

acyl-CoA dehydrogenase, in which one hydrogen atom each are removed from the α- and β-carbons.

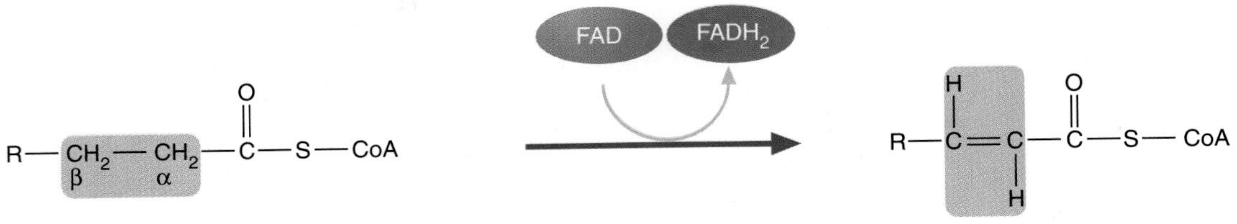

Acyl-CoA α,β-**Enoyl-CoA**

The electrons that are transferred to $FADH_2$ in this reaction are then donated to coenzyme Q, a component of the mitochondrial electron transport chain (ETC). The second reaction, catalyzed by enoyl-CoA hydrase, involves a hydration of the double bond between the α- and β-carbons.

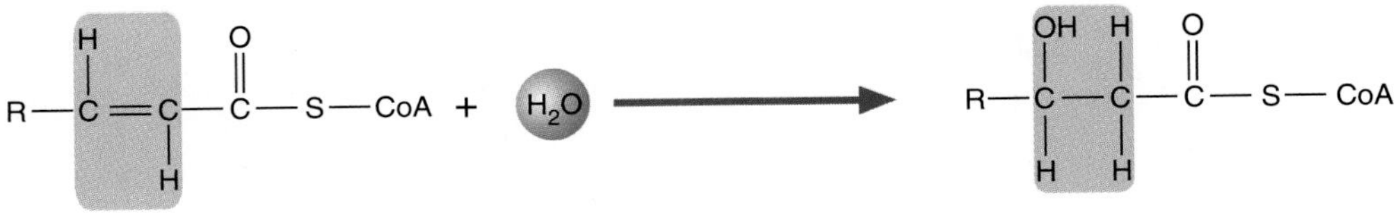

α,β-**Enoyl-CoA**   **3-Hydroxylacyl-CoA**

The β-carbon is now hydroxylated. In the next reaction this hydroxyl group is dehydrogenated. The production of a β-ketoacyl-CoA is catalyzed by 3-hydroxyacyl-CoA dehydrogenase.

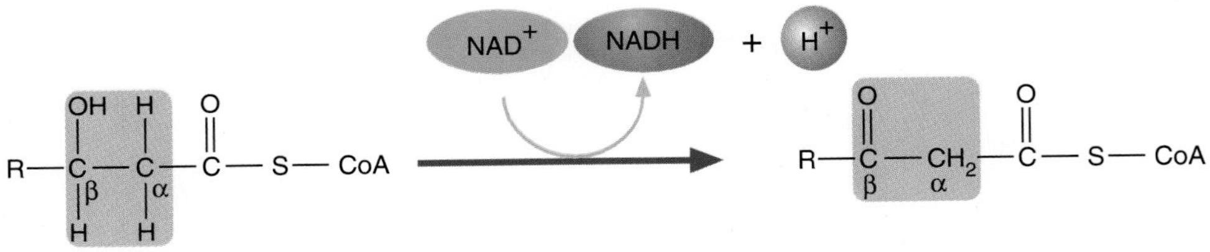

3-Hydroxylacyl-CoA β-**Ketoacyl-CoA**

The electrons transferred to NAD^+ are later donated to Complex I of the ETC. Finally, thiolase (sometimes referred to as β-ketoacyl-CoA thiolase) catalyzes a C_α-C_β cleavage.

β-**Ketoacyl-CoA**   **Acyl-CoA**   **Acetyl-CoA**

In this reaction, sometimes called a **thiolytic cleavage,** an acetyl-CoA molecule is released. The other product, an acyl-CoA, now contains two fewer C atoms.

The four steps just outlined constitute one cycle of β-oxidation. During each later cycle, a two-carbon fragment is removed. This process, sometimes called the β-*oxidation spiral,* continues until, in the last cycle, a four-carbon acyl-CoA is cleaved to form two acetyl-CoAs.

The following equation summarizes the oxidation of palmitoyl-CoA.

$$CH_3(CH_2)_{14}\overset{O}{\overset{\|}{C}}-S-CoA \; + \; 7\;FAD \; + \; 7\;NAD^+ \; + \; 7\;CoASH \; + \; 7\;H_2O \longrightarrow$$

$$8\;CH_3\overset{O}{\overset{\|}{C}}-S-CoA \; + \; 7\;FADH_2 \; + \; 7\;NADH^+ \; + \; 7\;H^+$$

The acetyl CoA molecules produced by fatty acid oxidation are converted via the *citric acid cycle* to CO_2 and H_2O as additional NADH and $FADH_2$ are formed. A portion of the energy released as NADH and $FADH_2$ are oxidized by the electron transport chain is later captured in ATP synthesis via *oxidative phosphorylation*. The complete oxidation of acetyl CoA is discussed in Chapter 11.

β-Oxidation in Peroxisomes   β-Oxidation of fatty acids also occurs within peroxisomes. In animals peroxisomal β-oxidation appears to shorten very long-chain fatty acids. The resulting medium-chain fatty acids are further degraded within mitochondria. In many plant cells, β-oxidation occurs predominantly in peroxisomes. (Fatty acids are not an important source of energy in most plant tissues. Although some plant mitochondria contain β-oxidation enzymes, this pathway is not considered to contribute to energy generation to any substantial degree.) In some germinating seeds, β-oxidation occurs in glyoxysomes. (Glyoxysomes are specialized peroxisomes that possess the glyoxylate cycle enzymes. Refer to p. 300–301.) The acetyl-CoA derived from glyoxysomal β-oxidation is converted to carbohydrate by the glyoxylate cycle and gluconeogenesis.

Peroxisomal membrane possesses an acyl-CoA ligase activity that is specific for very long-chain fatty acids. Mitochondria apparently cannot activate long-chain fatty acids such as tetracosanoic (24:0) and hexacosanoic (26:0). Peroxisomal carnitine acyltransferases catalyze the transfer of these molecules into peroxisomes, where they are oxidized to form acetyl-CoA and medium-chain acyl-CoA molecules (i.e., those possessing between 6 and 12 carbons). Medium-chain acyl-CoAs are further degraded via β-oxidation within mitochondria.

Although the reactions of peroxisomal β-oxidation are similar to those in mitochondria, there are some notable differences. First, the initial reaction in the peroxisomal pathway is catalyzed by a different enzyme. This reaction is a dehydration catalyzed by an acyl-CoA oxidase. The reduced coenzyme $FADH_2$ then donates its electrons directly to O_2 instead of UQ. This feature of peroxisomal β-oxidation is in sharp contrast to the mitochondrial pathway, which synthesizes ATP. The H_2O_2 produced when $FADH_2$ is oxidized is converted to H_2O by catalase. Second, the next two reactions in peroxisomal β-oxidation are catalyzed by two enzyme activities (enoyl-CoA hydrase and 3-hydroxyacyl CoA dehydrogenase) found on the same protein molecule. Finally, the last enzyme in the pathway (β-ketoacyl-CoA thiolase) has a different substrate specificity than its mitochondrial version. It does not efficiently bind medium-chain acyl-CoAs.

KEY *Concepts*

In β-oxidation, fatty acids are degraded by breaking the bond between the α- and β-carbon atoms. The ketone bodies are produced from excess molecules of acetyl-CoA.

QUESTION 10.2

In addition to β-oxidation, peroxisomes have other vital roles in lipid metabolism. For example, the synthesis of various ether-type lipids occurs within peroxisomes. In a rare autosomal recessive disease called *Zellweger syndrome,* affected individuals lack peroxisomes. Abnormalities in several organs (especially in brain, liver, and kidney) result in death in the first year of life. Because the absence of an organelle cannot be confirmed by microscopic methods, biochemical techniques must be used to diagnose Zellweger syndrome. (The organelle could be so altered by the genetic defect as to be undetectable.) Suggest in general terms several biochemical methods of diagnosing this malady.

The Ketone Bodies Most of the acetyl-CoA produced during fatty acid oxidation is used by the citric acid cycle or in isoprenoid synthesis (Section 10.3). Under normal conditions, fatty acid metabolism is so carefully regulated that only small amounts of excess acetyl-CoA are produced. In a process called **ketogenesis,** acetyl-CoA molecules are converted to acetoacetate, β-hydroxybutyrate, and acetone, a group of molecules called the **ketone bodies** (Figure 10.7).

Ketone body formation, which occurs within the matrix of liver mitochondria, begins with the condensation of two acetyl-CoAs to form acetoacetyl-CoA. Then acetoacetyl-CoA condenses with another acetyl-CoA to form β-hydroxy-β-methylglutaryl-CoA (HMG-CoA). In the next reaction, HMG-CoA is cleaved to form acetoacetate and acetyl-CoA. Acetoacetate is then reduced to form β-hydroxybutyrate. Acetone is formed by the spontaneous decarboxylation of acetoacetate when the latter molecule's concentration is high. (This condition, referred to as **ketosis,** occurs in uncontrolled diabetes, a metabolic disease discussed in Special Interest Box 15.1, and during starvation. In both of these conditions there is massive lipolysis.)

Several tissues, most notably cardiac and skeletal muscle, use ketone bodies to generate energy. During prolonged starvation (i.e., in the absence of sufficient glucose) the brain uses ketone bodies as an energy source. The mechanism by which acetoacetate and β-hydroxybutyrate are converted to acetyl-CoA is illustrated in Figure 10.8.

Eicosanoid Metabolism

As previously discussed, many important eicosanoids are derived from arachidonic acid. Almost all cellular arachidonic acid is stored in cell membranes as esters at C-2 of glycerol in phosphoglycerides. Release of arachidonic acid from membrane, considered to be the rate-limiting step in eicosanoid synthesis (Figure 10.9), results from binding an appropriate chemical signal to its receptor on a target cell plasma

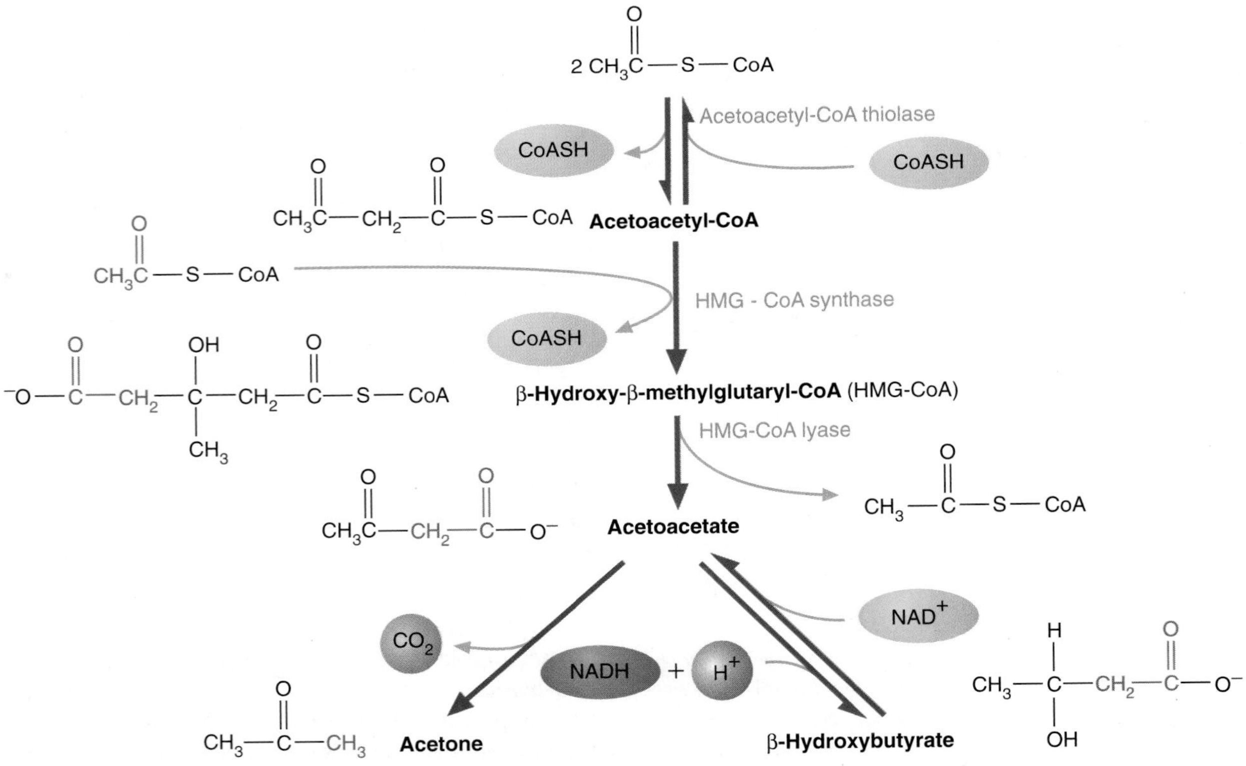

FIGURE 10.7

Ketone Body Formation.

Ketone bodies (acetoacetate, acetone, and β-hydroxybutyrate) are produced within the mitochondria when excess acetyl-CoA is available. Under normal circumstances, only small amounts of ketone bodies are produced.

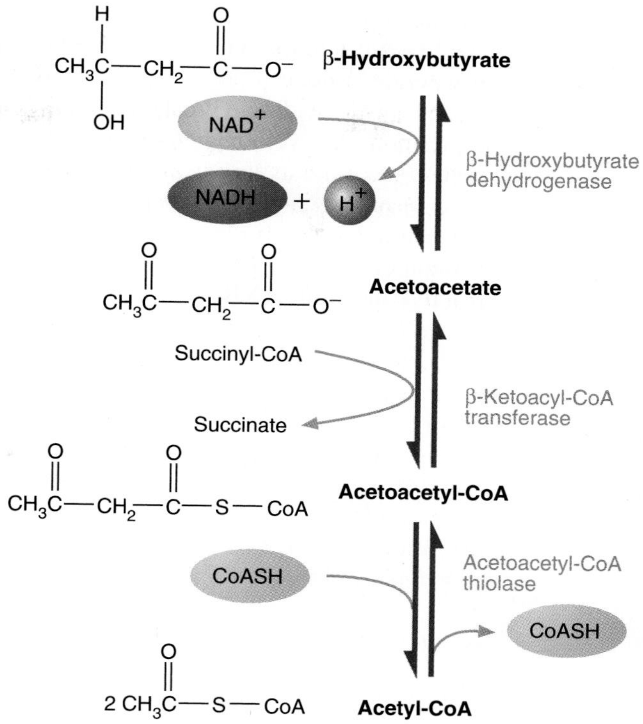

FIGURE 10.8

Conversion of Ketone Bodies to Acetyl-CoA.

Some organs (e.g., heart and skeletal muscle) can use ketone bodies as an energy source under normal conditions. During starvation the brain uses them as an important fuel source. Because liver does not have β-ketoacid-CoA transferase, it cannot use ketone bodies as an energy source. These reactions are reversible.

FIGURE 10.9

Synthesis of Selected Prostaglandins and Thromboxanes.

Each step is catalyzed by a cell-specific enzyme.

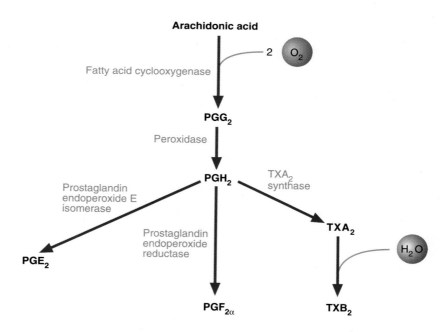

membrane. For example, the release of arachidonic acid in platelets is caused by binding thrombin, an enzyme that plays an important role in blood clotting. (Thrombin is a proteolytic enzyme that converts the soluble plasma protein fibrinogen into fibrin, which then forms an insoluble meshwork.) Platelet aggregation, triggered by the eicosanoid TXA_2, is an early critical step in the blood-clotting process.

Most often, the release of arachidonic acid is catalyzed by phospholipase A_2. Certain steroids that suppress inflammation inhibit phospholipase A_2. Phospholipase A_2 cleaves acyl groups from C-2 of a phosphoglyceride, thus forming a fatty acid and lysophosphoglyceride. Once they are released, arachidonic acid molecules may be converted (depending on both cell type and intracellular conditions) into a variety of eicosanoid molecules. Prostaglandin synthesis begins when cyclooxygenase converts arachidonic acid into PGG_2. (Aspirin inactivates cyclooxygenase by acetylating a critical serine residue in the enzyme.) Then the formation of PGH_2 from PGG_2 is catalyzed by peroxidase. (Prostaglandin endoperoxidase synthase, an ER enzyme, possesses both the cyclooxygenase and the peroxidase activities.) PGH_2 is a precursor of several eicosanoids. For example, PGE_2 and PGF_2 are formed from PGH_2 by the actions of prostaglandin endoperoxide E isomerase and prostaglandin endoperoxide reductase, respectively. In platelets and lung cells, TXA_2 synthase catalyzes the conversion of PGH_2 to TXA_2. Within seconds, TXA_2 is spontaneously hydrolyzed to the inactive molecule TXB_2.

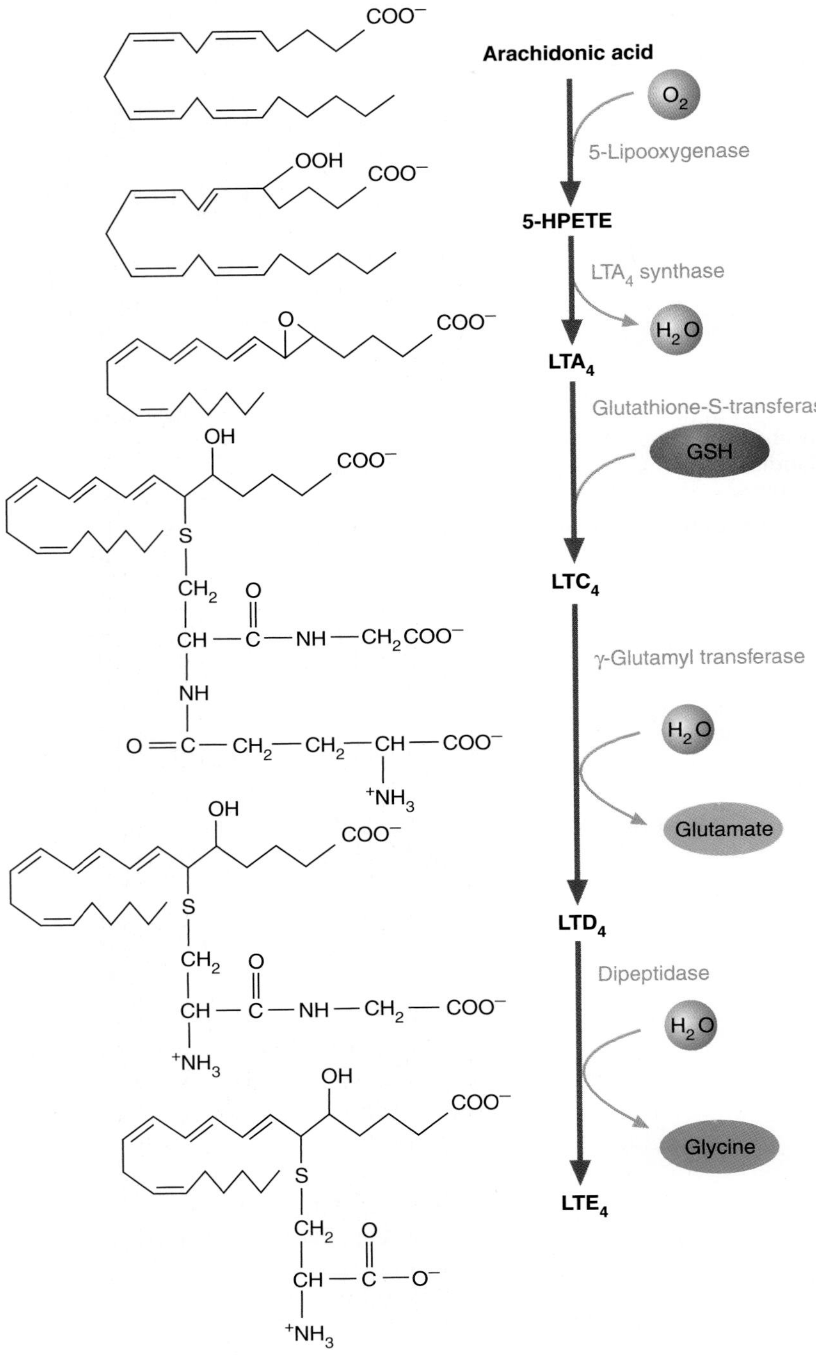

FIGURE 10.10

Synthesis of Selected Leukotrienes.

LTC_4, LTD_4, and LTE_4 are components of slow-reacting substance of anaphalaxis.

In a separate pathway, arachidonic acid is converted into the leukotrienes. Several enzymes, referred to as lipoxygenases, catalyze the addition of hydroperoxy groups to arachidonic acid. (Lipoxygenases are found in many mammalian tissues and in some plants.) The products of these reactions, called the monohydroperoxyeicosate-traenoic acids (HPETEs), are the direct precursors of the leukotrienes. For example, 5-lipooxygenase catalyzes the synthesis of 5-HPETE (Figure 10.10). 5-HPETE is then converted to LTA_4. (Note that LTA_4 possesses an epoxide. **Epoxides** are three-membered rings containing an ether functional group.) LTC_4 is formed by the addition of GSH. LTC_4 is converted to LTD_4 by the removal of glutamic acid. Finally, LTE_4 is formed when glycine is removed from LTD_4. The role of leukotrienes is unclear, although several are believed to act as chemoattractants or as intracellular signals.

QUESTION

10.3

Rheumatoid arthritis is an autoimmune disease in which the joints are chronically inflamed. (In **autoimmune diseases** the immune system fails to distinguish between self and nonself. For reasons that are not understood, specific lymphocytes are stimulated to produce antibodies, referred to as *autoantibodies*. These molecules bind to surface antigens on the patient's own cells as if they were foreign. Then the immune system attacks affected cells.) In rheumatoid arthritis, several types of white blood cells (e.g., lymphocytes and macrophages) infiltrate joint tissue and then rupture. When lysosomal enzymes leak, tissue is eroded. The inflammatory response is perpetuated by the release of eicosanoids by white blood cells. A variety of eicosanoids have been implicated. For example, macrophages are known to produce PGE_4, TXA_2, and several leukotrienes.

Currently, the treatment of rheumatoid arthritis consists of suppressing pain and inflammation. (The disease continues to progress despite treatment.) Because of its low cost and relative safety, aspirin plays an important role in the treatment of rheumatoid arthritis and other types of inflammation. Certain steroids are more potent than aspirin in reducing inflammation, that is, they immediately and dramatically reduce painful symptoms. However, steroids have serious side effects. For example, prednisone may depress the immune system as well as cause fat redistribution to the neck ("buffalo hump") and serious behavioral changes. For these and other reasons, prednisone is used to treat rheumatoid arthritis only when a patient does not respond to aspirin or similar drugs.

Review the effects of aspirin and steroids on eicosanoid metabolism. Suggest a reason why this information is relevant to the treatment of rheumatoid arthritis. Does it explain the difference between the effectiveness of aspirin and steroids in treating inflammation?

Fatty Acid Biosynthesis

Although fatty acid synthesis occurs within the cytoplasm of most animal cells, liver is the major site for this process. (Recall, for example, that liver produces VLDL.) Fatty acids are synthesized when the diet is low in fat and/or high in carbohydrate or protein. Most fatty acids are synthesized from dietary glucose. As discussed, glucose is converted to pyruvate in the cytoplasm. After entering the mitochondrion, pyruvate is converted to acetyl-CoA, which condenses with oxaloacetate, a citric acid cycle intermediate, to form citrate. When mitochondrial citrate levels are sufficiently high (i.e., cellular energy requirements are low), citrate enters the cytoplasm, where it is cleaved to form acetyl-CoA and oxaloacetate. The net reaction for the synthesis of palmitic acid from acetyl-CoA is as follows:

$$8\ \text{Acetyl-CoA} + 14\ \text{NADPH} + 14\ H^+ + 7\ \text{ATP} \longrightarrow$$

$$\text{Palmitate} + 14\ \text{NADP}^+ + 7\ \text{ADP}^+ + 7\ P_i + 8\ \text{CoASH} + 6\ H_2O$$

A relatively large quantity of NADPH is required in fatty acid synthesis. A substantial amount of NADPH is provided by the pentose phosphate pathway (p. 322). Reactions catalyzed by isocitrate dehydrogenase (p. 296) and malic enzyme (p. 303) provide smaller amounts.

The biosynthesis of fatty acids is outlined in Figure 10.11. At first glance, fatty acid synthesis appears to be the reverse of the β-oxidation pathway. For example, fatty

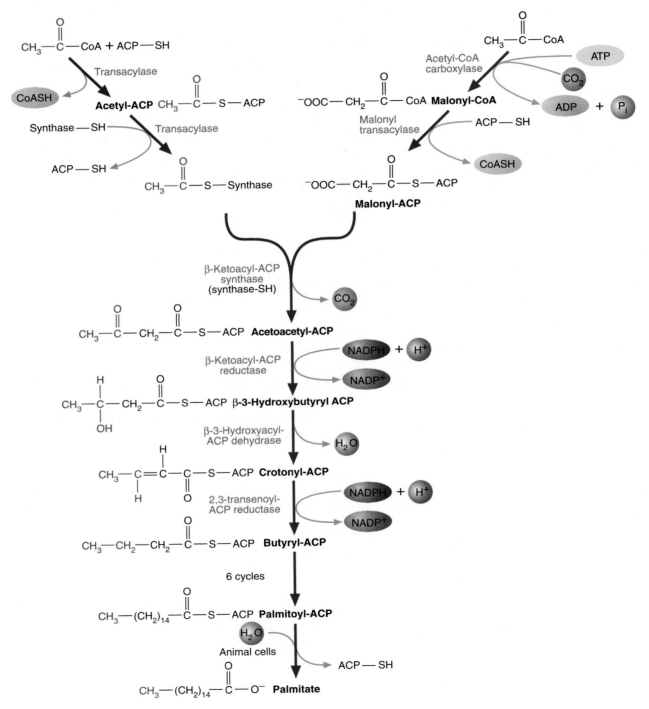

FIGURE 10.11

Fatty Acid Biosynthesis.

During each cycle of fatty acid biosynthesis, the molecule grows by two carbons. Most of the reactions in this pathway occur on a multienzyme complex.

acids are constructed by the sequential addition of two-carbon groups supplied by acetyl-CoA. Additionally, the same intermediates are found in both pathways (i.e., β-ketoacyl-, β-hydroxyacyl-, and α,β-unsaturated acyl groups).

Saturated fatty acids containing up to 16 carbon atoms (palmitate) are assembled in cytoplasm from acetyl-CoA. Depending on cellular conditions, the product of this process (palmitoyl-CoA) can be used directly in the synthesis of several types of lipid (e.g., triacylglycerol or phospholipids), or it can enter the mitochondrion. Several mitochondrial enzymes catalyze elongating and desaturating reactions (discussed below). ER possesses similar enzymes. A closer examination, however, reveals several notable differences between fatty acid synthesis and β-oxidation.

1. **Location.** Fatty acid synthesis occurs predominantly in the cytoplasm. (Recall that β-oxidation occurs within mitochondria and peroxisomes.)

2. **Enzymes.** The enzymes that catalyze fatty acid synthesis are significantly different in structure than those in β-oxidation. In eukaryotes, most of these enzymes are components of a multienzyme complex referred to as fatty acid synthase.

3. **Thioester linkage.** The intermediates in fatty acid synthesis are linked through a thioester linkage to **acyl carrier protein** (ACP), a component of fatty acid synthase. (Recall that acyl groups are attached to CoASH through a thioester linkage during β-oxidation.) Acyl groups are attached to both ACP and CoASH via a phosphopantetheine prosthetic group (Figure 10.12).

4. **Electron carriers.** In contrast to β-oxidation, which produces NADH and $FADH_2$, fatty acid synthesis consumes NADPH.

Fatty acid synthesis begins with the carboxylation of acetyl-CoA to form malonyl-CoA. (Acetyl-CoA carboxylation is considered an activating reaction. Activation is necessary in fatty acid synthesis because the condensation of acetyl groups is an endergonic reaction. As malonyl-CoA is decarboxylated during the condensation reaction, sufficient energy is generated to drive the process.) The carboxylation of acetyl-CoA to form malonyl-CoA (Figure 10.13), catalyzed by acetyl-CoA carboxylase, is the rate-limiting step in fatty acid synthesis. Acetyl-CoA carboxylase contains two subunits, each of which is bound to biotin. (Recall that biotin acts as a CO_2 carrier.) Fatty acid synthesis begins when acetyl-CoA carboxylase dimers aggregate to form high molecular filamentous polymers (four million to eight million D). Polymerization

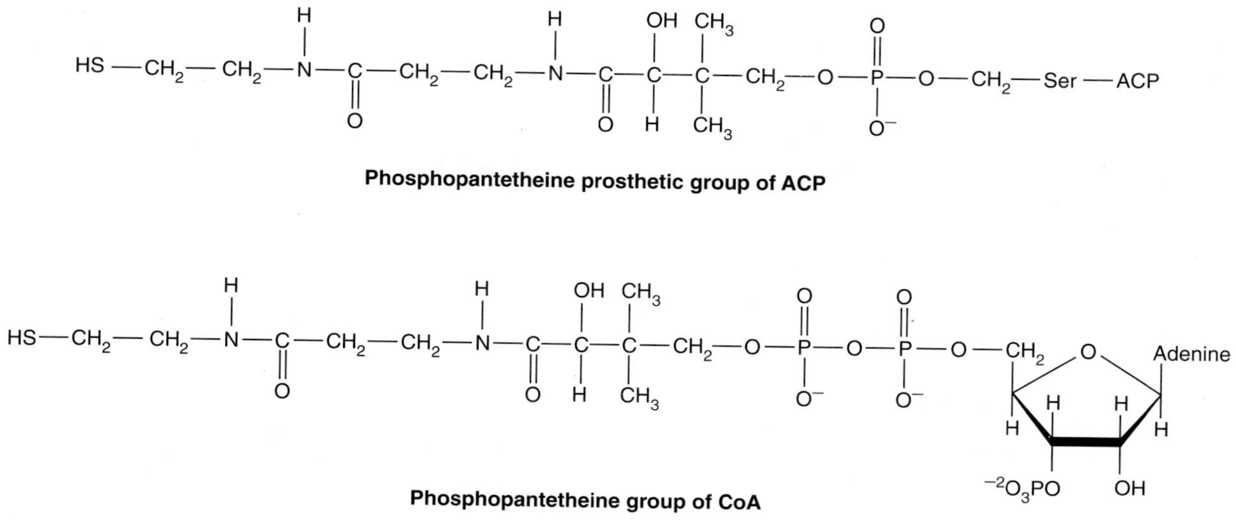

Phosphopantetheine prosthetic group of ACP

Phosphopantetheine group of CoA

FIGURE 10.12

Comparison of the Phosphopantetheine Group in Acyl Carrier Protein (ACP) and in Coenzyme A (CoASH).

Fatty acids are attached to this prosthetic group on ACP during fatty acid biosynthesis and on CoASH during β-oxidation.

begins when cytoplasmic citrate levels rise. Depolymerization occurs when malonyl-CoA or palmitoyl-CoA levels are high. Phosphorylation of acetyl-CoA carboxylase in response to the binding of glucagon or epinephrine also causes depolymerization. In contrast, insulin facilitates dimer aggregation (via phosphorylation at a separate site on the enzyme).

The remaining reactions in fatty acid synthesis take place on the fatty acid synthase multienzyme complex. This complex, the site of seven enzyme activities and ACP, is a 500-kD dimer. Because the enormous polypeptides in the dimer are arranged in a head-to-tail configuration, two fatty acids can be constructed simultaneously. A proposed mechanism for palmitate synthesis is shown in Figure 10.14.

During the first reaction on fatty acid synthase, acetyl transacylase catalyzes the transfer of the acetyl group from an acetyl-CoA molecule to the SH group of a cysteinyl residue of β-ketoacyl-ACP synthase. Malonyl-ACP is formed when malonyl transacylase transfers a malonyl group from malonyl-CoA to the SH group of the pantatetheine prosthetic group of ACP (reaction 2). Then β-ketoacyl-ACP synthase catalyzes a condensation reaction (reaction 3) in which acetoacetyl-ACP is formed (Figure 10.15).

During the next three steps, consisting of two reductions and a dehydration, the acetoacetyl group is converted to a butyryl group. (After each reaction on fatty acid synthase the product group is transferred to the next reaction site by the phosphopantetheine residue attached to ACP.) β-Ketoacyl-ACP reductase catalyzes the reduction of acetoacetyl-ACP to form β-hydroxybutyryl-ACP. β-Hydroxyacyl-ACP dehydrase later catalyzes a dehydration, thus forming crotonyl-ACP. Butyryl-ACP is produced when 2,3-trans-enoyl-ACP reductase reduces the double bond in crotonyl-

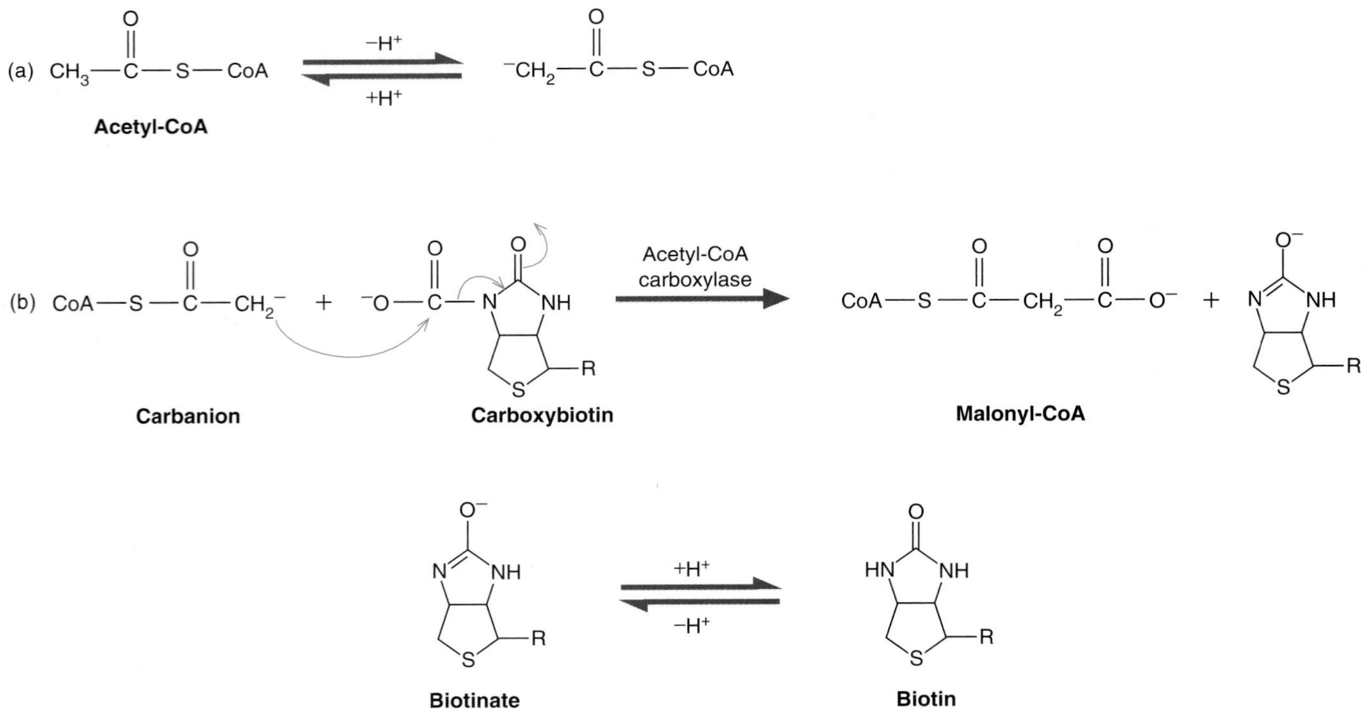

FIGURE 10.13

Synthesis of Malonyl-CoA.

(a) The reaction begins when acetyl-CoA forms a carbanion by losing a proton. (A *carbanion* is a molecule that contains a carbon atom bearing a negative charge.) (b) The carbanion attacks the carbon in carboxybiotin to yield malonyl-CoA and biotinate. Biotinate is reconverted to biotin by the addition of a proton. (Carboxybiotin is formed from biotin and bicarbonate in a reaction that is coupled to ATP hydrolysis.)

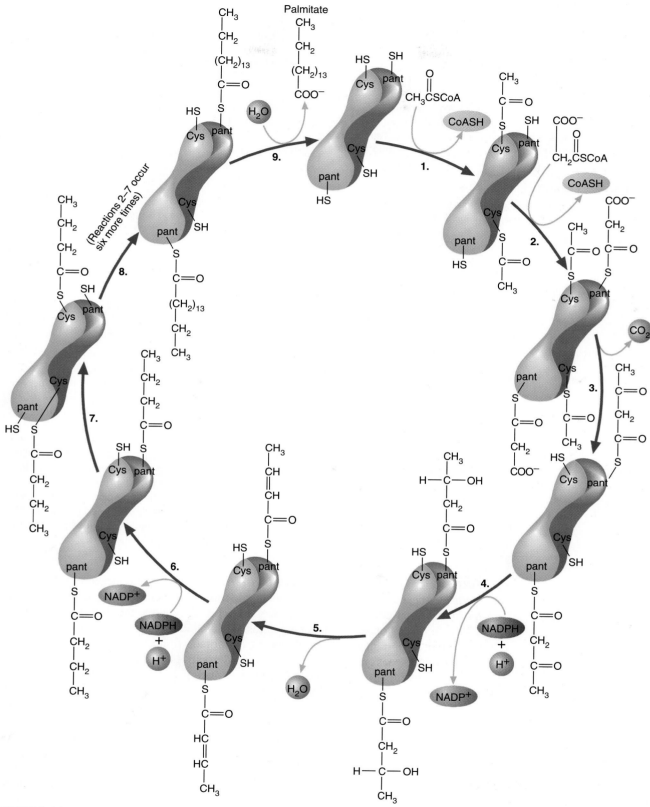

FIGURE 10.14

Fatty Acid Biosynthesis.

The bicolored structures represent the dimer of fatty acid synthase. Each component in the dimer possesses a Cys-SH residue belonging to β-ketoacyl ACP synthase and pant-SH, the pantetheine sulfhydryl group of ACP.

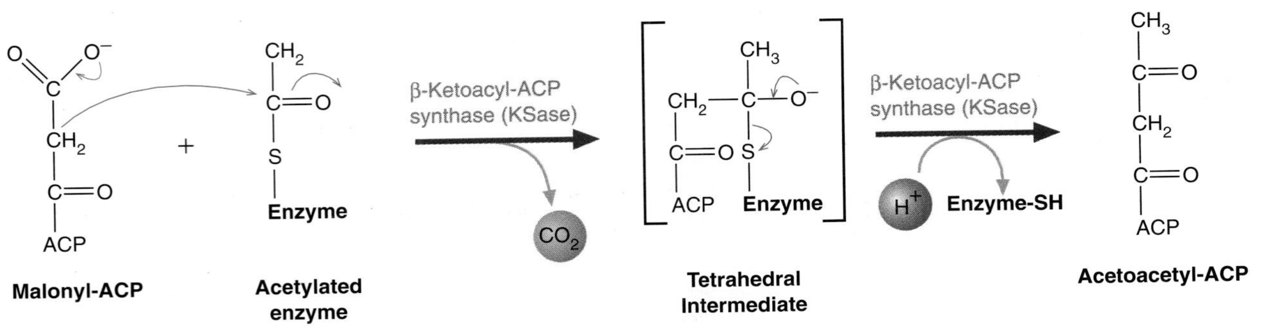

FIGURE 10.15

Formation of Acetoacetyl-ACP.

The acetyl group is shown bound to the enzyme β-ketoacyl-ACP synthase through a cysteine residue. The carbonyl group of the acetyl group is attacked by the central carbon on the malonyl group attached to ACP. Acetoacetyl-ACP is generated as the C—S bond is broken.

ACP. In the last step of the first cycle of fatty acid synthesis, the butyryl group is transferred from the pantetheine group to the cysteine residue of β-ketoacyl-ACP synthase. The newly freed ACP-SH group now binds to another malonyl group. The process is then repeated. Eventually, palmitoyl-ACP is synthesized. Now the palmitoyl group is released from fatty acid synthase when thioesterase converts it to palmitate.

QUESTION 10.4

Excessive consumption of fructose has been linked to a condition referred to as *hyper-triglyceridemia* (high blood levels of triacylglycerols). The most common source of fructose for most Americans is sucrose. (The fructose content of fresh fruits and vegetables is so low in comparison to many processed foods that it would be difficult to consume sufficient quantities to induce hypertriglyceridemia.) Sucrose is digested in the small intestine by the enzyme sucrase to yield one molecule each of fructose and glucose. Digestion is so rapid that the blood concentrations of these sugars become quite high. Recall that once it reaches the liver, fructose is converted to fructose-1-phosphate. It is now believed that fructose-1-phosphate promotes hexokinase D activity. (Apparently, fructose-1-phosphate binds to and inactivates a protein that depresses hexokinase D activity.) After reviewing fructose metabolism and fatty acid and triacylglycerol synthesis, suggest how hypertriglyceridemia might result from a diet that is rich in sucrose.

Fatty Acid Elongation and Desaturation Elongation and desaturation of fatty acids synthesized in cytoplasm or obtained from the diet are accomplished primarily by microsomal enzymes. (These processes occur only when the diet provides an inadequate supply of appropriate fatty acids.) Fatty acid elongation and desaturation are especially important in the regulation of membrane fluidity as well as the synthesis of the precursors for a variety of fatty acid derivatives, such as the eicosanoids. For example, myelination (a process in which a myelin sheath is formed around certain nerve cells) depends especially on microsomal fatty acid synthetic reactions. Very long chain saturated and monounsaturated fatty acids are important constituents of the cerebrosides and sulfatides found in myelin. Cells apparently regulate membrane fluidity by adjusting the types of fatty acids that are incorporated into membrane lipids. For example, in cold weather, more unsaturated fatty acids are incorporated. (Recall that unsaturated fatty acids have a lower freezing point than do saturated fatty acids.) If the diet does not provide a sufficient number of these molecules, fatty acid synthetic pathways are activated. Although elongation and desaturation are closely integrated processes, for the sake of clarity they are discussed separately.

Microsomal fatty acid chain elongation, which uses two-carbon units provided by malonyl-CoA, is a cycle of condensation, reduction, dehydration, and reduction reactions similar to those observed in cytoplasmic fatty acid synthesis. In contrast to the cytoplasmic process, the intermediates in the microsomal elongation process are CoA esters. These reactions can lengthen both saturated and unsaturated fatty acids. Reducing equivalents are provided by NADPH.

$$R-\overset{\overset{\displaystyle O}{\|}}{C}-S-CoA \quad + \quad {}^{-}O-\overset{\overset{\displaystyle O}{\|}}{C}-CH_2-\overset{\overset{\displaystyle O}{\|}}{C}-S-CoA \longrightarrow$$

$$\longrightarrow R-CH_2-CH_2-\overset{\overset{\displaystyle O}{\|}}{C}-S-CoA \quad + \quad CO_2 \quad + \quad CoASH$$

KEY *Concepts*

In animals, fatty acids are synthesized in the cytoplasm from acetyl-CoA and malonyl-CoA. Microsomal enzymes elongate and desaturate newly synthesized fatty acids as well as those obtained in the diet.

Acyl-CoA molecules are desaturated in microsomal membrane in the presence of NADH and O_2. All components of the desaturase system are integral membrane proteins that are apparently randomly distributed on the cytoplasmic surface of the ER. The association of cytochrome b_5 reductase (a flavoprotein), cytochrome b_5, and oxygen-dependent desaturases constitutes an electron transport system. This system efficiently introduces double bonds into long-chain fatty acids (Figure 10.16). Both the flavoprotein and cytochrome b_5 (found in a ratio of approximately 1:30) have hydrophobic peptides that anchor the proteins into the microsomal membrane. Animals typically have Δ^9, Δ^6, and Δ^5 desaturases that use electrons supplied by NADH via the electron transport system to activate the oxygen needed to create the double bond.

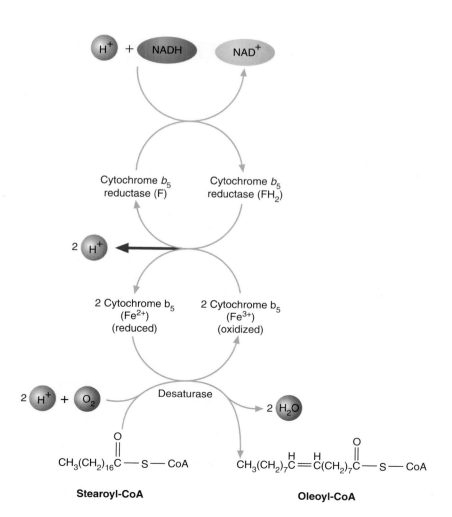

FIGURE 10.16

Desaturation of Stearoyl-CoA.

The desaturase uses electrons provided by an electron transport system composed of cytochrome b_5 reductase and cytochrome b_5 to activate the oxygen (not shown) needed to create the double bond. NADH is the electron donor.

Because elongation and desaturation systems are in close proximity to each other in microsomal membrane, a variety of long-chain polyunsaturated acids are typically produced. A prominent example of this interaction is the synthesis of arachidonic acid ($20:4^{\Delta5,8,11,14}$) from linoleic acid ($18:2^{\Delta9,12}$).

Fatty Acid Synthesis in Plants   Because of less intensive research efforts and several technical problems, plant fatty acid synthesis is less well understood than the animal process. It is known, however, that plant fatty acid synthesis has several notable features:

1. **Location.** Plant fatty acid synthesis appears to be limited to chloroplasts. A chloroplast isozyme of pyruvate dehydrogenase catalyzes the conversion of pyruvate to acetyl-CoA. Pyruvate is also derived from glycerate-3-phosphate, an intermediate in the **Calvin cycle,** a biosynthetic pathway in which plants incorporate CO_2 into sugar molecules. (The Calvin cycle is discussed in Chapter 12.)

2. **Metabolic control.** The regulation of fatty acid synthesis in plants is poorly understood. It remains unclear whether the reaction catalyzed by acetyl-CoA carboxylase is a rate-limiting step in plants, since malonyl-CoA is used in several other biosynthetic pathways (e.g., bioflavonoid synthesis). (Recall that this reaction is rate-limiting in animal cells.)

3. **Enzymes.** The structures of both acetyl-CoA carboxylase and fatty acid synthase in plants more closely resemble their counterparts in *Escherichia coli* than those in animal cells. For example, in *E. coli* and plants, each of the enzyme activities of fatty acid synthase are found on a separate protein.

Regulation of Fatty Acid Metabolism in Mammals

Because animals have such varying requirements for energy, the metabolism of fatty acids (*the* major energy source in animals) is carefully regulated (Figure 10.17). Both short- and long-term regulatory mechanisms are used. In most short-term regulation (measured in minutes) the activities of already existing molecules of key regulatory enzymes are modified by hormones. For example, glucagon or epinephrine (released when the body's energy reserves are low or when there is an increased energy requirement) stimulates the phosphorylation of several enzymes. When the adipocyte enzyme hormone-sensitive lipase is phosphorylated, it catalyzes the hydrolysis of triacylglycerol. (The release of norepinephrine from neurons in the sympathetic nervous system and growth hormone from the pituitary also activates hormone-sensitive lipase.) Subsequently, fatty acids are released into the blood. Hormones also regulate the use of fatty acids within tissues. For example, acetyl-CoA carboxylase is inhibited by glucagon binding. Then cellular malonyl-CoA concentration decreases, and fatty acid synthesis shuts down. Because malonyl-CoA inhibits carnitine acyltransferase I activity, fatty acids groups may now be transported into the mitochondria, where they are degraded to generate energy. The effect of insulin on fatty acid metabolism is opposite to that of glucagon and epinephrine. The secretion of insulin in response to high blood glucose levels promotes lipogenesis. Insulin promotes fatty acid synthesis by stimulating the phosphorylation of acetyl-CoA carboxylase (by a process that is independent of the cAMP-protein kinase mechanism). Simultaneous lipolysis is prevented by insulin's inhibition of the cAMP-mediated activation of protein kinase. This latter process leads to dephosphorylation (and therefore inactivation) of hormone-sensitive lipase.

Hormones are also involved in the long-term regulation of fatty acid metabolism. During hours or days, certain hormones alter the amounts of key enzymes. For example, when nutrients become plentiful after a long fast, insulin promotes the synthesis of the enzymes involved in lipogenesis (e.g., acetyl-CoA carboxylase and fatty acid synthase). Certain nutrients also affect enzyme concentrations. For example, starvation inhibits the synthesis of acetyl-CoA carboxylase and fatty acid synthase. In contrast, low-fat diets increase synthesis of these enzymes.

KEY *Concepts*

Fatty acid metabolism is highly regulated to ensure consistently adequate energy production. Hormones regulate the activity and the synthesis of key enzymes in both the synthetic and degradative pathways.

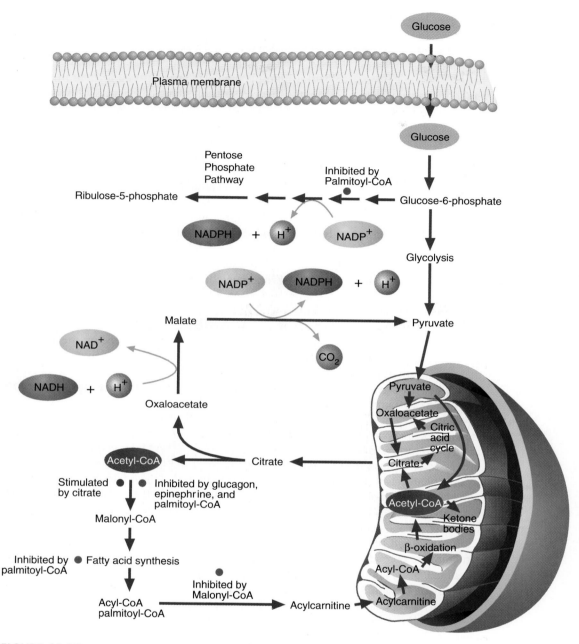

FIGURE 10.17

Intracellular Fatty Acid Metabolism.

Fatty acids are synthesized in cytoplasm from acetyl-CoA, formed from citrate. Citrate is produced from acetyl-CoA and oxaloacetate in the citric acid cycle, a reaction pathway in the mitochondrial matrix. Citrate is transferred to cytoplasm when β-oxidation is suppressed, i.e., when the cell needs little energy. Later it is cleaved to form oxaloacetate and acetyl-CoA. When the cell needs more energy, fatty acids are transported into the mitochondrion as acylcarnitine derivatives. Then acyl-CoA is degraded to acetyl-CoA via β-oxidation. (The further oxidation of acetyl-CoA to generate ATP is described in Chapter 11.) Note that the hormones glucagon and epinephrine and the substrates citrate, malonyl-CoA and palmitoyl-CoA are important regulators of fatty acid metabolism. Fatty acid metabolism and carbohydrate metabolism are interrelated. Pyruvate, the precursor of acetyl-CoA, is a product of glycolysis. A portion of the NADPH, the reducing agent required in fatty acid synthesis, is generated by several reactions in the pentose phosphate pathway. NADPH is also produced by converting malate, formed by the reduction of oxaloacetate, to pyruvate.

10.2 MEMBRANE LIPID METABOLISM

Phospholipid Metabolism

As described, the lipid bilayer of cell membranes is composed primarily of phospholipids and sphingolipids. After a discussion of the metabolism of these lipid classes, several aspects of membrane biogenesis are briefly described.

Most of the reactions involved in lipid biosynthesis appear to be located in the SER (although several enzyme activities have also been detected in the Golgi complex). Because each enzyme is a membrane associated protein with its active site facing the cytoplasm, the biosynthesis of phospholipid occurs at the interface of ER membrane and cytoplasm. The fatty acid composition of phospholipids changes somewhat after their synthesis. (Typically, unsaturated fatty acids replace the original saturated fatty acids incorporated during synthesis.) Most of this remodeling is accomplished by certain phospholipases and acyl transferases. Presumably, this process allows a cell to adjust the fluidity of its membranes.

The syntheses of phosphatidylethanolamine and phosphatidylcholine are similar (Figure 10.18). Phosphatidylethanolamine synthesis begins in the cytoplasm when ethanolamine enters the cell and is then immediately phosphorylated. Subsequently, phosphoethanolamine reacts with CTP (cytidine triphosphate) to form the activated intermediate CDP-ethanolamine. (Several nucleotides serve as high-energy carriers of specific molecules. CDP derivatives have an important role in the transfer of polar head groups in phosphoglyceride synthesis. Recall that UDP plays a similar role in glycogen synthesis.) CDP-ethanolamine is converted to phosphatidylethanolamine when it reacts with DAG. (This reaction is catalyzed by an enzyme on the endoplasmic reticulum.) As noted above, the biosynthesis of phosphatidylcholine is similar to that of phosphatidylethanolamine. The choline required in this pathway is obtained in the diet. However, phosphatidylcholine is also synthesized in the liver from phosphatidylethanolamine (Figure 10.19). Phosphatidylethanolamine is methylated in three steps by the enzyme phosphatidylethanolamine-N-methyltransferase to form the trimethylated product phosphatidylcholine. *S-Adenosylmethionine* (SAM) is the methyl donor in this set of reactions. (The role of SAM in cellular methylation processes is discussed in Chapter 13.)

Phosphatidylserine is generated in a reaction in which the ethanolamine residue of phosphatidylethanolamine is exchanged for serine (Figure 10.20). This reaction, which is catalyzed by an ER enzyme, is reversible. In mitochondria, phosphatidylserine is converted to phosphatidylethanolamine in a decarboxylation reaction.

Phospholipid turnover is rapid. (**Turnover** is the rate at which all molecules in a structure are degraded and replaced with newly synthesized molecules.) For example, in animal cells, approximately two cell divisions are required for the replacement of one-half of the total number of phospholipid molecules. Phosphoglycerides are degraded by the phospholipases. Each phospholipase, which catalyzes the cleavage of a specific bond in phosphoglyceride molecules, is named according to the bond cleaved. Phospholipases A_1 and A_2, which hydrolyze the ester bonds of phosphoglycerides at C-1 and C-2, respectively, contribute to the phospholipid remodeling previously described.

Sphingolipid Metabolism

Recall that sphingolipids in animals possess ceramide, a derivative of the amino alcohol sphingosine. The synthesis of ceramide begins with the condensation of palmitoyl-CoA with serine to form 3-ketosphinganine. This reaction is catalyzed by 3-ketosphinganine synthase, a pyridoxal-5′-phosphate-dependent enzyme. (Because pyridoxal-5′-phosphate plays an important role in amino acid metabolism, the biochemical function of this coenzyme is discussed in Chapter 13.) 3-Ketosphinganine is subsequently reduced by NADPH to form sphinganine. In a two-step process involving acyl-CoA and $FADH_2$, sphinganine is converted to ceramide. Sphingomyelin is formed when ceramide reacts with phosphatidylcholine. (In an alternative reaction, CDP-choline is used in place of phosphatidylcholine.) When ceramide reacts with UDP-glucose, glucosylceramide (a common cerebroside, sometimes referred to as glucosylcerebroside) is produced. Galactocerebroside, a precursor of other glycolipids,

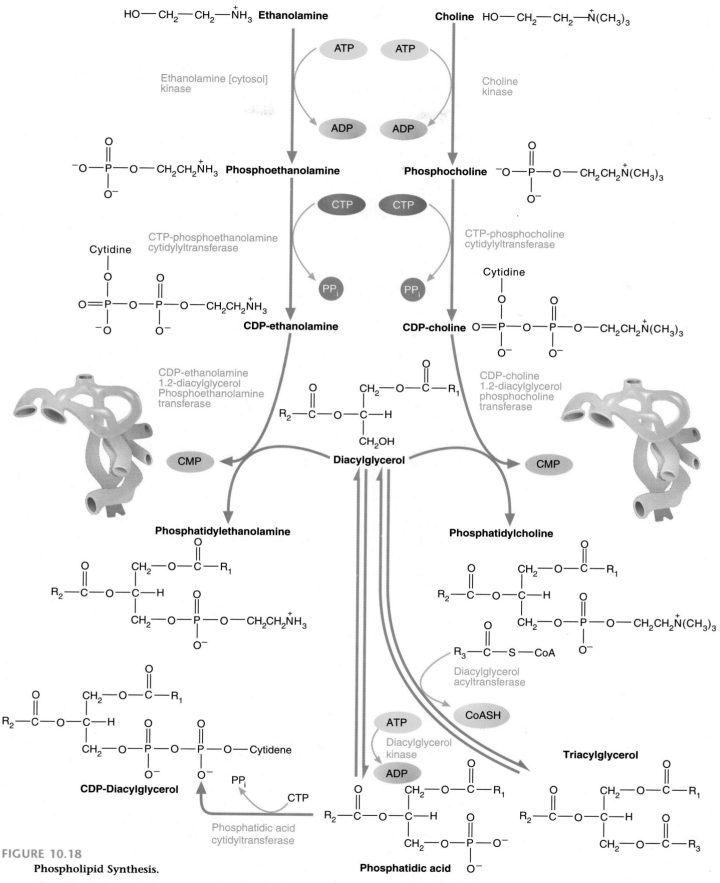

FIGURE 10.18

Phospholipid Synthesis.

After ethanolamine or choline has entered a cell, it is phosphorylated and converted to a CDP derivative. Then phosphatidylethanolamine or phosphatidylcholine is formed when diacylglycerol reacts with the CDP derivative. Triacylglycerol is produced if diacylglycerol reacts with acyl-CoA. CDP-Diacylglycerol, formed from phosphatidic acid and CTP, is a precursor of several phospholipids, e.g., phosphatidylglycerol and phosphatidylinositol.

FIGURE 10.19

Conversion of Phosphatidylethanolamine to Phosphatidylcholine.

Methylation reactions that use S-adenosylmethionine (SAM) are discussed in Chapter 13. (SAH is an abbreviation of S-adenosylhomocysteine.)

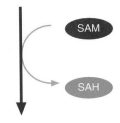
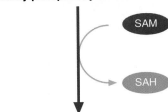

Phosphatidylethanolamine

SAM → SAH

N-Methylphosphatidylethanolamine

SAM → SAH

N,N-Dimethylphosphatidylethanolamine

SAM → SAH

Phosphatidylcholine

FIGURE 10.20

Phosphatidylserine Synthesis.

Phosphatidylserine can be synthesized from phosphatidylethanolamine in a reaction in which the polar head groups are exchanged. Phosphatidylethanolamine can also be synthesized from phosphatidylserine in a decarboxylation reaction. This reaction is an important source of ethanolamine in many eukaryotes.

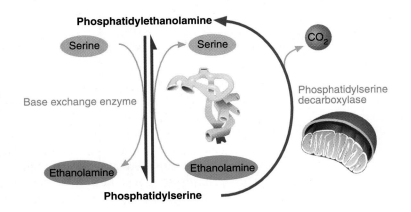

Because membranes are dynamic structures, the mechanism by which they are synthesized is complex. Currently, little is known about the synthesis of the membrane bilayer except for the following features: phospholipid translocation across membranes and the intracellular transfer of phospholipids between membranes.

If recently synthesized phospholipid molecules remained only on the cytoplasmic surface of the ER, a monolayer would form. Unassisted bilayer transfer of phospholipid, however, is extremely slow. (For example, half-lifes of eight days have been measured across artificial membrane.) A process known as *phospholipid translocation* is now believed to be responsible for maintaining the bilayer in membranes (Figure 10D). Transmembrane movement of phospholipid molecules (or flip-flop), which may occur in as little as 15 seconds, appears to be mediated by phospholipid translocator proteins. One protein (sometimes referred to as *flippase*) that transfers choline-containing phospholipids across the ER membrane has been identified. Because the hydrophilic polar head group of a phospholipid molecule is probably responsible for the low rate of spontaneous translocation, an interaction between flippase and polar head groups is believed to be involved in phosphatidylcholine transfer. Translocation results in a higher concentration of phosphatidylcholine on the lumenal side of the ER membrane than that of other phospholipids. This process is therefore partially responsible for the membrane asymmetry discussed in Chapter 9.

Two mechanisms have been proposed to explain the transport of phospholipids from the ER to other cellular membranes: protein-mediated transfer and a vesicular process. Several experiments have demonstrated that water-soluble proteins, known as *phospholipid exchange proteins,* can bind to specific phospholipid molecules and transfer them to another bilayer. Vesicular transport of phospholipids and membrane proteins in structures known as *transition vesicles* from the ER to the Golgi complex is not clearly understood. However, evidence of transfer of luminal material from the ER to the Golgi cisternae clearly supports vesicular transport.

FIGURE 10D

Translocation of Newly Synthesized Phospholipid.

Transfer of selected phospholipid molecules allows balanced growth of the bilayer.

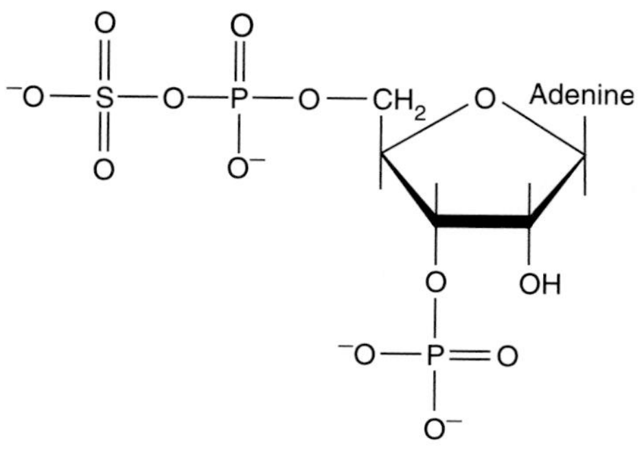

FIGURE 10.21

3'-Phosphoadenosine-5'-Phosphosulfate (PAPS).

PAPS is a high-energy sulfate donor.

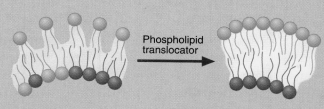

is synthesized when ceramide reacts with UDP-galactose. The sulfatides are synthesized when the galactocerebrosides react with the sulfate donor molecule *3'-phosphoadenosine-5'-phosphosulfate* (PAPS) (Figure 10.21). The transfer of sulfate groups is catalyzed by the microsomal enzyme sulfotransferase. Sphingolipids are degraded within lysosomes. Recall that specific diseases, called the sphingolipidoses (p. 227), result when enzymes required for degrading these molecules are missing or defective. The synthesis of sphingomyelin and glycosphingolipids is shown in Figure 10.22.

10.3 ISOPRENOID METABOLISM

Isoprenoids occur in all eukaryotes. Despite the astonishing diversity of isoprenoid molecules, the mechanisms by which different species synthesize them are similar. In fact, the initial phase of isoprenoid synthesis (the synthesis of isopentenyl pyrophosphate) appears to be identical in all of the species in which this process has been investigated. Figure 10.23 illustrates the relationships among the isoprenoid classes.

Because of its importance in human biology, cholesterol has received enormous attention from researchers. For this reason the metabolism of cholesterol is better understood than that of any other isoprenoid molecule.

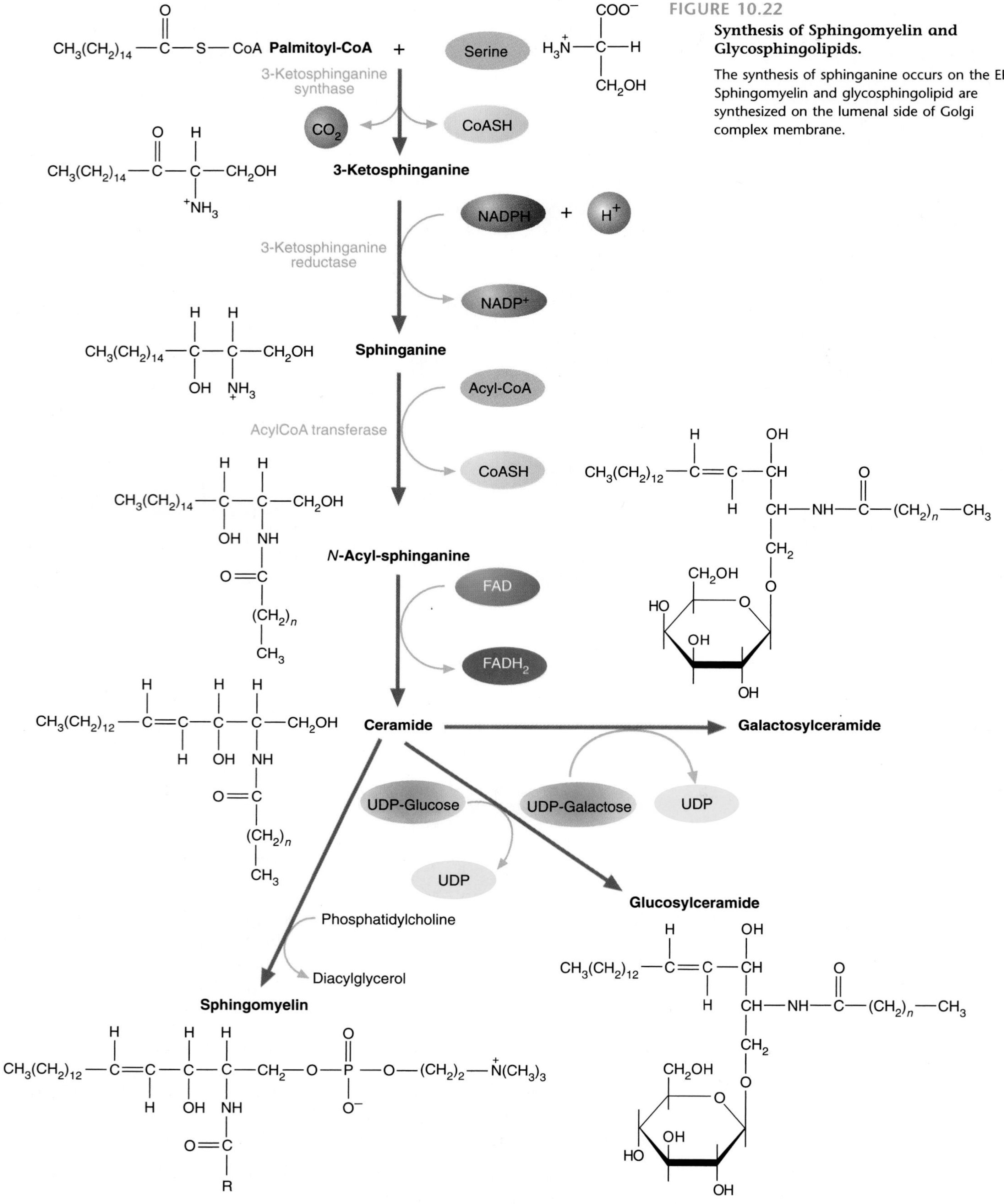

$CH_3(CH_2)_{14}$—C—S—CoA **Palmitoyl-CoA**   +   **Serine**

3-Ketosphinganine synthase

CO_2 CoASH

3-Ketosphinganine

3-Ketosphinganine reductase

NADPH + H$^+$

NADP$^+$

Sphinganine

Acyl-CoA

AcylCoA transferase

CoASH

N-Acyl-sphinganine

FAD

FADH$_2$

Ceramide   **Galactosylceramide**

UDP-Glucose   UDP-Galactose   UDP

UDP

Glucosylceramide

Phosphatidylcholine

Diacylglycerol

Sphingomyelin

FIGURE 10.22

Synthesis of Sphingomyelin and Glycosphingolipids.

The synthesis of sphinganine occurs on the ER. Sphingomyelin and glycosphingolipid are synthesized on the lumenal side of Golgi complex membrane.

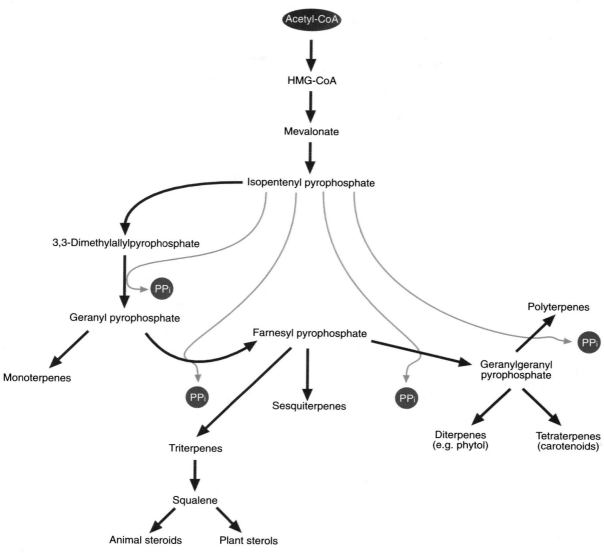

FIGURE 10.23

Isoprenoid Biosynthesis.

Isoprenoid biosynthetic pathways produce an astonishing variety of products in different cell types and different species. Despite this diversity, the beginning of isoprenoid biosynthesis appears to be identical in most of the species investigated (e.g., yeast, mammals, and plants).

Cholesterol Metabolism   The cholesterol that is used throughout the body is derived from two sources: diet and *de novo* synthesis. When the diet provides sufficient cholesterol, the synthesis of this molecule is depressed. (Recall that in normal individuals cholesterol delivered by LDL suppresses cholesterol synthesis.) Cholesterol biosynthesis is stimulated when the diet is low in cholesterol. As described previously, cholesterol is used as a cell membrane component and in the synthesis of important metabolites. An important mechanism for disposing of cholesterol is conversion to bile acids.

Cholesterol Synthesis   Although all tissues can make cholesterol (e.g., adrenal glands, ovaries, testes, skin, and intestine), most cholesterol molecules are synthesized in the liver. Cholesterol synthesis can be divided into three phases:

1. formation of HMG-CoA (β-hydroxy-β-methylglutaryl-CoA) from acetyl-CoA,
2. conversion of HMG-CoA to squalene, and
3. conversion of squalene to cholesterol.

FIGURE 10.24

Cholesterol Synthesis.

Several reactions occur in the cytoplasm. Enzyme key: 1 = HMG-CoA reductase, 2 = Squalene synthase, 3 = Squalene monooxygenase, 4 = 2,3-oxidosqualene lanosterol cyclase, 5 = Enzymes catalyzing 20 separate reactions.

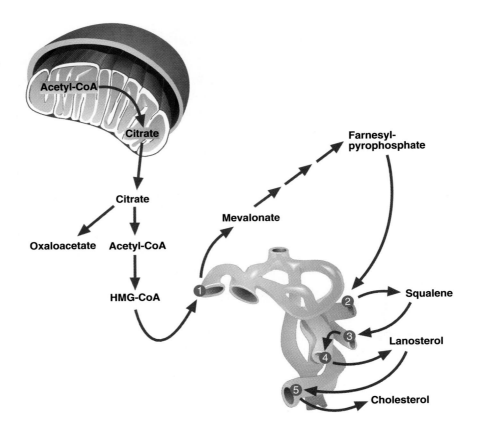

The first phase of cholesterol synthesis is a cytoplasmic process (Figure 10.24). (Recall that the initial substrate, acetyl-CoA, is produced in mitochondria from fatty acids or pyruvate. Also observe the similarity of the first phase of cholesterol synthesis to ketone body synthesis. Refer to Figure 10.7). The condensation of two acetyl-CoA molecules to form β-ketobutyryl-CoA (also referred to as acetoacetyl-CoA) is catalyzed by thiolase.

$$2 \; CH_3-\overset{\overset{\displaystyle O}{\|}}{C}-S-CoA \;\rightleftharpoons\; CH_3-\overset{\overset{\displaystyle O}{\|}}{C}-CH_2-\overset{\overset{\displaystyle O}{\|}}{C}-S-CoA \; + \; CoASH$$

Acetyl-CoA   **β-Ketobutyryl-CoA**

In the next reaction, β-ketobutyryl-CoA condenses with another acetyl-CoA to form β-hydroxy-β-methylglutaryl-CoA (HMG-CoA). This reaction is catalyzed by β-hydroxy-β-methylglutaryl-CoA synthase (HMG-CoA synthase).

$$CH_3-\overset{\overset{\displaystyle O}{\|}}{C}-CH_2-\overset{\overset{\displaystyle O}{\|}}{C}-S-CoA \; + \; CH_3-\overset{\overset{\displaystyle O}{\|}}{C}-S-CoA \;\longrightarrow\; {}^-O-\overset{\overset{\displaystyle O}{\|}}{C}-CH_2-\overset{\overset{\displaystyle CH_3}{\underset{\underset{\displaystyle OH}{|}}{C}}}{}-CH_2-\overset{\overset{\displaystyle O}{\|}}{C}-S-CoA \; + \; CoASH$$

β-Ketobutyryl-CoA   **Acetyl-CoA**   **HMG-CoA**

The second phase of cholesterol synthesis begins when HMG-CoA is reduced to form mevalonate. NADPH is the reducing agent.

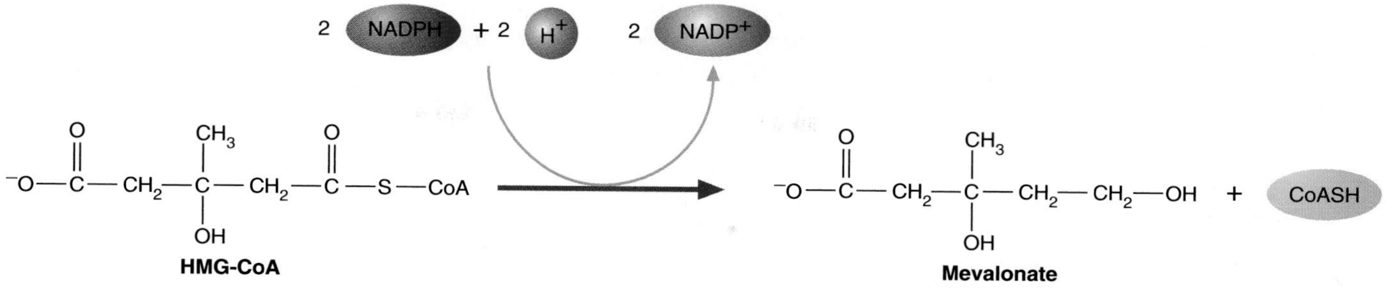

HMG-CoA

Mevalonate

HMG-CoA reductase, which catalyzes the latter reaction, is the rate-limiting enzyme in cholesterol synthesis. An accumulation of cholesterol in the cell either from endogenous synthesis or exogenously from LDL intake and degradation reduces the activity of HMG-CoA reductase in two ways: it inhibits HMG-CoA reductase synthesis and enhances degradation of existing enzyme. The activity and cellular concentration of HMG-CoA reductase, located on the cytoplasmic surface of the endoplasmic reticulum, is affected to varying degrees by the concentrations of intermediate products of the pathway (e.g., mevalonate, farnesol, squalene and 7-dehydrocholesterol). The precise mechanism by which this strategically important enzyme is regulated, however, remains unresolved.

In a series of cytoplasmic reactions, mevalonate is subsequently converted to farnesylpyrophosphate. Mevalonate kinase catalyzes the synthesis of phosphomevalonate. A second phosphorylation reaction catalyzed by phosphomevalonate kinase produces 5-pyrophosphomevalonate.

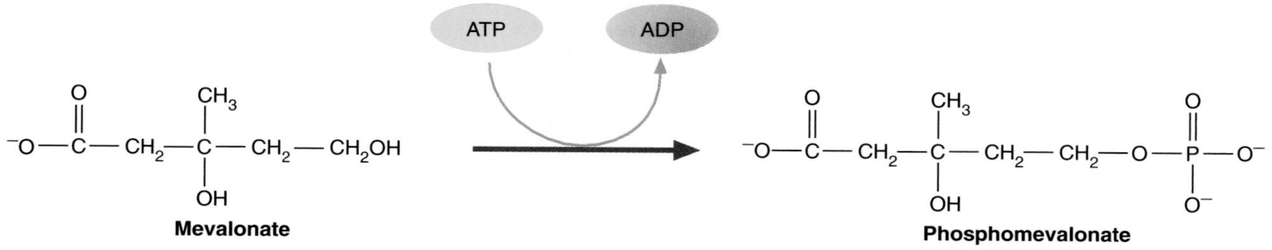

Mevalonate

Phosphomevalonate

5-Pyrophosphomevalonate

(Phosphorylation reactions significantly increase the solubility of these hydrocarbon molecules in the cytoplasm.) 5-Pyrophosphomevalonate is converted to isopentenyl pyrophosphate in a process involving a decarboxylation and a dehydration.

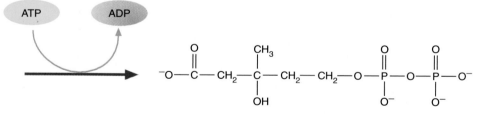

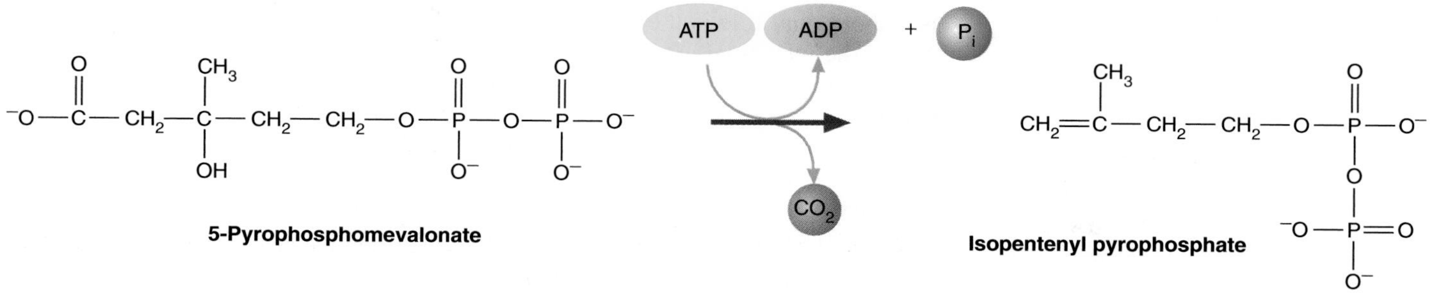

5-Pyrophosphomevalonate

Isopentenyl pyrophosphate

Isopentenyl pyrophosphate is next converted to its isomer dimethylallylpyrophosphate by isopentenyl pyrophosphate isomerase. (A $CH_2{=}CH{-}CH_2{-}$ group on an organic molecule is sometimes referred to as an *allyl group*.)

Isopentenylpyrophosphate   **Dimethylallylpyrophosphate**

Geranylpyrophosphate is generated during a condensation reaction between isopentenylpyrophosphate and dimethylallylpyrophosphate (Figure 10.25). Pyrophosphate is also a product of this reaction and two subsequent reactions. (Recall that reactions in which pyrophosphate is released are irreversible because of subsequent pyrophosphate hydrolysis.) Geranyl transferase catalyzes the condensation reaction between geranylpyrophosphate and isopentenylpyrophosphate that forms farnesylpyrophosphate. Squalene is synthesized when farnesyl transferase (a microsomal enzyme) catalyzes the condensation of two farnesylpyrophosphate molecules. (Farnesyl transferase is sometimes referred to as squalene synthase.) This reaction requires NADPH as an electron donor.

The last phase of the cholesterol biosynthetic pathway (Figure 10.26) begins by binding squalene to a specific cytoplasmic protein carrier called **sterol carrier protein.** The conversion of squalene to lanosterol occurs while the intermediates are bound to this protein. The enzyme activities required for the oxygen-dependent epoxide formation (squalene monooxygenase) and subsequent cyclization (2,3-oxidosqualene lanosterol cyclase) that result in lanosterol synthesis have been localized in microsomes. Squalene monooxygenase requires NADPH and FAD for activity. After its synthesis, lanosterol binds to a second carrier protein, to which it remains attached during the remaining reactions. All of the enzyme activities that catalyze the remaining 20 reactions needed to convert lanosterol to cholesterol are embedded in microsomal membranes. In a series of transformations involving NADPH as well as some oxygen, lanosterol is converted to 7-dehydrocholesterol. This product is then reduced by NADPH to form cholesterol.

Recall that cholesterol is a precursor for all steroid hormones as well as the bile salts. These processes are briefly outlined now. The metabolism of steroids is extremely complex and as yet incompletely understood. A variety of cells and organelles figure prominently in the production and processing of these powerful substances. The initial reaction in steroid hormone synthesis, the conversion of cholesterol to pregnenolone (Figure 10.27), is catalyzed by desmolase, a mitochondrial enzyme. (Desmolase is an enzyme complex composed of two hydroxylases, one of which is a cytochrome P_{450} enzyme. Cytochrome P_{450} is involved in reactions involving steroids as well as xenobiotics. Its properties are discussed in Chapter 19.) After its synthesis, pregnenolone is transported to the ER, where it is converted to progesterone. Pregnenolone and progesterone are precursors for all other steroid hormones (Figure 10.28). In addition to its precursor role, progesterone also acts as a hormone. Its primary hormonal role is the regulation of several physiological changes in the uterus. During the menstrual cycle, progesterone is produced by specialized cells within the ovary. During pregnancy the progesterone that is produced in large quantities by the placenta prevents uterine smooth muscle contractions.

The amount and types of steroids synthesized in a specific tissue are carefully regulated. Cells in each tissue are programmed during embryonic and fetal development to respond to chemical signals by inducing the synthesis of a unique set of specific enzymes. The most important chemical signals that are now believed to influence

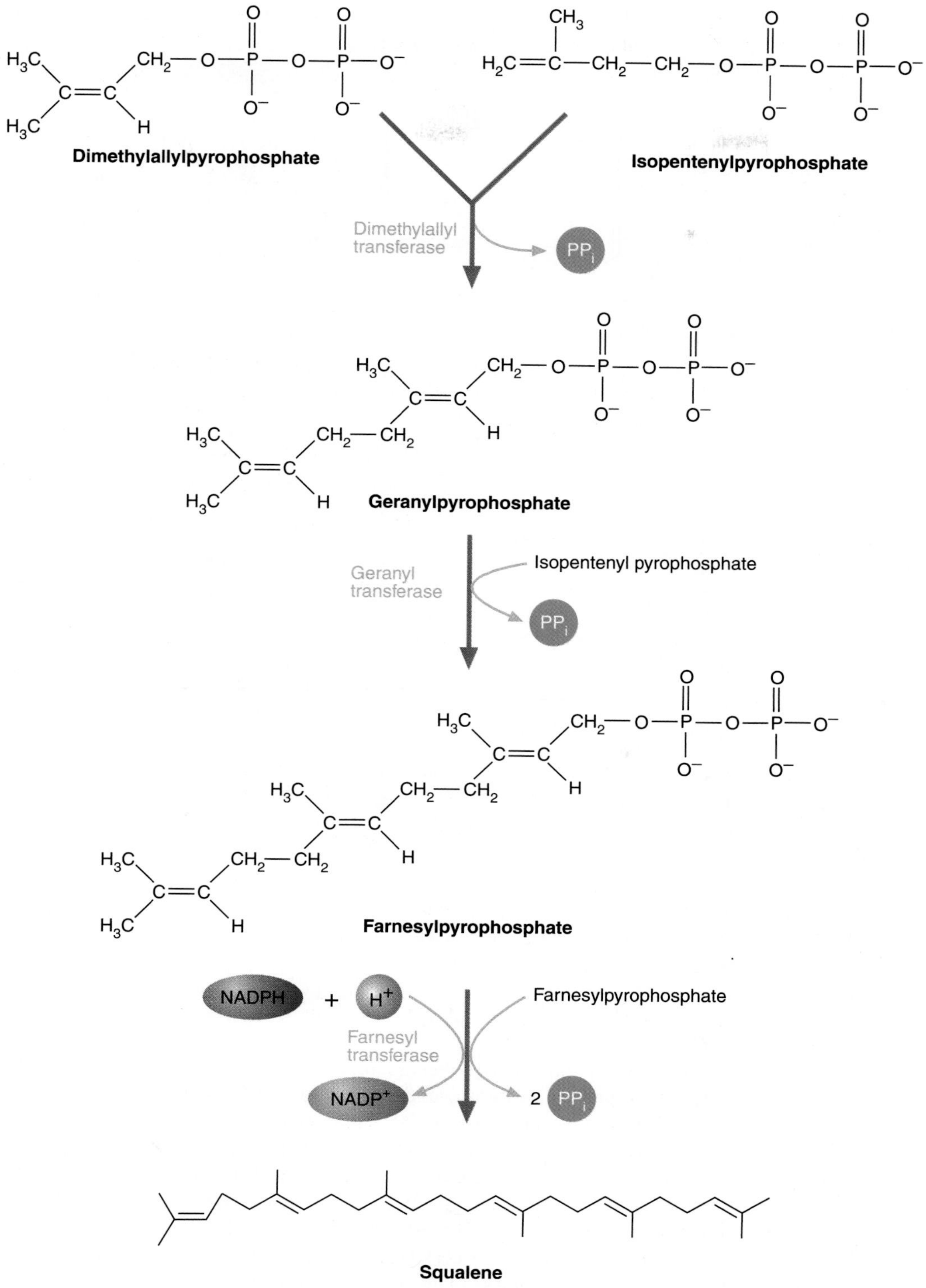

FIGURE 10.25

Synthesis of Squalene from Dimethylallylpyrophosphate and Isopentenylpyrophosphate.

The immediate precursor of squalene is farnesylpyrophosphate, which contains three C_5 isoprenoid groups.

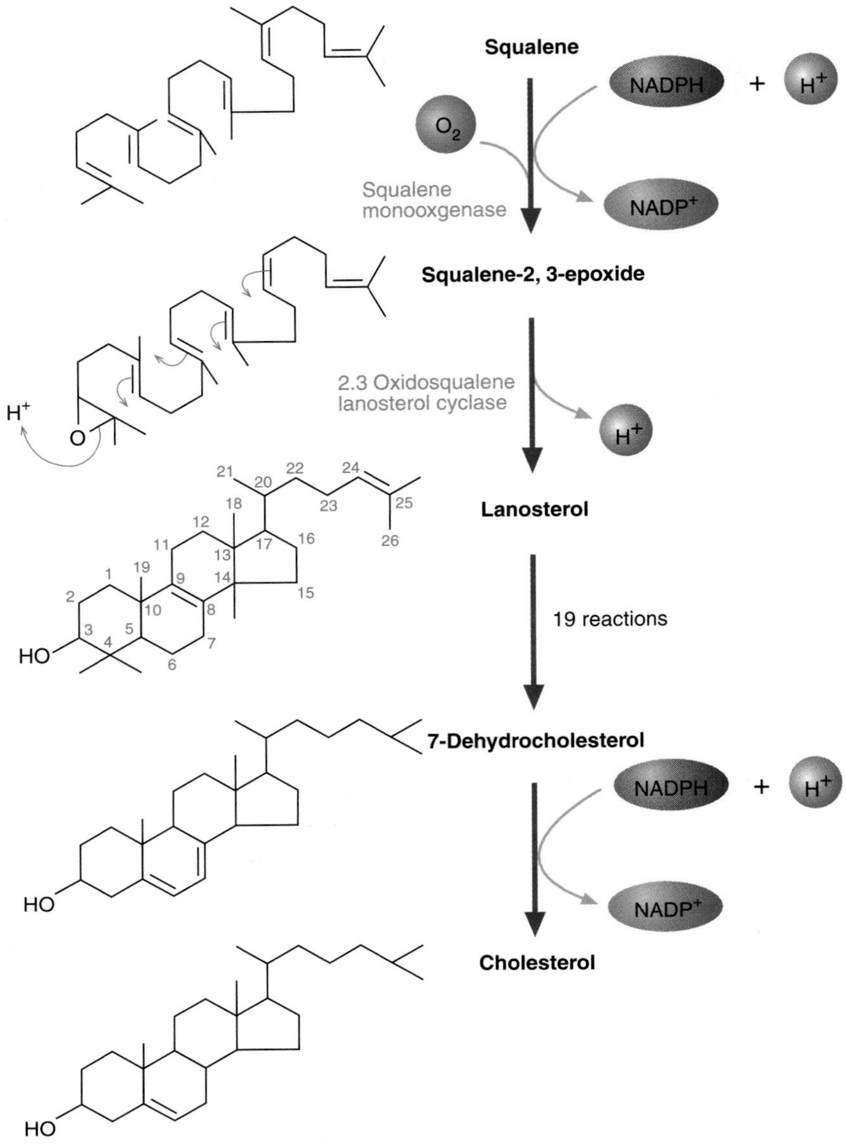

FIGURE 10.26

Synthesis of Cholesterol from Squalene.

This is the major route in mammals. In an alternative minor route, squalene is converted to desmosterol, which is then reduced to form cholesterol. The details of these and many reactions in the major route are poorly understood. (Desmosterol differs from cholesterol because of a C=C double bond between C-24 and C-25.)

steroid metabolism are peptide hormones secreted from the pituitary (a hormone-producing structure found in the brain) and several prostaglandins. (See Chapter 15 for a discussion of hormones and hormone action.) For example, *adrenocorticotropic hormone* (ACTH) is a peptide hormone secreted by the pituitary gland that stimulates adrenal steroid synthesis. One of the consequences of ACTH binding to adrenal cell receptors is an increased synthesis of 17-α-hydroxylase and 11-β-hydroxylase. In contrast, prostaglandin $F_{2\alpha}$ has been observed to inhibit the induction of progesterone synthesis in the ovaries. This latter process is stimulated by luteinizing hormone (LH), a protein produced by the pituitary. (The functional effect of prostaglandin $F_{2\alpha}$ on progesterone synthesis is still unclear.)

The enzymatic processes by which cholesterol is converted to biologically active steroids, as well as the means by which steroids are inactivated and prepared for excretion, compose an elaborate mechanism referred to as **biotransformation.** During biotransformation the same (or in some cases similar) enzymes are also used to solubilize hydrophobic xenobiotics so that they can be more easily excreted. An overview of biotransformation is presented in Chapter 19. The biotransformation of cholesterol into bile acids is outlined next.

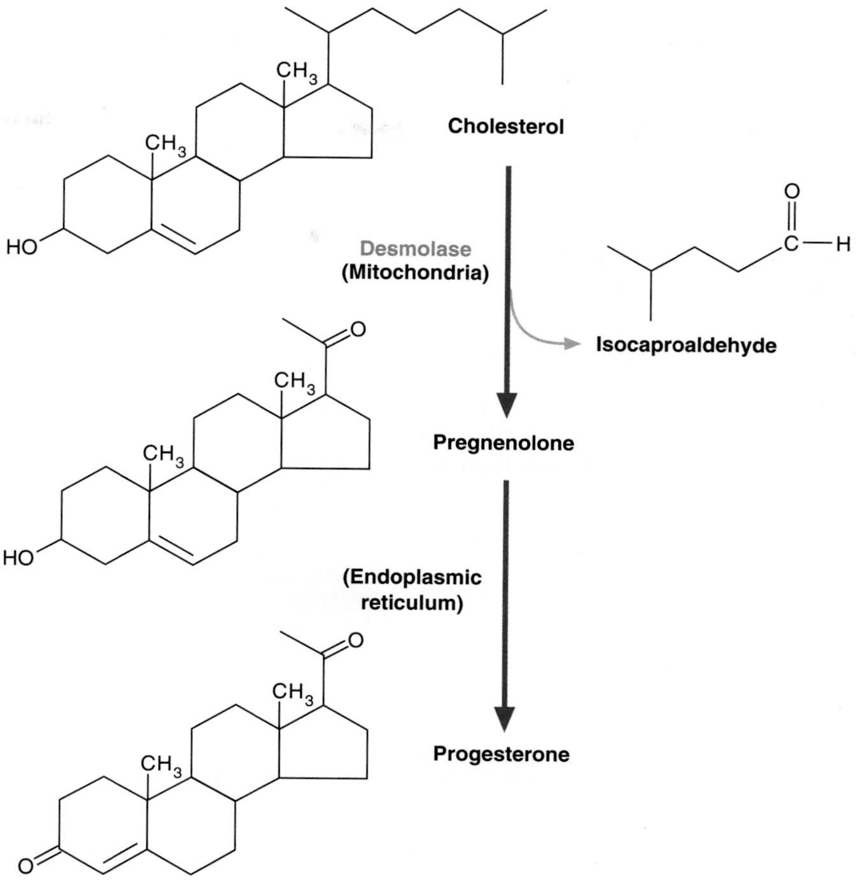

FIGURE 10.27

Progesterone Synthesis.

Pregnenolone is synthesized in the mitochondria. It is transported to the ER, where it is converted to progesterone. This latter process oxidizes the hydroxyl group and isomerizes a C=C double bond.

QUESTION 10.5

Cortisol (also referred to as *hydrocortisone*) is a potent glucocorticoid. (**Glucocorticoids** are hormones that promote carbohydrate, protein, and fat metabolism. For example, glucocorticoids stimulate gluconeogenesis, lipolysis, and an increased uptake of amino acids by the liver.) Cortisol also possesses a small amount of mineralocorticoid activity. (**Mineralocorticoids** regulate Na^+ and K^+ metabolism. For example, aldosterone, the most important mineralocorticoid in humans, induces the reabsorption of Na^+ from urine. It also promotes the secretion of K^+ and H^+ into urine.) For steroids to have either glucocorticoid or mineralocorticoid activity, they must possess a hydroxy group at C-11.

In *Addison's disease* an inadequate secretion of glucocorticoids and mineralocorticoids results in hypoglycemia, an imbalance in the body's Na^+ and K^+ concentrations, dehydration, and low blood pressure. In the past, individuals who had undiagnosed Addison's disease found that consumption of large amounts of licorice provided some relief from their symptoms. (Licorice, an extract of the plant *Glycyrrhiza glabra,* is used as a flavoring agent in candy and some medicines.) In these patients, licorice consumption resulted in sodium retention, hypokalemia (low blood K^+ concentration), and a rise in blood pressure. This effect was more pronounced when cortisol was administered.

Recently, it has been discovered that the active ingredient in licorice, called glycyrrhizic acid, inhibits 11-β-hydroxysteroid dehydrogenase, the enzyme that reversibly converts cortisol to cortisone, its inactive metabolite. Refer to the structure of cortisol (Figure 10.28), and deduce the structure of cortisone. Because cortisol is regarded as a glucocorticoid, why does glycyrrhizic acid consumption appear to affect mineral metabolism? Cortisone is often used to treat Addison's disease although it is physiologically inactive. Suggest a reason why its use is justified.

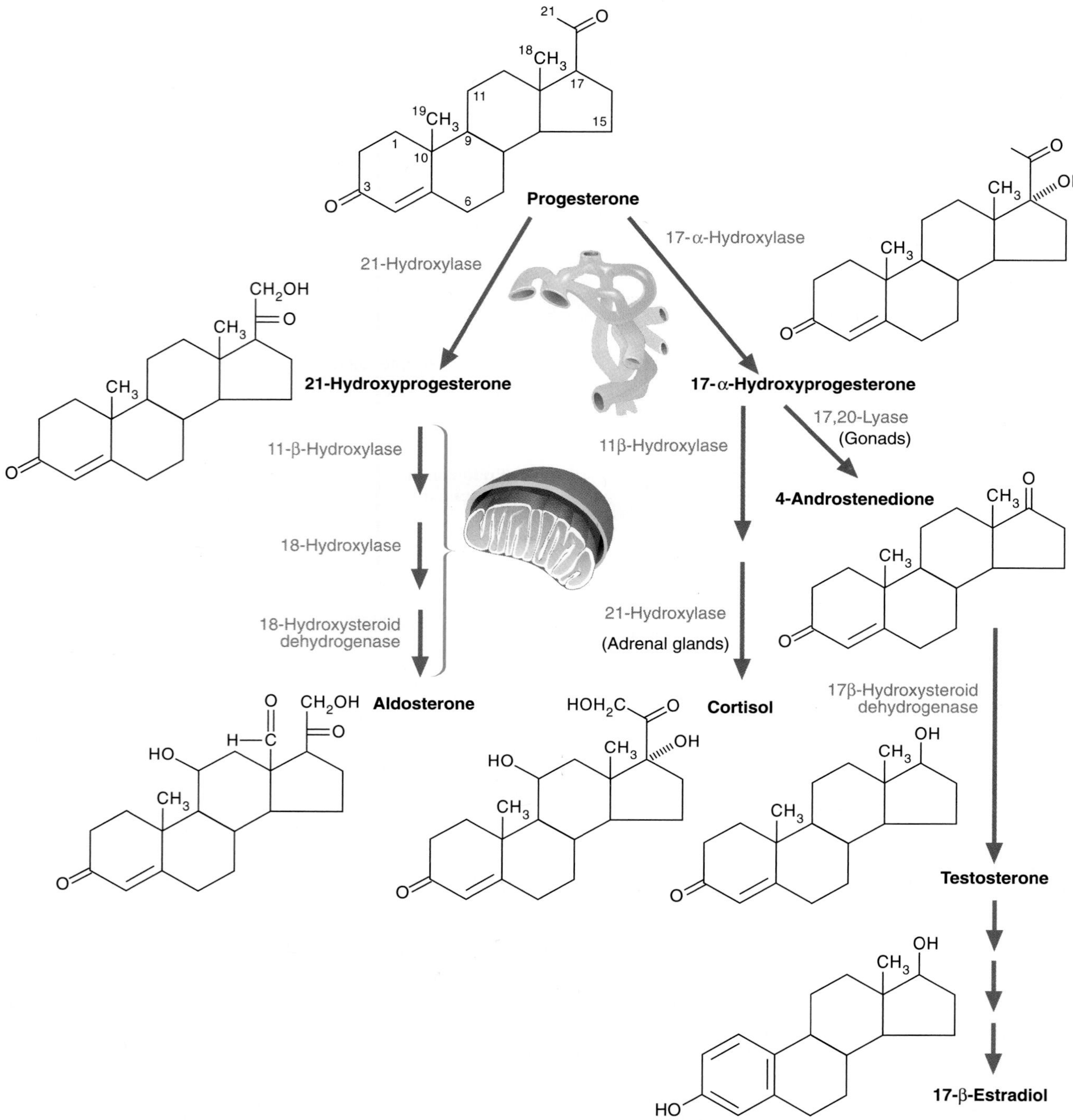

FIGURE 10.28
Synthesis of Selected Steroids.

The enzyme 17α-hydroxylase is found in all steroid-producing cells. Most other enzymes are tissue-specific.

Cholesterol Degradation Unlike many other types of biomolecules, cholesterol and other steroids cannot be degraded to smaller molecules. Instead, they are converted to derivatives, whose improved solubility properties allow their excretion. The most important mechanism for degrading and eliminating cholesterol is the synthesis of the bile acids. Bile acid synthesis, which occurs in the liver, is outlined in Figure 10.29. The conversion of cholesterol to 7-α-hydrocholesterol, catalyzed by cholesterol-7-hydroxylase (a microsomal enzyme), is the rate-limiting reaction in bile acid synthesis.

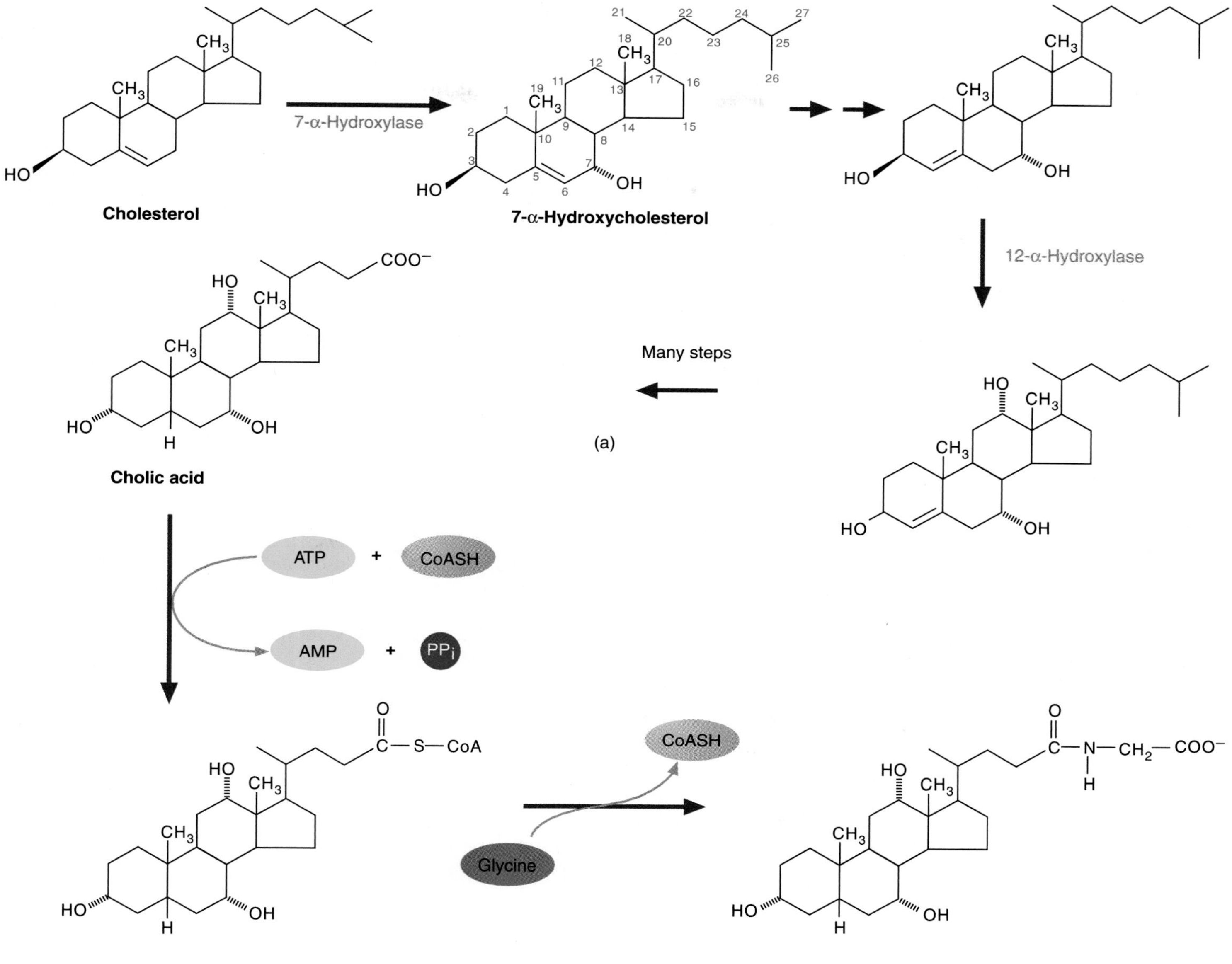

FIGURE 10.29

Synthesis of the Bile Acid Cholic Acid (a) and the Bile Salt Glycocholate (b).

Because of their higher solubility, bile acids primarily act as detergents during dietary fat digestion.

KEY *Concepts*

Cholesterol is synthesized from acetyl-CoA in a multistep pathway that occurs primarily in the liver. Small amounts of cholesterol are used to synthesize biologically powerful steroid hormones. Cholesterol is degraded primarily by conversion to the bile salts, which facilitate the emulsification and absorption of dietary fat.

In later reactions, the double bond at C-5 is rearranged and reduced, and an additional hydroxyl group is introduced. The products of this process, cholic acid and deoxycholic acid, are converted to bile salts by microsomal enzymes that catalyze conjugation reactions. (In **conjugation reactions** a molecule's solubility is increased by converting it into a derivative that contains a water-soluble group. Amides and esters are common examples of these conjugated derivatives.) Most bile acids are conjugated with glycine or taurine (Figure 10.30).

The bile salts are important components of *bile,* a yellowish green liquid produced by hepatocytes that aids in the digestion of lipids. In addition to bile salts, bile contains cholesterol, phospholipids, and bile pigments (bilirubin and biliverdin; see Chapter 19). (The bile pigments are degradation products of heme.) After it is secreted into the bile ducts and stored in the gallbladder, bile is used in the small intestine to

FIGURE 10.30
Glycine and Taurine.

$$H_3\overset{+}{N}—CH_2—\overset{\overset{\textstyle O}{\|}}{C}—O^-$$

Glycine

$$H_3N^+—CH_2—CH_2—SO_3^-$$

Taurine

enhance the absorption of dietary fat. Bile acts as an emulsifying agent, that is, it promotes the breakup of large fat droplets into smaller ones. Bile salts are also involved in the formation of so-called biliary micelles, which aid in absorbing fat and the fat-soluble vitamins (A, D, E, and K). Most bile salts are reabsorbed in the distal ileum (near the end of the small intestine). They enter the blood and are transported back to the liver, where they are resecreted into the bile ducts with other bile components. The biological significance of bile acid conjugation reactions appears to be that the conjugation process prevents premature absorption of bile acids in the biliary tract (the duct system and gallbladder) and small intestine. The reabsorption of bile salts in the distal ileum of the small intestine (necessary for effective recycling) is apparently triggered by the glycine or taurine signal. (It has been estimated that bile salt molecules are recycled 18 times before they are finally eliminated.)

QUESTION 10.6

The formation of gallstones (crystals that are usually composed of cholesterol and inorganic salts) within the gallbladder or bile ducts afflicts millions of people. Predisposing factors for this excruciatingly painful malady include obesity and infection of the gallbladder (*cholecystitis*). Because cholesterol is virtually insoluble in water, it is solubilized in bile by its incorporation into micelles composed of bile salts and phospholipids. Gallstones tend to form when cholesterol is secreted into bile in excessive quantities. Suggest a reason why an obese person is prone to gallstone formation. (Hint: HMG-CoA reductase activity is higher in obese individuals.)

Sterol Metabolism in Plants

Relatively little is known about plant sterols. (Most of the research effort in steroid metabolism has been expended in the investigation of steroid-related human diseases.) It appears, however, that the initial phase of plant sterol synthesis is very similar to that of cholesterol synthesis with the following exception. In plants and algae the cyclization of squalene-2,3-epoxide leads to the synthesis of *cycloartenol* instead of lanosterol. Many subsequent reactions in plant sterol pathways involve methylation reactions involving SAM. There appear to be two separate isoprenoid biosynthetic pathways in plant cells: the endoplasmic reticulum/cytoplasm pathway and a separate chloroplast pathway. The roles of these pathways in plant isoprenoid metabolism are still unclear.

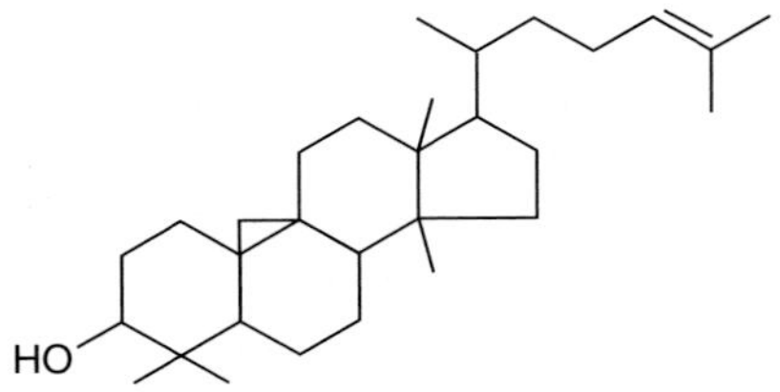

Cycloartenol

KEY WORDS

acyl carrier protein, *265*

autoimmune disease, *263*

biotransformation, *282*

Calvin cycle, *270*

conjugation reaction, *285*

epoxides, *263*

fatty acid–binding protein, *253*

glucocorticoid, *283*

ketogenesis, *260*

ketone body, *260*

ketosis, *260*

lipogenesis, *251*

lipolysis, *251*

mineralocorticoid, *283*

sterol carrier protein, *280*

thiolytic cleavage, *258*

turnover, *272*

SUMMARY

Acetyl-CoA plays a central role in most lipid-related metabolic processes. For example, acetyl-CoA is used in the synthesis of fatty acids. When fatty acids are degraded to generate energy, acetyl-CoA is the product.

Depending on the body's current energy requirements, newly digested fat molecules are used to generate energy or are stored within adipocytes. When the body's energy reserves are low, fat stores are mobilized in a process referred to as lipolysis. In lipolysis, triacylglycerols are hydrolyzed to fatty acids and glycerol. Glycerol is transported to the liver, where it can be used in lipid or glucose synthesis. Most fatty acids are degraded to form acetyl-CoA within mitochondria in a process referred to as β-oxidation. Peroxisomal β-oxidation appears to shorten very long fatty acids. Other mechanisms degrade odd-chain fatty acids and unsaturated fatty acids. When the product of fatty acid degradation (acetyl-CoA) is present in excess, ketone bodies are produced.

The first step in eicosanoid synthesis is the release of arachidonic acid from C-2 of glycerol in membrane phosphoglyceride molecules. Cyclooxygenase converts arachidonic acid into PGG_2, which is a precursor of the prostaglandins and the thromboxanes. The lipoxygenases convert arachidonic acid to the precursors of the leukotrienes.

Fatty acid synthesis begins with the carboxylation of acetyl-CoA to form malonyl-CoA. The remaining reactions of fatty acid synthesis take place on the fatty acid synthase multienzyme complex. Several enzymes are available to elongate and desaturate dietary and newly synthesized fatty acids.

After phospholipids are synthesized at the interface of the SER and the cytoplasm, they are often "remodeled," that is, their fatty acid composition is adjusted. The turnover (i.e., the degradation and replacement) of phospholipids, mediated by the phospholipases, is rapid.

Synthesis of the ceramide component of sphingolipids begins with the condensation of palmitoyl-CoA with serine to form 3-ketosphinganine. In a two-step process involving acyl-CoA and $FADH_2$, sphinganine (formed when 3-ketosphinganine is reduced by NADPH) is converted to ceramide. Sphingolipids are degraded within lysosomes.

Phospholipid translocator proteins, phospholipid exchange proteins, and transition vesicles are involved in the complicated process of membrane synthesis and delivery of membrane components to their cellular destinations.

Cholesterol synthesis can be divided into three phases: formation of HMG-CoA from acetyl-CoA, conversion of HMG-CoA to squalene, and conversion of squalene to cholesterol. Cholesterol is the precursor for all steroid hormones and the bile salts. Bile salts are used to emulsify dietary fat. They are the primary means by which the body can rid itself of cholesterol.

SUGGESTED READINGS

Drayr, J.-P., and Vamecq, J., The Gluconeogenicity of Fatty Acids in Mammals, *Trends Biochem. Sci.,* 14:478–479, 1989.

Goldstein, J. L., and Brown, M.S., Regulation of the Mevalonate Pathway, *Nature,* 343:425–430, 1990.

Gurr, M.I., and Harwood, J.L., *Lipid Biochemistry: An Introduction,* Chapman and Hall, London, 1991.

Hampton, R., Dimster-Denk, D., and Rine, J. The Biology of HMG-CoA Reductase: the Pros of Contra-regulation, *Trends. Biochem. Sci.,* 21:140–145, 1996.

Johnson, M., Carey, F., and McMillan, R.M., Alternative Pathways of Arachidonate Metabolism: Prostaglandins, Thromboxanes and Leukotrienes, *Essays Biochem.,* 19:40–141, 1983.

Masters, C., and Crane, D., The Role of Peroxisomes in Lipid Metabolism, *Trends Biochem. Sci.,* 9:314–319, 1984.

Weissman, G., Aspirin, *Sci. Amer.,* 264:84–90, 1991.

QUESTIONS

1. Define the following terms:
   a. *de novo*
   b. oil bodies
   c. β-oxidation
   d. turnover
   e. thiolytic cleavage
   f. autoantibodies
   g. ketone bodies
2. What is the function of each of the following substances?
   a. carnitine
   b. flippase
   c. thrombin
   d. thiolase
   e. desmolase
3. What are the differences between β-oxidation in mitochondria and in peroxisomes? What similarities are there between these processes?
4. Consuming aspirin shortly before vigorous exercise reduces aches and pains. Why is this so? Suggest a possible consequence of this behavior.
5. List three differences between fatty acid synthesis and β-oxidation.
6. How does fatty acid synthesis in plants differ from fatty acid synthesis in animals?
7. Explain how hormones act to modify the metabolism of fatty acids in both the short and long term. Give examples.
8. What is the difference between a steroid and a sterol?
9. Explain why the head group of a phospholipid is largely responsible for the slow rate of transmembrane migration.

10. Explain how conjugation increases the water solubility of water-insoluble compounds.

11. Show how the following fatty acid is oxidized:

CH₃CH₂CH₂CH(CH₃)—CH₂—C(=O)—OH

Indicate at which points α- and β-oxidation are carried out.

12. During periods of stress or fasting, blood glucose levels fall. In response, fatty acids are released by adipocytes. Explain how the drop in blood glucose triggers fatty acid release.

13. Butyric acid, a simple four-carbon fatty acid, is oxidized by β-oxidation. Calculate the number of FADH₂ and NADH molecules produced in this oxidation. How many acetyl-CoA molecules are also produced?

14. Insulin is released after carbohydrate intake. Describe two ways insulin acts to influence fatty acid metabolism.

15. β-Oxidation of naturally occurring monounsaturated fatty acids requires an additional enzyme. What is this enzyme and how does it accomplish its task?

16. The acyl-CoA dehydrogenase deficiency diseases are a group of inherited defects that impair the β-oxidation of fatty acids. Symptoms of the disease range from nausea and vomiting to frequent comas. Symptoms may be alleviated by eating regularly and avoiding periods of starvation (12 hours or more). Why does this simple procedure alleviate the symptoms?

17. There is an unusually high concentration of phosphatidylcholine on the lumenal side of the ER. What structural feature of phosphatidylcholine is responsible for this? Explain how this structural feature produces this effect.

18. Identify the hydrophobic and hydrophilic regions in the following molecule. How do you think it would orient itself in a membrane?

19. Phospholipases show an enhanced activity for a substrate above the critical micelle concentration. (The critical micelle concentration, or cmc, is that concentration of a lipid above which micelles begin to form.)
    a. What type of noncovalent interactions are possible between the lipid and the enzyme at this stage?
    b. What do these interactions suggest about the structure of phospholipases?

20. When the production of acetyl-CoA exceeds the body's capacity to oxidize it, acetoacetic acid, β-hydroxybutyrate, and acetone accumulate. When generated in large amounts, these substances can exceed the blood's buffering capacity. As the blood pH falls, the ability of red blood cells to carry oxygen is affected. Subsequently, the brain can be starved for oxygen, and a fatal coma can result. Explain how severe dieting can produce this condition.

21. Gaucher's disease is an inherited deficiency of β-glucocerebrosidase. Glucocerebroside is deposited in macrophages that die, releasing their contents into the tissues. Some affected individuals may have neurologic disorders while quite young; others may not show ill effects until much later in life. The disease may be detected by assaying white blood cells for the ability to hydrolyze the β-glycosidic bond of artificial substrates. Examine the following glucocerebroside and determine which bond would be cleaved by glucocerebrosidase.

CH_3 — $(CH_2)_{12}$ — CH=CH — CH(OH) — CH(NH—C(=O)—R) — CH₂—O

(glucose ring: CH₂OH, OH, OH, OH)

(structure for question 18:)

CH_3 — $(CH_2)_{15}CH_2$ — C(=O) — O — CH(—CH₂—O—C(=O)—CH₂(CH₂)₁₅—CH₃)(—CH₂—O—P(=O)(O⁻)—O—CH₂CH₂N⁺(CH₃)₃)—H

AEROBIC METABOLISM

Chapter Eleven

In aerobic cells most energy is generated within the mitochondrion. Dioxygen (O_2) is used to oxidize nutrient molecules.

OBJECTIVES

After you have studied this chapter, you should be able to answer these questions:

1. How is energy obtained from the breakdown of acetyl-CoA by the reactions of the citric acid cycle?

2. How is the energy released during the electron transport pathway captured and used to drive biosynthetic processes?

3. How are the pathways of energy metabolism interrelated?

4. How are the pathways of energy metabolism regulated so that each cell's energy requirements are consistently met.

5. Why do calorie-restricted diets promote obesity?

6. What are the toxic products of oxygen metabolism and how do they damage cells?

7. How are toxic oxygen metabolites formed and destroyed?

The use of oxygen by aerobic organisms provides enormous advantages in comparison to an anaerobic lifestyle because the aerobic oxidation of nutrients such as glucose and fatty acids yields a substantially greater amount of energy than does fermentation. Oxygen also facilitates reactions such as hydroxylations. Because oxygen is a highly reactive molecule, however, an inescapable price is paid for its use: toxic metabolites are formed. Research efforts reveal that aerobic cells have an array of mechanisms that protect against oxygen's deleterious effects. Numerous enzymes and antioxidant molecules usually prevent oxidative cell damage with extraordinary precision. Despite this high level of protection, however, oxygen metabolites sometimes cause serious damage to living cells. Abnormally stressful conditions can overwhelm an organism's antioxidant mechanisms. Disorders such as cancer, heart disease, and the nerve damage following spinal cord injuries are now known to be caused in part by oxygen metabolites.

The earth's ancient atmosphere is now believed to have consisted largely of methane, ammonia, hydrogen, nitrogen, carbon dioxide, and water vapor. The ancient sea contained simple organic molecules such as carboxylic and amino acids. These molecules, formed as electrical discharges and solar radiation produced chemical changes in the atmospheric gases, were a rich source of the building block and fuel molecules required to sustain emerging life forms. Eventually, these early living organisms (primordial prokaryotic cells) became so abundant that organic molecules were consumed faster than they were formed by natural forces. As supplies of preformed molecules dwindled, some organisms evolved new mechanisms for obtaining food. Eventually, one such mechanism, photosynthesis, had a dramatic and far-reaching effect on the global environment. Over three billion years ago, photosynthetic cells began to produce their own food by using light energy to transform CO_2 and H_2O into organic molecules. Dioxygen (O_2) is a by-product of this process. As photosynthesis occurred on an ever increasing scale, the oxygen content of the atmosphere increased. Because O_2 combines readily with other molecules (e.g., $4 NH_3 + 3 O_2 \rightarrow 2 N_2 + 6 H_2O$), the earth's atmosphere was gradually converted (over a billion-year time span) to one consisting principally of dinitrogen, water vapor, carbon dioxide, and oxygen. Most living organisms that arose under the reducing conditions of the primitive earth were unprepared for living in an oxidizing atmosphere. The species that survived the transition did so because they developed methods to protect themselves from oxygen's toxic effects. Their descendants, today's living organisms, use one of the following strategies. Obligate anaerobes, organisms that grow only in the absence of oxygen, avoid the gas by living in highly reduced environments such as soil. They use fermentative processes to satisfy their energy requirements. Aerotolerant anaerobes, which also depend on fermentation for their energy needs, possess detoxifying enzymes and antioxidant molecules that protect against oxygen's toxic products. Facultative anaerobes not only possess the mechanisms needed for detoxifying oxygen metabolites, they can also generate energy by using oxygen as an electron acceptor when the gas is present. Finally, obligate aerobes are highly dependent on oxygen for energy production. They protect themselves from its toxic effects with enzymes and antioxidant molecules.

Facultative anaerobes and obligate aerobes that use oxygen to generate energy employ the following biochemical processes: the citric acid cycle, the electron transport pathway, and oxidative phosphorylation. In eukaryotes these processes occur within the mitochondrion (Figure 11.1). This chapter begins with a detailed discussion of the citric acid cycle, a pathway in which acetyl-CoA is converted to CO_2 and H_2O as three molecules of NAD^+ and one molecule of FAD are reduced. This is followed by a discussion of the oxidation of reduced coenzymes by the electron transport chain

FIGURE 11.1

Aerobic Metabolism in the Mitochondrion.

In eukaryotic cells aerobic metabolism occurs within the mitochondrion. Acetyl CoA, the oxidation product of pyruvate, fatty acids, and certain amino acids (not shown), is oxidized by the reactions of the citric acid cycle within the mitochondrial matrix. The principal products of the cycle are the reduced coenzymes NADH and $FADH_2$ and CO_2. The high energy electrons of NADH and $FADH_2$ are subsequently donated to the electron transport chain (ETC), a series of electron carriers in the inner membrane. The terminal electron acceptor for the ETC is O_2. The energy derived from the electron transport mechanism drives ATP synthesis by creating a proton gradient across the inner membrane. The large folded surface of the inner membrane is studded with ETC complexes, numerous types of transport proteins, and ATP synthase, the enzyme complex responsible for ATP synthesis. The inner membrane is impermeable to most small molecules and ions.

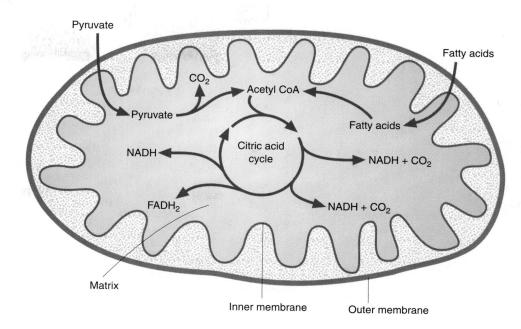

(ETC) and the mechanism by which the energy released by electron transport is captured by oxidative phosphorylation. Emphasis is placed on regulatory mechanisms. The chapter ends with a discussion of the formation of toxic oxygen products and the strategies that cells use to protect themselves.

11.1 CITRIC ACID CYCLE

The citric acid cycle (Figure 11.2) is a series of biochemical reactions aerobic organisms use to release chemical energy stored in acetyl CoA. Acetyl CoA is a product of catabolic reactions in carbohydrate, lipid and amino acid metabolism. It is synthesized from pyruvate (a partially oxidized product of the degradation of sugars and certain amino acids) in a series of reactions (discussed below). Acetyl CoA is also the product of β-oxidation of fatty acids (described in Chapter 10) and certain reactions in amino acid metabolism (Chapter 14). In the citric acid cycle, the carbon atoms derived from the acetyl group of acetyl CoA are oxidized to CO_2. The high-energy electrons removed from citric acid cycle intermediates are transferred to NAD^+ and FAD to form the reduced coenzymes NADH and $FADH_2$.

In the first reaction of the citric acid cycle, a two-carbon acetyl group condenses with a four-carbon molecule (oxaloacetate) to form a six-carbon molecule (citrate) (Figure 11.3). During the subsequent seven reactions, in which two CO_2 molecules are produced and four pairs of electrons are removed from carbon compounds, citrate is reconverted to oxaloacetate. During one step in the cycle, the high-energy molecule guanosine triphosphate (GTP) is produced during a substrate-level phosphorylation. The net reaction for the citric acid cycle is as follows:

$$\text{Acetyl-CoA} + 3 \text{ NAD}^+ + \text{FAD} + \text{GDP} + P_i + 2 \text{ H}_2\text{O} \rightarrow$$
$$2 \text{ CO}_2 + 3 \text{ NADH} + \text{FADH}_2 + \text{CoASH} + \text{GTP} + 3 \text{ H}^+$$

In addition to its role in energy production, the citric acid cycle plays another important role in metabolism. As discussed in the following text, cycle intermediates are substrates in a variety of biosynthetic reactions.

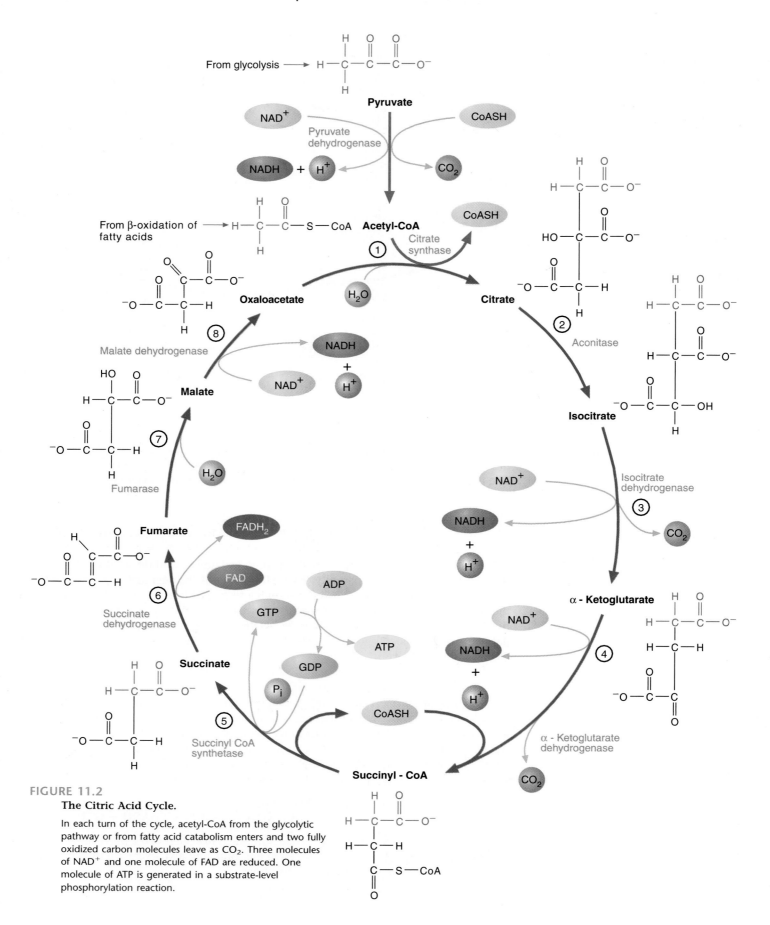

FIGURE 11.2

The Citric Acid Cycle.

In each turn of the cycle, acetyl-CoA from the glycolytic pathway or from fatty acid catabolism enters and two fully oxidized carbon molecules leave as CO_2. Three molecules of NAD^+ and one molecule of FAD are reduced. One molecule of ATP is generated in a substrate-level phosphorylation reaction.

FIGURE 11.3

Main Reactions of the Citric Acid Cycle.

Oxaloacetate, a four-carbon molecule, condenses with acetyl-CoA to form citrate, a six-carbon molecule. Then two molecules of CO_2 are formed. Three molecules of NADH, one molecule of $FADH_2$, and one molecule of GTP are also formed.

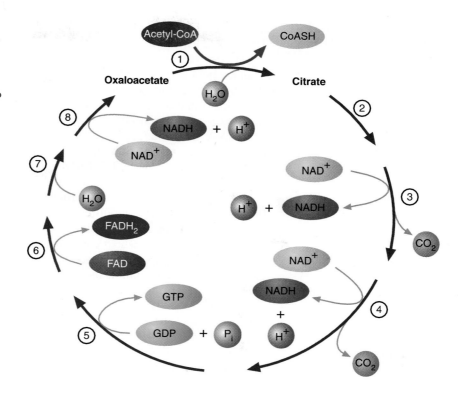

Conversion of Pyruvate to Acetyl-CoA

After its transport into the mitochondrial matrix, pyruvate is converted to acetyl-CoA in a series of reactions catalyzed by the enzymes in the pyruvate dehydrogenase complex. The net reaction, an oxidative decarboxylation, is as follows:

$$\text{Pyruvate} + NAD^+ + \text{CoASH} \rightarrow \text{Acetyl-CoA} + NADH + CO_2 + H_2O + H^+$$

Despite the apparent simplicity of this highly exergonic reaction ($\Delta G^{\circ\prime} = -8.0$ kcal/mole), its mechanism is one of the most complex known. The pyruvate dehydrogenase complex is a large multienzyme structure that contains three enzyme activities: pyruvate decarboxylase (E_1), dihydrolipoyl transacetylase (E_2), and dihydrolipoyl dehydrogenase (E_3). Each enzyme activity is present in multiple copies. Table 11.1 summarizes the number of copies of each enzyme and the required cofactors of *E. coli* pyruvate dehydrogenase.

In the first step, pyruvate decarboxylase catalyzes the decarboxylation of pyruvate. A nucleophile is formed when a basic residue of the enzyme extracts a proton from

TABLE 11.1

E. Coli Pyruvate Dehydrogenase Complex

Enzyme Activity	Function	No. of Copies per Complex*	Cofactors
Pyruvate Decarboxylase (E_1)	Decarboxylates pyruvate	24 (20–30)	TPP
Dihydrolipoyl Transacetylase (E_2)	Catalyzes transfer of acetyl group to CoASH	24 (60)	Lipoic acid, CoASH
Dihydrolipoyl Dehydrogenase (E_3)	Reoxidizes dihydrolipoamide	12 (20–30)	NAD^+, FAD

*The number of molecules of each enzyme activity found in mammalian pyruvate dehydrogenase is shown in parentheses.

the thiazole ring of thiamine pyrophosphate (TPP). The intermediate, hydroxyethyl-TPP (HETPP), forms after the nucleophilic thiazole ring attacks the carbonyl group of pyruvate (Figure 11.4).

In the next several steps, the hydroxyethyl group of HETPP is converted to acetyl-CoA by dihydrolipoyl transacetylase. Lipoic acid (Figure 11.5) plays a crucial role in this transformation. Lipoic acid is bound to the enzyme through an amide linkage with the ϵ-amino group of a lysine residue. It reacts with HETPP to form an acetylated lipoic acid and free TPP. The acetyl group is then transferred to the sulfhydryl group of coenzyme A. Subsequently, the reduced lipoic acid is reoxidized by dihydrolipoyl dehydrogenase. The FADH$_2$ is reoxidized by NAD$^+$ to form the FAD required for the oxidation of the next reduced lipoic acid residue.

The activity of pyruvate dehydrogenase is regulated by two mechanisms: product inhibition and covalent modification. The enzyme complex is inhibited directly by high concentrations of the reaction products acetyl-CoA and NADH. These molecules also activate a kinase, which converts the active pyruvate dehydrogenase complex to an inactive phosphorylated form. High concentrations of the substrates pyruvate, CoASH, and NAD$^+$ inhibit the activity of the kinase. The pyruvate dehydrogenase complex is reactivated by a dephosphorylation reaction catalyzed by a phosphoprotein phosphatase. The phosphoprotein phosphatase is activated when the mitochondrial ATP concentration is low.

Reactions of the Citric Acid Cycle

As previously mentioned, the citric acid cycle begins with the condensation of acetyl-CoA with oxaloacetate to form citrate.

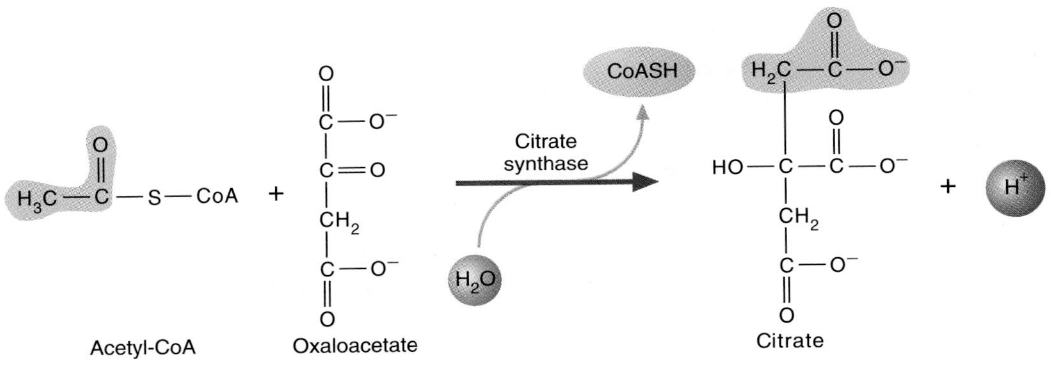

Acetyl-CoA   Oxaloacetate   Citrate

Note that this reaction is an aldol condensation. Because the standard free energy change is equal to -8 kcal/mole, citrate formation is highly exergonic.

In the next reaction of the cycle, citrate is reversibly converted to isocitrate by aconitase. During this isomerization reaction, an intermediate called *cis*-aconitate is formed by dehydration. The carbon-carbon double bond of *cis*-aconitate is then rehydrated to form the isomer isocitrate.

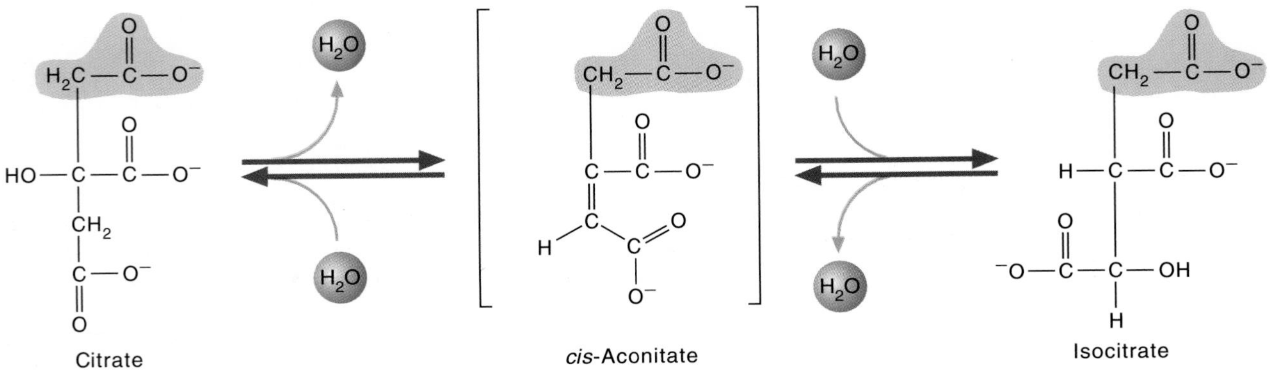

Citrate   *cis*-Aconitate   Isocitrate

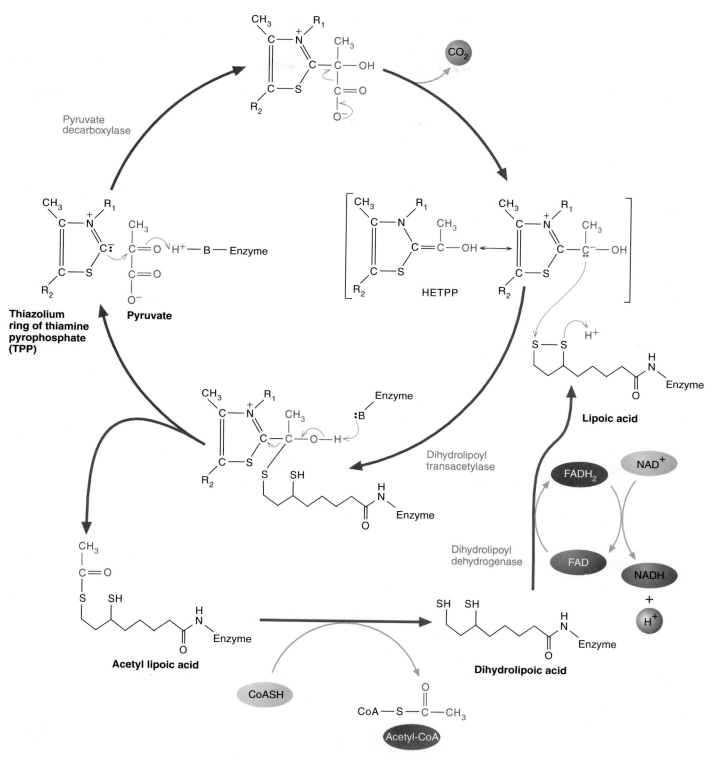

FIGURE 11.4

The Reactions Catalyzed by the Pyruvate Dehydrogenase Complex.

Pyruvate decarboxylase, which contains TPP, catalyzes the formation of the HETPP. Using lipoic acid as a cofactor, dihydrolipoyl transacetylase converts the hydroxyethyl group of HETPP to acetyl-CoA. Dihydrolipoyl dehydrogenase reoxidizes the reduced lipoic acid. (Refer to Figure 11.5 for the structure of lipoic acid.)

FIGURE 11.5

Lipoamide.

Lipoic acid is covalently bonded to the enzyme through an amide linkage with the ϵ-amino group of a lysine residue.

The oxidative decarboxylation of isocitrate, catalyzed by isocitrate dehydrogenase, occurs in two steps. First, isocitrate is oxidized to form oxalosuccinate, a transient intermediate.

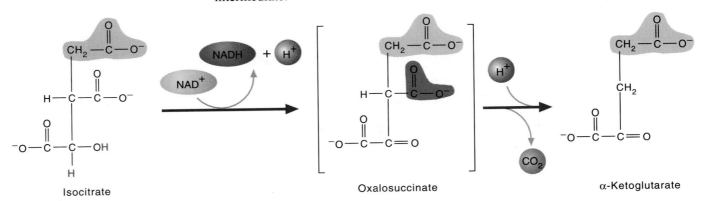

Isocitrate   Oxalosuccinate α-Ketoglutarate

Immediate decarboxylation of oxalosuccinate results in the formation of α-ketoglutarate, an α-keto acid. There are two forms of isocitrate dehydrogenase in mammals. The NAD$^+$-requiring isozyme is found only within mitochondria. The other isozyme, which requires NADP$^+$, is found in both the mitochondrial matrix and the cytoplasm. In some circumstances the latter enzyme is used within both compartments to generate NADPH, which is required in biosynthetic processes.

The conversion of α-ketoglutarate to succinyl-CoA is catalyzed by the enzyme activities in the α-ketoglutarate dehydrogenase complex.

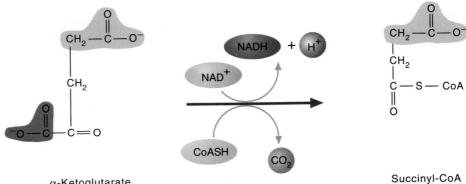

α-Ketoglutarate   Succinyl-CoA

This exergonic reaction ($\Delta G^{\circ\prime} = -8$ kcal/mole), an oxidative decarboxylation, is analogous to the conversion of pyruvate to acetyl-CoA catalyzed by pyruvate dehydrogenase. In both reactions, energy-rich thioester molecules are products, that is, acetyl-CoA and succinyl-CoA. Other similarities between the two multienzyme complexes are that the same cofactors (TPP, CoASH, lipoic acid, NAD^+, and FAD) are required and the same or similar allosteric effectors are inhibitors. In the case of α-ketoglutarate dehydrogenase, inhibition is promoted by succinyl-CoA, NADH, ATP, and GTP. An important difference between the two complexes is that the control mechanism of the α-ketoglutarate dehydrogenase complex does not involve covalent modification.

The cleavage of the thioester bond of succinyl-CoA to form succinate, catalyzed by succinate thiokinase, is coupled in mammals to the substrate-level phosphorylation of GDP. In many other organisms, ADP is phosphorylated instead.

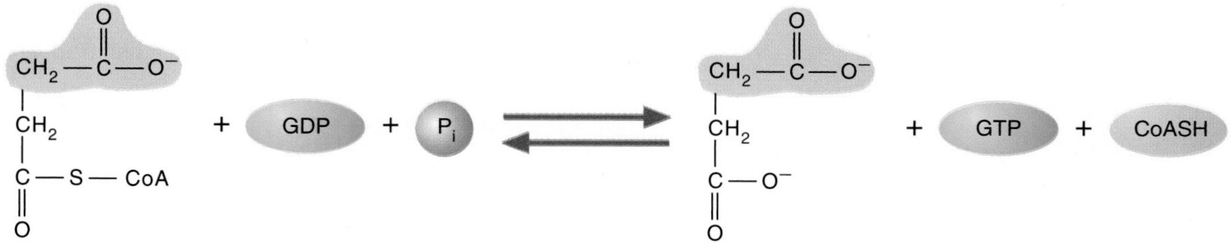

Succinyl-CoA   Succinate

ATP is synthesized in the following reaction catalyzed by nucleoside diphosphate kinase:

Succinate dehydrogenase catalyzes the oxidation of succinate to form fumarate.

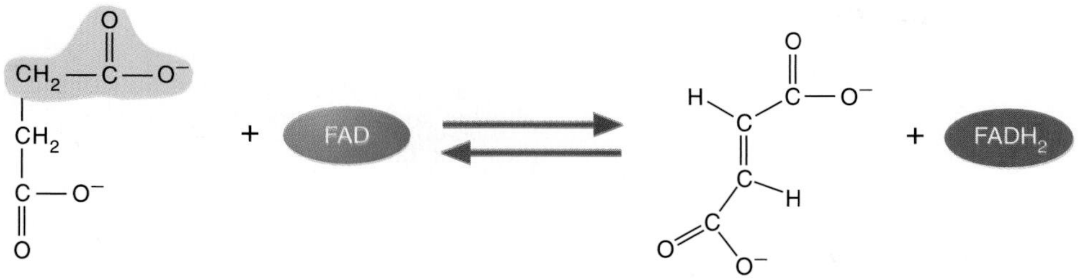

Succinate   Fumarate

KEY *Concepts*

The citric acid cycle begins with the condensation of a molecule of acetyl-CoA with oxaloacetate to form citrate, which is eventually reconverted to oxaloacetate. During this process, two molecules of CO_2, three molecules of NADH, one molecule of $FADH_2$, and one molecule of GTP are produced.

Unlike the other citric acid cycle enzymes, succinate dehydrogenase is not found within the mitochondrial matrix. Instead, it is tightly bound to the inner mitochondrial membrane. Succinate dehydrogenase is activated by high concentrations of succinate, ATP, and P_i and inhibited by oxaloacetate. Recall that the enzyme is also inhibited by malonate (p. 130), a structural analog of succinate. (This inhibitor was used by Hans Krebs in his pioneering work on the citric acid cycle.)

Fumarate is converted to L-malate in a reversible stereospecific hydration reaction catalyzed by fumarase (also referred to as fumarate hydratase).

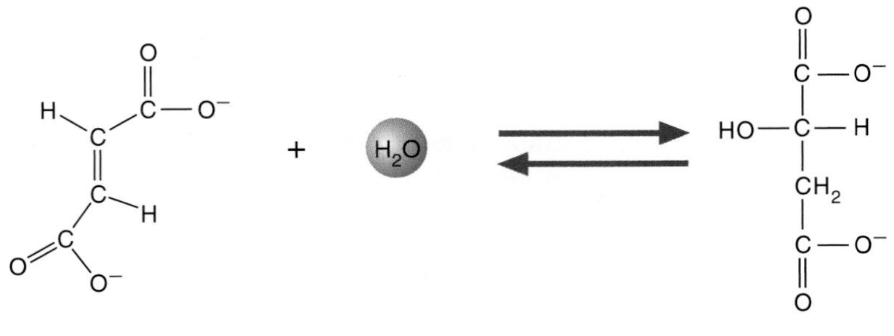

Fumarate   L-Malate

Finally, oxaloacetate is regenerated with the oxidation of L-malate.

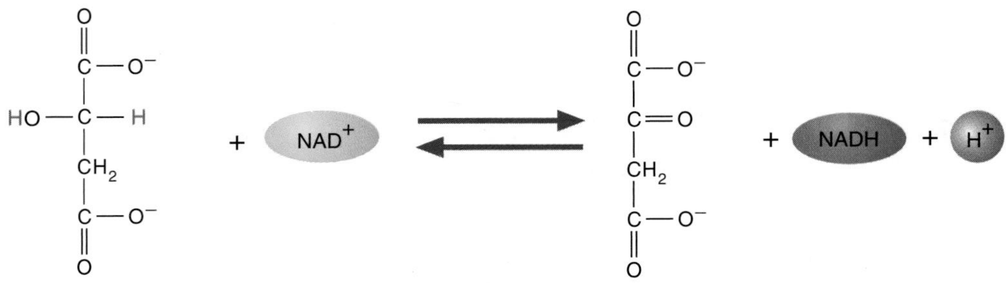

L-Malate   Oxaloacetate

Malate dehydrogenase uses NAD^+ as the oxidizing agent in a highly endergonic reaction ($\Delta G^{\circ\prime} = +7$ kcal/mole). The reaction is pulled to completion because of the removal of oxaloacetate in the next round of the cycle.

The Amphibolic Citric Acid Cycle

As previously described, amphibolic pathways can function in both anabolic and catabolic processes. The citric acid cycle is obviously catabolic, since acetyl groups are oxidized to form CO_2 and energy is conserved in reduced coenzyme molecules. The citric acid cycle is also anabolic, since several citric acid cycle intermediates are precursors in biosynthetic pathways (Figure 11.6). For example, oxaloacetate is used in both gluconeogenesis (Chapter 8) and amino acid synthesis (Chapter 13). α-Ketoglutarate also plays an important role in amino acid synthesis. The synthesis of porphyrins such as heme uses succinyl-CoA (Chapter 13). Finally, recall that the synthesis of fatty acids and cholesterol in the cytoplasm requires acetyl-CoA. Because acetyl-CoA cannot penetrate the inner mitochondrial membrane, it is converted to citrate. After its transport into the cytoplasm, citrate is cleaved to form acetyl-CoA and oxaloacetate by citrate lyase.

Anabolic processes drain the citric acid cycle of the molecules required to sustain its role in energy generation. Several reactions, referred to as **anaplerotic** reactions, replenish them. One of the most important anaplerotic reactions is catalyzed by pyruvate carboxylase. A high concentration of acetyl-CoA, an indicator of an insufficient oxaloacetate concentration, activates pyruvate carboxylase. As a result, oxaloacetate concentration increases. Any excess oxaloacetate that is not used within the citric acid cycle is used in gluconeogenesis. Other anaplerotic reactions include the synthesis of succinyl-CoA from certain fatty acids (Chapter 10) and certain amino acids (Chapter 14).

KEY *Concepts*

The citric acid cycle is an amphibolic pathway, that is, it plays a role in both anabolism and catabolism. The citric acid cycle intermediates used in anabolic processes are replenished by several anaplerotic reactions.

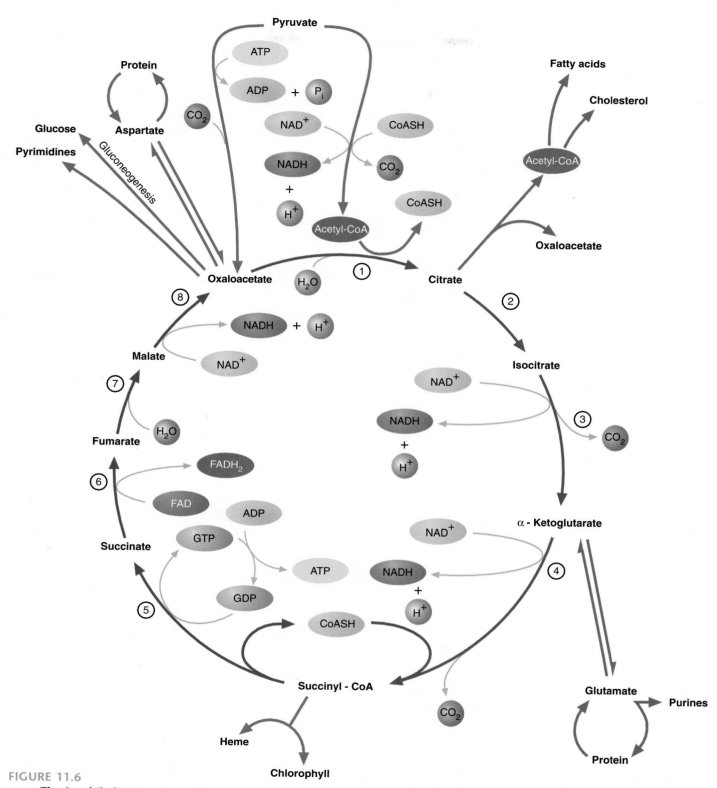

FIGURE 11.6

The Amphibolic Citric Acid Cycle.

The citric acid cycle operates in both
anabolic processes (e.g., the synthesis of
fatty acids, cholesterol, heme, and
glucose) and catabolic processes (e.g.,
amino acid degradation and energy
production).

QUESTION

11.1

In the past, mammals were considered unable to use fatty acids in gluconeogenesis. (Acetyl-CoA cannot be converted to pyruvate because the reaction catalyzed by pyruvate dehydrogenase is irreversible.) Recent experimental evidence indicates that certain unusual fatty acids (i.e., those with odd chains or two carboxylic acid groups) can be converted in small but measurable quantities to glucose. Recall that one molecule of propionyl-CoA is produced when each odd-carbon chain fatty acid is oxidized. Describe a possible biochemical pathway by which a liver cell might synthesize glucose from propionyl-CoA. (Hint: Refer to Figure 10B.) One of the products of the β-oxidation of dicarboxylic acids is succinyl-CoA. Propose a biochemical pathway for the conversion of the molecule illustrated in Figure 11.7 to glucose.

FIGURE 11.7
Adipic Acid.

$$HO-\overset{\overset{\displaystyle O}{\|}}{C}-(CH_2)_4-\overset{\overset{\displaystyle O}{\|}}{C}-OH$$

The Glyoxylate Cycle

Plants as well as some fungi, algae, protozoans, and bacteria can grow by using two-carbon compounds. (Molecules such as ethanol, acetate, and acetyl-CoA, derived from fatty acids, are the most common substrates.) The series of reactions responsible for this capability, referred to as the **glyoxylate cycle,** is a modified version of the citric acid cycle. In plants the glyoxylate cycle occurs in organelles called glyoxysomes (p. 29). In other eukaryotic organisms and in bacteria, glyoxylate enzymes occur in cytoplasm.

The glyoxylate cycle (Figure 11.8) consists of five reactions. The first two reactions (the synthesis of citrate and isocitrate) are familiar ones, since they also occur in the citric acid cycle. However, the formation of citrate from oxaloacetate and acetyl-CoA and the isomerization of citrate to form isocitrate are catalyzed by glyoxysome-specific isozymes. The next two reactions are unique to the glyoxylate cycle. Isocitrate is split into two molecules (succinate and glyoxylate) by isocitrate lyase. (This reaction is an aldol cleavage.) Succinate, a four-carbon molecule, is eventually con-

FIGURE 11.8

The Glyoxylate Cycle.

Using some of the enzymes of the citric acid cycle, the glyoxylate cycle converts two molecules of acetyl-CoA to one molecule of oxaloacetate. Both decarboxylation reactions of the citric acid cycle are bypassed.

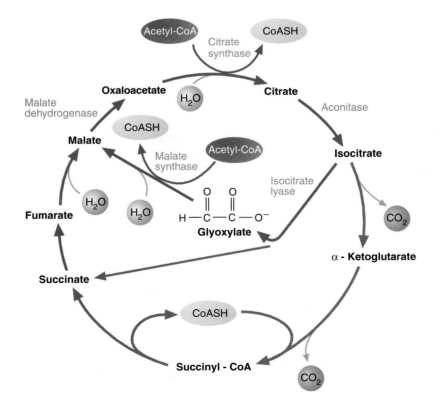

verted to malate by mitochondrial enzymes (Figure 11.9). The two-carbon molecule glyoxylate reacts with a second molecule of acetyl-CoA to form malate in a reaction catalyzed by malate synthase. The cycle is completed as malate is converted to oxaloacetate by malate dehydrogenase.

The glyoxylate cycle allows for the net synthesis of larger molecules from two-carbon molecules for the following reason. The decarboxylation reactions of the citric acid cycle, in which two molecules of CO_2 are lost, are bypassed. By using two molecules of acetyl-CoA, the glyoxylate cycle produces one molecule each of succinate and oxaloacetate. The succinate product is used in the synthesis of metabolically important molecules, the most notable of which is glucose. (In organisms, such as animals, that do not possess isocitrate lyase and malate synthase, gluconeogenesis always involves molecules with at least three carbon atoms. In these organisms there is no net synthesis of glucose from fatty acids.) The oxaloacetate product is used to sustain the glyoxylate cycle.

Citric Acid Cycle Regulation

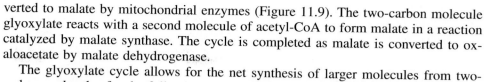

The citric acid cycle is precisely regulated so that the cell's energy and biosynthetic requirements are constantly met. Regulation is achieved primarily by the modulation of key enzymes and the availability of certain substrates. Because of its prominent role in energy production, the cycle also depends on a continuous supply of NAD^+, FAD, and ADP.

The citric acid cycle enzymes citrate synthase, isocitrate dehydrogenase, and α-ketoglutarate dehydrogenase are closely regulated because they catalyze reactions that represent important metabolic branch points (Figure. 11.10).

As previously described, citrate synthase, the first enzyme in the cycle, catalyzes the formation of citrate from acetyl-CoA and oxaloacetate. Because the concentrations

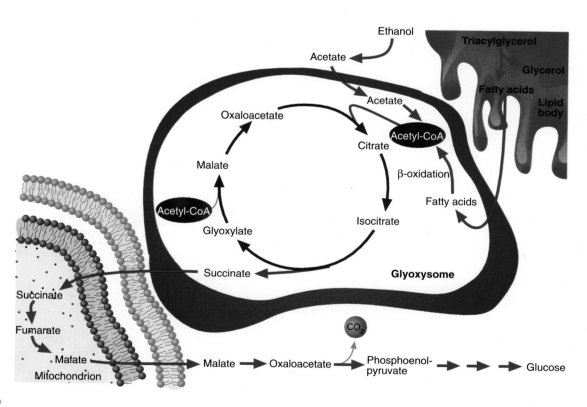

FIGURE 11.9

Role of the Glyoxylate Cycle in Gluconeogenesis.

The acetyl-CoA used in the glyoxylate cycle is derived from the breakdown of fatty acids (β-oxidation, see Chapter 10). In organisms with the appropriate enzymes, glucose can be produced from two-carbon compounds such as ethanol and acetate. In plants the reactions are localized within lipid bodies, glyoxysomes, mitochondria, and the cytoplasm.

FIGURE 11.10

Control of the Citric Acid Cycle.

The major regulatory sites of the cycle are indicated. Activators and inhibitors of regulated enzymes are shown in color.

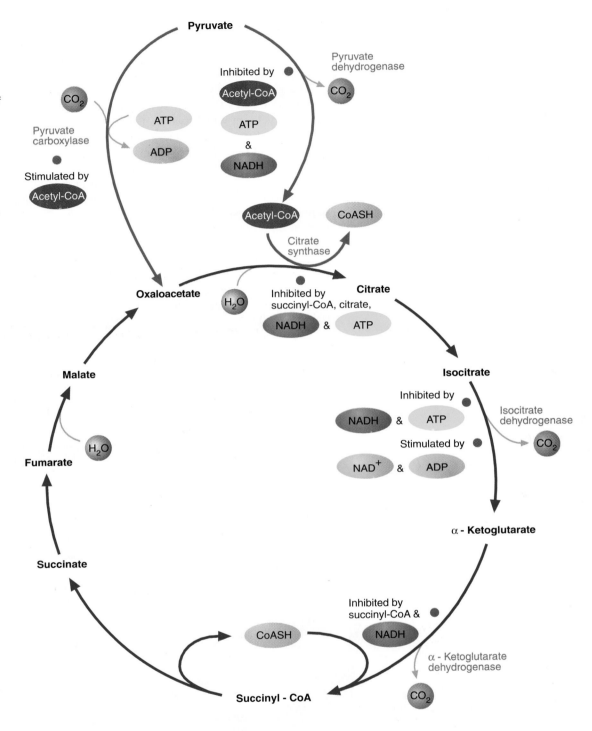

of acetyl-CoA and oxaloacetate are low in mitochondria in relation to the amount of the enzyme, any increase in substrate availability stimulates citrate synthesis. (Under these conditions the reaction is first order with respect to substrate. Therefore the rate of citrate synthesis is influenced by changes in concentrations of acetyl-CoA and oxaloacetate.) High concentrations of succinyl-CoA (a "downstream" intermediate product of the cycle) and citrate inhibit citrate synthase by acting as allosteric inhibitors. Other allosteric regulators of this reaction are NADH and ATP, whose concentrations reflect the cell's current energy status. A resting cell has high NADH/NAD$^+$ and ATP/ADP ratios. As a cell becomes metabolically active, NADH and ATP concentrations decrease. Consequently, key enzymes such as citrate synthase become more active.

Isocitrate dehydrogenase catalyzes the second closely regulated reaction in the cycle. Its activity is stimulated by relatively high concentrations of ADP and NAD$^+$ and inhibited by ATP and NADH. Isocitrate dehydrogenase is closely regulated because of its important role in citrate metabolism (Figure 11.11). As previously described, the conversion of citrate to isocitrate is reversible. An equilibrium mixture of the two molecules consists largely of citrate. (The reaction is driven forward because isocitrate is rapidly transformed to α-ketoglutarate.) Of the two molecules, only citrate can penetrate the mitochondrial inner membrane. (When a substantial number of citrate molecules move into cytoplasm, the cell's current requirement for energy is low.) As previously described, citrate transport transfers acetyl-CoA out of mitochondria. Once in the cytoplasm, citrate is cleaved by citrate lyase. The acetyl-CoA formed is used in several biosynthetic processes, such as fatty acid synthesis. Oxaloacetate is used in biosynthetic reactions, or it can be converted to malate. Malate either reenters the mitochondrion, where it is reconverted to oxaloacetate, or is converted in the cytoplasm to pyruvate by malic enzyme. Pyruvate then reenters the mitochondrion. Also a precursor of acetyl-CoA and oxaloacetate in the cytoplasm, citrate acts directly to regulate several cytoplasmic processes. Citrate is an allosteric activator of the first

FIGURE 11.11

Citrate Metabolism.

When citrate, a citric acid cycle intermediate, moves from the mitochondrial matrix into the cytoplasm, it is cleaved to form acetyl-CoA and oxaloacetate by citrate lyase. The citrate lyase reaction is driven by ATP hydrolysis. Most of the oxaloacetate is reduced to malate by malate dehydrogenase. Malate may then be oxidized to pyruvate and CO_2 by malic enzyme. The NADPH produced in this reaction is used in cytoplasmic biosynthetic processes, such as fatty acid synthesis. Pyruvate enters the mitochondria, where it may be converted to oxaloacetate or acetyl-CoA. Malate may also reenter the mitochondria, where it is reoxidized to form oxaloacetate.

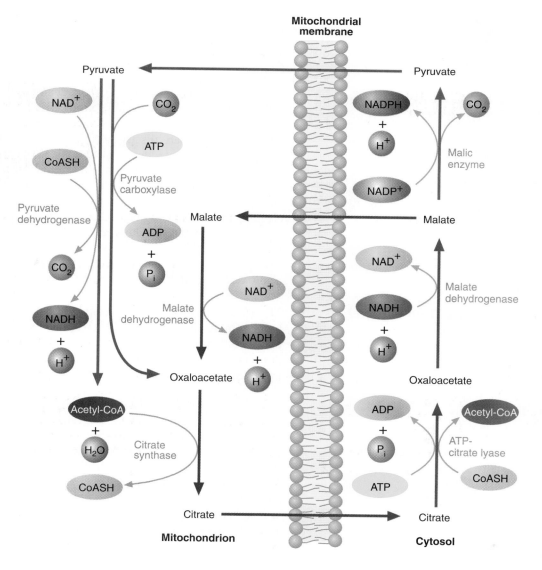

reaction of fatty acid synthesis. In addition, citrate metabolism provides some of the NADPH used in fatty acid synthesis. Finally, because citrate is an inhibitor of PFK-1, it inhibits glycolysis.

The activity of α-ketoglutarate dehydrogenase is strictly regulated because of the important role of α-ketoglutarate in several metabolic processes (e.g., amino acid metabolism). When a cell's energy stores are low, α-ketoglutarate dehydrogenase is activated and α-ketoglutarate is retained within the cycle at the expense of biosynthetic processes. As the cell's supply of NADH rises, the enzyme is inhibited, and α-ketoglutarate molecules become available for biosynthetic reactions.

Two enzymes outside the citric acid cycle profoundly affect its regulation. The relative activities of pyruvate dehydrogenase and pyruvate carboxylase determine the degree to which pyruvate is used to generate energy and biosynthetic precursors. For example, if a cell is using a cycle intermediate such as α-ketoglutarate in biosynthesis, the concentration of oxaloacetate falls and acetyl-CoA accumulates. Because acetyl-CoA is an activator of pyruvate carboxylase (and an inhibitor of pyruvate dehydrogenase), more oxaloacetate is produced from pyruvate, thus replenishing the cycle.

KEY *Concepts*

The citric acid cycle is closely regulated, thus ensuring that the cell's energy and biosynthetic needs are met. Allosteric effectors and substrate availability primarily regulate the enzymes citrate synthase, isocitrate dehydrogenase, α-ketoglutarate dehydrogenase, pyruvate dehydrogenase, and pyruvate carboxylase.

QUESTION 11.2

Fluoroacetate, $F\!-\!CH_2\!-\!COO^-$, is a toxic substance found in certain plants that grow in Australia and South Africa. Animals that ingest these plants die. Research indicates, however, that fluoroacetate is not poisonous by itself. Once it is consumed, fluoroacetate is converted into a toxic metabolite, fluorocitrate. In affected cells, citrate accumulates. Can you suggest how fluoroacetate is converted to fluorocitrate? Why are animals killed while the plants are left unaffected by fluoroacetate?

11.2 ELECTRON TRANSPORT

The mitochondrial electron transport chain (ETC), also referred to as the electron transport system, is a series of electron carriers that transfer the electrons derived from reduced coenzymes to oxygen. (There are other electron transport systems within cells. Several examples are discussed in Chapters 12 and 19.) During this transfer, a decrease in oxidation-reduction potential ($\Delta E^{\circ\prime}$) occurs. Recall that when NADH is the electron donor and oxygen is the electron acceptor, the change in potential is $+1.14$ V. This process, in which oxygen is used to generate energy from food molecules, is sometimes referred to as **aerobic respiration.** The energy released during electron transfer is coupled to several endergonic processes, the most prominent of which is ATP synthesis. Other processes driven by electron transport pump Ca^{2+} into the mitochondrial matrix and generate heat in brown adipose tissue (described on p. 310). Reduced coenzymes, derived from glycolysis, the citric acid cycle, and fatty acid oxidation, are the principal sources of electrons.

Electron Transport and Its Components

The components of the ETC in eukaryotes are located in the inner mitochondrial membrane (Figure 11.12). Most ETC components are organized into four complexes, each of which consists of several proteins and prosthetic groups. Each complex is briefly described below. The roles of two other molecules, coenzyme Q (ubiquinone, UQ) and cytochrome c (cyt c), are also described.

Complex I, also referred to as the NADH dehydrogenase complex, catalyzes the transfer of electrons from NADH to UQ. Composed of over two dozen different polypeptides, complex I is the largest protein component in the inner membrane. In addition to one molecule of FMN, the complex contains seven iron-sulfur centers (Figure 11.13). Iron-sulfur centers, which may consist of two or four iron atoms complexed with an equal number of sulfide ions, mediate one-electron transfer reactions. Proteins that contain iron-sulfur centers are often referred to as *nonheme iron proteins*. Although the structure and function of complex I are still poorly understood, it is believed that NADH reduces FMN to $FMNH_2$. Electrons are then transferred from $FMNH_2$ to an iron-sulfur center, one electron at a time. After transfer from one iron-sulfur center to another, the electrons are eventually donated to UQ (Figure 11.14).

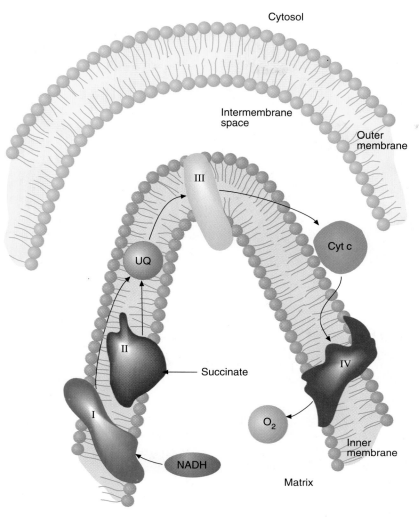

FIGURE 11.12

The Electron Transport Chain.

Complexes I and II transfer electrons from NADH and succinate to UQ. Complex III transfers electrons from UQH$_2$ to cytochrome c. Complex IV transfers electrons from cytochrome c to O$_2$. The arrows represent the flow of electrons.

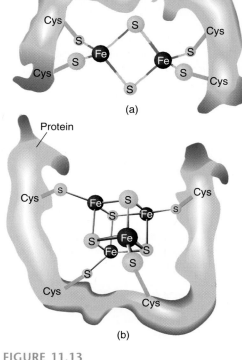

FIGURE 11.13

(a) 2 Fe, 2 S and (b) 4 Fe, 4 S Iron-Sulfur Centers.

The cysteine residues are part of a polypeptide.

KEY *Concepts*

The electron transport chain is a series of complexes consisting of electron carriers located in the inner mitochondrial membrane in eukaryotic cells.

Figure 11.15 illustrates the transfer of electrons through complex I. Electron transport is accompanied by the movement of protons from the matrix across the inner membrane and into the intermembrane space. The significance of this phenomenon for ATP synthesis is discussed below.

The succinate dehydrogenase complex (complex II) consists primarily of the citric acid cycle enzyme succinate dehydrogenase and two iron-sulfur proteins. Complex II mediates the transfer of electrons from succinate to UQ. The oxidation site for succinate is located on the larger of the iron-sulfur proteins. This molecule also contains a covalently bound FAD.

Complex III transfers electrons from reduced coenzyme Q (UQH$_2$) to cytochrome c. Because it contains two b-type cytochromes, one cytochrome c$_1$ (cyt c$_1$), and one iron-sulfur center, complex III is sometimes referred to as the cytochrome bc$_1$ complex. (The cytochromes are a series of membrane-bound electron transport proteins that contain a heme prosthetic group similar to those found in hemoglobin and myoglobin. Electrons are transferred one at a time as each oxidized iron atom (Fe^{3+}) is reversibly reduced to Fe^{2+}.) The movement of electrons from UQH$_2$ to cytochrome

FIGURE 11.14

Structure and Oxidation States of Coenzyme Q.

The length of the side chain varies among species. For example, some bacteria have six isoprene units, while mammals have ten. Some coenzyme Q molecules (ubiquinone) are functionally interchangeable.

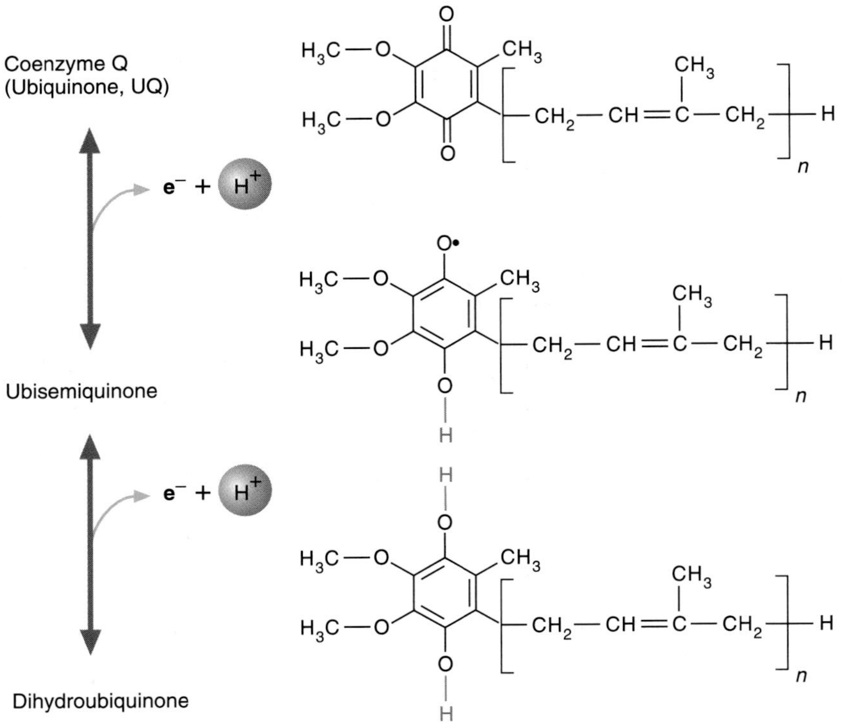

Coenzyme Q
(Ubiquinone, UQ)

e^- + H^+

Ubisemiquinone

e^- + H^+

Dihydroubiquinone

FIGURE 11.15

Transfer of Electrons Through Complex I of the Mitochondrial Electron Transport Chain.

Electron transfer begins with the reduction of FMN by NADH, a process that requires one proton from the matrix. $FMNH_2$ subsequently transfers a pair of electrons to six to eight Fe-S centers. (Because the path of the electrons is not known, only two Fe-S centers are shown.) The transfer of electrons to the first Fe-S center releases two protons into the intermembrane space. When UQ is reduced to UQH_2, two more protons are taken up from the matrix. Not shown in the figure is the transfer of one or two additional protons across the membrane. This latter process is not understood.

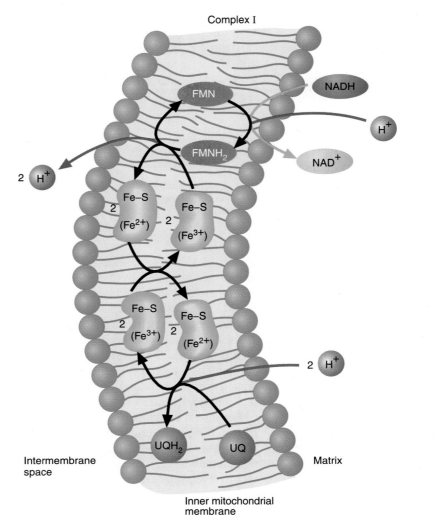

c is a complex, multistep process. Because UQ is lipid-soluble, it diffuses within the inner membrane between the electron donors in complex I or II and the electron acceptor in complex III. Electron transfer begins with the oxidation of UQH_2 by the iron-sulfur protein in complex III, which generates ubisemiquinone ($UQH \cdot$). Then the reduced iron-sulfur protein transfers an electron to cyt c_1, which transfers it to cyt c. See Figure 11.16 for additional details.

Cytochrome oxidase (complex IV) is a protein complex that catalyzes the four-electron reduction of O_2 to form H_2O. The membrane-spanning complex in mammals may contain between six and thirteen subunits, depending on species. It also contains two atoms of copper in addition to the heme iron atoms of cytochromes a and a_3. (The copper atoms alternate between the +1 and +2 oxidation states, Cu^{1+} and Cu^{2+}.) The iron atom of cyt a_3 is closely associated with a copper atom referred to as Cu_B. The other copper atom (Cu_A) is a short distance from the heme of cyt a. Cytochrome c, a protein that is loosely attached to the inner membrane on its outer surface, transfers electrons one at a time to cyt a and Cu_A. The electrons are then donated to cyt a_3 and Cu_B, which occur on the matrix (inner) side of the membrane. This electron shuttle allows four electrons and four protons to be delivered to the dioxygen molecule bound to cyt a_3-Fe^{2+}. Two water molecules are formed and leave the site.

$$O_2 + 4\,H^+ + 4\,e^- \rightarrow 2\,H_2O$$

During each sequential redox reaction in the ETC, an electron loses energy. During the oxidation of NADH there are three steps in which the change in reduction potential ($\Delta E^{\circ\prime}$) is sufficient for ATP synthesis. These steps, which occur within complexes I, III, and IV, are referred to as sites I, II, and III, respectively (Figure 11.17). Recent experimental evidence indicates that approximately 2.5 molecules of ATP are synthesized for each pair of electrons transferred between NADH and O_2 in the ETC. Approximately 1.5 molecules of ATP result from the transfer of each pair donated by the $FADH_2$ produced by succinate oxidation. The mechanism by which ATP synthesis is believed to be coupled to electron transport is described on page 309.

FIGURE 11.16

Electron Transport Through Complex III.

The black arrows represent the flow of electrons. The red arrows represent the path of UQ in its various oxidation states (the Q cycle) and of protons. UQH_2 is oxidized to UQ in two steps at an enzyme site adjacent to the intermembrane space. The first electron is transferred to the Fe-S protein. The second electron is transferred to cyt b. (Two molecules of UQH_2 undergo those reactions.) One of the two molecules of UQ produced diffuses to the site on the matrix side where it is reduced to form UQH_2. (The transfer of electrons from the two b cytochromes is inhibited by antimycin.) Once formed, the UQH_2 diffuses back to the oxidation site, where it joins the pool of UQH_2 coming from complexes I and II. The electrons transferred from UQH_2 to the Fe-S center then reduce cyt c. Four protons are released on the cytoplasmic side of the inner membrane.

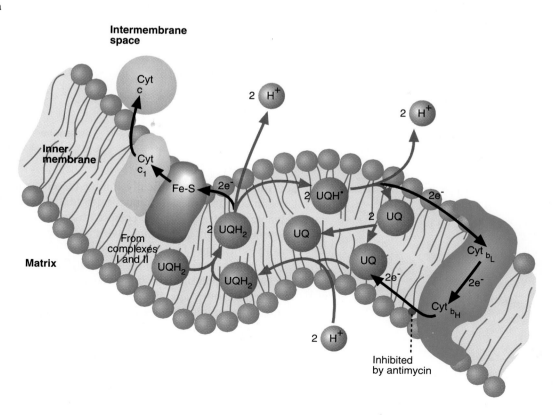

FIGURE 11.17

The Energy Relationships in the Mitochondrial Electron Transport Chain.

Relatively large decreases in free energy occur in three steps. During each of these steps (i.e., at sites I, II, and III), sufficient energy is released to account for the synthesis of ATP.

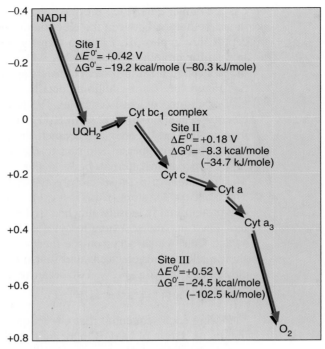

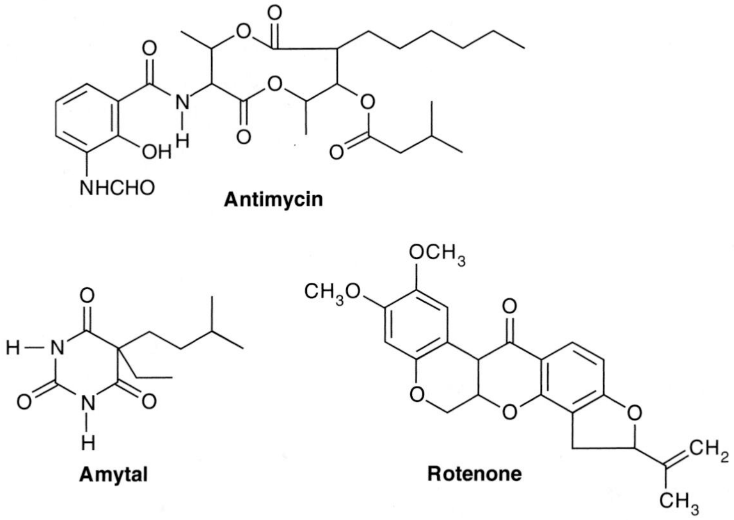

Site I
$\Delta E^{0'} = +0.42$ V
$\Delta G^{0'} = -19.2$ kcal/mole (-80.3 kJ/mole)

Cyt bc_1 complex

Site II
$\Delta E^{0'} = +0.18$ V
$\Delta G^{0'} = -8.3$ kcal/mole
(-34.7 kJ/mole)

Cyt c
Cyt a
Cyt a_3

Site III
$\Delta E^{0'} = +0.52$ V
$\Delta G^{0'} = -24.5$ kcal/mole
(-102.5 kJ/mole)

O_2

Electron flow ⟶

FIGURE 11.18

Several Inhibitors of the Mitochondrial Electron Transport Chain.

Antimycin blocks the transfer of electrons from the b cytochromes. Amytal and rotenone block NADH dehydrogenase.

Antimycin

Amytal

Rotenone

Electron Transport Inhibitors

Several molecules specifically inhibit the electron transport process (Figure 11.18). Used in conjunction with reduction potential measurements, inhibitors have been invaluable in determining the correct order of ETC components. In such experiments, electron transport is measured with an oxygen electrode. (Oxygen consumption is a sensitive measure of electron transport.) When electron transport is inhibited oxygen consumption is reduced or eliminated. ETC components on the oxygen side of the blockage become more oxidized because they are no longer able to accept electrons. Those on the other side become more reduced because they are unable to donate electrons. For example, antimycin A inhibits cyt b. If this inhibitor is added to a suspension of mitochondria, NAD^+, the flavins, and cyt b molecules become more reduced. The cytochromes c_1, c, and a become more oxidized. Other prominent examples of ETC inhibitors include rotenone and amytal, which inhibit NADH dehydrogenase (Complex I). Carbon monoxide (CO), azide (N_3^-), and cyanide (CN^-) inhibit cytochrome oxidase.

11.3 OXIDATIVE PHOSPHORYLATION

Oxidative phosphorylation, the process whereby the energy generated by the ETC is conserved by the phosphorylation of ADP to yield ATP, has been studied since the 1940s. The only type of phosphorylation reaction with which biochemists were then familiar was substrate-level phosphorylation (e.g., the oxidation of glyceraldehyde-3-phosphate, Section 8.3). It is not surprising, therefore, that the first mechanism proposed to explain the coupling of electron transport and ATP synthesis involved a high-energy intermediate. According to the *chemical coupling hypothesis,* a high-energy intermediate generated by the electron transport process is used in a second reaction to drive the formation of ATP from ADP and P_i. Despite research efforts that spanned several decades, the proposed intermediate has never been identified. In addition, the hypothesis could not account for several experimental findings. For example, it failed to explain how certain molecules, called uncouplers (described below), prevent ATP synthesis during electron transport. More important, however, the chemical coupling hypothesis did not explain why the entire inner mitochondrial membrane must be intact during ATP synthesis.

The Chemiosmotic Theory

In 1961, Peter Mitchell, a British biochemist, proposed a mechanism by which the free energy generated during electron transport drives ATP synthesis. Now widely accepted, Mitchell's model, referred to as the **chemiosmotic coupling theory** (Figure 11.19), has the following principal features:

1. As electrons pass through the ETC, protons are transported from the matrix and released into the intermembrane space. As a result, an electrical potential Ψ and a proton gradient ΔpH arise across the inner membrane. The electrochemical proton gradient is sometimes referred to as the **protonmotive force** Δp.
2. Protons, which are present in the intermembrane space in great excess, can pass through the inner membrane and back into the matrix down their concentration gradient only through special channels. (The inner membrane itself is impermeable to ions such as protons.) As protons pass through a channel, each of which contains an ATP synthase activity, ATP synthesis occurs.

Mitchell suggested that the free energy release associated with electron transport and ATP synthesis is coupled by the protonmotive force created by the ETC. (The term

FIGURE 11.19

The Chemiosmotic Theory.

The flow of electrons through the electron transport complexes is coupled to the flow of protons across the inner membrane from the matrix to the intermembrane space. This process raises the matrix pH. In addition, the matrix becomes negatively charged with respect to the intermembrane space. Protons flow passively into the matrix through a channel in the ATP synthase. This flow is coupled to ATP synthesis.

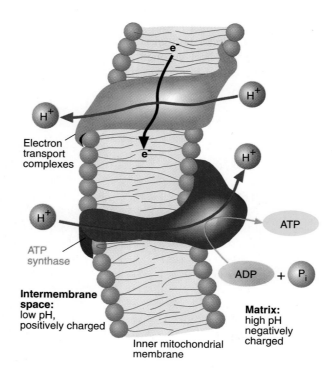

chemiosmotic emphasizes that chemical reactions can be coupled to osmotic gradients.)

Evidence Supporting the Chemiosmotic Theory   Evidence that supports the chemiosmotic theory includes the following:

1. Actively respiring mitochondria expel protons. The pH of a weakly buffered suspension of mitochondria measured by an electrode drops when O_2 is added. The typical pH gradient across the inner membrane is approximately 0.05 pH unit.

2. ATP synthesis stops when the inner membrane is disrupted. For example, although electron transport continues, ATP synthesis stops in mitochondria placed in a hypertonic solution. Mitochondrial swelling results in proton leakage across the inner membrane.

3. A variety of molecules that inhibit ATP synthesis are now known to specifically dissipate the proton gradient (Figure 11.20). According to the chemiosmotic theory, a disrupted proton gradient dissipates the energy derived from food molecules as heat. **Uncouplers** collapse the proton gradient by equalizing the proton concentration on both sides of membranes. (As they diffuse across the membrane, uncouplers pick up protons from one side and release them on the other.) **Ionophores** are hydrophobic molecules that dissipate osmotic gradients by inserting themselves into a membrane and forming a channel. For example, gramicidin is an antibiotic that forms a channel in membranes that allows the passage of H^+, K^+, and Na^+.

QUESTION 11.3	Dinitrophenol (DNP) is an uncoupler used as a diet aid in the 1920s, until several deaths occurred. Suggest why DNP consumption results in weight loss. The deaths caused by DNP were a result of liver failure. Explain. (Hint: Liver cells contain an extraordinarily large number of mitochondria.)

Brown Fat   Newborn babies, hibernating animals, and cold-adapted animals all require more heat production than is normally generated by metabolism. Recall from Chapter 4 that heat is a consequence of cellular reactions that create an ordered state. (The loss of heat, the most disorganized form of energy, increases the entropy of the surroundings.) Warm-blooded animals use this heat to maintain their body temperature. Under normal conditions, electron transport and ATP synthesis are tightly coupled, so heat production is kept to a minimum. In a specialized form of adipose tissue called *brown fat,* most of the energy produced by the mitochondrial ETC is not

FIGURE 11.20

Uncouplers.

(a) Dinitrophenol, (b) gramicidin A. Dinitrophenol diffuses across the membrane, picking up protons on one side and releasing them on the other. Gramicidin A forms an end-to-end dimer, which makes a pore that inserts itself into the membrane. From J.D. Rawn, *Biochemistry,* © 1989, p. 1039, fig. 31.18. Reprinted by permission of Prentice-Hall, Inc., Upper Saddle River, NJ.

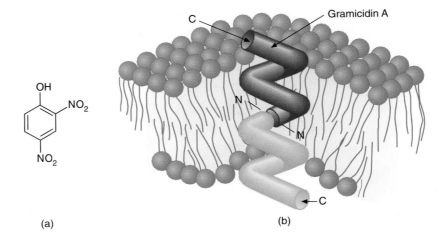

(a)   (b)

used to produce ATP. Instead, it is dissipated as heat. (This tissue has a brown appearance because of the large number of mitochondria it contains.) The mitochondrial inner membrane of brown fat cells is now known to contain a unique 32-kD protein called *uncoupling protein* or *thermogenin*. When the uncoupling protein is active, it dissipates the proton gradient by translocating protons. Uncoupling protein is activated when it binds to fatty acids. As the proton gradient decreases, large amounts of energy are dissipated as heat.

The entire process of heat generation from brown fat, called *nonshivering thermogenesis,* is regulated by norepinephrine. (In shivering thermogenesis, heat is produced by nonvoluntary muscle contraction.) Norepinephrine is a neurotransmitter released from specialized neurons that terminate in brown adipose tissue. Using cAMP as a second messenger, norepinephrine initiates a cascade mechanism that ultimately hydrolyzes fat molecules. As stated previously, the fatty acids produced activate the uncoupler protein. Fatty acid oxidation continues until the norepinephrine signal is terminated or the cell's fat reserves are depleted.

ATP Synthase

Early electron microscopic studies of mitochondria revealed the presence of numerous lollipop-shaped structures studding the inner membrane on its inner surface (Figure 11.21). Experiments begun in the early 1960s, using submitochondrial particles, revealed that each lollipop is a proton translocating ATP synthase. (*Submitochondrial particles,* or SMP, are generated by sonicating mitochondria. Figure 11.21 illustrates that SMP are inside out, that is, the lollipops project to the outside.) Further work showed that the ATP synthase (Figure 11.22) consists of two major components. The F_1 unit, the active ATPase, possesses five different subunits present in the ratio α_3, β_3, γ, δ, and ϵ. There are three nucleotide-binding catalytic sites on F_1. The F_0 unit, a transmembrane channel for protons, has three subunits present in the ratio a_1, b_2, and c_{12}.

It is currently believed that the translocation of three protons through the ATP synthase is required for each ATP molecule synthesized. (The transfer of an additional proton is required for the transport of ATP and OH^- out of the matrix in exchange for ADP and P_i.) It now appears, however, that the protonmotive force is not directly responsible for ATP synthesis but that ADP and P_i are spontaneously converted to

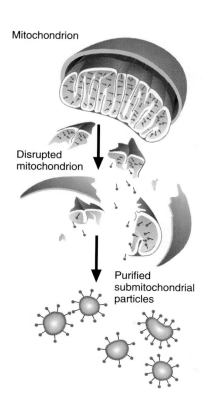

FIGURE 11.21

Submitochondrial Particles.

After disruption of mitochondria, the inner membrane fragments reseal to form submitochondrial particles.

Mitochondrion

Disrupted mitochondrion

Purified submitochondrial particles

FIGURE 11.22

ATP Synthase from *Escherichia Coli.*

The rotor consists of ϵ, γ, and c_{12} subunits. The stator consists of a_1, b_2, δ, α_3, and β_3 subunits. The molecular components of ATP synthase are well conserved among bacteria, plants, and animals.

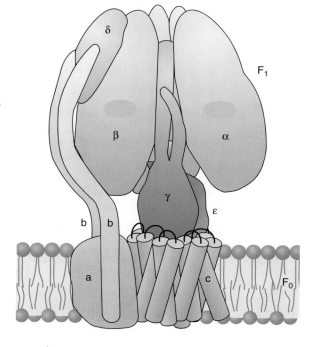

Control of Oxidative Phosphorylation

ATP within the catalytic sites. Recent evidence indicates that the energy-requiring step is the release of ATP from the enzyme. ATP synthase behaves as a rotating molecular machine in which the flow of protons releases ATP from the catalytic sites. The rotor (or revolving) component of the machine consists of subunits ϵ, γ, and c_{12}, whereas subunits a, b_2, δ, α_3, and β_3 comprise the stator (or stationary) component. As protons flow through F_0, a conformational change is translated into a relative rotation between the rotor and stator components, i.e., the globular portion of ATP synthase spins like a top. The rotation releases ATP from the enzyme's catalytic sites.

Control of oxidative phosphorylation allows a cell to produce only the precise amount of ATP required to sustain its activities. Recall that under normal circumstances, electron transport and ATP synthesis are tightly coupled. The value of the *P/O ratio* (the number of moles of P_i consumed for each oxygen atom reduced to H$_2$O) reflects the degree of coupling observed between electron transport and ATP synthesis. The measured maximum ratio for the oxidation of NADH is 2.5. The maximum P/O ratio for FADH$_2$ is 1.5.

The control of oxidative phosphorylation by ADP concentration is illustrated by the fact that mitochondria can oxidize NADH and FADH$_2$ only when ADP and P_i are present in sufficient concentration. If isolated mitochondria are provided with an oxidizable substrate (e.g., succinate), all of the ADP is eventually converted to ATP. At this point, oxygen consumption becomes greatly depressed. Oxygen consumption increases dramatically when ADP is supplied. The control of aerobic respiration by ADP is referred to as **respiratory control.** The formation of ATP appears to be directly related to the ATP mass action ratio ([ATP]/[ADP][P_i]). In other words, ATP synthase is inhibited by a high concentration of its product (ATP) and activated when ADP and P_i concentrations are high. The relative amounts of ATP and ADP within mitochondria are controlled largely by the two transport proteins in the inner membrane: the ATP-ADP translocator and the phosphate carrier.

The *ATP-ADP translocator* is a dimeric protein responsible for the 1:1 exchange of intramitochondrial ATP for the ADP produced in the cytoplasm. As previously described, there is a potential difference (inside negative) across the inner mitochondrial membrane. Because ATP molecules have one more negative charge than ADP molecules, the outer transport of ATP and inward transport of ADP is favored. The transport of HPO$_4^-$ into the matrix in exchange for OH$^-$ is mediated by a phosphate carrier (Figure 11.23). Of the four protons pumped out of the matrix for every ATP synthesized, three move back into the matrix and drive the phosphorylation of ADP. The translocation of the fourth proton drives the uptake of ADP and P_i.

FIGURE 11.23

The Phosphate/Hydroxide Exchange Protein.

The phosphate carrier transports H$_2$PO$_4^-$ across the inner mitochondrial membrane in exchange for an OH$^-$ moving in the opposite direction.

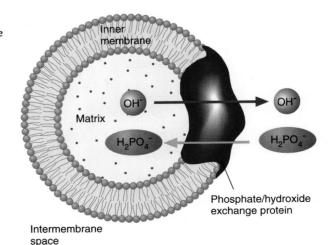

*The Complete Oxidation
of Glucose*

A summary of the sources of ATP produced from one molecule of glucose is provided in Table 11.2. ATP production from fatty acids, the other important energy source, is discussed on p. 316. Several aspects of this summary require further discussion. Recall that two molecules of NADH are produced during glycolysis. When oxygen is available, the oxidation of this NADH by the ETC is preferable (in terms of energy production) to lactate formation. The inner mitochondrial membrane, however, is impermeable to NADH. Animal cells have evolved several shuttle mechanisms to transfer electrons from cytoplasmic NADH to the mitochondrial ETC. The most prominent examples are the glycerol phosphate shuttle and the malate-aspartate shuttle.

In the **glycerol phosphate shuttle** (Figure 11.24a), DHAP, a glycolytic intermediate, is reduced by NADH to form glycerol-3-phosphate. This reaction is followed by the oxidation of glycerol-3-phosphate by mitochondrial glycerol-3-phosphate dehydrogenase. (The mitochondrial enzyme uses FAD as an electron acceptor.) Because glycerol-3-phosphate interacts with the mitochondrial enzyme on the outer face of the inner membrane, the substrate does not actually enter the matrix. The $FADH_2$ produced in this reaction is then oxidized by the ETC. FAD as an electron acceptor produces only 1.5 ATP per molecule of cytoplasmic NADH.

Although the **malate-aspartate shuttle** (Figure 11.24b) is a more complicated mechanism than the glycerol-phosphate shuttle, it is more energy efficient. The shuttle begins with the reduction of cytoplasmic oxaloacetate to malate by NADH. After its transport into the mitochondrial matrix, malate is reoxidized. The NADH produced is then oxidized by the ETC. For the shuttle to continue, oxaloacetate must be returned to the cytoplasm. Because the inner membrane is impermeable to oxaloacetate, it is converted to aspartate in a transamination reaction (Chapter 14) involving glutamate. Both products of this reaction, aspartate and α-ketoglutarate, are then

TABLE 11.2

Summary of ATP Synthesis from the Oxidation of One Molecule of Glucose

	NADH	FADH$_2$	ATP
Glycolysis (cytoplasm)			
Glucose → glucose-6-phosphate			−1
Fructose-6-phosphate → fructose-1,6-bisphosphate			−1
Glyceraldehyde-3-phosphate → Glycerate-1,3-bisphosphate	+2		
Glycerate-1,3-bisphosphate → glycerate-3-phosphate			+2
Phosphoenolpyruvate → pyruvate			+2
Mitochondrial Reactions			
Pyruvate → acetyl-CoA	+2		
Citric acid cycle			
Oxidation of isocitrate, α-ketoglutarate, and malate	+6		
Oxidation of succinate		+2	
GDP → GTP			+1.5*
Oxidative Phosphorylation			
2 Glycolytic NADH			+4.5† (3)‡
2 NADH (pyruvate to acetyl-CoA)			+5
6 NADH (citric acid cycle)			+15
2 FADH$_2$ (citric acid cycle)			+3
			31 (29.5)

*This number reflects the price of transport into the cytoplasm.
†Assumes the aspartate-malate shuttle.
‡Assumes the glycerol phosphate shuttle.

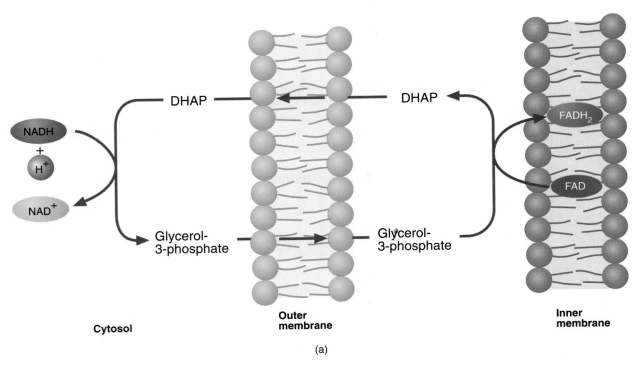

Cytosol

Outer
membrane

Inner
membrane

(a)

FIGURE 11.24

Shuttle Mechanisms that Transfer Electrons from Cytoplasmic NADH to the Respiratory Chain.

(a) The glycerol-3-phosphate shuttle. DHAP is reduced to form glycerol-3-phosphate. Glycerol-3-phosphate is reoxidized by mitochondrial glycerol-3-phosphate dehydrogenase. (b) The aspartate-malate shuttle. Oxaloacetate is reduced by NADH to form malate. Malate is transported into the mitochondrial matrix, where it is reoxidized to form oxaloacetate and NADH. Because oxaloacetate cannot penetrate the inner membrane, it is converted to aspartate in a transamination involving glutamate. Two inner membrane carriers are required for this shuttle mechanism: the glutamate-aspartate transport protein and the malate-α-ketoglutarate transport protein.

transported into the cytoplasm. Oxaloacetate and glutamate are regenerated in a transamination reaction. In addition to the cytoplasmic and mitochondrial enzymes, the shuttle requires two inner membrane carriers: the glutamate-aspartate transport protein and the malate-α-ketoglutarate transport protein. The glutamate-aspartate transporter requires moving a proton into the matrix. Therefore, the net ATP synthesis using this mechanism is somewhat reduced. Instead of generating 2.5 molecules of ATP for each NADH molecule, the yield is approximately 2.25 molecules of ATP.

One final issue concerned with ATP synthesis from glucose remains. Recall that two molecules of ATP are produced in the citric acid cycle (from GTP). The price for their transport into the cytoplasm, where they will be used, is the uptake of two protons into the matrix. Therefore the total amount of ATP produced from a molecule of glucose is reduced by about 0.5 molecule of ATP.

Depending on the shuttle used, the total number of molecules of ATP produced per molecule of glucose varies (approximately) from 29.5 to 31. Assuming that the average amount of ATP produced is 30 molecules, the net reaction for the complete oxidation of glucose is as follows:

$$C_6H_{12}O_6 + 6\ O_2 + 30\ ADP + 30\ P_i \rightarrow 6\ CO_2 + 6\ H_2O + 30\ ATP$$

The number of ATP molecules generated during the complete oxidation of glucose is in sharp contrast to the 2 molecules of ATP formed by glycolysis. Quite obviously, organisms that use oxygen to oxidize glucose have a substantial advantage.

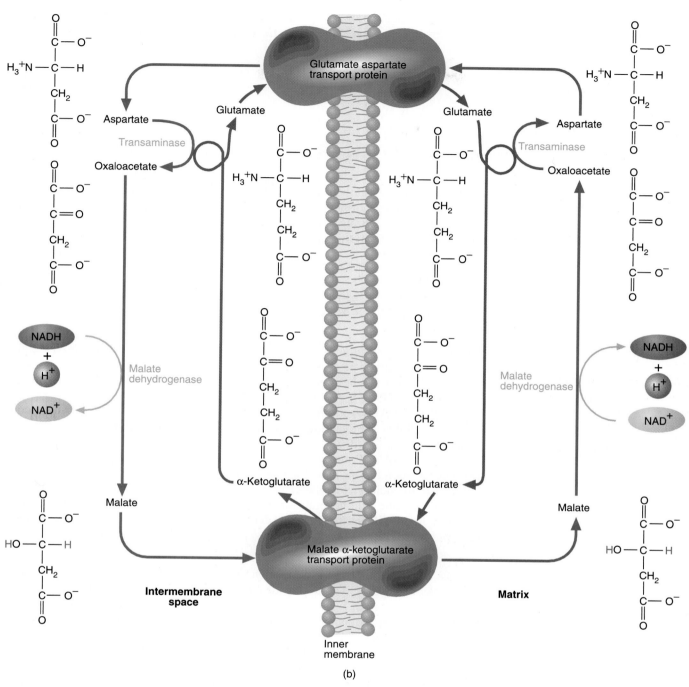

QUESTION

11.4

Traditionally, the oxidation of each NADH and FADH$_2$ by the ETC was believed to result in the synthesis of three molecules of ATP and two molecules of ATP, respectively. As noted, recent measurements, which have considered such factors as proton leakage across the inner membrane, have reduced these values somewhat. Using the earlier values, calculate the number of ATP molecules generated by the aerobic oxidation of a glucose molecule. First assume that the glycerol phosphate shuttle is operating, then assume that the malate-aspartate shuttle is transferring reducing equivalents into the mitochondrion.

The Complete Oxidation of a Fatty Acid

The aerobic oxidation of a fatty acid also generates a large number of ATP molecules. The following equation summarizes the oxidation of palmitoyl-CoA.

$$CH_3(CH_2)_{14}\overset{O}{\underset{\|}{C}}-S-CoA \; + \; 7 \; FAD \; + \; 7 \; NAD^+ \; + \; 7 \; CoASH \; + \; 7 \; H_2O \longrightarrow$$

$$8 \; CH_3\overset{O}{\underset{\|}{C}}-S-CoA \; + \; 7 \; FADH_2 \; + \; 7 \; NADH^+ \; + \; 7 \; H^+$$

As previously described, the oxidation of each $FADH_2$ during electron transport and oxidative phosphorylation yields approximately 1.5 molecules of ATP. Similarly, the oxidation of each NADH yields approximately 2.5 molecules of ATP. Therefore the yield of ATP from the oxidation of palmitoyl-CoA to form CO_2 and H_2O is calculated as follows:

7 $FADH_2$	× 1.5 ATP/$FADH_2$	=	10.5 ATP
7 NADH	× 2.5 ATP/NADH	=	17.5 ATP
8 Acetyl-CoA	× 10 ATP/acetyl-CoA	=	80 ATP
			108 ATP

Recall that the formation of palmitoyl-CoA from palmitic acid uses two ATP equivalents. (The synthesis of ATP from AMP involves two sequential phosphorylation reactions.) The net synthesis of ATP per molecule of palmitoyl-CoA is therefore 106 molecules of ATP.

The yield of ATP from the oxidations of palmitic acid and glucose can be compared. Recall that the total number of ATP molecules produced per glucose molecule is approximately 31. If glucose and palmitic acid molecules are compared in terms of the number of ATP molecules produced per carbon atom, palmitic acid is a superior energy source. The ratio for glucose is 31/6 or 5.2 ATP molecules per carbon atom. Palmitic acid yields 106/16 or 6.6 ATP molecules per carbon atom. The oxidation of palmitic acid generates more energy than that of glucose because palmitic acid is a more reduced molecule. (Glucose with its six oxygen atoms is a partially oxidized molecule.)

11.4 OXIDATIVE STRESS

Oxygen is not essential to generate energy; many living organisms (all of which are anaerobic bacteria) use glycolysis to provide all their energy needs. Why then is oxygen used by the vast majority of living organisms to extract energy from organic molecules? In addition to the large amounts of energy generated by using this gaseous substance, it is readily available and easily distributed within organisms. (Oxygen diffuses rapidly into and out of cells because it is soluble in the nonpolar lipid core of membranes.)

Reactive Oxygen Species

As previously mentioned, the advantages of using oxygen are linked to a dangerous property it possesses. A number of effects including cancer, myocardial infarct (Special Interest Box 11.3), inflammation, and aging, have been associated with the formation of derivatives of oxygen referred to as **reactive oxygen species (ROS).** Examples of ROS include the superoxide radical, hydrogen peroxide, the hydroxyl radical, and singlet oxygen. Because ROS are so reactive, they can seriously damage living cells if formed in significant amounts. ROS formation is usually kept to a minimum by cellular antioxidant defense mechanisms. Under certain conditions, however, referred to collectively as **oxidative stress,** some damage may occur. This damage results primarily from enzyme inactivation, polysaccharide depolymerization, DNA breakage, and membrane destruction. Examples of circumstances that may cause serious oxidative damage include the overconsumption of certain drugs or exposure to certain environmental contaminants. Several types of cells are now known to produce large quantities of ROS normally. Scavenger cells such as macrophages and neutrophils continuously search the body for microorganisms and damaged cells. In an oxygen-consuming process referred to as the *respiratory burst* (described below), ROS are generated and used to kill and dismantle these cells.

Formation of ROS

The properties of oxygen are directly related to its molecular structure. The diatomic oxygen molecule is a diradical. A **radical** is an atom or group of atoms that contain one or more unpaired electrons. Dioxygen is a diradical because it possesses two unpaired electrons. For this and other reasons, when it reacts, dioxygen can accept only one electron at a time.

Recall that during mitochondrial electron transport, H_2O is formed as a consequence of the sequential transfer of four electrons to O_2. During this process, several ROS are formed. Cytochrome oxidase (as well as other oxygen-activating proteins) traps these reactive intermediates within its active site until all four electrons have been transferred to oxygen. However, ROS may occasionally leak out of active sites before they can be completely reduced. Under normal circumstances, cellular antioxidant defense mechanisms minimize any damage. ROS are also formed during nonenzymatic processes. For example, exposure to UV light and ionizing radiation cause ROS formation.

The first ROS formed during the reduction of oxygen is the superoxide radical (O_2^-). O_2^- acts as a nucleophile and (under specific circumstances) as either an oxidizing agent or a reducing agent. Because of its solubility properties, O_2^- causes considerable damage to the phospholipid components of membranes. When it is generated in an aqueous environment, O_2^- reacts with itself to produce O_2 and hydrogen peroxide (H_2O_2).

$$2 H^+ + 2 O_2^- \rightarrow O_2 + H_2O_2$$

H_2O_2 is not a radical because it does not have any unpaired electrons. The limited reactivity of H_2O_2 allows it to cross membranes and become widely dispersed. The subsequent reaction of H_2O_2 with Fe^{2+} (or other transition metals) results in the production of the hydroxyl radical $(\cdot OH)$, a highly reactive species.

$$Fe^{2+} + H_2O_2 \rightarrow Fe^{3+} + \cdot OH + OH^-$$

The hydroxyl radical, which is highly reactive, diffuses only a short distance before it reacts with whatever biomolecule it collides with. Radicals such as the hydroxyl radical are especially dangerous because they can initiate an autocatalytic radical chain reaction (Figure 11.25). Singlet oxygen $(^1O_2)$, a highly excited state created when dioxygen absorbs sufficient energy to shift an unpaired electron to a higher orbital, can be formed from superoxide.

$$2 O_2^- + 2 H^+ \rightarrow H_2O_2 + {}^1O_2$$

or from peroxides.

$$2 ROOH \rightarrow 2 ROH + {}^1O_2$$

Because it is a powerful oxidant, singlet oxygen is even more reactive than the hydroxyl radical, although it is not a radical.

As mentioned, ROS are generated during several other cellular activities besides the reduction of O_2 to form H_2O. These include the metabolism of xenobiotics (Chapter 19) and the respiratory burst (Figure 11.26) in white blood cells. In addition, electrons often leak from the electron transport pathways in mitochondria and the endoplasmic reticulum (refer to Chapter 19) to form superoxide by combining with O_2.

To protect themselves from oxidative stress, several antioxidant defense mechanisms have developed in living organisms. These mechanisms employ several metalloenzymes and antioxidant molecules.

KEY *Concepts*

ROS form because oxygen is reduced by accepting one electron at a time. ROS formation is a normal by-product of metabolism and the result of conditions such as exposure to radiation.

EXERCISE, DIET, AND NUTRIENT METABOLISM

What do all calorie-restricted diet plans have in common? For most people (as many as 95% of those who attempt weight loss by dieting) these plans fail to achieve long-term weight loss. So, despite several decades of effort and billions of dollars spent, Americans are heavier today than ever before.

In the past, the concept of dieting was based on the premise that excess weight results primarily from eating too much. Exercise has always been acknowledged to promote weight loss, but the major emphasis of dieting strategies was on calorie restriction. Physiological and biochemical research now appears to indicate that this view is not only flawed, but actually counterproductive, for severe calorie restriction actually promotes obesity. (For example, most people who lose substantial weight not only gain the lost weight back but they also gain additional pounds.)

It is becoming increasingly clear why calorie-restricted diets are so ineffective. When individuals diet (i.e., they consume fewer calories than they need to sustain their bodily functions and physical activity), their brains apparently interpret the lower food intake as the beginning of a famine. To prevent starvation the body adapts by reducing its *basal metabolic rate (BMR)*. (The BMR is a measure of the energy required to support essential life-sustaining metabolic processes. BMR is determined when a person is at rest after an all-night fast.) This change, which is still poorly understood, is mediated in part by reducing the serum levels of thyroid hormone. (Thyroid hormone increases the metabolic rate by increasing O_2 consumption and the rate of synthesis of numerous enzymes.) This ability to adapt to low food intake had survival value because humans were often confronted by famines.

Much of the early weight loss that results from calorie restriction is actually a loss of muscle. The breakdown of muscle protein serves two purposes. First, because liver glycogen is quickly depleted during a diet, hepatic gluconeogenesis becomes increasingly important in providing blood glucose. (Water loss also accounts for some of the weight loss in the early phase of dieting. Recall that glycogen binds a relatively large amount of water.) Amino acids become an important substrate for this process. Second, because of its high consumption of ATP during muscle contraction and its relatively large mass, skeletal muscle is a metabolically demanding tissue. If its size can be reduced, the body can conserve vital energy reserves to outlive the famine. After a weight-loss goal is achieved, the individual's BMR is lower than before, and the primary user of energy, skeletal muscle, is significantly depleted. (Dieters who have lost significant amounts of weight often appear gaunt.) Weight, in the form of fat in adipose tissue, is regained because several centers in the brain (e.g., the hypothalamus) promote behavior (e.g., intake of high-calorie foods and reduction in physical activity) that reestablishes a normal amount of energy reserves.

(Normal body weight as determined by the brain is often referred to as the *set point*. It is now believed that an individual's set point is determined by genetic factors as well as several physiological factors related to the level of physical activity.)

Current research indicates that a specific form of exercise (i.e., endurance training) combined with a balanced diet is required for achieving and maintaining a healthful body weight. (A balanced diet is composed of a wide variety of nutritious foods. A diet that consists largely of vegetables, fruits, whole grains, and lean meat as well as fish is low in fat and simple sugars and high in fiber.) A significant amount of the insight gained in obesity research has been achieved by observing the effect of exercise on specific aspects of metabolism, such as the levels of certain enzymes and membrane proteins in muscle cells. Endurance training promotes a greater reliance by the body on using fat as a fuel. Endurance training has also been proven to lower the risk of developing cardiovascular disease and reduce the incidence of degenerative complications of diabetes.

Energy Sources for Exercising Muscle

There are two metabolic strategies for supplying the energy requirements of contracting muscles: anaerobic metabolism and aerobic metabolism. The term *anaerobic exercise* describes short intense bursts of physical activity. Glycolysis generates the ATP required to drive muscle contraction. At such high levels of metabolism, muscle uses oxygen faster than it can be supplied by the cardiovascular system. As soon as the oxygen supply is depleted, lactate levels begin to increase. Muscle contraction can continue only until muscle lactic acid levels rise to a level that causes muscle fatigue. (When lactic acid levels reach a certain threshold value, muscle cells become unresponsive to neural stimulation.) When muscle contraction occurs at a pace for which adequate amounts of oxygen can be supplied, the exercise is said to be *aerobic*. When exercise is aerobic, the activity can be sustained long enough to induce substantial changes in metabolism. Among these are an increased BMR and resistance to fatigue.

When adequate oxygen is available, muscle uses two primary fuels: glucose and fatty acids. During rest or low-intensity exercise, energy is provided by small amounts of both glucose and fatty acids. A small amount of glucose is required in part because it is converted to oxaloacetate, needed in the citric acid cycle. As physical activity intensifies, the release of fatty acids from adipose tissue fat stores increases. (This process is mediated by the sympathetic nervous system, which stimulates the adrenal gland to secrete epinephrine and norepinephrine. Recall that these hormones activate hormone-sensitive lipase.) Eventually, muscle glycogen is depleted, and glucose derived from hepatic glycogen is used. If intense physical activity lasts long enough, the body's glycogen reserves become almost totally depleted. Then muscle cells must depend on fatty acid oxidation to supply the energy for muscle contraction. However, because glucose is primarily responsible for maintaining a large supply of citric acid cycle intermediates, the capacity of the muscle to generate energy then drops to 60% of its previous level. In endurance training, exercise is performed at a submaximal level, that is, when glucose is available to supplement fatty acid oxidation. Following a description of the practical features of endurance training, the metabolic changes induced by endurance training are discussed.

Endurance Training

Endurance training incorporates the following features:

1. The exercise is performed regularly, for example, every day or every other day.

2. A large mass of muscle must be used to perform the exercise. Exercises such as jogging, walking, bicycling, and cross-country skiing are excellent choices because a large amount of muscle is used. The more muscle is trained, the larger the amount of fat that can be oxidized.

3. The exercise must be nonstop. To induce a "training effect" (i.e., a change in muscle metabolism), the heart rate must be elevated above a specific threshold for a minimum time. This threshold value is determined largely by a person's age. For example, a 20-year-old's minimum training heart rate is approximately 130 beats per minute, whereas 104 is sufficient for an individual over 60. These values are 65% of the age-adjusted maximum heart rate. (Maximum heart rate is calculated by subtracting a person's age from 220.)

4. The exercise must be performed for a minimum time. The time required depends on the type of exercise chosen. For example, between 30 and 45 minutes (depending on speed) are required for a walking program, while 20 minutes of jogging are sufficient to induce a training effect.

5. The exercise must be aerobic. In other words, it must not be so intense that breathing is difficult. An aerobic workout can also be achieved by monitoring breathing or heart rate. Breathing should not be labored. The exercise should not cause the heart rate to exceed 80% of the age-adjusted maximum heart rate. For a 20-year-old the maximum heart rate during exercise should not exceed 160 (i.e., 80% of 200, the maximum heart rate).

The Training Effect

The endurance training of skeletal muscle is now known to induce a number of metabolic adaptations that enhance the body's capacity to burn fat. (The precise mechanisms by which these changes are effected is not understood.) These adaptations also serve to conserve muscle glycogen, an important consideration if both fatty acids and glucose are to be degraded simultaneously. The most obvious adaptations occur in skeletal muscle.

1. The number of mitochondria per muscle fiber increases.

This increase is partially responsible for making muscle cells more responsive to metabolic regulators. For example, muscle glycogen is conserved in part because of the inhibition of PFK-1 by various metabolic regulators produced during fatty acid oxidation, such as citrate. The larger number of mitochondria increase the efficiency with which these regulators can inhibit glycolytic enzymes.

2. Fatty acids are more efficiently degraded because of increased synthesis of molecules that facilitate fatty acid transport and oxidation. The concentration of citric acid cycle and β-oxidation enzymes increases as well as the components of the ETC. In addition, the capacity of the muscle cell to remove fatty acids from blood and to transport them into mitochondria increases. For example, increases in the synthesis of fatty acid transporter proteins and fatty acid–binding proteins, as well as carnitine and carnitine acyltransferase, have been observed.

3. The vascularization of muscle tissue increases. As the number of capillaries increases, the transit time for blood through muscle increases (i.e., there is increased resistance to flow because of a greater surface area for the exchange of nutrients). The exchange of nutrients and waste products between the blood and muscle fibers is more efficient.

Endurance training also has a noteworthy effect on fat metabolism. In sedentary individuals, especially those who have undergone calorie-restricted diets, adipocytes are resistant to the stimulation by the sympathetic nervous system that accompanies physical exercise. For reasons that are still poorly understood, endurance training results in an increased sensitivity of adipocytes to these hormones. This is an important point, because the amount of fatty acids transported into muscle cells and used to drive muscle contraction is directly related to their concentration in blood. Trained muscle can generate more energy by oxidizing ketone bodies.

Getting Started

After many years of research, it has become clear that the success of an exercise program in achieving and maintaining weight loss is directly related to motivation. After years spent in a sedentary lifestyle people find that initiating and sustaining an exercise program is difficult. Weeks, sometimes months, of consistent effort are required to effect the metabolic changes described above. Despite the mild discomfort of this period, endurance training in combination with a healthful diet is the only proven method of permanent weight control. In addition, the other benefits of this activity (e.g., improvements in cardiovascular fitness) are well worth such an investment of time and effort. For some individuals with a history of cardiovascular disease or joint damage caused by years of obesity, any exercise program should be monitored by a physician. The risks of exercise for these individuals should be weighed against the serious risks of a sedentary lifestyle.

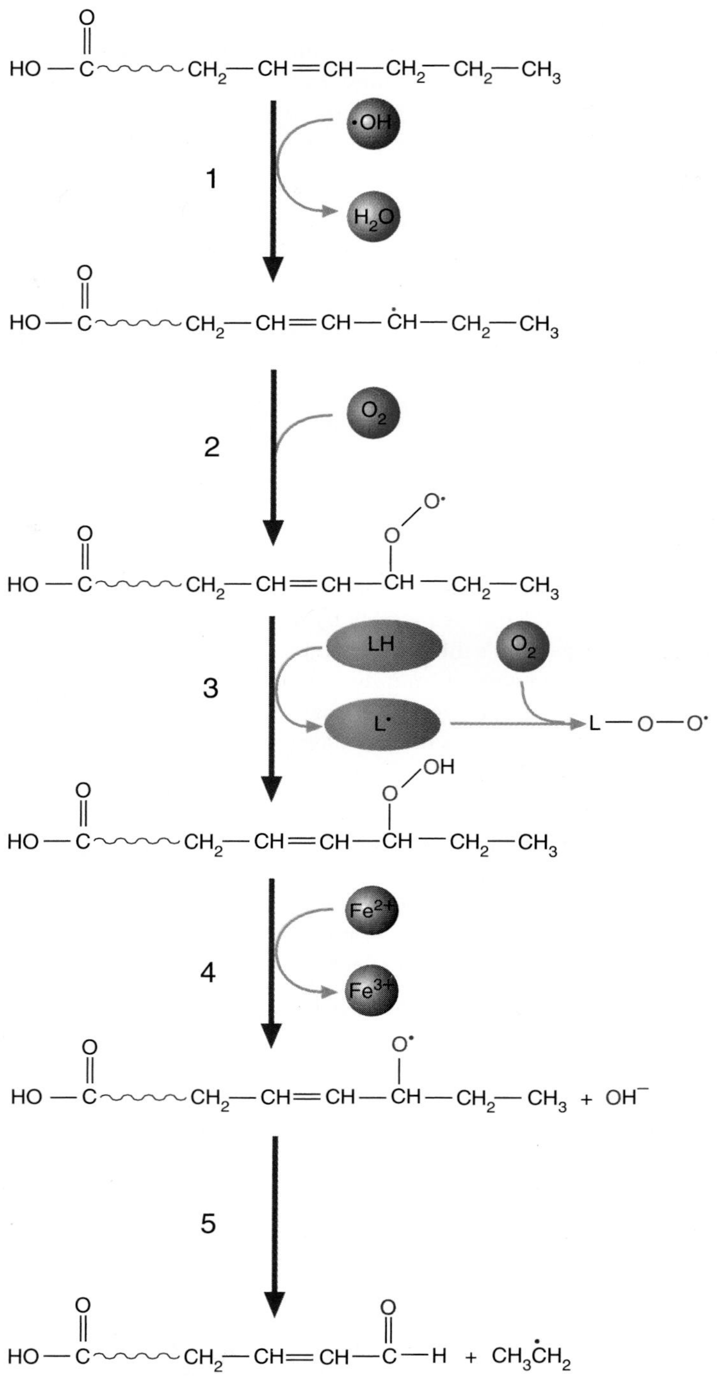

FIGURE 11.25

Radical Chain Reaction.

Step 1: Lipid peroxidation reactions begin after a hydrogen atom is extracted from an unsaturated fatty acid (LH → L·). Step 2: The lipid radical (L·) then reacts with O_2 to form a peroxyl radical (L· + O_2 → L—O—O·). Step 3: The radical chain reaction begins when the peroxyl radical extracts a hydrogen atom from another fatty acid molecule (L—O—O· + L'H → L—O—O—H + L'·). Step 4: The presence of a transition metal such as Fe^{2+} initiates further radical formation (L—O—O—H + Fe^{2+} → LO· + HO^- + Fe^{3+}). Step 5: One of the most serious consequences of lipid peroxidation is the formation of aldehydes, which involves a radical cleavage reaction. The chain reaction continues as the free radical product then reacts with a nearby molecule.

Antioxidant Enzyme Systems

The major enzymatic defenses against oxidative stress are provided by superoxide dismutase, catalase, and glutathione peroxidase. The wide distribution of these enzymatic activities underscores the ever present problem of oxidative damage.

The superoxide dismutases (SOD) are a class of enzyme that catalyze the formation of H_2O_2 and O_2 from the superoxide radical.

$$2\ O_2^- + 2\ H^+ \rightarrow H_2O_2 + O_2$$

There are two major forms of SOD. In humans the Cu-Zn isozyme occurs in cytoplasm. A Mn-containing isozyme is found in the mitochondrial matrix.

FIGURE 11.26

The Respiratory Burst.

The respiratory burst provides a dramatic example of the destructiveness of ROS. Within seconds after a phagocytic cell binds to a bacterium (or other foreign structure), its oxygen consumption increases nearly 100-fold. During endocytosis the bacterium is incorporated into a large vesicle called a phagosome. Phagosomes then fuse with lysosomes to form phagolysosomes. Two destructive processes then ensue: the respiratory burst and digestion by lysosomal enzymes. The respiratory burst is initiated when NADPH oxidase converts O_2 to O_2^- Two molecules of O_2^- combine in a reaction catalyzed by SOD (superoxide dismutase) to form H_2O_2. H_2O_2 is next converted to several types of bacteriocidal (bacteria-killing) molecules by myeloperoxidase (MPO), an enzyme found in abundance in phagocytes. For example, MPO catalyzes the oxygenation of halide ions (e.g., Cl^-) to form hypohalides. Hypochloride (the active ingredient in household bleach) is extremely bacteriocidal. In the presence of Fe^{2+}, O_2^- and H_2O_2 react to form $\cdot OH$ and $^1O_2$ (singlet oxygen), both of which are extremely reactive. Some researchers consider the formation of $\cdot OH$ and $^1O_2$ to be controversial. After the bacterial cell disintegrates, lysosomal enzymes digest the fragments that remain.

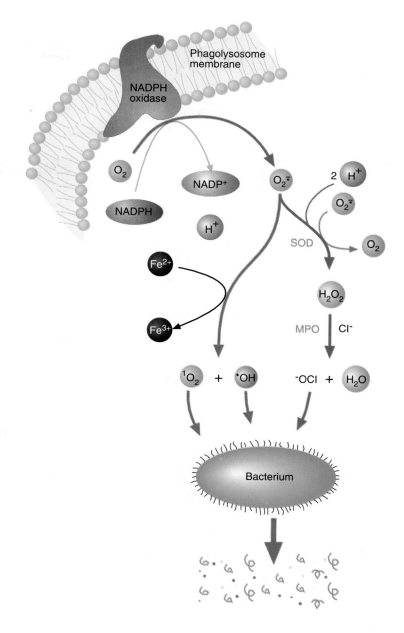

Relatively large amounts of H_2O_2 are generated within peroxisomes as a by-product in various oxidative reactions.

$$RH_2 + O_2 \rightarrow R + H_2O_2$$

Catalase is a heme-containing enzyme that uses H_2O_2 to oxidize other substrates.

$$RH_2 + H_2O_2 \rightarrow R + 2\ H_2O$$

When H_2O_2 is present in excessive amounts, catalase converts it to water.

$$2\ H_2O_2 \rightarrow 2\ H_2O + O_2$$

Glutathione peroxidase, a selenium-containing enzyme, is a key component in an enzymatic system most responsible for controlling cellular peroxide levels. Recall that this enzyme catalyzes the reduction of a variety of substances by the reducing agent

FIGURE 11.27

The Glutathione Redox Cycle.

Glutathione peroxidase utilizes GSH to reduce the peroxides generated by cellular aerobic metabolism. GSH is regenerated from its oxidized form, GSSG, by glutathione reductase. NADPH, the reducing agent in this reaction, is supplied by the pentose phosphate pathway as well as several other reactions.

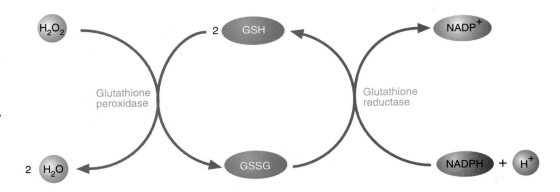

GSH (Section 5.2). In addition to reducing H_2O_2 to form water, glutathione peroxidase transforms organic peroxides into alcohols.

$$2 \text{ GSH} + R\text{—}O\text{—}O\text{—}H \rightarrow G\text{—}S\text{—}S\text{—}G + R\text{—}OH + H_2O$$

Several ancillary enzymes support glutathione peroxidase function (Figure 11.27). GSH is regenerated from GSSG by glutathione reductase.

$$G\text{—}S\text{—}S\text{—}G + \text{NADPH} + H^+ \rightarrow 2 \text{ GSH} + \text{NADP}^+$$

The NADPH required in the reaction is provided primarily by several reactions of the pentose phosphate pathway (described below). Recall that NADPH is also produced by the reactions catalyzed by isocitrate dehydrogenase and malic enzyme.

QUESTION

11.5   Selenium is generally considered a toxic element. (It is the active component of loco weed.) However, there is growing evidence that selenium is also an essential trace element. Because glutathione peroxidase activity is essential to protect red blood cells against oxidative stress, a selenium deficiency can damage red blood cells. Although sulfur is in the same chemical family as selenium, it cannot be substituted. Can you explain why? (Hint: Selenium is more easily oxidized than sulfur.) Would sulfur or selenium be a better scavenger for oxygen when this gas is present in trace amounts?

11.6   Ionizing radiation is believed to damage living tissue by producing hydroxyl radicals. Drugs that protect organisms from radiation damage usually have —SH groups. Unfortunately, they must be taken *before* radiation exposure. How do such drugs protect against radiation damage? Can you suggest any type of nonsulfhydryl group-containing molecule that would protect against hydroxyl radical–induced damage?

The Pentose Phosphate Pathway

A substantial amount of the NADPH required for reductive processes such as lipid biosynthesis and antioxidant mechanisms is supplied by the **pentose phosphate pathway.** (For this reason this pathway is most active in the cells in which relatively large amounts of lipids are synthesized, for example, adipose tissue, adrenal cortex, mammary glands, and the liver. It is also quite active in cells that are at high risk for oxidative damage, such as red blood cells.) The pentose phosphate pathway, sometimes referred to as the hexose monophosphate shunt, is an alternative pathway for glucose oxidation in which no ATP is generated. In addition to NADPH production, it is also responsible for the synthesis of ribose-5-phosphate, a component of nucleotides and nucleic acids. In plants, the pentose phosphate pathway is involved in the synthesis of glucose during the dark reactions of photosynthesis (Chapter 12).

The pentose phosphate pathway occurs in two phases: oxidative and nonoxidative. In the oxidative phase of the pathway, the conversion of glucose-6-phosphate to ribulose-5-phosphate is accompanied by the production of two molecules of NADPH.

In the nonoxidative phase, ribulose-5-phosphate is primarily converted to either ribose-5-phosphate or xylulose-5-phosphate.

The oxidative phase of the pentose phosphate pathway consists of three reactions (Figure 11.28a). In the first reaction, glucose-6-phosphate dehydrogenase (G-6-PD) catalyzes the oxidation of glucose-6-phosphate. 6-Phosphogluconeolactone and NADPH are products in this reaction. 6-Phosphogluconeolactone is then hydrolyzed to produce 6-phosphogluconate. A second molecule of NADPH is produced during the oxidative decarboxylation of 6-phosphogluconate, a reaction that yields ribulose-5-phosphate.

The nonoxidative phase of the pathway begins with the conversion of ribulose-5-phosphate to ribose-5-phosphate by ribulose-5-phosphate isomerase or to xylulose-5-phosphate by ribulose-5-phosphate epimerase. During the remaining reactions of the pathway (Figure 11.28b), transketolase and transaldolase catalyze the interconversions of trioses, pentoses, and hexoses. Transketolase is a TPP-requiring enzyme that transfers two-carbon units from a ketose to an aldose. Two reactions are catalyzed by transketolase. In the first reaction, the enzyme transfers, a two-carbon unit from xylulose-5-phosphate to ribose-5-phosphate, yielding glyceraldehyde-3-phosphate and sedoheptulose-7-phosphate. In the second transketolase-catalyzed reaction, a two-carbon unit from another xylulose-5-phosphate molecule is transferred to erythrose-4-phosphate to form a second molecule of glyceraldehyde-3-phosphate and fructose-6-phosphate. (Erythrose-4-phosphate is used by some organisms to synthesize aromatic amino acids.) Transaldolase transfers three-carbon units from a ketose to an aldose. In the reaction catalyzed by transaldolase, a three-carbon unit is transferred from sedoheptulose-7-phosphate to glyceraldehyde-3-phosphate. The products formed are fructose-6-phosphate and erythrose-4-phosphate. The result of the nonoxidative phase of the pathway is the synthesis of ribose-5-phosphate and the glycolytic intermediates glyceraldehyde-3-phosphate and fructose-6-phosphate.

The pentose phosphate pathway is regulated to meet the cell's moment-by-moment requirements for NADPH and ribose-5-phosphate. The oxidative phase is very active in cells such as red blood cells or hepatocytes in which demand for NADPH is high. (See Special Interest Box 11.2.) In contrast, the oxidative phase is virtually absent in cells (e.g., muscle cells) that synthesize little or no lipid. G-6-PD catalyzes the rate-limiting step in the pentose phosphate pathway. Its activity is inhibited by NADPH and stimulated by GSSG and glucose-6-phosphate. In addition, diets high in carbohydrate increase the synthesis of both G-6-PD and phosphogluconate dehydrogenase.

KEY *Concepts*

The major enzymatic defenses against oxidative stress are superoxide dismutase, catalase, and glutathione peroxidase. The pentose phosphate pathway produces NADPH and ribose-5-phosphate and several glycolytic intermediates.

QUESTION 11.7

In some regions where malaria is endemic (e.g., the Middle East), fava beans are a staple food. Fava beans are now known to contain two β-glycosides called vicine and convicine.

Vicine   Convicine

It is believed that the aglycone components of these substances, called divicine and isouramil, respectively, can oxidize GSH. Individuals who eat fresh fava beans are protected to a certain extent from malaria. A condition known as *favism* results when some G-6-PD-deficient individuals develop a severe hemolytic anemia after eating the beans. Explain why.

FIGURE 11.28

The Pentose Phosphate Pathway.

(a) The oxidative phase; (b) the nonoxidative phase.

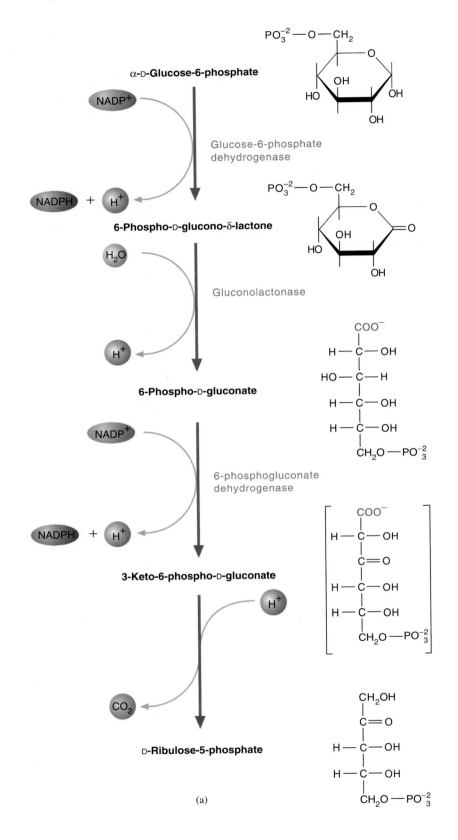

(a)

FIGURE 11.28
(Continued)

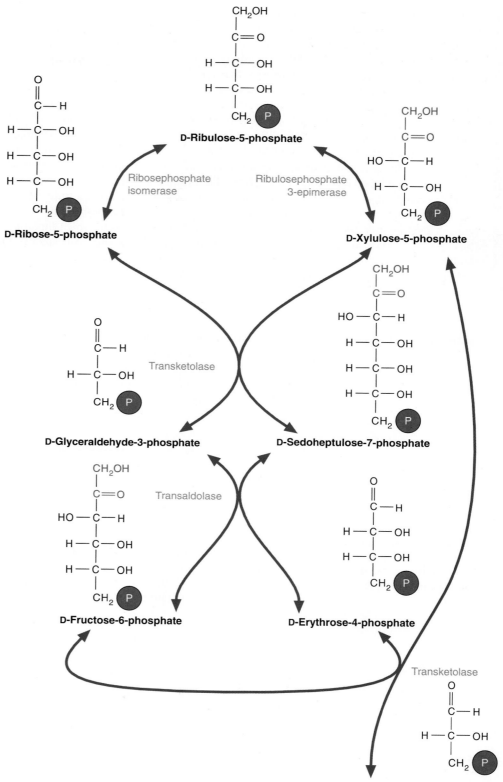

(b)

GLUCOSE-6-PHOSPHATE DEHYDROGENASE DEFICIENCY

Because of their role in oxygen transport, red blood cells are especially prone to oxidative stress. Recall, for example, that H_2O_2 oxidizes the iron in hemoglobin to form methemoglobin. In addition, lipid peroxidation makes cell membranes fragile. When such cells pass through narrow blood vessels, they may rupture. If the oxidative stress is severe, hemolytic anemia results. Fortunately, red blood cells are usually well-protected. A vital component of this protection is a highly active pentose phosphate pathway. The major role of NADPH produced by the oxidative reactions of the pathway is to reduce GSSG to GSH.

In *glucose-6-phosphate dehydrogenase deficiency,* the red blood cell's capacity to protect itself from oxidative stress is reduced. Affected individuals produce low amounts of NADPH because they possess a defective enzyme. (There are over 100 known variants of the G-6-PD gene. The capacity to produce NADPH therefore varies widely among G-6-PD-deficient individuals.) A lower than normal NADPH concentration impairs the individual's capacity to generate GSH.

Under normal conditions, many carriers of the mutant gene are asymptomatic. However, any additional oxidative stress can have serious consequences. For example, administration of the antimalarial drug primaquine to G-6-PD-deficient individuals results in hemolytic anemia. The drug kills the malarial parasite *Plasmodium* because it stimulates the production of hydrogen peroxide. The resultant lowering of NADPH and GSH levels in red blood cells (that already have lower than normal amounts) causes the lysis of the red cell membrane. G-6-PD-deficient individuals are resistant to malaria. (*Plasmodium* is especially sensitive to oxidizing conditions, so any circumstance that lowers cellular antioxidant capacity inhibits the infection.) It is not surprising, therefore, that G-6-PD deficiency is one of the most common human genetic anomalies. In geographic areas in which malaria is endemic (e.g., the Mediterranean and Middle East regions), individuals who possess the defective enzyme are less likely to die of the disease than those who do not. (Recall that sickle-cell trait also confers resistance to malaria.)

Antioxidant Molecules

Living organisms use antioxidant molecules to protect themselves from radicals. Some prominent antioxidants include GSH, α-tocopherol (vitamin E), ascorbic acid (vitamin C), and β-carotene (Figure 11.29).

α-*Tocopherol,* a potent radical scavenger, belongs to a class of compounds referred to as *phenolic antioxidants.* Phenols are effective antioxidants because the radical products of these molecules are resonance-stabilized and thus relatively stable.

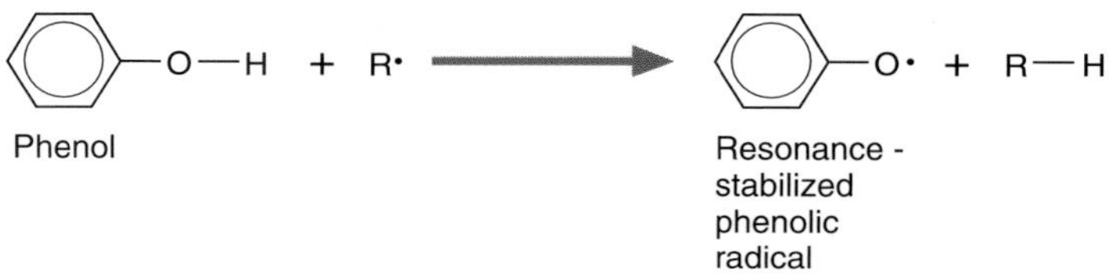

Phenol   Resonance-stabilized phenolic radical

KEY *Concepts*

Antioxidant molecules protect cell components from oxidative damage. Prominent antioxidants include GSH and the dietary components α-tocopherol, β-carotene, and ascorbic acid.

Because vitamin E (found in vegetable and seed oils, whole grains, and green, leafy vegetables) is lipid-soluble, it plays an important role in protecting membranes from lipid peroxyl radicals.

β-*Carotene,* found in yellow-orange and dark green fruits and vegetables such as carrots, sweet potatoes, broccoli, and apricots, is a member of a class of plant pig-

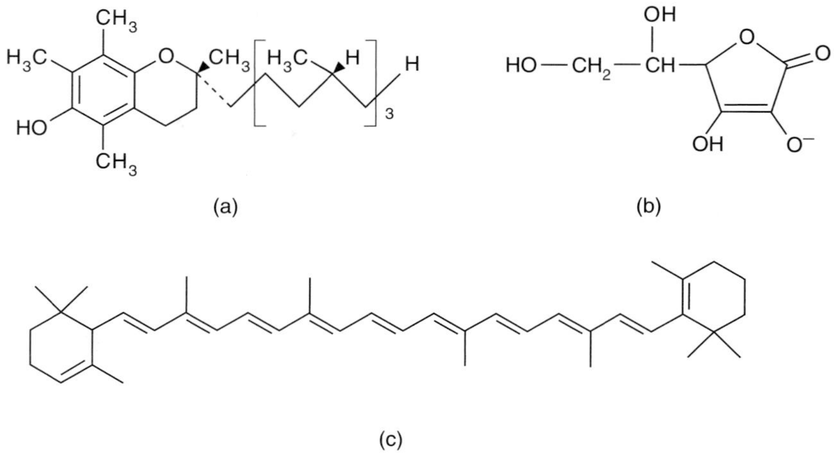

(a)   (b)

FIGURE 11.29

Selected Antioxidant Molecules.

(a) α-Tocopherol (vitamin E),
(b) ascorbate (vitamin C), and
(c) β-carotene.

(c)

ment molecules referred to as the *carotenoids*. In plant tissue the carotenoids absorb some of the light energy used to drive photosynthesis and protect against the ROS that form at high light intensities. In animals, β-carotene is a precursor of retinol (vitamin A) and an important antioxidant in membranes. (Retinol is a precursor of retinal, the light-absorbing pigment found in the rod cells of the retina.)

Ascorbic acid has been shown to be an efficient antioxidant. Present largely as ascorbate, this water-soluble molecule scavenges a variety of ROS within the aqueous compartments of cells and in extracellular fluids. Ascorbate is reversibly oxidized as follows.

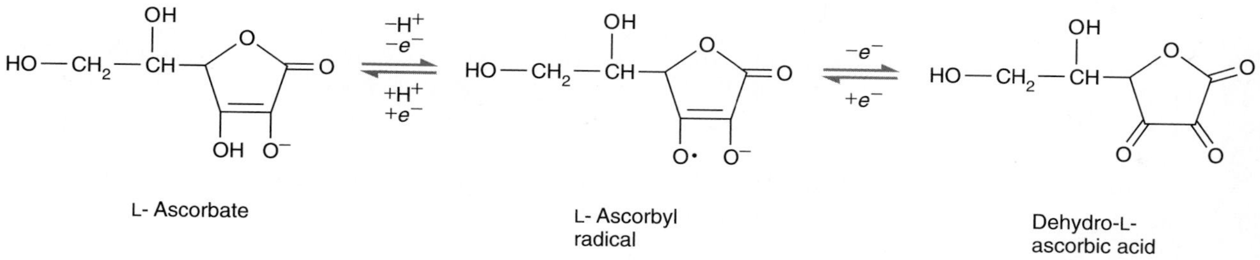

L- Ascorbate   L- Ascorbyl radical   Dehydro-L-ascorbic acid

Ascorbate protects membranes through two mechanisms. First, ascorbate reacts with peroxyl radicals formed in the cytoplasm before they can reach the membrane, thereby preventing lipid peroxidation. Second, ascorbate enhances the antioxidant activity of vitamin E by regenerating reduced α-tocopherol from the α-tocopheroxyl radical (Figure 11.30). Ascorbate is then regenerated by reacting with GSH.

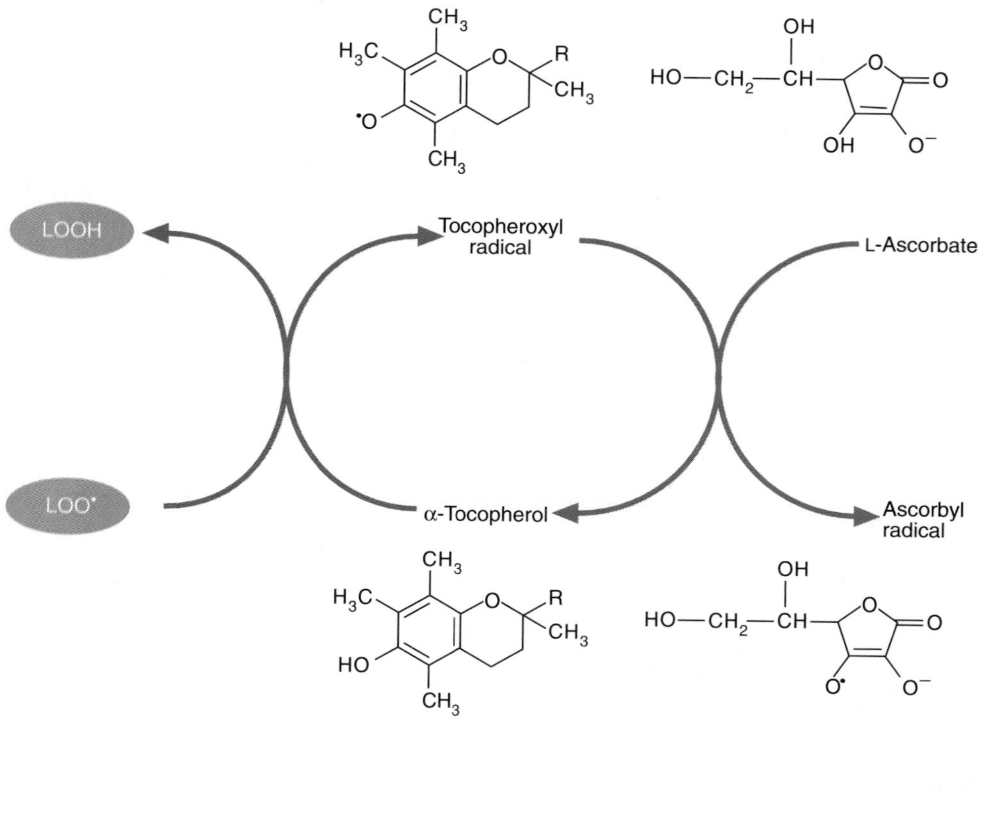

FIGURE 11.30

Regeneration of α-Tocopherol by L-Ascorbate.

L-Ascorbate, a water-soluble molecule, protects membranes from oxidative damage by regenerating α-tocopherol from α-tocopheroxyl radical. The ascorbyl radical formed in this process is reconverted to L-ascorbate during a reaction with GSH.

ISCHEMIA AND REPERFUSION

The tissue damage that occurs during a myocardial infarct or a cerebrovascular accident (stroke) is caused by *ischemia*, a condition in which there is inadequate blood flow. Heart attacks and strokes are usually caused by atherosclerosis accompanied by blood clot formation in an essential artery. (In atherosclerosis, soft masses of fatty material called *plaques* are formed in the linings of blood vessels.) Unlike skeletal muscle, which is fairly resistant to ischemic injury, heart and brain are extremely sensitive to hypoxic (low-oxygen) conditions. For example, significant brain damage occurs if the brain is deprived of oxygen for more than a few minutes. The stimulation of anaerobic glycolysis, which leads to lactate production and acidosis, is an early response of cells to ischemia. Because energy production by glycolysis is inefficient, ATP levels begin to fall. Without sufficient ATP, cells cannot maintain appropriate intracellular ion concentrations. As osmotic pressure increases, affected cells swell and leak their contents. (Recall that the leakage of specific enzymes into blood is used to diagnose heart and liver damage.) Similarly, lysosomal enzymes begin to leak from lysosomes. Because lysosomal enzymes are active only at low pH values, their presence in an increasingly acidic cytoplasm eventually results in the hydrolysis of cell components. If an oxygen supply is not soon reestablished, affected cells are irreversibly damaged.

The reoxygenation of an ischemic tissue, a process referred to as *reperfusion*, can be a life-saving therapy. For example, using streptokinase to remove artery-occluding clots in heart attack patients, accompanied by administration of oxygen, has been a remarkably successful life-saving strategy. However, depending on the duration of the hypoxic episode, the reintroduction of oxygen to ischemic tissue may also result in further damage. Recent research with antioxidants reveals that ROS are largely responsible for reperfusion-initiated cell damage. The exact mechanism by which reperfusion causes ROS production is still unclear. However, there are several likely possibilities. For example, the leakage of electrons from swollen mitochondria may result in ROS formation. In addition, the release of iron from cell components such as myoglobin, which can result from ROS-inflicted damage, can cause additional production of $\cdot$OH. Finally, the acidosis caused by lactate accumulation in compromised heart muscle cells unload abnormally high amounts of oxygen from hemoglobin. This latter condition greatly facilitates further ROS synthesis. Currently, ROS-quenching enzymes and antioxidant molecules are being investigated for clinical use.

QUESTION 11.8 BHT (butylated hydroxytoluene) is widely used as a food preservative. Quercitin is an example of a large group of potent antioxidants found in fruits and vegetables called the flavonoids.

BHT

Quercitin

What structural characteristic of these molecules is responsible for their antioxidant properties?

KEY WORDS

aerobic respiration, *304*

anaplerotic, *298*

chemiosmotic coupling theory, *309*

glycerol phosphate shuttle, *313*

glyoxylate cycle, *300*

ionophore, *310*

malate-aspartate shuttle, *313*

oxidative phosphorylation, *309*

oxidative stress, *316*

pentose phosphate pathway, *322*

protonmotive force, *309*

radical, *317*

reactive oxygen species (ROS), *316*

respiratory control, *312*

uncoupler, *310*

SUMMARY

Aerobic organisms have an enormous advantage over anaerobic organisms, that is, a greater capacity to obtain energy from organic food molecules. To use oxygen to generate energy requires the following biochemical pathways: the citric acid cycle, the electron transport pathway, and oxidative phosphorylation.

The citric acid cycle is a series of biochemical reactions that eventually completely oxidize organic substrates, such as glucose and fatty acids, to form CO_2, H_2O, and the reduced coenzymes NADH and $FADH_2$. Pyruvate, the product of the glycolytic pathway, is converted to acetyl-CoA, the citric acid cycle substrate.

In addition to its role in energy generation, the citric acid cycle also plays an important role in several biosynthetic processes, such as gluconeogenesis, amino acid synthesis, and porphyrin synthesis.

The glyoxylate cycle, found in plants and some fungi, algae, protozoans, and bacteria, is a modified version of the citric acid cycle in which two-carbon molecules, such as acetate, are converted to precursors of glucose.

The NADH and $FADH_2$ molecules produced in glycolysis, the β-oxidation pathway, and the citric acid cycle generate usable energy in the electron transport pathway. The pathway consists of a series of redox carriers that receive electrons from NADH and $FADH_2$. At the end of the pathway the electrons, along with protons, are donated to oxygen to form H_2O.

During the oxidation of NADH, there are three steps in which the energy loss is sufficient to account for ATP synthesis. These steps, which occur within complexes I, III, and IV, are referred to as sites I, II, and III, respectively.

Oxidative phosphorylation is the mechanism by which electron transport is coupled to the synthesis of ATP. According to the chemiosmotic theory, the creation of a proton gradient that accompanies electron transport is coupled to ATP synthesis.

The complete oxidation of a molecule of glucose results in the synthesis of 29.5 to 31 molecules of ATP, depending on whether the glycerol phosphate shuttle or the malate-aspartate shuttle transfers electrons from cytoplasmic NADH to the mitochondrial ETC.

The use of oxygen by aerobic organisms is linked to the production of ROS. ROS form because the diradical oxygen molecule accepts electrons one at a time. Examples of ROS include the superoxide radical, hydrogen peroxide, the hydroxyl radical, and singlet oxygen. The danger from the highly reactive ROS is usually kept to a minimum by cellular antioxidant defense mechanisms.

SUGGESTED READINGS

Chans, S.I., and Li, P.M., Cytochrome Oxidase: Understanding Nature's Design of a Proton Pump, *Biochem.*, 29(1):1–12, 1990.

Denton, R.M., and Halestrap, A.P., Regulation of Pyruvate Metabolism in Mammalian Tissues, *Essays Biochem.*, 15:37–77, 1979.

Gibble, G.W., Fluoroacetate Toxicity, *J. Chem. Ed.*, 50:460–462, 1973.

Hinkle, P.C., Kumar, A., Resetar, A., and Harris, D.L., Mechanistic Stoichiometry of Mitochondrial Oxidative Phosphorylation, *Biochem.*, 30:3576–3582, 1991.

Hinkle, P.C., and McCarty, R.E., How Cells Make ATP, *Sci. Amer.*, 238(3):104–123, 1978.

Junge, W., Lill, H., and Engelbrecht, S., ATP Synthase: An Electrochemical Transducer with Rotatory Mechanics, *Trends Biochem. Sci.* 22(11):420–423, 1997.

Krebs, H.A., The History of the Tricarboxylic Cycle, *Perspect. Biol. Med.*, 14:154–170, 1970.

Kornberg, H.L., Tricarboxylic Acid Cycles, *Bioessays*, 7:236–238, 1987.

Mitchell, P., Keilin's Respiratory Chain Concept and Its Chemiosmotic Consequences, *Science*, 206:1148–1159, 1979.

Nicholls, D.G., and Ferguson, S.J., *Bioenergetics 2*, Academic Press, London, 1992.

Nicholls, D.G., and Rial, E., Brown Fat Mitochondria, *Trends Biochem.* Sci., 9:489–491, 1984.

Rice-Evans, C.A., Miller, N.J., and Paganga, G., Antioxidant Properties of Phenolic Compounds, *Trends Plant Sci.* 2(4):152–159, 1997.

Sies, H. (Ed.), *Oxidative Stress*, Academic Press, London, 1985.

QUESTIONS

1. Define the following terms:
   a. aerotolerant anaerobes
   b. anaplerotic reactions
   c. glyoxysomes
   d. chemical coupling hypothesis
   e. chemiosmotic coupling theory
   f. ionophore
   g. respiratory control
   h. ischemia
   i. aerobic respiration

2. Describe the transformation of the earth's atmosphere that occurred about three billion years ago.

3. Describe the lifestyles of the following organisms:
   a. obligate anaerobes
   b. aerotolerant anaerobes
   c. facultative anaerobes
   d. obligate aerobes

4. A runner needs a tremendous amount of energy during a race. Explain how the use of ATP by contracting muscle affects the citric acid cycle.

5. Describe two important roles of the citric acid cycle.

6. Acetyl-CoA is manufactured in the mitochondria and used in the cytoplasm to synthesize fatty acids. However, acetyl-CoA cannot penetrate the mitochondrial membrane. How is this problem solved?

7. Describe the glyoxylate cycle. How is it used to synthesize complex molecules from two-carbon molecules?

8. Which of the following conditions indicates a low cell energy status? What biochemical reaction(s) does each condition affect?
   a. high $NADH/NAD^+$ ratio
   b. high ADP/ATP ratio
   c. high acetyl-CoA concentration
   d. low citrate concentration
   e. high succinyl-CoA concentration

9. List three functions of citric acid in the cytoplasm.

10. What are the principal sources of electrons for the electron transport pathway?

11. Describe the processes that are believed to be driven by mitochondrial electron transport?

12. Describe the principal features of the chemiosmotic theory.

13. The chemical coupling hypothesis failed to explain why mitochondrial membrane must be intact during ATP synthesis. How does the chemiosmotic theory account for this phenomenon?

14. How does dinitrophenol inhibit ATP synthesis?

15. Four protons are required to drive the phosphorylation of ADP. Account for the function of each proton in this process.

16. Describe the roles of the pentose phosphate pathway.

17. List several reasons why oxygen is widely used in energy production.

18. Which of the following are reactive oxygen species? Why is each ROS dangerous?
    a. O_2
    b. OH^-
    c. $RO\cdot$
    d. O_2^-
    e. CH_3OH
    f. $^1O_2$

19. Describe the types of cellular damage produced by ROS.

20. Describe the enzymatic activities used by cells to protect themselves from oxidative damage.

21. Give an example of a genetic defect that has survival value. Describe how this defect contributes to survival.

22. Biochemical pathways may seem excessively complex. Organisms sometimes use several reaction pathways to make relatively simple molecules. Can you suggest how these phenomena contribute to the efficient control of metabolic processes?

23. How could the following molecules be synthesized by using intermediates in the citric acid cycle?
    a. aspartic acid
    b. alanine

24. $^{14}CH_3—COOH$ is fed to microorganisms during an experiment. Trace the ^{14}C label through the citric acid cycle.

25. Ethanol is oxidized in the liver to form acetate, which is converted to acetyl-CoA. Determine how many molecules of ATP are produced from one mole of ethanol. Note that two moles of NADH are produced when ethanol is oxidized to form acetate.

PHOTOSYNTHESIS

Chapter Twelve

Evolution of Oxygen by an Aquatic Plant.

OBJECTIVES

After you have studied this chapter, you should be able to answer these questions:

1. What is the principal role of chlorophyll in photosynthesis?

2. What are antenna pigments and what role do they play in photosynthesis?

3. What are photosystems and what functions do they serve in photosynthesizing cells?

4. What similarities are there between aerobic respiration and the light reactions of photosynthesis? What differences are there?

5. What is the Z scheme?

6. What is the water-oxidizing clock?

7. What is the metabolic relationship between the light reactions and the light-independent reactions of photosynthesis?

8. How is CO_2 incorporated into carbohydrate molecules?

9. What is photorespiration and how do some plants avoid it?

10. What is the key regulatory enzyme in photosynthesis and how is its activity controlled?

Without question, photosynthesis is the most important biochemical process on the earth. With a few minor exceptions, photosynthesis is the only mechanism by which an external source of energy is harnessed by the living world. As with other energy-yielding processes, photosynthesis involves oxidation-reduction reactions. Water is the source of electrons and protons that reduce CO_2 to form organic compounds. This chapter is devoted to a discussion of the principles of photosynthetic processes. The relationship between photosynthetic reactions and the structure of chloroplasts as well as the relevant properties of light are emphasized.

As living organisms became abundant on the primitive earth, their consumption of the organic nutrients produced by geochemical processes outpaced production. Developing an alternative source of organic molecules that were useful as the energy sources and the raw materials required for biosynthetic processes became critical for survival. The abundant CO_2 in the earth's early atmosphere was a natural carbon source for organic synthesis. (Much of this CO_2 was generated during the anaerobic degradation of organic nutrients by living organisms.) However, CO_2 is an oxidized, low-energy molecule. For this reason the processes by which CO_2 is incorporated into organic molecules require energy and reducing power. (The formation of carbon-carbon bonds requires free energy now provided by ATP hydrolysis. Reducing power is required because a strong electron donor must provide the high-energy electrons needed to convert CO_2 to a CH_2O unit once the former has been incorporated into an organic molecule.)

The evolution of photosynthetic mechanisms (referred to as **photosystems**) provided both the energy and the reducing power for organic synthesis. The organisms that possessed them had a definite survival advantage, since they no longer depended on an uncertain supply of preformed organic nutrients. These primitive organisms are believed to have been similar to modern green sulfur bacteria. Green sulfur bacteria possess a photosystem that uses light energy to drive a relatively simple electron transport process. As a pigment molecule in the photosystem absorbs light energy, an electron is energized and then donated to the first of several electron acceptors. Eventually, two light-excited electrons are donated to NAD^+, thus forming the reducing agent NADH. (Modern green sulfur bacteria may also use NADPH as a reducing agent in photosynthesis. In more advanced species, NADPH is used exclusively as the reducing agent.) The membrane component that mediates the conversion of light energy into chemical energy is a pigment-protein complex referred to as a **reaction center.** The electrons removed from the reaction center are replaced when oxidized components of the reaction center strip electrons from H_2S, thus generating S. As electrons flow through the photosystem, protons are pumped across the membrane, thus creating an electrochemical gradient. ATP is synthesized as protons move back into the cell through an ATP synthase. This photosystem therefore provides the bacterial cell with the NADH and ATP that incorporate CO_2 into organic molecules. However, H_2S is not normally produced in large quantities and most H_2S molecules are produced in relatively isolated areas. (Recall, for example, the hydrothermal vents described in Special Interest Box 4.1.)

The next critical step in the evolution of life was a photosystem that could remove and use electrons from H_2O. Because water is abundant, photosynthetic organisms could now penetrate and occupy vast new areas on the planet. Water-based photosynthesis also had a profound impact on other organisms. As photosynthetic organisms proliferated, they provided a new and richer supply of organic molecules for other life forms. Their most significant contribution, however, was the accumulation of gaseous oxygen in the atmosphere. As previously described, some organisms (i.e., those that survived this period) adapted to these changing conditions by evolving mechanisms that protected them against oxygen's toxic effects. Eventually, organisms began to use oxygen to generate energy.

Because the electrons removed from H_2O have a more positive redox potential than those in H_2S, more elaborate and sophisticated mechanisms are required for the production of ATP and NADPH. In this chapter the principles of this process, that is, photosynthesis in plants and algae, are described. The discussion begins with a detailed view of chloroplast structure. After a brief review of the relevant properties of light, the reactions that constitute modern photosynthesis will be described. These include the light reactions and the light-independent reactions. During the light reactions, electrons are energized and eventually used in the synthesis of both ATP and NADPH. These molecules are then used in the light-independent reactions (often referred to as the dark reactions) to drive the synthesis of carbohydrate. Several photosynthetic variations, referred to as C4 metabolism and crassulacean acid metabolism, are also discussed. The chapter ends with a discussion of several mechanisms that control photosynthesis in plants.

KEY *Concepts*

Incorporating CO_2 into organic molecules requires energy and reducing power. In photosynthesis, both of these requirements are provided by a complex process driven by light energy.

12.1 CHLOROPHYLL AND CHLOROPLASTS

The essential feature of photosynthesis is the absorption of light energy by specialized pigment molecules (Figure 12.1). The **chlorophylls** are green pigment molecules that resemble heme. *Chlorophyll a* plays the principal role in eukaryotic photosynthesis, since its absorption of light energy directly drives photochemical events. *Chlorophyll b* acts as a light-harvesting pigment by absorbing light energy and passing it on to chlorophyll a. The orange-colored **carotenoids** are isoprenoid molecules that either function as light-harvesting pigments (e.g., lutein, a xanthophyll) or protect against ROS (e.g., β-carotene).

In plants and algae, photosynthesis takes place within specialized organelles called chloroplasts (Chapter 2, p. 30). Chloroplasts resemble mitochondria in several respects. First, both organelles have an outer and an inner membrane with different permeability characteristics (Figure 12.2). The outer membrane of each organelle is highly permeable, while the inner membrane possesses specialized carrier molecules that regulate molecular traffic. Second, the chloroplast inner membrane encloses an inner space, referred to as the **stroma,** that resembles the mitochondrial matrix. The stroma possesses a variety of enzymes (e.g., those that catalyze the light-independent reactions and starch synthesis), DNA, and ribosomes. There are also notable differences between the organelles. For example, chloroplasts are substantially larger than mitochondria. Although their shapes and sizes vary, many plant mitochondria are rod-shaped structures approximately 1500 nm long and 500 nm wide. Many chloroplasts are spheroid-shaped with lengths from 4000 to 6000 nm and widths of approximately 2000 nm. The reasons for this range of sizes are unknown. In addition, chloroplasts possess a distinct third membrane, referred to as the **thylakoid membrane,** that forms an intricate set of flattened vesicles. As described previously, thylakoid membrane is folded into a series of the disklike vesicular structures called **grana.** Each *granum* is a stack of several flattened vesicles. The internal compartment created by the formation of grana is referred to as the **thylakoid space.** The thylakoid membrane that interconnects the grana is called the **stromal lamellae.** Adjacent layers of membrane that fit closely together within each granum are said to be appressed. The stromal lamellae are nonappressed.

The pigments and proteins responsible for the light-dependent reactions of photosynthesis are found within thylakoid membrane. Most of these molecules are organized into the working units of photosynthesis.

1. **Photosystem I.** Photosystem I (PSI), which energizes and transfers the electrons that eventually are donated to $NADP^+$, is a large membrane-spanning protein-pigment complex composed of several polypeptides. The largest of these are two nearly identical 83 kD subunits designated A and B. Although it possesses over 200 molecules of chlorophyll a, the essential role of photosystem I (the donation of energized electrons to a series of electron carriers within thylakoid membrane) is performed by two special chlorophyll a molecules that reside within the reaction center. These molecules, referred to as a special pair, are located in the core complex of PSI, the AB dimer. Because they absorb light at 700 nm,

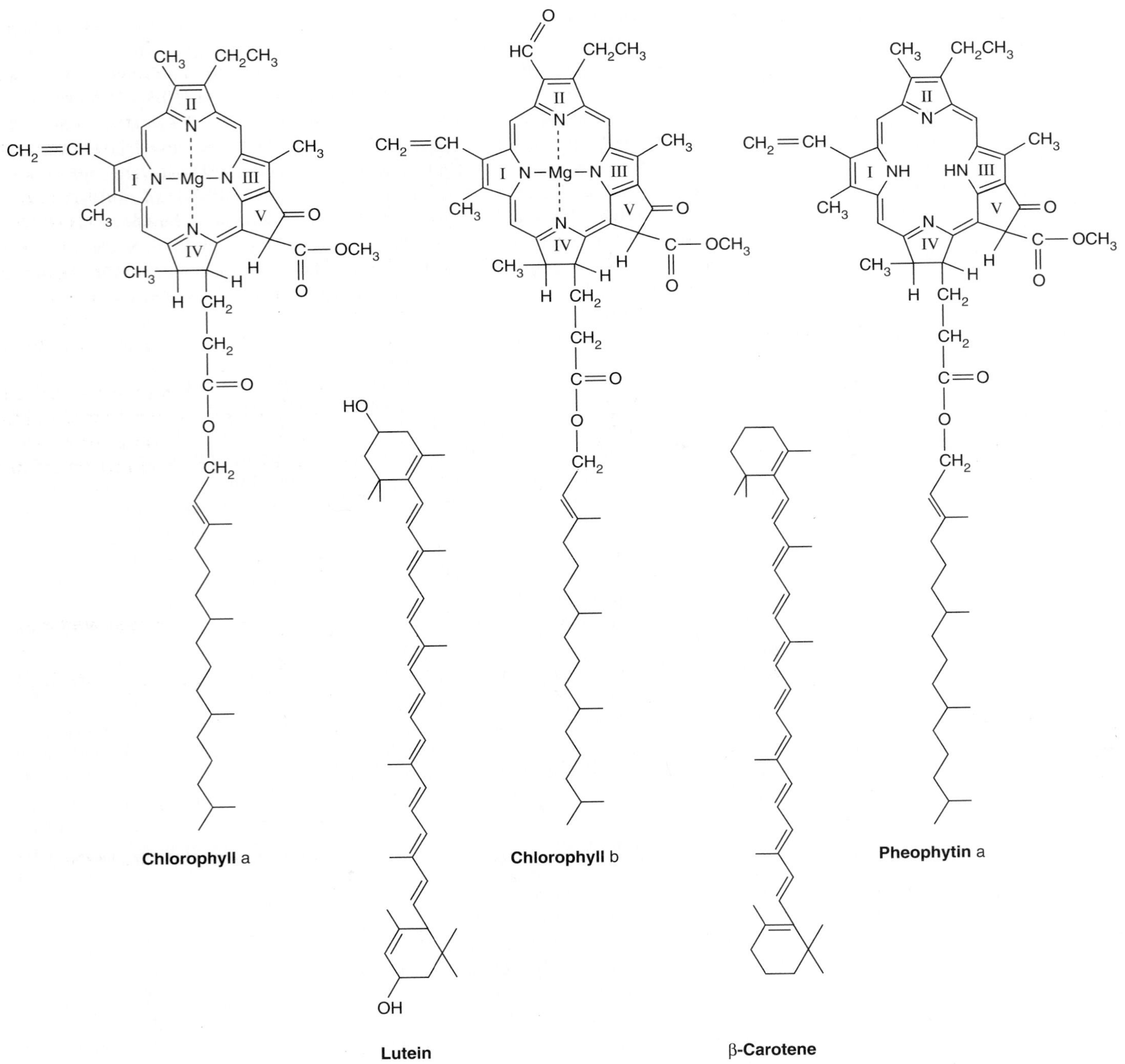

FIGURE 12.1

Pigment Molecules Used in Photosynthesis.

Chlorophylls a and b are found in almost all photosynthesizing organisms. They possess a complex cyclic structure (called a porphyrin) with a magnesium atom at its center. Chlorophyll a possesses a methyl group attached to ring II of the porphyrin, whereas chlorophyll b has an aldehyde group attached to the same site. Pheophytin a is similar in its structure to chlorophyll a. The magnesium atom is replaced by two protons. Chlorophylls a and b and pheophytin a all possess a phytol chain esterified to the porphyrin. The phytol chain extends into and anchors the molecule to the membrane. Lutein and β-carotene are the most abundant carotenoids in thylakoid membranes.

FIGURE 12.2

Details of Chloroplast.

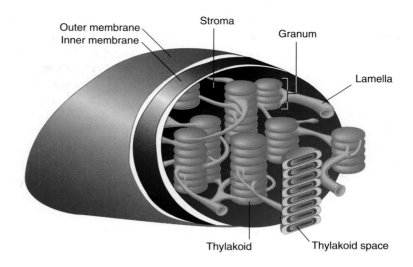

the special pair within photosystem I is sometimes referred to as P700. In addition to the special pair, the AB dimer contains a series of single electron carriers: A_0, A_1 and F_x. A_0 is a specific chlorophyll a molecule that accepts an energized electron from P700 and transfers it to A_1. A_1, which has been identified as phylloquinone (vitamin K_1), is a ubiquinone-like molecule that is sometimes abbreviated as Q. The electron is then transferred from A_1 to F_x, an 4Fe-4S center. Subsequently, the electron is donated to F_A and F_B, two 4Fe-4S centers in an adjacent 9 kD protein. Chlorophyll a molecules other than the special pair, as well as small amounts of chlorophyll b and carotenoids that occur in photosystem I, act as antenna pigments. **Antenna pigments** absorb light energy and transfer it to the reaction center. (This phenomenon is described more fully in Section 12.2.) Most PSI complexes are located in nonappressed thylakoid membrane, i.e., membrane that is directly exposed to the stroma.

2. **Photosystem II.** The function of photosystem II is to oxidize water molecules and donate energized electrons to electron carriers that eventually reduce photosystem I. Photosystem II is a large membrane-spanning protein-pigment complex now believed to possess at least 23 components. The most prominent of these is the reaction center, a protein-pigment complex composed of two polypeptide subunits known as D_1 (33 kD) and D_2 (31 kD) (the D_1/D_2 dimer), cytochrome b_{559}, and a special pair of chlorophyll a molecules (referred to as P680) that absorb light at 680 nm. The oxygen-evolving component of PSII (located on D1) includes a manganese cluster and a tyrosine residue (tyr[161]), often referred to as Y_z. The manganese cluster, which is involved in the splitting of water molecules, is shielded from the thylakoid space by a 33 kD peripheral protein. Also associated with the D_1/D_2 dimer are several electron acceptors. Pheophytin a is a chlorophyll-like pigment that accepts an electron from P680. This electron is donated in turn to two forms of plastoquinone (PQ), an ubiquinone-like molecule. Q_A is a plastoquinone that is permanently bound to D_2, whereas Q_B is reversibly bound to D_1. Several hundred antenna pigment molecules are also associated with the reaction center. A fraction of these light-harvesting molecules are associated with 43 kD and 47 kD proteins that are components of the core structure of PSII. The preponderance of accessory pigment molecules as well as several proteins, however, belong to a detachable unit, referred to as *light-harvesting complex II (LHCII)*. LHCII consists of a transmembrane protein that binds numerous chlorophyll a and chlorophyll b molecules as well as carotenoids. LCHII units are a major component of thylakoid membrane. They possess approximately half of the chloroplast's chlorophyll content as well as a substantial percentage of photosynthesis-related protein. (PSI also possesses a light-harvesting complex, but

it is, as yet, poorly characterized.) Most PSII units are found in thylakoid membrane within grana, that is, in appressed membrane, not exposed to stroma.

3. **Cytochrome b_6f complex.** Cytochrome b_6f complex, found throughout the thylakoid membrane, is similar in structure and function to the cytochrome bc_1 complex in mitochondrial inner membrane. (Recall that cytochrome bc_1 complex is involved in transferring electrons from UQ to cytochrome c in the mitochondrial ETC and pumping protons across the inner membrane.) The cytochrome b_6f complex plays a critical role in the transfer of electrons from PSII to PSI. An iron-sulfur site on the complex accepts electrons from the membrane-soluble electron carrier plastoquinone and donates them to a small water-soluble copper-containing protein called plastocyanin. The mechanism that transports electrons from reduced plastoquinone (PQH_2) though the cytochrome b_6f complex appears to be similar to the Q cycle in mitochondria (Figure 11.16).

4. **ATP synthase.** The chloroplast ATP synthase, also referred to as *CF_0CF_1 ATP synthase,* is structurally similar to the mitochondrial ATP synthase. The CF_0 component is a membrane-spanning protein complex that contains a proton-conducting channel. The CF_1 head piece, which projects into the stroma, possesses an ATP-synthesizing activity. Although the actual mechanism of chloroplast ATP synthesis is not known, it is clear that a transmembrane proton gradient produced during light-driven electron transport drives ADP phosphorylation. The synthesis of each ATP molecule is believed to require pumping approximately three protons across the membrane into the thylakoid space. The ATP synthase is found in thylakoid membrane that is directly in contact with the stroma.

The arrangement of these working units in thylakoid membrane is illustrated in Figure 12.3.

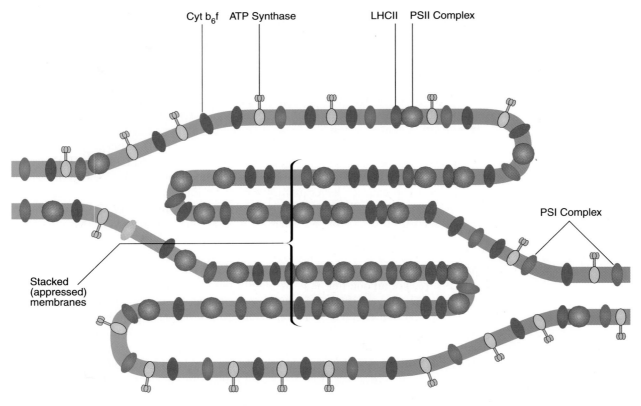

FIGURE 12.3

The Working Units of Photosynthesis.

PSI are most abundant in the unstacked stromal lamellae. In contrast, PSII are located primarily in the stacked regions of thylakoid membrane. Cytochrome b_6f is found in both areas of thylakoid membrane. The ATP synthase is found only in thylakoid membrane that is directly in contact with the stroma.

12.2 LIGHT

The sun emits energy in the form of electromagnetic radiation, which travels through space in waves (Figure 12.4). (The types of electromagnetic radiation, referred to as the electromagnetic spectrum, are shown in Figure 12.5.) The most relevant portion of the electromagnetic spectrum used in photosynthesis is visible light.

Many of the properties of light are explained by its wave behavior. Energy waves are described by the following terms:

1. **Wavelength.** A wavelength λ is the distance from the crest of one wave to the crest of the next wave.
2. **Amplitude.** Amplitude a is the height of a wave. The intensity of electromagnetic radiation (e.g., the brightness of light) is proportional to a^2.
3. **Frequency.** Frequency v is the number of waves that pass a point in space per second.

For each type of radiation the wavelength multiplied by the frequency equals the velocity c of the radiation.

$$\lambda v = c$$

This equation rearranges to

$$\lambda = c/v$$

The wavelength therefore depends on both the frequency and the velocity of the wave.

In addition to behaving as a wave, visible light (and other types of electromagnetic radiation) can also be considered to consist of particles. When light interacts with

FIGURE 12.4

Properties of Waves.

(a) A wavelength λ is the distance between two consecutive peaks in a wave. The amplitude a or height of a wave is related to the intensity of electromagnetic radiation. (b) Frequency is the number of waves that pass a point in space per second. Radiation with the shortest wavelength has the highest frequency.

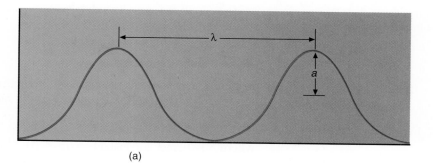

(a)

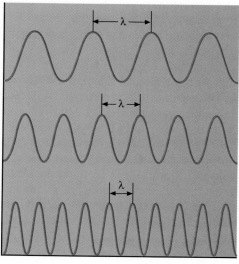

(b)

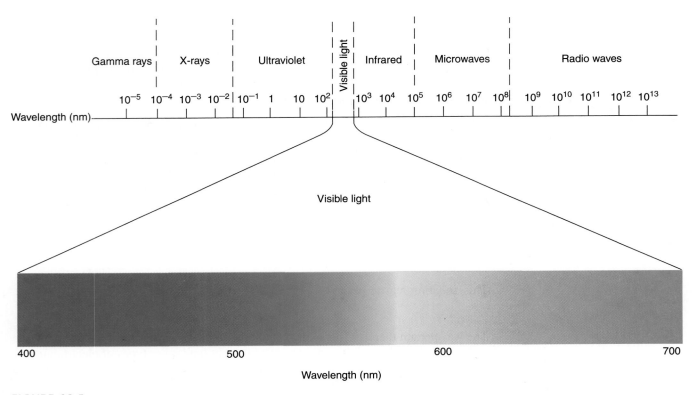

FIGURE 12.5

The Electromagnetic Spectrum.

Gamma rays, which have short wavelengths, have high energy. At the other end of the spectrum, the radio waves (long wavelengths) have low energy. Visible light is the portion of the spectrum to which the visual pigments in the retina of eyes are sensitive. Pigment molecules in chloroplasts are also sensitive to portions of the visible spectrum.

matter, it does so in discrete packets of energy called photons. The energy ϵ of a photon is proportional to the frequency of the radiation.

$$\epsilon = h\upsilon$$

where h is Planck's constant (6.63×10^{-34} J · s).

High-frequency radiation such as gamma rays and X-rays, which have short wavelengths, are high energy. In contrast, low-frequency radiation such as radio waves has long wavelengths and therefore low energy. Note that a photon's energy is inversely proportional to its wavelength.

According to the quantum theory, radiant energy can be absorbed or emitted only in specific quantities called quanta. When a molecule absorbs a quantum of energy, an electron is promoted from its ground state orbital (the lowest energy level) to a higher orbital (Figure 12.6). For absorption to occur, the energy difference between the two orbitals must exactly equal the energy of the absorbed photon. Complex molecules often absorb at several wavelengths. For example, chlorophyll absorbs light in the violet-blue end of the visible spectrum, as well as in the red end of the spectrum. Chlorophyll appears green because it reflects light in the green portion. (The absorption spectrum of chlorophyll and several other pigment molecules is discussed in Appendix B, Supplement 12.1.) Molecules that absorb electromagnetic energy possess structures called chromophores. In **chromophores** electrons move easily to higher energy levels when energy is absorbed. Visible chromophores typically possess extended chains of conjugated double bonds. Molecules with a small number of conjugated double bonds or isolated double bonds absorb energy in the ultraviolet portion of the electromagnetic spectrum. In other words, without the resonance stabilization pro-

FIGURE 12.6

An Electron Absorbing Light.

If a molecule absorbs a photon, an electron becomes excited and moves to a higher orbital. There is usually no change in the spin of the excited electron. As long as the spins of the two unpaired electrons remain antiparallel, the molecule is said to be in an excited singlet state. An excited molecule can return to its ground state by releasing energy as fluorescence or heat. In addition, the energy may be transferred to another molecule, or the energized electron itself may be donated to another molecule.

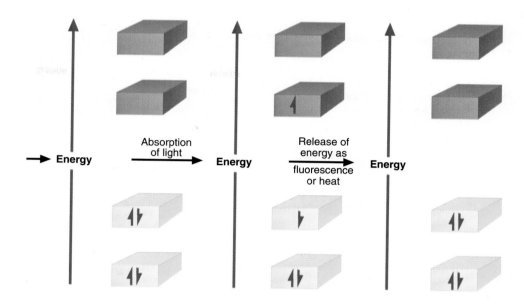

vided by a sufficient number of double bonds, the absorption of considerable energy is required for an electron to reach a higher orbital.

Once an electron is excited, it can return to its ground state in several ways:

1. **Fluorescence.** In **fluorescence** a molecule's excited state decays as it emits a photon. Because some energy is lost during the transition, a fluorescently emitted photon has a longer wavelength than that of the photon originally absorbed. Fluorescent decay can occur as quickly as 10^{-15} s. (Although various chlorophylls absorb light energy throughout the visible spectrum, they emit only photons with the lowest energy possible, that is, red light.)

2. **Resonance energy transfer.** In **resonance energy transfer,** the excitation energy is transferred to a neighboring molecule through interaction between adjacent molecular orbitals.

3. **Oxidation-reduction.** An excited electron is transferred to a neighboring molecule. An excited electron occupies a normally unoccupied orbital and is bound less tightly than when it occupies a normally filled orbital. A molecule with an excited electron is a strong reducing agent. It returns to its ground state by reducing another molecule.

4. **Radiationless decay.** The excited molecule decays to its ground state by converting the excitation energy into heat.

Of all these responses to energy absorption, the most important in photosynthesis are resonance energy transfer and oxidation-reduction. Resonance energy transfer plays a critical role in harvesting light energy by accessory pigment molecules (Figure 12.7). Eventually, the energy absorbed and transmitted by light-harvesting complexes reaches the reaction center chlorophyll molecules. When these molecules become excited, they can lose an electron to a specific acceptor molecule (Figure 12.8). P700 passes electrons to ferredoxin (an iron-sulfur protein), and P680 passes them to pheophytin a. (Once P700 and P680 have been oxidized, they are referred to as P700* and P680*.) The electron hole left in the reaction center chlorophyll molecules is filled by an electron from a donor molecule. Plastocyanin and water play this role in PSI and PSII, respectively. Fluorescence also plays a role in photosynthesis when light absorption exceeds the capacity of the photosystems to transfer energy. Then photons are reemitted by a protective mechanism.

FIGURE 12.7

Resonance Energy Transfer.

Energy flows through a light-harvesting complex. (a) Once a photon is absorbed by a molecule in a light-harvesting complex, it migrates randomly through the complex by resonance energy transfer. The energy is donated from one antenna molecule to another (yellow hexagons) until it is trapped by a reaction center (dark green hexagons) or it is reemitted (b). The reaction center traps the excitation energy because its lowest excited state has a lower energy than the antenna molecules have.

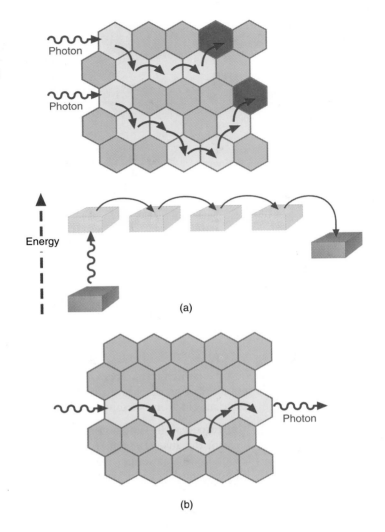

(a)

(b)

FIGURE 12.8

Electron Transfer.

Electron transfer initiates photosynthesis. When light energy is absorbed by a chlorophyll complex (Chl), an electron is transferred to an acceptor (A). The oxidized chlorophyll complex (Chl*) extracts an electron from a donor (D).

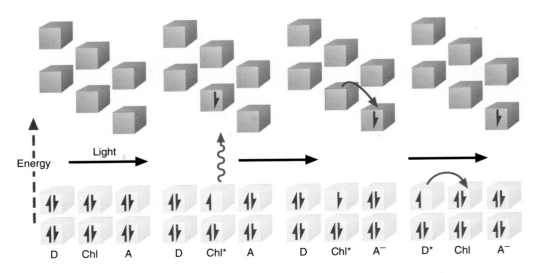

12.3 LIGHT REACTIONS

As described previously, the **light reactions** are a mechanism by which electrons are energized and subsequently used in ATP and NADPH synthesis. In O_2-evolving species both photosystems I and II are required. (In other species only photosystem I is used.) During photosynthesis, the two photosystems, which are connected in series, couple the light-driven oxidation of water molecules to the reduction of $NADP^+$. The overall reaction is

$$2 \text{ NADP}^+ + 2 \text{ H}_2\text{O} \rightleftarrows 2 \text{ NADPH} + \text{O}_2 + 2 \text{ H}^+$$

Since the standard reduction potentials for the half-reactions are

$$O_2 + 4\ e^- + 4\ H^+ \rightleftarrows 2\ H_2O \qquad E_0' = +0.816 \text{ V}$$

and

$$NADP^+ + H^+ + 2\ e^- \rightleftarrows NADPH \qquad E_0' = -0.320 \text{ V}$$

the redox potential for the process is -1.136 V. The minimum free energy change for this process (calculated using the equation $\Delta G_0' = -nF\ \Delta E_0'$; Section 4.2) is approximately 104.7 kcal (438 kJ) per mole of O_2 generated. In comparison, a mole of photons of 700 nm light provides approximately 40.6 kcal (170 kJ). Experimental observations have revealed that the absorption of 8 or more photons (i.e., 2 photons per electron) are required for each O_2 generated. Consequently, a total of 325 kcal (1360 kJ) (i.e., 8 times 40.6 kcal) are absorbed for each mole of O_2 produced. This energy is more than sufficient to account for reducing $NADP^+$ and establishing the proton gradient for ATP synthesis.

The process of light-driven photosynthesis begins with the excitation of PSII by light energy. One electron at a time is transferred to a chain of electron carriers that connects the two photosystems. As electrons are transferred from PSII to PSI, protons are pumped across the thylakoid membrane from the stroma into the thylakoid space. ATP is synthesized as protons flow back into the thylakoid space through the ATP synthase. When P700 absorbs an additional photon it releases an energized electron. (This electron is immediately replaced by an electron provided by PSII.) The newly energized electron is passed through a series of iron-sulfur proteins and a flavoprotein to $NADP^+$, the final electron acceptor. This sequence, referred to as the **Z scheme,** is outlined in Figure 12.9. A more detailed view of the Z scheme is shown in Figure 12.10.

Photosystem II and Oxygen Generation

When LHCII absorbs a photon, its energy is transferred as previously described to P680. Then an energized electron is donated to *pheophytin a,* a molecule similar to chlorophyll in its structure. Reduced pheophytin a passes this electron to Q_A (plastoquinone). When a second electron is transferred from P680 and two protons are

FIGURE 12.9

The Z Scheme.

Two photons must be absorbed to drive an electron from H_2O to $NADP^+$. The black arrows represent the flow of electrons. The vertical arrangement of the components is according to their E_0' values. The most negative values are at the top of the figure. As electrons move from one carrier to the next, they lose energy, that is, their E_0' values become less negative.

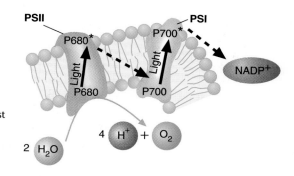

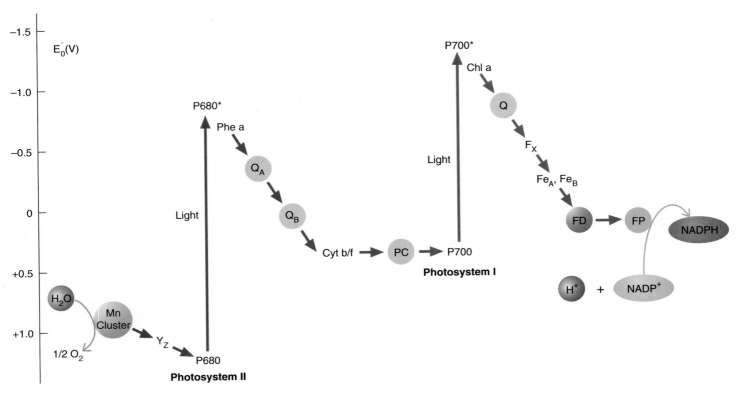

FIGURE 12.10

More Details of the Z Scheme.

The flow of electrons from photosystem II to photosystem I drives the transport of protons into the thylakoid lumen. Electron transfer through the iron-sulfur proteins Fe$_A$ and Fe$_B$ is not understood. E_0' values are approximate.

transferred from the stroma, reduced plastoquinone (Q_AH_2), also referred to as plastoquinol is formed.

Plastoquinone	Plastoquinol

One at a time, Q_AH_2 donates its two electrons to Q_B. Two additional stromal protons are used in the reduction of Q_B. Reduced Q_B (Q_BH_2) donates its electrons to the cytochrome b$_6$f complex. Finally, the cytochrome b$_6$f complex donates its electrons to plastocyanin (PC), a mobile peripheral membrane protein. PC, a single-electron carrier, then transfers these electrons to P700 in PSI.

Recall that electrons transferred from P680 are replaced when H_2O is oxidized. An *oxygen-evolving complex,* composed in part of the manganese cluster and tyrosine residue mentioned above, is responsible for the transfer of electrons from H_2O to P680*. The evolution of one O_2 requires splitting two H_2O, which releases four protons and four electrons. Experimental evidence indicates that H_2O is converted to O_2 by a mechanism referred to as the *water-oxidizing clock* (Figure 12.11). The O_2-evolving complex has five oxidation states: S_0, S_1, S_2, S_3 and S_4. S_0 is the most reduced state and S_4 is the most oxidized state of the complex. It is now believed that the Mn clus-

FIGURE 12.11

FIGURE 12.11

The Water-Oxidizing Clock.

The O_2-evolving apparatus has five oxidation states. Four electrons are removed for every O_2 that is evolved. Each electron is sequentially transferred to a tyrosine residue (Y_Z) and then to P680. Protons are released during several steps in the cycle.

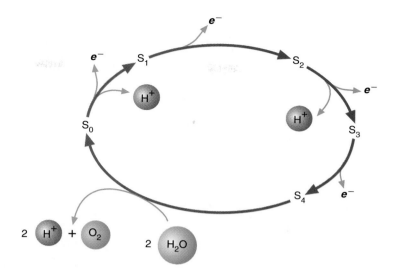

ter, a group of four manganese atoms bound near the PSII reaction center, is mostly responsible for these transitions. As the oxygen-evolving complex abstracts electrons and protons from H_2O, it cycles through the five oxidation states. The electrons are transferred one at a time to a tyrosine residue on the D_1 polypeptide and then to the P680 reaction center. The protons released in the process remain in the thylakoid lumen, where they contribute to the pH gradient that drives ATP synthesis.

QUESTION 12.1

Excessive amounts of light can depress photosynthesis. Recent research indicates that PSII is extremely vulnerable to light damage. Plants often survive this damage because they possess efficient repair systems. It now appears that cells delete and resynthesize damaged components and recycle undamaged ones. For example, the D_1 polypeptide, apparently the most vulnerable component of PSII, is rapidly replaced after it is damaged. After reviewing the role of PSII, suggest the proximate cause of light-induced damage of D_1. (Hint: The D_1/D_2 dimer binds two molecules of β-carotene.)

Photosystem I and NADPH Synthesis

As described, the absorption of a photon by P700 leads to the release of an energized electron. This electron is then passed through a series of electron carriers, the first of which is a chlorophyll a molecule. As the electron is donated sequentially to phylloquinone and to several iron-sulfur proteins (the last of which is ferredoxin), it is moved from the lumenal surface of the thylakoid membrane to its stromal surface. Ferredoxin, a mobile, water-soluble protein, then donates each electron to a flavoprotein called ferredoxin-NADP oxidoreductase. The flavoprotein uses a total of two electrons and a stromal proton to reduce $NADP^+$ to NADPH. The transfer of electrons from ferredoxin to $NADP^+$ is referred to as the *noncyclic electron transport pathway*. In some species (e.g., algae), electrons can return to PSI by way of a *cyclic electron transport pathway* (Figure 12.12). In this process, which typically occurs when a chloroplast has a high $NADPH/NADP^+$ ratio, no NADPH is produced. Instead, electrons are used to pump additional protons across the thylakoid membrane. Consequently, additional molecules of ATP are synthesized.

QUESTION 12.2

Because PSII and PSI operate in series, photosynthesis is efficient when they receive light at equal rates. Recent research indicates that phosphorylation may help balance the activities of the two photosystems. Apparently, it does so by affecting the capacity of LHCII to interact with PSII. The unphosphorylated form of LHCII binds to PSII. If PSII is more active than PSI, plastoquinone becomes reduced. Reduced plastoquinone activates a kinase, which then phosphorylates LHCII. After the phosphorylated LHCII detaches from

PSII, electron transport through PSII slows. (In this process, some appressed layers of thylakoid membrane are converted into nonappressed layers.) When plastoquinone becomes reoxidized, the kinase becomes inactive. Then the phosphate groups on LHCII are removed by a phosphatase allowing it to interact with PSII. Review the effects of covalent modification on proteins (Chapter 5) and suggest why phosphorylation has this effect on photosynthesis.

FIGURE 12.12

The Cyclic Electron Transport Pathway.

A Q cycle similar to that observed in the electron transport pathway that links PSI and PSII is believed to be responsible for pumping two protons across the thylakoid membrane for each electron transported.

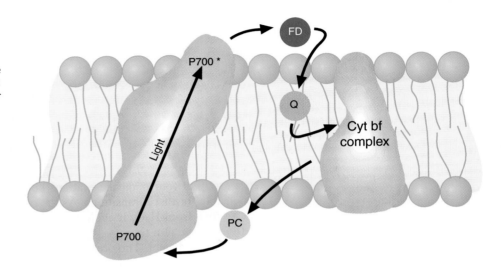

Photophosphorylation

During photosynthesis light energy captured by an organism's photosystems is transduced (i.e., converted from one form to another) into ATP phosphate bond energy. This conversion is referred to as **photophosphorylation.** It is apparent from the preceding discussions that there are many similarities between mitochondrial and chloroplast ATP synthesis. For example, many of the same molecules and terms that are encountered in aerobic respiration (Chapter 11) are also relevant to discussions of photosynthesis. Additionally, in both organelles, electron transport is used to induce a proton gradient, which in turn drives ATP synthesis. Although there is a variety of differences between aerobic respiration and photosynthesis, the essential difference between the two processes is the conversion of light energy into redox energy by chloroplasts. (Recall that mitochondria produce redox energy by extracting high-energy electrons from food molecules.) Another critical difference involves the permeability characteristics of mitochondrial inner membrane and thylakoid membrane. In contrast to the inner membrane, the thylakoid membrane is permeable to Mg^{2+} and Cl^-. Therefore Mg^{2+} and Cl^- move across the thylakoid membrane as electrons and protons are transported during the light reaction. The electrochemical gradient across the thylakoid membrane that drives ATP synthesis therefore consists mainly of a proton gradient that may be as great at 3.5 pH units.

Although experimental observations of chloroplasts played a critical role in the development of the chemiosmotic theory, several issues related to chloroplast ATP synthesis remain unclear. The most prominent of these is the $H^+/2\ e^-$ ratio. Note that in the light reactions outlined in Figure 12.13, a total of six H^+ ions are transported for each pair of electrons. Because the movement of three H^+ ions through the ATP synthase is required for the synthesis of one ATP molecule, two ATP molecules are produced for every NADPH molecule that is synthesized. However, in some circumstances (e.g., high light intensity) a ratio closer to $4\ H^+/2\ e^-$ has been observed. With this ratio, approximately 1.3 ATP molecules are synthesized for each NADPH molecule. The reason for the ratio reduction is unknown. Some recent experimental evidence indicates that under high light intensity the cytochrome b_6f complex fails to pump the additional two protons.

KEY *Concepts*

Eukaryotic photosynthesizing cells possess two photosystems, PSI and PSII, which are connected in series in a mechanism referred to as the Z scheme. The water-oxidizing clock component of PSII generates O_2. The protons are used in the synthesis of ATP in a chemiosmotic mechanism. PSI is responsible for the synthesis of NADPH.

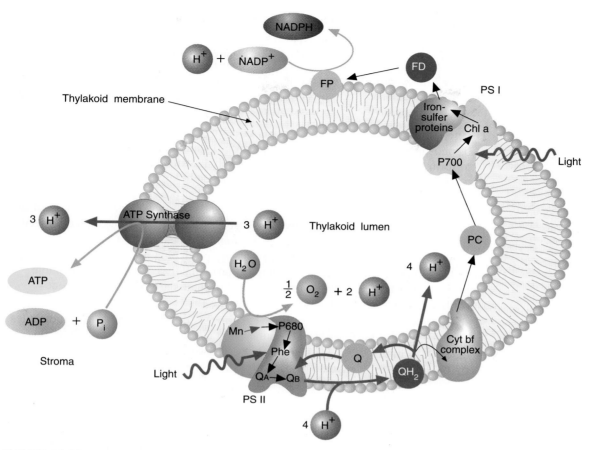

FIGURE 12.13

The Light Reactions.

The transfer of two electrons from PSII to PSI translocates four protons. This proton movement probably occurs in a Q cycle that resembles the one described for mitochondrial electron transport. Two additional protons are released as H_2O is oxidized to O_2. One additional proton is removed from the stroma for each $NADP^+$ reduced to NADPH by PSI. The flow of protons back into the stroma through the ATPase drives the synthesis of ATP.

QUESTION 12.3

A variety of herbicides kill plants by inhibiting photosynthetic electron transport. Atrazine, a triazine herbicide, blocks electron transport between Q_A and Q_B in PSII. DCMU (3-(3,4-dichlorophenyl)-1,1-dimethylurea) also blocks electron flow between the two molecules of plastoquinone. Paraquat is a member of a family of compounds called the bipyridylium herbicides. Paraquat is reduced by PSI but is easily reoxidized by O_2 in a process that produces superoxide and hydroxyl radicals. Plants die because their cell membranes are destroyed by radicals. Of the herbicides just discussed, determine which, if any, are most likely to be toxic to humans and other animals. What specific damage may occur?

12.4 THE LIGHT-INDEPENDENT REACTIONS

The incorporation of CO_2 into carbohydrate by eukaryotic photosynthesizing organisms, a process that occurs within chloroplast stroma, is often referred to as the **Calvin cycle.** Because the reactions of the Calvin cycle can occur without light if sufficient ATP and NADPH are supplied, they have often been called the dark reactions. The name *dark reactions* is somewhat misleading, however. The Calvin cycle reactions typically occur only when the plant is illuminated, since ATP and NADPH are produced by the light reactions. Therefore **light-independent reactions** is a more appropriate term. Because of the types of reactions that occur in the Calvin cycle, it is also referred to as the *reductive pentose phosphate cycle* (RPP cycle) and the *photosynthetic carbon reduction cycle* (PCR cycle).

The Calvin Cycle

The net equation for the Calvin cycle (Figure 12.14) is

$$3\ CO_2 + 6\ NADPH + 9\ ATP \rightarrow$$
$$\text{Glyceraldehyde-3-Phosphate} + 6\ NADP^+ + 9\ ADP + 8\ P_i$$

For every three molecules of CO_2 that are incorporated into carbohydrate molecules, there is a net gain of one molecule of glyceraldehyde-3-phosphate. The reactions of the cycle can be divided into three phases:

1. **Carbon fixation.** This phase consists of a single reaction. Ribulose-1,5-bisphosphate carboxylase catalyzes the carboxylation of ribulose-1,5-bisphosphate to form two molecules of glycerate-3-phosphate. Ribulose-1,5-bisphosphate carboxylase, a complex molecule composed of eight large subunits (L) (56 kD) and eight small subunits (S) (14 kD), is the pacemaker enzyme of the Calvin cycle. Its activity is regulated by CO_2, O_2, Mg^{2+}, and pH, as well as other metabolites. Each L subunit contains an active site that binds substrate. The catalytic activity of the L subunits is enhanced by the S subunits. Because the CO_2 fixation reaction is extremely slow, plants compensate by producing a large number of copies of the enzyme, which often constitutes approximately half of a leaf's soluble protein. (For this reason, ribulose-1,5-bisphosphate carboxylase is often described as the world's most abundant enzyme.)

2. **Reduction.** The next phase of the cycle consists of two reactions. Six molecules of glycerate-3-phosphate are phosphorylated at the expense of six ATP molecules to form glycerate-1,3-bisphosphate. The latter molecules are then reduced by NADP-glyceraldehyde-3-phosphate dehydrogenase to form six molecules of glyceraldehyde-3-phosphate. These reactions are similar to reactions encountered in gluconeogenesis. Unlike the dehydrogenase in gluconeogenesis, the Calvin cycle enzyme uses NADPH as a reducing agent.

3. **Regeneration.** As noted previously, the net production of fixed carbon in the Calvin cycle is one molecule of glyceraldehyde-3-phosphate. The other five glyceraldehyde-3-phosphate molecules are processed in the remainder of the Calvin cycle reactions to regenerate three molecules of ribulose-1,5-bisphosphate. Two molecules of glyceraldehyde-3-phosphate are isomerized to form dihydroxyacetone phosphate. One dihydroxyacetone molecule combines with a third glyceraldehyde-3-phosphate molecule to form fructose-1,6-bisphosphate. The latter molecule is then hydrolyzed to fructose-6-phosphate. Fructose-6-phosphate subsequently combines with a fourth molecule of glyceraldehyde-3-phosphate to form xylulose-5-phosphate and erythrose-4-phosphate. Erythrose-4-phosphate combines with dihydroxyacetone phosphate to form sedoheptulose-1,7-bisphosphate, which is then hydrolyzed to form sedoheptulose-7-phosphate. The fifth molecule of glyceraldehyde-3-phosphate combines with sedoheptulose-7-phosphate to form ribose-5-phosphate and a second molecule of xylulose-5-phosphate. Ribose-5-phosphate and both molecules of xylulose-5-phosphate are separately isomerized to ribulose-5-phosphate. In the last step, three molecules of ribulose-5-phosphate are phosphorylated at the expense of three ATP molecules to form three molecules of ribulose-1,5-bisphosphate. The remaining molecule of glyceraldehyde-3-phosphate is either used within the chloroplast in starch synthesis or exported to the cytoplasm, where it may be used in the synthesis of sucrose or other metabolites.

QUESTION 12.4   Many of the reactions in the regeneration phase of the Calvin cycle are similar to reactions encountered in previous chapters in this textbook. Review the reactions in this phase of the Calvin cycle and determine which reactions they resemble. (Hint: Review Chapters 8 and 11.)

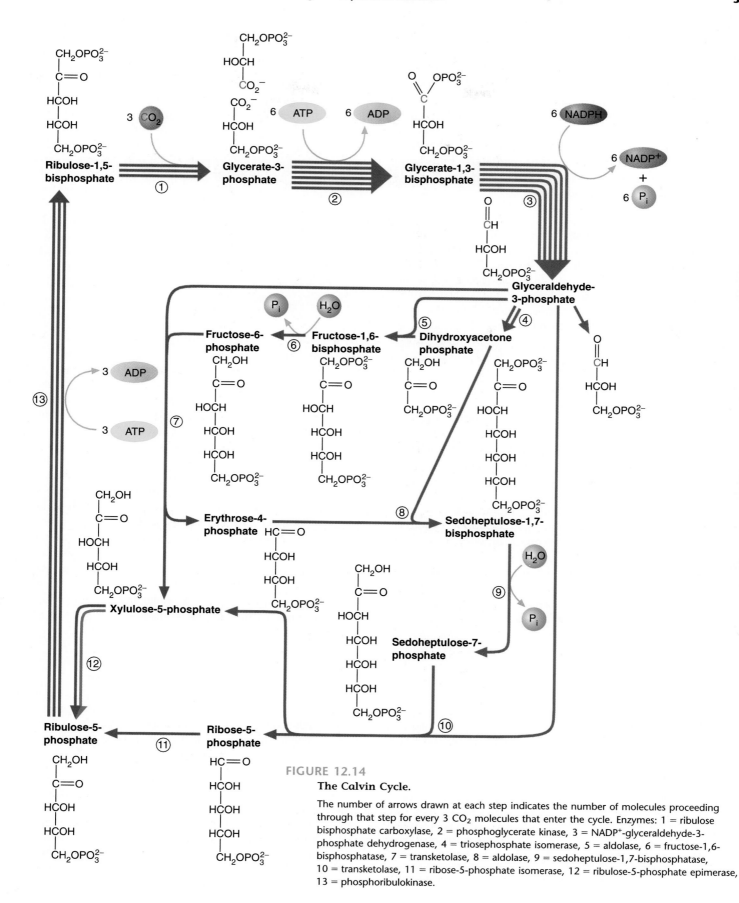

FIGURE 12.14

The Calvin Cycle.

The number of arrows drawn at each step indicates the number of molecules proceeding through that step for every 3 CO_2 molecules that enter the cycle. Enzymes: 1 = ribulose bisphosphate carboxylase, 2 = phosphoglycerate kinase, 3 = NADP+-glyceraldehyde-3-phosphate dehydrogenase, 4 = triosephosphate isomerase, 5 = aldolase, 6 = fructose-1,6-bisphosphatase, 7 = transketolase, 8 = aldolase, 9 = sedoheptulose-1,7-bisphosphatase, 10 = transketolase, 11 = ribose-5-phosphate isomerase, 12 = ribulose-5-phosphate epimerase, 13 = phosphoribulokinase.

STARCH AND SUCROSE METABOLISM

Because glyceraldehyde-3-phosphate and dihydroxyacetone phosphate are readily interconverted, these two molecules (referred to as the **triose phosphates**) are both considered to be Calvin cycle products. The synthesis of triose phosphate is sometimes referred to as the *C3 pathway*. Plants that produce triose phosphates during photosynthesis are called *C3 plants*. Triose phosphate molecules are used by plant cells in such biosynthetic processes as the formation of polysaccharides, fatty acids, and amino acids. Initially, most triose phosphate is used in the synthesis of starch and sucrose. The metabolism of each of these molecules is briefly discussed below.

Starch Metabolism

During very active periods of photosynthesis, triose phosphates are converted to starch. Under normal conditions, approximately 30% of the CO_2 fixed by leaves is incorporated into starch, which is stored as water-insoluble granules. During a subsequent dark period, most chloroplast starch is degraded and converted to sucrose. Sucrose is then exported to storage organs and rapidly growing tissues. In these tissues (e.g., tubers and seeds), most sucrose molecules are used to synthesize starch, which is stored primarily within a specialized plastid called an *amyloplast*.

Triose phosphates retained within the chloroplast are converted to fructose-6-phosphate by aldolase and fructose-1,6-bisphosphatase. Glucose-1-phosphate, the starting material for starch synthesis, is produced from fructose-6-phosphate by phosphoglucoisomerase and phosphoglucomutase. The conversion of glucose-1-phosphate to ADP-glucose by ADP-glucose pyrophosphorylase is the rate-limiting step in starch synthesis. ADP-glucose is incorporated into starch by starch synthase. Like glycogen synthase (p. 187), starch synthase adds each monosaccharide unit to a preexisting polysaccharide chain. The $\alpha(1,6)$ branch points of amylopectin are introduced by branching enzyme.

Several enzymes contribute to starch breakdown. Both α- and β-amylases cleave $\alpha(1,4)$ glycosidic bonds. β-Amylase catalyzes the successive removal of maltose units from the nonreducing ends of starch chains. Maltose is degraded to form glucose by

α-glucosidase. Glucose-1-phosphate is the product when $\alpha(1,4)$ glycosidic bonds at nonreducing ends are broken by starch phosphorylase. Branch points in starch are removed by debranching enzyme. The products of starch digestion, glucose and glucose-1-phosphate, are then converted to triose phosphate and exported to the cytoplasm. In photosynthesizing cells, most triose phosphate is converted to sucrose.

Sucrose imported from leaves is the substrate for most of the starch synthesis in nonphotosynthesizing cells. Sucrose is converted into fructose and UDP-glucose in a reversible reaction catalyzed by sucrose synthase. Fructose is then converted to glucose-1-phosphate by hexokinase and phosphoglucomutase. UDP-glucose is converted to glucose-1-phosphate by UDP-glucose pyrophosphorylase. The conversion of sucrose to two molecules of glucose-1-phosphate is a cytoplasmic process. After its transport into an amyloplast, glucose-1-phosphate is used in starch synthesis. (Smaller amounts of the glycolytic intermediates glyceraldehyde-3-phosphate and dihydroxyacetone phosphate are also transported into amyloplasts and used in starch synthesis.)

Sucrose Metabolism

Sucrose has several important roles in plants. First, sucrose accounts for a large portion of the CO_2 absorbed during photosynthesis. Second, most of the carbon translocated throughout plants is in the form of sucrose. Finally, sucrose is an important energy storage form in many plants.

Sucrose is synthesized in the cytoplasm (Figure 12A). After their transport from chloroplasts, triose phosphates are converted to fructose-1,6-bisphosphate and subsequently to glucose-6-phosphate. The latter molecule is converted to glucose-1-phosphate by phosphoglucomutase. UDP-glucose (formed by glucose-1-phosphate uridyltransferase from glucose-1-phosphate) and fructose-6-phosphate combine to form sucrose-6-phosphate. Sucrose-6-phosphate synthesis is catalyzed by sucrose phosphate synthase. Sucrose phosphatase catalyzes the hydrolysis of sucrose-6-phosphate to form sucrose and P_i. The free energy change of the latter reaction ($\Delta G^{\circ\prime} = -18.4$ kJ/mole) ensures that sucrose production continues in sucrose-storing tissues.

12.5 When plant cells are illuminated, their cytoplasmic ATP/ADP and NADH/NAD ratios rise significantly. The following shuttle mechanism is believed to contribute to the transfer of ATP and reducing equivalents from the chloroplast into the cytoplasm. Once dihydroxyacetone phosphate is transported from the stroma into the cytoplasm, it is converted to glyceraldehyde-3-phosphate and then to glycerate-1,3-bisphosphate. (This reaction is the reverse of the reaction in which glyceraldehyde-3-phosphate is formed during carbon fixation.) In the cytoplasmic reaction, the reducing equivalents are donated to NAD^+ to form NADH. In a later reaction, glycerate-1,3-bisphosphate is converted to glycerate-3-phosphate with the concomitant production of one molecule of ATP. Glycerate-3-phosphate is then transported back into the chloroplast, where it is reconverted to glyceraldehyde-3-phosphate.

This shuttle somewhat depresses mitochondrial respiration processes. Review the regulation of aerobic respiration (Chapter 11) and suggest how photosynthesis suppresses this aspect of mitochondrial function.

FIGURE 12A

Synthesis of Sucrose.

Dihydroxyacetone phosphate is transported from the stroma into the cytoplasm, where it is used to synthesize sucrose.

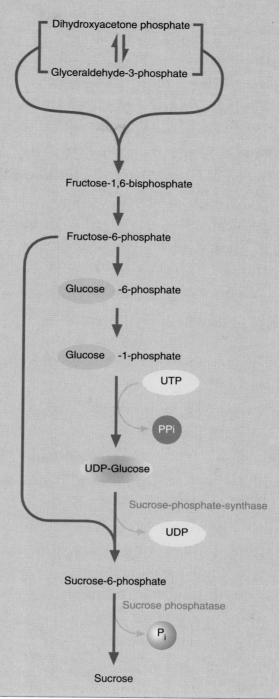

Photorespiration

Photorespiration is perhaps the most curious feature of photosynthesis. In this light-dependent process, oxygen is consumed, and CO_2 is liberated by plant cells that are actively engaged in photosynthesis. Photorespiration is a multistep mechanism initiated by ribulose bisphosphate carboxylase. In addition to its carboxylation function, this enzyme also possesses an oxygenase activity. (For this reason the name *ribulose-1,5-bisphosphate carboxylase-oxygenase,* or *rubisco,* is sometimes used.) Because the enzyme's active site binds to both CO_2 and O_2, these substrates compete.

In the oxygenation reaction, ribulose-1,5-bisphosphate is converted to glycolate-2-phosphate and glycerate-3-phosphate (Figure 12.15). In a complex series of reactions, glycolate-2-phosphate is oxidized by O_2. Ultimately, the glycerate-3-phosphate produced by this pathway is used to produce (via the Calvin cycle) ribulose-1,5-bisphosphate. Photorespiration is a wasteful process. It loses fixed carbon (as CO_2), and consumes both ATP and NADH.

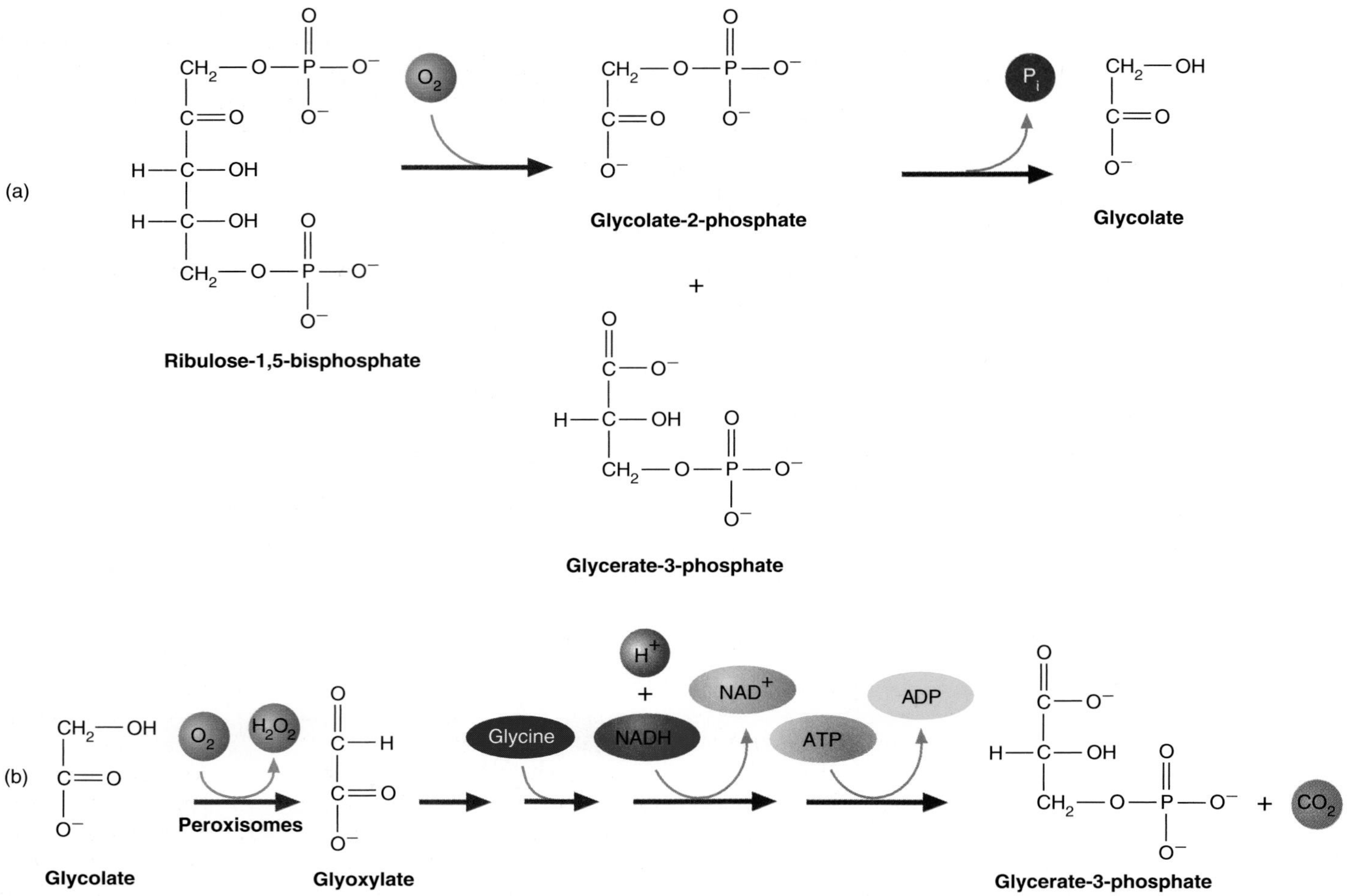

(a)

Ribulose-1,5-bisphosphate

Glycolate-2-phosphate

Glycolate

+

Glycerate-3-phosphate

(b)

Glycolate   **Glyoxylate**   **Glycerate-3-phosphate**

Peroxisomes

FIGURE 12.15

Photorespiration.

Photorespiration is a complex multistep process catalyzed by enzymes in several cellular compartments. (a) The synthesis of glycolate occurs within the stroma. (b) Glycolate is converted to glyoxylate within peroxisomes. In a complex series of reactions that occur in peroxisomes, mitochondria and cytoplasm, glyoxylate is converted to glycerate-3-phosphate and CO_2.

KEY *Concepts*

The Calvin cycle is a series of light-independent reactions in which CO_2 is incorporated into organic molecules. The Calvin cycle reactions occur in three phases: carbon fixation, reduction, and regeneration. Photorespiration is a wasteful process in which photosynthesizing cells evolve CO_2.

The rate of photorespiration depends on several parameters. These include the concentrations of CO_2 and O_2 to which photosynthesizing cells are exposed. Photorespiration is depressed by CO_2 concentrations above 0.2%. (Because photorespiration and photosynthesis occur concurrently, CO_2 is released during CO_2 fixation. When the rates of CO_2 release and fixation are equal, the *CO$_2$ compensation point* has been reached. The lower the CO_2 compensation point, the less photorespiration takes place. Many C3 plants have CO_2 compensation points between 0.02% and 0.03% of CO_2 in the air near photosynthesizing cells.) In contrast, high O_2 concentrations and high temperatures promote photorespiration. Consequently, this process is favored when plants are exposed to high temperatures and any condition that causes low CO_2 and/or high O_2 concentrations. For example, photorespiration is a serious problem for C3 plants in hot, dry environments. To conserve water, these plants close their stomata, thus reducing the CO_2 concentration within leaf tissue. (*Stomata* are pores on the surface of leaves. When they are open, CO_2, O_2, and H_2O vapor can readily diffuse down the concentration gradients between the leaf's interior and the external environment.) In addition, as photosynthesis continues, O_2 levels increase. Depending on the severity of the circumstances, from 30% to 50% of a plant's yield of fixed carbon may be lost. This effect can be serious because several C3 plants (e.g., soybeans and oats) are major food crops.

The purpose of photorespiration is still unknown. Because the ribulose-1,5-bisphosphate carboxylases of all photosynthetic organisms so far investigated possess oxygenase activity, it is currently believed that the enzyme's structure may make photorespiration necessary. Nevertheless, two types of photosynthesizing plants have developed elaborate mechanisms to suppress photorespiration. These mechanisms, referred to as C4 metabolism and crassulacean acid metabolism, are described in Special Interest Box 12.2.

12.5 REGULATION OF PHOTOSYNTHESIS

The regulation of photosynthesis is complex. This is unavoidable, since plants must adapt to a wide variety of environmental conditions. Although the control of most photosynthetic processes is far from being completely understood, several control features are well-established. Most of these processes are directly or indirectly controlled by light. After a brief description of general light-related effects, the control of the activity of ribulose-1,5-bisphosphate carboxylase, the key regulatory enzyme in photosynthesis, is discussed.

Light Control of Photosynthesis

Investigations of photosynthesis are complicated by several factors. The most prominent of these is that the photosynthetic rate depends on temperature and cellular CO_2 concentration as well as on light. Nevertheless, numerous investigations have firmly established light as an important regulator of most aspects of photosynthesis. This is not surprising, considering light's role in driving photosynthesis.

Many of the effects of light on plants are mediated by changes in the activities of key enzymes. Because plant cells possess enzymes that operate in several competing pathways (i.e., glycolysis, pentose phosphate pathway, and the Calvin cycle), careful metabolic regulation is critical. Light assists in this regulation by activating certain photosynthetic enzymes and deactivating several enzymes in degradative pathways. Among the light-activated enzymes are ribulose-1,5-bisphosphate carboxylase, $NADP^+$-glyceraldehyde-3-phosphate dehydrogenase, fructose-1,6-bisphosphatase, sedoheptulose-1,7-bisphosphatase, and phosphoribulokinase. Examples of light-inactivated enzymes include phosphofructokinase and glucose-6-phosphate dehydrogenase.

Light affects enzymes by indirect mechanisms. Among the best researched are the following:

1. **pH.** Recall that during the light reactions, protons are pumped across the thylakoid membrane from the stroma into the thylakoid space. As the pH of the stroma increases from 7 to approximately 8, the activities of several enzymes are affected. For example, the pH optimum of ribulose-1,5-bisphosphate carboxylase is 8.

2. **Mg^{2+}.** Several photosynthetic enzymes are activated by Mg^{2+}. Light induces an increase in the stromal Mg^{2+} concentration. (Recall that Mg^{2+} moves across thylakoid membrane into the stroma during the light reactions.)

3. **The ferredoxin-thioredoxin system.** Thioredoxins are small proteins that transfer electrons from reduced ferredoxin to certain enzymes (Figure 12.16). (Recall that ferredoxin is an electron donor in PSI.) When exposed to light, PSI reduces ferredoxin, which then reduces ferredoxin-thioredoxin reductase (FTR), an iron-sulfur protein that mediates the transfer of electrons between ferredoxin and thioredoxin. Reduced thioredoxins activate several enzymes (e.g., fructose-1,6-bisphosphatase, sedoheptulose-1,7-bisphosphatase, and phosphoribulokinase) and inactivate others (e.g., $NADP^+$-glyceraldehyde-3-phosphate dehydrogenase).

4. **Phytochrome.** Phytochrome is a 120-kD protein that possesses a chromophore (Figure 12.17). Phytochrome exists in two forms: P_r and P_{fr}. P_r, the inactive blue form, absorbs red light (670 nm). The absorption of longer wavelengths (720 nm) (i.e., far red light) converts P_r to P_{fr}, the active green form. (In the dark, P_{fr} decays back to P_r.) Phytochrome apparently mediates hundreds of plant responses to light, many of which are initiated by changes in intracellular Ca^{2+} levels.

ALTERNATIVES TO C3 METABOLISM

C4 Metabolism

C4 plants are found primarily in the tropics. Such plants include sugar cane and maize (corn). They have been assigned the name "C4 plants" because a four-carbon molecule (oxaloacetate) plays a prominent role in a biochemical pathway that avoids photorespiration. This pathway is called the *C4 pathway* or the *Hatch-Slack pathway* (after its discoverers).

C4 plants possess two types of photosynthesizing cells in their leaves: mesophyll cells and bundle sheath cells. (In C3 plants, photosynthesis occurs in mesophyll cells.) Most mesophyll cells in both plant types are positioned so that they are in direct contact with air when the leaf's stomata are open. In C4 plants, CO_2 is captured in specialized mesophyll cells that incorporate it into oxaloacetate (Figure 12B). Phosphoenolpyruvate carboxylase (PEP carboxylase) catalyzes this reaction. Oxaloacetate is then reduced to malate. Once formed, malate diffuses into bundle sheath cells. (As their name implies, bundle sheath cells form a layer around vascular bundles, which contain phloem and xylem vessels.) Within bundle sheath cells, malate is decarboxylated to pyruvate in a reaction that reduces $NADP^+$ to NADPH. The pyruvate product of this latter reaction diffuses back to a mesophyll cell, where it can be reconverted to PEP. Although this reaction is driven by the hydrolysis of one molecule of ATP, there is a net cost of two ATP molecules. An additional ATP molecule is required to convert the AMP product to ADP so that it can be rephosphorylated during photosynthesis. This circuitous process delivers CO_2 and NADPH to the chloroplasts of bundle sheath cells, where ribulose-1,5-bisphosphate carboxylase and the other enzymes of the Calvin cycle use them to synthesize triose phosphates.

C4 metabolism is important because it assimilates CO_2 when it is advantageous for the plant. In hot environments, C4 plants open their stomata only at night after the air temperature decreases and the risk of water loss is low. The CO_2 enters through the stomata and is immediately incorporated into oxaloacetate. The next morning, when light becomes available to drive photosynthesis, the CO_2 released within bundle sheath cells is fixed into sugar molecules as photosynthesis provides adequate amounts of ATP and NADPH to drive this process. Because the concentration of CO_2 within bundle sheath cells is significantly higher than that of O_2, photorespiration is virtually eliminated. Consequently, the net rate of photosynthesis in C4 plants can be at least one-third higher than that of C3 plants. Agricultural scientists are investigating the use of genetic engineering to introduce the C4 pathway into C3 plants.

Crassulacean Acid Metabolism

Crassulacean acid metabolism (CAM) is another mechanism by which certain plants avoid photorespiration. CAM plants, most of which are succulents (e.g., cacti), typically grow in regions of high light intensity and very limited water supply. (CAM is named for the Crassulaceae, a group of plants in which this pathway was first investigated.) They employ a strategy similar to that used by C4 plants. At night, when their stomata are open, CO_2 is incorporated into oxaloacetate by PEP carboxylase. After malate is formed, it is stored within the vacuole until photosynthesis begins the next morning, when CO_2 is regenerated.

The most significant difference between C4 metabolism and CAM is the way in which PEP carboxylation is separated from the Calvin cycle. Recall that in C4 metabolism the two processes are spatially separated (i.e., two cell types are used). In CAM the processes are temporally separated within mesophyll cells. In other words, during daylight hours, CO_2 is regenerated from malate that was synthesized during the night.

Examples of phytochrome-mediated processes include seed germination, stem elongation, flower production, and the differentiation of chloroplasts from proplastids. (Several of these processes are also known to be promoted by specific plant hormones.) Phytochrome has specific effects on photosynthetic processes. These include controlling the rate of synthesis of the small subunit of ribulose-1,5-bisphosphate carboxylase and positioning chloroplasts within photosynthesizing cells.

Control of Ribulose-1,5-Bisphosphate Carboxylase

When most plants are exposed to moderate to high illumination, an increase in CO_2 accelerates the plant's photosynthetic rate. In other words, when adequate light is available, the overall rate of photosynthesis depends on the rate of CO_2 fixation. As described, ribulose-1,5-bisphosphate carboxylase (rubisco) is the key regulatory en-

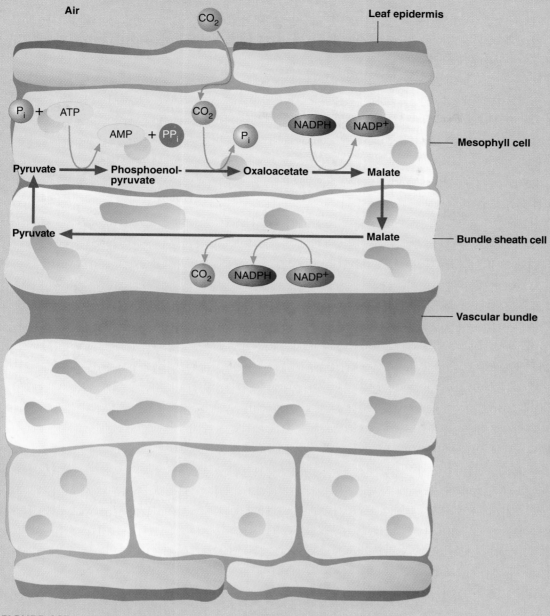

Air

CO₂

Leaf epidermis

Mesophyll cell

P_i + ATP

CO₂

NADPH   NADP⁺

AMP + PP$_i$

P_i

Pyruvate → Phosphoenol-pyruvate → Oxaloacetate → Malate

Pyruvate ← Malate

Bundle sheath cell

CO₂   NADPH   NADP⁺

Vascular bundle

FIGURE 12B

C4 Metabolism.

In the C4 pathway, mesophyll cells, which are in direct contact with the air space in the leaf, take up CO_2 and use it to synthesize oxaloacetate, which is then reduced to malate. (Some C4 plants synthesize aspartate instead of malate.) Malate then diffuses to bundle sheath cells, where it is reconverted to pyruvate. The CO_2 released in this reaction is used in the Calvin cycle. Pyruvate returns to the mesophyll.

zyme in CO_2 fixation. That its activity is regulated at several levels and by a variety of factors and metabolites attests to the critical role that this enzyme plays in photosynthesis.

The genes that code for ribulose-1,5-bisphosphate carboxylase are found within the chloroplast (the L subunit) and the nucleus (the S subunit). The light-activated synthesis of both subunits is believed to be partially mediated by phytochrome. Once the S subunit is transported from the cytoplasm into the chloroplast, both subunits are processed and combined to form the L_8S_8 holoenzyme. Another protein produced in the cytoplasm, called the *large subunit-binding protein*, apparently helps form the holoenzyme. When light is no longer available, the synthesis of both subunits is rapidly depressed.

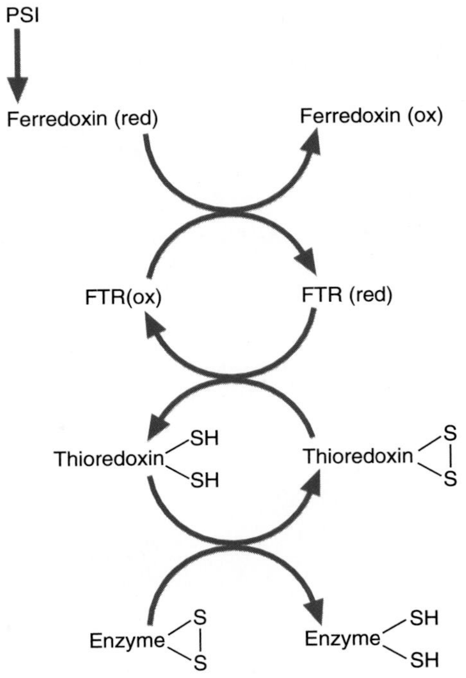

FIGURE 12.16

The Ferredoxin-Thioredoxin System.

Using the light energy captured by PSI, energized electrons are donated to ferredoxin. Electrons donated by ferredoxin to FTR (ferredoxin-thioredoxin reductase) are used to reduce the disulfide bridge of thioredoxin. Thioredoxin then reduces the disulfide bridges of susceptible enzymes. Some enzymes are activated by this process, while others are inactivated.

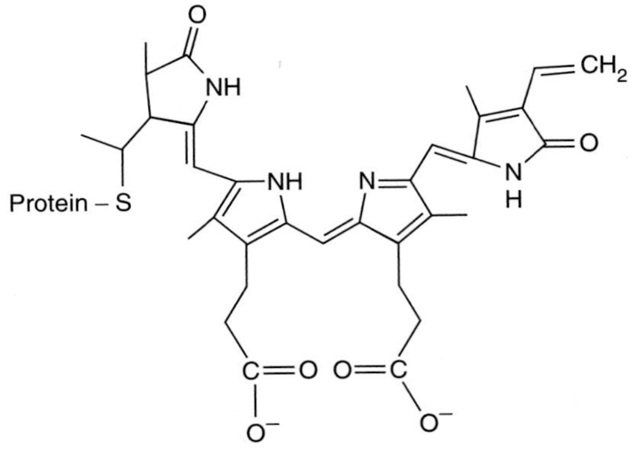

FIGURE 12.17

Phytochrome.

The absorption of light changes the arrangement of conjugated double bonds in the molecule.

The activity of rubisco is affected by several metabolic signals. The enzyme becomes more active as the pH of the stroma rises and the transport of Mg^{2+} that accompanies pH change increases. Rubisco is also activated by high concentrations of NADPH and fructose-6-phosphate. Without light, the stromal pH falls and the concentrations of activating metabolites decrease. In addition, the enzyme is inhibited when ribulose-1,5-bisphosphate binds to the active site in the L subunit (see below). Rubisco is also inhibited by fructose-1,6-bisphosphate, a molecule whose concentration is regulated by the ferredoxin-thioredoxin system. Inhibition by this metabolite occurs in the following manner. The enzyme fructose-1,6-bisphosphate phosphatase, which converts fructose-1,6-bisphosphate to fructose-6-phosphate, is inactive in the dark. When plant cells are exposed to light, the ferredoxin-thioredoxin system converts the oxidized (inactive) enzyme to its reduced (active) version. Consequently, fructose-6-phosphate concentration rises. When light is no longer available, the reduced phosphatase is reoxidized (inactivated). In the absence of an active phosphatase, the concentration of the inhibitor fructose-1,6-bisphosphate rises.

O
‖
CH_2—O—P—O⁻
|
O
‖
O⁻
|
HO—C—C—O⁻
|
H—C—OH
|
H—C—OH
|
CH_2OH

D-Carboxyarabinitol-1-phosphate

KEY *Concepts*

Light is the principal regulator of photosynthesis. Light affects the activities of regulatory enzymes in photosynthetic processes by indirect mechanisms, which include changes in pH, Mg^{2+} concentration, the ferredoxin-thioredoxin system, and phytochrome.

Photosynthetic Studies

In addition to fructose-1,6-bisphosphate, the activity of rubisco is also inhibited by gluconate-6-phosphate (an intermediate in the pentose phosphate pathway), O_2 (see the discussion of photorespiration), and (in some plants) carboxyarabinitol-1-phosphate. The mechanisms by which these molecules affect the enzyme's activity are not well understood. For example, the pathway by which carboxyarabinitol-1-phosphate (CA1P) is synthesized is unknown. It is known, however, that CA1P is made in large amounts in the dark. It is rapidly degraded when plant cells are illuminated.

Because rubisco is an allosteric enzyme, its activation is cooperative. Each L subunit of the enzyme is activated by binding a CO_2 to the ϵ-amino group of a lysine residue. To form this carbamate group (lysine—NH—CO_2^-) requires Mg^{2+}. (Binding Mg^{2+} is believed to stabilize the carbamate group. Mg^{2+} is also required to bind substrate molecules.) The activation reaction, which depends on ATP, is apparently facilitated by an enzyme called rubisco activase. (The activity of rubisco activase is another link between light and photosynthesis, since the concentration of ATP rises when plant cells are illuminated.) Rubisco activase activates the L subunit by promoting release of ribulose-1,5-bisphosphate from the active site, thus allowing the carbamoylation reaction to proceed. The carbamoylation of one L subunit results in a conformational change that helps activate additional L subunits.

Considering the above discussions related to photosynthesis control, it becomes obvious that the effects of light on plant function are complicated. Photosynthetic components are regulated by overlapping mechanisms. When parameters other than light are considered (e.g., CO_2 concentration and temperature), it becomes apparent that the metabolic control systems of plants are extraordinarily sophisticated. Despite considerable research efforts, many aspects of controlling function in plants remain to be elucidated. This endeavor is clearly important, since greater insight into photosynthetic production is critical for improving crop yields and developing plant resources in the future.

Despite the complexity of photosynthesis, many aspects of this life-sustaining process are now understood. Technologies whose use contributed (and continue to contribute) to this research effort include spectroscopy, photochemistry, X-ray crystallography, and radioactive tracers. The principles of these technologies and examples of their use are briefly discussed in Appendix B, Supplement 12.1. **Information provided in this discussion is necessary for answering several end-of-chapter questions.**

KEY WORDS

antenna pigment, *335*	crassulacean acid metabolism, *352*	photophosphorylation, *344*	stromal lamella, *333*
C4 metabolism, *352*		photorespiration, *349*	thylakoid membrane, *333*
Calvin cycle, *345*	fluorescence, *339*	photosystem, *332*	thylakoid space, *333*
carotenoid, *333*	granum, *333*	reaction center, *332*	triose phosphate, *348*
chlorophyll, *333*	light-independent reaction, *345*	resonance energy transfer, *339*	Z scheme, *341*
chromophore, *338*	light reaction, *341*	stroma, *333*	

SUMMARY

In plants, photosynthesis takes place in chloroplasts. Chloroplasts possess three membranes. The outer membrane is highly permeable, whereas the inner membrane possesses a variety of carrier molecules that regulate molecular traffic into and out of the chloroplast. A third membrane, called the thylakoid membrane, forms an intricate series of flattened vesicles called grana.

Photosynthesis consists of two major phases: the light reactions and the light-independent reactions. During the light reactions, water is oxidized, O_2 is evolved, and the ATP and NADPH required to drive carbon fixation are produced. The major working units of the light reactions are photosystems I and II, cytochrome b_6f complex, and the ATP synthase. During the light-independent reactions, CO_2

is incorporated into organic molecules. The first stable product of carbon fixation is glycerate-3-phosphate. The Calvin cycle is composed of three phases: carbon fixation, reduction, and regeneration.

Most of the carbon incorporated during the Calvin cycle is used initially to synthesize starch and sucrose, both of which are important energy sources. Sucrose is also important because it is used to translocate fixed carbon throughout the plant.

Photorespiration is a process whereby O_2 is consumed and CO_2 is released from plants. Its role in plant metabolism is not understood, since it is apparently a wasteful process. C4 plants and CAM plants, which grow in relatively stringent environments, have developed biochemical and anatomical mechanisms for suppressing photorespiration.

Because plants must adapt to environmental conditions, the regulation of photosynthesis is complex. Several features of this control are now well established. Light is an important regulator of most aspects of photosynthesis. Many of the effects of light are mediated by changes in the activities of key enzymes. The mechanisms by which light effects these changes include changes in pH, Mg^{2+} concentration, the ferredoxin-thioredoxin system, and phytochrome. The most important enzyme in photosynthesis is ribulose-1,5-bisphosphate carboxylase. Its activity is highly regulated. Light activates the synthesis of both types of the enzyme's subunits. In addition, its activity is affected by allosteric effectors.

SUGGESTED READINGS

Barber, J., and Andersson, B., Too Much of a Good Thing: Light Can Be Bad for Photosynthesis, *Trends Biochem. Sci.,* 17:61–66, 1992.

Furuya, M., and Schafer, E., Photoperception and Signalling of Induction Reactions by Different Phytochromes, *Trends Plant Sci.,* 1(9):301–307, 1996.

Govindjee, and Coleman, W.J., How Plants Make Oxygen, *Sci. Amer.,* 262:50–58, 1990.

Heber, U., and Krause, G.H., What Is the Physiological Role of Photorespiration? *Trends Biochem. Sci.,* 5:32–34, 1980.

Hoganson, C.W., and Babcock, G.T., A Metalloradical Mechanism for the Generation of Oxygen from Water in Photosynthesis, *Science,* 277:1953–1956, 1997.

Huber, R., A Structural Basis of Light Energy and Electron Transfer in Biology, *Eur. J. Biochem.,* 187:283–305, 1990.

Prince, R.C., Photosynthesis: the Z-Scheme Revised, *Trends Biochem Sci.,* 21(4):121–122, 1996.

Rogner, M., Boekema, E.J., and Barber, J., How Does Photosystem 2 Split Water? The Structural Basis of Efficient Energy Conversion, *Trends Biochem. Sci.,* 21(2):44–49, 1996.

Rutherford, A.W., Photosystem III, the Water-Splitting Enzyme, *Trends Biochem. Sci.,* 14:227–232, 1989.

Schnapf, J.L., and Baylor, D.A., How Photoreceptor Cells Respond to Light, *Sci. Amer.,* 256(4):40–47, 1987.

Youvan, D.C., and Marrs, B.L., Molecular Mechanism of Photosynthesis, *Sci. Amer.,* 256(6):42–48, 1987.

QUESTIONS

1. Define the following terms:
    a. photosystem
    b. reaction center
    c. light reaction
    d. dark reaction
    e. chloroplast
    f. photorespiration
2. What was the most significant contribution of early photosynthetic organisms to the earth's environment?
3. List the three primary photosynthetic pigments and describe the role each plays in photosynthesis.
4. List five ways in which chloroplasts resemble mitochondria.
5. Excited molecules can return to the ground state by several means. Describe each briefly. Which of these processes are important in photosynthesis? Describe how they function in a living organism.
6. What is the final electron acceptor in photosynthesis?
7. What reactions occur during the light reactions of photosynthesis?
8. What reactions occur during the light-independent reactions of photosynthesis?
9. Why is the oxygen-evolving system referred to as a clock?
10. The statement has been made that the more extensively conjugated a chromophore is, the less energy a photon needs to excite it. What is conjugation and how does it contribute to this phenomenon?
11. The chloroplast has a highly organized structure. How does this structure help make photosynthesis possible?
12. Increasing the intensity of the incident light but not the energy increases the rate of photosynthesis. Why is this so?
13. Both oxidative phosphorylation and photophosphorylation trap energy in high-energy bonds. How are these processes different? How are they the same?
14. What is the Z scheme of photosynthesis? How are the products of this reaction used to fix carbon dioxide?
15. Without carbon dioxide, chlorophyll fluoresces. How does carbon dioxide prevent this fluorescence?
16. Where does carbon dioxide fixation take place in the cell?
17. In C3 plants, high concentrations of oxygen inhibit photosynthesis. Why is this so?
18. Generally, increasing the concentration of carbon dioxide increases the rate of photosynthesis. What conditions could prevent this effect?
19. It has been suggested that chloroplasts, like mitochondria, evolved from living organisms. What features of the chloroplast suggest that this is true?
20. Explain why photorespiration is repressed by high concentrations of carbon dioxide.

21. Why does exposing C3 plants to high temperatures raise the carbon dioxide compensation point?

22. Certain herbicides act by promoting photorespiration. These herbicides are lethal to C3 plants but do not affect C4 plants. Why is this so?

23. Corn, a major grain, is a C4 plant, and many weeds in temperate climates are C3 plants. Therefore the herbicides described in Question 22 are widely used. What effect is likely if these materials are not degraded before they wash into the ocean?

24. If one plots the rate of photosynthesis versus the incident wavelength of light, an action spectrum is obtained. How can the action spectrum provide information about the nature of the light-absorbing pigments involved in photosynthesis?

25. Using the action spectrum for photosynthesis in Question 24, determine what wavelengths of light appear to be optimal for photosynthesis.

26. What is the Emerson enhancement effect? How was it used to demonstrate the existence of two different photosystems? (Hint: Refer to Supplement 12.1, Appendix B.)

27. Explain the following observation. When a photosynthetic system is exposed to a brief flash of light, no oxygen is evolved. Only after several bursts of light is oxygen evolved.

NITROGEN METABOLISM I: SYNTHESIS

Chapter Thirteen

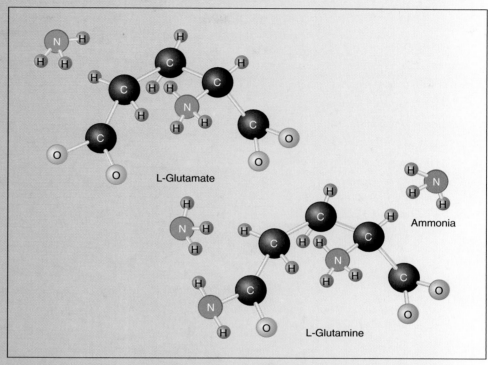

Glutamate, glutamine, and ammonia are among the most important molecules in nitrogen metabolism.

OBJECTIVES

After you have studied this chapter, you should be able to answer these questions:

1. What is nitrogen fixation and why is it an important process in the biosphere?
2. What are nonessential and essential amino acids?
3. What is the amino acid pool?
4. What role do transamination reactions play in amino acid metabolism?
5. What are the amino acid families? On what basis are amino acids classified into these families?
6. What is one-carbon metabolism?
7. What role do folic acid and S-adenosylmethionine play in one-carbon metabolism?
8. What are the biogenic amines?
9. What role do the glutathione-S-transferases play in metabolism?
10. How are the nucleotides synthesized?
11. How are the biosynthetic pathways for heme and chlorophyll similar and how are they different?

Nitrogen is an essential element found in proteins, nucleic acids, and myriad other bio-molecules. Despite the important role that it plays in living organisms, biologically useful nitrogen is scarce. Although nitrogen gas (N_2) is plentiful in the atmosphere, it is almost chemically inert. Therefore, converting N_2 to a useful form requires a major expenditure of energy. Only a few organisms can perform this function, referred to as nitrogen fixation. Certain microorganisms (all of which are bacteria or cyanobacteria) can reduce N_2 to form NH_3 (ammonia). Plants and microorganisms can absorb NH_3 and NO_3^- (nitrate), the oxidation product of ammonia. Both molecules are then used to synthesize nitrogen-containing biomolecules. (The conversion of NH_3 to NO_3^- in a process referred to as nitrification is also performed by certain microorganisms.) Animals cannot synthesize the nitrogen-containing molecules they require from NH_3 and NO_3^-. Instead, they must acquire "organic nitrogen," primarily amino acids, from dietary sources (i.e., plants and plant-eating animals). In a complex series of reaction pathways, animals then use amino acid nitrogen to synthesize important metabolites. This chapter describes the synthesis of the major nitrogen-containing molecules (e.g., amino acids and nucleotides) as well as a small selection of molecules that represent the rich diversity of metabolites that contain this critically important element.

Nitrogen is found in an astonishingly vast array of biomolecules. Examples of major nitrogen-containing metabolites include amino acids, nitrogenous bases, porphyrins, and several lipids. In addition, miscellaneous nitrogen-containing metabolites that are required in smaller amounts (e.g., the biogenic amines and glutathione) are also critically important in the metabolism of many eukaryotes.

As previously mentioned, the incorporation of nitrogen into organic molecules begins with the fixation (reduction) of N_2 by prokaryotic microorganisms. The organisms that fix nitrogen are found in many environments. For example, some organisms live in symbiotic relationships within the roots of certain plants (e.g., *Rhizobium* is found within the root nodule cells of leguminous plants such as peas and beans). Other nitrogen-fixing organisms are found in marine and fresh water, in hot springs, or within the guts of certain animals. Plants such as corn depend on absorbing NH_3 and NO_3^-, which is synthesized by soil bacteria or provided by artificial fertilizers. Because little fixed nitrogen is usually available to plants, nitrogen supply is often the limiting factor in plant growth and development. However plants acquire NH_3, whether by nitrogen-fixation, absorption from the soil, or by reduction of absorbed NO_3^-, it is assimilated by conversion into the amide group of glutamine (p. 368). Then this "organic nitrogen" is transferred to other carbon-containing compounds to produce the amino acids used by the plant to synthesize nitrogenous molecules (e.g., proteins, nucleotides and nucleic acids). Organic nitrogen, primarily in the form of amino acids, then flows throughout the ecosystem as plants are consumed by animals and decomposing microorganisms. The complex process that transfers nitrogen throughout the living world is referred to as the *nitrogen cycle.*

Organisms other than plants vary widely in their capacity to synthesize amino acids from metabolic intermediates and fixed nitrogen. For example, many microorganisms can produce all the amino acids they need. In contrast, animals can synthesize only about half the amino acids they require. The **nonessential amino acids (NAA)** are synthesized from readily available metabolites. The amino acids that must be provided in the diet are referred to as **essential amino acids (EAA).**

Following the digestion of dietary protein in the body's digestive tract, free amino acids are transported across intestinal mucosal cells and into the blood. Because most diets do not provide amino acids in the proportions that the body requires, their concentrations must be adjusted by metabolic mechanisms. The amino acids released to the blood in the intestine already show some changes in their relative concentrations. For example, alanine levels are higher and glutamate and glutamine levels are lower than those found in the predigested protein. Further changes occur when the blood reaches the liver, where the fate of each amino acid is determined. Excessive amounts of all NAA and most EAA are degraded. The concentrations of certain EAA, referred to as the **branched chain amino acids (BCAA),** remain unchanged. (The BCAA are leucine, isoleucine, and valine.) Therefore blood leaving the liver after a protein-rich meal is enriched in BCAA because of selective degradation of excessive amounts of other amino acids. Apparently, BCAA represent a major transport form of amino nitrogen from the liver to other tissues, where they are used in the synthesis of the NAA required for protein synthesis, as well as various amino acid derivatives.

Transamination reactions dominate amino acid metabolism. In these reactions, catalyzed by a group of enzymes referred to as the *aminotransferases* or *transaminases,* α-amino groups are transferred from an α-amino acid to a α-keto acid.

| Acceptor keto acid | Donor amino acid | New keto acid | New amino acid |

(In α-*keto acids* a carbonyl group is directly adjacent to the carboxyl group. Examples include α-ketoglutarate and pyruvate.) Because transamination reactions are readily reversible, they play important roles in both the synthesis and degradation of the amino acids.

After a discussion of nitrogen fixation, the essential features of amino acid biosynthesis are described. This is followed by descriptions of the biosynthesis of selected nitrogen-containing molecules. A special emphasis is placed on the anabolic pathways of the nucleotides. In the following chapter (Chapter 14) the flow of nitrogen atoms is traced through several catabolic pathways to the nitrogenous waste products excreted by animals.

13.1 NITROGEN FIXATION

Several circumstances limit the amount of usable nitrogen available in the biosphere. Because of the chemical stability of the atmospheric gas dinitrogen (N_2), its reduction to form NH_3 (referred to as **nitrogen fixation**) requires a large energy input. For example, at least 16 ATP are required to reduce 1 N_2 to 2 NH_3. In addition, only a few prokaryotic species can "fix" nitrogen. The most prominent of these are several species of free-living bacteria (e.g., *Azotobacter vinelandii* and *Clostridium pasteurianum*), the cyanobacteria (e.g., *Nostoc muscorum* and *Anabaena azollae*), and symbiotic bacteria (e.g., several species of *Rhizobium*). (Symbiotic organisms form mutualistic, that is, mutually beneficial, relationships with host plants or animals. *Rhizobium* species, for example, infect the roots of leguminous plants such as soybeans and alfalfa. See Special Interest Box 13.1.)

All species that can fix nitrogen possess the *nitrogenase complex.* Its structure, similar in all species so far investigated, consists of two proteins called nitrogenase and nitrogenase reductase. Nitrogenase (220 kD), also referred to as *Fe-Mo protein,* is a heterodimer that contains two molybdenum (Mo) atoms, between 28 and 32 iron atoms, and four polypeptide subunits. It catalyzes the reaction $N_2 + 6H^+ + 6e \rightarrow 2NH_3$. Nitrogenase reductase (60 kD) (also referred to as *Fe protein*) is a dimer containing identical subunits.

NITROGEN FIXATION AND AGRICULTURE

Farmers have used crop rotation utilizing legumes for centuries to preserve soil fertility. (In crop rotation, fields are alternately used to grow nitrogen-demanding crops such as corn and the nitrogen-producing legumes.) Despite its importance in food production, the origin of this practice remains obscure. It was not until 1888 that bacteria (later called *Rhizobium*) were discovered within the root nodules of leguminous plants. Subsequent patent applications for a legume inoculant, filed by both British and American companies, were an early milestone in commercial attempts to improve agriculture.

As nitrogen's role in soil fertility became known, commercially produced fertilizers came into use. Guano was one of the most popular commercial fertilizers until the early twentieth century. (Guano is a rich source of nitrate. Found in large deposits on islands off the coast of Chile and Peru, guano is essentially the manure of sea birds.) Commercial production of artificial fertilizers became possible in the early twentieth century. Between 1907 and 1909 the German chemist Fritz Haber developed a method of producing ammonia from N_2 and H_2 at high temperature and pressure, referred to as the Haber process. The original Haber process used H_2 obtained from the production of coke (coal heated so intensely that most gases have been removed) and N_2 derived from the fractional distillation of liquid air. The hydrogen and nitrogen gases were heated at 550°C at a pressure of 2 atm in the presence of an iron catalyst. The ammonia produced was oxidized to form nitrate, which was used as a fertilizer. The higher temperatures and pressures (i.e.,

700°C and 1000 atm) used in modern industrial nitrogen fixation have significantly improved the efficiency of the Haber process.

Unfortunately, modern improvements in the efficiency of the Haber process have significantly raised the costs of fertilizer production. Because of an increasing world population (projected to be six billion by the year 2000), there is ever increasing pressure to make fertilizer production more cost-effective. Because of high cost and dwindling fossil fuel supplies associated with fertilizer production, alternative technologies are being explored. One possible answer to future fixed nitrogen requirements is the use of DNA technology to produce plant species with their own nitrogen-fixing capability. Although scientists have been working for over a decade on this project, success has not yet been achieved. Biological nitrogen fixation is unexpectedly complex. For example, nitrogen fixation by the bacterium *Klebsiella* requires the functions of at least 18 genes (referred to as nif genes). In addition to coding for the protein components of nitrogenase, these genes also code for molecules used in electron transport, metal processing, and coordination of the components into a functioning nitrogen-fixing system. Despite these difficulties, any success in this endeavor (e.g., producing plants that can fix their own nitrogen or develop symbiotic relationships with nitrogen-fixing bacteria) will reduce dependence on expensive artificial fertilizers. The consequences in terms of food production and food prices will be revolutionary.

Despite considerable research, nitrogen fixation is not yet completely understood. However, several aspects of nitrogen fixation have been elucidated (Figure 13.1). NADH (or NADPH) is the ultimate source of the electrons required in the reduction of dinitrogen. The reduced coenzyme molecules donate electron pairs to the iron-sulfur protein ferredoxin, which then transfers them to nitrogenase reductase. The hydrolysis of at least 4 molecules of ATP then facilitates the transfer of each electron pair from nitrogenase reductase to nitrogenase. A total of eight electrons are transferred per N_2 reduced. Ammonia is produced when 6 electrons are transferred from nitrogenase to N_2. Once it is synthesized, ammonia is translocated out of the bacte-

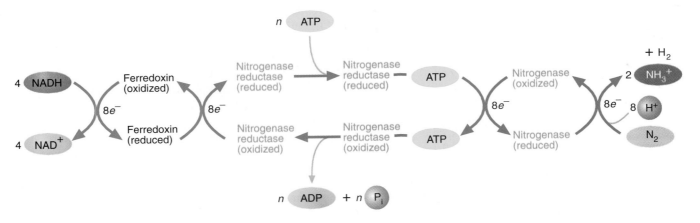

FIGURE 13.1

The Flow of Electrons in Nitrogen Fixation.

rial cell into symbiotic host cells, where it is used in glutamine synthesis (Section 13.2). The remaining two electrons reduce 2 H^+ to form H_2. (In the presence of ATP and the absence of N_2, significant amounts of H_2 are evolved. It has been estimated, for example, that U.S. soybean crops infected with Rhizobium japonicum produce billions of cubic feet of H_2 annually.)

Both components of the nitrogenase complex are irreversibly inactivated by O_2. Nitrogen-fixing organisms solve this problem in several ways. Anaerobic organisms such as *Clostridium* grow only in anaerobic soil, while many of the cyanobacteria produce specialized nitrogenase-containing cells called heterocysts. The thick cell walls of heterocysts isolate the enzyme from atmospheric oxygen. Legumes produce an oxygen-binding protein called *leghemoglobin,* which traps oxygen before it can interact with the nitrogenase complex.

13.2 AMINO ACID BIOSYNTHESIS

Living organisms differ in their capacity to synthesize the amino acids required for protein synthesis. While plants and many microorganisms can produce all their amino acids from readily available precursors, other organisms must obtain some preformed amino acids from their environment. For example, mammalian tissues can synthesize NAA (Table 13.1) by relatively simple reaction pathways. In contrast, EAA must be obtained from the diet because mammals lack the long and complex reaction pathways required for their synthesis.

Amino Acid Metabolism Overview

Amino acids serve a number of functions. Although the most important role of amino acids is the synthesis of proteins, they are also the principal source of the nitrogen atoms required in various synthetic reaction pathways. In addition, the nonnitrogen parts of amino acid (referred to as carbon skeletons) are a source of energy, as well as precursors, in several reaction pathways. Therefore an adequate intake of amino acids, in the form of dietary protein, is essential for an animal's proper growth and development.

Dietary protein sources differ widely in their proportions of the EAA. In general, complete proteins (those containing sufficient quantities of EAA) are of animal origin (e.g., meat, milk, and eggs). Plant proteins often lack one or more EAA. For example, gliadin (wheat protein) has insufficient amounts of lysine, and zein (corn protein) is low in both lysine and tryptophan. Because plant proteins differ in their amino acid compositions, plant foods can provide a high-quality source of essential amino acids only if they are eaten in appropriate combinations. One such combination includes beans (low in methionine) and cereal grains (low in lysine).

TABLE 13.1

The Essential and Nonessential Amino Acids in Humans

Essential	Nonessential
Isoleucine	Alanine
Leucine	Arginine[*]
Lysine	Asparagine
Methionine	Aspartate
Phenylalanine	Cysteine
Threonine	Glutamate
Tryptophan	Glutamine
Valine	Glycine
	Histidine[*]
	Proline
	Serine
	Tyrosine

[*]Amino acids that are essential for infants.

The amino acid molecules that are immediately available for use in metabolic processes are referred to as the **amino acid pool.** In animals, amino acids in the pool are derived from the breakdown of both dietary and tissue proteins. Excreted nitrogenous products such as urea and uric acid are output from the pool. Amino acid metabolism is a complex series of reactions in which the amino acid molecules required for the syntheses of proteins and metabolites are continuously being synthesized and degraded. Depending on current metabolic requirements, certain amino acids are synthesized or interconverted and then transported to tissue, where they are used. When nitrogen intake (primarily amino acids) equals nitrogen loss, the body is said to be in *nitrogen balance.* This is the condition of healthy adults. In *positive nitrogen balance,* a condition that is characteristic of growing children, pregnant women, and recuperating patients, nitrogen intake exceeds nitrogen loss. The excess of nitrogen is retained as the amount of tissue proteins being synthesized exceeds the amount being degraded. *Negative nitrogen balance* exists when an individual cannot replace nitrogen losses with dietary sources. *Kwashiorkor* ("the disease the first child gets when the second is on the way") is a form of malnutrition caused by a prolonged insufficient intake of protein. Its symptoms include growth failure, apathy, ulcers, liver enlargement, and diarrhea, as well as decreased mass and function of the heart and kidneys. Prevalent in Africa, Asia, and Central and South America, kwashiorkor can be treated by feeding high-quality protein-rich foods such as milk, eggs, and meat.

Transport of amino acids into cells is mediated by specific membrane-bound transport proteins, several of which have been identified in mammalian cells. They differ in their specificity for the types of amino acids transported and in whether the transport process is linked to the movement of Na^+ across the plasma membrane. (Recall that the gradient created by the active transport of Na^+ can move molecules across membrane. Na^+-dependent amino acid transport is similar to that observed in the glucose transport process illustrated in Figure 9.28.) For example, several Na^+-dependent transport systems have been identified within the lumenal plasma membrane of enterocytes. Na^+-independent transport systems are responsible for transporting amino acids across the portion of enterocyte plasma membrane in contact with blood vessels. The γ-glutamyl cycle (Section 13.3) is believed to assist in transporting some amino acids into specific tissues (i.e., brain, intestine, and kidney).

Reactions of Amino Groups

Once amino acid molecules enter cells, the amino groups are available for synthetic reactions. This metabolic flexibility is effected primarily by transamination reactions in which amino groups are transferred from an α-amino acid to an α-keto acid. However, another class of reactions, in which NH_4^+ is directly incorporated into certain amino acid molecules, also occurs. These reaction types are discussed next.

Transamination Eukaryotic cells possess a large variety of aminotransferases. Found within both the cytoplasm and mitochondria, these enzymes possess two types of specificity: (1) the type of α-amino acid that donates the α-amino group and (2) the α-keto acid that accepts the α-amino group. Although the aminotransferases vary widely in the type of amino acids they bind, most of them use glutamate as the amino group donor.

Acceptor α-keto acid   Glutamate   New amino acid α-Ketoglutarate

Because glutamate is produced when α-ketoglutarate (a citric acid cycle intermediate) accepts an amino group, these two molecules (referred to as the *α-ketoglutarate/ glutamate pair*) have a strategically important role in both amino acid metabolism and metabolism in general. Two other such pairs have important functions in metabolism. In addition to its role in transamination reactions, the *oxaloacetate/aspartate* pair is involved in the disposal of nitrogen in the urea cycle (Chapter 14). One of the most important functions of the *pyruvate/alanine pair* is in the alanine cycle (Figure 8.15). Because α-ketoglutarate and oxaloacetate are citric acid cycle intermediates, transamination reactions often represent an important mechanism for meeting the energy requirements of cells.

Transamination reactions require the coenzyme pyridoxal-5′-phosphate (PLP), which is derived from pyridoxine (vitamin B$_6$). PLP is also required in numerous other reactions of amino acids. Examples include racemizations, decarboxylations and several side chain modifications. (**Racemizations** are reactions in which L- and D-amino acids are interconverted.) The structures of the vitamin and its coenzyme form are illustrated in Figure 13.2.

PLP is bound in the enzyme active site by a Schiff base formed from the aldehyde group of PLP and the ϵ-amino group of a lysine residue (an aldimine).

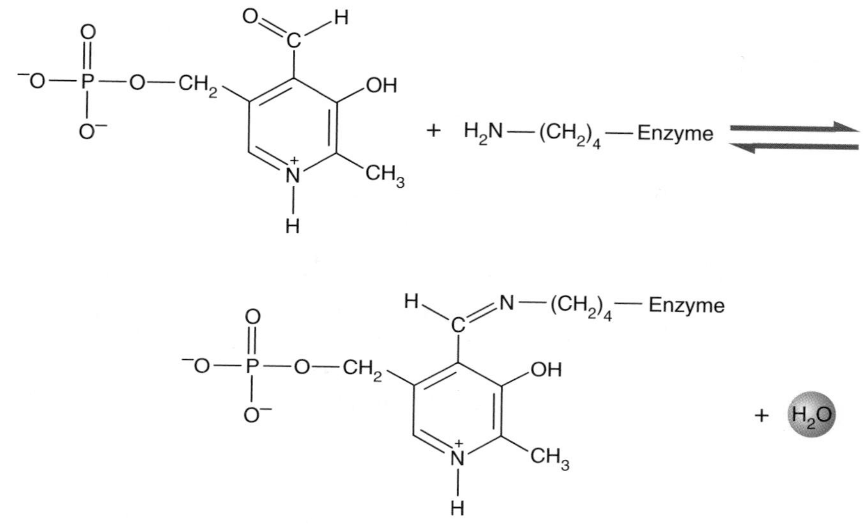

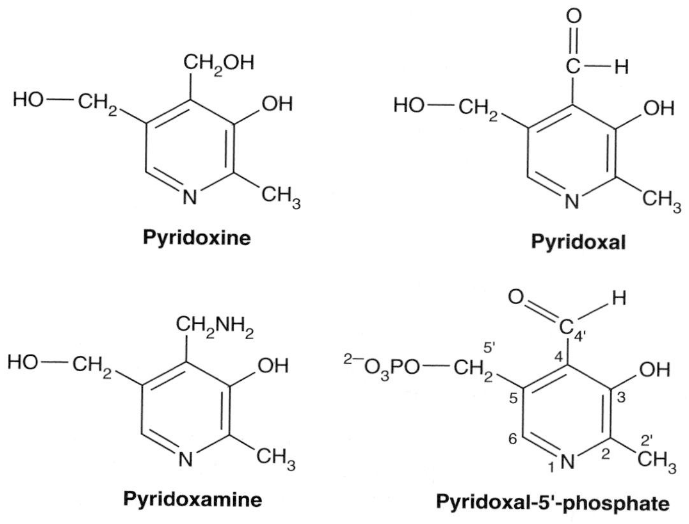

FIGURE 13.2

Vitamin B$_6$.

Vitamin B$_6$ includes (a) pyridoxine, (b) pyridoxal, and (c) pyridoxamine. (Pyridoxine is found in leafy green vegetables. Pyridoxal and pyridoxamine are found in animal foods such as fish, poultry, and red meat.) The biologically active form of vitamin B$_6$ is (d) pyridoxal-5′-phosphate.

Pyridoxine

Pyridoxal

Pyridoxamine

Pyridoxal-5′-phosphate

Additional stabilizing forces include ionic interactions between amino acid side chains and PLP's pyridinium ring and phosphate group. The positively charged pyridinium ring also functions as an electron sink, stabilizing negatively charged reaction intermediates.

Amino acid substrates become bound to PLP via the α-amino group in an imine exchange reaction. Then one of three bonds of the α-carbon atom is selectively broken in the active sites in each type of PLP-dependent enzyme.

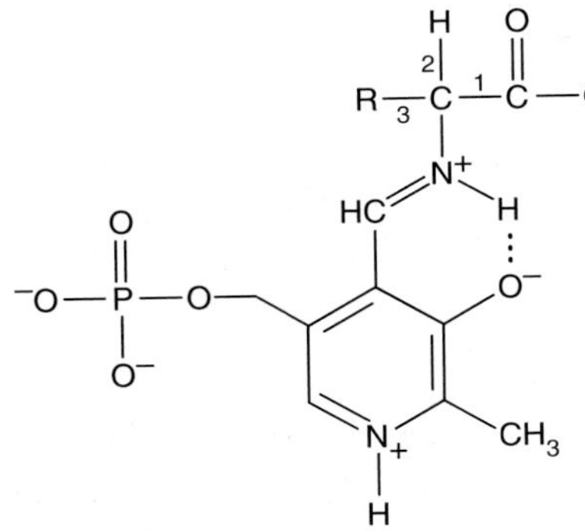

This selectivity is accomplished by orienting the amino acid substrate so that the bond to be broken lies perpendicular to the plane of the pyridinium ring. If bond 1 is broken, an α-decarboxylation occurs. Racemizations or transaminations occur when bond 2 is broken. Side-chain decarboxylations and several types of eliminations break bond 3.

Despite the apparent simplicity of the transamination reaction, the mechanism is quite complex. The reaction begins with the formation of a Schiff base between PLP and the α-amino group of an α-amino acid (Figure 13.3). When the α-hydrogen atom is removed by a general base in the enzyme active site, a resonance-stabilized intermediate forms. With the donation of a proton from a general acid and a subsequent hydrolysis, the newly formed α-keto acid is released from the enzyme. A second α-keto acid then enters the active site and is converted into an α-amino acid in a reversal of the reaction process that has just been described. Transamination reactions are examples of a reaction mechanism referred to as a *bimolecular ping-pong reaction*. The mechanism is so named because the first substrate must leave the active site before the second one can enter.

Because transamination reactions are reversible, it is theoretically possible for all amino acids to be synthesized by transamination. However, experimental evidence indicates that there is no net synthesis of an amino acid if its α-keto acid precursor is not independently synthesized by the organism. For example, alanine, aspartate, and glutamate are nonessential for animals because their α-keto acid precursors (i.e., pyruvate, oxaloacetate, and α-ketoglutarate) are readily available metabolic intermediates. Because the reaction pathways for synthesizing molecules such as phenylpyruvate, α-keto-β-hydroxybutyrate, and imidazolepyruvate do not occur in animal cells, phenylalanine, threonine, and histidine must be provided in the diet. (Reaction pathways that synthesize amino acids from metabolic intermediates, not only by transamination, are referred to as *de novo* pathways.)

Direct Incorporation of Ammonium Ions into Organic Molecules   There are two principal means by which ammonium ions are incorporated into amino acids and eventually other metabolites: (1) reductive amination of α-keto acids and (2) formation of the amides of aspartic and glutamic acids.

FIGURE 13.3

The Transamination Mechanism.

The donor amino acid forms a Schiff's base with pyridoxal phosphate within the enzyme's active site. After a proton is lost, a carbanion forms and is resonance-stabilized by interconversion to a quinonoid intermediate. After an enzyme-catalyzed proton transfer and a hydrolysis, the α-keto product is released. A second α-keto acid then enters the active site. This acceptor α-keto acid is converted to an α-amino acid product as the mechanism just described is reversed.

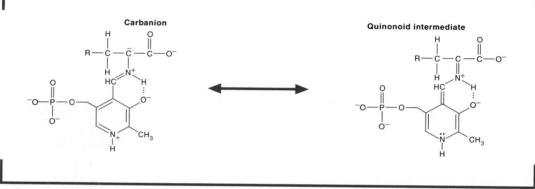

Glutamate dehydrogenase, an enzyme found in both the mitochondria and cytoplasm of eukaryotic cells as well as in some bacterial cells, catalyzes the direct amination of α-ketoglutarate.

$$^-O-\overset{\overset{\textstyle O}{\|}}{C}-CH_2-CH_2-\overset{\overset{\textstyle O}{\|}}{C}-\overset{\overset{\textstyle O}{\|}}{C}-O^- \; + \; NH_4^+ \; + \; NADH \; + \; H^+ \rightleftharpoons$$

α-Ketoglutarate

$$^-O-\overset{\overset{\textstyle O}{\|}}{C}-CH_2-CH_2-\overset{\overset{\textstyle H}{|}}{\underset{\underset{\textstyle +NH_3}{|}}{C}}-\overset{\overset{\textstyle O}{\|}}{C}-O^- \; + \; NAD^+ \; + \; H_2O$$

Glutamate

The primary function of this enzyme in eukaryotes appears to be catabolic (i.e., a means of producing NH_4^+ in preparation for nitrogen excretion). However, the reaction is reversible. When excess ammonia is present, the reaction is driven toward glutamate synthesis.

Ammonium ions are also incorporated into cell metabolites by the formation of glutamine, the amide of glutamate.

KEY *Concepts*
_____

In transamination reactions, amino groups are transferred from one carbon skeleton to another. In reductive amination, amino acids are synthesized by the incorporating of NH_4^+ into α-keto acids. Ammonium ions are also incorporated into cellular metabolites by the amination of glutamate to form glutamine.

$$^-O-\overset{\overset{\textstyle O}{\|}}{C}-CH_2-CH_2-\overset{\overset{\textstyle H}{|}}{\underset{\underset{\textstyle +NH_3}{|}}{C}}-\overset{\overset{\textstyle O}{\|}}{C}-O^- \; + \; ATP \; + \; NH_4^+ \longrightarrow$$

Glutamate

$$H_2N-\overset{\overset{\textstyle O}{\|}}{C}-CH_2-CH_2-\overset{\overset{\textstyle H}{|}}{\underset{\underset{\textstyle +NH_3}{|}}{C}}-\overset{\overset{\textstyle O}{\|}}{C}-O^- \; + \; ADP \; + \; P_i$$

Glutamine

The brain, a rich source of the enzyme glutamine synthase, is especially sensitive to the toxic effects of NH_4^+. Brain cells convert NH_4^+ to glutamine, a neutral, nontoxic molecule. Glutamine is then transported to the liver, where the production of nitrogenous waste occurs.

 In plants, the pathway by which most NH_4^+ is incorporated into organic molecules requires two enzymes: glutamine synthase and glutamate synthase. After NH_4^+ is incorporated into glutamine by glutamine synthase, the amide amino group is transferred to the 2-keto group of α-ketoglutarate by glutamate synthase. The two electrons required in this reaction are provided by reduced ferredoxin in some plant tissues (e.g., leaves) and NADPH in other tissues (e.g., roots and germinating seeds).

$$H_2N-\overset{\overset{\textstyle O}{\|}}{C}-CH_2-CH_2-\overset{\overset{\textstyle H}{|}}{\underset{\underset{\textstyle +NH_3}{|}}{C}}-\overset{\overset{\textstyle O}{\|}}{C}-O^- \; + \; ^-O-\overset{\overset{\textstyle O}{\|}}{C}-CH_2-CH_2-\overset{\overset{\textstyle O}{\|}}{C}-\overset{\overset{\textstyle O}{\|}}{C}-O^- \; + \; 2\,e^-$$

Glutamine   **α-Ketoglutarate**

$$\longrightarrow \; 2\;^-O-\overset{\overset{\textstyle O}{\|}}{C}-CH_2-CH_2-\overset{\overset{\textstyle H}{|}}{\underset{\underset{\textstyle +NH_3}{|}}{C}}-\overset{\overset{\textstyle O}{\|}}{C}-O^-$$

Glutamate

One of the two glutamate products of this reaction is then used as a substrate in a glutamine synthase-catalyzed reaction. Consequently, one molecule of glutamate is produced for every NH_4^+ that enters the process.

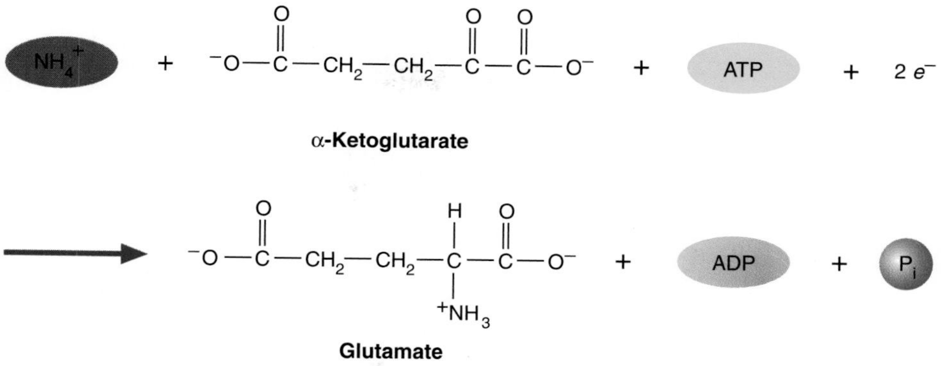

α-Ketoglutarate

Glutamate

Synthesis of the Amino Acids

The amino acids differ from other classes of biomolecules in that each member of this class is synthesized by a unique pathway. Despite the tremendous diversity of amino acid synthetic pathways, they have one common feature. The carbon skeletons of each amino acid are derived from commonly available metabolic intermediates. Thus in animals, all NAA molecules are derivatives of either glycerate-3-phosphate, pyruvate, α-ketoglutarate, or oxaloacetate. Tyrosine, synthesized from the essential amino acid phenylalanine, is an exception to this rule.

On the basis of the similarities in their synthetic pathways, the amino acids can be grouped into six families: glutamate, serine, aspartate, pyruvate, the aromatics, and histidine. The amino acids in each family are ultimately derived from one precursor molecule. In the discussions of the amino acid synthesis that follow, the intimate relationship between amino acid metabolism and several other metabolic pathways is apparent. Amino acid biosynthesis is outlined in Figure 13.4.

The Glutamate Family The glutamate family includes—in addition to glutamate— glutamine, proline, and arginine. As described, α-ketoglutarate may be converted to glutamate by reductive amination and by transamination reactions involving a number of amino acids. Although the relative contribution of these reactions to glutamate synthesis varies with cell type and metabolic circumstances, transamination appears to play a major role in the synthesis of most glutamate molecules in eukaryotic cells. In addition to serving as a component of proteins and as a precursor for other amino acids, glutamate is also used in the central nervous system as an excitatory neurotransmitter. (The binding of *excitatory neurotransmitters* to certain receptors on nerve cell membrane promotes membrane depolarization.)

The conversion of glutamate to glutamine, catalyzed by glutamine synthase, takes place in a number of mammalian tissues (liver, brain, kidney, muscle, and intestine). BCAA (branched chain amino acids) are an important source of amino groups in glutamine synthesis. As mentioned, blood leaving the liver is selectively enriched in BCAA. Much more BCAA are taken up by peripheral tissues than is needed for protein synthesis. The amino groups of the BCAA may be used primarily for the synthesis of nonessential amino acids. In addition to its role in protein synthesis, glutamine is the amino group donor in numerous biosynthetic reactions (e.g., purine, pyrimidine, and amino sugar syntheses) and, as was previously mentioned, as a safe storage and transport form of NH_4^+. Glutamine is therefore a major metabolite in living organisms. Other functions of glutamine vary, depending on the cell type being considered. For example, in the kidney and small intestine, glutamine is a major source of energy. In the small intestine, approximately 55% of glutamine carbon is oxidized to CO_2.

Proline is a cyclized derivative of glutamate. As shown in Figure 13.5, a γ-glutamyl-phosphate intermediate is reduced to glutamic-γ-semialdehyde. The enzyme

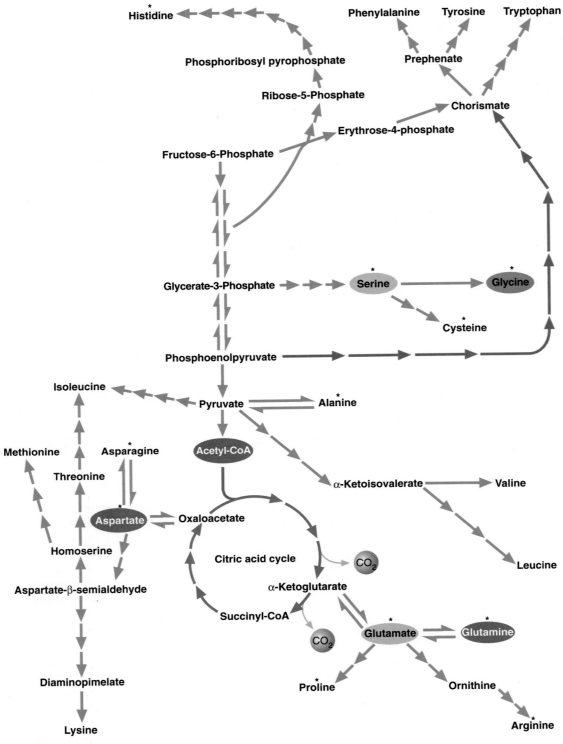

FIGURE 13.4

Biosynthesis of the Amino Acids.

Intermediates in the central metabolic pathways provide the carbon skeleton precursor molecules required for the synthesis of each amino acid. The number of reactions in each pathway is indicated. The nonessential amino acids for mammals are indicated by asterisks. (In mammals, tyrosine can be synthesized from phenylalanine.)

FIGURE 13.5

Biosynthesis of Proline and Arginine from Glutamate.

Proline is synthesized from glutamate in three steps. The second step is a spontaneous cyclization reaction. In arginine synthesis the acetylation of glutamate prevents the cyclization reaction. In mammals the reactions that convert ornithine to arginine are part of the urea cycle.

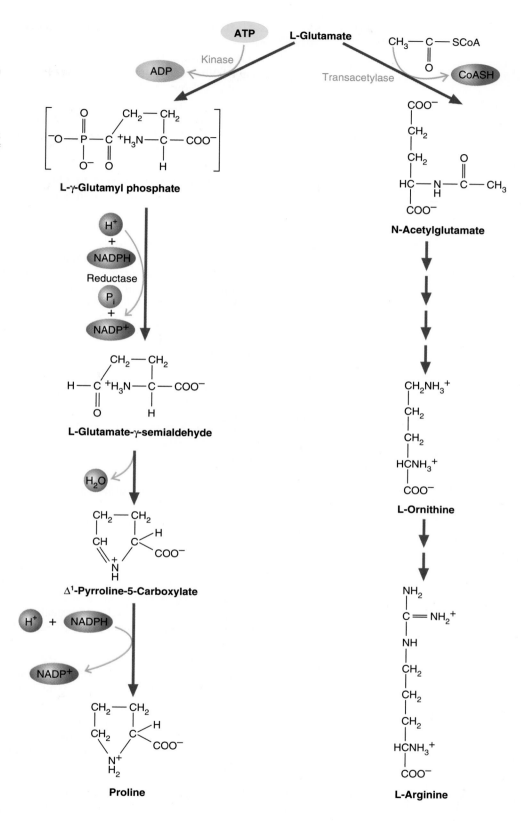

catalyzing the phosphorylation of glutamate (γ-glutamyl kinase) is regulated by negative feedback inhibition by proline. Glutamate-γ-semialdehyde cyclizes spontaneously to form Δ^1-pyrroline-5-carboxylate. Δ^1-Pyrroline-5-carboxylate reductase catalyzes the reduction of Δ^1-pyrroline-5-carboxylate to form proline. The interconversion of Δ^1-pyrroline-5-carboxylate and proline may act as a shuttle mechanism to transfer reducing equivalents derived from the pentose phosphate pathway into mitochondria. This process may partially explain the high turnover of proline in many cell types. Proline can also be synthesized from ornithine, a urea cycle intermediate. (The urea cycle is a pathway in which urea, the principal mammalian nitrogenous waste product, is produced. Urea synthesis is discussed in Chapter 14.) The enzyme catalyzing ornithine's conversion to glutamate-γ-semialdehyde, ornithine aminotransferase, is found in relatively high concentration in cells (e.g., fibroblasts) where the demand for proline incorporation into collagen is high.

Glutamate is also a precursor of arginine. Arginine synthesis begins with the acetylation of the α-amino group of glutamate. N-acetylglutamate is then converted to ornithine in a series of reactions that include a phosphorylation, a reduction, a transamination, and a deacetylation (removal of an acetyl group). The subsequent reactions in which ornithine is converted to arginine are part of the urea cycle. In infants, in whom the urea cycle is insufficiently functional, arginine is an essential amino acid.

The Serine Family The members of the serine family—serine, glycine, and cysteine—derive their carbon skeletons from the glycolytic intermediate glycerate-3-phosphate. The members of this group play important roles in numerous anabolic pathways. Serine is a precursor of ethanolamine and sphingosine. Glycine is used in the purine, porphyrin, and glutathione synthetic pathways. Together, serine and glycine contribute to a series of biosynthetic pathways that are referred to collectively as one-carbon metabolism (discussed in Section 13.3). Cysteine plays a significant role in sulfur metabolism (Chapter 14).

Serine is synthesized in a direct pathway from glycerate-3-phosphate that involves dehydrogenation, transamination, and hydrolysis by a phosphatase (Figure 13.6). Cellular serine concentration controls the pathway through feedback inhibition of phosphoglycerate dehydrogenase and phosphoserine phosphatase. The latter enzyme catalyzes the only irreversible step in the pathway. A high-protein diet also inhibits serine synthesis.

The conversion of serine to glycine consists of a single complex reaction catalyzed by serine hydroxymethyltransferase, a pyridoxal phosphate–requiring enzyme. During the reaction, which is an aldol cleavage, serine binds to pyridoxal phosphate. The reaction yields glycine and a chemically reactive formaldehyde group that is transferred to tetrahydrofolate (THF) to form N^5,N^{10}-methylene tetrahydrofolate. (The coenzyme tetrahydrofolate is discussed in Section 13.3.) Serine is the major source of glycine. Smaller amounts of glycine can be derived from choline, when the latter molecule is present in excess. The synthesis of glycine from choline consists of two dehydrogenations and a series of demethylations. Glycine acts as an inhibitory neurotransmitter within the central nervous system. (When *inhibitory neurotransmitters* bind to nerve cell receptors, which are usually linked to chloride channels, the membrane becomes hyperpolarized. Because the inside of the membrane is more negative in hyperpolarized neurons than it is in resting neurons, action potentials are unlikely.)

Cysteine synthesis is a primary component of sulfur metabolism. The carbon skeleton of cysteine is derived from serine (Figure 13.7). In animals the sulfhydryl group is transferred from methionine by way of the intermediate molecule homocysteine. (In many other organisms, H_2S is the source of the sulfhydryl group.) Both enzymes involved in the conversion of serine to cysteine (cystathionine synthase and γ-cystathionase) require pyridoxal phosphate.

FIGURE 13.6

Biosynthesis of Serine and Glycine.

Serine inhibits 3-phosphoglycerate dehydrogenase, the first reaction in the pathway.

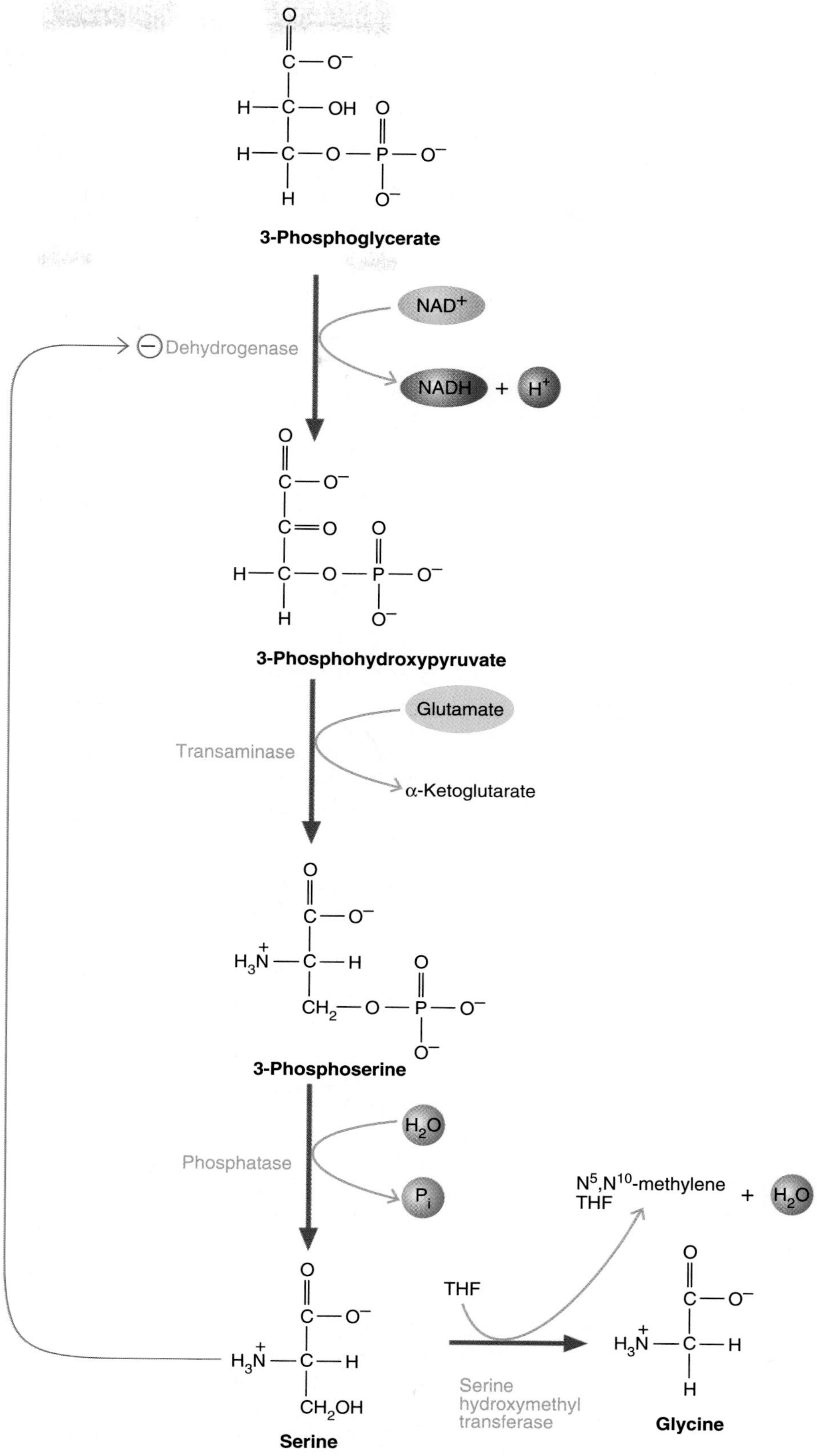

3-Phosphoglycerate

NAD$^+$

⊖ Dehydrogenase

NADH + H$^+$

3-Phosphohydroxypyruvate

Transaminase

Glutamate

α-Ketoglutarate

3-Phosphoserine

Phosphatase

H$_2$O

P$_i$

Serine

THF

Serine hydroxymethyl transferase

N^5,N^{10}-methylene THF + H$_2$O

Glycine

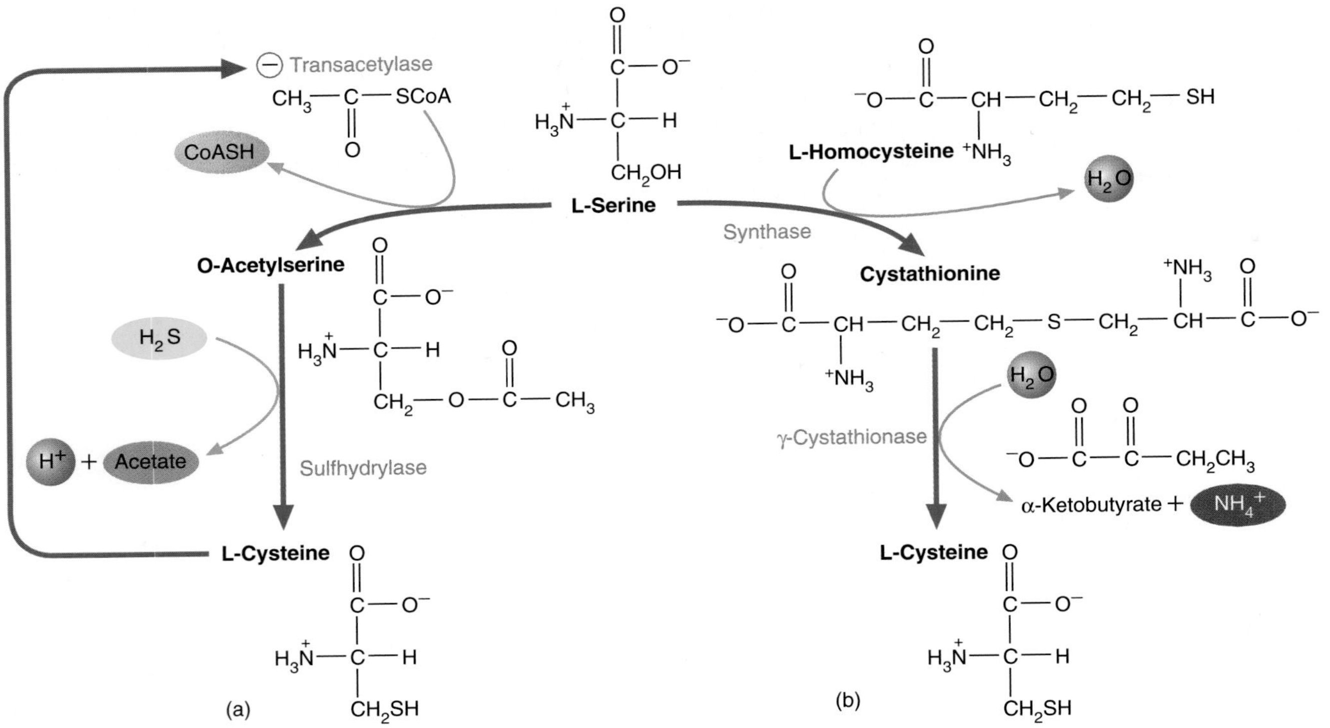

FIGURE 13.7

Biosynthesis of Cysteine.

(a) In plants and some bacteria, cysteine is synthesized in a two-step pathway. Serine is acetylated by serine acetyltransferase. The acetyl group is then displaced in a reaction with H_2S. (b) In animals serine condenses with homocysteine (derived from methionine) to form cystathionine. γ-Cystathionase catalyzes the cleavage of cystathionine to yield cysteine, α-ketobutyrate, and NH_4^+.

Aspartate Family   Aspartate, the first member of the aspartate family of amino acids, is derived from oxaloacetate in a transamination reaction.

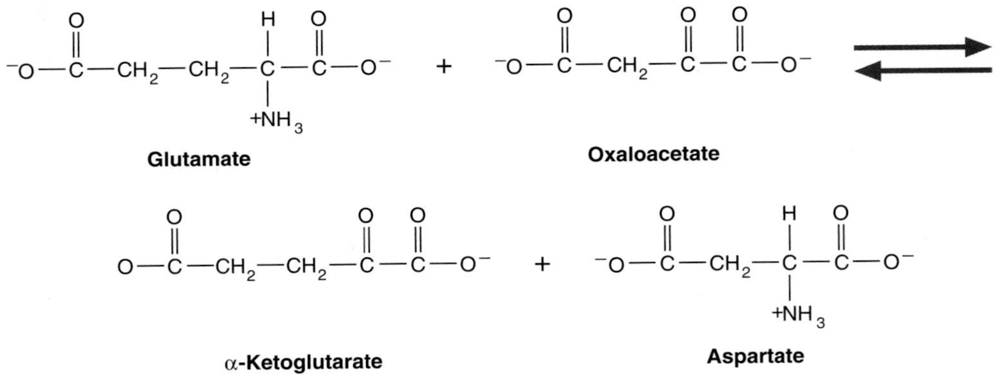

Aspartate transaminase (AST) (also known as glutamic oxaloacetic transaminase, or GOT), the most active of the aminotransferases, is found in most cells. Because AST isozymes occur in both mitochondria and the cytoplasm and the reaction that it catalyzes is reversible, this enzymatic activity significantly influences the flow of carbon and nitrogen within the cell. For example, excess glutamate is converted via AST to aspartate. Aspartate is then used as a source of both nitrogen (for urea formation) and the citric acid cycle intermediate fumarate. Aspartate is also an important precursor in nucleotide synthesis.

The aspartate family also contains asparagine, lysine, methionine, and threonine. Threonine contributes to the reaction pathway in which isoleucine is synthesized. The synthesis of isoleucine, often considered to be a member of the pyruvate family, is discussed on p. 377.

Asparagine, the amide of aspartate, is not formed directly from aspartate and NH_4^+. Instead, the amino group of glutamine is transferred by transamination during an ATP-requiring reaction catalyzed by asparagine synthase.

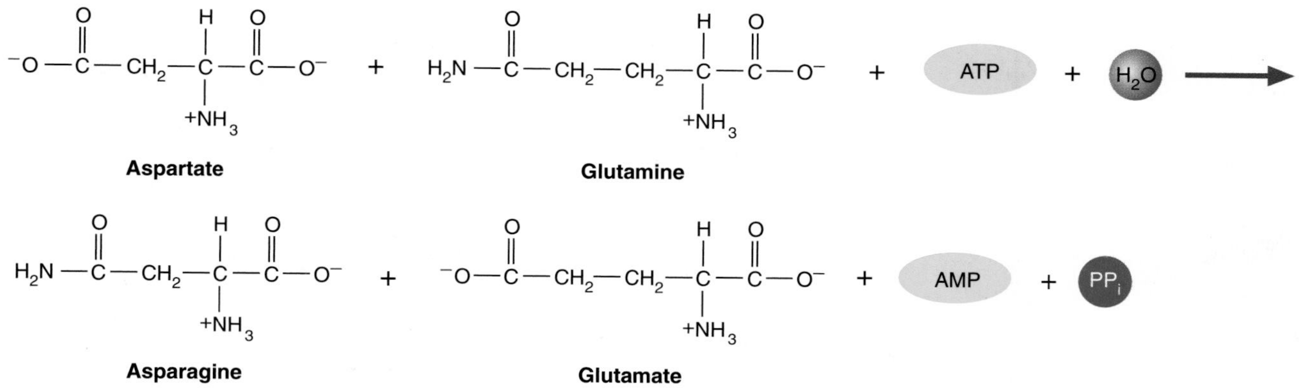

Aspartate   **Glutamine**

Asparagine   **Glutamate**

The synthesis of the other members of the aspartate family (Figure 13.8) is initiated by aspartate kinase (often referred to as aspartokinase) in an ATP-requiring reaction in which the side chain carboxyl group is phosphorylated. Aspartate-β-semialdehyde, produced by the NADPH-dependent reduction of β-aspartylphosphate, represents an important branch point in plant and bacterial amino acid synthesis. The semialdehyde can either react with pyruvate to form dihydropicolinic acid (a precursor of lysine) or be reduced to homoserine. Lysine is synthesized from dihydropicolinic acid in a series of reactions that is still poorly characterized. Homoserine also occurs at a branch point. It is the precursor in the synthesis of both methionine and threonine.

The Pyruvate Family The pyruvate family consists of alanine, valine, leucine, and isoleucine. Alanine is synthesized from pyruvate in a single step.

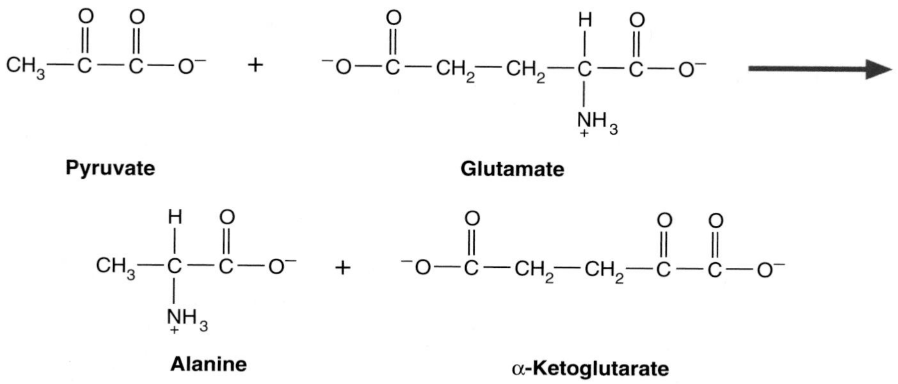

Pyruvate   **Glutamate**

Alanine   **α-Ketoglutarate**

Although the enzyme that catalyzes this reaction, alanine aminotransferase, has cytoplasmic and mitochondrial forms, the majority of its activity has been found in the cytoplasm. Recall that the alanine cycle (Chapter 8) contributes to the maintenance of blood glucose. It has recently become apparent that the BCAA are the ultimate source of many of the amino groups transferred from glutamate in the alanine cycle.

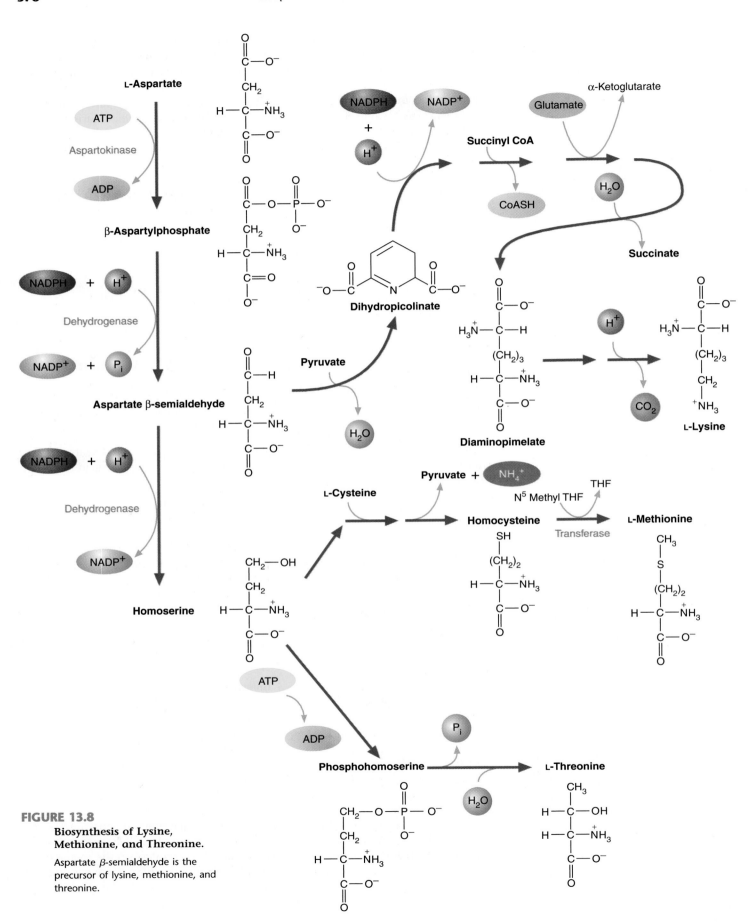

FIGURE 13.8

Biosynthesis of Lysine, Methionine, and Threonine.

Aspartate β-semialdehyde is the precursor of lysine, methionine, and threonine.

The synthesis of valine, leucine, and isoleucine from pyruvate is illustrated in Figure 13.9. Valine and isoleucine are synthesized in parallel pathways with the same four enzymes. Valine synthesis begins with the condensation of pyruvate with hydroxyethyl-TPP (a decarboxylation product of a pyruvate-thiamine pyrophosphate intermediate) catalyzed by acetohydroxy acid synthase. The α-acetolactate product is then reduced to form α,β-dihydroxyisovalerate. Valine is produced in a subsequent transamination reaction. (α-Ketoisovalerate is also a precursor of leucine.) Isoleucine synthesis also involves hydroxyethyl-TPP, which condenses with α-ketobutyrate to form α-aceto-α-hydroxybutyrate. (α-Ketobutyrate is derived from L-threonine in a deamination reaction catalyzed by threonine deaminase.) α,β-Dihydroxy-β-methylvalerate, the reduced product of α-aceto-α-hydroxybutyrate, subsequently loses an H_2O molecule, thus forming α-keto-β-methylvalerate. Isoleucine is then produced during a transamination reaction. In the first step of leucine biosynthesis from α-ketoisovalerate, acetyl-CoA donates a two-carbon unit. Leucine is formed after isomerization, reduction, and transamination.

The Aromatic Family The aromatic family of amino acids includes phenylalanine, tyrosine, and tryptophan. Of these, only tyrosine is considered to be nonessential in mammals. Either phenylalanine or tyrosine is required for the synthesis of dopamine, epinephrine, and norepinephrine, an important class of biologically potent molecules referred to as the **catecholamines** (Special Interest Box 13.2). Tryptophan is a precursor in the synthesis of NAD, NADP, and the neurotransmitter serotonin.

The benzene ring of the aromatic amino acids is formed by the *shikimate pathway.* The carbons in the benzene ring are derived from erythrose-4-phosphate and phosphoenolpyruvate. These two molecules condense to form 2-keto-3-deoxyarabinoheptulosinate-7-phosphate, a molecule that is subsequently converted to chorismate in a series of reactions that are outlined in Figure 13.10. Chorismate is the branch point in the syntheses of various aromatic compounds.

Figure 13.11 illustrates the syntheses of phenylalanine, tyrosine, and tryptophan from chorismate. (Chorismate is also a precursor in the synthesis of the aromatic rings in the mixed terpenoids, e.g., the tocopherols, the ubiquinones, and plastoquinone.)

Tyrosine is not an essential amino acid in animals because it is synthesized from phenylalanine in a hydroxylation reaction. The enzyme involved, phenylalanine-4-monoxygenase, requires the coenzyme tetrahydrobiopterin (Section 13.3), a folic acid–like molecule derived from GTP. Because this reaction also is a first step in phenylalanine catabolism, it is discussed further in Chapter 14.

Histidine Histidine is considered to be nonessential in heatlhy human adults. In human infants and many animals, histidine must be provided by the diet. Because of its unique chemical properties, histidine contributes substantially to protein structure and function. Recall, for example, that histidine residues bind heme prosthetic groups in hemoglobin. In addition, histidine often acts as a general acid during enzyme-catalyzed reactions. Of all the amino acids, histidine's biosynthesis is the most unusual. Histidine is synthesized from phosphoribosylpyrophosphate (PRPP), ATP, and glutamine (Figure 13.12). Synthesis begins with the condensation of PRPP with ATP to form phosphoribosyl-ATP. Phosphoribosyl-ATP is then hydrolyzed by phosphoribosyl-ATP pyrophosphorylase to phosphoribosyl-AMP. In the next step, a hydrolytic reaction opens the adenine ring. After an isomerization and the transfer of an amino group from glutamine, imidazole glycerol phosphate is synthesized. (The other product of the latter reaction, 5'-phosphoribosyl-4-carboxamide-5-aminoimidazole, is used in the synthesis of purine nucleotides. See Section 13.3.) Histidine is produced from imidazole glycerol phosphate in a series of reactions that include a dehydration, a transamination, a phosphorolysis, and an oxidation.

KEY*Concepts*

There are six families of amino acids: glutamate, serine, aspartate, pyruvate, the aromatics, and histidine. The nonessential amino acids are derived from precursor molecules available in many organisms. The essential amino acids are synthesized from metabolites produced only in plants and some microorganisms.

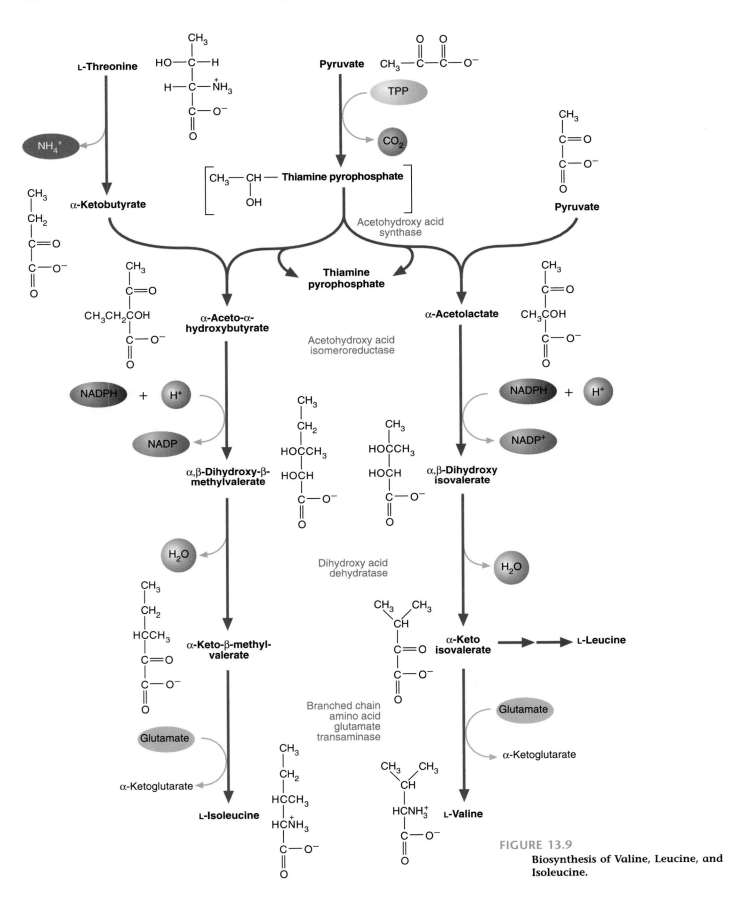

FIGURE 13.9

Biosynthesis of Valine, Leucine, and Isoleucine.

FIGURE 13.10

Chorismate Biosynthesis.

Chorismate is an intermediate in the shikimate pathway.

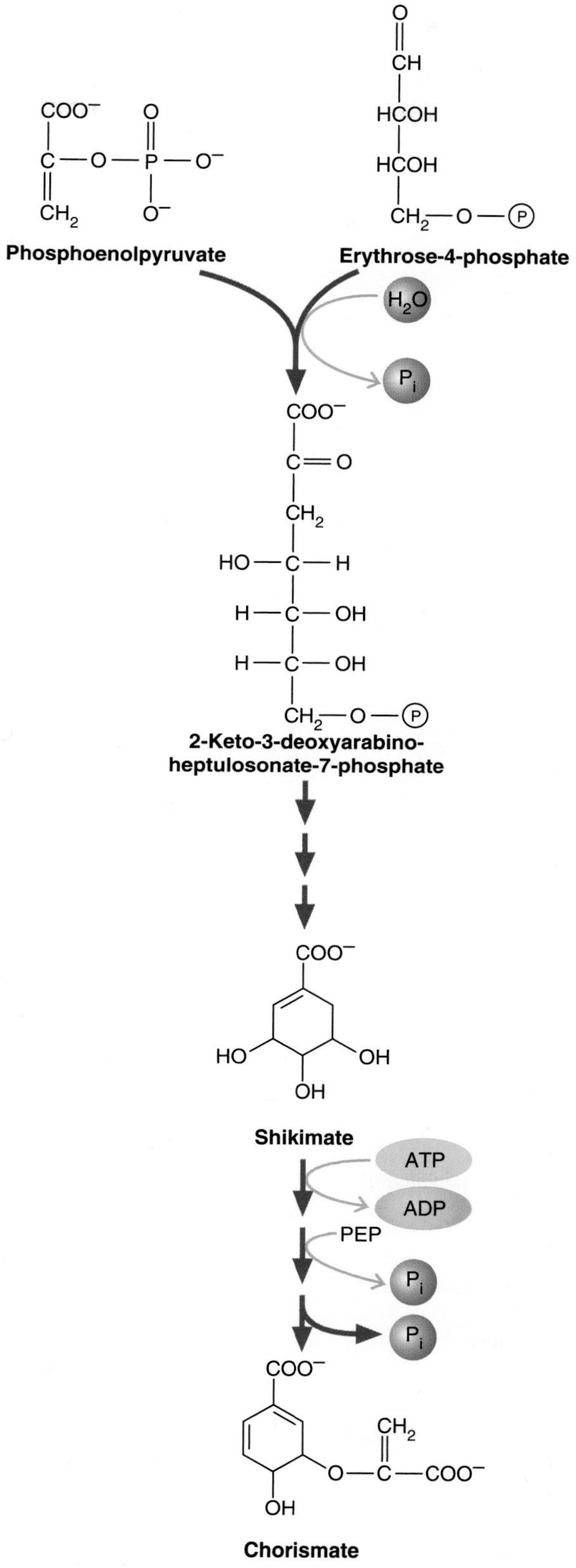

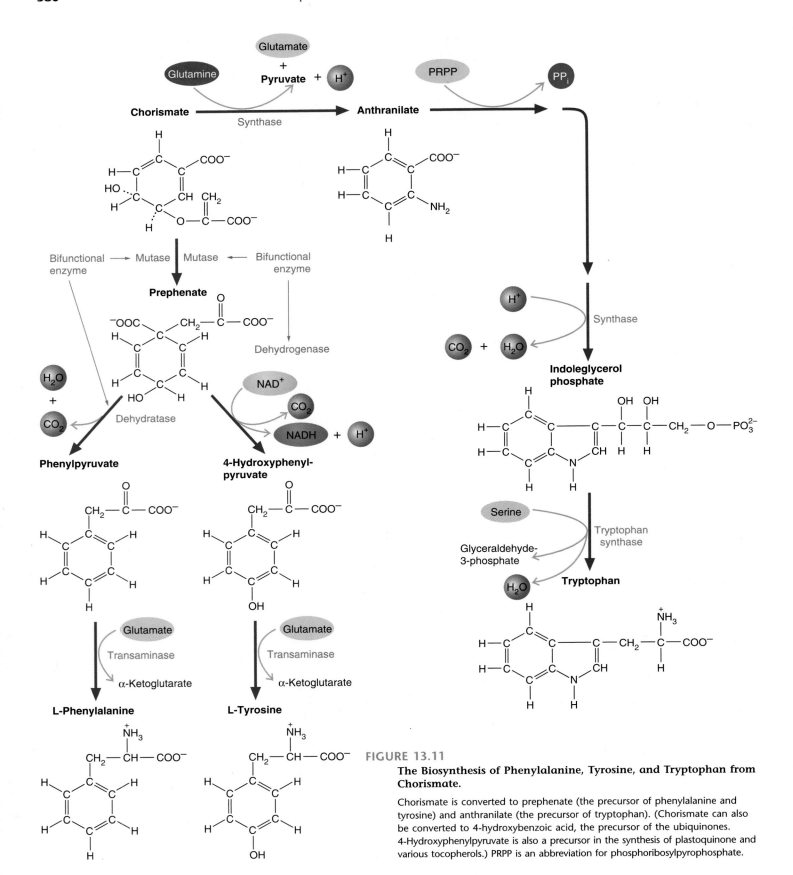

FIGURE 13.11

The Biosynthesis of Phenylalanine, Tyrosine, and Tryptophan from Chorismate.

Chorismate is converted to prephenate (the precursor of phenylalanine and tyrosine) and anthranilate (the precursor of tryptophan). (Chorismate can also be converted to 4-hydroxybenzoic acid, the precursor of the ubiquinones. 4-Hydroxyphenylpyruvate is also a precursor in the synthesis of plastoquinone and various tocopherols.) PRPP is an abbreviation for phosphoribosylpyrophosphate.

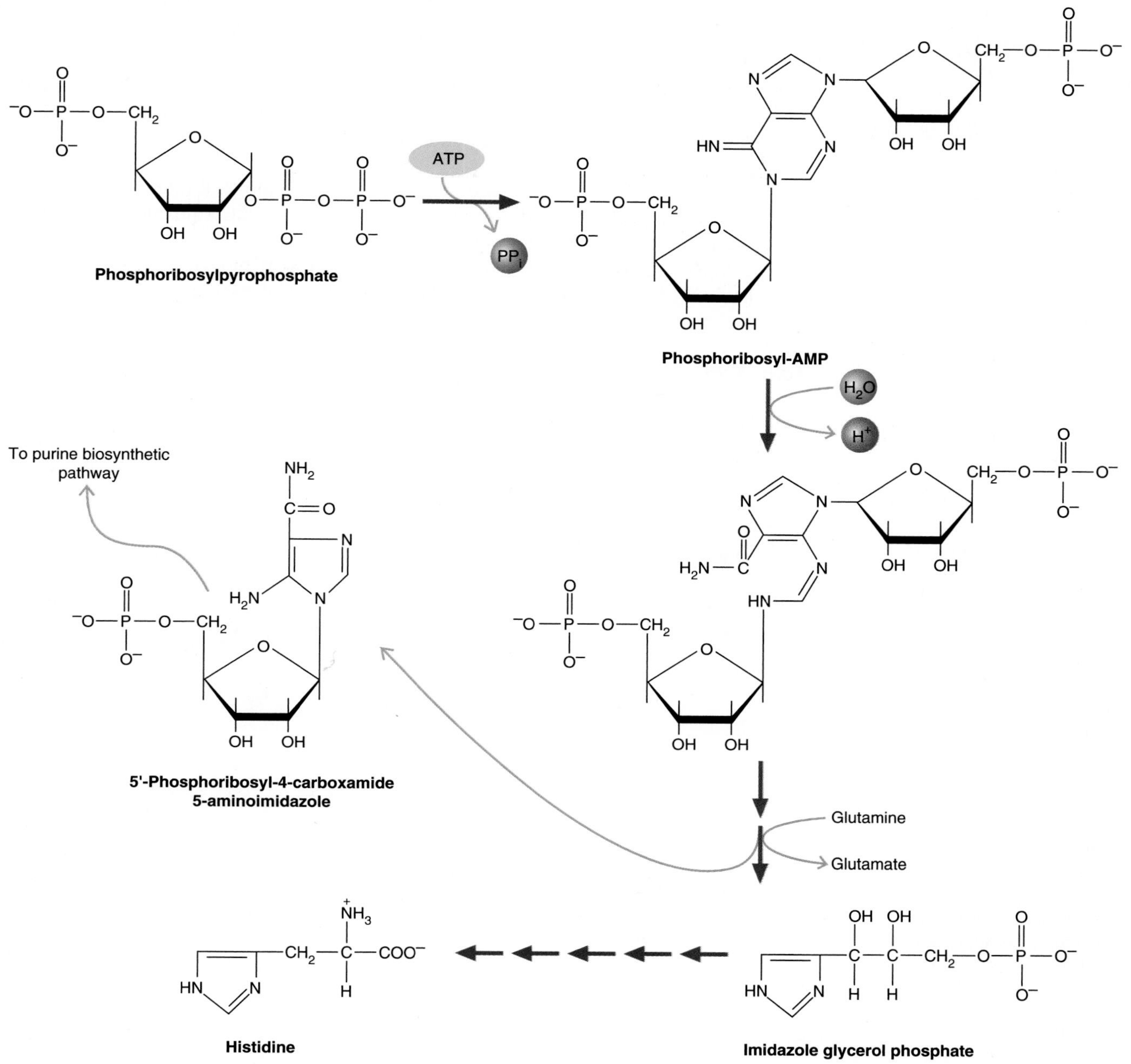

FIGURE 13.12

Histidine Biosynthesis.

The ATP used in the first reaction in the pathway is regenerated when 5-phosphoribosyl-4-carboxamide-5-aminoimidazole (released in a subsequent reaction) is diverted into the purine nucleotide biosynthetic pathway.

13.3 BIOSYNTHETIC REACTIONS OF AMINO ACIDS

As described, amino acids are precursors of many physiologically important nitrogen-containing molecules, in addition to serving as building blocks for polypeptides. In the following discussion the syntheses of several examples of these molecules (e.g., neurotransmitters, glutathione, alkaloids, nucleotides, and heme) are described. Because many of these processes involve the transfer of carbon groups, this section begins with a brief description of one-carbon metabolism.

One-Carbon Metabolism

Carbon atoms have several oxidation states. Those of biological interest are found in methanol, formaldehyde, and formate. Table 13.2 lists the equivalent one-carbon groups that are actually involved in synthetic reactions.

The most important carriers of one-carbon groups in biosynthetic pathways are folic acid and S-adenosylmethionine. The metabolism of each are described briefly. (The function of biotin, a carrier of CO_2 groups, is discussed in Section 8.3.)

Folic Acid   Folic acid, also known as folate or folacin, is a B vitamin. Its structure consists of a pteridine nucleus and para-aminobenzoic acid, linked to one or more glutamic acid residues (Figure 13.13). Once it is absorbed by the body, folic acid is converted by dihydrofolate reductase to the biologically active form, tetrahydrofolic

TABLE 13.2
One-Carbon Groups

Oxidation Level	Methanol (most reduced)	Formaldehyde	Formate (most oxidized)
One-carbon group	Methyl ($-CH_3$)	Methylene ($-CH_2-$)	Formyl ($-CHO$)
			Methenyl ($-CH=$)

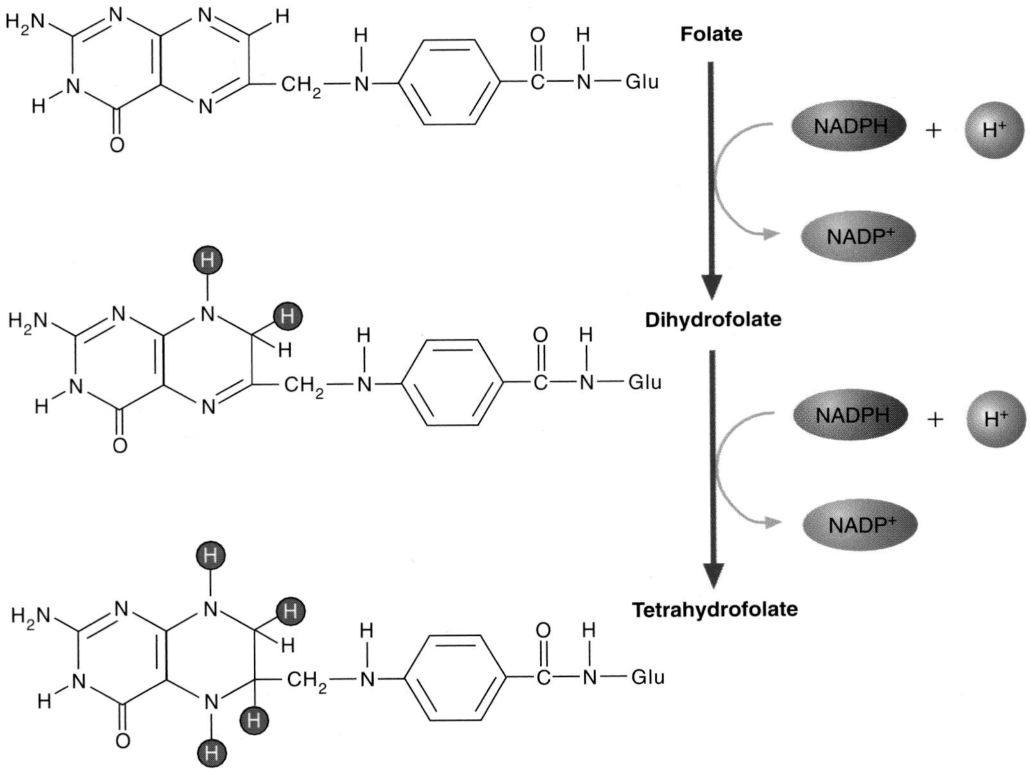

FIGURE 13.13
Biosynthesis of Tetrahydrofolate (THF).

The vitamin folic acid (folate) is converted to its biologically active form by two successive reductions of the pteridine ring. Both reactions are catalyzed by dihydrofolate reductase.

acid (THF). The carbon units carried by THF (i.e., methyl, methylene, methenyl, and formyl groups) are bound to N^5 and/or N^{10} of the pteridine ring. Figure 13.14 illustrates the interconversions of the one-carbon units carried by THF, as well as their origin and metabolic fate. A substantial amount of one-carbon units enter the THF

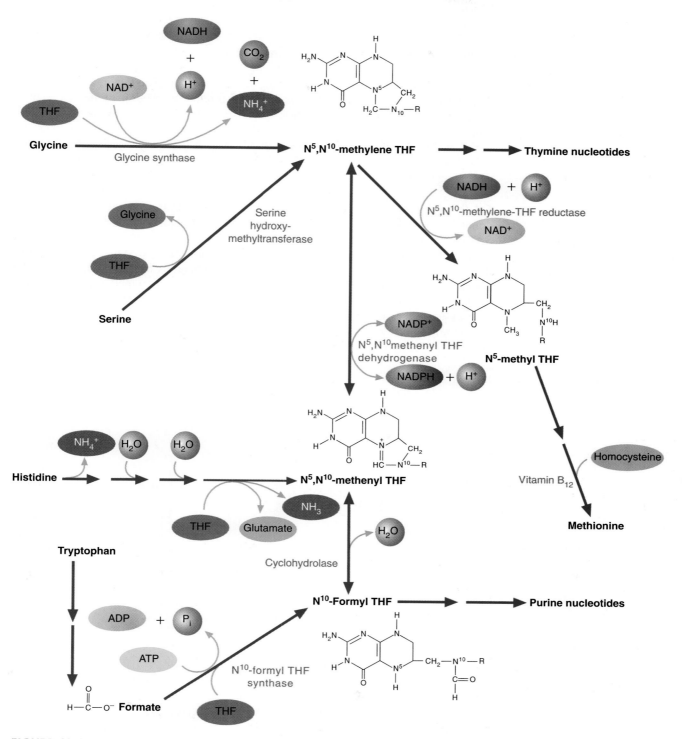

FIGURE 13.14

Structures and Enzymatic Interconversions of THF Coenzymes.

The THF coenzymes play a critical role in one-carbon metabolism. The interconversions of the coenzymes are reversible except for the conversion of N^5,N^{10}-methylene THF to N^5-methyl THF.

pool as N^5,N^{10}-methylene THF, produced during the conversion of serine to glycine and the cleavage of glycine (catalyzed by glycine synthase).

In Figure 13.14, vitamin B_{12} is required for the N^5-methyl THF-dependent conversion of homocysteine to methionine. **Vitamin B_{12}** (cobalamin) is a complex, cobalt-containing molecule synthesized only by microorganisms (Figure 13.15). (During the purification of cobalamin, a cyanide group attaches to cobalt.) Animals obtain it from intestinal flora and by consuming foods derived from other animals (e.g., liver, eggs, shrimp, chicken, and pork). A deficiency of vitamin B_{12} results in **pernicious anemia.** In addition to low red blood cell counts, the symptoms of this malady include weakness and various neurological disturbances. Pernicious anemia is most often caused by decreased secretion of intrinsic factor, a glycoprotein secreted by stomach cells, which is required for the absorption of the vitamin in the intestine. Vitamin B_{12} absorption can also be inhibited by several gastrointestinal disorders, such as coeliac disease or tropical sprue, both of which damage the lining of the intestine. A reduction in vitamin B_{12} absorption has also been observed in the presence of intestinal overgrowths of microorganisms induced by antibiotic treatments.

S-Adenosylmethionine   S-Adenosylmethionine (SAM) is the major methyl group donor in one-carbon metabolism. Formed from methionine and ATP (Figure 13.16), SAM contains an "activated" methyl thioether group, which can be transferred to a variety of acceptor molecules (Table 13.3). S-Adenosylhomocysteine is a product in these reactions. The loss of free energy that accompanies S-adenosylhomocysteine formation makes the methyl transfer irreversible. SAM acts as a methyl donor in many transmethylation reactions, some of which occur in the synthesis of phospholipids, several neurotransmitters, and glutathione.

The importance of SAM in metabolism is reflected in the several mechanisms that provide for the synthesis of sufficient amounts of its precursor, methionine, when the

KEY *Concepts*

Tetrahydrofolate, the biologically active form of folic acid, and S-adenosyl methionine are important carriers of single carbon atoms in a variety of synthetic reactions.

FIGURE 13.15

Structure of Cyanocobalamin (Vitamin B_{12}).

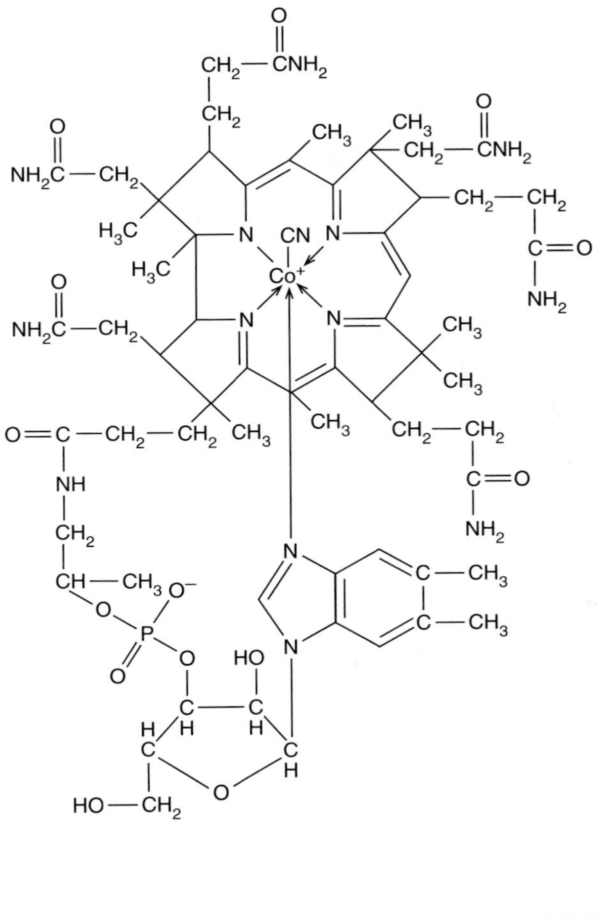

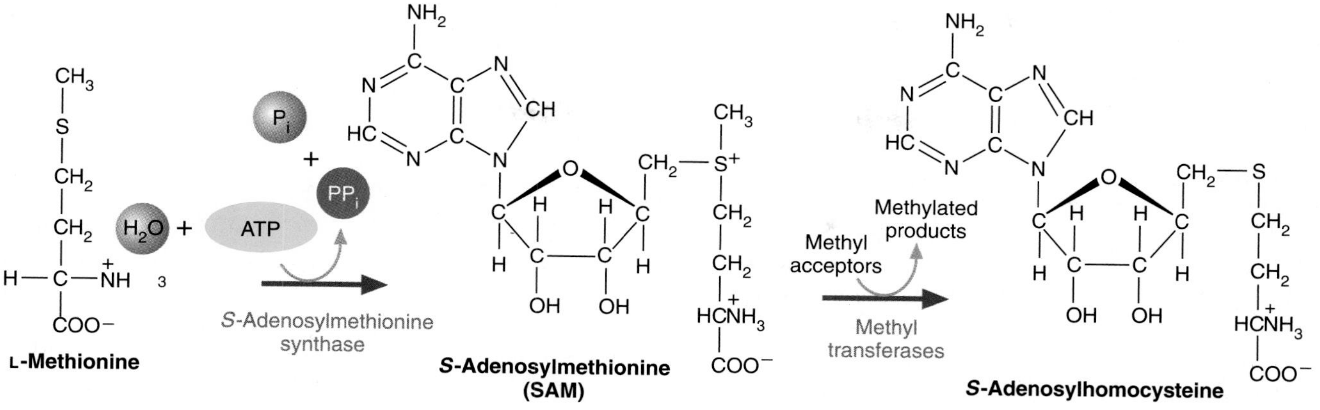

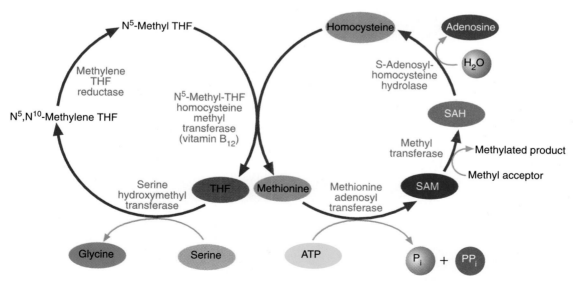

FIGURE 13.16

The Formation of S-Adenosylmethionine.

One of the principal functions of SAM is to serve as a methylating agent.

TABLE 13.3

Examples of Transmethylation Acceptors and Products

Methyl Acceptors	Methylated Product
Phosphatidylethanolamine	Phosphatidylcholine
Norepinephrine	Epinephrine
Guanidinoacetate	Creatine
γ-Aminobutyric acid	Carnitine

latter molecule is temporarily absent from the diet. For example, choline is used as a source of methyl groups to convert homocysteine into methionine. Homocysteine can also be methylated in a reaction utilizing N^5-methyl THF. This latter reaction is a bridge between the THF and SAM pathways (Figure 13.17).

FIGURE 13.17

The Tetrahydrofolate and S-Adenosylmethionine Pathways.

The THF and SAM pathways intersect at the reaction, catalyzed by N^5-methyl THF homocysteine methyltransferase, in which homocysteine is converted to methionine.

NEUROTRANSMITTERS

More than 30 different substances have been proven or proposed to act as neurotransmitters. **Neurotransmitters** are either excitatory or inhibitory in nature. As noted, excitatory neurotransmitters (e.g., glutamate and acetylcholine) promote the depolarization of the membrane in another cell (either another neuron or an effector cell, such as a muscle cell). If the second (postsynaptic) cell is a neuron, the wave of depolarization (referred to as an action potential) triggers the release of neurotransmitter molecules as it reaches the end of the axon. (Most neurotransmitter molecules are stored in numerous membrane-enclosed *synaptic vesicles.*) When the action potential reaches the nerve ending, the neurotransmitter molecules are released by exocytosis into the synapse. If the postsynaptic cell is a muscle cell, sufficient release of excitatory neurotransmitter molecules results in muscle contraction. Inhibitory neurotransmitters (e.g., glycine) make the membrane potential in the postsynaptic cell even more negative, that is, they inhibit the formation of an action potential.

A significant percentage of neurotransmitter molecules are either amino acids or amino acid derivatives (Table 13.4). The latter class is often referred to as the **biogenic amines.** After

TABLE 13.4

Amino Acid and Amine Neurotransmitters

Amino Acids	Amines
Glycine	Norepinephrine*
Glutamate	Epinephrine*
γ-Aminobutyric acid (GABA)	Dopamine*
	Serotonin
	Histamine

*These molecules are referred to as the catecholamines.

a brief discussion of several biogenic amines, the properties of nitric oxide, a newly recognized neurotransmitter, is described.

γ-Aminobutyric Acid

γ-Aminobutyric acid (GABA) acts as an inhibitory neurotransmitter in the central nervous system. The binding of GABA to its receptor increases the nerve cell membrane's permeability to chloride ions. (The benzodiazopines, a class of tranquilizers that alleviates

FIGURE 13A

Biosynthesis of the Catecholamines.

Dopamine, norepinephrine, and epinephrine act as neurotransmitters and/or hormones. (PNMT is an abbreviation for phenylethanolamine-N-methyltransferase.)

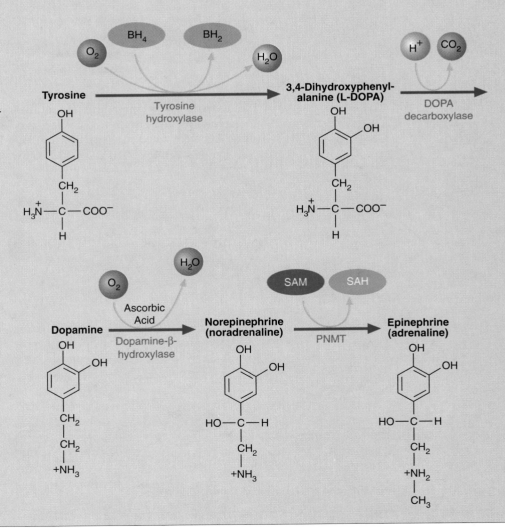

anxiety and aggressive behavior, have been shown to enhance GABA's ability to increase membrane conductance of chloride.)

GABA is produced by the decarboxylation of glutamate. The reaction is catalyzed by glutamate decarboxylase, which is a pyridoxal phosphate–requiring enzyme:

Glutamate → **γ-Aminobutyric acid (GABA)**

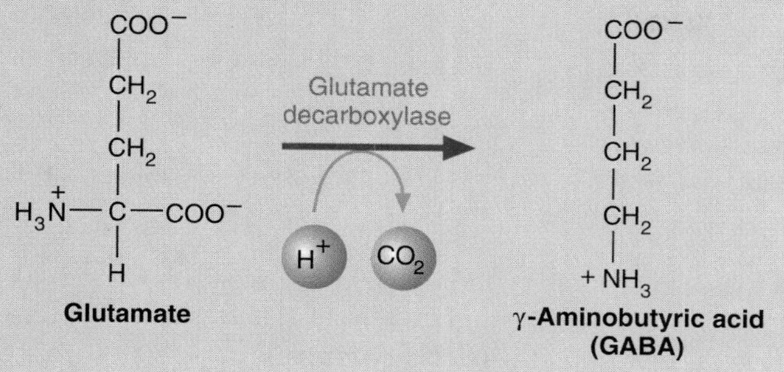

The Catecholamines

The *catecholamines* (dopamine, norepinephrine, and epinephrine) are derivatives of tyrosine. Dopamine (D) and norepinephrine (NE) are used in the brain as excitatory neurotransmitters. Outside the central nervous system, NE and epinephrine (E) are released primarily from the adrenal medulla, as well as the peripheral nervous system. Because both NE and E regulate aspects of metabolism, they are often considered hormones.

The first, and rate-limiting, step in catecholamine synthesis is the hydroxylation of tyrosine to form 3,4-dihydroxyphenylalanine (DOPA) (Figure 13A). Tyrosine hydroxylase, the mitochondrial enzyme that catalyzes the reaction, requires a cofactor known as *tetrahydrobiopterin* (BH_4). (BH_4, a folic acid–like molecule, is an essential cofactor in the hydroxylation of aromatic amino acids. BH_4 is regenerated from its oxidized metabolite, BH_2, by reduction with NADPH.)

the central nervous system. Deficiency in dopamine production has been found to be associated with Parkinson's disease, a serious degenerative neurological disorder. (See Special Interest Box 13.3.) The precursor L-DOPA is used to alleviate the symptoms of Parkinson's disease because dopamine cannot penetrate the blood-brain barrier. (The blood-brain barrier protects the brain from toxic substances. Many polar molecules and ions cannot move from blood capillaries, while most lipid-soluble substances readily pass across. The blood-brain barrier consists of connective tissue and specialized cells called astrocytes that envelope the capillaries.) Once L-DOPA is transported into appropriate nerve cells, it is converted to dopamine.

Norepinephrine is synthesized from tyrosine in the chromaffin cells of the adrenal medulla in response to fright, cold, and exercise as well as low levels of blood glucose. NE acts to stimulate the degradation of triacylglycerol and glycogen. It also increases cardiac output and blood pressure. The hydroxylation of dopamine to produce NE is catalyzed by the copper-containing enzyme dopamine-β-hydroxylase, an oxidase that requires the antioxidant ascorbic acid for full activity.

As described, the secretion of epinephrine in response to stress, trauma, extreme exercise, or hypoglycemia causes a rapid mobilization of energy stores, that is, glucose from the liver and fatty acids from adipose tissue. The reaction in which NE is methylated to form E is mediated by the enzyme phenylethanolamine-N-methyltransferase (PNMT). Although the enzyme occurs predominantly in the chromaffin cells of the adrenal medulla, it is also found in certain portions of the brain where E functions as a neurotransmitter. Recent evidence indicates that both E and NE are present in several other organs (e.g., liver, heart, and lung). Bovine PNMT is a monomeric protein (30 kD) that uses SAM as a source of methyl groups.

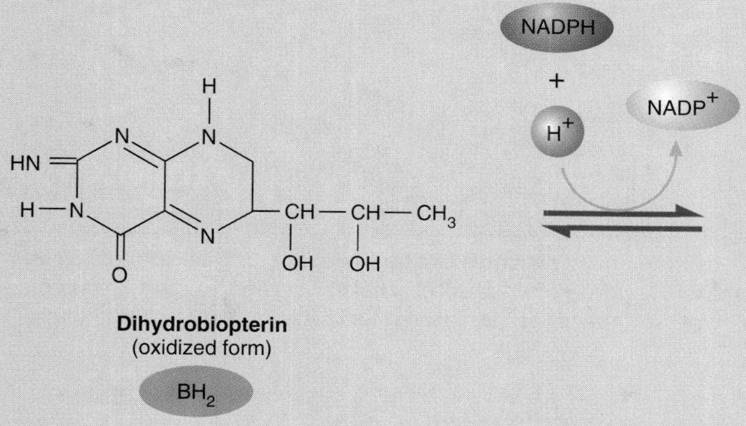

Dihydrobiopterin (oxidized form) — BH_2

Tetrahydrobiopterin (reduced form) — BH_4

Tyrosine hydroxylase uses BH_4 to activate O_2. One oxygen atom is attached to tyrosine's aromatic ring, while the other atom oxidizes the coenzyme. DOPA, the product of the reaction, is used in the synthesis of the other catecholamines.

DOPA decarboxylase, a pyridoxal phosphate–requiring enzyme, catalyzes the synthesis of dopamine from DOPA. Dopamine is produced in neurons found in certain structures in the brain. It is believed to exert an inhibitory action within

Serotonin

Serotonin is found in various cells within the central nervous system, where it inhibits feeding. (Serotonin has been implicated in human eating disorders such as anorexia nervosa, bulimia, and the carbohydrate craving associated with seasonal affective disorder (SAD). SAD is a clinical depression triggered by the decreased daylight in autumn and winter.) Additionally, serotonin appears to affect mood,

Tryptophan   **5-Hydroxytryptophan**   **5-Hydroxytryptamine (serotonin)**

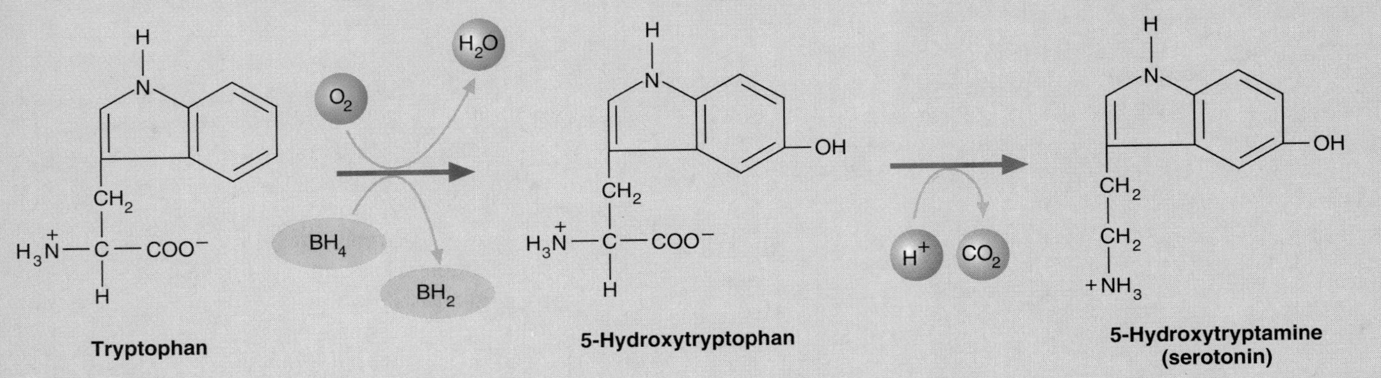

temperature regulation, pain perception, and sleep. The hallucinogenic drug LSD (lysergic acid diethylamide) apparently competes with serotonin for specific brain cell receptors. Serotonin is also found in the gastrointestinal tract, blood platelets, and mast cells.

Tryptophan hydroxylase uses O_2 and the electron donor BH_4 to hydroxylate C—5 of tryptophan. The product, called 5-hydroxytryptophan, then undergoes a decarboxylation catalyzed by 5-hydroxytryptophan decarboxylase, a pyridoxal phosphate-requiring enzyme. Serotonin, often referred to as 5-hydroxytryptamine, is the product of this reaction.

Histamine

Histamine, an amine produced in numerous tissues throughout the body, has complex physiological effects. It is a mediator of allergic and inflammatory reactions, a stimulator of gastric acid production, and a neurotransmitter in several areas of the brain. Histamine is formed by the decarboxylation of L-histidine in a reaction catalyzed by histidine decarboxylase, a pyridoxal phosphate–requiring enzyme.

Histidine   **Histamine**

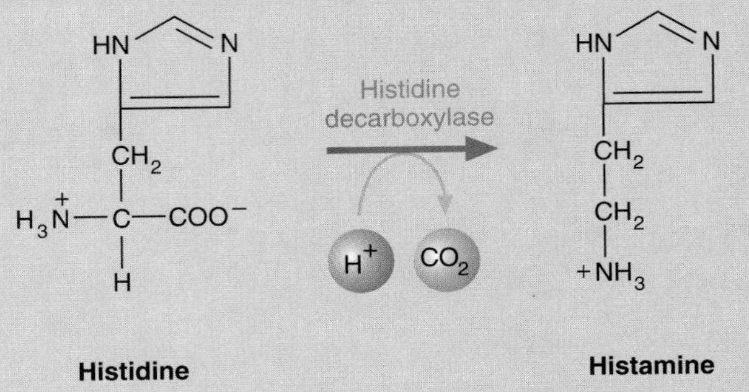

Nitric Oxide

Nitric oxide (NO) is a highly reactive gas. Because of its free radical structure, NO has been regarded, until recently, primarily as a contributing factor in the destruction of the ozone layer in the earth's atmosphere and as a precursor of acid rain.

13.1

Amethopterin, also referred to as **methotrexate,** is a structural analog of folate. (**Analogs** are compounds that closely resemble other molecules.) Methotrexate has been used to treat several types of cancer. It has been especially successful in childhood leukemia.

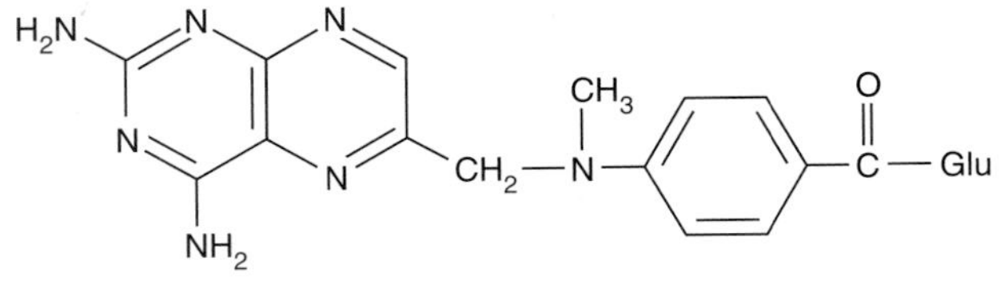

Amethopterin (methotrexate)

Using your knowledge of cell biology and biochemistry, suggest a biochemical mechanism that explains why amethopterin is effective against cancer. (Hints: Compare the structures of folate and methotrexate. Review Figure 13.13.)

Recent research has revealed, however, that NO is produced in at least a dozen cell types found in the mammalian body (e.g., in blood vessels, liver, several gastrointestinal organs, and the brain). Physiological functions in which NO is now believed to play a role include the dilation of blood vessels, the inhibition of platelet aggregation, and the destruction of foreign or damaged cells by macrophages. NO is also produced in many areas of the brain, where its formation has been linked to the neurotransmitter function of glutamate.

Glutamate is released from neurons in response to the synthesis of the second messenger molecule cGMP. (The conversion of GTP to cGMP is catalyzed by the enzyme guanylate cyclase.) When glutamate binds to a certain class of receptors, a transient flow of Ca^{2+} through the postsynaptic membrane triggers

the synthesis of NO. The enzyme nitric oxide synthase, which catalyzes the conversion of arginine to citrulline and NO, requires NADPH and O_2 (Figure 13B). (The activation of nitric oxide synthase by Ca^{2+} is mediated by *calmodulin*, a multipurpose calcium-binding protein.) Once it is synthesized, NO diffuses back to the presynaptic cell, where it activates guanylate cyclase. (Because of its reactive nature, NO is rapidly inactivated by oxidation to nitrite and nitrate.) Consequently, NO acts as a so-called retrograde neurotransmitter, that is, it promotes a cycle in which glutamate is released from the presynaptic neuron and then binds to and promotes action potentials in the postsynaptic neuron. This potentiating mechanism is now believed to play a role in learning and memory formation, as well as other functions in mammalian brain.

FIGURE 13B

Synthesis of Nitric Oxide.

Nitric oxide synthase in some cells (e.g., certain neurons) is activated by Ca^{2+}. In other cells the nitric oxide synthase activity is activated by cytokines. (Cytokines are a group of polypeptides produced by certain immune system cells.)

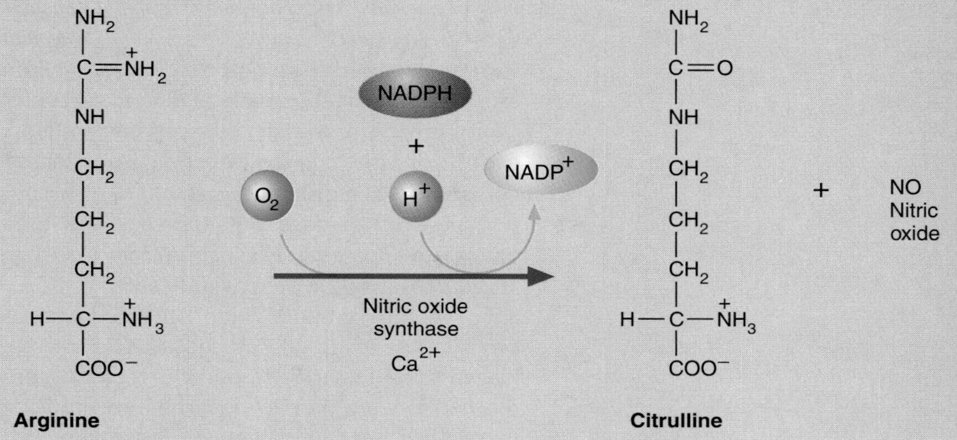

Arginine + O_2 + H^+ (NADPH → $NADP^+$) → Citrulline + NO Nitric oxide

Nitric oxide synthase Ca^{2+}

QUESTION

13.2 Melatonin is a hormone derived from serotonin. It is produced in the brain's light-sensitive pineal gland. The pineal's secretion of melatonin is depressed by nerve impulses that originate in the retina of the eye and other light sensitive tissue in the body in response to light. Pineal function is involved in *circadian rhythms,* patterns of activity associated with light and dark, such as sleep/wake cycles. In many mammals the functioning of the pineal gland also regulates seasonal cycles of fertility and infertility. (Melatonin inhibits the secretion of certain hormones from the hypothalamus and pituitary that stimulate the functioning of the ovaries and testes.) For example, in some species (e.g., deer) the males are fertile only in early spring, ensuring that newborn animals will be mature enough to survive the next winter.

After serotonin is produced within the pineal gland, it is converted to 5-hydroxy-N-acetyltryptamine by N-acetyl transferase. 5-Hydroxy-N-acetyltryptamine is then methylated by O-methyl transferase. SAM is the methylating agent. With this information, draw the synthetic pathway of melatonin.

13.3   Auxins are a class of plant growth regulators. Synthesized in meristematic (actively growing) tissue in response to light, auxin molecules diffuse to nearby cells. Depending on several parameters (e.g., concentration and location), auxins may either stimulate or inhibit cell growth. For example, auxins stimulate growth in the main shoot but inhibit growth in lateral shoots. Indole-3-acetic acid (IAA), the best-characterized auxin, is derived from tryptophan.

$$CH_2COOH$$

Auxin
(indole acetic acid)

Compare the structure of IAA with that of tryptophan. Suggest two reaction types involved in IAA synthesis.

13.4   It now appears that oxidative stress is a causative factor in the brain damage associated with stroke, head trauma, and several age-related neurological disorders (e.g., Parkinson's disease and Huntington's disease). There is also compelling evidence that excitatory neurotransmitters such as glutamate contribute to the brain damage. In a destructive mechanism that is not yet completely understood, glutamate can act as an excitotoxin; that is, excessive release of glutamate excites nearby neurons to death. Under normal circumstances, neurons are saved from such damage by glutamate transporters, which remove glutamate from the extracellular space. The excitotoxic effect of glutamate apparently results from the disabling of glutamate transporters. NO or its oxidized derivative the peroxynitrite anion ($ONOO^-$), is believed to actually cause the damage. (The peroxynitrite anion is formed when NO reacts with $O_2^{\overline{\cdot}}$. Once it forms, the peroxynitrite anion quickly decomposes to $\cdot OH$ and NO_2.) Use your knowledge of oxidative stress to suggest how $O_2^{\overline{\cdot}}$ might be produced within nerve cells. What defense mechanisms are probably available to protect the brain from oxidative damage? Why might they provide inadequate protection after a stroke or head trauma? Considering NO's role in normal glutamate function in the brain, why is this molecule such a lethal factor in excessive oxidative stress?

Glutathione

The nitrogen-containing molecule glutathione (γ-glutamylcysteinylglycine) is the most common intracellular thiol. (Its concentration in mammalian cells varies from 0.5 to 10 mM.) The functions of glutathione (GSH) include involvement in DNA and RNA synthesis as well as the synthesis of certain eicosanoids and other biomolecules. (In many of these processes, GSH acts as a reducing agent. As such, it maintains the sulfhydryl groups of enzymes and other molecules in a reduced state.) In addition to protecting cells from radiation, oxygen toxicity, and environmental toxins, GSH also promotes amino acid transport. After a brief discussion of its synthesis, the transport role of GSH is described. This is followed by a discussion of an interesting class of enzymes called the glutathione-S-transferases.

GSH is synthesized in a pathway composed of two reactions. In the first reaction, γ-glutamylcysteine synthase catalyzes the condensation of glutamate with cysteine (Figure 13.18). γ-Glutamylcysteine, the product of this reaction, then combines with glycine to form GSH.

Transport   Transport of GSH out of cells appears to serve several functions. Among these are (1) transfer of the sulfur atoms of cysteine between cells, (2) protection of the plasma membrane from oxidative damage, and (3) transfer to membrane-bound γ-glutamyl transpeptidase leading to the formation of γ-glutamyl amino acid derivatives. The latter process, which initiates the γ-glutamyl cycle, occurs in brain, intestine, pancreas, liver, and kidney cells.

The γ-glutamyl cycle provides for the active transport of several amino acids, especially cysteine and methionine, as well as GSH itself. Some researchers consider

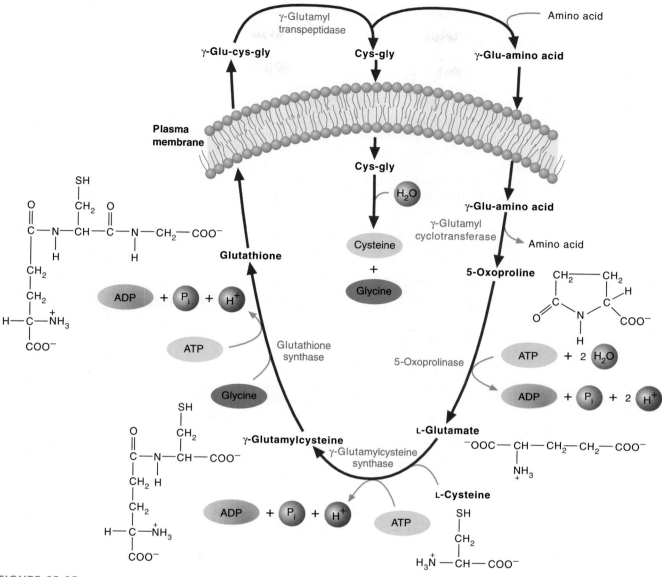

FIGURE 13.18

The γ-Glutamyl Cycle.

The functions of the γ-glutamyl cycle are described in the text. (1) Glutathione is excreted from the cell. γ-Glutamyltranspeptidase converts GSH to a γ-Glu-amino acid derivative and Cys-Gly. (2) The γ-glu-amino acid is transported into the cell, where it is converted to 5-oxoproline and the free amino acid. 5-Oxoproline is eventually reconverted to GSH. (3) Cys-Gly is transported into the cell, where (4) it is hydrolyzed to cysteine and glycine.

the amino acid transport function of GSH controversial. An alternative view is that the γ-glutamyl cycle generates a signal that activates amino acid uptake. 5-Oxoproline has been proposed as such a signal.

Several human diseases are associated with deficiencies in GSH metabolism. One of the most notable is *GSH synthase deficiency* (also known as *5-oxoprolinuria*). GSH synthase deficiency is characterized by a severe acidosis, hemolysis (red blood cell destruction), and central nervous system damage. Because of the enzyme deficiency, the concentration of γ-glutamylcysteine increases. This molecule is then converted to 5-oxoproline and cysteine by γ-glutamyl cyclotransferase. The production of 5-oxoproline soon exceeds the capacity of oxoprolinase to convert it to glutamate. Consequently, the concentration of 5-oxoproline in blood and urine begins to rise.

Parkinson's disease, formerly known as "paralysis agitans," is a movement disorder caused by damage to brain structures called the basal ganglia and substantia nigra. Symptoms of Parkinson's disease, most commonly observed in adults past 40 years of age, include tremor, skeletal muscle rigidity, and difficulty in initiating movement. Although the ultimate cause of this malady is not understood, that certain neurons within the substantia nigra cannot produce the neurotransmitter dopamine is believed to be primarily responsible. (Dopamine produced within the substantia nigra normally acts to inhibit neural activities within the basal ganglia.) As stated, because dopamine does not cross the protective blood-brain barrier, the precursor molecule L-DOPA (also known as levodopa) is used to treat Parkinson's patients.

In the late 1970s a substantial clue to the cause of the nerve cell destruction in Parkinson's disease was provided by young drug addicts using the synthetic heroin substitute MPPP (1-methyl-4-phenyl-4-proprionoxypiperidine) (Figure 13C). Several unfortunate individuals, later found to have consumed MPPP, were diagnosed with Parkinson disease despite their youth and lack of a family history of the disease. Considerable research revealed that under certain reaction conditions the synthesis of MPPP produces a toxic by-product called MPTP (1-methyl-4-phenyl-1,2,3,6-tetrahydropyridine). Once it has been consumed, MPTP is converted to MPP^+ (1-methyl-4-phenylpyridinium) in the brain by the enzyme monoamine oxidase. After its synthesis, MPP^+ is transported by a dopamine-specific transport mechanism into certain neurons. Although the mechanism by which nerve cells are destroyed by MPP^+ is not completely understood, it appears that one of its effects is to inhibit NADH dehydrogenase, a component of the mitochondrial electron transport complex.

FIGURE 13C

Formation of MPP+, a Neurotoxin.

MPTP is a toxic by-product formed during the synthesis of MPPP. MPP^+, formed in the brain from MPTP, destroys certain neurons.

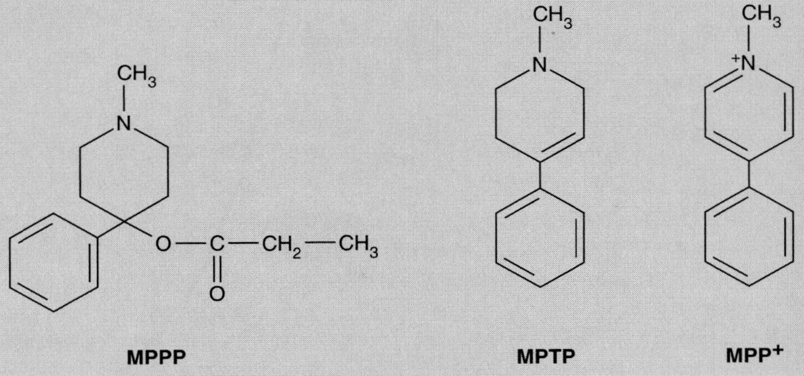

MPPP   **MPTP**   **MPP⁺**

KEY *Concepts*

Glutathione (GSH), the most common intracellular thiol, is involved in many cellular activities. In addition to reducing sulfhydryl groups, GSH protects cells against toxins and promotes the transport of some amino acids.

Glutathione-S-Transferases   As mentioned, GSH contributes to the protection of cells from environmental toxins. GSH does so by reacting with a large variety of foreign molecules to form GSH conjugates (Figure 13.19). The bonding of these substances with GSH, which prepares them for excretion, may be spontaneous, or it may be catalyzed by the GSH-S-transferases (also known as the *ligandins*). Before their excretion in urine, GSH conjugates are usually converted to mercapturic acids by a series of reactions initiated by γ-glutamyltranspeptidase.

Alkaloids

The **alkaloids** are a large, heterogeneous group of basic nitrogen-containing molecules produced in the leaves, seeds, or bark of some plants. They are derived from α-amino acids (or closely related molecules) in complex and poorly understood pathways. Although many alkaloids have profound physiological properties when consumed by animals, their roles in plants are relatively obscure. Because they often have bitter tastes or are poisonous, alkaloids may protect plants against herbivores, insects and microbes. Several examples of this group of natural products are illustrated in Figure 13.20.

Alkaloids are classified according to their heterocyclic rings. For example, cocaine, a central nervous system stimulant (Chapter 19), and atropine, a muscle relaxant, are examples of the *tropane alkaloids*. Nicotine, the addictive and toxic component of tobacco, is an example of the *pyridine alkaloids*. (Nicotine is an effective insecticide.) The addictive components of opium (codeine and morphine) are examples of the *isoquinoline alkaloids*.

Nucleotides

Nucleotides are complex nitrogen-containing molecules required for cell growth and development. Not only are nucleotides the building blocks of the nucleic acids, they

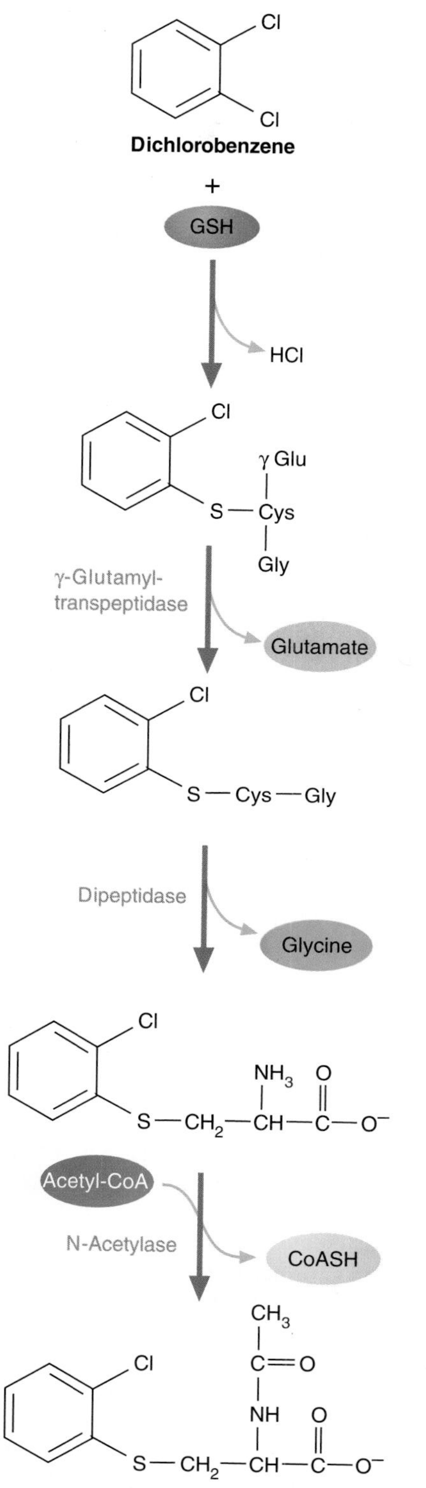

FIGURE 13.19

Formation of a Mercapturic Acid Derivative of a Typical Organic Contaminant.

GSH-S-transferase catalyzes the synthesis of a GSH derivative of dichlorobenzene.

FIGURE 13.20

Several Alkaloids.

Over 5000 alkaloids have been isolated from plants. Their roles in plants are usually unknown. Alkaloids are often physiologically potent molecules in animals.

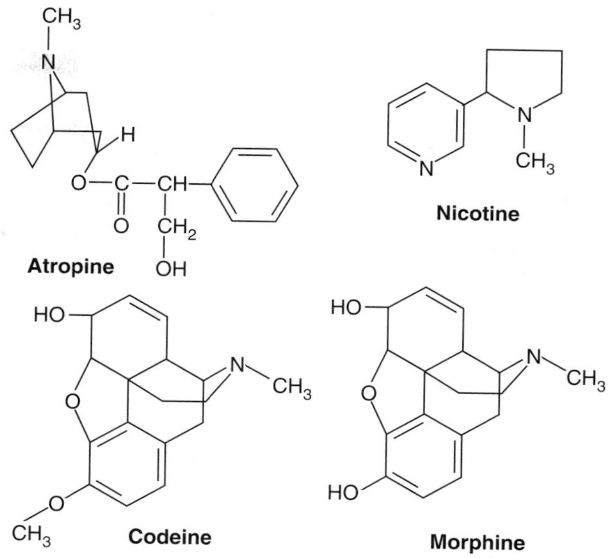

also play several essential roles in energy transformation and regulate many metabolic pathways. As described, each nucleotide is composed of three parts: a nitrogenous base, a pentose sugar, and one or more phosphate groups. The nitrogenous bases are derivatives of either purine or pyrimidine, which are planar heterocyclic aromatic amines.

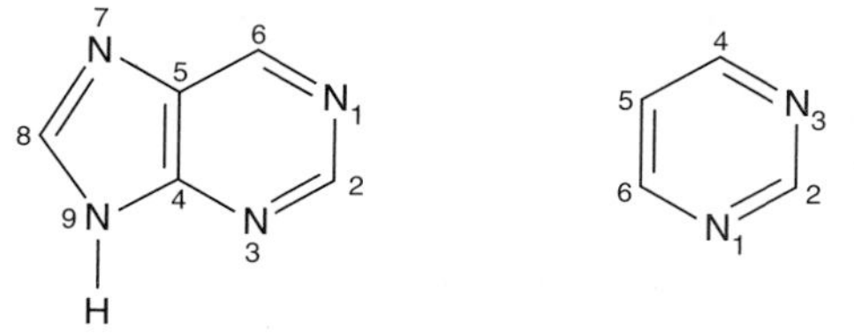

Common naturally occurring **purines** include adenine, guanine, xanthine and hypoxanthine; thymine, cytosine and uracil are common **pyrimidines** (Figure 13.21). Because of their aromatic structures, the purines and pyrimidines absorb UV light. At pH 7, this absorption is especially strong at 260 nm. Purine and pyrimidine bases have tautomeric forms, i.e., they undergo spontaneous shifts in the position of hydrogen atoms or double bonds. This property is especially important because the precise location of hydrogen atoms on the oxygen and nitrogen atoms affects the interaction of bases in nucleic acid molecules. Adenine and cytosine have both amino and imino forms; guanine, thymine and uracil have both keto (lactam) and enol (lactim) forms (Figure 13.22). At physiological pH the amino and keto forms are the most stable.

When a purine or pyrimidine base is linked through a β-N-glycosidic linkage to C-1 of a pentose sugar, the molecule is called a **nucleoside.** Two types of sugar are found in nucleosides: ribose or deoxyribose. Ribose-containing nucleosides with adenine, guanine, cytosine and uracil are referred to as adenosine, guanosine, cytidine and uridine, respectively. When the sugar component is deoxyribose, the prefix *deoxy* is used. For example, the deoxy nucleoside with adenine is called deoxyadenosine. Because the base thymine usually occurs only in deoxyribonucleosides, deoxythymidine is called thymidine. Possible confusion in the identification of atoms

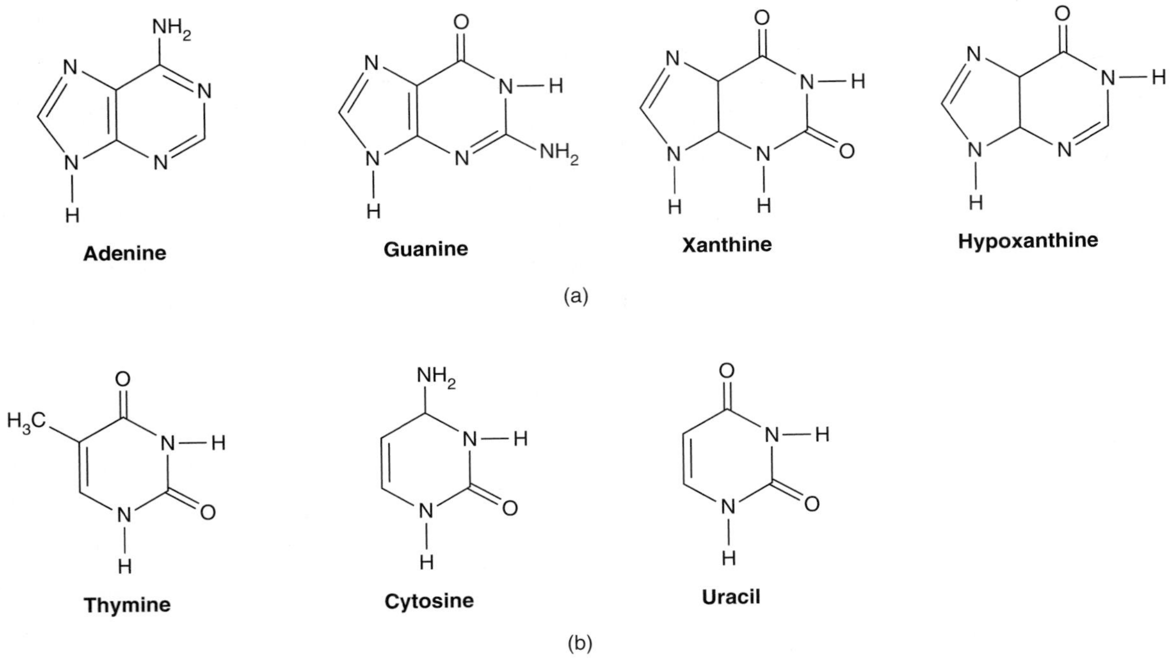
Adenine
Guanine
Xanthine
Hypoxanthine
(a)
Thymine
Cytosine
Uracil
(b)

FIGURE 13.21
The Most Common Naturally Occurring Purines (a) and Pyrimidines (b).

in the base and sugar components of nucleosides is avoided by using a superscript prime to denote the atoms in the sugar.

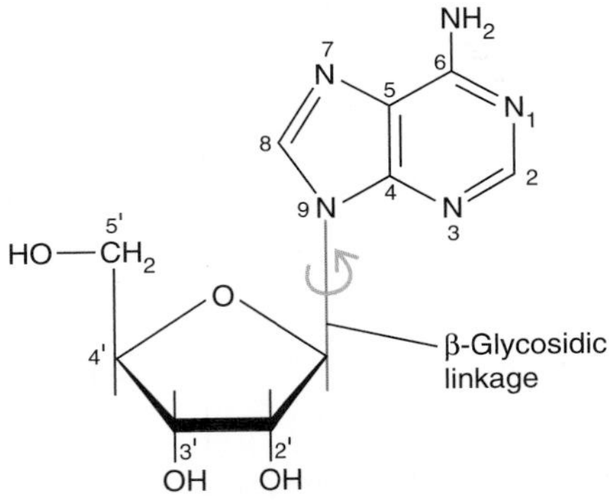
β-Glycosidic linkage

Rotation around the N-glycosidic bond of nucleosides creates two conformations: *syn* and *anti.* Purine nucleosides occur as either syn or anti forms. In pyrimidine nucleosides the anti conformation predominates because of steric hinderance between the pentose sugar and the carbonyl oxygen at C-2.

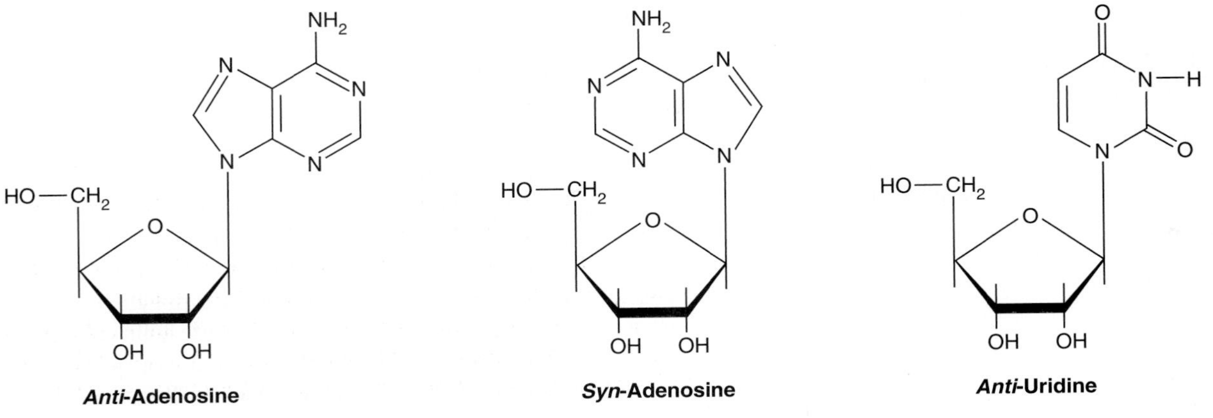
Anti-Adenosine
Syn-Adenosine
Anti-Uridine

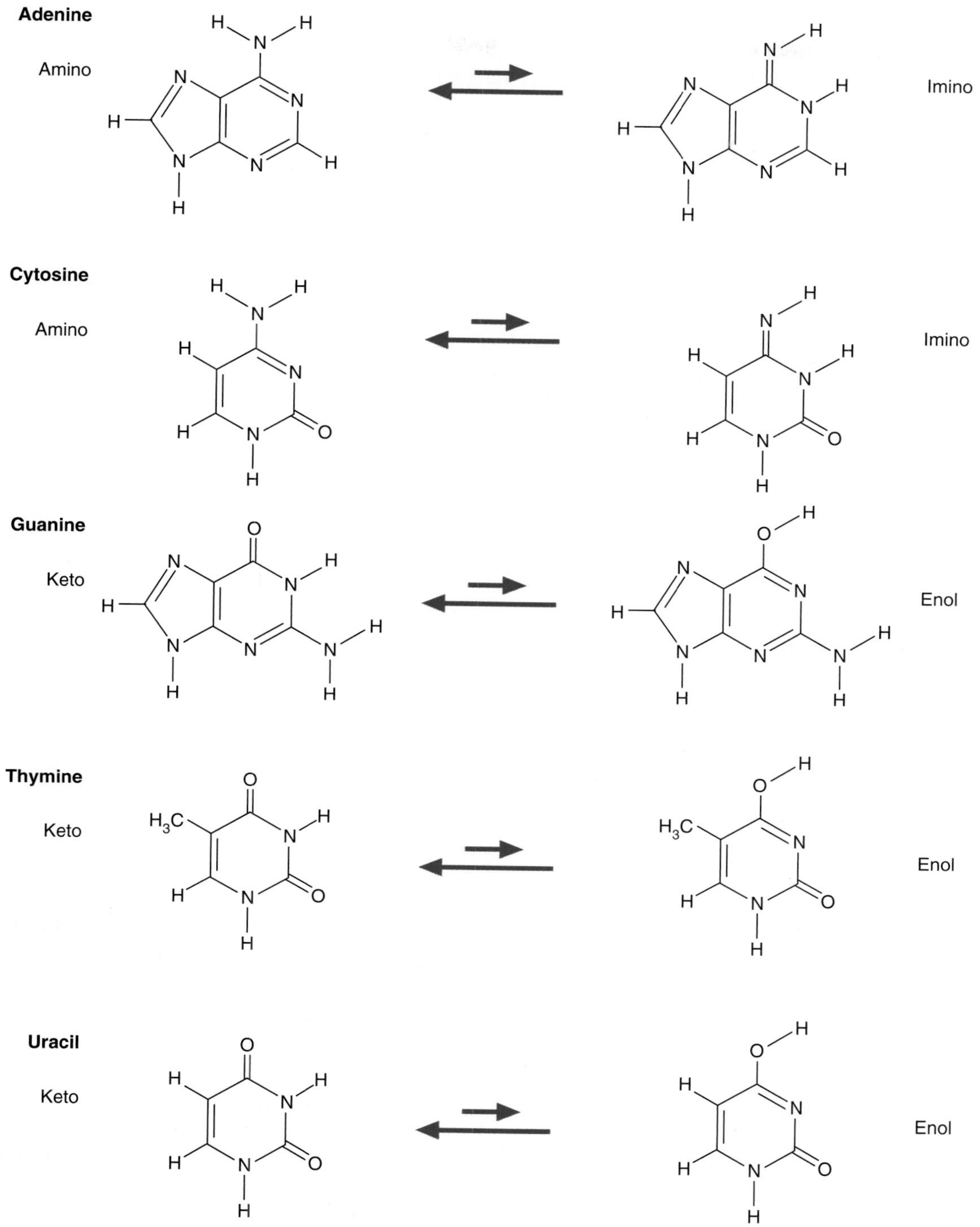

FIGURE 13.22

Tautomers of Adenine, Cytosine, Guamine, Thymine, and Uracil.

At physiological pH the amino and keto tautomers of nitrogenous bases are the predominant forms.

Nucleotides are nucleosides in which one or more phosphate groups are bound to the sugar (Figure 13.23). Most naturally occurring nucleotides are 5′-phosphate esters. If one phosphate group is attached at the 5′-carbon of the sugar, the molecule is named as a nucleoside monophosphate, e.g., adenosine-5′-monophosphate (AMP). Nucleoside di- and triphosphates contains two and three phosphate groups, respectively. Phosphate groups make nucleotides strongly acidic. (Protons dissociate from the phosphate groups at physiological pH.) Because of their acidic nature, nucleotides may also be named as acids. For example, AMP is often referred to as adenylic acid or adenylate. Nucleoside di- and triphosphates form complexes and Mg^{2+}. In nucleoside triphosphates such as ATP, Mg^{2+} can form α,β and β,γ complexes.

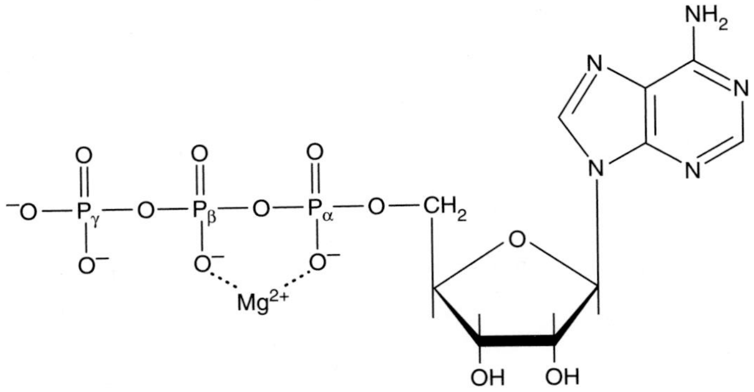

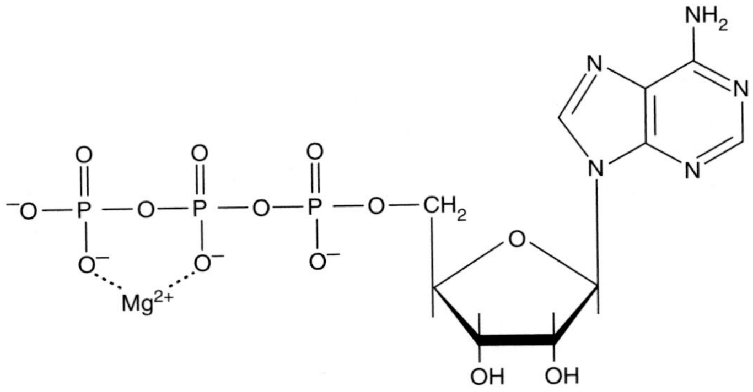

Purine and pyrimidine nucleotides can be synthesized in *de novo* and salvage pathways. These pathways are described below.

Purine Nucleotides   The *de novo* synthesis of purine nucleotides begins with the formation of 5-phospho-α-D-ribosyl-1-pyrophosphate (PRPP) catalyzed by ribose-5-phosphate pyrophosphokinase (PRPP synthetase).

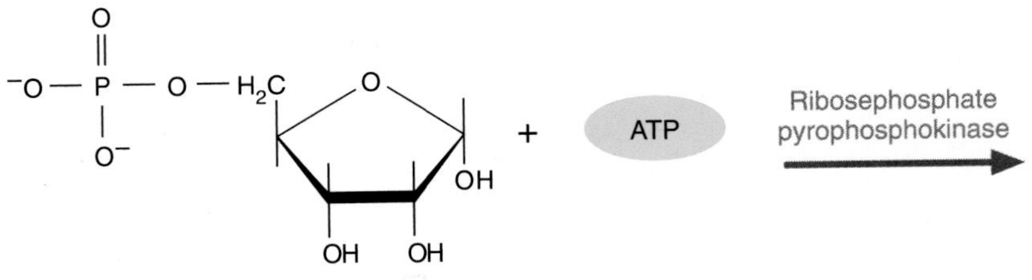

α-D-ribose-5-phosphate

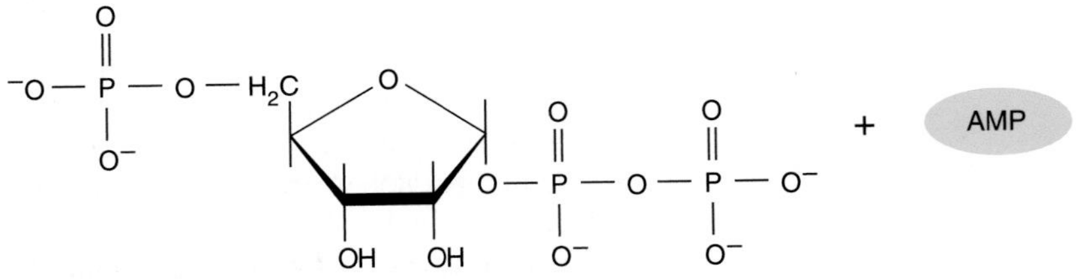

5-Phospho-α-D-ribosyl-1-pyrophosphate (PRPP)

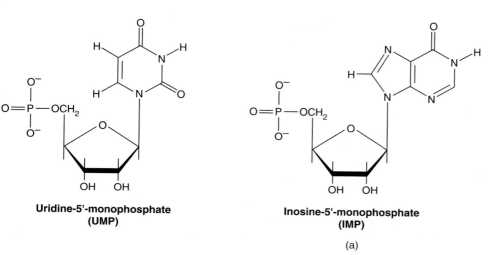

Adenosine-5'-monophosphate (AMP)

Guanosine-5'-monophosphate (GMP)

Cytidine-5'-monophosphate (CMP)

Uridine-5'-monophosphate (UMP)

Inosine-5'-monophosphate (IMP)

(a)

FIGURE 13.23

Common Ribonucleotides (a) and Deoxyribonucleotides (b, page 398).

The names of nucleotides containing both deoxyribose and thymine do not have the prefix *deoxy*. Inosine-5'-monophosphate (IMP) is an intermediate in purine nucleotide synthesis. The base component of IMP is hypoxanthine.

(The substrate for this reaction, α-D-ribose-5-phosphate, is a product of the pentose phosphate pathway.) Figure 13.24 illustrates the initial phase in the pathway by which PRPP is converted to inosine monophosphate (inosinate), the first purine nucleotide. The process begins with the displacement of the pyrophosphate group of PRPP by the amide nitrogen of glutamine in a reaction catalyzed by glutamine PRPP amidotransferase. This reaction is the committed step in purine synthesis. The product formed is 5-phospho-β-D-ribosylamine.

Once 5-phospho-β-D-ribosylamine is formed, the building of the purine ring structure begins. Phosphoribosylglycinamide synthase catalyzes the formation of an amide bond between the carboxyl group of glycine and the amino group of 5-phospho-β-D-ribosylamine. In eight subsequent reactions the first purine nucleotide IMP is formed. Other precursors of the base component of IMP (hypoxanthine) include CO_2, aspartate, and N^{10}-formyl THF. This pathway requires the hydrolysis of four ATP molecules.

The conversion of IMP to either adenosine monophosphate (AMP or adenylate) or guanosine monophosphate (GMP or guanylate) requires two reactions (Figure 13.25).

FIGURE 13.23
Continued.

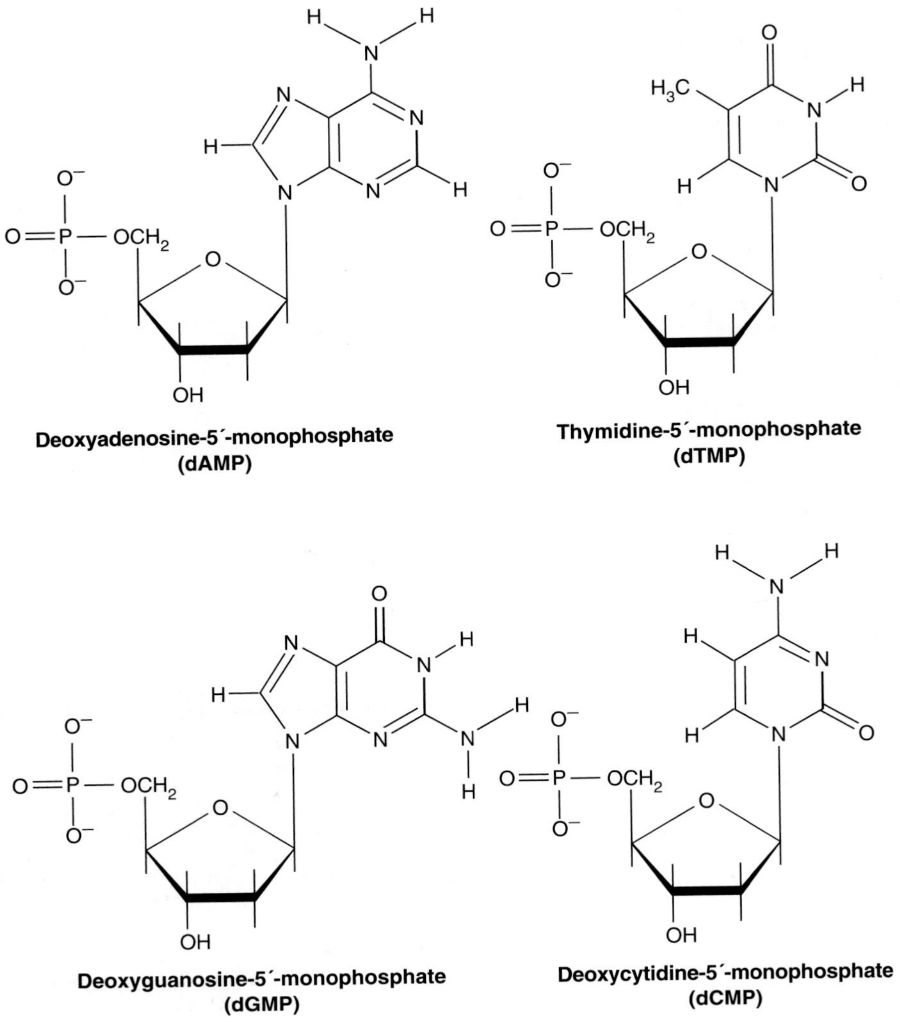

Deoxyadenosine-5′-monophosphate
(dAMP)

Thymidine-5′-monophosphate
(dTMP)

Deoxyguanosine-5′-monophosphate
(dGMP)

Deoxycytidine-5′-monophosphate
(dCMP)

(b)

AMP differs from IMP in only one respect: An amino group replaces a keto group. The amino nitrogen provided by aspartate becomes linked to IMP in a GTP-requiring reaction catalyzed by adenylosuccinate synthase. In the next step the product adenylosuccinate eliminates fumarate to form AMP. (The enzyme that catalyzes this reaction also catalyzes a similar step in IMP synthesis.) The conversion of IMP to GMP begins with a dehydrogenation utilizing NAD^+ catalyzed by IMP dehydrogenase. The product is referred to as xanthosine monophosphate (XMP). XMP is then converted to GMP by the donation of an amino nitrogen by glutamine in an ATP-requiring reaction catalyzed by GMP synthase.

Nucleoside triphosphates are the most common nucleotide used in metabolism. They are formed in the following manner. Recall that ATP is synthesized from ADP and P_i during certain reactions in glycolysis and aerobic metabolism. ADP is synthesized from AMP in a reaction catalyzed by adenylate kinase.

$$AMP + ATP \rightarrow 2\ ADP$$

Other nucleoside triphosphates are synthesized in ATP-requiring reactions catalyzed by a series of nucleoside monophosphate kinases.

$$NMP + ATP \rightleftarrows NDP + ADP$$

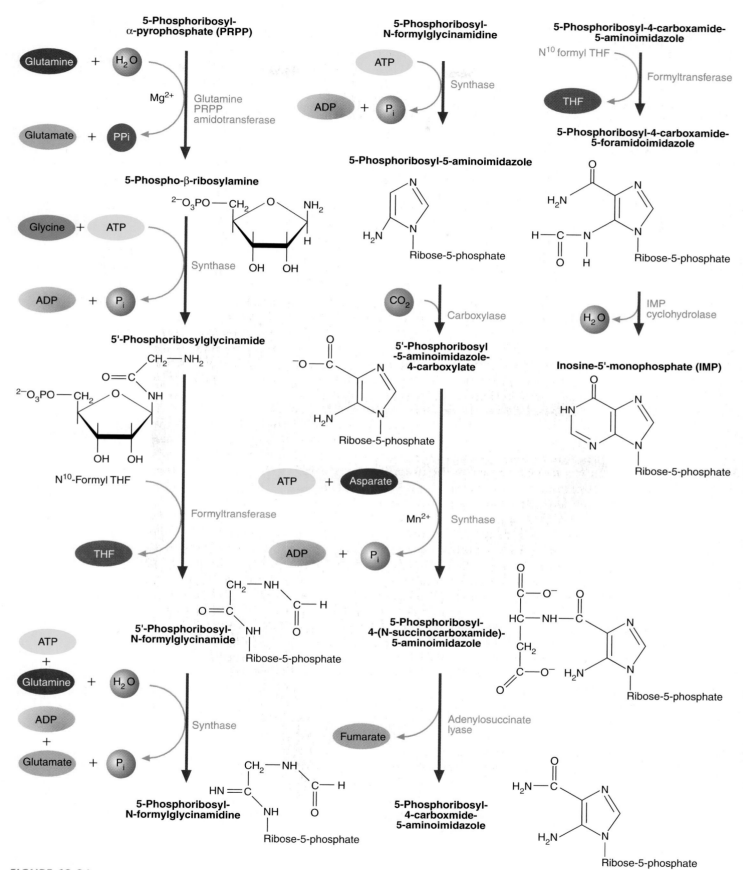

FIGURE 13.24

Synthesis of Inosine-5′-Monophosphate

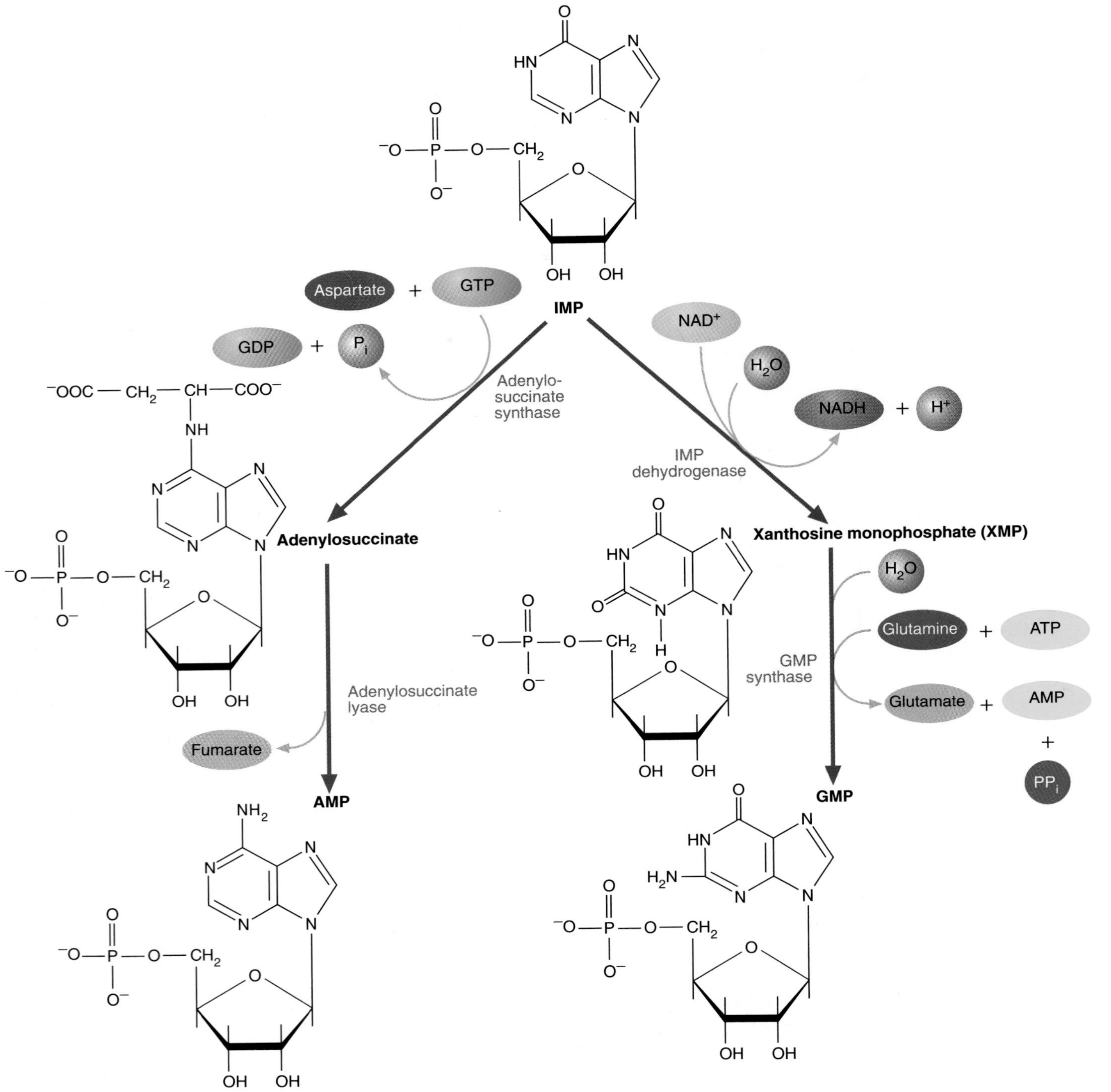

FIGURE 13.25

Biosynthesis of AMP and GMP from IMP.

AMP formation requires GTP and GMP formation requires ATP.

Nucleoside diphosphate kinase catalyzes the formation of nucleoside triphosphates.

$$N_1DP + N_2TP \rightleftarrows N_1TP + N_2DP$$

where N_1 and N_2 are purine or pyrimidine bases.

In the purine salvage pathway, purine bases obtained from the normal turnover of cellular nucleic acids or (to a lesser extent) from the diet are reconverted into nucleotides. Because the *de novo* synthesis of nucleotides is metabolically expensive (i.e., relatively large amounts of phosphoryl bond energy are used), many cells have mechanisms to retrieve purine bases. Hypoxanthine-guaninephosphoribosyltransferase (HGPRT) catalyzes nucleotide synthesis using PRPP and either hypoxanthine or guanine. The hydrolysis of pyrophosphate makes these reactions irreversible.

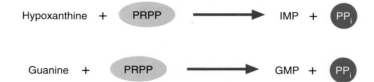

Hypoxanthine + PRPP $\longrightarrow$ IMP + PP$_i$

Guanine + PRPP $\longrightarrow$ GMP + PP$_i$

 Deficiency of HGPRT causes *Lesch-Nyhan syndrome,* a devastating disease characterized by excessive production of uric acid (Section 14.3) and certain neurological symptoms (self-mutilation, involuntary movements, and mental retardation). Affected children appear normal at birth but begin to deteriorate at about three to four months of age.

Adenine phosphoribosyltransferase (ARPT) catalyzes the transfer of adenine to PRPP, thus forming AMP.

Adenine + PRPP $\longrightarrow$ AMP + PP$_i$

The relative importance of the *de novo* and salvage pathways is unclear. However, the severe symptoms of hereditary HGPRT deficiency indicate that the purine salvage pathway is vitally important. In addition, investigations of purine nucleotide synthesis inhibitors for treating cancer indicate that both pathways must be inhibited for significant tumor growth suppression.

 The regulation of purine nucleotide biosynthesis is summarized in Figure 13.26. The pathway is controlled to a considerable degree by PRPP availability. Several products of the pathway inhibit both ribose-5-phosphate pyrophosphokinase and glutamine-PRPP

FIGURE 13.26

Purine Nucleotide Biosynthesis Regulation.

Feedback inhibition is indicated by red arrows. The stimulation of AMP synthesis by GTP and GMP synthesis by ATP ensures a balanced synthesis of both families of purine nucleotides.

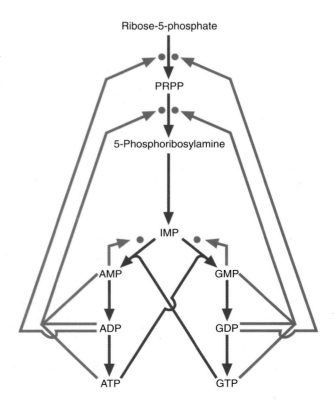

amidotransferase. The combined inhibitory effect of the end products is synergistic (i.e., the net inhibition is greater than the inhibition of each nucleotide acting alone). At the IMP branch point, both AMP and GMP regulate their own syntheses by feedback inhibition of adenylosuccinate synthase and IMP dehydrogenase. The hydrolysis of GTP drives the synthesis of adenylosuccinate, while ATP drives XMP synthesis. This reciprocal arrangement is believed to facilitate the maintenance of appropriate cellular concentrations of adenine and guanine nucleotides.

Pyrimidine Nucleotides   In pyrimidine nucleotide synthesis the pyrimidine ring is assembled first and then linked to ribose phosphate. The carbon and nitrogen atoms in the pyrimidine ring are derived from bicarbonate, aspartate, and glutamine. Synthesis begins with the formation of carbamoyl phosphate in an ATP-requiring reaction catalyzed by the cytoplasmic enzyme carbamoyl phosphate synthetase II (Figure 13.27). (Carbamoyl phosphate synthetase I is a mitochondrial enzyme involved in the urea cycle, described in Chapter 14.) One molecule of ATP provides a phosphate group, while the hydrolysis of another ATP drives the reaction. Carbamoyl phosphate next reacts with aspartate to form carbamoyl aspartate. The closure of the pyrimidine ring is then catalyzed by dihydroorotase. The product, dihydroorotate, is then oxidized to form orotate. Dihydroorotate dehydrogenase, the enzyme that catalyzes this reaction, is a flavoprotein associated with the inner mitochondrial membrane. (The NADH produced in this reaction donates its electrons to the electron transport complex.) After its transport into the cytoplasm, orotate is converted by orotate pyrophosphoribosyl transferase to orotidine-5′-phosphate (OMP), the first nucleotide in the pathway, by reacting with PRPP. Uridine-5′-phosphate (UMP) is produced when OMP is decarboxylated in a reaction catalyzed by OMP decarboxylase. Both orotate pyrophosphoribosyl transferase and OMP decarboxylase activities occur on a protein referred to as UMP synthase. (In a rare genetic disease called *orotic aciduria,* there is excessive urinary excretion of orotic acid because UMP synthase is defective. Symptoms include retarded growth as well as anemia. Treatment with a combination of pyrimidine nucleotides, which inhibit the production of orotate and provide the building blocks for nucleic acid synthesis, reverses the disease process.) UMP is a precursor for the other pyrimidine nucleotides. Two sequential phosphorylation reactions form UTP, which then accepts an amide nitrogen from glutamine to form CTP.

Deoxyribonucleotides   All of the nucleotides discussed so far are ribonucleotides, molecules that are principally used as the building blocks of RNA, as nucleotide derivatives of molecules such as sugars, or as energy sources. The nucleotides required for DNA synthesis, the 2′-deoxyribonucleotides, are produced by reducing ribonucleoside diphosphates (Figure 13.28). The electrons used in the synthesis of 2′-deoxyribonucleotides are ultimately donated by NADPH. *Thioredoxin,* a low-molecular-weight protein (13 kD) with two sulfhydryl groups, mediates the transfer of hydrogen atoms from NADPH to ribonucleotide reductase, the enzyme that catalyzes the reduction of ribonucleotides to form deoxyribonucleotides. The regeneration of reduced thioredoxin is catalyzed by thioredoxin reductase.

Regulation of ribonucleotide reductase is complex. The binding of dATP (deoxyadenosine triphosphate) to a regulatory site on the enzyme decreases catalytic activity. The binding of deoxyribonucleoside triphosphates to several other enzyme sites alters substrate specificity so that there are differential increases in the concentrations of each of the deoxyribonucleotides. This latter process balances the production of the 2′-deoxyribonucleotides required for cellular processes, especially that of DNA synthesis.

The deoxyuridylate (dUMP) produced by dephosphorylation of the dUDP product of ribonucleotide reductase is not a component of DNA, but its methylated derivative deoxythymidylate (dTMP) is. The methylation of dUMP is catalyzed by thymidylate synthase, which utilizes N^5,N^{10}-methylene THF. As the methylene group is transferred, it is reduced to a methyl group, while the folate coenzyme is oxidized to form

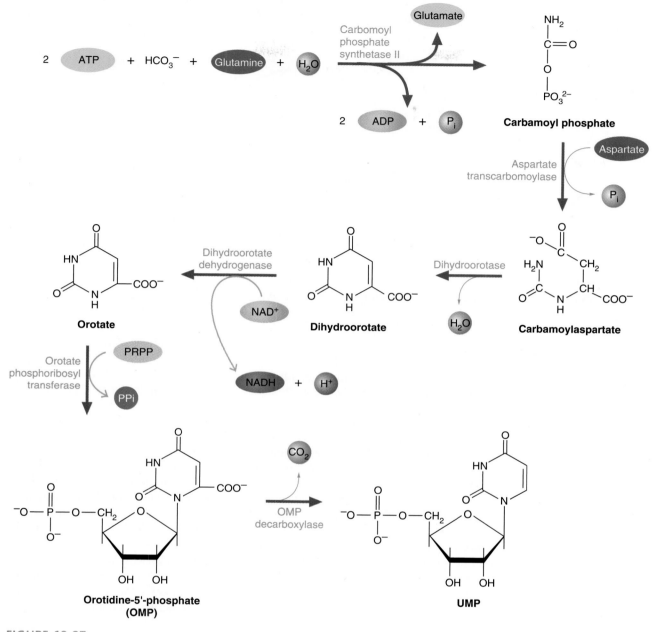

dihydrofolate. THF is regenerated from dihydrofolate by dihydrofolate reductase and NADPH. (This reaction is the site of action of some anticancer drugs, such as methtrexate.) Deoxyuridylate can also be synthesized from dCMP by deoxycytidylate deaminase.

The pyrimidine salvage pathway, which uses preformed pyrimidine bases from dietary sources or from nucleotide turnover, is of minor importance in mammals. In bacteria, phosphoriboyltransferase catalyzes the synthesis of nucleotides from PRPP and either uracil or thymine. In an alternative pathway found in both bacteria and some higher animals, uridine phosphorylase catalyzes the synthesis of uridine from uracil and ribose-1-phosphate. UMP is produced in a later reaction between uridine and ATP, catalyzed by uridine kinase. Similar enzymes that catalyze salvage reactions for other pyrimidine nucleotides have also been identified.

KEY *Concepts*

Nucleotides are the building blocks of the nucleic acids. They also regulate metabolism as well as transfer energy. The purine and pyrimidine nucleotides are synthesized in both *de novo* and salvage pathways.

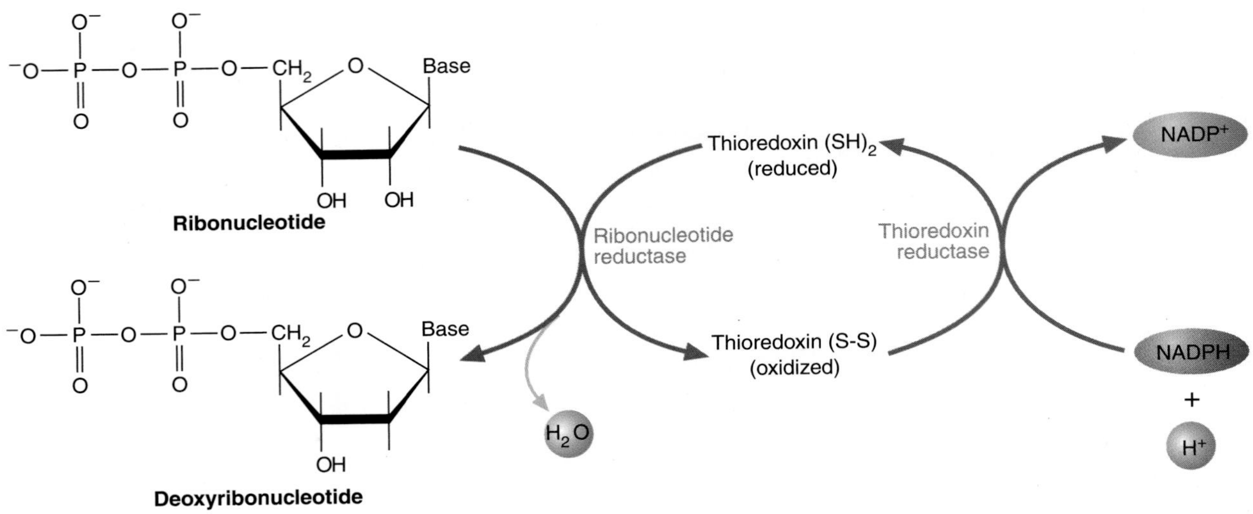

FIGURE 13.28

Deoxyribonucleotide Biosynthesis.

Electrons for the reduction of ribonucleotides ultimately come from NADPH. Thioredoxin, a small protein with two thiol groups, mediates the transfer of electrons from NADPH to ribonucleotide reductase.

In mammals, carbamoyl phosphate synthetase II is the key regulatory enzyme in the biosynthesis of pyrimidine nucleotides. The enzyme is inhibited by UTP, the product of the pathway, and stimulated by purine nucleotides. In many bacteria, aspartate carbamoyl transferase is the key regulatory enzyme. It is inhibited by CTP and stimulated by ATP.

Heme

Heme, one of the most complex molecules synthesized by mammalian cells, has an iron-containing porphyrin ring. As described previously, heme is an essential structural component of hemoglobin, myoglobin, and the cytochromes. Almost all aerobic cells synthesize heme because it is required for the cytochromes of the mitochondrial ETC. The heme biosynthetic pathway is especially prominent in liver, bone marrow, and intestine cells and in reticulocytes (the nucleus-containing precursor cells of red blood cells). Heme is synthesized from the relatively simple components glycine and succinyl-CoA.

In the first step of heme synthesis, glycine and succinyl-CoA condense to form δ-aminolevulinate (ALA) (Figure 13.29). This reaction, which requires pyridoxal phosphate, is the rate-controlling step in porphyrin synthesis. ALA synthase, a mitochondrial enzyme, is allosterically inhibited by *hemin,* an Fe^{3+}-containing derivative of heme. (When a red blood cell's porphyrin concentration exceeds that of globin, heme accumulates and is subsequently oxidized to hemin.) Hemin production also decreases synthesis of ALA synthase. In the next step of porphyrin synthesis, two molecules of ALA condense to form porphobilinogen. Porphobilinogen synthase, which catalyzes this reaction, is a zinc-containing enzyme that is extremely sensitive to heavy-metal poisoning. (See Special Interest Box 13.4.) Uroporphyrinogen synthase catalyzes the symmetric condensation of four porphobilinogen molecules. An additional protein is also required in this reaction. Uroporphyrinogen III cosynthase alters the specificity of uroporphyrinogen synthase so that the asymmetric molecule uroporphyrinogen III is produced. When four CO_2 molecules are removed, catalyzed by uroporphyrinogen decarboxylase, coproporphyrinogen is synthesized. This reaction is followed by the removal of two additional CO_2 molecules, thus forming protoporphyrinogen IX. Oxidation of the porphyrin ring's methylene groups forms protoporphyrin IX, the direct precursor of heme. The final step in the synthesis of heme (also called *protoheme IX*) is the insertion of Fe^{2+}, a reaction that occurs spontaneously but is accelerated by ferrochelatase.

KEY *Concepts*

Heme is an iron- and nitrogen-containing porphyrin ring system synthesized from glycine and succinyl-CoA. Protoporphyrin IX, the precursor of heme, is also a precursor of the chlorophylls.

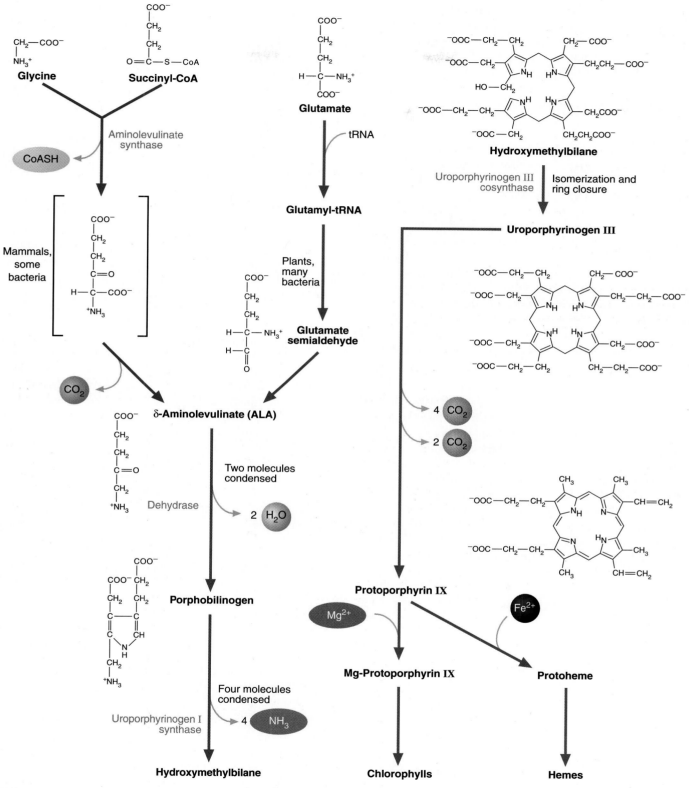

FIGURE 13.29

Heme and Chlorophyll Biosynthesis.

In plants and some bacterial species, ALA is synthesized from glutamate in a process involving glutamyl tRNA. Refer to Chapter 16 for a discussion of tRNA (transfer RNA).

LEAD POISONING

The heavy metal lead is absorbed primarily in the gastrointestinal and respiratory tracts and deposited in soft tissues (e.g., kidney, liver, and central nervous system) and in bone. (In affected individuals, approximately 95% of lead is sequestered in bone.) Symptoms of lead intoxication, which vary according to the degree of exposure, include anorexia, muscle pain and weakness, abdominal pain, infertility, stillbirth, and encephalopathy. (*Lead encephalopathy* is a disorder of the brain's cerebral cortex. It is characterized by clumsiness, headache, irritability, insomnia, mental retardation, and, in extreme cases, hallucinations and paralysis.) Possible renal effects of lead exposure include chronic nephritis (an inflammatory condition), as well as disturbances in the kidney's capacity to reabsorb nutrients such as amino acids, glucose, and phosphate. Despite intensive research efforts, federal regulations, and increasing public awareness, lead poisoning (also referred to as "plumbism") remains a serious public health problem. For example, children in impoverished neighborhoods in large cities are still at high risk for high blood levels of lead. Lead-based paint is consumed by childern because of its sweet taste. In addition, the soil in these areas often has lead levels substantially above acceptable standards. (These high levels are no doubt due in part to lead compounds formerly used in gasoline.)

Lead and Heme Synthesis

Lead is toxic largely because it forms bonds with the sulfhydryl groups of proteins. Any protein with free sulfhydryl groups is therefore vulnerable. Among the best-researched examples of lead-sensitive biomolecules are several enzymes that catalyze reactions in heme biosynthesis. Inhibition of porphobilinogen synthase by lead occurs with relatively low lead levels. Therefore, detecting its substrate (ALA) in the urine serves is early warning of lead intoxication. The inhibition of ferrochelatase is a more reliable indicator of a serious lead exposure. In acute lead poisoning (due to accidental ingestion of relatively large amounts of lead compounds), its substrate (protoporphyrin IX) accumulates in tissues. In chronic lead poisoning (a slow, progressive process), protoporphyrin IX complexed with zinc appears in blood. (Because of its high affinity for zinc, protoporphyrin IX forms complexes with this metal when ferrochelatase is inhibited.) Because zinc protoporphyrin in blood is easily measured, its detection is a valuable diagnostic tool.

Lead Poisoning: An Ancient Heritage

Since ancient times the soft grayish-blue metal called lead has been extremely useful. Because it resists corrosion and can be easily shaped, lead has many commercial and industrial applications. For example, lead alloys have long been used in plumbing and shipbuilding. Additionally, several lead compounds have vibrant colors and have been valued as components of paint and cosmetics. However, lead is highly toxic. First used at least 8000 years ago (probably in areas near the Aegean Sea), lead soon became a source of economic strength in the ancient world. For this reason, lead poisoning may have been one of the earliest occupational diseases. However, plumbism was not limited to artisans and metalworkers. Because lead containers stored and preserved wine and foods, and lead pipes transported water, the wealthy were also at high risk. The decline of the Roman Empire has been blamed in part on the effects of lead-contaminated wine and food (e.g., insanity and infertility) on the Roman aristocracy.

Although several ancient physicians were aware that lead was harmful, it was not until the Industrial Revolution in Europe and America that any sustained attention was paid to lead poisoning. Numerous observations of sterility, miscarriages, stillbirths, and premature delivery in both female leadworkers and the wives of male leadworkers resulted, by the end of the nineteenth century, in the removal of female workers from the industry. In the twentieth century, improvements in testing techniques as well as an awakening social conscience have significantly reduced lead exposure. The most serious (and obvious) effects of lead toxicity are now rarely observed. However, lead is believed to be responsible for more subtle injuries. For example, in one controversial hypothesis, some cases of renal disease and hypertension are linked to mild lead exposure. In addition, several researchers have associated intellectual dullness and lowered IQ scores to relatively low levels of lead exposure.

Protoporphyrin IX is also a precursor of the chlorophylls. After magnesium (Mg^{3+}) is incorporated, the enzyme Mg-protoporphyrin methylesterase catalyzes the addition of a methyl group to form Mg-protoporphyrin IX monomethylester. This molecule is then converted to chlorophyll in several light-induced reactions.

KEY WORDS

SUMMARY

Nitrogen is an essential element in living systems, since it is found in proteins, nucleic acids, and myriad other biomolecules. Biologically useful nitrogen, a scarce resource, is produced in a process referred to as nitrogen fixation. Only a few organisms are capable of fixing nitrogen to form ammonia and nitrate, the oxidation product of ammonia.

Organisms vary widely in their ability to synthesize amino acids. Some organisms (e.g., plants and some microorganisms) can produce all required amino acid molecules from fixed nitrogen. Animals can produce only some amino acids. Nonessential amino acids are produced from readily available precursor molecules, while essential amino acids must be acquired in the diet.

Two types of reactions play prominent roles in amino acid metabolism. In transamination reactions, new amino acids are produced when α-amino groups are transferred from donor α-amino acids to acceptor α-keto acids. Because transamination reactions are reversible, they play an important role in both amino acid synthesis and degradation. Ammonium ions can also be directly incorporated into amino acids and eventually other metabolites.

On the basis of the biochemical pathways in which they are synthesized, the amino acids can be divided into six families: glutamate, serine, aspartate, pyruvate, aromatics, and histidine.

Amino acids are precursors of many physiologically important biomolecules. Many of the processes that synthesize these molecules involve the transfer of carbon groups. Because many of these transfers involve one-carbon groups (e.g., methyl, methylene, methenyl, and formyl), the overall process is referred to as one-carbon metabolism. S-Adenosyl methionine (SAM) and tetrahydrofolate (THF) are the most important carriers of one-carbon groups.

Molecules derived from amino acids include several neurotransmitters (e.g., GABA, the catecholamines, serotonin, histamine, and nitric oxide) and hormones (e.g., indole acetic acid). Glutathione is an example of an amino acid derivative that plays an essential role in cells. Alkaloids are a diverse group of basic nitrogen-containing molecules. The role of alkaloids in the plants that produce them is poorly understood. Several alkaloid molecules have profound physiological effects on animals. The nucleotides, molecules that serve as the building blocks of the nucleic acids (as well as energy sources and metabolic regulators), possess heterocyclic nitrogenous bases as part of their structures. These bases, called the purines and the pyrimidines, are derived from various amino acid molecules. Heme is an example of a complex heterocyclic ring system that is derived from glycine and succinyl-CoA. The biosynthetic pathway that produces heme is similar to the one that produces chlorophyll in plants.

SUGGESTED READINGS

Coyle, J.T., and Puttfarcken, P., Oxidative Stress, Glutamate and Neurodegenerative Disorders, *Science,* 262:689–695, 1993.

Lancaster, J.R., Nitric Oxide in Cells, *Amer. Sci.,* 80(3):248–259, 1992.

Lea, P.J., and Miflin, B.J., Transport and Metabolism of Asparagine and Other Nitrogen Compounds Within the Plant, in Miflen, B.J. (Ed.), *The Biochemistry of Plants,* Vol. 5, pp. 569–607, Academic Press, New York, 1980.

Meister, A., and Anderson, M.E., Glutathione, *Ann. Rev. Biochem.,* 52:711–760, 1983.

Orme-Johnson, W.H., Molecular Basis of Biological Nitrogen Fixation, *Ann. Rev. Biophys. Biophys. Chem.,* 14:419–459, 1985.

Reichard, P., and Ehrenberg, A., Ribonucleotide Reductase—A Radical Enzyme, *Science,* 221:514–519, 1983.

Weaver, L.M. and Herrmann, K.M., Dynamics of the Skikimate Pathway in Plants, *Trends Plant Sci.,* 2(9):346–351, 1997.

Wedein, R.P., *Poison in the Pot: The Legacy of Lead,* Southern Illinois University Press, Carbondale and Edwardville, 1984.

QUESTIONS

1. Define the following terms:
   a. essential amino acid
   b. nitrogen balance
   c. *de novo* pathway
   d. biogenic amine
   e. excitotoxin
   f. retrograde neurotransmitter

2. Why are transamination reactions important in both the synthesis and degradation of amino acids?

3. Give two reasons why nitrogen compounds are limited in the biosphere.

4. Nitrogenase complexes are irreversibly inactivated by oxygen. Explain how nitrogen-fixing bacteria solve this problem.

5. Using reaction equations, illustrate how α-ketoglutarate is converted to glutamate. Name the enzymes and cofactors required.

6. In PLP-catalyzed reactions, the pyridinium ring acts as an electron sink. Describe this process.

7. In PLP-catalyzed reactions, would a benzene ring be a more or less effective electron sink than the pyridinium ring?

8. The concentration in the blood of which of the following amino acids is not affected by passage through the liver?
   a. alanine
   b. isoleucine
   c. phenylalanine
   d. valine
   e. glycine
   f. proline

9. What are the two major classes of neurotransmitters? How do their modes of action differ? Give an example of each type of neurotransmitter.

10. Illustrate the pathways to synthesize the following amino acids:
    a. glutamine
    b. methionine
    c. threonine

11. When susceptible people consume monosodium glutamate, they experience several extremely unpleasant symptoms, such as increased blood pressure and body temperature. Using your knowledge of glutamate activity, explain these symptoms.

12. Determine the synthetic family to which each of the following amino acids belongs:
    a. alanine
    b. phenylalanine
    c. methionine
    d. tryptophan
    e. histidine
    f. serine

13. What are the two most important carriers in one-carbon metabolism? Give examples of their roles in metabolism.

14. Glutathione is an important intracellular thiol. List five functions of glutathione in the body.

15. Radiation exerts part of its damaging effects by causing the formation of hydroxyl radicals. Write a reaction equation to explain how glutathione acts to protect against this form of radiation damage.

16. Shown below are six compounds. Indicate which are nucleosides, nucleotides, or purine or pyrimidine bases.

(a)

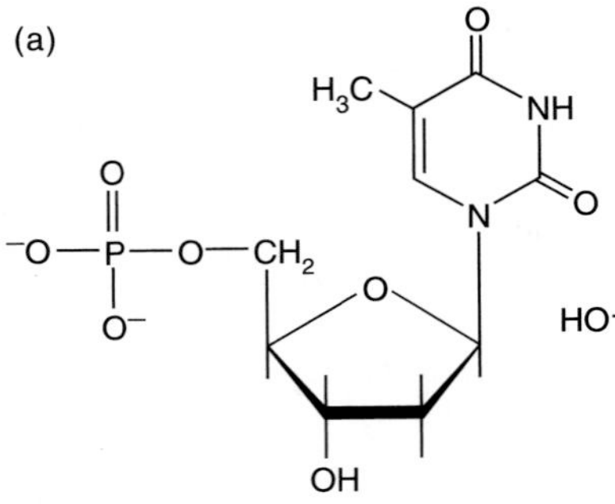

(b)

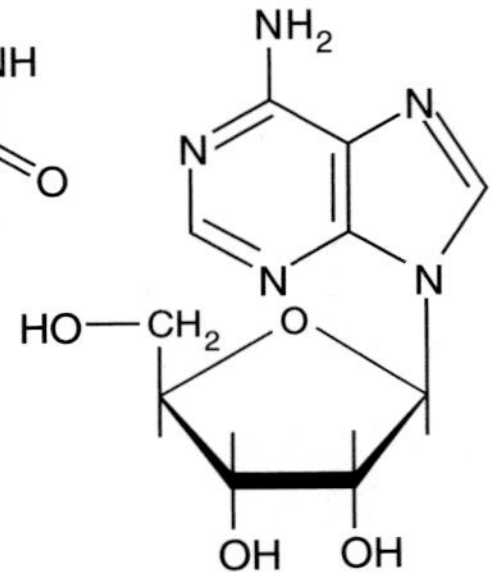

(c)

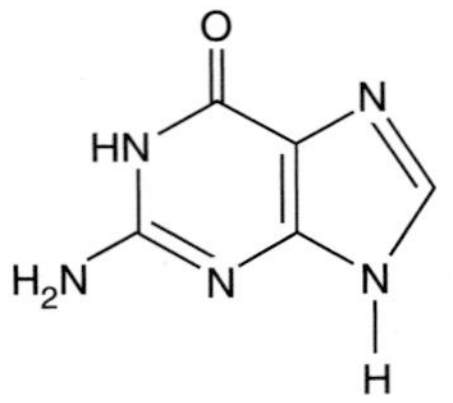

(d)

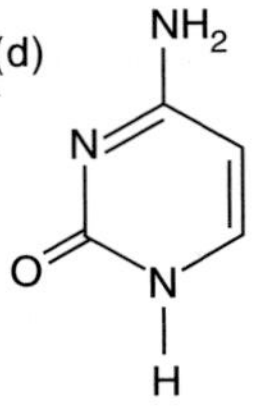

(e)

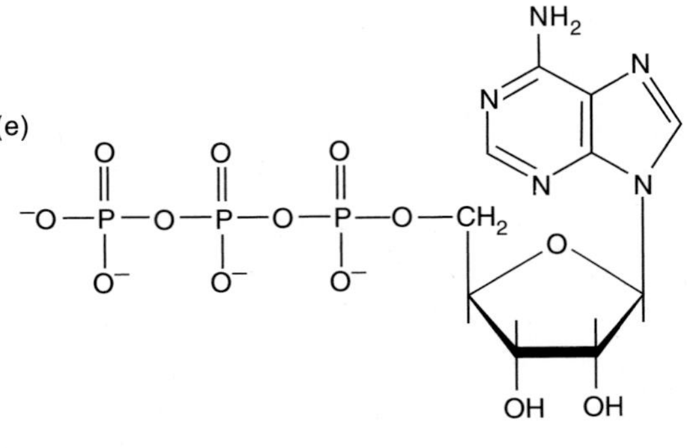

17. In pyrimidine nucleosides the anti conformation predominates because of steric interactions with pentose. Do the purine nucleosides have similar interactions?

18. Describe the steric interactions that determine the conformations that pyrimidine nucleosides assume.

19. Outline the reactions involved during the assembly of the purine ring. Include structures in your answer.

20. Referring to Question 19, calculate the number of ATP molecules that are required to synthesize a purine. Referring to the purine salvage pathway, calculate the number of ATP molecules that are required to prepare the same molecule. How many ATP molecules are saved by the purine salvage pathway?

21. In pyrimidine nucleotide synthesis, the carbon and nitrogen atoms are derived from bicarbonate, aspartate, and glutamine. Devise a simple experiment to prove the source of the nitrogen atoms. Do not forget to take into account the nitrogen exchange in amino acids.

22. List ten essential amino acids in humans. Why are they essential?

23. Most amino acids can be readily interconverted with the corresponding α-keto acid. This is not true of lysine. When lysine is deaminated, a Schiff base is produced. Explain how this product is formed.

24. Explain how the γ-glutamyl cycle acts to transport amino acids across a membrane. How does the location of the γ-glutamyl transpeptidase help drive this process?

25. By definition, essential amino acids are not synthesized by an organism. Arginine is classified as an essential amino acid in children even though it is part of the urea cycle. Explain.

26. The amino acids glutamine and glutamate are central to amino acid metabolism. Explain.

27. A mutant bacterium is unable to synthesize glycine. What intermediate in purine biosynthesis will accumulate?

28. If an organism is incubated with $^{14}CH_2(OH)CH(NH_2)COOH$, what positions in the purine ring will be labeled?

29. Indicate the source of each carbon and nitrogen in the pyrimidine ring.

30. Transamination reactions have been described as ping-pong reactions. Using the reaction of alanine with α-ketoglutarate, indicate how this ping-pong reaction works.

31. Pyridoxal phosphate acts as an intermediate carrier of amino groups during transamination reactions. Write a series of reactions to show the role of pyridoxal phosphate in the reaction of alanine and α-ketoglutarate.

32. What is the biologically active form of folic acid? How is it formed?

33. In PLP-catalyzed reactions, the bond broken in the substrate molecule must be perpendicular to the plane of the pyridinium ring. Considering the bonds present in this ring, describe why this arrangement stabilizes the carbanion.

NITROGEN METABOLISM II: DEGRADATION

Chapter Fourteen

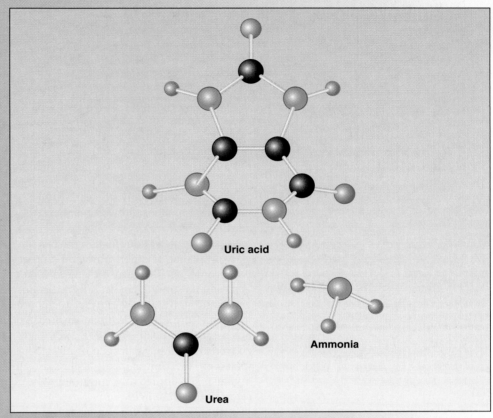

Uric acid, urea, and ammonia are among the most common nitrogenous waste molecules.

OBJECTIVES

After you have studied this chapter, you should be able to answer these questions:

1. How do various animal species dispose of nitrogenous waste?
2. What role does protein turnover play in cellular metabolism?
3. How are proteins targeted for degradation?
4. How are amino groups removed from amino acids?
5. How is waste nitrogen incorporated into urea molecules?
6. How are the citric acid cycle and the urea cycle interrelated?
7. By what biochemical pathways are the carbon skeletons of the amino acids degraded into common metabolic intermediates?
8. What role does the destruction of neurotransmitters play in the functioning of neurons and muscle cells?
9. How are dietary nucleic acids degraded?
10. What degradation products result from the catabolism of the purine and pyrimidine bases?

The metabolism of nitrogen-containing molecules such as the proteins and the nucleic acids differs significantly from that of carbohydrates and lipids. Whereas the latter molecules can be stored and mobilized as needed for biosynthetic reactions or for energy generation, there is no nitrogen-storing molecule. (One exception to this rule is storage protein in seeds.) Organisms must constantly replenish their supply of usable nitrogen to replace organic nitrogen that is lost in catabolism. For example, animals must have a steady supply of amino acids in their diets to replace the nitrogen excreted as urea, uric acid, and other nitrogenous waste products.

Despite their apparent stability, most living cells are constantly undergoing renovation. To maintain, repair, and/or reproduce themselves, cells must acquire nutrients from their environment. As described, nutrients are used both as building block molecules and as the energy sources required to drive cellular processes. One of the most obvious aspects of cellular renovation is the turnover of protein and nucleic acids, the macromolecules that are most responsible for the complexity of modern living processes. Additionally, the capacity to both synthesize and degrade these molecules in a timely manner allows organisms to respond to physiological and environmental cues.

Because of the continual synthesis and degradation of proteins and nucleic acids, as well as that of other nitrogen-containing molecules, nitrogen atoms flow through living organisms. Because of the difficulty and expense encountered in fixing nitrogen (and its subsequent scarcity), it is not surprising that living organisms recycle organic nitrogen into a variety of metabolites before it is eventually reconverted to its inorganic form. In some organisms (e.g., plants and microorganisms) this process does not begin until death. Then *decomposers,* microorganisms that inhabit soil and water, convert the organic nitrogen of all dead organisms to ammonia. Ammonia, or its oxidized products nitrate and nitrite may subsequently be absorbed and used by nearby organisms. Alternatively, nitrate may be converted to atmospheric nitrogen, a process referred to as *denitrification*

Animals, which typically have a more energetic and aggressive lifestyle, appear to be more wasteful of organic nitrogen. To maintain metabolic flexibility, animals have had to develop mechanisms for disposing of excess and toxic nitrogen-containing molecules (i.e., amino acids and nucleotides), that are not immediately required in cellular processes. Such molecules are converted into nitrogenous waste. Although many variations are observed among species, the following generalizations can be made. The nitrogen in amino acids is removed by deamination reactions and converted to ammonia. Because this latter molecule is so toxic, it must be detoxified and/or excreted as fast as it is generated. Because of its solubility, many aquatic animals can excrete ammonia itself, which dissolves in the surrounding water and is quickly diluted. (These organisms are referred to as *ammonotelic.*) Terrestrial animals, which must conserve body water, convert ammonia to molecules that can be excreted without a large loss of water. Mammals, for example, convert ammonia to urea. (Urea-producing organisms are referred to as *ureotelic.*) Other animals, such as birds, certain reptiles, and insects, which have even more stringent water conservation problems, are called *uricotelic* because they convert ammonia to uric acid. In many animals (e.g., humans and birds), uric acid is also the nitrogenous waste product of purine nucleotide catabolism.

Because nitrogen catabolic pathways are similar in many organisms and most research efforts in nitrogen catabolism have concentrated on mammals, the mammalian pathways are the focus of this chapter. The chapter begins with a discussion of the pathways that degrade amino acids to form ammonia and the carbon skeletons used in anabolic and catabolic processes. This is followed by a discussion of urea synthesis. The chapter ends with descriptions of the degradation of several amine neurotransmitters and the nucleotides.

14.1 AMINO ACID CATABOLISM

Despite the complexity of amino acid degradative pathways, the following general statements can be made. The catabolism of the amino acids usually begins by removing the amino group. Amino groups can then be disposed of in urea synthesis. The carbon skeletons produced from the standard amino acids are then degraded to form seven metabolic products: acetyl-CoA, acetoacetyl-CoA, pyruvate, α-ketoglutarate, succinyl-CoA, fumarate, and oxaloacetate. Depending on the animal's current metabolic requirements, these molecules are used to synthesize fatty acids or glucose or to generate energy. Amino acids degraded to form acetyl-CoA or acetoacetyl-CoA are referred to as **ketogenic** because they can be converted to either fatty acids or ketone bodies. The carbon skeletons of the **glucogenic** amino acids, which are degraded to pyruvate or a citric acid cycle intermediate, can then be used in gluconeogenesis. After discussions of deamination pathways and urea synthesis, the pathways that degrade carbon skeletons are described.

Deamination

The removal of the α-amino group from amino acids involves two types of biochemical reaction: transamination and oxidative deamination. Both reactions have been described (Section 13.2). (Recall that transamination reactions occupy important positions in nonessential amino acid synthesis.) Because these reactions are reversible, amino groups are easily shifted from abundant amino acids and used to synthesize those that are scarce. Amino groups become available for urea synthesis when amino acids are in excess. Urea is synthesized in especially large amounts when the diet is high in protein or when there is massive breakdown of protein, for example, during starvation.

In muscle, excess amino groups are transferred to α-ketoglutarate to form glutamate.

$$\alpha\text{-Ketoglutarate} + \text{L-Amino acid} \rightleftharpoons \text{L-Glutamate} + \alpha\text{-Keto acid}$$

The amino groups of glutamate molecules are transported in blood to the liver by the alanine cycle (Figure 8.15).

$$\text{Pyruvate} + \text{L-Glutamate} \rightleftharpoons \text{L-Alanine} + \alpha\text{-Ketoglutarate}$$

In the liver, glutamate is formed as the reaction catalyzed by alanine transaminase is reversed. The oxidative deamination of glutamate yields α-ketoglutarate and NH_4^+.

In most extrahepatic tissues, the amino group of glutamate is released via oxidative deamination as NH_4^+. Ammonia is carried to the liver as the amide group of glutamine. The ATP-requiring reaction in which glutamate is converted to glutamine is catalyzed by glutamine synthetase.

$$\text{L-Glutamate} + NH_4^+ + \text{ATP} \rightarrow \text{L-Glutamine}$$

After its transport to the liver, glutamine is hydrolyzed by glutaminase to form glutamate and NH_4^+. An additional NH_4^+ is generated as glutamate dehydrogenase converts glutamate to α-ketoglutarate.

$$\text{L-Glutamine} + H_2O \rightarrow \text{L-Glutamate} + NH_4^+$$
$$\text{L-Glutamate} + H_2O + NAD^+ \rightarrow \alpha\text{-Ketoglutarate} + \text{NADH} + H^+ + NH_4^+$$

Most of the ammonia generated in amino acid degradation is produced by the oxidative deamination of glutamate. Additional ammonia is produced in several other reactions catalyzed by the following enzymes:

1. **L-Amino acid oxidases.** Small amounts of ammonia are generated by various L-amino acid oxidases, found in liver and kidney, that require a flavin mononucleotide (FMN) coenzyme. FMN is regenerated from $FMNH_2$ by reacting with O_2 to form H_2O_2.

PROTEIN TURNOVER

The cellular concentration of each type of protein is a consequence of a balance between its synthesis and its degradation. Although it appears to be wasteful, the continuous degradation and resynthesis of proteins, a process referred to as **protein turnover,** serve several purposes. First of all, metabolic flexibility is afforded by relatively quick changes in the concentrations of key regulatory enzymes, peptide hormones, and receptor molecules. Protein turnover also protects cells from the accumulation of abnormal proteins. Finally, numerous physiological processes are just as dependent on timely degradative reactions as they are on synthetic ones. Prominent examples include eukaryotic cell cycle control and antigen presentation. The progression of eukaryotic cells through the phases of the cell cycle (Section 17.1) is regulated by the precisely timed synthesis and degradation of a class of proteins called the *cyclins.* In *antigen presentation* certain immune system cells (e.g., macrophages) engulf foreign or abnormal substances. Most of the molecules that can elicit an immune response, referred to as *antigens,* are polypeptides or proteins. Partially degraded antigen is then transferred to the macrophage's plasma membrane where it is used to activate certain T lymphocytes (T cells) via cell-cell interactions. T cells are the principal regulators of the body's immune response.

Proteins differ significantly in their turnover rates, which are measured in half-lives. (A *half-life* is the time required for 50% of a specified amount of a protein to be degraded.) Proteins that play structural roles typically have long half-lives. For example, some connective tissue proteins (e.g., the collagens) often have half-lives that are measured in years. In contrast, the half-lives of regulatory enzymes are typically measured in minutes. Several selected examples are listed in Table 14.1.

Despite considerable research, the mechanisms of protein turnover are still unclear. However, several aspects of this process are now known. Proteins are degraded by proteolytic enzymes found throughout the cell. These include the cytoplasmic Ca^{2+}-activated calpains and the lysosomal cathepsins. In addition, ubiquination is now believed to have a major role in protein turnover. In **ubiquination,** illustrated in Figure 14A, several molecules of a small 76-residue eukaryotic protein called **ubiquitin** are covalently attached to some proteins destined for degradation. Once a protein is ubiquinated, it is degraded by a multisubunit proteolytic complex called a **proteosome.** Because ubiquitin molecules are not degraded in this process, they then become available for new rounds of protein degradation.

Ubiquitin, found in several cellular compartments (e.g., cytoplasm and the nucleus), belongs to a class of protein referred to as *stress proteins.* Stress proteins, also called **heat shock proteins** (hsp), are so named because their syntheses are accelerated (and in some cases initiated) when cells encounter stress. (The name "heat shock protein" is misleading, since a variety of stressful conditions besides elevated temperature induce their synthesis.) Other stress proteins act as **molecular chaperones,** that is, they promote protein folding (Special Interest Box 18.2).

The mechanisms that target protein for destruction by ubiquination or by other degradative processes are not fully un-

TABLE 14.1
Human Protein Half-Lives

Protein	Approximate Value of Half-Life (h)
Ornithine decarboxylase	0.5
Tyrosine aminotransferase	2
Tryptophan oxygenase	2
PEP carboxykinase	5
Arginase	96
Aldolase	118
Glyceraldehyde-3-phosphate dehydrogenase	130
Cytochrome c	150
Hemoglobin	2880

2. **Serine and threonine dehydratases.** Serine and threonine are not substrates in transamination reactions. Their amino groups are removed by the pyridoxal phosphate-requiring hepatic enzymes serine dehydratase and threonine dehydratase. The carbon skeleton products of these reactions are pyruvate and α-ketobutyrate, respectively.

3. **Bacterial urease.** A major source of ammonia in liver (approximately 25%) is produced by the action of certain bacteria in the intestine that possess the enzyme urease. Urea present in the blood circulating through the lower digestive tract diffuses across cell membranes and into the intestinal lumen. Once urea is hydrolyzed by bacterial urease to form ammonia, the latter substance diffuses back into the blood, which transports it to the liver.

Urea Synthesis

In ureotelic organisms the urea cycle disposes of approximately 90% of surplus nitrogen. As shown in Figure 14.1, urea is formed from ammonia, CO_2, and aspartate in a cyclic pathway referred to as the **urea cycle.** Because the urea cycle was dis-

derstood. However, the following features of proteins appear to mark them for destruction:

1. **N-terminal residues.** The N-terminal residue of a protein is partially responsible for its susceptibility to degradation. For example, proteins with methionine or alanine N-terminal residues have substantially longer half-lives than do those with leucine or lysine.

2. **Peptide motifs.** Proteins with certain homologous sequences are rapidly degraded. For example, proteins that have extended sequences containing proline, glutamate, serine, and threonine have half-lives of less than two hours. (PEST sequences are named for the one-letter abbreviations for these amino acids. See Table 5.1.) The *cyclin destruction box* is a set of homologous nine-residue sequences near the N-terminus of cyclins that ensure rapid ubiquination.

3. **Oxidized residues.** Oxidized amino acid residues (i.e., residues that are altered by oxidases or by attack by ROS) promote protein degradation.

FIGURE 14A

Ubiquination of Protein.

Three enzymes are involved in preparing ubiquitin for its role in protein degradation. In the first step, an activating enzyme E_1 forms a thiol ester with ubiquitin. (This reaction is driven by the hydrolysis of ATP to AMP.) Ubiquitin is then transferred from E_1 to E_2. Ubiquitin may be transferred from E_2 (ubiquitin-conjugating enzyme) directly to a target protein. Often E_2 is a substrate for a ubiquitin targeting protein called E_3, which identifies specific proteins to be degraded. In this latter process ubiquitin is transferred from E_2 to E_3 and then to the target protein. Ubiquitin is attached to target proteins via a covalent bond between the C terminal glycine of ubiquitin and the ϵ side chain amino group of the lysine residues of the targeted protein. Most cells possess a single type of E_1 and numerous families of E_2 and E_3.

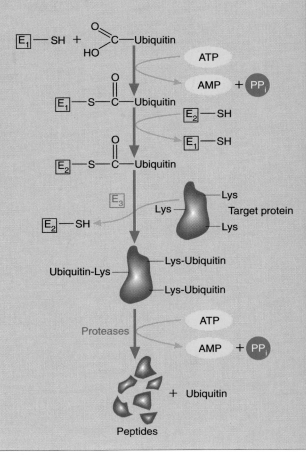

covered by Hans Krebs and Kurt Henseleit, it is often referred to as the **Krebs urea cycle** or the *Krebs-Henseleit cycle.*

Urea synthesis, which occurs in hepatocytes, begins with the formation of carbamoyl phosphate. The substrates for this reaction, catalyzed by carbamoyl phosphate synthetase I, are NH_4^+ and HCO_3^-. (The nitrogen source for carbamoyl phosphate synthetase II, the enzyme involved in pyrimidine synthesis, is glutamine.)

Because two molecules of ATP are required in carbamoyl phosphate synthesis, this reaction is essentially irreversible. (One is used to activate HCO_3^-. The second molecule is used to phosphorylate carbamate.) Carbamoyl phosphate subsequently reacts with ornithine to form citrulline. This reaction, catalyzed by ornithine transcarbamoylase, is driven to completion because phosphate is released from carbamoyl phosphate. (Recall from Table 4.2 that carbamoyl phosphate has a high phosphate group transfer potential.) Once it is formed, citrulline is transported to the cytoplasm, where it reacts with aspartate to form arginosuccinate. (The α-amino group of aspartate, formed from oxaloacetate by transamination reactions in the liver, provides the

413

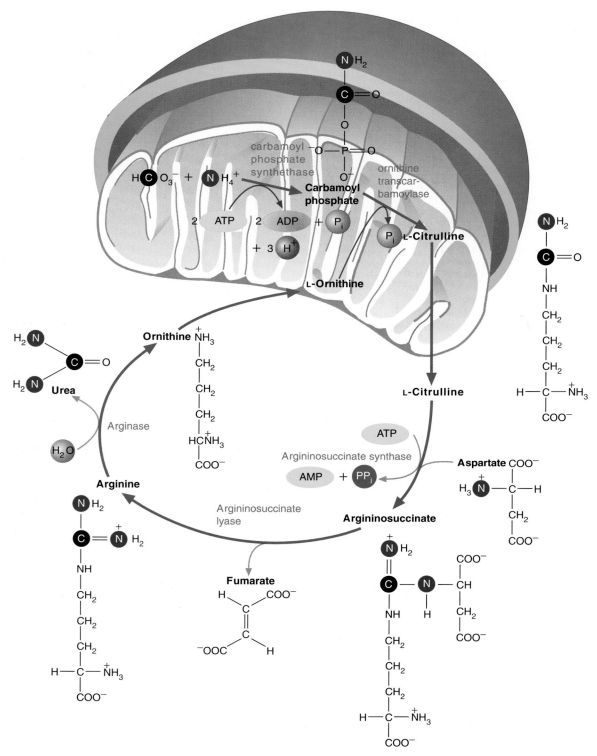

FIGURE 14.1

The Urea Cycle.

The urea cycle converts NH_4^+ to urea, a less toxic molecule. The sources of the atoms in urea are shown in color. Citrulline is transported across the inner membrane by a carrier for neutral amino acids. Ornithine is transported in exchange for H^+ or citrulline. Fumarate is transported back into the mitochondrial matrix (for reconversion to malate) by carriers for α-ketoglutarate or tricarboxylic acids.

second nitrogen that is ultimately incorporated into urea.) This reaction, which is catalyzed by arginosuccinate synthase, is reversible. It is pulled forward by the cleavage of pyrophosphate by pyrophosphatase. Arginosuccinate lyase subsequently cleaves arginosuccinate to form arginine (the immediate precursor of urea) and fumarate. In the final reaction of the urea cycle, arginase catalyzes the synthesis of urea. Once it forms, urea diffuses out of the hepatocytes and into the bloodstream. It is ultimately eliminated in the urine by the kidney. Ornithine, the other product of this reaction, is transported into the mitochondrial matrix, thus enabling the urea cycle to continue. Because arginase is found in significant amounts only in the livers of ureotelic animals, urea is produced only in this organ.

After its transport back into the mitochondrial matrix, fumarate is hydrated to form malate, a component of the citric acid cycle. The oxaloacetate product of the citric acid cycle can be used in energy generation, or it can be converted to glucose or aspartate. The relationship between the urea cycle and the citric acid cycle, often referred to as the **Krebs bicycle,** is outlined in Figure 14.2.

The net reaction for the urea cycle is

$$CO_2 + NH_4^+ + \text{aspartate} + 3\ ATP + 2\ H_2O \rightarrow$$
$$\text{urea} + \text{fumarate} + 2\ ADP + 2\ P_i + AMP + PP_i + 5\ H^+$$

Four high-energy phosphates are consumed in the synthesis of one molecule of urea. Two molecules each of ATP are required to regenerate ATP from AMP and two molecules of ATP from two molecules of ADP.

Control of the Urea Cycle

KEY *Concepts*

Urea is synthesized from ammonia, CO_2, and aspartate. The urea cycle is carefully regulated to prevent hyperammonemia.

Because ammonia is so toxic (see Special Interest Box 14.2), it is not surprising that the urea cycle is subject to stringent regulation. There are long- and short-term regulatory mechanisms. The levels of all five urea cycle enzymes are altered by variations in dietary protein consumption. Within several days after a significant dietary change, there are twofold to threefold changes in enzyme levels. Several hormones (e.g., glucagon and the glucocorticoids) are believed to be involved in the altered rates of enzyme synthesis.

The urea cycle enzymes are controlled in the short term by the concentrations of their substrates. Carbamoyl phosphate synthetase I is also allosterically activated by *N-acetylglutamate*. This latter molecule is a sensitive indicator of the cell's glutamate

FIGURE 14.2

The Krebs Bicycle.

The aspartate used in urea synthesis is generated from oxaloacetate, a citric acid cycle intermediate.

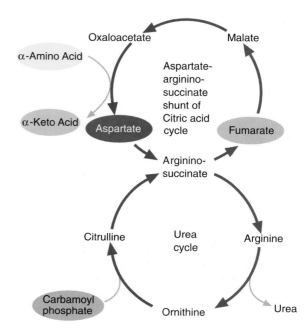

HYPERAMMONEMIA

Hyperammonemia is a condition in which the concentration of NH_4^+ is excessive (i.e., greater than 60 μM) in blood. Elevated concentrations of ammonia are serious; the consequences of **ammonia intoxication** include lethargy, tremors, slurred speech, blurred vision, protein-induced vomiting, coma, and death.

Hyperammonemia may be caused by genetic defects or cirrhosis of the liver. In congenital (inherited) hyperammonemia, a relatively rare condition, one or more of the urea cycle enzymes are missing or defective. Complete absence of a urea cycle enzyme is fatal soon after birth. Brain damage can be minimized in infants who have partial deficiencies in urea synthesis if aggressive therapy is initiated immediately after birth. (Therapy consists of diets with severe restrictions on protein intake.) In cirrhosis, loss of liver function is devastating due to widespread inflammation and necrosis (cell death). It is most commonly caused by prolonged, excessive consumption of ethanol.

(Refer to Special Interest Box 19.1.) Less common causes of cirrhosis include prolonged exposure to toxic chemicals such as carbon tetrachloride, hepatitis (an inflammation of the liver that is often caused by viral infections) and amebiasis (an infection with parasitic amebas).

Because most of the symptoms of ammonia intoxication are manifested in brain tissue, ammonia is considered a neurotoxic agent. Although significant research effort has been devoted to elucidating the effects of ammonia on brain cells, the mechanism of the damage is still unclear. As described previously, depletion of glutamate via the glutamine synthase–catalyzed reaction may contribute to brain cell damage. (Because glutamate and its derivative GABA are neurotransmitters, alterations in their concentrations are potentially toxic.) Depletion of α-ketoglutarate, a citric acid cycle intermediate, has also been implicated. Other toxic effects of ammonia on the brain may include inhibition of amino acid transport and the Na^+-K^+ATPase.

---

concentration. (Recall that a significant amount of NH_4^+ is derived from glutamate.) N-acetylglutamate is produced from glutamate and acetyl-CoA in a reaction catalyzed by N-acetylglutamate synthase.

14.1   In some clinical circumstances, patients with hyperammonemia are treated with antibiotics. Suggest a rational basis for this therapy.

Catabolism of Amino Acid Carbon Skeletons

The α-amino acids can be grouped into classes according to their end products. As previously mentioned, these are acetyl-CoA, acetoacetyl-CoA, pyruvate, and several citric acid cycle intermediates. Each group is briefly discussed below. The degradation pathways for the 20 α-amino acids found in proteins are outlined in Figure 14.3.

Amino Acids Forming Acetyl-CoA   In all, ten α-amino acids yield acetyl-CoA. This group is further divided according to whether or not pyruvate is an intermediate in acetyl-CoA formation. (Recall that pyruvate is converted to acetyl-CoA by the pyruvate dehydrogenase complex.) The amino acids whose degradation involves pyruvate are alanine, serine, glycine, cysteine, and threonine. The other five amino acids converted to acetyl-CoA by pathways not involving pyruvate are lysine, tryptophan, tyrosine, phenylalanine, and leucine. The two reaction sequences are outlined in Figures 14.4 and 14.5.

1. **Alanine.** Recall that the reversible transamination reaction involving alanine and pyruvate is an important component of the alanine cycle discussed previously (Section 8.4).

2. **Serine.** As described, serine is converted to pyruvate by serine dehydratase.

3. **Glycine.** Glycine can be converted to serine by serine hydroxymethyltransferase. (The hydroxymethyl group is donated by N^5,N^{10}-methylene THF as described in Section 13.3) Then serine is converted to pyruvate, as previously described. Most glycine molecules, however, are degraded to CO_2, NH_4^+, and a methylene group removed by THF. The enzyme involved is glycine synthase (also referred to as glycine cleavage enzyme), which requires NAD^+.

FIGURE 14.3

Degradation of the 20 α-Amino Acids Found in Proteins.

The α-amino groups are removed early in the catabolic pathways. Carbon skeletons are converted to common metabolic intermediates.

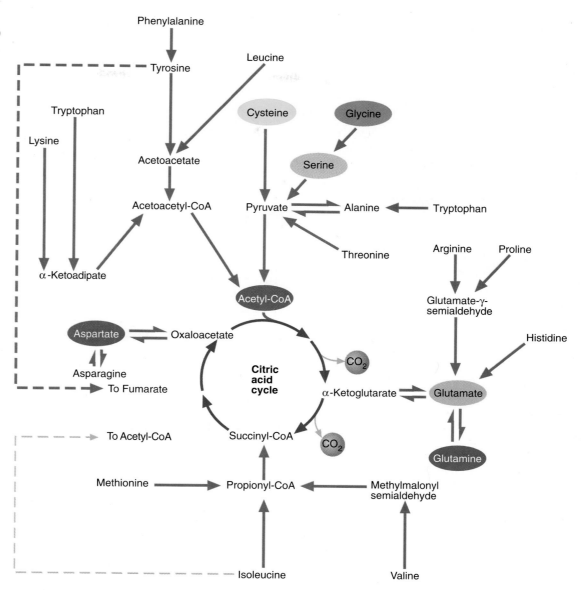

4. **Cysteine.** In animals, cysteine is converted to pyruvate by several pathways. In the principal pathway, the conversion occurs in three steps. Initially, cysteine is oxidized to cysteine sulfate. Pyruvate is produced after a transamination and a desulfuration reaction.

5. **Threonine.** In the major degradative pathway, threonine is oxidized by threonine dehydrogenase to form α-amino-β-ketobutyrate. The latter molecule is metabolized further to form lactate via pyruvate, or it can be cleaved by α-amino-β-ketobutyrate lyase to form acetyl-CoA and glycine. As previously discussed, glycine is converted to acetyl-CoA via pyruvate. Alternatively, threonine can be degraded to α-ketobutyrate by threonine dehydratase and subsequently to propionyl-CoA. Propionyl-CoA is then converted to succinyl-CoA (see p. 257).

6. **Lysine.** Lysine is converted to α-ketoadipate in a series of reactions that include two oxidations, removal of the side chain amino group, and a transamination. Acetoacetyl-CoA is produced in a further series of reactions that involve several oxidations, a decarboxylation, and a hydration.

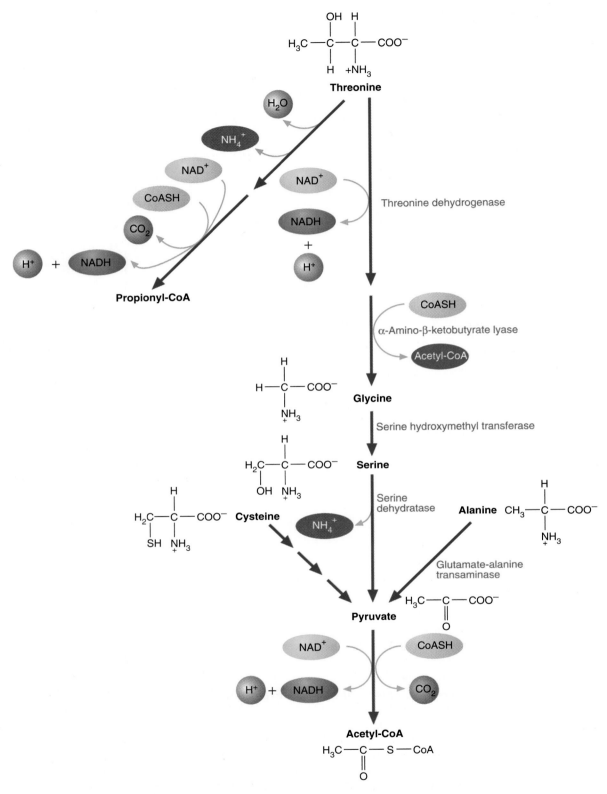

FIGURE 14.4

The Catabolic Pathways of Threonine, Glycine, Serine, Cysteine, and Alanine.

Pyruvate is an intermediate in the conversion of these amino acids to acetyl-CoA.

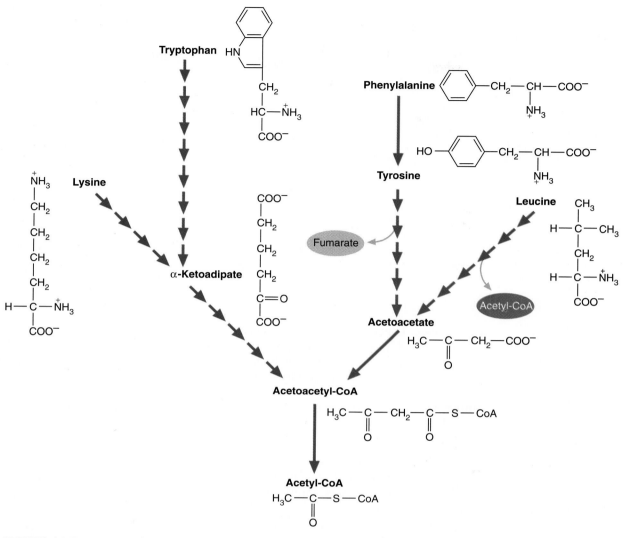

FIGURE 14.5

The Catabolic Pathways of Lysine, Tryptophan, Phenylalanine, Tyrosine, and Leucine.

These pathways are long and complex. The number of reactions in each segment is indicated by the number of arrows.

Acetoacetyl-CoA can be converted to acetyl-CoA in a reaction that is the reverse of a step in ketone body formation.

7. **Tryptophan.** Tryptophan is converted to α-ketoadipate in a long, complex series of eight reactions, which also yield formate and alanine. Acetyl-CoA is synthesized from α-ketoadipate as described above. The alanine produced in this pathway is converted to acetyl-CoA via pyruvate.

8. **Tyrosine.** Tyrosine catabolism begins with a transamination and a hydroxylation. Homogentisate is synthesized in the latter reaction, catalyzed by the ascorbate-requiring enzyme parahydroxyphenylpyruvate dioxygenase. Homogentisate is converted to maleylacetoacetate by homogentisate oxidase. Acetoacetate and fumarate are then generated in isomerization and hydration reactions.

9. **Phenylalanine.** Phenylalanine is converted to tyrosine by phenylalanine hydroxylase in a reaction illustrated in Figure 14.6. As was just described, tyrosine is degraded to form acetoacetate and fumarate.

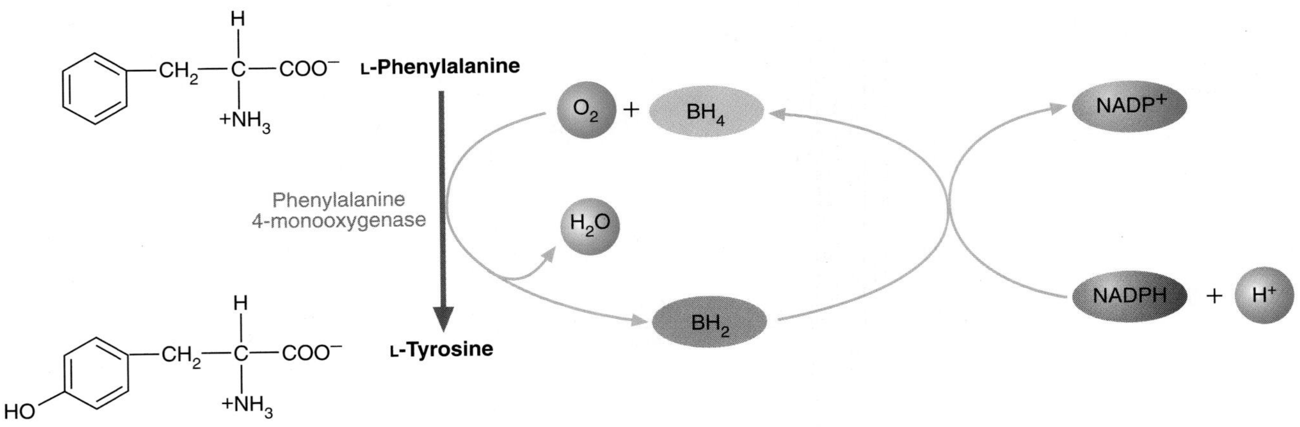

FIGURE 14.6

The Conversion of Phenylalanine to Tyrosine.

The reaction catalyzed by phenylalanine-4-monooxygenase is irreversible. The electrons required for the hydroxylation of phenylalanine are carried to O_2 from NADPH by tetrahydrobiopterin.

10. **Leucine.** Leucine, one of the branched chain amino acids, is converted to HMG-CoA in a series of reactions that include a transamination, two oxidations, a carboxylation, and a hydration. HMG-CoA is then converted to acetyl-CoA and acetoacetate by HMG-CoA lyase.

Amino Acids Forming α-Ketoglutarate   Five amino acids (arginine, proline, histidine, glutamate, and glutamine) are degraded to α-ketoglutarate. An outline of their catabolism is illustrated in Figure 14.7. Each pathway is briefly described.

1. **Glutamate and glutamine.** Glutamine is converted to glutamate by glutaminase. As described previously, glutamate is converted to α-ketoglutarate by glutamate dehydrogenase or by transamination.

2. **Arginine.** Recall that arginine is cleaved by arginase to form ornithine and urea. In a subsequent transamination reaction, ornithine is converted to glutamate-γ-semialdehyde. Glutamate is then produced as glutamate-γ-semialdehyde is hydrated and oxidized. As described, α-ketoglutarate is produced by a transamination reaction or by oxidative deamination.

3. **Proline.** Proline catabolism begins with an oxidation reaction that produces Δ^1-pyrroline. The latter molecule is converted to glutamate-γ-semialdehyde by a hydration reaction. Glutamate is then formed by another oxidation reaction.

4. **Histidine.** Histidine is converted to glutamate in four reactions: a nonoxidative deamination, two hydrations, and the removal of a formamino group (NH=CH—) by THF.

Amino Acids Forming Succinyl-CoA   Succinyl-CoA is formed from the carbon skeletons of methionine, isoleucine, and valine. An outline of the reactions that degrade these amino acids is illustrated in Figure 14.8.

1. **Methionine.** Methionine degradation begins with the formation of S-adenosylmethionine, which is followed by a demethylation reaction, as described (Figure 13.16). S-Adenosylhomocysteine, the product of the latter reaction, is hydrolyzed to adenosine and homocysteine. Homocysteine then combines with serine to yield cystathionine. Cysteine, α-ketobutyrate, and NH_4^+ result from the cleavage of cystathionine. α-Ketobutyrate is then converted to propionyl-CoA by α-ketoacid dehydrogenase. Propionyl-CoA is converted to succinyl-CoA in three steps. The

FIGURE 14.7

The Catabolic Pathways of Arginine, Proline, Histidine, Glutamine, and Glutamate.

All these amino acids are eventually converted to α-ketoglutarate.

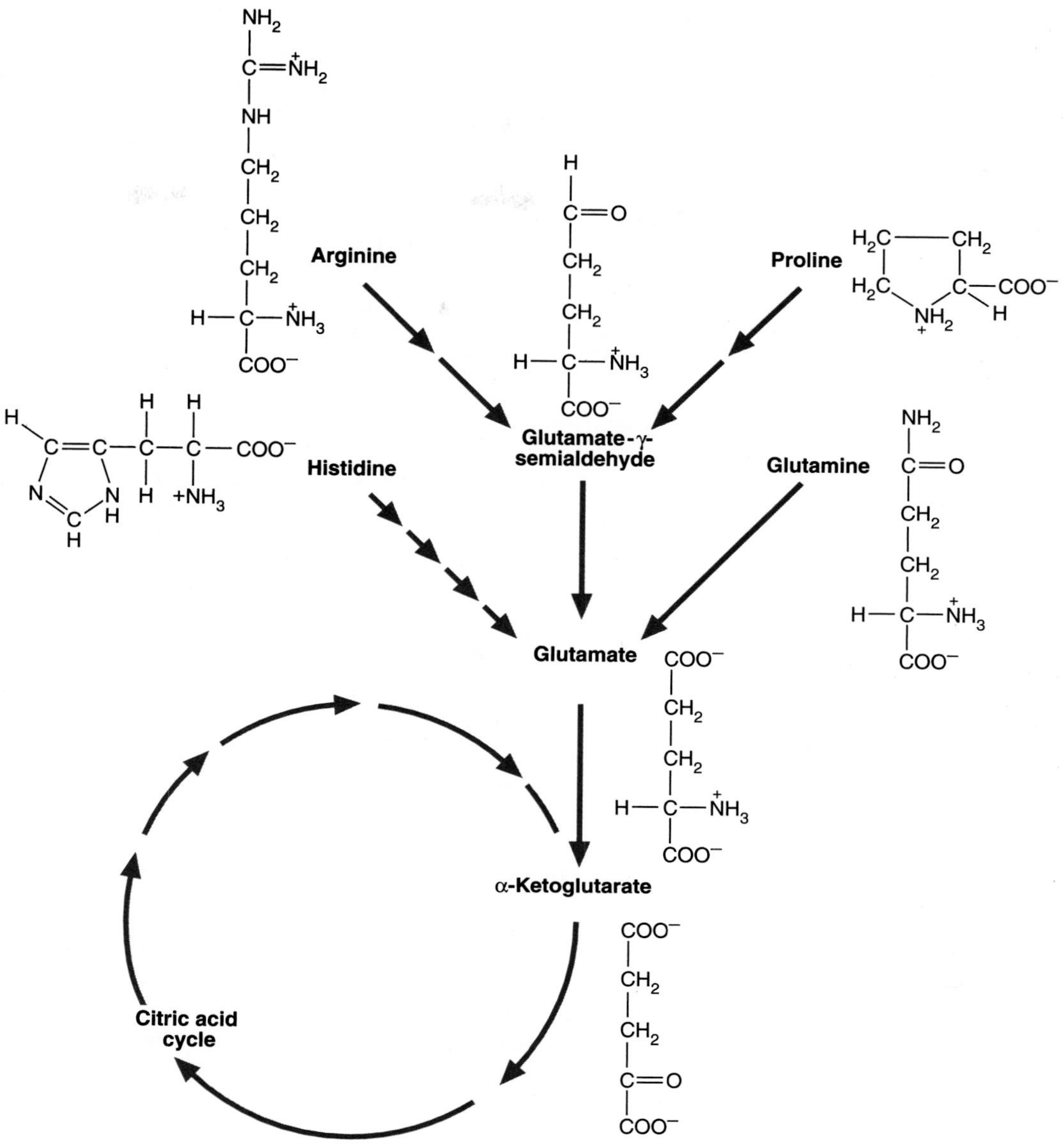

Arginine

Proline

Glutamate-γ-semialdehyde

Histidine

Glutamine

Glutamate

α-Ketoglutarate

Citric acid cycle

enzyme that catalyzes the last of these, methylmalonyl-CoA mutase, requires methylcobalamin. The conversion of methionine to cysteine is sometimes referred to as the **transulfuration pathway** (Figure 14.9). A substantial amount of the sulfate produced from cysteine degradation is excreted in urine. Sulfate is also used in the synthesis of sulfatides and proteoglycans. Additionally, molecules such as steroids and certain drugs are excreted as sulfate esters (Chapter 19).

2. **Isoleucine and valine.** The first four reactions in the degradation of isoleucine and valine are identical. Initially, both amino acids undergo transamination reactions to form α-keto-β-methylvalerate and α-ketoisovalerate, respectively. This is followed by the formation of CoA derivatives, as well as oxidative decarboxylation, oxidation, and dehydration reactions. The product of the isoleucine pathway is then hydrated, dehydrogenated, and cleaved to form acetyl-CoA and propionyl-CoA. In the valine degradative pathway the α-keto acid intermediate is converted into propionyl-CoA after a double bond is hydrated and CoA is removed by hydrolysis. After the formation of an aldehyde by the oxidation of the

KEY Concepts

Amino acid carbon skeletons can be degraded into one or more of several metabolites. These include acetyl-CoA, acetoacetyl-CoA, α-ketoglutarate, succinyl-CoA, and oxaloacetate.

FIGURE 14.8

The Catabolic Pathways of Methionine, Isoleucine, and Valine.

Propionyl-CoA and L-methylmalonyl-CoA are intermediates in the conversion of these amino acids to succinyl-CoA. Methylmalonyl-CoA mutase is a vitamin B$_{12}$-requiring enzyme.

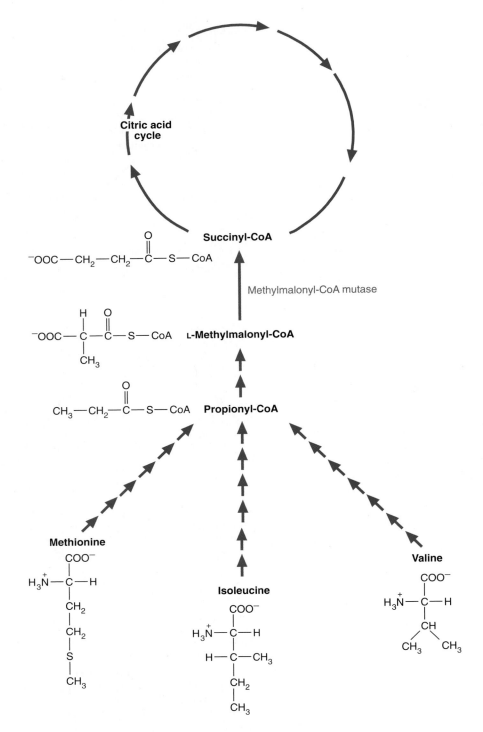

hydroxyl group, propionyl-CoA is produced as a new thioester is formed during an oxidative decarboxylation.

Amino Acids Forming Oxaloacetate   Both aspartate and asparagine are degraded to form oxaloacetate. Aspartate is converted to oxaloacetate with a single transamination reaction. Asparagine is initially hydrolyzed to yield aspartate and NH$_4^+$ by asparaginase.

DISORDERS OF AMINO ACID CATABOLISM

Defects in amino acid catabolism were among the first genetic diseases to be recognized and investigated by medical scientists. These "inborn errors of metabolism" result from **mutations** (permanent changes in genetic information, i.e., DNA structure). Most commonly, in the genetic diseases related to amino acid metabolism, the defective gene codes for an enzyme. The metabolic blockage that results from such a deficit disrupts what are ordinarily highly coordinated cellular and organismal processes, producing abnormal amounts and/or types of metabolites. Because these metabolites (or their heightened concentrations) are often toxic, permanent damage or death ensues. Several of the most commonly observed inborn errors of amino acid metabolism are discussed below.

Alkaptonuria, caused by a deficiency of homogentisate oxidase, was the first disease to be linked to genetic inheritance involving a single enzyme. In 1902, Archibald Garrod proposed that a single inheritable unit (later called a gene) was responsible for the urine in alkaptonuric patients turning black. Large quantities of homogentisate, the substrate for the defective enzyme, are excreted in urine. Homogentisate turns black when it is oxidized as the urine is exposed to air. Although black urine appears to be an essentially benign (if somewhat disconcerting) condition, alkaptonuria is not innocuous, since alkaptonuric patients develop arthritis in later life. In addition, pigment accumulates gradually and unevenly darkens the skin.

Albinism is an example of a genetic defect with serious consequences. The enzyme tyrosinase is deficient. Consequently, *melanin,* a black pigment found in skin, hair, and eyes, is not produced. It is formed from tyrosine in several cell types, for example, the melanocytes in skin. In such cells, tyrosinase converts tyrosine to DOPA and DOPA to dopaquinone. A large number of molecules of the latter product, which is highly reactive, condense to form melanin. Because of the lack of pigment, affected individuals (called albinos) are extremely sensitive to sunlight. In addition to their susceptibility to skin cancer and sunburn, they often have poor eyesight.

Phenylketonuria, caused by a deficiency of phenylalanine hydroxylase, is one of the most common genetic diseases associated with amino acid metabolism. If this condition is not identified and treated immediately after birth, mental retardation and other forms of irreversible brain damage occur. This damage results mostly from the accumulation of phenylalanine. (The actual mechanism of the damage is not understood.) When it is present in excess, phenylalanine undergoes transamination to form phenylpyruvate, which is also converted to phenyllactate and phenylacetate. Large amounts of these molecules are excreted in the urine. Phenylacetate gives the urine its characteristic musty odor. Phenylketonuria is treated with a low phenylalanine diet.

In *maple syrup urine disease,* also called *branched chain ketoaciduria,* the α-keto acids derived from leucine, isoleucine, and valine accumulate in large quantities in blood. Their presence in urine imparts a characteristic odor that gives the malady its name. All three α-keto acids accumulate because of a deficient branched chain α-keto acid dehydrogenase complex. (This enzymatic activity is responsible for the conversion of the α-keto acids to their acyl-CoA derivatives.) If left untreated, affected individuals experience vomiting, convulsions, severe brain damage, and mental retardation. They often die before one year of age. As with phenylketonuria, treatment consists of rigid dietary control.

Deficiency of methylmalonyl-CoA mutase results in *methylmalonic acidemia,* a condition in which methylmalonate accumulates in blood. The symptoms are similar to those of maple syrup urine disease. Methylmalonate may also accumulate because of a deficiency of adenosylcobalamin or weak binding of this coenzyme by a defective enzyme. Some affected individuals respond to injections of large daily doses of vitamin B_{12}.

QUESTIONS

14.2   Taurine is a sulfur-containing amine synthesized from cysteine. With the exception of its incorporation in bile salts, taurine's physiological role is still poorly understood. However, several pieces of information suggest that taurine is an important metabolite. For example, taurine is found in brain tissue in large amounts. In addition, domestic cats have recently been observed to develop congestive heart failure if fed a taurine-free diet. (Cats cannot synthesize taurine. For this reason they must consume meat in their diet. Cats that are fed vegetarian diets soon become listless and will eventually die prematurely.) In most animals, taurine is synthesized from cysteine sulfinate (the oxidation product of cysteine) in two reactions: a decarboxylation followed by an oxidation of the sulfinate group ($-SO_2^-$) to form sulfonate ($-SO_3^-$). With this information, determine the biosynthetic pathway for taurine. (Hint: The structure of taurine is illustrated in Chapter 10 on p. 286. Also refer to Fig. 14.9 on the next page.)

14.3   Some amino acids are classified as both ketogenic and glucogenic. Review the amino acid catabolic pathways and determine which amino acids belong to both categories.

FIGURE 14.9

FIGURE 14.9

The Transulfuration Pathway.

The sulfur atom of methionine becomes the sulfur atom of cysteine. The sulfate generated in cysteine catabolism is excreted or used in several biosynthetic or catabolic pathways. The transulfuration and methylation pathways are intimately related.

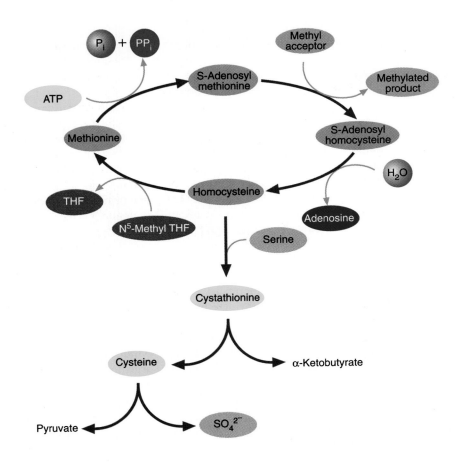

14.2 DEGRADATION OF SELECTED NEUROTRANSMITTERS

The previous discussion of amino acid catabolic disorders indicates that catabolic processes are just as important for the proper functioning of cells and organisms as are anabolic processes. This is no less true for molecules that act as neurotransmitters. To maintain precision in information transfer, neurotransmitters are usually quickly degraded or removed from the synaptic cleft. An extreme example of enzyme inhibition illustrates the importance of neurotransmitter degradation. Recall that acetylcholine is the neurotransmitter that initiates muscle contraction. Shortly afterwards, the action of acetylcholine is terminated by the enzyme acetylcholinesterase. (Acetylcholine must be destroyed rapidly so that muscle can relax before the next contraction.) Acetylcholinesterase is a serine esterase that hydrolyzes acetylcholine to acetate and choline. Serine esterases have catalytic mechanisms similar to those of the serine proteases (discussed in Section 6.4). Both types of enzymes are irreversibly inhibited by DFP (diisopropylfluorophosphate). Exposure to DFP causes muscle paralysis because acetylcholinesterase is irreversibly inhibited. With each nerve impulse, more acetylcholine molecules enter the neuromuscular synaptic cleft. The accumulating acetylcholine molecules repetitively bind to acetylcholine receptors. The overstimulated muscle cells soon become paralyzed (nonfunctional). Affected individuals suffocate because of paralyzed respiratory muscles.

The catecholamines epinephrine, norepinephrine, and dopamine are inactivated by oxidation reactions catalyzed by monoamine oxidase (MAO) (Figure 14.10). Because MAO is found within nerve endings, catecholamines must be transported out of the synaptic cleft before inactivation. (The process by which neurotransmitters are transported back into nerve cells so that they can be reused or degraded is referred to as *reuptake*.) Catecholamines are also inactivated in methylation reactions catalyzed by catechol-O-methyltransferase in which a variety of methylated products are produced.

After its reuptake into nerve cells, serotonin is degraded in a two-step pathway (Figure 14.11). In the first reaction, serotonin is oxidized by MAO. The product,

FIGURE 14.10

Inactivation of the Catecholamines.

Monoamine oxidase is a flavoprotein that catalyzes the oxidative deamination of amines to form the corresponding aldehydes. O_2 is the electron acceptor, and NH_3 and H_2O_2 are the other products.

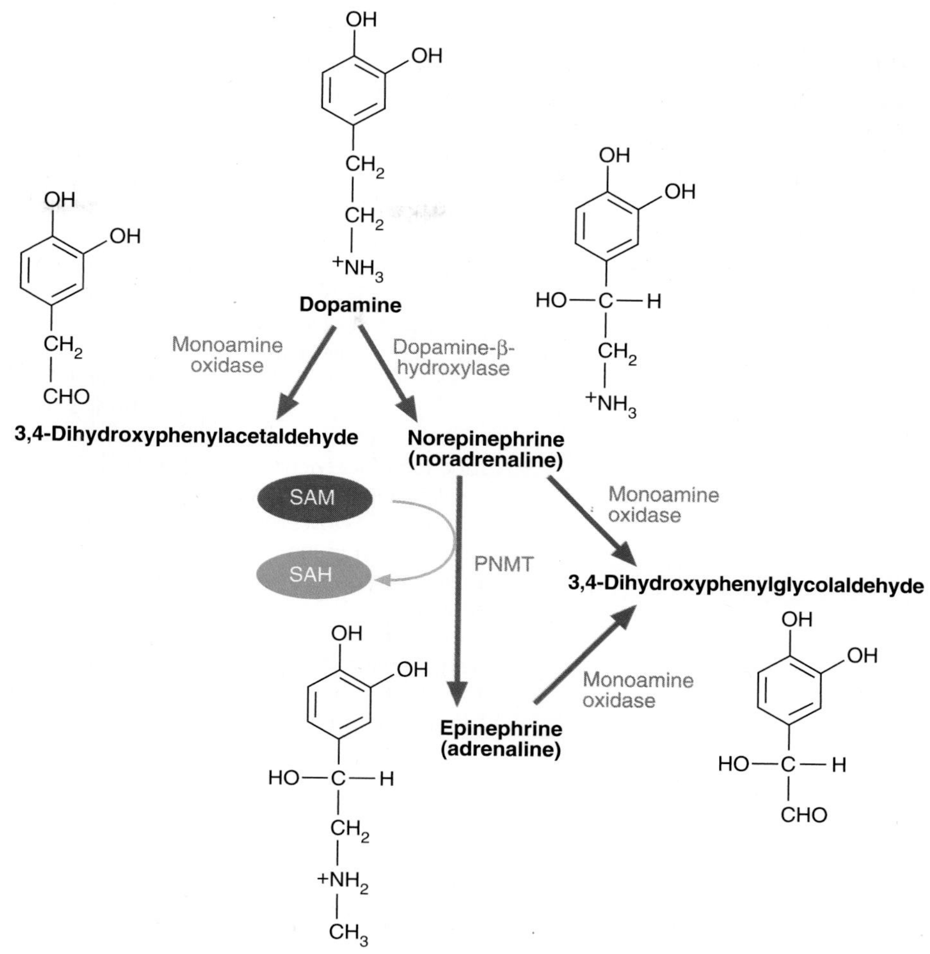

FIGURE 14.11

Degradation of Serotonin.

In the major catabolic pathway, serotonin is deaminated and oxidized to form 5-hydroxyindole-3-acetaldehyde. The latter molecule is then further oxidized to form 5-hydroxyindole-3-acetate.

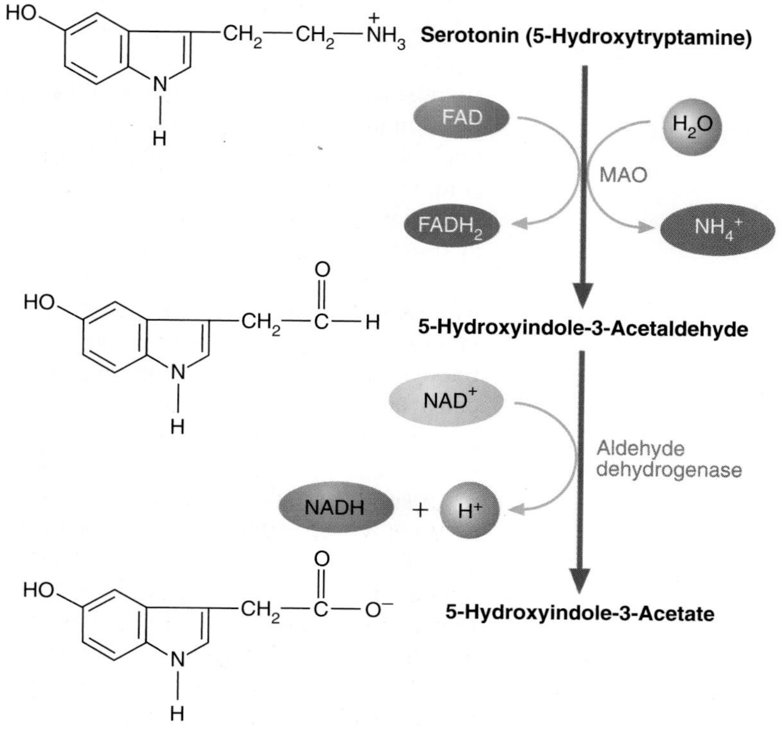

14.4   A variety of medical conditions are currently being treated with medications that block the biological activity or the metabolism of neurotransmitters. The term *antagonist* is used to describe molecules that block the biological actions of normal neurotransmitters. For example, certain drug molecules used to treat hypertension (high blood pressure) antagonize the action of the catecholamines. (The binding of catecholamines to specific receptor molecules in the cardiovascular system constricts blood vessels.) Another interesting example is certain medications that treat *obsessive compulsive disorder* (OCD). (OCD is a condition characterized by the persistent intrusion of unwanted and disturbing thoughts and/or the compulsive performance of certain acts such as handwashing.) For unknown reasons, serotonin reuptake inhibitors have been remarkably effective in improving patient symptoms.

    Myasthenia gravis is treated with drugs that inhibit acetylcholinesterase, the enzyme that degrades acetylcholine. *Myasthenia gravis* is an autoimmune disease in which autoantibodies bind to and initiate the destruction of the acetylcholine receptor in skeletal muscle cell membranes. Gradually, the number of functional acetylcholine receptors is reduced. This condition is characterized by muscle weakness and fatigability. Eventually, patients develop difficulty in speaking and swallowing. However, a short time after consuming reversible cholinesterase inhibitors (e.g., neostigmine or physostigmine), patients experience significant improvement in their symptoms. Based on your knowledge of the action of acetylcholine, can you suggest how anticholinesterase drugs achieve this short-term clinical improvement? (Hint: For a muscle cell to contract, a threshold number of acetylcholine receptors must bind acetylcholine. In normal individuals this number of receptors is significantly lower than the number of receptors in muscle cell membrane. Also note that the productive binding and unbinding of a neurotransmitter to its receptor are often rapid.)

5-hydroxyindole-3-acetaldehyde, is then further oxidized by aldehyde dehydrogenase to form 5-hydroxyindole-3-acetate.

14.3 NUCLEOTIDE DEGRADATION

In most living organisms, purine and pyrimidine nucleotides are constantly degraded. In animals, degradation occurs because of the normal turnover of nucleic acids and nucleotides as well as the digestion of dietary nucleic acids. During digestion, nucleic acids are hydrolyzed to oligonucleotides by enzymes called **nucleases. (Oligonucleotides** are defined as short nucleic acid segments containing fewer than 50 nucleotides.) Enzymes that are specific for breaking internucleotide bonds in DNA are called *deoxyribonucleases* (DNases); those that degrade RNA are called *ribonucleases* (RNases). Once formed, oligonucleotides are further hydrolyzed by various *phosphodiesterases,* a process that produces a mixture of mononucleotides. *Nucleotidases* remove phosphate groups from nucleotides, yielding nucleosides. These latter molecules are hydrolyzed by *nucleosidases* to free bases and phosphorylated sugars, which are then absorbed. Alternatively, nucleosides may be absorbed by intestinal enterocytes.

Generally speaking, dietary purine and pyrimidine bases are not used in significant amounts to synthesize cellular nucleic acids. Instead, they are degraded within enterocytes. Purines are degraded to uric acid in humans and birds. Pyrimidines are degraded to β-alanine or β-aminoisobutyric acid, as well as NH_3 and CO_2. In contrast to the catabolic processes for other major classes of biomolecules (e.g., sugars, fatty acids, and amino acids), purine and pyrimidine catabolism does not result in ATP synthesis. The major pathways for the degradation of purine and pyrimidine bases are described next.

Purine Catabolism

Purine nucleotide catabolism is outlined in Figure 14.12. As noted in the figure, there is some variation in the specific pathways used by different organisms or tissues to degrade AMP. In muscle, for example, AMP is initially converted to IMP by AMP deaminase (also referred to as adenylate aminohydrolase). IMP is subsequently hydrolyzed to inosine by 5'-nucleotidase. In most tissues, however, AMP is hydrolyzed by 5'-nucleotidase to form adenosine. Adenosine is then deaminated by adenosine deaminase (also called adenosine aminohydrolase) to form inosine.

Purine nucleoside phosphorylase converts inosine, guanosine, and xanthosine to hypoxanthine, guanine, and xanthine, respectively. (The ribose-1-phosphate formed during these reactions is reconverted to PRPP by phosphoribomutase.) Hypoxanthine is oxidized to xanthine by xanthine oxidase, an enzyme that contains molybdenum, FAD, and two different Fe-S centers. (Xanthine oxidase-catalyzed reactions produce O_2^- in addition to forming H_2O_2.) Guanine is deaminated to xanthine by aminohydrolase. Xanthine molecules are further oxidized to uric acid by xanthine oxidase.

Several diseases result from defects in purine catabolic pathways. *Gout,* which is often characterized by high blood levels of uric acid and recurrent attacks of arthritis, is caused by several metabolic abnormalities. (See Special Interest Box 14.4.) Two different immunodeficiency diseases are now known to result from defects in purine catabolic reactions. In *adenosine deaminase deficiency,* large concentrations of dATP inhibit ribonucleotide reductase. Consequently, DNA synthesis is depressed. For reasons that are not yet clear, this metabolic distortion is observed primarily in the T and

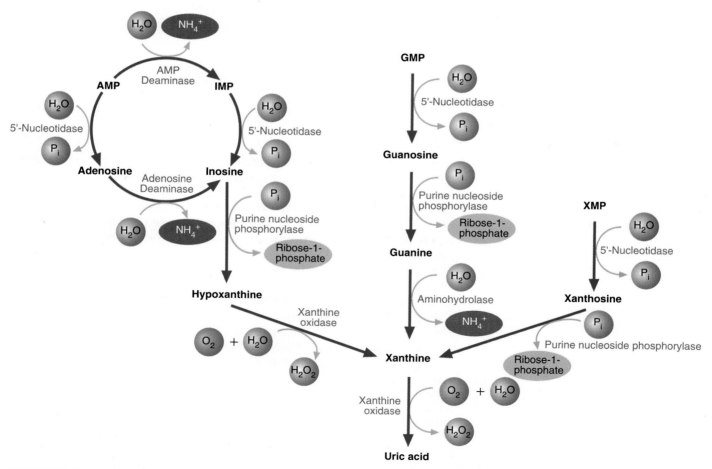

FIGURE 14.12

Purine Nucleotide Catabolism.

Ribose-1-phosphate is released in AMP, GMP, and XMP catabolism. Xanthine oxidase-catalyzed reactions generate O_2^-.

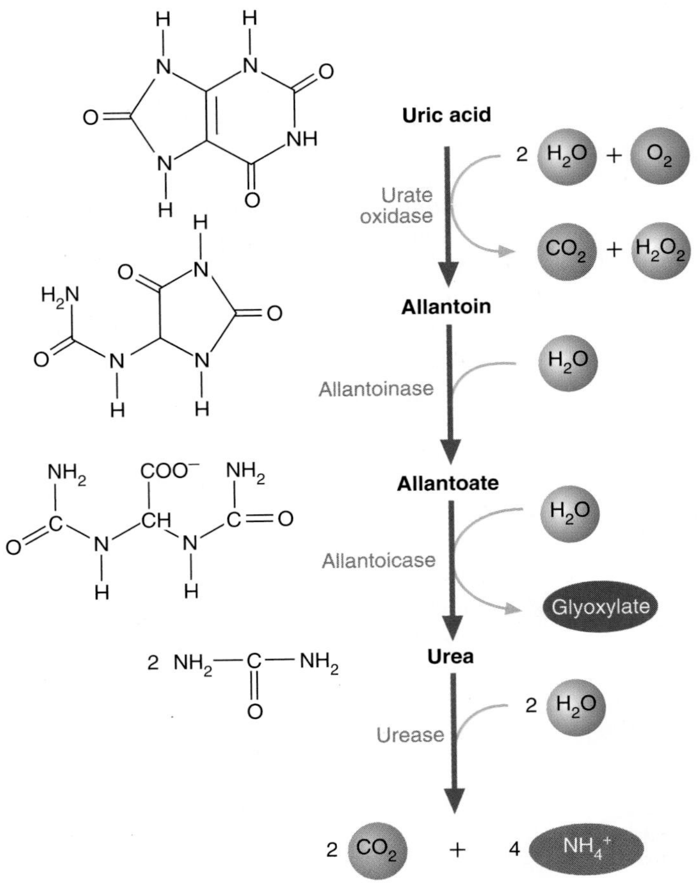

FIGURE 14.13

FIGURE 14.13

Uric Acid Catabolism.

Many animals possess enzymes that allow them to convert uric acid to other excretory products.

B lymphocytes. (*T lymphocytes,* or **T cells,** bear antibody-like molecules on their surfaces. They bind to and destroy foreign cells in a process referred to as **cellular immunity.** *B lymphocytes,* or **B cells,** produce antibodies that bind to foreign substances, thereby initiating their destruction by other immune system cells. The production of antibodies by B cells is referred to as the **humoral immune response.**) Children with adenosine deaminase deficiency usually die before the age of two because of massive infections. In *purine nucleoside phosphorylase deficiency* levels of purine nucleotides are high and synthesis of uric acid decreases. High levels of dGTP are apparently responsible for the impairment of T cells that is characteristic of this malady.

Many animals degrade uric acid further (Figure 14.13). Urate oxidase converts uric acid to allantoin, an excretory product in many mammals. Allantoinase catalyzes the hydration of allantoin to form allantoate, which is excreted by bony fish. Other fish as well as amphibians produce allantoicase, which splits allantoic acid into glyoxylate and urea. Finally, marine invertebrates degrade urea to NH_4^+ and CO_2 in a reaction catalyzed by urease.

QUESTION 14.5

One of the more fascinating aspects of biochemistry is that living organisms use the same molecule for different purposes. Two interesting examples are allantoin and allantoate. As previously mentioned, these molecules serve as nitrogenous waste in several animal groups. Some plant species (i.e., certain legumes such as soybeans and snapbeans) begin to synthesize allantoin and allantoate once they become infected by nitrogen-fixing bacteria. Both molecules, referred to as the *ureides,* are nitrogen transport compounds. (Other legumes, such as peas and alfalfa, use asparagine for nitrogen transport whether they are infected or not.) Once they are synthesized, the ureides are transported through the xylem to the leaves. In leaves, the nitrogen is released and used primarily in amino acid synthesis. Allantoate is degraded to glyoxylate, four molecules of NH_4^+, and two molecules of CO_2 by three reactions that are not yet completely characterized. Based on the information provided in Chapter 13 and this chapter, trace the transport of NH_4^+ in root nodules to its incorporation into amino acids in leaves. Assume that the same or similar enzymes are used to synthesize allantoin and allantoate as observed in animals.

Pyrimidine Catabolism

In humans the purine ring cannot be degraded. This is not true for the pyrimidine ring. An outline of the pathway for pyrimidine nucleotide catabolism is illustrated in Figure 14.14.

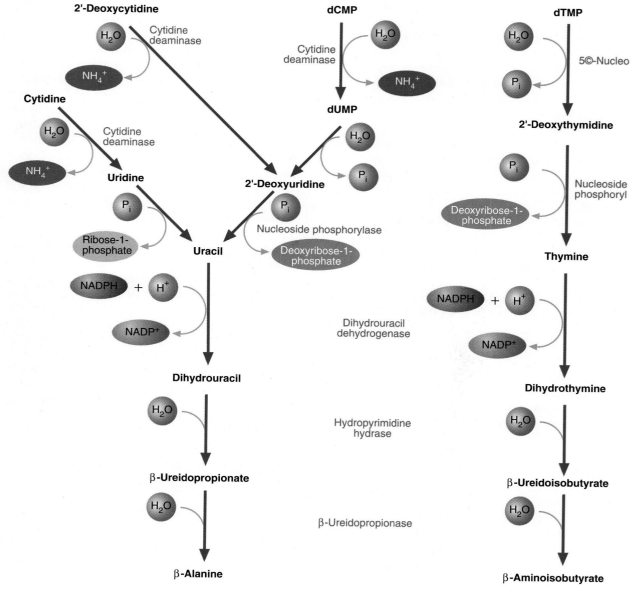

FIGURE 14.14

Degradation of Pyrimidine Bases.

Both uracil and thymine are degraded to β-alanine and β-aminoisobutyrate, respectively, in parallel pathways. The entire pathway is present in mammalian liver.

GOUT

Gout is a disorder in which sodium urate crystals are deposited in and around joints. (The name "gout" is derived from *gutta*, the Latin word for "drop." According to ancient belief, a poisonous substance falls drop by drop into joints.) This deposition, which occurs because of **hyperuricemia** (high blood levels of uric acid), causes a form of arthritis (joint inflammation). The initial attacks of gouty arthritis are usually acute (sudden) and most frequently affect the big toe, although other joints in the foot or leg may also be involved. The inflammation caused by urate crystal deposition attracts white blood cells, which engulf the crystals. Further tissue destruction is caused when urate crystals disrupt the lysosomal membranes in the white blood cells. In addition, visible structures called tophi (urate crystal "stones") may form near joints and cause grotesque deformities. The deposition of urate crystals within the kidney causes impaired renal function.

There are two forms of gout: primary and secondary. *Primary gout* is most often caused by genetic defects in purine metabolism. For example, several variants of ribose-5-phosphate pyrophosphokinase are not effectively regulated by allosteric inhibitors (e.g., P_i, GDP, or ADP). Consequently, PRPP concentrations rise, causing the increased synthesis of purine nucleotides. (Recall that PRPP concentration is an important regulator of purine nucleotide synthesis.) The overproduction of purine nucleotides then leads to increased uric acid synthesis. HGPRT deficiency also causes hyperuricemia because of decreased salvage of purine bases. Hyperuricemia can also be caused by genetic defects in other pathways. For example, in glucose-6-phosphatase deficiency, hypoglycemia develops in affected individuals because they cannot produce blood glucose from glucose-6-phosphate. Consequently, high liver concentrations of glucose-6-phosphate stimulate the synthesis of ribose-5-phosphate and PRPP.

Secondary (or acquired) *gout* is caused by seemingly unrelated disorders. These conditions may cause hyperuricemia by either overproduction of uric acid or its undersecretion by the kidneys. For example, the massive cell destruction that occurs in leukemia or because of the use of certain anticancer drugs results in an overproduction of uric acid. Hyperuricemia also results when certain drugs interfere with the renal secretion of uric acid into the urine. Patients with lead poisoning are also likely to develop gout because of renal damage (see below).

Gout is treated with diet as well as with several drugs. Dietary control (i.e., reduced consumption of food that is rich in nucleic acids such as liver and sardines) depresses uric acid synthesis in some individuals who are susceptible to primary gout. Allopurinol and colchicine are often used in gout therapy. Because allopurinol inhibits xanthine oxidase, it depresses uric acid synthesis. (Allopurinol is converted to alloxanthine by xanthine oxidase. Alloxanthine acts as a competitive inhibitor of the enzyme.) Hypoxanthine and xanthine, whose levels increase with allopurinol treatment, are easily excreted because of their solubility properties. In addition, the conversion of allopurinol to allopurinol ribonucleotide by HGPRT reduces PRPP levels. This circumstance depresses purine nucleotide synthesis. Colchicine, an alkaloid that is known to disrupt microtubules, reduces joint inflammation. It is currently believed that colchicine acts against inflammation by disrupting white blood cell activity.

Saturnine Gout

In years past, gout was associated with rich diets and especially with excessive consumption of alcoholic beverages. In recent years this association has been discounted because so many individuals lead overindulgent lives without developing gout. However, recent clinical research as well as some historical detective work indicates that the old connection between gout and alcoholic beverages may have been accurate.

Until well into the nineteenth century many bottles of wine and other alcoholic beverages were likely to be contaminated with lead. For example, the large-scale consumption of port wine by the English gentry during the eighteenth century is now believed to have been largely responsible for a gout epidemic that occurred among this population. (Port wines were imported from Portugal. To maximize their profits, Portuguese exporters added lead salts, which are very effective preservatives. In recent years, port wine bottles from this era were tested and found to contain large amounts of lead.) Similarly, in the past, rum was often stored in containers lined with lead-containing glazes.

The term *saturnine gout* reflects the connection made between gout and lead exposure by several nineteenth century physicians. (The medieval alchemists believed that the planet Saturn had leadlike properties.) Proving the connection has been more difficult. Because bone is the major reservoir for lead (both calcium and lead are divalent), chronic lead exposure may often not be easily diagnosed. Lead can be transferred in small amounts from bone to tissues such as the kidney over long periods of time. Consequently, tissue damage may continue for years after the original lead exposure. Long before tissue damage becomes obvious, blood lead levels have returned to near normal values. Saturnine gout is now believed to be caused by hyperuricemia from kidney damage. Although the kidney damage is irreversible, further damage can be avoided by removing lead from the body with chelation therapy. A chelating agent such as ethylenediaminetetracetic acid (EDTA) binds to lead with a higher affinity than it does to calcium. (Chelating agents are molecules with carboxylate groups that bind to metal cations. EDTA binds to metals with two or more positive charges.) Because lead-EDTA chelate is soluble, it is excreted in the urine.

Before they can be degraded, cytidine and deoxycytidine are converted to uridine and deoxyuridine, respectively, by deamination reactions catalyzed by cytidine deaminase. Similarly, deoxycytidylate (dCMP) is deaminated to form deoxyuridylate (dUMP). The latter molecule is then converted to deoxyuridine by 5′-nucleotidase. Uridine and deoxyuridine are then further degraded by nucleoside phosphorylase to form uracil. Thymine is formed from thymidylate (dTMP) by the sequential actions of 5′-nucleotidase and nucleoside phosphorylase.

Uracil and thymine are converted to their end products, β-alanine and β-aminoisobutyrate, respectively, in parallel pathways. In the first step, uracil and thymine are reduced by dihydrouracil dehydrogenase to their corresponding dihydro derivatives. As these latter molecules are hydrolyzed, the rings open, yielding β-ureidopropionate and β-ureidoisobutyrate, respectively. Finally, β-alanine and β-aminoisobutyrate are produced in deamination reactions catalyzed by β-ureidopropionase.

In several conditions, β-aminoisobutyrate is produced in such large quantities that it appears in urine. Among these are a genetic predisposition for slow β-aminoisobutyrate conversion to succinyl-CoA and diseases that cause massive cell destruction, such as leukemia. Because it is soluble, excess β-aminoisobutyrate does not cause problems comparable to those observed in gout.

QUESTION 14.6

The products of pyrimidine base catabolism, β-alanine and β-aminoisobutyrate, can be further degraded to acetyl-CoA and succinyl-CoA, respectively. Can you suggest the type of reactions required to accomplish these transformations?

KEY WORDS

ammonia intoxication, *416*	hyperammonemia, *416*	mutation, *423*	transulfuration pathway, *421*
B cell, *428*	hyperuricemia, *430*	nuclease, *426*	ubiquination, *412*
cellular immunity, *428*	ketogenic, *411*	oligonucleotide, *426*	ubiquitin, *412*
glucogenic, *411*	Krebs bicycle, *415*	protein turnover, *412*	urea cycle, *412*
heat shock protein, *412*	Krebs urea cycle, *413*	proteosome, *412*	
humoral immune response, *428*	molecular chaperones, *412*	T cell, *428*	

SUMMARY

Animals are constantly synthesizing and degrading nitrogen-containing molecules such as proteins and nucleic acids. Protein turnover is believed to provide cells with metabolic flexibility, protection from accumulations of abnormal proteins, and the timely destruction of proteins during developmental processes. Ubiquitin is one stress protein that plays an important role in targeting proteins for destruction.

In general, amino acid degradation begins with deamination. Most deamination is accomplished by transamination reactions, which are followed by oxidative deaminations that produce ammonia. Although most deaminations are catalyzed by glutamate dehydrogenase, other enzymes also contribute to ammonia formation. Ammonia is prepared for excretion by the enzymes of the urea cycle. Aspartate and CO_2 also contribute atoms to urea.

Amino acids are classified as ketogenic or glucogenic on the basis of whether their carbon skeletons are converted to fatty acids or to glucose. Several amino acids can be classified as both ketogenic and glucogenic because their carbon skeletons are precursors for both fat and carbohydrates.

The degradation of neurotransmitters is critical to the proper functioning of information transfer in animals. The amine neurotransmitters such as acetylcholine, the catecholamines, and serotonin are among the best-researched examples.

The turnover of nucleic acids is accomplished by several types of enzymes. The nucleases degrade the nucleic acids to oligonucleotides. (The deoxyribonucleases degrade DNA; the ribonucleases degrade RNA.) The phosphodiesterases convert the oligonucleotides to mononucleotides. By removing phosphate groups, the nucleotidases convert nucleotides to nucleosides. The nucleosidases hydrolyze nucleosides to form free bases and phosphorylated sugars. Dietary nucleic acids are generally degraded in the intestine and are not used in salvage pathways. Cellular purines are converted to uric acid. Many animals degrade uric acid further because they produce enzymes that are not present in primates. Pyrimidine bases are degraded to either β-alanine or β-aminoisobutyrate.

SUGGESTED READINGS

Bachmair, A., Finley, D., and Varshavsky, A., *In vivo* Half-Life of a Protein Is a Function of Its Amino-Terminal Residue, *Science*, 234:179–186, 1986.

Hershko, A., The Ubiquitin Pathway for Protein Degradation, *Trends Biochem. Sci.,* 16(7):265–268, 1991.

Hilt, W. and Wolf, D.H., Proteasomes: Destruction as a Programme, *Trends Biochem. Sci.,* 21(3):96–102, 1996.

Holmes, F.L., Hans Krebs and the Discovery of the Ornithine Cycle, *Fed. Proc.,* 39:216–225, 1980.

Rechsteiner, M. and Rogers, S.W., PEST Sequences and Regulation by Proteolysis, *Trends Biochem. Sci.,* 21(7):267–271, 1996.

Wellner, D., and Meister, A., A Survey of Inborn Errors of Amino Acid Metabolism and Transport in Man, *Ann. Rev. Biochem.,* 50:911–968, 1981.

Winkler, R.G., Blevins, D.G., Polacco, J.C., and Randall, D.D., Ureide Catabolism in Nitrogen-Fixing Legumes, *Trends Biochem. Sci.,* 13:97–100, 1988.

QUESTIONS

1. Define the following terms:
   a. denitrification
   b. ammonotelic
   c. protein turnover
   d. ubiquination

2. What are the major molecules used to excrete nitrogen?

3. What are three purposes served by protein turnover?

4. What are the structural features of proteins that mark them for destruction?

5. What are the seven metabolic products produced by the degradation of amino acids?

6. Indicate which of the following amino acids are ketogenic and which are glucogenic:
   a. tyrosine
   b. lysine
   c. glycine
   d. alanine
   e. valine
   f. threonine

7. Describe how each of the following amino acids is degraded:
   a. lysine
   b. glutamate
   c. glycine
   d. aspartate
   e. tyrosine
   f. alanine

8. Mammals excrete most nitrogen atoms as urea. The process requires the expenditure of considerable amounts of ATP energy. Why is it not practical to excrete nitrogen as ammonia as some aquatic species do? What toxic effects would this have on a mammal?

9. In humans the purine ring cannot be degraded. How is it excreted? What reactions are involved?

10. The urea cycle occurs partially in the cytosol and partially in the mitochondria. Discuss the urea cycle reactions with reference to their cellular locations.

11. Describe how the glucose-alanine cycle acts to transport ammonia to the liver.

12. In individuals with PKU, is tyrosine an essential amino acid?

13. Urea formation is energetically expensive, requiring the expenditure of four moles of ATP per mole of urea formed. However, NADH is produced when fumarate is reconverted to aspartate. How many ATP molecules are produced by the mitochondrial oxidation of the NADH? What is the net ATP requirement for urea synthesis?

14. Describe the Krebs bicycle. What compound links the citric acid and urea cycles?

15. Describe how increasing concentrations of ammonia stimulate the formation of N-acetylglutamate and turn on the urea cycle.

16. Explain how a defective enzyme in the urea cycle can produce high levels of ammonia.

17. PKU can be caused by deficiencies in phenylalanine hydroxylase and by enzymes catalyzing the formation and regeneration of 5,6,7,8-tetrahydrobiopterin. How can this second defect cause the symptoms of PKU?

18. Individuals who cannot produce 5,6,7,8-tetrahydrobiopterin must be supplied with L-dopa and 5-hydroxytryptophan, metabolic precursors to norepinephrine and serotonin. Why does supplying of 5,6,7,8-tetrahydrobiopterin have no effect?

19. In their *in vitro* studies using liver slices, Krebs and Henseleit observed that urea formation was stimulated by the addition of ornithine, citrulline, and arginine. Other amino acids had no effect. Explain these observations.

20. Specify which type of carbon unit is transferred by each of the following compounds:
    a. N^5,N^{10}-methylene THF
    b. serine
    c. choline
    d. S-Adenosylmethionine

21. Caffeine, a purine alkaloid found in chocolate, coffee, and tea, is excreted as uric acid. Using your knowledge of the metabolism of other purine compounds, suggest how caffeine is metabolized.

22. Individuals suffering from gout should not drink coffee or tea. Why is this so?

23. Most amino acids are degraded in the liver. This is not true of the branched chain amino acids. Where are they primarily degraded?

24. Describe how a protein is targeted for degradation.

25. Some animals living in a fluid medium excrete nitrogen as ammonia. Land animals, which conserve water, excrete urea and uric acid. Why does the excretion of these molecules aid in water conservation?

INTEGRATION OF METABOLISM

Chapter Fifteen

Food supplies the human body with the nourishment required to sustain living processes. Complex regulatory mechanisms ensure that the demands of each cell for energy and metabolites are consistently met.

OBJECTIVES

After you have studied this chapter, you should be able to answer these questions:

1. *What are the metabolic contributions of each major organ in mammals?*
2. *Why do mammals need to consume food only intermittently?*
3. *What is the role of hormones in the feeding-fasting cycle?*
4. *What specific effects do glucagon and insulin have on metabolism?*
5. *What is the hormone cascade system and how is it controlled?*
6. *How are the structures of the hypothalamus and pituitary gland related to the functioning of the processes they regulate?*
7. *What role does the down-regulation of receptors play in metabolism?*
8. *What diseases do overproduction and underproduction of hormones cause?*
9. *What are the two major forms of diabetes mellitus?*
10. *What roles do growth factors play in animals?*
11. *What are the major second messengers and how do they mediate hormone messages?*
12. *What are the major plant hormones?*

Previous chapters deal with several important topics, for example, the metabolism of car-bohydrates, lipids, amino acids, and other molecules. However, the whole is not just the sum of its parts. Multicellular organisms are extraordinarily complex, more so than their components would suggest. This chapter takes a wider view of functioning of the mam-malian body. Initially, the division of labor that allows the sophisticated functioning of the multicellular body is considered. This is followed by a discussion of the feeding-fasting cy-cle, a complex multiorgan process. Hormones and growth factors, the major tools of intercellular communication, and their mechanisms of action will then be described. This chapter also includes a discussion of diabetes mellitus, a disease that has widespread metabolic effects.

It should now be evident that the maintenance of living processes in multicellular organisms is a complicated business. Recall that, despite changes in their internal and external environments, these organisms must constantly sustain adequate (if not optimal) operating conditions as they simultaneously engage in growth and repair activities. To accomplish these functions, the anabolic and catabolic reaction pathways that use carbohydrates, lipids, and proteins as energy sources and biosynthetic precursors must be precisely regulated. As described, multicellular organisms can efficiently exploit their environment because of the division of labor among their constituent cells, tissues, and organs. Mammals, the most carefully investigated group of multicellular organisms, have a sophisticated and mutually beneficial division of labor. Each organ performs specific functions to serve both the short- and long-term interests of the body.

The operation of such a complex system as the body is maintained by a continuous flow of information among its parts. A simple system for information transfer is composed of a stimulus sent by a sender, a message carrier (or messenger), and a receiver. In such a system, only one response to the signal is possible. Physiological systems, however, are extraordinarily complex and require finely modulated responses to complex stimuli. In addition, for coordinated functioning, each body part must also receive information about events in other parts. Because multicellular organisms are hierarchical organizations of cells, tissues, organs, and organ systems, it is not surprising that a large number of signals, message carriers, and receivers are required. In the mammalian body, much information transfer is accomplished by hormones. As described below, these messenger molecules are arranged in complex hierarchies that allow for a high degree of sophisticated regulation.

In this chapter the focus of the discussion is the integration of the major metabolic processes in mammals (Figure 15.1). The chapter begins with a description of the metabolic contributions of several major organs. This is followed by a discussion of the feeding-fasting cycle, which illustrates several important control mechanisms. Subsequently, the major mammalian hormones and their mechanisms of action are described. The chapter ends with a brief description of plant hormones.

15.1 THE DIVISION OF LABOR

Each organ in the mammalian body has several roles that contribute to the individual's function. For example, some organs are consumers of energy so that they may perform certain energy-driven tasks (e.g., muscle contraction). Other organs, such as those in the digestive tract, are responsible for efficiently supplying energy-rich nutrient molecules for use elsewhere. The roles of several organs in relation to their metabolic contributions are discussed below.

Small Intestine

The most obvious role of the small intestine is the digestion of nutrients such as carbohydrates, lipids, and proteins into molecules that are small enough to be absorbed (sugars, fatty acids, glycerol, and amino acids). Nutrient absorption by the enterocytes

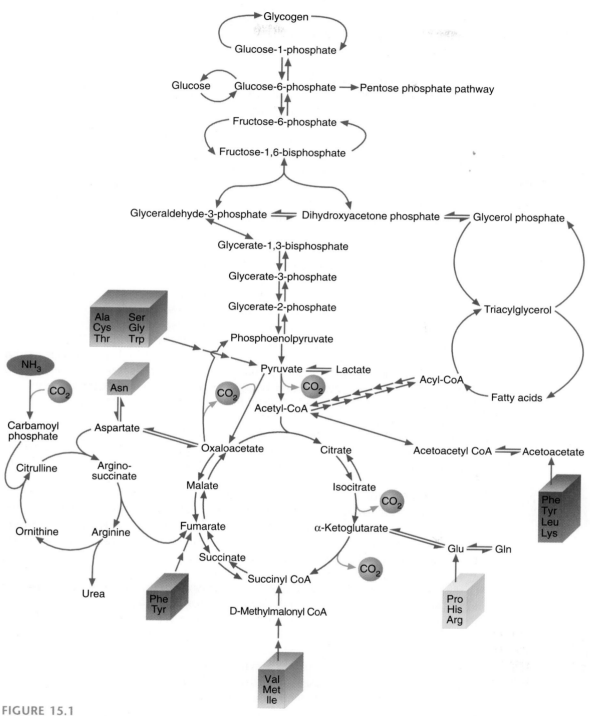

FIGURE 15.1

Nutrient Metabolism in Mammals.

Despite the variability of the mammalian diet, these organisms usually provide their cells with adequate nutrients. Control mechanisms that regulate biochemical pathways are responsible for this phenomenon.

of the small intestine is an extremely vital and complicated process that involves numerous enzymes and transport mechanisms. As described, enterocytes then transport these molecules (as well as water, minerals, vitamins, and other substances) into the blood and lymph, which carries them throughout the body.

The small intestine also appears to be a major site for glutamine metabolism. Glutamine is a significant energy source for enterocytes. (During the digestive process, enterocytes obtain glutamine from degraded dietary protein. Under fasting conditions, glutamine is acquired from arterial blood.) In addition to metabolizing glutamine to produce CO_2 and H_2O, enterocytes also convert glutamine to several intermediate products. Some glutamine molecules are converted to Δ^1-pyrroline-5-carboxylate. The latter molecule is then converted to proline. Other products of glutamine metabolism include lactate, citrate, ornithine, and citrulline. These intermediates are processed further in other organs, such as the kidney or the liver.

Liver

The liver performs a stunning variety of metabolic activities. In addition to its key roles in carbohydrate, lipid, and amino acid metabolism, the liver monitors and regulates the chemical composition of blood and synthesizes several plasma proteins. The liver distributes several types of nutrients to other parts of the body. Because of its metabolic flexibility, the liver reduces the fluctuations in nutrient availability caused by drastic dietary changes and intermittent feeding and fasting. For example, a sudden shift from a high-carbohydrate diet to one that is rich in proteins increases (within hours) the synthesis of the enzymes required for amino acid metabolism. Finally, the liver plays a critically important protective role in processing foreign molecules (Chapter 19).

Muscle

Skeletal muscle is specialized to perform intermittent mechanical work. As described previously, the energy sources that provide ATP for muscle contraction depend on the degree of muscular activity and the physical status of the individual. During fasting and prolonged starvation, some skeletal muscle protein is degraded to provide amino acids (e.g., alanine) to the liver for gluconeogenesis.

In contrast to skeletal muscle, cardiac muscle must continuously contract to sustain blood flow throughout the body. To maintain its continuous operation, cardiac muscle relies mainly on fatty acids. (It is not surprising, therefore, that cardiac muscle is densely packed with mitochondria.) It can also use other energy sources, such as glucose, ketone bodies, pyruvate, and lactate. (Ketone bodies are used principally during starvation.) Unlike skeletal muscle, cardiac muscle is strictly an energy-consuming tissue.

Adipose Tissue

As described, the role of adipose tissue is primarily the storage of energy in the form of triacylglycerols. Depending on current physiological conditions, adipocytes convert excess nutrient molecules (i.e., fatty acids and glucose) to fat deposits or degrade fat to generate energy-rich fatty acids and glycerol. Recall that these metabolic activities are regulated by several hormones (i.e., insulin, glucagon, and epinephrine).

Brain

The brain ultimately directs most metabolic processes in the body. Sensory information from numerous sources is integrated in several areas in the brain. These areas then direct the activities of the motor neurons that innervate muscles and glands. Much of the body's hormonal activity is controlled either directly or indirectly by the hypothalamus and the pituitary gland (see Section 15.3).

Like the heart, the brain does not provide energy to other organs or tissues. Under normal conditions, the brain uses glucose as its sole fuel. Because it stores little glycogen, the brain is highly dependent on a continuous supply of glucose in the blood. During prolonged starvation, the brain can adapt to using ketone bodies as an energy source.

Kidney

The kidney has several important functions that contribute significantly to maintaining a stable internal environment. These include

1. The filtration of blood plasma, which results in the excretion of water-soluble waste products (e.g., urea and certain foreign compounds),
2. Reabsorption of electrolytes, sugars, and amino acids from the filtrate,
3. The regulation of blood pH, and
4. The regulation of the body's water content.

Considering the functions of the kidney, it is not surprising that much of the energy generated in this organ is consumed by transport processes. Energy is provided largely by fatty acids and glucose. Under normal conditions, the small amounts of glucose formed by gluconeogenesis are used only within certain kidney cells. The rate of gluconeogenesis increases during starvation and acidosis. The kidney uses glutamine and glutamate (via glutaminase and glutamate dehydrogenase, respectively) to generate ammonia, which is used in pH regulation. (Recall that NH_3 reversibly combines with H^+ to form NH_4^+.)

Each organ in the mammalian system contributes to the body's overall function.

15.2 THE FEEDING-FASTING CYCLE

Despite their consistent requirements for energy and biosynthetic precursor molecules, mammals consume food only intermittently. This is possible because of elaborate mechanisms for storing and mobilizing energy-rich molecules derived from food. The changes in the status of various biochemical pathways during transitions between feeding and fasting illustrate metabolic integration and the profound regulatory influence of hormones. Substrate concentrations are also an important factor in metabolism. In discussions of the feeding-fasting cycle, the terms "postprandial" and "postabsorptive" are often used. In the **postprandial** state, which occurs directly after a meal has been digested and absorbed, blood nutrient levels are elevated above those in the fasting phase. During the **postabsorptive** state, for example, after an overnight fast, nutrient levels in blood are low.

The Feeding Phase

As the feeding phase begins, food is propelled along the gastrointestinal tract by muscular contractions. As it moves through the organs, food is broken into smaller particles and exposed to enzymes. Ultimately, the products of digestion (consisting largely of sugars, fatty acids, glycerol, and amino acids) are absorbed by the small intestine and secreted into the blood and lymph. This phase is regulated by interactions between enzyme-producing cells of the digestive organs, the nervous system, and several hormones. The nervous system is responsible for the waves of smooth muscle contraction that propel food along the tract, as well as regulating the secretions of several digestive structures (e.g., salivary and gastric glands). Hormones such as gastrin, secretin, and cholecystokinin also contribute to the digestive process. (See Table 15.1 in Section 15.3.) They do so by stimulating the secretion of enzymes or digestive aids such as bicarbonate and bile.

The early postprandial state is illustrated in Figure 15.2. As described, sugars and amino acids are absorbed and transported by the portal blood to the liver. Most lipid molecules are transported from the small intestine in lymph as chylomicrons. Chylomicrons pass into the bloodstream, which carries them to tissues such as muscle and adipose tissue. After most triacylglycerol molecules have been removed from chylomicrons, these structures, now referred to as *chylomicron remnants,* are then taken up by the liver. The phospholipid, protein, cholesterol, and few remaining triacylglycerol molecules are then degraded or reused. For example, cholesterol is used to synthesize bile acids, and fatty acids are used in new phospholipid synthesis. Phospholipids, as well as other newly synthesized lipid and protein molecules, are then incorporated into lipoproteins for export to other tissues.

As glucose moves through the blood from the small intestine to the liver, the β-cells within the pancreas are stimulated to release insulin. (High blood glucose and insulin levels depress glucagon secretion by the pancreatic α-cells. The opposing effects of insulin and glucagon on glucose and fat metabolism are illustrated in Figure 15.3.) Insulin release triggers several processes that ensure the storage of nutrients. These include glucose uptake by muscle and adipose tissue, glycogenesis in liver and muscle, fat synthesis in liver and adipocytes, and gluconeogenesis (using excess dietary amino acids). Recall that in the liver, most glycogen and fatty acids are synthesized from three-carbon molecules such as lactate, not directly from blood glucose. In addition, insulin also influences amino acid metabolism. For example, insulin promotes the transport of amino acids into cells (especially liver and muscle cells). In general, insulin stimulates protein synthesis in most tissues.

FIGURE 15.2
The Early Postprandial State.

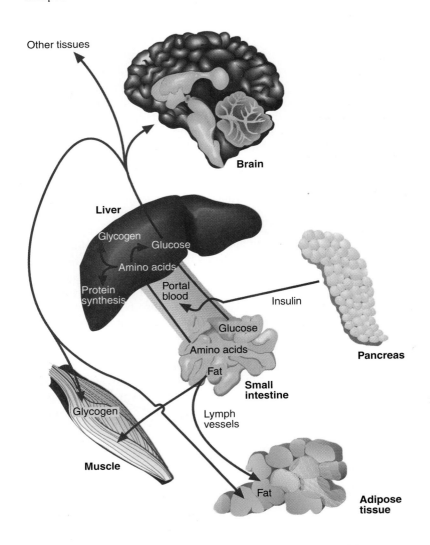

Although the effects of insulin on postprandial metabolism are profound, other factors (e.g., substrate supply and allosteric effectors) also affect the rate and degree to which these processes occur. For example, elevated levels of fatty acids in blood promote lipogenesis in adipose tissue. Regulation by several allosteric effectors further ensures that competing pathways do not occur simultaneously, for example, in many cell types fatty acid synthesis is promoted by citrate (an activator of acetyl-CoA carboxylase), while fatty acid oxidation is depressed by malonyl-CoA (an inhibitor of carnitine acyltransferase I activity). The control of fatty acid metabolism is described in Section 10.1).

The Fasting Phase

The early postabsorptive state (Figure 15.4) of the feeding-fasting cycle begins as the nutrient flow from the intestine diminishes. As blood glucose and insulin levels fall back to normal, glucagon is released. Glucagon acts to prevent hypoglycemia by promoting glycogenolysis and gluconeogenesis in liver. Decreased insulin also promotes lipolysis and the release of amino acids such as alanine and glutamine from muscle. (Recall that several tissues use fatty acids in preference to glucose. Glycerol and alanine are substrates for gluconeogenesis, and glutamine is an energy source for enterocytes.)

If a fast becomes prolonged (e.g., overnight), several metabolic strategies maintain blood glucose levels. Increased mobilization of fatty acids from adipose tissue during the postabsorptive state is stimulated by norepinephrine. These fatty acids provide an alternative to glucose for muscle. (Reduced skeletal muscle consumption of

FIGURE 15.3

The Opposing Effects of Insulin and Glucagon on Blood Glucose Levels.

In general, insulin promotes anabolic processes (e.g., fat synthesis, glycogenesis, and protein synthesis). Glucagon raises blood glucose levels by promoting glycogenolysis in liver and protein degradation in muscle. It also promotes lipolysis.

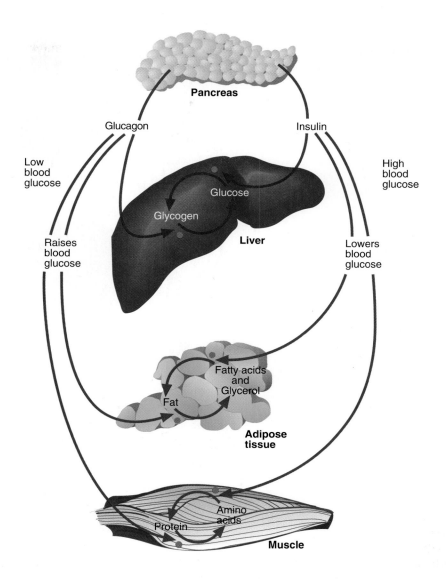

KEY *Concepts*

During the feeding phase, food is consumed, digested, and absorbed. Absorbed nutrients are then transported to the organs, where they are either used or stored. During fasting, several metabolic strategies maintain blood glucose levels.

15.3 INTERCELLULAR COMMUNICATION

glucose spares its use for brain. Recall that glucose is normally the only fuel source in brain.) In addition, the action of glucagon increases gluconeogenesis, using amino acids derived from muscle. (During fasting, insulin levels decline significantly.)

Under conditions of extraordinarily prolonged fasting (starvation), the body makes metabolic changes to ensure that adequate amounts of blood glucose are available to sustain energy production in the brain and other glucose-requiring cells. Additionally, fatty acids from adipose tissue and ketone bodies from liver are mobilized to sustain the other tissues. Because glycogen is depleted after several hours of fasting, gluconeogenesis plays a critical role in providing sufficient glucose. During early starvation, large amounts of amino acids from muscle are used for this purpose. However, after several weeks, the breakdown of muscle protein declines significantly because the brain is using ketone bodies as a fuel source. Figure 15.5 illustrates the plasma levels of glucose, fatty acids, and ketone bodies as starvation proceeds for several days.

Hormones are synthesized and secreted by specialized cells and exert biochemical effects on target cells. When these target cells are distant from the hormone-secreting cells, the hormones are referred to as **endocrine** hormones. (Recall that paracrine hormones exert their effects on nearby cells.) Some hormones exert very specific effects

FIGURE 15.4
The Early Postabsorptive State.

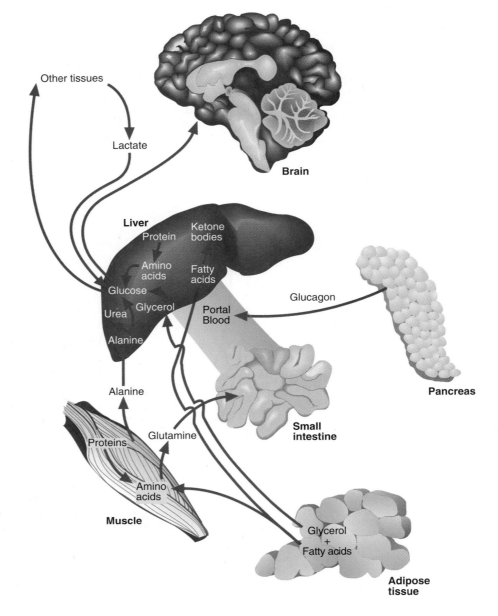

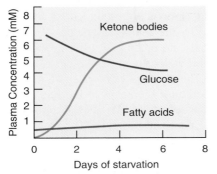

FIGURE 15.5

The Plasma Levels of Fatty Acids, Glucose, and Ketone Bodies During the Early Days of Starvation.

The concentrations of fatty acids and ketones increase. In contrast, glucose levels decrease.

on one type of target cell; other hormones act on a variety of target cells. For example, thyroid-stimulating hormone (TSH) stimulates follicular cells in the thyroid gland to release T_3 (triiodothyronine) and T_4 (thyroxine) (Figure 15.6). In contrast, T_3 and T_4 stimulate a variety of cellular reactions in numerous cell types. Some hormones have different effects on different target cells. For example, among the many physiological effects of the thyroid hormones are the stimulation of glycogenolysis in liver cells and glucose absorption in the small intestine.

Control of physiological responses often involves several hormones. In some systems, two or more hormones act in opposition to each other (e.g., insulin and glucagon in the regulation of blood glucose). In other control systems, several hormones act in information hierarchies. This section begins with a description of the best-researched example of such a hierarchy, referred to as a hormone cascade mechanism. This is followed by a discussion of growth factors, specialized proteins that stimulate cell division in susceptible cells.

Sensitive techniques are now available to detect and measure hormones. The most common of these, radioimmunoassays and ELISA (enzyme-linked immunosorbent assays), are described in Appendix B, Supplement 15.

FIGURE 15.6

Structure of the Thyroid Hormones T$_3$ and T$_4$.

Triiodothyronine (T$_3$)

Thyroxine (T$_4$)

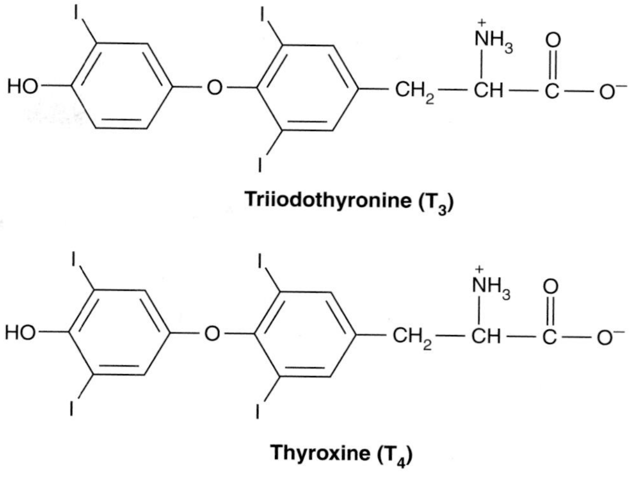

The Hormone Cascade System

The list of molecules now recognized as hormones has grown astonishingly large. Table 15.1 contains a small selection of the better-known mammalian hormones.

In mammals, most metabolic activities are controlled to one degree or another by hormones. The synthesis and secretion of many hormones are regulated by a complex cascade mechanism and ultimately controlled by the central nervous system. In this system, outlined in Figure 15.7, sensory signals are received by the hypothalamus, an area in the brain that links the nervous and endocrine systems. (In addition to regulating hormone production, nerve cells in the hypothalamus monitor and/or regulate vital body functions, such as body temperature, blood pressure, water balance, and body weight. The hypothalamus is also associated with certain behaviors, for example, anger, sexual arousal, and feelings of pain and pleasure.) Once it is appropriately stimulated, the hypothalamus induces the secretion of several hormones produced by the anterior lobe of the pituitary gland. (The pituitary gland, which is attached to the hypothalamus by the pituitary stalk, consists of two distinct parts: the anterior lobe, or adenohypophysis, and the posterior lobe, or neurohypophysis.) The hypothalamus does so by synthesizing a series of specific peptide-releasing hormones. Releasing hormones then pass into a specialized capillary bed referred to as the hypothalamohypophyseal portal system, which transports them directly to the adenohypophysis. Each releasing hormone stimulates specific cells to synthesize and secrete one or more types of hormone. For example, the tripeptide thyrotropin-releasing hormone (TRH) stimulates the secretion of TSH and prolactin. (Prolactin promotes milk production in new mothers. Ordinarily, prolactin release is prevented by other hormonal and neural factors.) The hormones of the anterior pituitary are sometimes referred to as tropic ("to turn" or "to change"), because they stimulate the synthesis and release of hormones from other endocrine glands. For example TSH stimulates the thyroid gland to release the thyroid hormones T$_3$ and T$_4$.

The anatomy and function of the posterior pituitary differ from those of the anterior lobe. The hormones secreted by the neurohypophysis (i.e., the peptides vasopressin and oxytocin; see Section 5.2) are actually synthesized in separate types of neurons that originate in the hypothalamus. After their synthesis, both hormones are packaged with associated proteins called neurophysins into secretory granules. The granules then migrate down the axons into the posterior lobe. They are secreted into the bloodstream when an action potential reaches the nerve endings and initiates exocytosis. Oxytocin is secreted as a response to nerve impulses initiated by the stretching of the uterus late in pregnancy and suckling during breast feeding. Vasopressin (antidiuretic hormone) is secreted in response to neural signals from specialized cells called osmoreceptors, which are sensitive to changes in blood osmolality.

Animals employ several mechanisms to prevent excessive hormone synthesis and release. The most prominent of these is feedback inhibition. The hypothalamus and anterior pituitary are controlled by the target cells they regulate. For example, TSH

TABLE 15.1

Selected Mammalian Hormones

Source	Hormone	Function
Hypothalamus	Gonadotropin–releasing hormone* (GnRH)	Stimulates LH and FSH secretion
	Corticotropin–releasing hormone* (CRH)	Stimulates ACTH secretion
	Growth hormone-releasing hormone* (GHRH)	Stimulates GH secretion
	Somatostatin*	Inhibits GH and TSH secretion
	Thyrotropin-releasing hormone* (TRH)	Stimulates TSH and prolactin secretion
Pituitary	Luteinizing hormone* (LH)	Stimulates cell development and synthesis of sex hormones in ovaries and testes
	Follicle-stimulating hormone* (FSH)	Promotes ovulation and estrogen synthesis in ovaries and sperm development in testes
	Corticotropin* (ACTH) (adrenocorticotropic hormone)	Stimulates steroid synthesis in adrenal cortex
	Growth hormone* (GH)	General anabolic effects in many tissues
	Thyrotropin* (TSH) (thyroid-stimulating hormone)	Stimulates thyroid hormone synthesis
	Prolactin	Stimulates milk production in mammary glands and assists in the regulation of the male reproductive system
	Oxytocin*	Uterine contraction and milk ejection
	Vasopressin*	Blood pressure and water balance
Gonads	Estrogens† (estradiol)	Maturation and function of reproductive system in females
	Progestins† (progesterone)	Implantation of fertilized eggs and maintenance of pregnancy
	Androgens† (testosterone)	Maturation and function of reproductive system in males
Adrenal cortex	Glucocorticoids† (cortisol, corticosterone)	Diverse metabolic effects as well as inhibiting the inflammatory response
	Mineralocorticoids† (aldosterone)	Mineral metabolism
Thyroid	Triiodothyronine‡ (T3) Thyroxine‡ (T4)	General stimulation of many cellular reactions
Gastrointestinal tract	Gastrin*	Stimulates secretion of stomach acid and pancreatic enzymes
	Secretin*	Regulates pancreatic exocrine secretions
	Cholecystokinin*	Stimulates secretion of digestive enzymes and bile
	Somatostatin*	Inhibits secretion of gastrin and glucagon
Pancreas	Insulin*	General anabolic effects including glucose uptake and lipogenesis
	Glucagon*	Glycogenolysis and lipolysis
	Somatostatin*	Inhibits the secretion of glucagon

*Peptide or polypeptide.
†Steroid.
‡Amino acid derivative.

release by the anterior pituitary is inhibited when blood levels of T_3 and T_4 rise (Figure 15.8). The thyroid hormones inhibit the responsiveness of TSH synthesizing cells to TRH. In addition, several tropic hormones inhibit the synthesis of their releasing factors.

Target cells also possess mechanisms that protect against overstimulation by hormones. In a process referred to as **desensitization,** target cells adjust to changes in

FIGURE 15.7

The Hormone Cascade System.

Hormone synthesis and secretion are ultimately controlled by the central nervous system. Neurons in the hypothalamus either stimulate or inhibit hormone synthesis and/or secretion in the pituitary. Pituitary hormones initiate metabolic activities in their target cells.

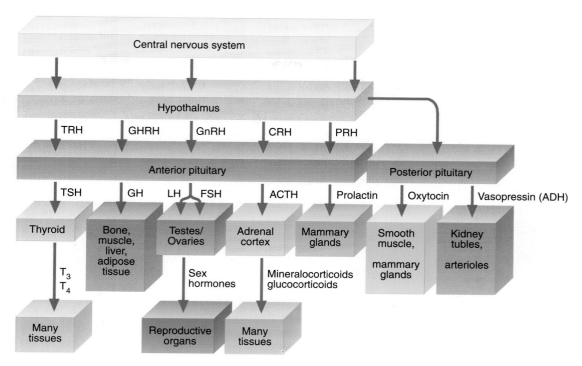

FIGURE 15.8

Feedback Inhibition.

The secretion of thyroxine from the thyroid is stimulated by TSH. The secretion of TSH is in turn stimulated by TRH. As blood levels of thyroxine rise, TRH secretion is inhibited.

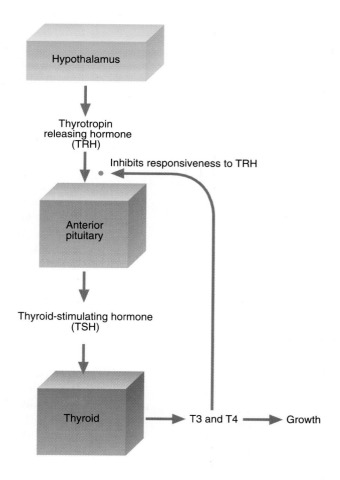

Many of the hormones in the mammalian body are controlled by a complex cascade mechanism and ultimately regulated by the central nervous system.

stimulation levels by decreasing the number of cell surface receptors or by inactivating those receptors. The reduction in cell surface receptors in response to stimulation by specific hormone molecules is called **down-regulation.** In down-regulation, receptors are internalized by endocytosis. Depending on cell type and several metabolic factors, the receptors may eventually be recycled to the cell surface or be degraded. If degraded, proteins must be synthesized to replace receptors. Some disease states are caused by or associated with target cell insensitivity to specific hormones. For example, some cases of diabetes are associated with **insulin resistance,** caused by a decrease in functional insulin receptors. (Diabetes is discussed in Special Interest Box 15.2.)

QUESTION 15.1

The thymus gland is a bilobed organ found just above the heart. It promotes the differentiation of certain lymphocytes to form T cells. (Recall that T cells confer cellular immunity.) In addition, the thymus gland secretes several thymic hormones that stimulate T cell function after these cells have left the thymus gland.

Physical and emotional stress are now known to depress cellular immunity. For example, the risk of developing cancer and severe infections increases with the duration of chronic stress. Although the mechanism by which stress induces this change in function is not clearly understood, elevated blood levels of glucocorticoids (e.g., cortisol) are believed to play an important role. Thymic hormone secretion is depressed when glucocorticoid levels are high. In addition, circadian rhythms alter thymic hormone and glucocorticoid levels. (Thymic hormone levels are higher and glucocorticoid levels are lower at night than during the day.) Assuming that glucocorticoids depress thymic hormone secretion, outline (in general terms) how thymic hormone is affected by stressful situations and by light/dark cycles. What other hormones are involved in both of these processes? (Hint: Review melatonin synthesis, outlined in Chapter 13.)

Growth Factors

The survival of multicellular organisms requires that cell growth and cell division (mitosis) be rigorously controlled. The conditions that regulate these processes have not yet been completely resolved. However, a variety of hormonelike polypeptides and proteins, called **growth factors** (or **cytokines**), are now believed to regulate the growth, differentiation, and proliferation of various cells. Often, the actions of several growth factors are required to promote cellular responses. Growth factors differ from hormones in that they are synthesized by a variety of cell types rather than by specialized glandular cells. Examples of mammalian growth factors include epidermal growth factor (EGF), platelet-derived growth factor (PDGF), and the somatomedins. Similar molecules such as the interleukins, which promote cell proliferation and differentiation within the immune system, are also considered cytokines. Several growth-suppressing molecules have also been characterized. The mechanisms by which the cytokines exert their effects are not clearly understood. The few aspects of cytokine action that have been identified resemble some of those observed with hormones (Section 15.4). Although some growth factors are found in circulating blood, most have paracrine and/or autocrine activity.

Epidermal growth factor (EGF) (6.4 kD), one of the first cellular growth factors identified, is a **mitogen** (a stimulator of cell division) for a large number of epithelial cells, such as epidermal and gastrointestinal lining cells. EGF triggers cell division when it binds to plasma membrane EGF receptors, which are transmembrane tyrosine kinases structurally similar to insulin receptors.

Platelet-derived growth factor (PDGF) (31 kD) is secreted by blood platelets during the clotting reaction. Acting with EGF, PDGF stimulates mitosis in fibroblasts and other nearby cells during wound healing. PDGF also promotes collagen synthesis in fibroblasts.

The **somatomedins** are a group of polypeptides that mediate the growth-promoting actions of GH. Produced in the liver as well as a variety of other tissue cells (e.g., muscle, fibroblasts, bone, and kidney) when GH binds to its cell surface receptor, the somatomedins are the major stimulators of growth in animals. In contrast to other tis-

Growth factors are a group of hormonelike polypeptides and proteins that promote cell growth as well as cell division and differentiation.

HORMONE-RELATED DISEASES

Because hormones are so influential in the regulation of metabolic processes, it is not surprising that there are numerous hormone-related diseases. In general, such diseases are caused by either overproduction or underproduction of a specific hormone or the insensitivity of target cells.

Hormone Overproduction

The oversecretion of hormone molecules is most often caused by a tumor. Several types of pituitary tumor cause endocrine diseases. For example, one of the most common causes of Cushing's disease is an abnormal proliferation of ACTH-producing cells. *Cushing's disease* is characterized by obesity, hypertension, and elevated blood glucose levels. Patients with Cushing's disease develop a characteristic appearance: a puffy "moon face" and a "buffalo hump" caused by fat deposits between the shoulders. Occasionally, Cushing's disease is caused by adrenocortical tumors.

Depending on the patient's age, pituitary tumors that develop from somatotroph cells (the cells that synthesize GH) cause either gigantism or acromegaly. *Gigantism,* marked by pronounced growth of the long bones, is caused by excessive secretion of growth hormone (GH) during childhood. In adulthood, excessive GH production causes *acromegaly,* in which connective tissue proliferation and bone thickening result in coarsened and exaggerated facial features as well as enlarged hands and feet.

Not all hypersecretion diseases are caused by tumors. For example, *Graves' disease,* the most common type of hyperthyroidism, is an autoimmune disease. For unknown reasons, autoantibodies are produced that bind to TSH receptors in the thyroid gland. (These antibodies are called *long-acting thyroid stimulators,* or LATS.) The resulting excessive production of thyroid hormone causes *thyrotoxicosis,* characterized by *goiter* (an enlarged thyroid gland) and *exophthalmos* (abnormal eyeball protrusion). Affected individuals become more sensitive to catecholamines. (Apparently, the number of catecholamine receptors is increased in the heart and the central nervous system.) During stressful situations (e.g., surgery or myocardial infarction) when large amounts of catecholamines are released, a life-threatening "thyroid storm" may occur. The consequences of a thyroid storm include agitation, delirium, coma, and heart failure.

Hormone Underproduction

Inadequate hormone production has a variety of causes. The most common are the autoimmune destruction of hormone-producing cells, genetic defects, and an inadequate supply of precursor molecules.

As described, in Addison's disease adrenal cortex function is inadequate. The most common cause of Addison's disease is the autoimmune destruction of the adrenal gland. (The major cause of adrenal insufficiency before the 1920s was tuberculosis of the adrenal gland.) Addison's disease also results from prolonged glucocorticoid therapy. (Recall that certain glucocorticoids are used as anti-inflammatory drugs.)

Hypothyroidism (thyroid hormone deficiency) may result from autoimmune disease (*Hashimoto's disease*) or from deficient synthesis of TSH or TRH (thyroid-stimulating hormone–releasing factor). Because adequate ingestion of iodine is a prerequisite for thyroid hormone synthesis, iodine deficiency also causes hypothyroidism. In children, thyroid hormone deficiency (called *cretinism*) causes depressed growth and mental retardation. Severe hypothyroidism in adults (*myxedema*) results in symptoms such as edema (abnormal fluid accumulation) and goiter. Hypothyroidism is usually treated with hormone replacement therapy.

Growth hormone deficiency may be hereditary or a consequence of a pituitary tumor or head trauma. Congenital GH deficiency results in shortened stature (*dwarfism*). Currently, affected children are treated with commercially produced GH, a recombinant DNA product. In *Laron's dwarfism,* exogenous GH has no effect on the patient's cells because of a defective GH response mechanism. (The somatomedins, which mediate the effects of GH within cells, are discussed in Section 15.3.)

Diabetes insipidus is characterized by the passage of copious amounts of very dilute urine. It is caused by either an inadequate synthesis of vasopressin or the failure of the kidney to respond to vasopressin. The most common causes of diabetes insipidus are tumors and surgical procedures in which the neurohypophyseal nerve tracts are cut. In several forms of kidney disease, the organ's capacity to respond to vasopressin is compromised.

---

sues, the somatomedins are secreted by the liver into blood. In addition to stimulating cell division, the somatomedins promote (but to a lesser degree) the same metabolic processes as does the hormone insulin (e.g., glucose transport and fat synthesis). For this reason the two somatomedins found in humans have been renamed **insulinlike growth factors I and II** (IGF-I and IGF-II). Like other polypeptide growth factors, the somatomedins trigger intracellular processes by binding to cell surface receptors. Not surprisingly, somatomedin receptors are also tyrosine kinases.

Interleukin-2 (IL-2) (13 kD) is a member of a group of cytokines that regulate the immune system in addition to promoting cell growth and differentiation. IL-2 is secreted by T cells after they have been activated by binding to a specific antigen presenting cell. These cells are also stimulated to produce IL-2 receptors. The binding of IL-2 to these receptors stimulates cell division so that numerous identical T cells

DIABETES MELLITUS

Diabetes mellitus is a group of devastating metabolic diseases caused by insufficient insulin synthesis, increased insulin destruction or ineffective insulin action. All of its metabolic effects result when the body's cells fail to acquire glucose from the blood. The metabolic imbalances that occur have serious, if not life threatening, consequences (Figure 15A). In insulin-dependent diabetes mellitus (IDDM), also called type I diabetes, inadequate amounts of insulin are secreted because the β-cells of the pancreas have been destroyed. Because IDDM usually occurs before the age of 20, it has (until recently) been referred to as juvenile-onset diabetes. Noninsulin-dependent diabetes mellitus (NIDDM), also called type II or adult-onset diabetes, is caused by the insensitivity of target tissues to insulin. Although these forms of diabetes share some features, they differ significantly in others. Once quite rare, diabetes is now the third leading cause of death in the United States, where it afflicts at least 5% of the population.

The most obvious symptom of diabetes is **hyperglycemia** (high blood glucose levels), caused by inadequate cellular uptake of glucose. Among diabetics the severity of hyperglycemia may vary considerably. Because the kidney's capacity to reabsorb glucose from the urinary filtrate is limited, glucose appears in the urine **(glucosuria).** (The kidneys filter blood and then reabsorb substances such as glucose, amino acids, and ions from the filtrate.) Glucosuria results in **osmotic diuresis,** a process in which an excessive loss of water and electrolytes (Na^+, K^+

FIGURE 15A

The Metabolic Consequences of Insulin Deficiency or Resistance.

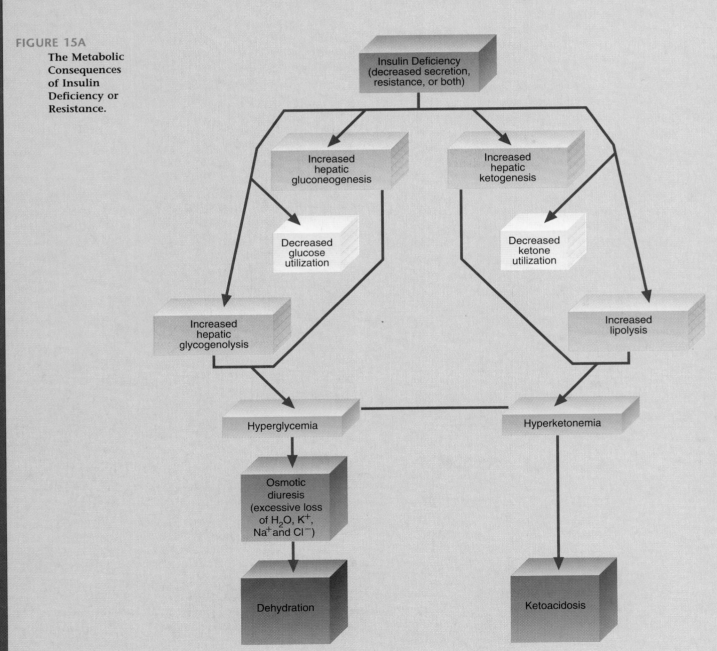

and Cl⁻) is caused by the presence of solute in the filtrate. In severe cases, despite the consumption of large volumes of fluid, patients may become dehydrated.

Without insulin to regulate fuel metabolism, its three principal target tissues (liver, adipose tissue, and muscle) fail to absorb nutrients appropriately. Instead, these tissues function as if the body were undergoing starvation. In the liver, gluconeogenesis accelerates because large amounts of amino acids are mobilized from muscle and synthesis of gluconeogenic enzymes (Section 8.4) increases. Glycogenolysis (ordinarily suppressed by insulin) produces additional glucose. The liver delivers these glucose molecules to an already hyperglycemic bloodstream. Increased lipolysis (Section 10.1) in adipose tissue (caused by the unopposed action of glucagon) releases large quantities of fatty acids into blood. In the liver, because these molecules are degraded by β-oxidation in combination with low concentrations of oxaloacetate (caused by excessive gluconeogenesis), large amounts of acetyl CoA, the substrate for forming ketone bodies, are produced. Fatty acids not used to generate energy or ketone bodies are used in VLDL synthesis. This process causes hyperlipoproteinemia (high blood concentrations of lipoproteins) because lipoprotein lipase synthesis is depressed when insulin is lacking. In effect, many of the cells of diabetics "starve in the midst of plenty." In severe cases, despite increased appetite and food consumption, the failure to use glucose effectively eventually causes body weight changes.

Insulin-Dependent Diabetes

In most cases of insulin-dependent diabetes, the insulin-producing β-cells have been destroyed by the immune system. Although the symptoms of IDDM often manifest themselves abruptly, it now appears that β-cell destruction is caused by an inflammatory process over several years. The symptoms are not obvious until virtually all insulin-producing capacity is destroyed. As in other inflammatory and autoimmune processes, β-cell destruction is initiated when an antibody binds to a cell surface antigen. One of the most common autoantibodies found in type I diabetes is now believed to bind specifically to an antigen with glutamate decarboxylase activity. (Recall that glutamate decarboxylase catalyzes the synthesis of GABA from glutamate.) Autoantibodies to insulin and the tyrosine phosphatase IA-2 have also been detected. (Tyrosine phosphatase IA-2, one of several enzymes that remove phosphate groups from specific phosphotyrosine residues in key regulatory proteins, is found only in brain and insulin-producing cells in the pancreas.) The significance of this phenomenon is unknown.

The most serious acute symptom of type I diabetes is **ketoacidosis.** Elevated concentrations of ketones in the blood **(ketosis)** and low blood pH along with hyperglycemia cause excessive water losses. (The odor of acetone on a patient's breath is characteristic of ketoacidosis.) Ketoacidosis and dehydration, if left untreated, can lead to coma and death. IDDM patients are treated with injections of insulin obtained from animals or from recombinant DNA technology. Before Frederick Banting and Charles Best discovered insulin in 1922, most type I diabetics died within a year after being diagnosed. Although exogenous insulin prolongs life, it is not a cure. Most diabetics have a shortened life span because of the long-term complications of their disease (discussed below).

Many researchers now believe that IDDM is caused by both genetic and environmental factors. Although the precise cause is still unknown, individuals who have inherited certain genetic markers are at high risk for developing the disease. Certain HLA antigens, that is, HLA-DR3 and HLA-DR4, are found in a large majority of type I diabetics. The *HLA* or *histocompatibility antigens,* found on the surface of most of the body's cells, play an important role in determining how the immune system reacts to foreign substances or cells. If a type I patient with both HLA-DR3 and HLA-DR4 antigens has an identical twin, the twin's risk for developing the disease is approximately 50%. That the risk is not 100% suggests that some environmental exposure is required for developing type I diabetes.

Noninsulin-Dependent Diabetes

Noninsulin-dependent diabetes is a milder disease than the insulin-dependent form. Its onset is slow, often occurring after the age of 40. In contrast to type I patients, most individuals with type II diabetes have normal or often elevated blood levels of insulin. For a variety of reasons, type II patients are resistant to insulin. The most common cause of insulin resistance is the down-regulation of insulin receptors. (Defective insulin receptors or improper insulin receptor processing also causes type II diabetes in some patients.) Approximately 85% of type II diabetics are obese. Because obesity itself promotes tissue insensitivity to insulin, individuals who are prone to this form of diabetes are at risk for the disease when they gain weight.

Treatment of NIDDM usually consists of diet control and exercise. Often, obese patients become more sensitive to insulin (i.e., there is an up-regulation of insulin receptors) when they lose weight. Because sustained muscular activity increases the uptake of glucose without requiring insulin, exercise also decreases hyperglycemia. In some cases, oral hypoglycemic drugs are used. It is believed that these molecules promote additional insulin release from the pancreas.

When the failure of type II diabetic patients to control hyperglycemia is accompanied by other serious medical conditions (e.g., renal insufficiency, myocardial infarction, or infections), a serious metabolic state referred to as **hyperosmolar hyperglycemic nonketosis** (HHNK) can result. (Ketoacidosis is rare in type II diabetes.) Because of the additional metabolic stress, insulin resistance is exacerbated, and blood glucose levels rise. The patient can then become severely dehydrated. The resulting lower blood volume depresses renal function, which results in further increases in blood glucose concentrations. Eventually, the patient becomes comatose. Because the onset is slow, it may not be recognized until the patient is severely dehydrated. (This is especially true for elderly diabetics, who often have a depressed thirst mechanism.) For this reason, HHNK is often more life-threatening than ketoacidosis.

Long-Term Complications of Diabetes

Despite the efforts of physicians and patients to control the symptoms of diabetes, few diabetics avoid the long-term consequences of their disease. Diabetics are especially prone to develop kidney failure, myocardial infarction, stroke, blindness, and neuropathy. (In *diabetic neuropathy,* nerve damage causes the loss of sensory and motor functions.) In addition, circulatory problems often cause gangrene, which leads to tens of thousands of amputations annually.

Most diabetic complications stem from damage to the vascular system. For example, damaged capillaries in the eye and kidney lead to blindness and kidney damage, respectively. Similarly, the accelerated form of atherosclerosis found in diabetics leads to serious cases of myocardial infarction and stroke. It is now believed that most of this damage is initiated by hyperglycemia. High blood glucose levels promote the nonenzymatic glycosylation of protein molecules, a process that is referred to as the Maillard reaction. The **Maillard reaction** (named for the French chemist who discovered it in 1912) is initiated when a sugar aldehyde or ketone group condenses with a free amino group to form a Schiff base (Figure 15B). The Schiff base rearranges to form a stable ketoamine called the **Amadori product.** Amadori products degrade into reactive carbonyl-containing products. These reactive products react with free amino groups as well as other amino acid side chains on nearby proteins to form a complex series of cross-linkages and adducts. (An **adduct** is the product of an addition reaction. In addition reactions, two molecules react to form a third molecule.) When

levels of blood glucose are high, glycosylation end products accumulate and cause extensive damage throughout the cardiovascular system. When cells in the lining of blood vessels are damaged, a repair process involving macrophages and growth factors is initiated that inadvertently promotes atherosclerosis. The capacity of affected blood vessels to nourish nearby tissue is eventually compromised.

Diabetics are also damaged by another consequence of hyperglycemia: increased metabolism by the sorbitol pathway. In some cells that do not require insulin for glucose uptake (e.g., Schwann cells in peripheral nerve and ocular tissues such as lens epithelial cells), the hexose enters by facilitated transport down its concentration gradient. Glucose is then converted to sorbitol (p. 159) by the NADPH-requiring enzyme aldose dehydrogenase. The oxidation of some sorbitol molecules to form fructose by sorbitol dehydrogenase is coupled to the reduction of NAD^+. Increased concentrations of NADH promote the reduction of pyruvate to form lactate. The accumulation of sorbitol and the redox changes caused by sorbitol oxidation are associated with several pathological changes in diabetics, e.g., nerve damage and cataract formation. The physiological significance of the sorbitol pathway in normal cells is unknown.

FIGURE 15B

The Maillard Reaction.

Any amino group-containing molecule can undergo the Maillard reaction, so nucleotides and amines are also susceptible. Proteins, however, are most susceptible to this process.

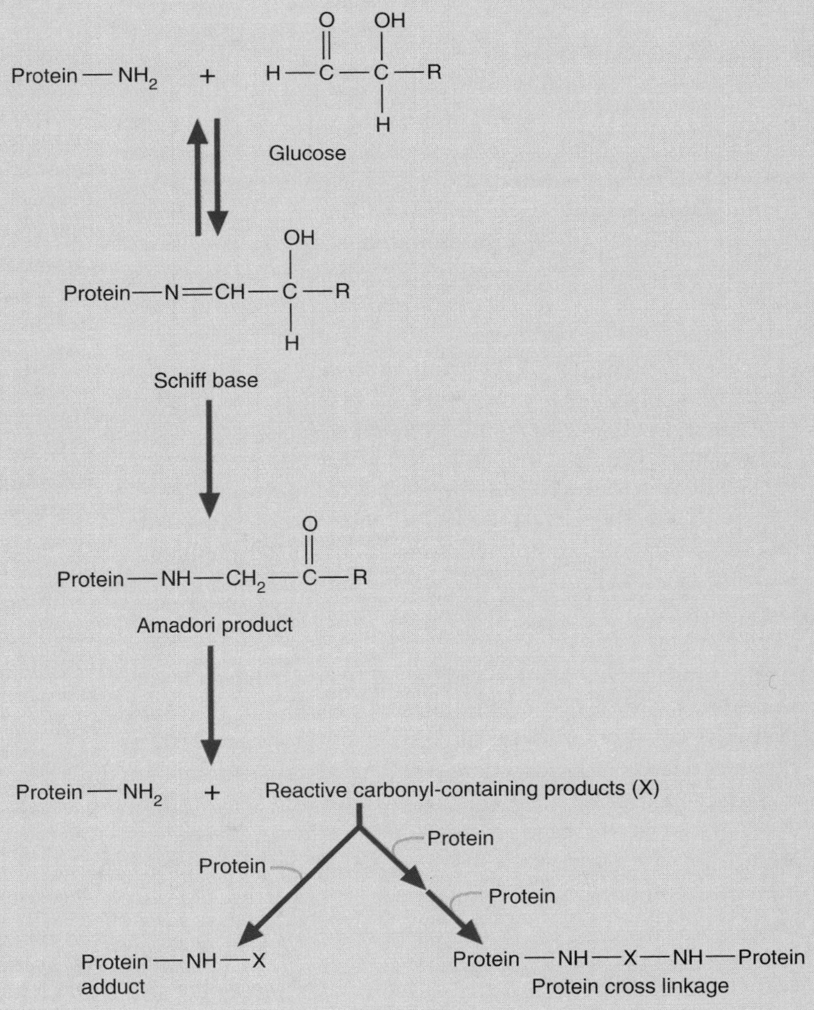

are produced. This process, as well as other aspects of the immune response, continues until the antigen is eliminated from the body.

Several cytokines are growth inhibitors. The **interferons** are a group of polypeptides produced by a variety of cells in response to several stimuli, such as antigens, mitogens, viral infections, and certain tumors. The type I interferons protect cells from viral infection by stimulating the phosphorylation and inactivation of a protein factor required to initiate protein synthesis. Type II interferons, produced by T lymphocytes, inhibit the growth of cancerous cells in addition to having several immunoregulatory effects. As their names imply, the **tumor necrosis factors** (TNF) are toxic to tumor cells. Both TNF-α (produced by antigen-activated phagocytic white blood cells) and TNF-β (produced by activated T cells) suppress cell division. They may also have a role in regulating several developmental processes.

QUESTION	15.2	The term *diabetes* derived from the Greek word *diabeinein* ("to go to excess"), was first used by Aretaeus (A.D. 81–138) to identify a group of symptoms that included intolerable thirst and "a liquefaction of the flesh and limbs into urine." After reviewing Special Interest Box 15.2, explain the physiological and biochemical basis for Aretaeus's findings.

15.4 MECHANISMS OF HORMONE ACTION

As discussed, hormones typically initiate their actions within target cells by binding to a receptor. Water-soluble hormone molecules (e.g., polypeptides, proteins, and epinephrine) bind to receptor molecules on the outer surface of the target cell's plasma membrane. This binding process, which is reversible, triggers a mechanism that initiates a phosphorylation cascade either directly (e.g., insulin) or indirectly via second messenger molecules (e.g., glucagon). As a result, the activities of specific enzymes and/or membrane transport mechanisms are altered. Examples of well-researched second messengers include cyclic AMP (cAMP), cyclic GMP (cGMP), the phosphatidylinositol-4,5-bisphosphate derivatives diacylglycerol (DAG) and inositol triphosphate (IP$_3$), and calcium ions. In contrast, the lipid-soluble steroid and thyroid hormones enter target cells, where they bind to specific receptor molecules. Each hormone-receptor complex then binds to specific regions of the target cell's DNA. Such binding activates genes and often leads to the synthesis of new proteins.

In this section, each type of hormone mechanism is briefly described. This is followed by a discussion of the insulin receptor, a critically important cellular component.

The Second Messengers

When a hormone molecule binds to a plasma membrane receptor, an intracellular signal called a second messenger is generated. The second messenger actually delivers the hormonal message. This process, referred to as **signal transduction,** also amplifies the original signal. In other words, a few hormone molecules initiate a mechanism that converts thousands of substrate molecules to product. The following second messengers are described below: cAMP, cGMP, DAG, IP$_3$, and Ca^{2+}.

cAMP Recall that cAMP is generated from ATP by adenylate cyclase when a hormone molecule binds to its receptor. The interaction between the receptor and adenylate cyclase is mediated by a G protein (Figure 15.9). (**G proteins** are so named because they bind guanine nucleotides. A variety of G proteins have been characterized. The G protein that stimulates cAMP synthesis when hormones such as glucagon, TSH, and epinephrine bind is referred to as G$_s$. G$_i$, discussed below, inhibits adenylate cyclase and therefore decreases cAMP levels.) In their unstimulated state, G proteins bind GDP. As a consequence of hormone binding and the resulting conformational change, the receptor interacts with a nearby G$_s$ protein. As G$_s$ binds to the receptor, GDP dissociates. Then the binding of GTP to G$_s$ allows one of its subunits to interact and stimulate adenylate cyclase. (G proteins usually contain three subunits: α, β,

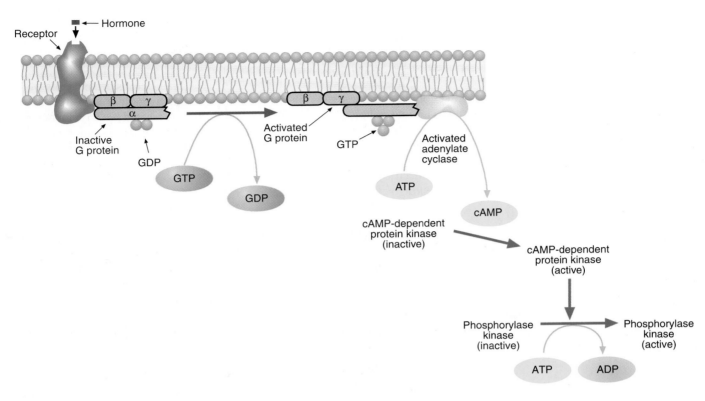

FIGURE 15.9

The Adenylate Cyclase Second Messenger System that Controls Glycogenolysis.

When the receptor is unoccupied, the G protein α subunit has GDP bound and is complexed with the β and γ subunits. The binding of hormone activates the receptor and leads to replacement of GDP with GTP. The activated subunit interacts with and activates adenylate cyclase. The cAMP produced binds to cAMP-dependent protein kinase. Eventually, the α subunit is inactivated as GTP is hydrolyzed to GDP.

and γ. The α subunit binds guanine nucleotides and has GTPase activity. The membrane-associated β and γ units bind reversibly with the α subunit as adenylate cyclase is alternately activated and deactivated.) The cAMP molecules formed by adenylate cyclase diffuse into the cytoplasm, where they bind to and activate cAMP-dependent protein kinase. Active protein kinase then phosphorylates and thereby alters the catalytic activity of key regulatory enzymes. Adenylate cyclase remains active only as long as it interacts with the α subunit of G_s. Once GTP has been hydrolyzed to GDP, the α subunit dissociates from adenylate cyclase and reassociates with the β and γ subunits. Subsequently cAMP is inactivated by phosphodiesterase.

The target proteins affected by cAMP depend on the cell type. In addition, several hormones may activate the same G protein. Therefore different hormones may elicit the same effect. For example, glycogen degradation in liver cells is initiated by both epinephrine and glucagon.

Some hormones inhibit adenylate cyclase activity. Such molecules depress cellular protein phosphorylation reactions because their receptors interact with G_i protein. When G_i is activated, its α subunit binds to and depresses adenylate cyclase activity. For example, because its receptors in adipocytes are associated with G_i, PGE_1 (prostaglandin E_1) depresses lipolysis. (Recall that lipolysis is stimulated by glucagon and epinephrine.)

cGMP   Although cGMP is synthesized in almost all animal cells, its role in cellular metabolism is still relatively undefined. However, the following information is known.

cGMP is synthesized from GTP by guanylate cyclase. Two types of guanylate cyclase are involved in signal transduction. In one type, which is membrane-bound, the extracellular domain of the enzyme is a hormone receptor. The other type is a cytoplasmic enzyme.

Two types of molecule are now known to activate membrane-bound guanylate cyclase: atrial natriuretic peptide and bacterial enterotoxin. *Atrial natriuretic factor* (ANF) is a peptide that is released from heart atrial cells in response to increased blood volume. The biological effects of ANF, that is, lowering of blood pressure via vasodilation and diuresis (increased urine excretion), appear to be mediated by cGMP. cGMP activates the phosphorylating enzyme protein kinase G. The role of this enzyme in mediating ANF effects is still unclear. ANF activates guanylate cyclase in several cell types. In one type, those in the kidney's collecting tubules, ANF-stimulated cGMP synthesis increases renal excretion of Na^+ and water.

The binding of *enterotoxin* (produced by several bacterial species) to another type of guanylate cyclase found in the plasma membrane of intestinal cells causes diarrhea. For example, one form of traveler's diarrhea is caused by a strain of *E. coli* that produces *heat stable enterotoxin*. The binding of this toxin to an enterocyte plasma membrane receptor linked to guanylate cyclase triggers excessive secretion of electrolytes and water into the lumen of the small intestine.

Cytoplasmic guanylate cyclase possesses a heme prosthetic group. The enzyme is activated by Ca^{2+}, so any rise in cytoplasmic Ca^{2+} causes cGMP synthesis. This guanylate cyclase activity is activated by NO (Special Interest Box 13.2). Some evidence suggests that binding NO to the heme group activates the enzyme. In several cell types (e.g., smooth muscle cells), cGMP stimulates the functioning of ion channels.

The Phosphatidylinositol Cycle and Calcium   The phosphatidylinositol cycle (Figure 15.10) mediates the actions of hormones and growth factors. Examples include acetylcholine (e.g., insulin secretion in pancreatic cells), vasopression, TRH, EGF, and PDGF. Phosphatidylinositol-4,5-bisphosphate (PIP_2) is cleaved by phospholipase C to form the second messengers DAG (diacylglycerol) and IP_3 (inositol-1,4,5-triphosphate). Phospholipase C is activated by a G protein activated when hormone binds to a membrane receptor. Several types of G protein may be involved in the phosphatidylinositol cycle. For example, G_Q (shown in Figure 15.10) mediates the actions of vasopressin.

The DAG product of the phospholipase C-catalyzed reaction activates protein kinase C. Several protein kinase C activities have been identified. Depending on the cell, activated protein kinase C phosphorylates specific regulatory enzymes, thereby activating or inactivating them.

Once it is generated, IP_3 diffuses to the calcisome (SER), where it binds to a receptor. The IP_3 receptor is a calcium channel. Cytoplasmic calcium levels then rise as calcium ions flow through the activated open channel. Recent evidence indicates that the IP_3-stimulated calcium signal is potentiated for a brief time by the release of another signal (as yet uncharacterized) that activates a plasma membrane calcium channel. Recall that calcium ions are involved in the regulation of a large number of cellular processes (Special Interest Box 8.2). Because calcium levels are still relatively low even when the calcium release mechanism has been activated (approximately 10^{-6} M), the calcium-binding sites on calcium-regulated proteins must have a high affinity for the ion. Several calcium-binding proteins modulate the activity of other proteins in the presence of calcium. Calmodulin, a well-researched example of the calcium-binding proteins, mediates many calcium-regulated reactions. In fact, calmodulin is a regulatory subunit for some enzymes (e.g., phosphorylase kinase, which converts phosphorylase b to phosphorylase a in glycogen metabolism).

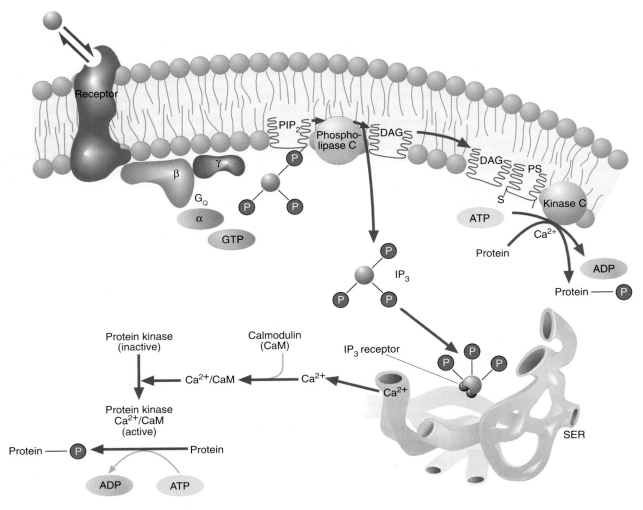

FIGURE 15.10

The Phosphatidylinositol Pathway.

The binding of certain hormones to their receptor activates the α subunit of a G protein. The α subunit then activates phospholipase C that cleaves IP_3 from PIP_2, leaving DAG in the membrane. DAG acting with phosphatidylserine (PS) and Ca^{2+} activates protein kinase C, which subsequently phosphorylates key regulators. IP_3 binds to receptors on the SER, liberating Ca^{2+}, which then activates another group of kinases.

QUESTION

15.3   Cholera toxin causes a massive diarrhea that is life-threatening because it prevents the α subunit of G_s from converting GTP to GDP. Describe why this inhibition leads to the diarrhea. (Hint: Refer to Special Interest Box 5.1.)

15.4   *Angina pectoris* is a very painful symptom of coronary artery disease. In this condition, which is caused by the narrowing of the heart's coronary arteries, insufficient oxygen flow during exertion causes chest pain that radiates to the neck and left arm. Traditionally, angina pectoris has been treated with nitroglycerin. It is now believed that nitroglycerin (Figure 15.11) relieves the pain because it acts to promote vasodilation (increased

FIGURE 15.11

Nitroglycerin.

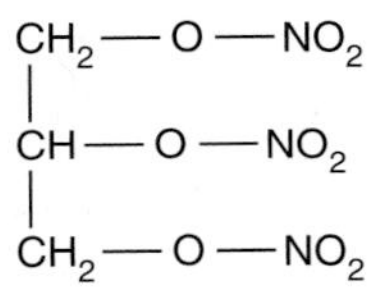

$$CH_2 - O - NO_2$$
$$|$$
$$CH - O - NO_2$$
$$|$$
$$CH_2 - O - NO_2$$

blood vessel diameter) throughout the body. Vasodilation, which results from smooth muscle relaxation, reduces blood pressure, which in turn reduces the heart's work load. Suggest how nitroglycerin relieves angina pectoris. (Hint: Cytoplasmic calcium ions stimulate muscle contraction.)

15.5

Cancer often results from a multistage process involving an initiating event (mediated by a viral infection or a carcinogenic chemical) followed by exposure to tumor promotors. Tumor promotors, a group of molecules that stimulate cell proliferation, cannot induce tumor formation by themselves. The phorbol esters, found in croton oil (obtained from the seeds of the croton plant, *Croton tiglium*), are potent tumor promotors. (Other examples of tumor promotors include asbestos and several components of tobacco smoke.) In one of the tumor-promoting actions of the phorbol esters, these molecules mimic the actions of DAG. In contrast to DAG, the phorbol esters are not easily disposed of. Explain the possible biochemical consequences of phorbol esters in an "initiated" cell. What enzyme is activated by both DAG and phorbol esters?

Steroid and Thyroid Hormone Mechanisms

Because the steroid and thyroid hormones are hydrophobic molecules, they are transported in the blood to their target cells bound to several types of protein. Examples of steroid transport proteins include transcortin (also called corticosteroid-binding globulin), androgen-binding protein, sex hormone–binding protein, and albumin. In addition to albumin, the thyroid hormones are also transported by thyroid-binding globulin and thyroid-binding prealbumin.

Once they reach their target cells, both types of hormone dissociate from their transport proteins and apparently diffuse through the plasma membrane. (An active transport process for thyroid hormones has been proposed.) After hormone molecules bind to their intracellular receptors, each hormone-receptor complex binds to specific DNA segments called **hormone response elements** (HRE). The binding of the hormone-receptor complex to an HRE either enhances or diminishes the transcription of a specific gene. (Transcription, the first major step in the expression of a gene, is described in Chapter 17. Gene expression often results in protein synthesis.) Several HREs can bind to the same hormone-receptor complex.

In the absence of hormone, several types of receptor have been observed to form complexes with other proteins. For example, unoccupied glucocorticoid receptors are found in the cytoplasm bound to hsp90. (Recall that hsp is the abbreviation for heat shock proteins. Refer to Special Interest Box 14.1 for a brief discussion of heat shock proteins.) The binding of hsp90 to the glucocorticoid receptor prevents the inappropriate binding of the unoccupied receptor to DNA. When corticosterone binds to the receptor, a change in the receptor's conformation makes it dissociate from hsp90. Then two hormone-bound receptors associate to form a functional complex that then moves into the nucleus, where it binds to its HRE.

Once thyroid hormone enters a cell, it binds temporarily with a specific cytoplasmic protein. Thyroid hormone molecules migrate to the nucleus and mitochondria, where they bind to receptors. In the nucleus the binding of thyroid hormone initiates the transcription of genes that play crucial roles in a variety of cellular processes, such as those that code for growth hormone and Na^+-K^+ ATPase. In mitochondria, thyroid hormones promote oxygen consumption and increased fatty acid oxidation. (The mechanism by which this latter process occurs is not understood.)

The Insulin Receptor

The insulin receptor is a member of a family of cell surface receptors for various anabolic polypeptides, for example, EGF, PDGF, and IGF-I. Although there are several structural differences among this group (Figure 15.12), they do possess the following structural features in common: an external domain that binds specific extracellular ligands, a transmembrane segment, and a cytoplasmic catalytic domain with tyrosine kinase activity. When a ligand binds to the external domain, a conformational change in the receptor protein activates the tyrosine kinase domain. The tyrosine kinase

PLANT HORMONES

For a variety of technical reasons, plant biochemical research is more difficult and therefore less well understood than that of animals. Consequently, relatively little is known about plant hormones. The most important hormones that have so far been investigated include auxins, gibberellins, cytokinins, abscisic acid, and ethylene (Figure 15C).

Auxins

The **auxins** (See p. 389) are a class of plant growth regulators. The most important auxin is indole acetic acid (IAA). Auxins are found throughout most plant tissues, although most IAA synthesis occurs in actively growing tissues. Most plants synthesize IAA from tryptophan by the indole-3-pyruvate pathway. Tryptophan aminotransferase catalyzes a transamination reaction, producing indole-3-pyruvic acid and glutamate. Indole-3-pyruvic acid is then converted to indole-3-acetaldehyde by indole pyruvate decarboxylase. De-pending on the plant species, IAA is synthesized from indole-3-acetaldehyde by either an NAD-dependent indoleacetaldehyde dehydrogenase or an oxygen-requiring indoleacetaldehyde oxidase. IAA is inactivated by several oxidases as well as by photooxidation.

The mechanisms by which auxins regulate plant growth are not fully understood. It is currently believed that membrane-bound auxins promote the transfer of protons out of cells. Apparently, decreased extracellular pH loosens and expands the cell wall. Auxins also increase nucleic acid and protein synthesis.

Gibberellins

The **gibberellins** are a large class of tetracyclic diterpenes, known primarily for their effect on stem growth. However, they are apparently involved in nearly all of the processes influenced by auxins. In some plants, application of gibberellins prompts flowering and seed production. The gibberellins are synthesized

FIGURE 15C

Common Plant Hormones.

The roles of these molecules in plant development are still poorly understood.

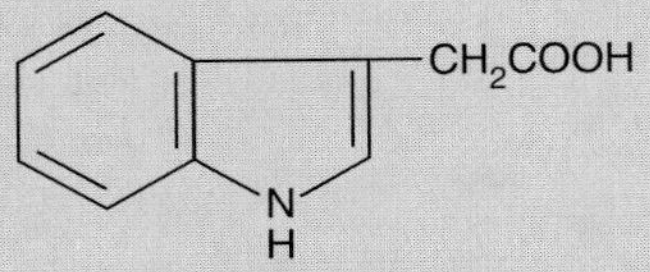

Auxin
(indole acetic acid)

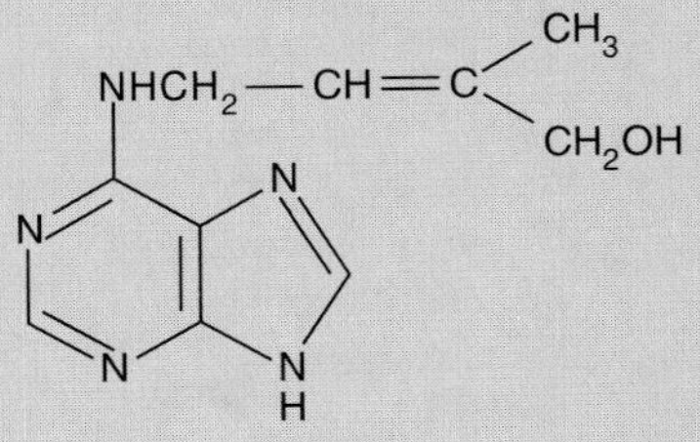

Cytokinin
(zeatin)

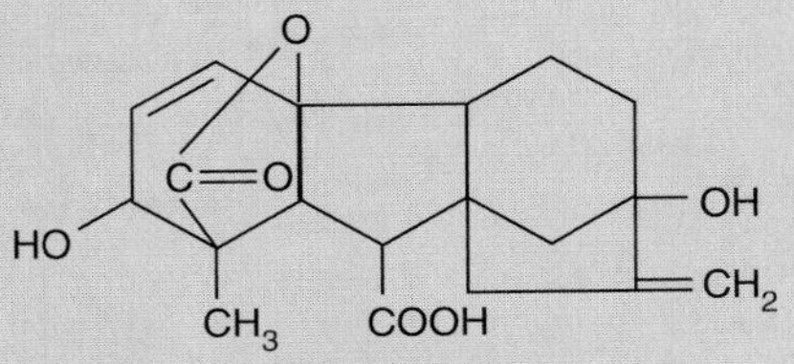

Gibberellin
(gibberellic acid)

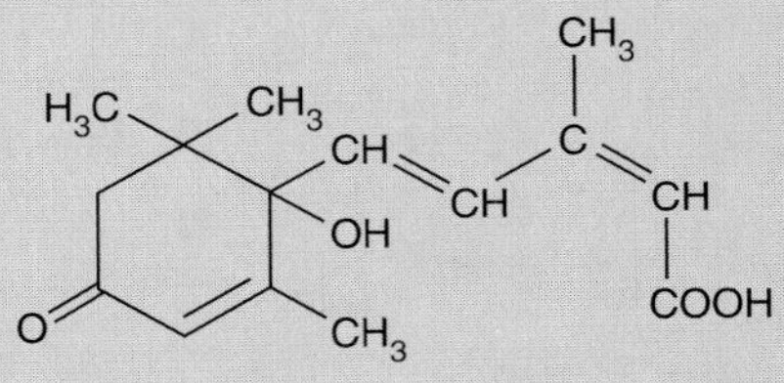

Abscisic acid

$H_2C = CH_2$

Ethylene

activity initiates a phosphorylation cascade that begins with an autophosphorylation of the tyrosine kinase domain. Because most research efforts have been devoted to the insulin receptor, its structure and proposed functions are described next.

The insulin receptor is a transmembrane glycoprotein composed of two types of subunits connected by disulfide bridges. Two large α subunits (130 kD) extend extracellularly, where they form the insulin-binding site. Each of the two β subunits (90 kD) contains a transmembrane segment and a tyrosine kinase domain.

Insulin binding activates receptor tyrosine kinase activity and causes a phosphorylation cascade that modulates various intracellular proteins. For example, insulin

from geranylgeranyl pyrophosphate in a complex reaction pathway that has been studied mostly in the fungus *Gibberella fujikuroi*. (In the 1920s, Japanese scientists discovered that this fungus causes rice seedlings to become abnormally tall. Later, the active substance was isolated and named "gibberellin.") The mode of action of the gibberellins is not understood, although they are known to induce nucleic acid and protein synthesis.

Cytokinins

The **cytokinins**, similar to adenine in structure, are found in greatest abundance in meristem and other developing tissues. Along with the auxins and gibberellins, they promote growth and developmental processes (e.g., stem and root growth) as well as chloroplast development and delay the aging of leaves.

Abscisic Acid

Although abscisic acid is found in most plant tissue, it is especially common in fleshy fruits. Abscisic acid antagonizes many growth processes. For example, it prevents fruit seeds from germinating while the fruit remains on the plant. Apparently, this hormone also protects leaves from excessive water loss by promoting the transport of K^+ out of guard cells. (Guard cells open stomata when they are swollen and close stomata when they become contracted.) As K^+ and water flow out of the guard cells, the stomata close, thus protecting the plant from further transpiration water losses.

Ethylene

Ethylene, a gas, is produced by plant tissues such as fruits, flowers, seeds, and leaves. Its synthesis from methionine is outlined in Figure 15D. Although ethylene is involved in longitudinal growth processes and flower senescence, it is widely used by commercial growers to accelerate ripening of fruit. Its mechanism of action is unknown.

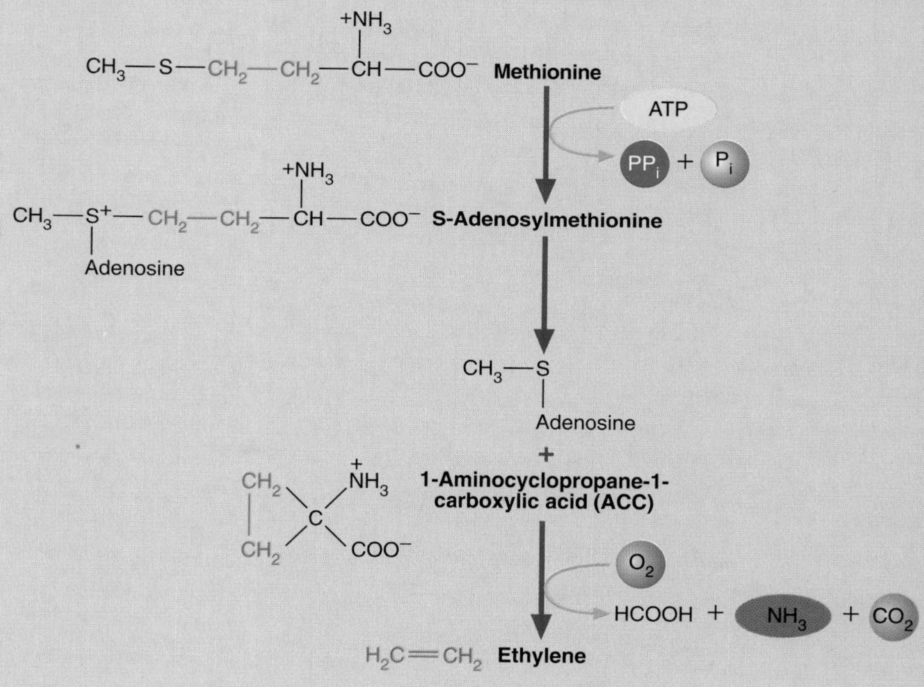

FIGURE 15D

Synthesis of Ethylene.

The immediate precursor of ethylene, 1-amino-cyclopropane-1-carboxylic acid, is synthesized from SAM in response to auxins and bruising or other types of stress.

binding inhibits hormone-sensitive lipase in adipocytes. It apparently does so by activating a phosphatase that dephosphorylates the lipase. In addition, several models of insulin action suggest that several second messengers are employed, for example, inositol monophosphate or DAG, in activating protein kinase C.

Insulin binding appears to initiate a phosphorylation cascade that induces the transfer of several types of protein to the cell surface. Examples of these molecules include the glucose transporter and the receptors for LDL and IGF-II. The movement of these molecules to the plasma membrane in the postabsorptive phase of the feeding-fasting cycle promotes the cell's acquisition of nutrients as well as growth-promoting signals.

FIGURE 15.12

Selected Receptors.

All three receptors possess an extracellular ligand-binding domain, a transmembrane segment, and an intracellular tyrosine kinase-containing domain. Each catalytic domain (colored dark green) is about 250 amino acid residues long. The amino acid sequences of the tyrosine kinase domains are similar to each other.

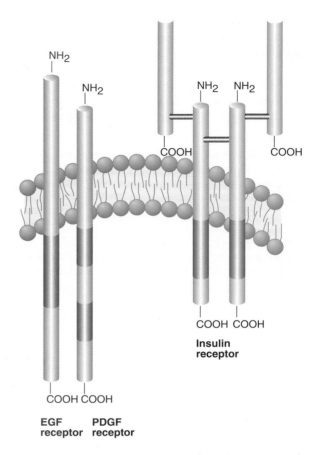

KEY WORDS

adduct, *448*

Amadori product, *448*

auxin, *454*

cytokine, *444*

cytokinin, *455*

desensitization, *442*

down-regulation, *444*

endocrine, *439*

epidermal growth factor, *444*

G protein, *449*

gibberellin, *454*

glucosuria, *446*

growth factor, *444*

hormone response element, *453*

hyperglycemia, *446*

hyperosmolar hyperglycemic nonketosis, *447*

insulin resistance, *444*

insulinlike growth factor, *445*

interferon, *449*

interleukin-2, *445*

ketoacidosis, *447*

ketosis, *447*

Maillard reaction, *448*

mitogen, *444*

osmotic diuresis, *446*

platelet-derived growth factor, *444*

postabsorptive, *437*

postprandial, *437*

signal transduction, *449*

somatomedin, *444*

tumor necrosis factor, *449*

SUMMARY

Multicellular organisms require sophisticated regulatory mechanisms to ensure that all their cells, tissues, and organs are cooperating. For example, the feeding-fasting cycle illustrates how a variety of organs contribute to the acquisition of food molecules as well as their use.

Hormones are molecules organisms use to convey information to other cells. When target cells are distant from the hormone-producing cell, such molecules are called endocrine hormones. To ensure proper control of metabolism, the synthesis and secretion of many mammalian hormones are regulated by a complex cascade mechanism ultimately controlled by the central nervous system. In addition, a negative feedback mechanism precisely controls various hormone syntheses. A variety of diseases are caused by either over-production or underproduction of a specific hormone or by the insensitivity of target cells.

Growth factors (or cytokines) are a group of polypeptides that regulate the growth, differentiation, or proliferation of various cells. They differ from hormones in that they are often produced by a variety of cell types rather than by specialized glandular cells.

Hormones and growth factors usually initiate their effects in a target cell by binding to a specific receptor. Water-soluble hormones typically bind to a receptor on the surface of the target cell. They alter the activities of several enzymes and/or transport mechanisms. Second messenger molecules such as cAMP, cGMP, IP_3, DAG, and Ca^{2+} often mediate a hormone's or growth factor's message. In general, the hydrophobic steroid and thyroid hormones bind to an intracellular receptor that binds to a DNA sequence referred to as a hormone response element.

Several plant hormones have been identified, for example, auxins, gibberellins, cytokinins, abscisic acid, and ethylene. Because of technical obstacles, little is known about their action.

SUGGESTED READINGS

Becker, A.B., and Roth, R.A., Insulin Receptor Structure and Function in Normal and Pathological Conditions, *Ann. Rev. Med.,* 41:99–115, 1990.

Brownlee, M., Vlassara, H., and Cerami, A., Nonenzymatic Glycosylation and the Pathogenesis of Diabetes Complications, *Ann. Inter. Med.,* 101:527–537, 1984.

Hartmann, F., and Plauth, M., Intestinal Glutamine Metabolism, *Metabolism,* 38(8)(Suppl 1):18–24, 1989.

Krebs, E. G., Role of the Cyclic AMP-Dependent Protein Kinase in Signal Transduction, *J. Am. Med. Assoc.,* 262:1815–1818, 1989.

Lienhard, G.E., Slot, J.W., James, D.E., and Mueckler, M.M., How Cells Absorb Glucose, *Sci. Amer.,* 266(1):86–91, 1992.

Linder, M.E., and Gilman, A.G., G Proteins, *Sci. Amer.,* 271(1):56–65, 1992.

Maclaren, N.K., and Atkinson, M.A., Insulin-dependent Diabetes Mellitus: The Hypothesis of Molecular Mimicry Between Islet Cell Antigens and Microorganisms, *Mol. Med. Today* 3(2):76–83, 1997.

Rasmussen, H., The Cycling of Calcium as an Intracellular Messenger, *Sci. Amer.,* 261(4):66–73, 1989.

Smith, K.A., Interleukin-2, *Sci. Amer.,* 262(3):50–57, 1990.

Uvnas-Moberg, K., The Gastrointestinal Tract in Growth and Reproduction, *Sci. Amer.,* 261(1):78–83, 1989.

QUESTIONS

1. Define the following terms:
   a. ketoacidosis
   b. hyperlipoproteinemia
   c. neurophysin
   d. growth factor
   e. chylomicron remnant
   f. down-regulation
   g. G protein

2. Which organ carries out the following activities?
   a. pH regulation
   b. gluconeogenesis
   c. absorption
   d. neural and endocrine integration
   e. lipogenesis
   f. urea synthesis

3. State the action of each of the following hormones:
   a. corticotropin
   b. insulin
   c. glucagon
   d. oxytocin
   e. LH
   f. GnRF
   g. somatostatin
   h. vasopressin
   i. FSH

4. Sustained metabolism of fats takes place only in the presence of carbohydrate. Explain.

5. What are the general functions of hormones in the body?

6. The hypothalamus and the pituitary are two endocrine glands located near one another in the brain. The secretions of the hypothalamus exert a powerful effect on the pituitary, yet if the pituitary is surgically transported to a remote location such as the kidney, the secretions of the hypothalamus have no effect. Suggest a reason for this observation.

7. Explain how a second messenger works. Why use a second messenger rather than simply relying on the original hormone to produce the desired effect?

8. NADH is an important reducing agent in cellular catabolism, whereas NADPH is an important reducing agent in anabolism. Review previous chapters and show how the synthesis and degradation of these two molecules are interconnected.

9. Dieters frequently fast in an attempt to reduce their weight. During these fasts, they frequently lose considerable muscle mass rather than fat. Why is this so?

10. You are being stalked by a large tiger. Explain how your metabolism responds to help you escape.

11. During fasting in humans, virtually all the glucose reserves are consumed in the first day. The brain requires glucose to function and adjusts only slowly to other energy sources. Explain how the body supplies the glucose required by the brain.

12. In uncontrolled diabetes levels of hydrogen ions are elevated. Explain how these ions are generated.

13. Why is it important for hormones to act at low concentrations and be metabolized quickly?

14. Hormones are often synthesized and stored in an inactive form within secretory vesicles. Secretion usually occurs only when the hormone-producing cell is stimulated. Explain the advantages that this process has over making the hormone molecules as they are needed.

15. Briefly discuss the major classes of second messenger that are now recognized.

16. Describe the general modes of hormone action.

17. What are phorbol esters and how do they promote tumors?

18. Briefly describe the five major classes of plant hormones.

19. Steroid hormones are often present in cells in low concentrations. This makes them difficult to isolate and identify. It is sometimes easier to isolate the proteins to which they bind by

using affinity chromatography. (Refer to Appendix B, Supplement 5.) Explain how you would use this technique to isolate a protein suspected of steroid hormone binding.

20. After about 6 weeks of fasting, the production of urea is decreased. Explain.

21. Extreme thirst is a characteristic symptom of diabetes. Explain.

22. During periods of prolonged exercise, muscles burn fat released from adipocytes in addition to glucose. Explain how the need for additional fatty acids by muscle is communicated to the adipocytes.

23. Bodybuilders often take anabolic steroids to increase their muscle mass. How do these steroids achieve this effect? (Common side effects of anabolic steroid abuse include heart failure, violent behavior, and liver cancer.)

24. During periods of starvation, the brain converts from using glucose to using ketone bodies instead. What survival advantages does this have for the animal?

25. What are four metabolic functions of the kidney? How are these affected by diabetes mellitus?

26. During periods of fasting, some muscle protein is depleted. How is this process initiated and what happens to the amino acids in these proteins?

27. The kidney has an unusually large demand for glutamine and glutamate. How could the metabolism of these compounds help maintain pH balance?

28. Hemoglobin molecules exposed to high levels of glucose are converted to glycosylated products. The most common, referred to as hemoglobin A_{1C} (HbA$_{1C}$), contains a β-chain glycosylated adduct. Because red blood cells last about three months, HbA$_{1C}$ concentration is a useful measure of a patient's blood sugar control. In general terms, describe why and how HbA$_{1C}$ forms.

NUCLEIC ACIDS

Chapter Sixteen

When James Watson (left) and Francis Crick discovered the structure of DNA in 1953, they were research students at the Henry Cavendish Laboratory of Cambridge University.

OBJECTIVES

After you have studied this chapter, you should be able to answer these questions:

1. What is the central dogma?
2. How is DNA damaged?
3. What historical developments led to the discovery of DNA structure?
4. What information did Watson and Crick use to determine DNA structure?
5. What are the structural differences among A-DNA, B-DNA, H-DNA, and Z-DNA?
6. What is the structural significance of DNA supercoiling?
7. What are nucleosomes?
8. What are the three most prominent forms of RNA?
9. What properties of viruses make them useful research tools for biochemists?
10. What types of genomes can viruses possess?
11. What is the difference between lytic and lysogenic viruses?
12. After many years of research and billions of dollars spent, why is AIDS still considered an incurable disease?

The determination of the structure of deoxyribonucleic acid by James Watson and Francis Crick in the early 1950s was the culmination of research that had begun almost a century before. The consequences of their seminal work are still unfolding as the genetic information of various organisms, including humans, is gradually being revealed. The structure and functioning of DNA and its companion molecule, RNA (ribonucleic acid), which have proven to be astonishingly complex, continue to fascinate researchers and students alike. Indeed, their manipulation has proven to be both intellectually and (occasionally) financially rewarding.

All biological processes are profoundly dependent on information. For example, in the previous chapter (Chapter 15) it was noted that chemical signals such as hormones and growth factors direct and regulate metabolic activities in multicellular organisms. But what mechanism directs the synthesis of these molecules or, for that matter, the synthesis of all the other molecules and supramolecular structures in living organisms? The complexity of living processes must require a large amount of information and this information must be stored and then retrieved in an orderly and precise manner. This latter concept is never more clear than when organismal growth and development are observed. Growing children and germinating seeds, for example, obviously develop according to a predetermined plan. Off-spring normally progress through predictable developmental stages that vary little from one generation to the next. Although the progeny of sexually reproducing species are not identical to their parents, the inheritance of many traits (e.g., hair and eye color in animals and flower color in plants) is clearly identifiable.

For countless centuries, humans have observed inheritance patterns without understanding the mechanisms that transmit physical traits and developmental processes from parent to offspring. Many human cultures have used such observations to improve their economic status, as in the breeding of domesticated animals or seed crops. It was not until the nineteenth century that the scientific investigation of inheritance, now referred to as **genetics,** began. By the beginning of the twentieth century, scientists generally recognized that physical traits are inherited as discrete units (later called **genes**) and that chromosomes within the nucleus are the repositories of genetic information. Eventually, the chemical structure of chromosomes was elucidated, and (after many decades of investigation) deoxyribonucleic acid (DNA) was identified as the genetic information. A complete set of this information in an organism is now referred to as its **genome.** Most eukaryotic cells possess a *diploid* genome, that is, one that consists of two copies of each type of chromosome. A *haploid* genome, which often occurs in gametes (sex cells, i.e., eggs and sperm), consists of one copy of each type of chromosome.

The transformation of the life sciences that followed the successful determination of DNA structure by James Watson and Francis Crick in 1953 was monumental. Every biological phenomenon was eventually reexamined in the light of the newly emerging science devoted to nucleic acid function called **molecular biology.** During the next few decades, in one of the most fascinating and complex investigations of the twentieth century, the molecular biologists formulated a general outline of biological inheritance. This work revealed the following principles:

1. The information encoded within DNA, which directs the functioning of living cells and is transmitted to offspring, consists of a specific sequence of nitrogenous bases. The physiological and genetic function of DNA requires the synthesis of relatively error-free copies. DNA synthesis involves the complementary pairing of nucleotide bases.

2. The mechanism that uses genetic information to direct cellular processes involves the synthesis of another type of nucleic acid, ribonucleic acid (RNA). RNA synthesis occurs by complementary pairing of ribonucleotide bases with the bases in a DNA molecule.

3. Several types of RNA are involved in the synthesis of the enzymes, structural proteins, and other polypeptides required for organismal function.

These processes are summarized by the following sequence:

DNA ⟶ RNA ⟶ Protein

This sequence has been referred to as the "central dogma of molecular biology" because it describes the flow of genetic information from DNA through RNA and eventually to proteins.

NUCLEIC ACID TECHNOLOGY	The chemical and physical properties of the nucleic acids are routinely exploited in a rapidly expanding set of techniques used in their isolation, purification, characterization, and manipulation. Refer to Appendix B, Supplements 16 and 17, for discussions of the most commonly used methods in nucleic acid technology. These techniques are referred to throughout this chapter and Chapter 17.

Because of the informational nature of genetic processes, some descriptive terms have been borrowed from other information sciences. For example, DNA synthesis is often referred to as **replication** (to copy or duplicate). Similarly, the process in which DNA is used to synthesize RNA is called **transcription,** and protein synthesis is referred to as **translation.** Several other terms (e.g., genetic code, codon, and anticodon), defined in Chapter 17, come from the language of information transfer.

This chapter, which focuses on the structure of the nucleic acids, begins with a description of DNA structure and the investigations that led to its discovery. This is followed by a discussion of current knowledge of DNA and chromosome structure, as well as the structure and roles of the several forms of RNA. The chapter ends with the description of several viruses, which are noncellular parasites composed of nucleic acid and proteins. In the following chapter (Chapter 17), several aspects of nucleic acid synthesis and function (i.e., DNA replication and transcription) are discussed. Protein synthesis is described in Chapter 18.

16.1 DNA

The structure of DNA consists of two polynucleotide strands wound around each other to form a right-handed double helix (Figure 16.1). (The double helix of DNA is so distinctive that this molecule is often referred to as the *double helix.*) In each polynucleotide the mononucleotides are linked to each other by 3′,5′-phosphodiester bonds (Figure 16.2). These bonds join the 5′-hydroxyl group of the deoxyribose of one nucleotide to the 3′-hydroxyl group of the sugar unit of another nucleotide through a phosphate group. The antiparallel orientation of the two polynucleotide strands allows hydrogen bonds to form between the nitrogenous bases that are oriented toward the helix interior. There are two types of base pairs (bp) in DNA: (1) The purine adenine pairs with the pyrimidine thymine, and (2) the purine guanine pairs with the pyrimidine cytosine. Because each base pair is oriented at an angle to the long axis of the helix, the overall structure of DNA resembles a twisted staircase. The dimensions of crystalline DNA have been precisely measured.

1. One turn of the double helix spans 3.4 nm and consists of approximately 10.4 base pairs. (Changes in pH and salt concentrations affect these values slightly.)
2. The diameter of the double helix is 2.4 nm. Note that the double helix has sufficient space in the interior only for base pairing a purine and a pyrimidine. Pairing two pyrimidines would be too narrow, while pairing purines would be too wide.
3. The distance between adjacent base pairs is 0.34 nm. The dimensions of both types of base pairs are illustrated in Figure 16.2.

As befits its role in living processes, DNA is a relatively stable molecule. Several types of noncovalent bonding contribute to this stability. Hydrophobic interactions

FIGURE 16.1

Two Models of DNA Structure.

(a) The DNA double helix is represented as a spiral ladder. The sides of the ladder represent the sugar-phosphate backbones. The rungs represent the pairs of bases. (b) In a space-filling model, the sugar-phosphate backbones are represented by strings of colored spheres. The base pairs consist of horizontal arrangements of dark blue spheres. Wide and narrow grooves are created by twisting the two strands around each other.

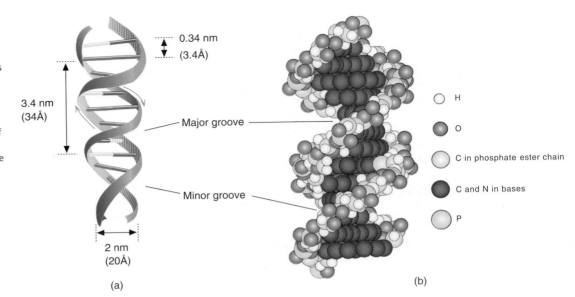

FIGURE 16.2

DNA Structure.

In this short segment of DNA, the bases are shown in orange and the sugars are blue. Each base pair is held together by either two or three hydrogen bonds. The two polynucleotide strands are antiparallel. Because of base pairing, the order of bases in one strand determines the order of bases along the other.

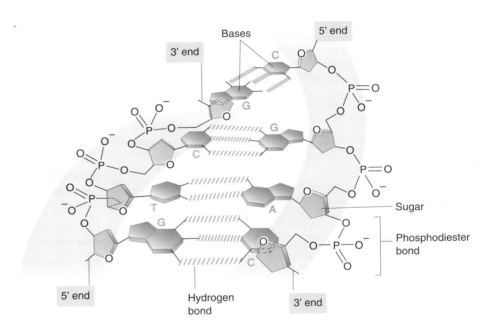

between the stacked base pairs in the double helix play an important role in stabilizing DNA. DNA's external surface, referred to as the *sugar-phosphate backbone,* is hydrophilic and therefore solvated by water. In addition, hydrogen bonding between complementary bases promotes stability as well as provides a mechanism for accurate pairings between the bases.

DNA Structure: The Nature of Mutation

DNA is vulnerable to certain types of structural change. This complex organic molecule is constantly exposed to solvent collisions, thermal fluctuations, and other spontaneous disruptive processes. In addition, a wide variety of xenobiotics, both natural and human-made, are known to alter DNA structure. Several examples of well-researched mutagenic factors are discussed below.

As described, tautomeric shifts are spontaneous changes in nucleotide base structure that interconvert amino and imino groups as well as keto and enol groups. Usually tautomeric shifts have little effect on DNA structure. However, if tautomers form during DNA replication, base mispairings may result. For example, the imino form of adenine will not base pair with thymine. Instead, it forms a base pair with cyto-

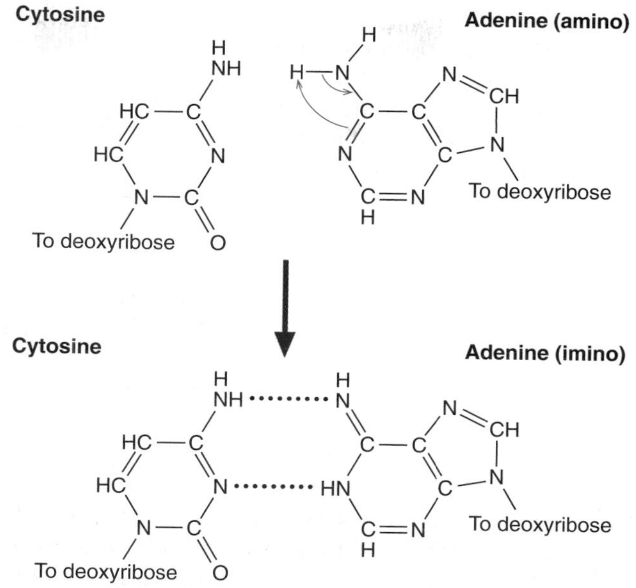

FIGURE 16.3

A Tautomeric Shift Causes a Transition Mutation.

As adenine undergoes a tautomeric shift, its imino form can base pair with cytosine. The transition shows up in the second generation of DNA replication when cytosine base pairs with guanine. In this manner an A-T base pair is replaced by a C-G base pair.

sine (Figure 16.3). If this pairing is not corrected immediately, a transition mutation results. In a **transition mutation** a pyrimidine base is substituted for another pyrimidine, or a purine is substituted for another purine. Transition mutations are **point mutations,** DNA base sequence changes that involve a single base pair.

Several spontaneous hydrolytic reactions also cause DNA damage. For example, it has been estimated that several thousand purine bases are lost daily from the DNA in each human cell. In depurination reactions the N-glycosyl linkage between a purine base and deoxyribose is cleaved. The protonation of N-3 and N-7 of guanine promotes hydrolysis. If a purine base is not replaced before the next round of DNA replication, a point mutation will result. Similarly, bases can be spontaneously deaminated. For example, the deaminated product of cytosine converts via a tautomeric shift to uracil. Eventually, what should be a CG base pair is converted to an AT base pair. (Uracil is similar in structure to thymine.)

Some ionizing radiation (e.g., UV, X-rays and γ-rays) can alter DNA structure. Low radiation levels may cause mutation; high levels can be lethal. Radiation-induced damage, caused by either a free radical mechanism (either abstraction of hydrogen atoms or the creation of ·OH, the hydroxyl radical), includes strand breaks, DNA-protein crosslinking, ring openings and base modifications. The hydroxyl radical, formed by the radiolysis of water, as well as oxidative stress (see Section 11.4), is known to cause some strand breakage and numerous base modifications (e.g., thymine glycol, 5-hydroxymethyl uracil and 8-hydroxymethyl uracil).

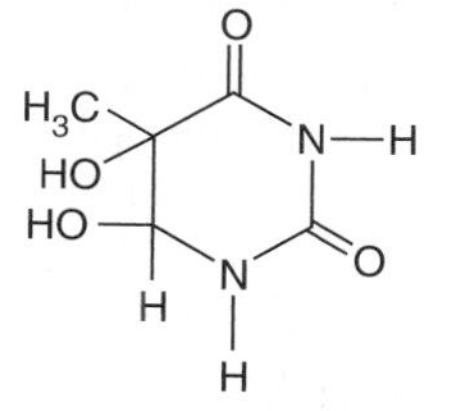

Thymine glycol 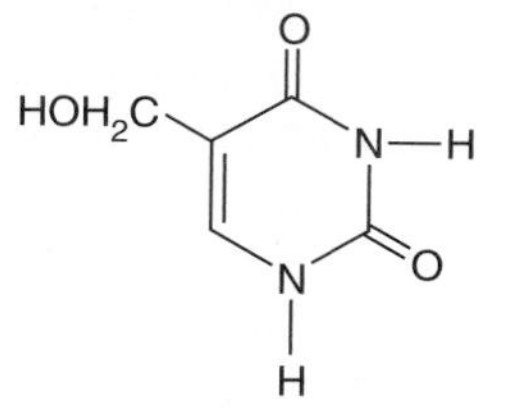   **5-Hydroxymethyl uracil** 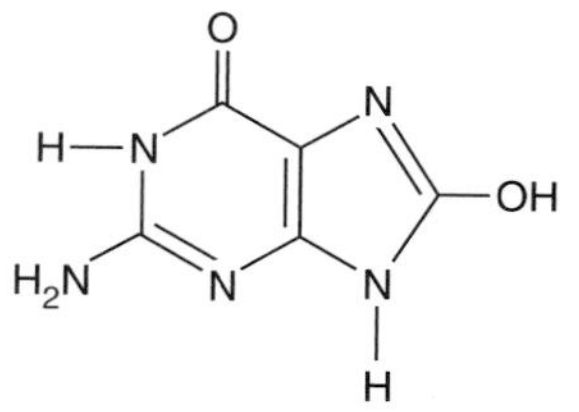   **8-Hydroxyguanine**

The most common UV-induced product is pyrimidine dimers (Figure 16.4). The helix distortion that results from helix formation stalls DNA synthesis.

A large number of xenobiotics can damage DNA. The most important of these molecules belong to the following classes:

1. **Base analogs.** Because their structures are similar to the normal nucleotide bases, **base analogs** can be inadvertently incorporated into DNA. For example,

FIGURE 16.4

Thymine Dimer Structure.

Adjacent thymines form dimers with high efficiency after absorbing UV light.

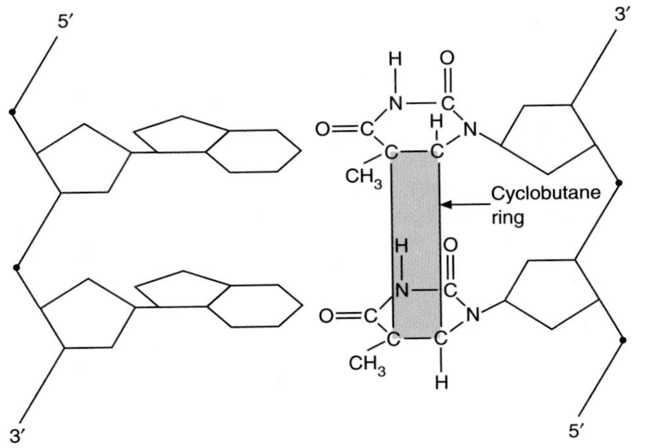

caffeine is a base analog of thymine. Because it can base pair with guanine, caffeine incorporation can cause a transition mutation.

2. **Alkylating agents. Alkylation** is a process in which electrophilic ("electron-loving") substances attack molecules that possess an unshared pair of electrons. When electrophiles react with such molecules, they usually add carbon-containing alkyl groups. Adenine and guanine are especially susceptible to alkylation, although thymine and cytosine can also be affected.

   DNA repair mechanisms (Section 17.1) usually eliminate the damaged nucleotide. Because the repair process can insert either purine or pyrimidine nucleotides, a transversion mutation may occur. (In a **transversion mutation,** another type of point mutation, a pyrimidine is substituted for a purine or vice versa.) Alkylations can also promote tautomer formation, which may result in transition mutations. Examples of alkylating agents include dimethylsulfate and dimethylnitrosamine. Mitomycin C is a bifunctional alkylating agent. It can prevent DNA synthesis by crosslinking guanine bases.

3. **Nonalkylating agents.** A variety of chemicals other than the alkylating agents can modify DNA structure. Nitrous acid (HNO_2), derived from the nitrosamines as well as from sodium nitrite ($NaNO_2$) deaminates bases. HNO_2 converts adenine, guanine, and cytosine to hypoxanthine, xanthine, and uracil, respectively. The aromatic polycyclic hydrocarbons (e.g., benzo[a]pyrene) are also mutagenic. Once consumed, these molecules can be converted to highly reactive derivatives by biotransformation reactions (Chapter 19). The reactive derivatives can then form adducts of most bases. Damage occurs primarily because this chemical modification prevents base pairing.

4. **Intercalating agents.** Certain planar molecules can distort DNA because they insert themselves (intercalate) between the stacked base pairs of the double helix. Either adjacent base pairs are deleted or new base pairs are inserted. If not corrected, deletions or insertions cause so-called frame-shift mutations (described in Chapter 18). In addition, chromosomes may break. The acridine dyes are examples of intercalating agents. Quinacrine is an acridine dye used to treat malaria and intestinal tapeworms.

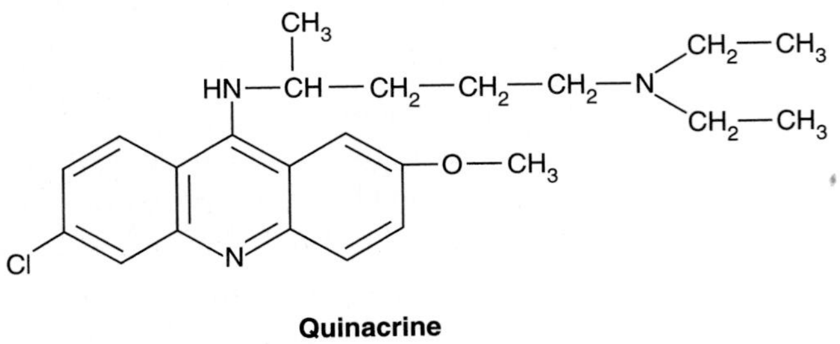

Quinacrine

16.1

The accumulation of oxidative DNA damage now appears to be a major cause of aging in mammals. Animals that have high metabolic rates (i.e., use large amounts of oxygen) or excrete large amounts of modified bases in the urine typically have shorter life spans. The excretion of relatively large amounts of oxidized bases indicates a reduced capacity to prevent oxidative damage. Despite substantial evidence that oxygen radicals damage DNA, the actual radicals that cause the damage are still not clear. In addition to the hydroxyl radical, suggest other possible culprits. Some tissues sustain more oxidative damage than others. For example, the human brain is believed to sustain more oxidative damage than most other tissues during an average life span. Suggest two reasons for this phenomenon.

DNA Structure: From Mendel's Garden to Watson and Crick

To modern eyes, the structure of DNA is both elegant and obvious. DNA is now a cultural icon, synonymous with the concept of information storage and retrieval. As mentioned, the correct structure of DNA was proposed in 1953 by James Watson and Francis Crick. The investigation that led to this remarkable discovery is instructive for several reasons. First, as often happens in scientific research, the road to the elucidation of DNA structure was long, frustrating, and tortuous. Living organisms are so complex that discerning any aspect of their function is extraordinarily difficult. Adding to this obstacle is the propensity of scientists (as well as other humans) to reject or ignore new information that does not fit comfortably with currently popular ideologies. This latter problem is probably unavoidable, since the scientific method requires a certain degree of skepticism. (How does one differentiate, for example, between breakthrough concepts and erroneous ideas?) However, skepticism can often be confused with an unimaginative adherence to the status quo. Albert Szent-Gyorgyi (Nobel Prize in Physiology or Medicine, 1937), who identified ascorbic acid as vitamin C and made significant contributions to the elucidation of muscle contraction and the citric acid cycle, once remarked, "Discovery consists in seeing what everyone else has seen and thinking what nobody else has thought."

A second, more concrete reason for the length of the discovery process is that the development of new concepts often requires the integration of information from several scientific disciplines. For example, the DNA model was based on discoveries in descriptive and experimental biology, genetics, organic chemistry, and physics. Significant scientific advancements are usually made by imaginative and industrious individuals who have the good fortune to work when sufficient information and technology are available for solving the scientific problems that interest them. The most talented of these investigators often help create new technologies.

The scientific revolution that eventually led to the DNA model began quietly in the abbey garden of an obscure Austrian monk named Gregor Mendel. Mendel discovered the basic rules of inheritance by cultivating pea plants. In 1865, Mendel published the results of his breeding experiments in the *Journal of the Brunn Natural History Society*. Although he sent copies of this publication to eminent biologists throughout Europe, his work was ignored until 1900. In that year, several botanists independently rediscovered Mendel's paper and recognized its significance. This long delay was due largely to the descriptive nature of nineteenth century biology; few biologists were familiar with the mathematics that Mendel used to analyze his data. By 1900, many biologists were trained in mathematics. In addition, this latter group of scientists had a frame of reference for Mendel's principles, since many of the details of meiosis, mitosis, and fertilization had become common knowledge.

Amazingly, the substance that constitutes the inheritable units that Mendel referred to in his work was being investigated almost simultaneously. The discovery of "nuclein," later renamed nucleic acid, was reported in 1869 by Friedrich Miescher, a Swiss pathologist. Working with the nuclei of pus cells, Miescher extracted nuclein and discovered that it was acidic and contained a large amount of phosphate. (Although Joseph Lister published his findings on antiseptic surgery in 1867, hospitals continued to be a rich source of pus for many years to come.) Interestingly, Miescher came to believe (erroneously) that nuclein was a phosphate storage compound.

The chemical composition of DNA was determined largely by Albrecht Kossel between 1882 and 1897 and P.A. Levene in the 1920s as suitable analytical techniques were developed. Unfortunately, Levene mistakenly believed that DNA was a small and relatively simple molecule. His concept, referred to as the *tetranuceotide hypothesis,* significantly retarded further investigations of DNA. Instead, proteins (the other major component of nuclei) were viewed as the probable carrier of genetic information. (By the end of the nineteenth century it was commonly accepted that the nucleus contains the genetic information.)

While geneticists focused on investigating the mechanisms of heredity and chemists elucidated the structures of nucleic acid components, microbiologists developed methods of studying bacterial cultures. In 1928, while investigating a deadly epidemic of pneumonia in Britain, Fred Griffith performed a remarkable series of experiments with two strains of pneumococcus. One bacterial strain, called the smooth form (or type S) because it is covered with a polysaccharide capsule, is pathogenic. The rough form (or type R) lacks the capsule and is nonpathogenic. Griffith observed that mice inoculated with a mixture of live R and heat-killed S bacteria died. He was amazed when live S bacteria were isolated from the dead mice. Although this transformation of R bacteria into S bacteria was confirmed in other laboratories, Griffith's discovery was greeted with considerable skepticism. (The concept of transmission of genetic information between bacterial cells was not accepted until the 1950s.) In 1944, Oswald Avery and his colleagues Colin MacLeod and Maclyn McCarty reported their painstaking isolation and identification of the transforming agent in Griffith's experiment as DNA. Not everyone accepted this conclusion because their DNA sample had a trace of protein impurities. Avery and McCarty later demonstrated that the digestion of DNA by deoxyribonuclease (DNase) inactivated the transforming agent. (Eventually, it was determined that the DNA that transforms R pneumococcus into the S form codes for an enzyme required to synthesize the gelatinous polysaccharide capsule. This capsule protects the bacteria from the animal's immune system and increases adherence and colonization of host tissues.)

Another experiment that confirmed DNA as the genetic material was reported by Alfred Hershey and Martha Chase in 1952. Using T2 bacteriophage, Hershey and Chase demonstrated the separate functions of viral nucleic acid and protein. (*Bacteriophage,* sometimes called *phage,* is a type of virus that infects bacteria.) When T2 phage infects an *Escherichia coli* cell, the bacterium is directed to synthesize several hundred new viruses. Within 30 minutes after infection the cell dies as it bursts open, thus releasing the viral progeny. In the first phase of their experiment, Hershey and Chase incubated bacteria infected with T2 phage in a culture medium containing ^{35}S (to label protein) and ^{32}P (to label DNA). In the second phase the radioactively labeled virus was harvested and allowed to infect nonlabeled bacteria. Immediately afterwards, the infected bacterial culture was subjected to shearing stress in a Waring blender. This treatment removed the phage from its attachment sites on the external surface of the bacterial cell wall. After separation from empty viral particles by centrifugation, the bacteria were analyzed for radioactivity. The cells were found to contain ^{32}P (thus confirming the role of DNA in "transforming" the bacteria into virus producers), whereas most of the ^{35}S remained in the supernatant. In addition, samples of the labeled infected bacteria produced some ^{32}P-labeled viral progeny.

By the early 1950s it had become clear that DNA was the genetic material. Because researchers also recognized that genetic information was critical for all living processes, determining the structure of DNA became an obvious priority. Linus Pauling (California Institute of Technology), and Maurice Wilkins and Rosalind Franklin (King's College, London), as well as Watson and Crick (Cambridge University), were all working toward this goal. The structure proposed by Watson and Crick in the April 25, 1953, issue of *Nature* was based on their scale model.

Considering how the scientific community responded to other concepts involving DNA as the genetic material, their acceptance of the Watson-Crick structure was unusually rapid. In 1962, Watson, Crick, and Wilkins were awarded the Nobel Prize in chemistry.

KEY *Concepts*

The model of DNA structure proposed by James Watson and Francis Crick in 1953 was based on information derived from the efforts of many individuals.

FIGURE 16.5

X-Ray Diffraction Study of DNA by Rosalind Franklin and R. Gosling.

Note the symmetry of the X-ray.

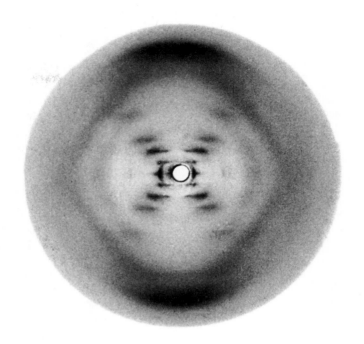

The information used to construct this model included the following:

1. The chemical structures and molecular dimensions of deoxyribose, the nitrogenous bases, and phosphate.
2. The 1:1 ratios of adenine:thymine and guanine:cytosine in the DNA isolated from a wide variety of species investigated by Erwin Chargaff (1948–1952). (This relationship is sometimes referred to as **Chargaff's rules.)**
3. Superb X-ray diffraction studies performed by Rosalind Franklin (Figure 16.5) indicating that DNA is a symmetrical molecule and probably a helix.
4. The diameter and pitch of the helix estimated by Wilkins and his colleague Alex Stokes from other X-ray diffraction studies.
5. The recent demonstration by Linus Pauling that protein, another complex class of molecule, could exist in a helical conformation.

DNA Structure: Variations on a Theme

The structure discovered by Watson and Crick, referred to as **B-DNA,** represents the sodium salt of DNA under highly humid conditions. DNA can assume different conformations because deoxyribose is flexible and the C^1-N-glycosidic linkage rotates. (Recall that furanose rings have a puckered conformation.

When DNA becomes partially dehydrated, it assumes the A form (Figure 16.6). In **A-DNA** the base pairs are no longer at right angles to the helical axis. Instead, they tilt 20° away from the horizontal. In addition, the distance between adjacent base pairs is slightly reduced, with 11 bp per helical turn instead of the 10.4 bp found in the B form. Each turn of the double helix occurs in 2.5 nm, instead of 3.4 nm, and the molecule's diameter swells to approximately 2.6 nm from the 2.4 nm observed in B-DNA. The A form of DNA is observed when it is extracted with solvents such as ethanol. The significance of A-DNA under cellular conditions is that the structure of RNA duplexes and RNA/DNA duplexes formed during transcription resemble the A-DNA structure.

The Z form of DNA (named for its "zigzag" conformation) radically departs from the B form. **Z-DNA** ($D = 1.8$ nm), which is considerably slimmer than B-DNA ($D = 2.4$ nm), is twisted into a left-handed spiral with 12 bp per turn. Each turn of Z-DNA occurs in 4.5 nm, compared with 3.4 nm for B-DNA. DNA segments with alternating purine and pyrimidine bases (especially CGCGCG) are most likely to adopt a Z configuration. In Z-DNA, the bases stack in a left-handed staggered dimeric

FIGURE 16.6

A-DNA, B-DNA, and Z-DNA.

Because DNA is a flexible molecule, it can assume different conformation forms depending on its base pair sequence and/or isolation conditions. Each molecular form in the figure possesses the same number of base pairs.

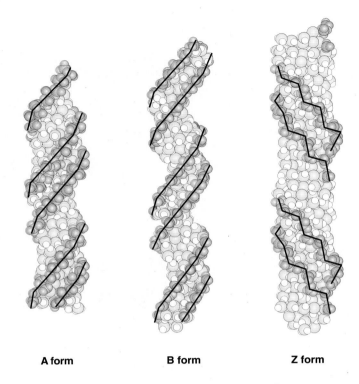

A form   B form   Z form

pattern, which gives the DNA a zigzag appearance and its flattened, nongrooved surface. Regions of DNA rich in GC repeats are often regulatory, binding specific proteins that initiate or block transcription. Certain physiologically relevant processes such as methylation and negative supercoiling (discussed on p. 470) stabilize the Z form.

Segments of DNA molecules have been observed to have several higher-order structures. Examples include cruciforms, triple helices, and supercoils. Each is briefly described below.

As their name implies, **cruciforms** (Figure 16.7) are crosslike structures. They are likely to form when a DNA sequence contains a palindrome. (A **palindrome** is defined as a sequence that provides the same information whether it is read forward or backward, e.g., "MADAM, I'M ADAM.") In contrast to language palindromes, the "letters" in a biological palindrome must form complementary pairs, that is, A aligns with T and C aligns with G. The DNA sequences that form palindromes, which may consist of several bases or thousands of bases, are called *inverted repeats*. In one proposed mechanism, cruciform formation begins with a small bubble, or *protocruciform*, and progresses as intrastrand base pairing occurs. The mechanism by which bubble formation is initiated is unknown. The function of cruciforms is unclear but is believed to be associated with the binding of various proteins to DNA. DNA palindromes also play a role in the function of an important class of enzymes called the restriction enzymes (Appendix B, Supplement 17).

In certain circumstances (e.g., low pH) a DNA sequence containing a long segment consisting of a polypurine strand hydrogen-bonded to a polypyrimidine strand can form a triple helix (Figure 16.8). The formation of the *triple helix*, also referred to as **H-DNA**, depends on the formation of nonconventional base pairing (**Hoogsteen base pairing**), which occurs without disrupting the Watson-Crick base pairs. The significance of H-DNA is not understood. However, H-DNA may have a role in genetic recombination (Section 17.1).

Packaging large DNA molecules to fit into cells requires DNA supercoiling (discussed on page 470). To undergo supercoiling, DNA molecules must be nicked and then either over-wound or under-wound before resealing. Small changes in DNA shape depend on

FIGURE 16.7

Cruciforms.

Cruciforms form because of palindrome sequences.

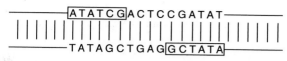

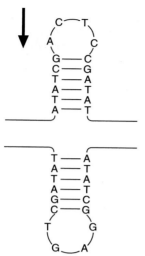

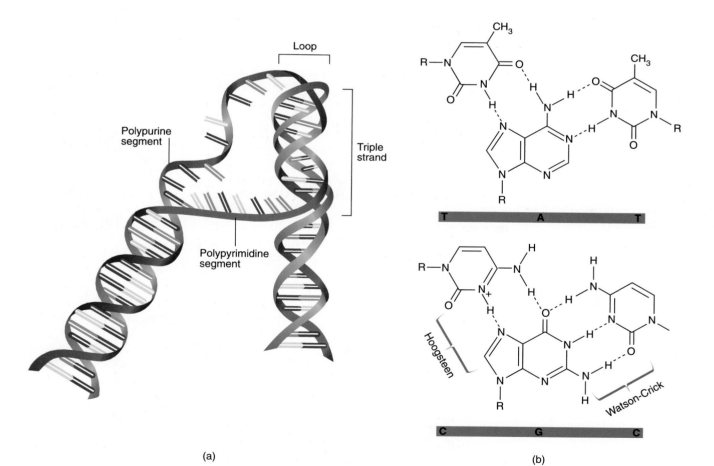

Loop

Polypurine segment

Polypyrimidine segment

Triple strand

(a)

(b)

FIGURE 16.8

H-DNA.

(a) DNA sequences with long segments such as (A-G)n bonded to (T-C)n can form H-DNA. (b) H-DNA formation depends on the formation of nonconventional (Hoogsteen) base pairing.

sequence. For example, four sequential AT pairs produce a bend in the molecule. Significant bending or wrapping around associated proteins, however, requires supercoiling.

DNA Supercoiling

DNA supercoiling, once considered an artifact of DNA extraction techniques, is now known to facilitate several biological processes. Examples include packaging DNA into a compact form, as well as replicating and transcribing DNA (Chapter 17). Because DNA supercoiling is a dynamic three-dimensional process the information that two-dimensional illustrations can convey is limited. To understand supercoiling, therefore, consider the following thought experiment. A long, linear DNA molecule is laid on a flat surface. After the ends are brought together, they are sealed to form an unpuckered circle (Figure 16.9a). Because this molecule is sealed without under- or overwinding, the helix is said to be relaxed and remains flat on a surface. The properties of such closed helical circles are described by several quantitative parameters. The number of times each strand crosses the other is called the **linking number** L. The number of helical turns is defined as twist T. For example, if there are 260 bp in a relaxed, circular DNA molecule, then $L = 25$ (i.e., 260/10.4) and $T = 25$. A DNA molecule's linking number can change only if a covalent bond in the backbone of one or both strands is broken and resealed. Twist can vary freely. If the relaxed circular DNA molecule is held and twisted a few times, it takes the shape shown in Figure 16.9b. When this twisted molecule is laid back on the flat surface, it rotates to eliminate the twist. However, consider what happens if this molecule is cut before it is twisted. (The enzymes that perform this function in living cells, called topoisomerases, are discussed in Chapter 17.) In addition to changing the linking number, this latter operation also changes another parameter, the writhing number W. (The **writhing,** or supercoiling, **number** is defined as the number of superhelical turns in a molecule's three-dimensional conformation. As with T values, the W number of a molecule can vary.) The relationship between these parameters is described by the following equation:

$$L = T + W$$

FIGURE 16.9

Linear and Circular DNA and DNA Winding.

(a) The linking number of this relaxed circular DNA molecule is 10. (b) When a relaxed DNA molecule is twisted, it reverts to its flat structure once it is released.

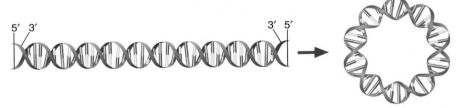

Linear double-stranded DNA molecule   Circular DNA molecule

(a)

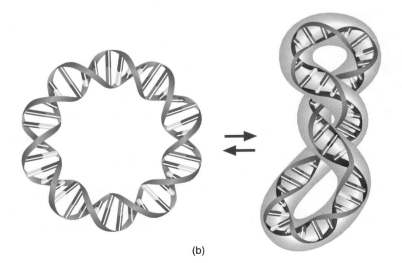

(b)

FIGURE 16.10

Supercoils.

Supercoils occur in two major forms: (a) toroidal and (b) interwound.

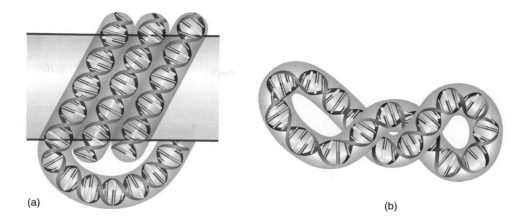

(a)

(b)

When L changes from its relaxed value, the strain introduced into the molecule is partitioned between T and W values. The dynamic changes in shape that occur during replication and transcription show constantly changing values for T and W. In the stable forms of packaged DNA, the T value is fixed to produce stable Watson and Crick base pairing and relieve all the strain by supercoiling.

When DNA is under-wound (T is greater than L), it twists to the left to relieve strain and *negative supercoiling results* (W is negative). When DNA is over-wound (L is greater than T), it twists to the left to relieve strain and *positive supercoiling* results (W is positive). When DNA is negatively supercoiled, it usually winds around itself to form an interwound supercoil (Figure 16.10). Positively supercoiled DNA is usually found where DNA coils around a protein core to form a toroidal supercoil.

When a negatively supercoiled DNA molecule is forced to lie in a plane, T decreases to equal L. The strain relieved by the formation of the negative supercoiling is reintroduced (Figure 16.11). During replication, topoisomerases nick the DNA to relieve torsional strain so that replication can proceed. DNA negative supercoiling also explains the propensity of certain DNA sequences to form cruciforms and H-DNA. Finally, the above discussion also applies to the linear DNA molecules found in the nuclei of eukaryotic cells. As is discussed in the next section, such molecules are constrained by their attachment to nuclear scaffolds, which are structural components of chromosomes.

Chromosomes and Chromatin

DNA, which contains the genes (the units of heredity), is packaged into structures called chromosomes. As it was originally defined, the term **chromosome** referred only to the dense, darkly staining structures visible within eukaryotic cells during meiosis or mitosis. However, this term is now also used to describe the DNA molecules that occur in prokaryotic cells. The physical structure and genetic organization of prokaryotic and eukaryotic chromosomes are significantly different.

In prokaryotes such as *E. coli,* a chromosome is a circular DNA molecule that is extensively looped and coiled (Figure 16.12). The *E. coli* chromosome is now believed to consist of a supercoiled DNA that is complexed with an RNA-protein core. In addition, the polyamines (polycationic molecules such as spermidine and spermine) may also assist in attaining the chromosome's highly compressed structure.

$$H_3\overset{+}{N}-CH_2-CH_2-CH_2-CH_2-\overset{+}{N}H_2-CH_2-CH_2-CH_2-\overset{+}{N}H_3$$
$$\textbf{Spermidine}$$

$$H_3\overset{+}{N}-CH_2-CH_2-CH_2-\overset{+}{N}H_2-CH_2-CH_2-CH_2-CH_2-\overset{+}{N}H_2-CH_2-CH_2-CH_2-\overset{+}{N}H_3$$
$$\textbf{Spermine}$$

(When the positively charged polyamines bind to the negatively charged DNA backbone they overcome the charge repulsion between adjacent DNA coils.) Investigations of the *E. coli* chromosome, which contains over three million bp (enough base pairs to account for about 3000 average size genes of 1000 bp each), have revealed

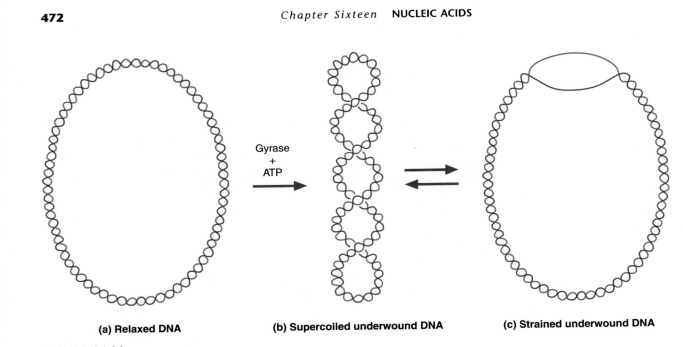

(a) Relaxed DNA   **(b) Supercoiled underwound DNA**   **(c) Strained underwound DNA**

Gyrase
+
ATP

FIGURE 16.11

The Effect of Strain on a Circular DNA Molecule.

Breakage and reformation of phosphodiester linkages allow the conversion of a relaxed circular form (a) to the negatively supercoiled form (b). The strain in the negatively supercoiled molecule can be relieved by some degree of strand separation (c).

FIGURE 16.12

The *E. Coli* Chromosome.

The circular *E. coli* chromosome is complexed with an RNA-protein core. Because the chromosome (3×10^6 bp) is highly supercoiled, the entire chromosome complex measures only 2μm across.

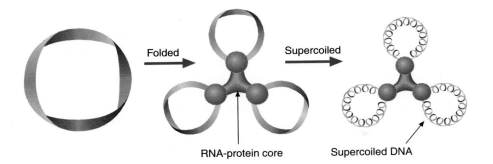

Folded   Supercoiled

RNA-protein core   Supercoiled DNA

that it is efficiently organized with little evidence of any nonfunctional base sequences.

In comparison to prokaryotes, the eukaryotes possess genomes which are extraordinarily complex. Depending on species, the chromosomes of eukaryotes vary in both length and number. For example, humans possess 23 pairs of chromosomes with a total of approximately three billion bp. The fruit fly *Drosophila melanogaster* has four chromosome pairs with 180 million bp, while corn (*Zea mays*) has ten chromosome pairs with a total of 6.6 billion bp. In addition, each eukaryotic chromosome possesses two unique structural elements: centromeres and telomeres. A *centromere* is a specific AT-rich DNA sequence associated with nonhistone proteins to form the kinetochore, which interacts with the spindle fibers during cell division. *Telomeres* are CCCA repetitive regions of DNA at the ends of the chromosomes that prevent loss of coding sequences during DNA replication.

Each eukaryotic chromosome consists of a single linear DNA molecule complexed with histones to form **nucleohistone.** (Small amounts of nonhistone proteins as well as RNA and polyamines may also affect DNA packaging.) The histones are a group of small basic proteins found in all eukaryotes. Consisting of five major classes (H1, H2A, H2B, H3, and H4), the histones are amazingly similar in their primary structure among eukaryotic species. Histones from different species as well as different phases of the cell cycle do differ, however, in the degree to which they undergo various chemical modifications (e.g., phosphorylation, acetylation, methylation, and

ADP-ribosylation). The significance of these modifications is currently under investigation. The binding of histones to DNA results in the formation of **nucleosomes,** which are the structural units of eukaryotic chromosomes.

Nucleosomes and Chromatin

When eukaryotic cells are not undergoing cell division, the chromosomes partially decondense to form **chromatin.** In electron micrographs, chromatin appears to have a beaded appearance. Each of these "beads" is a nucleosome, which is composed of a positively supercoiled segment of DNA forming a toroidal coil around an octameric histone core (two copies each of H2A, H2B, H3 and H4. See Figure 16.13b). Approximately 140 bp are in contact with each histone octamer. An additional 60 bp of spacer (or linker) DNA connect adjacent nucleosomes (Figure 16.13a). One molecule of histone H1 also occurs in each nucleosome, although the exact location is still unclear (Figure 16.13c). H1 is believed to facilitate coiling of the beaded fiber into higher-order structures.

As chromatin is compacted to form chromosomes, the nucleosomes are coiled into a higher order of structure referred to as the *30-nm fiber* (Figure 16.14). The 30-nm fiber is further coiled to form *200-nm filaments.* The three-dimensional structure of the 200-nm filaments is unclear but is believed to contain numerous supercoiled loops attached to a central protein complex referred to as a nuclear scaffold (Figure 16.15). Although the entire structural organization of eukaryotic chromosomes has not yet been elucidated, it is likely to consist of multiple levels of supercoiling.

The organization of genetic information in eukaryotic chromosomes has proven to be substantially more complex than in prokaryotes. The two types of genome differ in several respects:

1. **Genome size.** Eukaryotic genomes are substantially larger than those of prokaryotes. In contrast to the estimate of 3000 genes in *E. coli,* for example, estimates of human genes range between 50,000 and 100,000.

FIGURE 16.13

Chromatin and Nucleosome.

(a) Nucleosomes are connected by linker DNA. (b) Each nucleosome core is composed of a histone octomer, around which is wrapped one and three-quarters turn of DNA. (c) A proposed structure of a nucleosome. The H1 histone aids in stabilizing the wrapping of DNA around the histone octomer. *Source:* (b) and (c) From Devlin, *Textbook of Biochemistry with Clinical Correlations,* 1992. Copyright © 1992, Wiley-Liss. Reprinted by permission of Wiley-Liss, Inc., a subsidiary of John Wiley & Sons, Inc.

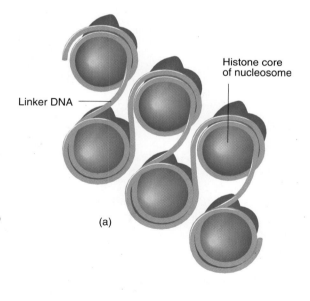

Linker DNA

Histone core of nucleosome

(a)

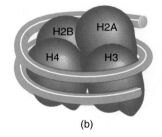

(b)

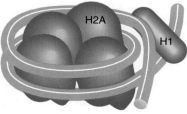

(c)

FIGURE 16.14

Chromatin.

Nuclear chromatin contains many levels of coiled structure.

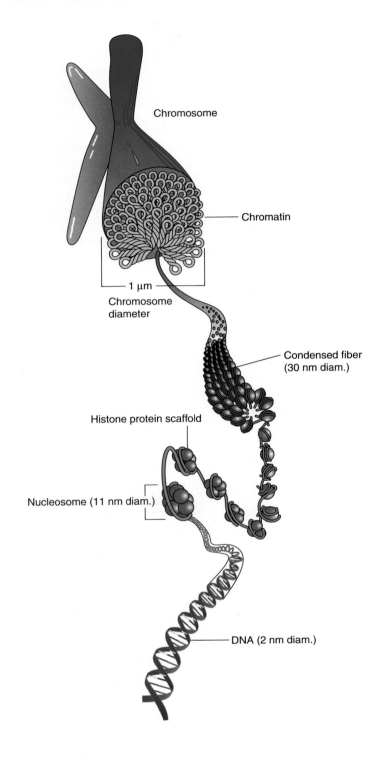

Chromosome

Chromatin

1 μm
Chromosome diameter

Condensed fiber (30 nm diam.)

Histone protein scaffold

Nucleosome (11 nm diam.)

DNA (2 nm diam.)

2. **Coding functions.** The majority of DNA sequences in eukaryotes do not appear to have coding functions (i.e., they do not specify the primary structure of proteins). Although the functions of these noncoding sequences are unknown, some of them probably have regulatory or structural roles. Especially noteworthy are the so-called repetitive sequences. These segments consist of short or moderately long sequences that are repeated hundreds or thousands of times, either in tandem or interspersed throughout the genome. For example, one group of short interspersed repeats containing 300 bp, called the Alu family, occurs approximately 500,000 times in the haploid human genome.

FIGURE 16.15

Chromatin.

In one proposal for the structure of 200-nm filaments, the 30-nm fiber is looped and attached to a nuclear scaffold composed of protein.

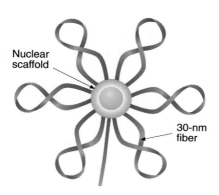

Nuclear scaffold

30-nm fiber

3. **Coding continuity.** Most eukaryotic genes investigated so far are discontinuous. Noncoding sequencs (called **introns** or intervening sequences) are interspersed between sequences called **exons** (expressed sequences), which code for a gene product (either a polypeptide or a type of RNA).

Organelle DNA

Mitochondria and chloroplasts are semiautonomous organelles, that is, they possess DNA as well as their own version of protein-synthesizing machinery. These organelles, both of which reproduce by binary fission, also require a substantial contribution of proteins and other molecules that are coded for by the nuclear genome. For example, mitochondrial DNA (mtDNA) codes for several types of RNA and certain inner membrane proteins. The remainder of mitochondrial proteins are synthesized in the cytoplasm and transported into the mitochondria. Similarly, the chloroplast genome codes for several types of RNA as well as certain proteins, many of which are directly associated with photosynthesis. The activities of nuclear and organelle genomes are highly coordinated. Consequently, their individual contributions to organelle function are often difficult to discern. Because mitochondria and chloroplasts are now believed to be the descendants of free-living bacteria, it is not surprising that they are susceptible to the actions of antibiotics (e.g., chloramphenicol and erythromycin) if their concentrations are sufficiently high. Many of these molecules are (or have been) used in clinical practice because they inhibit some aspect of bacterial genome function.

16.2 RNA

Ribonucleic acid is a class of polynucleotides, nearly all of which are involved in some aspect of protein synthesis. RNA molecules are synthesized in a process referred to as **transcription.** During transcription, new RNA molecules are produced by a mechanism similar to DNA synthesis, that is, through complementary base pair formation. The sequence of bases in RNA is therefore specified by the base sequence in one of the two strands in DNA. For example, the DNA sequence 5′-CCGATTACG-3′ is transcribed into the RNA sequence 3′-GGCUAAUGC-5′. (Complementary DNA and RNA sequences are antiparallel.)

RNA molecules differ from DNA in the following ways:

1. The sugar moiety of RNA is ribose instead of deoxyribose in DNA.
2. The nitrogenous bases in RNA differ somewhat from those observed in DNA. Instead of thymine, RNA molecules use uracil. In addition, the bases in some RNA molecules are modified by a variety of enzymes (e.g., methylases, thiolases, and deaminases).
3. In contrast to the double helix of DNA, RNA exists as a single strand. For this reason, RNA can coil back on itself and form unique and often quite complex three-dimensional structures (Figure 16.16). The shape of these structures is determined by complementary base pairing by specific RNA sequences, as well as by base stacking. In addition, the 2′-OH of ribose can form hydrogen bonds with nearby molecular groups. Because RNA is single stranded, Chargaff's rules do not apply. An RNA molecule's contents of A and U, as well as C and G, are usually not equal.

FIGURE 16.16

Secondary Structure of RNA.

(a) Many different types of secondary structures occur in RNA molecules. (b) A hairpin structure.

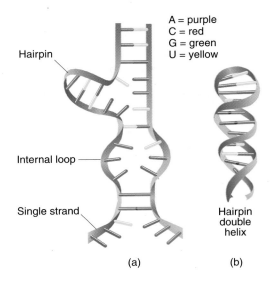

Hairpin

Internal loop

Single strand

Hairpin double helix

(a)   (b)

The most prominent types of RNA are transfer RNA, ribosomal RNA, and messenger RNA. The structure and function of each of these molecules is discussed below. Examples of other less abundant (but no less important) types of RNA (heterogeneous RNA and small nuclear RNA) are also described.

Transfer RNA

Transfer RNA (tRNA) molecules transport amino acids to ribosomes for assembly into proteins. Comprising about 15% of cellular RNA the average length of a tRNA molecule is 75 nucleotides. Because each tRNA molecule binds to a specific amino acid, cells possess at least one type of tRNA for each of the 20 amino acids commonly found in protein. The three-dimensional structure of tRNA molecules, which resembles that of a warped cloverleaf (Figure 16.17), results primarily from extensive intrachain base pairing. tRNA molecules contain a variety of modified bases. Examples include pseudouridine, 4-thiouridine, 1-methylguanosine, and dihydrouridine.

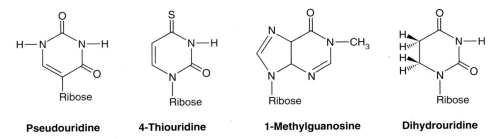

Pseudouridine   **4-Thiouridine**   **1-Methylguanosine**   **Dihydrouridine**

The structure of tRNA allows it to perform two critical functions involving the most obviously important structural components: the 3'-terminus and the anticodon loop. The *3'-terminus* forms a covalent bond to a specific amino acid. The *anticodon loop* contains a three-base-pair sequence that allows the tRNA to align its attached amino acid properly during protein synthesis. (This process is discussed in Chapter 18.) tRNAs also possess three other prominent structural features, referred to as the D loop, the TΨC loop, and the variable loop. (Ψ is an abbreviation for the modified base pseudouridine.) The function of these structures is unknown, but they are presumably related to the alignment of tRNA within the ribosome and/or the binding of a tRNA to the enzyme that catalyzes the attachment of the appropriate amino acid. The *D loop* is so named because it contains dihydrouridine. Similarly, the TΨC loop contains the base sequence thymine, pseudouridine, and cytosine. tRNAs can be classified on the basis of the length of their *variable loop*. The majority (approximately 80%) of tRNAs have variable loops with four to five nucleotides, while the others have variable loops with as many as 20 nucleotides.

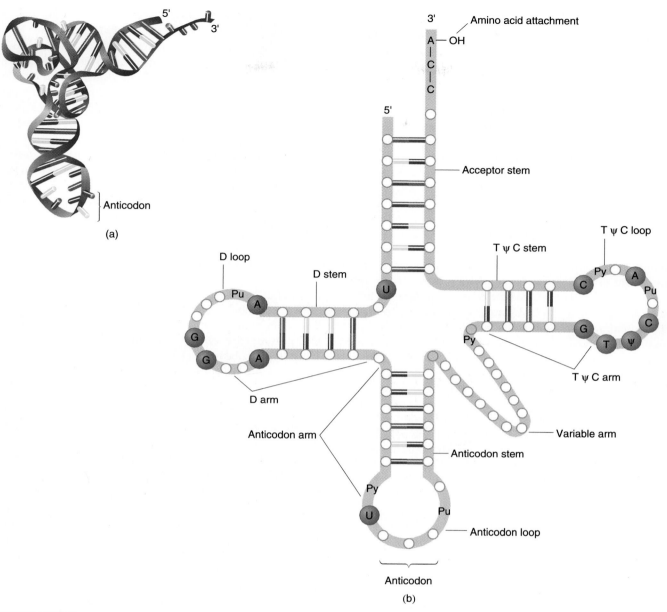

FIGURE 16.17

Transfer RNA.

(a) Three-dimensional structure of a tRNA molecule. (b) A tRNA molecule. The positions of invariant bases and bases that seldom vary are indicated.

Ribosomal RNA

Ribosomal RNA (rRNA) is the most abundant form of RNA in living cells. (In most cells, rRNA constitutes approximately 80% of the total RNA.) The secondary structure of rRNA is extraordinarily complex (Figure 16.18). Although there are species differences in the primary nucleotide sequences of rRNA, the overall three-dimensional structure of this class of molecules is conserved. As its name suggests, rRNA is a component of ribosomes.

As described, ribosomes are cytoplasmic structures that synthesize proteins. (Because they are composed of both protein and rRNA, the ribosomes are sometimes described as ribonucleoprotein bodies.) The ribosomes of prokaryotes and eukaryotes are similar in shape and function, although they differ in size and their chemical composition. Both types of ribosome consist of two subunits of unequal size, which are usually referred to in terms of their S values. (S is an abbreviation for the Svedberg (or sedimentation) unit, which is a measure of sedimentation velocity in a centrifuge.

ANCIENT DNA . . .

Evolutionary biology is essentially a historical science. Ever since the publication of *On the Origin of Species* by Charles Darwin in 1859, biologists have attempted to reconstruct the events and processes that gave rise to modern organisms by investigating fossils and the comparative anatomy of modern species. *Fossils,* the preserved part of ancient organisms, have been used to trace the lineage of modern organisms. For example, fossilized skeletal remains have allowed paleontologists to trace human lineage back about three million years. Most fossils were formed when recently deceased organisms were exposed to environmental conditions that slowed the process of decomposition. They were covered with sediment (fine soil particles suspended in water) or became embedded in bogs, tar pits, amber (a polymerized resin derived from plant essential oils), or ice. Arid desert conditions also promoted fossil formation.

Comparative anatomical studies have provided a wealth of information concerning the relationships of modern species. Consider, for example, that the similar structures of the forelimbs of most vertebrates suggest a common history for these species. Species with greater structural similarities (e.g., humans and chimpanzees) are more closely related than those with obvious dissimilarities (e.g., whales and birds). As nucleic acid and protein sequencing techniques have become available, however, the structures of these molecules have provided more precise information concerning the relationships of modern species. For example, by using DNA and protein sequence studies, molecular evolutionary rates have been calculated by detecting changes in the DNA base sequences or the polypeptide amino acid sequences of different species. This information in conjunction with fossil evidence, has been used to make estimates, referred to as an *evolutionary clock,* of the time required for evolutionary changes. In addition, DNA sequence information has provided an extraordinarily promising mechanism for comparing the genetic instructions of all existing species. Unfortunately, there are severe limitations on the conclusions that molecular paleontologists can infer from the study of modern DNA sequences, since they cannot check these sequences against an historical record. Or can they?

Although well-preserved nucleated cells have been observed in specimens dating as far back as 1912, DNA recovery became feasible only in the 1980s. The first successful extractions of ancient DNA (aDNA) took advantage of **cloning,** a recombinant DNA technique (described in Appendix B, Supplement 17), in which bacteria are used to generate a large number of copies of specific DNA sequences. These sequences are then investigated (via DNA sequencing and hybridization techniques) in terms of their relationship to comparable sequences of modern species. For example, in 1984, DNA was successfully cloned from a preserved quagga (an extinct animal that resembled both horses and zebras). Later investigations of quagga DNA confirmed its close similarity to both horse and zebra DNA.

Although DNA cloning was an important breakthrough, its use is awkward when applied to fossilized specimens. The principal reason is that cloning requires larger amounts of DNA than are often present in fossils. (This is an important consideration because interesting genes, such as those that code for protein, generally appear in only two copies per cell.) Investigations of aDNA seemed as elusive as ever in the early 1980s, despite their obvious advantages. However, this situation changed dramatically in 1985, when a new technique, called **PCR (polymerase chain reaction),** became available. By using PCR (described in Appendix B, Supplement 17) as many as one billion copies of DNA sequences can be produced in a test tube. Because PCR is extraordinarily sensitive (a single DNA molecule can be amplified), it seemed made to order for studies of aDNA.

Since PCR has been applied to aDNA investigations, the molecular paleontologists have extracted DNA fragments from a wide variety of fossils, artifacts, and museum specimens. For example, DNA sequences have been isolated from such disparate sources as amber-embedded insects (over 100 million years old), fossil herbarium specimens (millions of years old), and Egyptian mummies (over 6000 years old). Comparisons of these and other DNA sequences to those of modern species have provided important information concerning how populations change over time and how much time has elapsed since species shared a common ancestor (i.e., the evolutionary clock).

Although aDNA studies can provide invaluable information, they also have serious limitations.

Because sedimentation velocity depends on the molecular weight as well as the shape of a particle, S values are not necessarily additive.) Prokaryotic ribosomes (70 S) are composed of a 50 S subunit and a 30 S subunit, whereas ribosomes of eukaryotes (80 S) contain a 60 S subunit and a 40 S subunit.

Several different kinds of rRNA and protein are found in each type of ribosomal subunit. The large ribosomal subunit of *E. coli,* for example, contains 5 S and 23 S rRNAs and 34 polypeptides. The small ribosomal subunit of *E. coli* contains a 16 S rRNA and 21 polypeptides. A typical large eukaryotic ribosomal subunit contains three rRNAs (5 S, 5.8 S, and 28 S) and 49 polypeptides; the small subunit contains an 18 S rRNA and approximately 30 polypeptides. The functions of the rRNA and polypeptides in ribosomes are poorly understood and are being investigated.

1. **Microbial contamination.** Because aDNA is usually obtained from specimens that have been taken from nonsterile environments (e.g., soil) or handled by previous investigators (e.g., museum curators), their analysis must be handled carefully. For example, investigators should handle specimens with sterile gloves and decontaminate their surfaces with exposure to high-intensity UV light. When analyzing aDNA, the presence of contaminating DNA sequences must be ruled out.

2. **PCR inhibitors.** Substances such as heme and the cytochromes (or their decomposition products), which inhibit the enzyme used in PCR, often survive a DNA purification protocol. Such inhibition can sometimes be prevented by adding bovine serum albumin (which binds to a variety of inhibitors) or by diluting the sample.

3. **PCR contaminants.** Because PCR is such a sensitive method, the presence of even a single DNA contaminant can compromise an experiment. For this reason, specimen extraction should be carried out in stringently clean labs that are distant from areas where PCR is carried out (to prevent cross-contamination.)

4. **aDNA condition.** Because of long exposure to various environmental conditions, aDNA is often fragmented and partially decomposed. Studying it is therefore similar to investigating partially degraded fossils, and paleontologists seldom find complete fossilized organisms.

. . . AND NOT-SO-ANCIENT DNA

The identification techniques used in aDNA investigations are similar to those used with DNA samples of more recent vintage. Such methods are especially valuable in diagnostic clinical medicine and forensic investigations.

In diagnostic clinical medicine, DNA identification has been applied to investigations of patient tissue specimens obtained during biopsies or autopsies. Because of imaginative adaptations of DNA extraction techniques, the DNA contained in slides and preserved tissue specimens can now be investigated for evidence of infectious or genetic diseases. For example, *in situ* DNA hybridization is an ultrasensitive technique in which specific DNA probes are applied directly to tissue embedded in paraffin. (Paraffin-embedded tissue specimens are used in microscopic studies. Such slides can be stored indefinitely.) This technique recently confirmed the presence of the bacterium *Helicobacter pylori* in preserved specimens taken from the stomachs of ulcer patients. (*Helicobacter pylori* is now believed to be a causative agent in most cases of gastric ulcer, as well as in stomach cancer.) *In situ* hybridization, as well as PCR combined with Southern blotting, has detected viral sequences (e.g., HIV) in preserved specimens. (These and other biochemical tests now indicate that some patients died of HIV in the United States at least as early as the 1950s, although AIDS was not recognized until 1981.)

DNA techniques are also used in forensic investigations of both recent and old crimes. Because DNA persists for many years in dried biological specimens (e.g., blood and semen) as well as in bone, these materials can be used as evidence. DNA fingerprinting now often provides decisive information in court cases. **DNA fingerprinting,** a variation of Southern blotting, compares the banding patterns of DNA from different individuals, for example, crime scene specimen DNA with that of suspects. The entire genome in each sample is isolated and treated with a restriction enzyme. Because of genetic variations, the DNA from each individual fragments differently. (Such genetic differences are called restriction fragment length polymorphisms, or RFLPs.) After the restriction fragments are separated according to size on agarose gel electrophoresis and transferred to nitrocellulose filter paper, they are exposed to radiolabeled probes. Because the sizes of the fragments that bind to these probes differ from one individual to the next, banding patterns have been used successfully to either convict or acquit crime suspects. Because related individuals have similar RFLP patterns, DNA fingerprinting has also been used to identify crime victims. For example, comparisons of the DNA extracted from the bones of extensively decomposed crime victims with that of putative relatives has resulted in positive identifications.

Messenger RNA

As its name suggests, **messenger RNA** (mRNA) is the carrier of genetic information from DNA for the synthesis of protein. mRNA molecules, which typically constitute approximately 5% of cellular RNA, vary considerably in size. For example, mRNA from *E. coli* varies from 500 to 6000 nucleotides.

Prokaryotic mRNA and eukaryotic mRNA differ in several respects. First, many prokaryotic mRNAs are *polycistronic*, that is, they contain coding information for several polypeptide chains. In contrast, eukaryotic mRNA typically codes for a single polypeptide and is therefore referred to as *monocistronic*. (A **cistron** is a DNA sequence that contains the coding information for a polypeptide as well as several signals that are required for ribosome function.) Second, prokaryotic and eukaryotic mRNAs are processed differently. In contrast to prokaryotic mRNAs, which are translated into protein by

FIGURE 16.18

rRNA Structure.

Although their sequences differ, the three-dimensional structures of these 16-S-like rRNAs from (a) *E. coli* and (b) *Saccharomyces cervisiae* (yeast) appear remarkably similar.

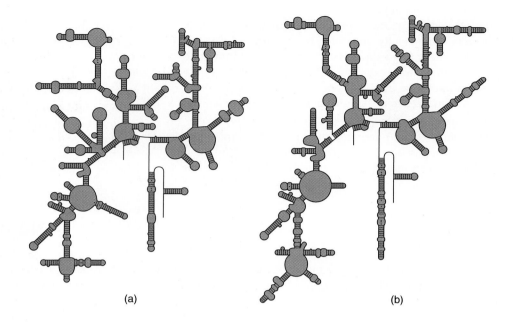

(a)   (b)

ribosomes immediately after they are synthesized, eukaryotic mRNAs are modified extensively. These modifications include capping (linkage of 7-methylguanosine to the 5′-terminal residue), splicing (removal of introns), and the attachment of an adenylate polymer referred to as a poly A tail. (Each of these processes is described in Chapter 17.)

Heterogeneous RNA and Small Nuclear RNA

Heterogeneous RNA (hnRNA) and small nuclear RNA (snRNA) play complementary roles in eukaryotic cells. **Heterogeneous nuclear RNA** molecules are the primary transcripts of DNA and are the precursors of mRNA. HnRNA is processed by splicing and modifications to form mRNA. **Splicing** is the enzymatic removal of the introns from the primary transcripts. A class of **small nuclear RNA** molecules (containing between 90 and 300 nucleotides), which are complexed with several proteins to form **small nuclear ribonucleoprotein particles** (snRNP or snurps), are involved in splicing activities as well as other forms of RNA processing.

QUESTION 16.2

When a gene is transcribed, only one DNA strand acts as a template for the synthesis of the RNA molecule. This strand is referred to as a **sense** strand; the nontranscribed strand is called the **antisense** strand. (The strand of a DNA segment that acts as the sense strand differs from one gene to the next.) Investigations of this aspect of nucleic acid metabolism have revealed that bacteria and viruses use the synthesis of so-called antisense RNAs to control certain aspects of cell metabolism. When an antisense RNA is produced (by transcription from the antisense DNA strand), it binds specifically (through complementary base pairing) to the corresponding mRNA. This binding prevents polypeptide synthesis from the mRNA.

Because mRNA–antisense RNA binding is so specific, antisense molecules are considered promising research tools. Numerous investigators are using antisense RNA molecules to study eukaryotic function by selectively turning on and off the activities of specific genes. This so-called "reverse genetics" is also useful in medical research. Although serious problems have been encountered in antisense research (e.g., the inefficient insertion of oligonucleotides into living cells and high manufacturing costs), antisense technology has already provided valuable insight into the mechanisms of several diseases (e.g., cancer and viral infections).

Consider the following sense DNA sequence:

5′-GCATTCGAATTGCAGACTCCTGCAATTCGGCAAT-3′

Determine the sequence of its complementary strand. Then determine the mRNA and antisense RNA sequences. (Recall that in RNA structure, U is substituted for T. So A in a DNA strand is paired with a U as RNA is synthesized.)

16.3 VIRUSES

Viruses lack most of the properties that distinguish life from nonlife. For example, viruses cannot carry on metabolic activities on their own. Yet under the appropriate conditions they can wreak havoc on living organisms. Often described as obligate, intracellular parasites, viruses can also be viewed as mobile genetic elements because of their structure, that is, each consists of a piece of nucleic acid enclosed within a protective coat. Once a virus has infected a host cell, its nucleic acid can hijack the cell's nucleic acid and protein-synthesizing machinery. As viral components accumulate, complete new viral particles are produced and then released from the host cell. In many circumstances, so many new viruses are produced that the host cell lyses (ruptures). Alternatively, the viral nucleic acid may insert itself into a host chromosome, resulting in transformation of the cell. (See Special Interest Box 16.2.)

Viruses have fascinated biochemists ever since their existence was suspected near the end of the nineteenth century. Driven in large part because of the role of viruses in numerous diseases, viral research has benefited biochemistry enormously. Because a virus subverts normal cell function to produce new virus, a viral infection can provide unique insight into cellular metabolism. For example, the infection of animal cells has provided invaluable and relatively unambiguous information about the mechanisms that glycosylate newly synthesized proteins. In addition, several eukaryotic genetic mechanisms have been elucidated with the aid of viruses and/or viral enzymes. Viral research has also provided substantial information concerning genome structure and carcinogenesis (the mechanisms by which normal cells are transformed into cancerous cells). Finally, viruses have been invaluable in the development of recombinant DNA technology.

The Structure and Function of Viruses

An enormous number of viruses have been identified since 1892, when the Russian researcher Dmitri Ivanovski first isolated the tobacco mosaic virus. Because their origins and evolutionary history is unclear, the scientific classification of viruses has been difficult. Often, viruses have been assigned to groups according to such properties as their microscopic appearance (e.g., rhabnoviruses have a bullet-shaped appearance), the anatomic structures where they were first isolated (e.g., adenoviruses were discovered in the adenoids, a type of lymphoid tissue), or the symptoms they produce in a host organism (e.g., the herpes viruses cause rashes that spread). In recent years, scientists have attempted to develop a systematic classification system based primarily on viral structure, although several other factors are also important (e.g., host and disease caused).

Viruses occur in a bewildering array of sizes and shapes. Virions (complete viral particles) range from 10 nm to approximately 400 nm in diameter. Although most viruses are too small to be seen with the light microscope, a few (e.g., the pox viruses) can be visualized because they are as large as the smallest bacteria.

Simple virions are composed of a *capsid* (a protein coat made of interlocking protein molecules called capsomeres), which encloses nucleic acid. (The term *nucleocapsid* is often used to describe the complex formed by the capsid and the nucleic acid.) Most capsids are either helical or icosahedral. (Icosahedral capsids are 20-sided structures composed of triangular capsomeres.) The nucleic acid component of virions is either DNA or RNA. Although most viruses possess double-stranded DNA (dsDNA) or single-stranded RNA (ssRNA), examples with single-stranded DNA (ssDNA) and double-stranded RNA (dsRNA) genomes have also been observed. There are two types of ssRNA genomes. A *positive-sense* RNA genome [(+)-ssRNA] acts as a giant mRNA, that is, it directs the synthesis of a long polypeptide that is cleaved and processed into smaller molecules. A *negative-sense* RNA genome [(−)-ssRNA] is complementary in base sequence to the mRNA that directs the synthesis of viral proteins. Viruses that employ (−)-ssRNA genomes must provide an enzyme, referred to as a transcriptase, that synthesizes the mRNA.

In more complex viruses, the nucleocapsid is surrounded by a membrane envelope, which usually arises from the host cell nuclear or plasma membranes. Envelope proteins, coded for by the viral genome, are inserted into the envelope membrane

VIRAL "LIFESTYLES"

Despite the diversity in the structures of viruses and the types of host cell that are infected, there are several basic steps in the life cycle of all viruses: infection (penetration of the virion or its nucleic acid into the host cell), replication (expression of the viral genome), maturation (assembly of viral components into virions), and release (the emission of new virions from the host cell). Because viruses usually possess only enough genetic information to specify the synthesis of their own components, each type must exploit some of the normal metabolic reactions of its host cell to complete the life cycle. For this reason there are numerous variations on these basic steps. This point can be illustrated by comparing the life cycles of two well-researched viruses: the T4 bacteriophage and the human immunodeficiency virus (HIV).

Bacteriophage T4

The T4 bacteriophage (Figure 16A) is a large virus with an icosahedral head and a long, complex tail similar in structure to T2 (p. 466). The head contains dsDNA, and the tail attaches to the host cell and injects the viral DNA into the host cell.

The life cycle of T4 (Figure 16B) begins with adsorbing the virion to the surface of an *E. coli* cell. Because the bacterial cell wall is rigid, the entire virion cannot penetrate into the cell's interior. Instead, the DNA is injected by flexing and constricting the tail apparatus. Once the DNA has entered the cell, the infective process is complete, and the next phase (replication) begins.

Within two minutes after the injection of T4 phage DNA into an *E. coli* cell, synthesis of host DNA, RNA, and protein stops and phage mRNA synthesis begins. Phage mRNA codes for the synthesis of capsid proteins and some of the enzymes required for the replication of the viral genome and the assembly of virion components. In addition, other enzymes are synthesized that weaken the host cell's cell wall, so that new phage can be released for new rounds of infection. Approximately 22 minutes after viral DNA (vDNA) is injected, the host cell, now filled with several hundred new virions, lyses. Upon release, the virions infect nearby bacteria, thus initiating new infections.

FIGURE 16A

The T4 Bacteriophage.

The DNA genome of the T4 bacteriophage induces the host cell to synthesize about 30 proteins.

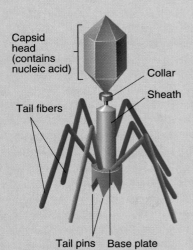

Capsid head (contains nucleic acid)

Collar

Sheath

Tail fibers

Tail pins   Base plate

Bacteriophage that initiate this so-called **lytic cycle** are referred to as *virulent* because they destroy their host cells. Many phage, however, do not initially kill their hosts. So-called *temperate* or *lysogenic* phage integrate their genome into that of the host cell. (The term **lysogeny** describes a condition in which the phage genome is integrated into a host chromosome.) The integrated viral genome (called the *prophage*) is copied along with host DNA during cell division for an indefinite period of time. Occasionally, lysogenic phage can enter a *lytic* phase. Certain external conditions, such as UV or ionizing radiation, activate the prophage, which directs the synthesis of new virions. Sometimes, a lysing bacterial cell releases a few virions that contain some bacterial DNA instead of phage genetic material. When such a virion infects a new host cell, this DNA is introduced into the host genome. This process is referred to as **transduction.**

HIV

The human immunodeficiency virus (HIV) is the causative agent of acquired immune deficiency syndrome (AIDS). AIDS is a lethal condition because HIV destroys the body's immune system, rendering it defenseless against disease-causing organisms (e.g., bacteria, protozoa, and fungi, as well as other viruses) in addition to some forms of cancer.

HIV is an enveloped virus (Figure 16C) that belongs to a unique group of RNA viruses called the retroviruses. (**Retroviruses** are so named because they contain an enzymatic activity referred to as reverse transcriptase, which synthesizes a DNA copy of a ssRNA genome.) HIV contains a cylindrical core within its capsid. In addition to two copies of its (+)-ssRNA genome, the core contains several enzymes: reverse transcriptase, ribonuclease, integrase, and protease. The RNA molecules are coated with multiple copies of two low-molecular-weight proteins, p7 and p9. (The numbers in these and other protein names indicate their kilodalton mass; for example, p7 is a protein with a 7-kD mass.) The bullet-shaped core itself is composed of hundreds of copies of p24, while copies of another protein form an inner lining of the viral envelope. The envelope of HIV contains two major viral proteins, gp120 and gp41, in addition to host proteins.

HIV infection occurs because of direct exposure of an individual's bloodstream to the body fluids of an infected person. Most HIV is transmitted through sexual contact, blood transfusions, and perinatal transmission from mother to child. Once HIV enters the body, it is believed to infect cells that bear the CD4 antigen on their plasma membranes. The principal group of cells that are attacked by HIV are the T-4 helper lymphocytes of the immune system. T-4 helper cells play a critical role in regulating the activities of other immune system cells. It is now known that T-cell infection requires the interaction of the gp120-CD4 complex with a chemokine receptor. (Chemokines are immune system chemotactic agents. They stimulate T cells by binding to receptors on the T cell plasma membrane.) In the early stages of infection, the CCR5 receptor helps HIV enter T cells. Later the CXCR4 receptor is used. (Recent evidence suggests that humans with two copies of a defective CCR5 gene, a

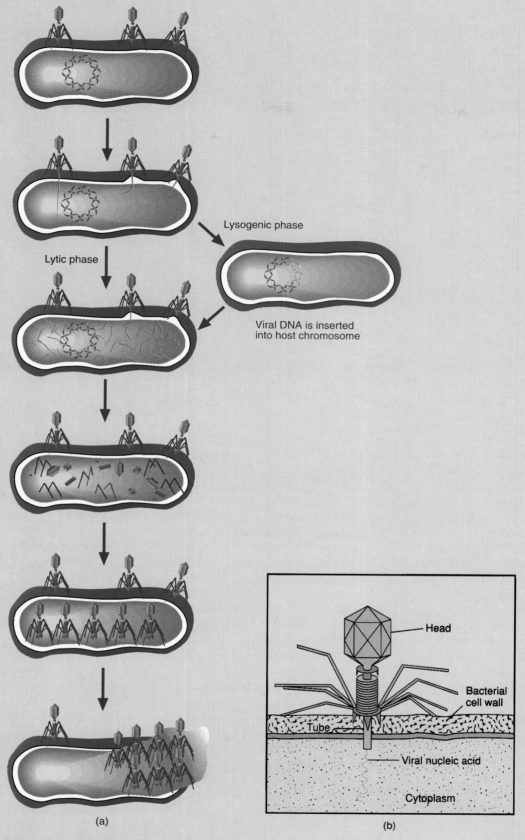

Lysogenic phase

Lytic phase

Viral DNA is inserted
into host chromosome

(a)

Head

Bacterial
cell wall

Tube

Viral nucleic acid

Cytoplasm

(b)

FIGURE 16B

Life Cycle of the T4 Bacteriophage.

(a) Events in the life cycle of T4. Once the host cell is infected, the virus can enter into the lytic phase immediately, or it may become lysogenic.
(b) Penetration of the bacterial cell wall by a bacteriophage.

FIGURE 16C

HIV.

The RNA genome is contained within a bullet-shaped capsid. Surrounding the capsid is an envelope with spikes. Changes in the chemical composition of the envelope, caused by mutations in the HIV genome, have made developing an HIV vaccine extraordinarily difficult.

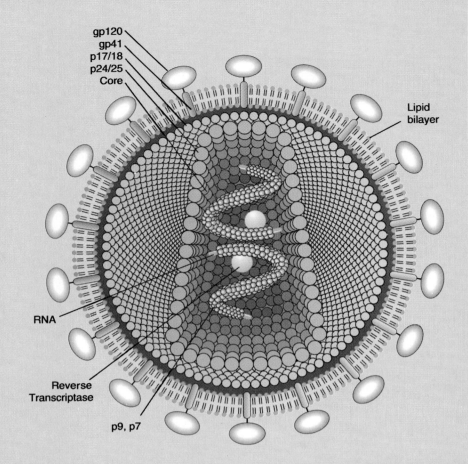

gp120
gp41
p17/18
p24/25
Core

Lipid bilayer

RNA

Reverse Transcriptase

p9, p7

relatively small portion of the population, are resistant to HIV infection.) Other cells that are known to be infected by HIV include some intestinal and nervous system cells.

Once the gp120 envelope protein of HIV binds to the CD4 antigen and chemokine receptor on the T cell, the viral envelope fuses with the host cell's plasma membrane. The two RNA strands are released into the cytoplasm. Reverse transcriptase, a heterodimer with several enzymatic activities, then catalyzes the synthesis of a ssDNA using the vRNA as a template. The heterodimer's RNase activity then degrades the vRNA. The same protein produces a double-stranded vDNA by forming a complementary strand of the ssDNA. Viral integrase integrates the vDNA into a host cell chromosome. The proviral DNA remains latent until the specific infected T cell is activated. The proviral DNA can then direct the cell to synthesize viral components. Newly synthesized viruses bud from the infected cell. Eventually, HIV-infected cells die. Several mechanisms are believed to contribute to cell death. These include the following:

1. The budding of a virus from the cell membrane may tear the membrane and cause massive leakages that cannot be repaired.

2. Massive release of new virus from a cell, directed by the provirus, may so deplete the cell that it disintegrates.

3. The binding of cell surface gp120 molecules to CD4 receptors on nearby healthy cells leads to the formation of large, nonfunctional multinucleated cell masses called *syncytia*.

HIV infection progresses through several stages, which may vary considerably in length among individuals. Initial symptoms, which usually occur soon after the initial exposure to the virus and last for several weeks, include fever, lethargy, headache, and other neurological complaints, diarrhea, and lymph node enlargement. (Antibodies to HIV are detectable during this period.) Exaggerated versions of these symptoms, referred to as the AIDS-related complex (ARC), may often recur. Eventually, the immune system becomes so compromised that the individual becomes susceptible to serious opportunistic diseases and is said to have developed AIDS. The time required for the development of AIDS may vary from two years to eight or ten years. For reasons that are not understood, a few patients do not develop AIDS even after 15 years of HIV infection. (It has recently been suggested that some of these individuals are infected with attenuated HIV variants.) Some of the most common AIDS-related diseases include *pneumocystis carinii* pneumonia, cryptococcal meningitis (inflammation of membranes that cover the brain and spinal cord), toxoplasmosis (brain lesions, heart and kidney damage, and fetal abnormalities), cytomegalovirus infections (pneumonia, kidney and liver damage, and blindness), and tuberculosis. HIV infection is also associated with several types of cancer, the most common of which is a rare skin cancer called Kaposi's sarcoma.

Because there is no cure for AIDS, treatment seeks to suppress symptoms (e.g., antibiotics for the infections) and slow viral reproduction. The drug AZT (azidothymidine), for example, is a nucleotide base analog that inhibits the vDNA synthesis catalyzed by reverse transcriptase. Indinavir is an example of the protease inhibitors, a class of drugs that prevent processing of viral protein. Because the viral genome mutates frequently (i.e., its surface antigens become altered), developing an AIDS vaccine is believed to take years.

FIGURE 16.19

Representative Viruses.

(a) Pox virus, (b) rhabdovirus, (c) mumps virus, (d) flexible-tailed bacteriophage, (e) herpes virus, (f) papilloma (wart) virus.

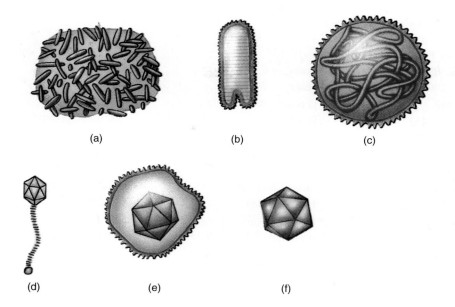

(a)   (b)   (c)

(d)   (e)   (f)

during virion assembly. Proteins that protrude from the surface of the envelope, called spikes, are believed to mediate the attachment of the virus to the host cell. Representative viruses are illustrated in Figure 16.19.

QUESTION

16.3

Recall that according to the central dogma, the flow of genetic information is from DNA to RNA and then to protein. Retroviruses are an exception to this rule. The alterations of the central dogma that are observed in retroviruses and other RNA viruses can be illustrated as follows:

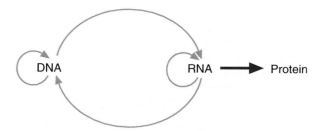

DNA   RNA ——▶ Protein

Compare this illustration with that of the original central dogma (p. 461). Describe in your own words the implications of each component of these figures.

16.4

HIV screening tests detect the presence of HIV antibodies in blood serum. In the most common HIV test, an ELISA kit that contains HIV antigens detects these antibodies. Because of the life-altering significance of a positive HIV test, the presence of HIV infection is confirmed by an additional, more expensive test. In most laboratories this confirming test is a Western blot analysis. Western blot analysis is similar to Southern blotting (refer to Appendix B, Supplement 16) with the following exceptions. Western blotting detects the presence of specific proteins instead of specific DNA sequences. In HIV analysis, these proteins are antibodies to gp41 and gp120. If there is antigen-antibody binding on a gel, it is detected with labeled "antihuman" antibodies. (Antihuman antibodies bind specifically to certain sites, other than the antigen-binding sites, that occur on all human antibodies.) After reviewing the discussion of ELISA (Appendix B, Supplement 15), describe how an HIV ELISA is conducted. Referring to the discussion of Southern blotting, describe in general terms how a Western blot analysis for HIV infection is performed. (Hint: An HIV Western blot begins by separating known HIV antigens on agarose gel electrophoresis.)

KEY WORDS

SUMMARY

The information required for directing all living processes is stored in the nucleotide sequences of DNA. DNA is composed of two antiparallel polynucleotide strands wound around each other to form a right-handed double helix. The deoxyribose-phosphodiester bonds form the backbones of the double helix, while the nucleotide bases project to its interior. The nucleotide base pairs form because of hydrogen bonding between certain bases: adenine and thymine, and cytosine and guanine. Mutations are changes in DNA structure, which may be caused by collisions with solvent molecules, thermal fluctuations, ROS, radiation, or xenobiotics.

DNA can have several conformations depending on the nucleotide sequence. In addition to the classical structure determined by Watson and Crick (B-DNA), A-DNA, H-DNA, and Z-DNA have also been observed. DNA supercoiling is now known to be a critical feature of several biological processes, such as DNA replication and transcription.

Each eukaryotic chromosome is composed of nucleohistone, a complex formed by winding a single DNA molecule around a histone octomer to form a nucleosome. The DNA of mitochondria and chloroplasts is similar to the chromosomes found in prokaryotes.

The forms of RNA that occur in cells, that is, transfer, ribosomal, messenger, heterogeneous, and small nuclear RNAs, are all involved in the synthesis of proteins. RNA differs from DNA in that it contains ribose (instead of deoxyribose), has a somewhat different base composition, and is usually single-stranded. Transfer RNA molecules bind to specific amino acids and transport them to the ribosome, where they are properly aligned during protein synthesis. The ribosomal RNAs are components of ribosomes. Messenger RNA contains within its nucleotide sequence the coding instructions for synthesizing a specific polypeptide. Heterogeneous nuclear RNA is the original transcript produced by complementary base pairing from a DNA template. It is then processed to form mRNA. The small nuclear RNAs are involved in splicing activities during mRNA synthesis.

Viruses are obligate intracellular parasites. Although they are acellular and cannot carry out metabolic activities on their own, viruses can wreak havoc on living organisms. Each type of virus infects a specific type of host (or small set of hosts). A virus does so because it can either inject its genome into the host cell or gain entrance for the entire viral particle. Each virus has the capacity to use the host cell's metabolic processes to manufacture new copies of itself, called virions. Viruses possess dsDNA, ssDNA, dsRNA, or ssRNA genomes.

HIV is a retrovirus that causes AIDS. Retroviruses are a class of RNA viruses that possess a reverse transcriptase activity that converts their RNA genome to a DNA molecule. This vDNA is then inserted into the host cell genome, causing a permanent infection. Eventually, HIV infection destroys the immune system of infected individuals.

SUGGESTED READINGS

Calladine, C.R., and Drew, H.R., *Understanding DNA: The Molecule and How It Works,* Academic Press, San Diego, 1992.

Frank-Kamenetskii, M.D., and Mirkin, S.M., Triple Helix DNA Structures, *Ann. Rev. Biochem.,* 64:65–95, 1995.

Gallo, R.C., and Montagnier, L., The Chronology of AIDS Research, *Nature,* 326(6112):435–436, 1987.

Herrmann, B., and Hummel, S. (Eds.), *Ancient DNA,* Springer-Verlag, New York, 1994.

Julian, M.M., Women in Crystallography, in Kass-Simon, G., and Farnes, P. (Eds.), *Women of Science: Righting the Record,* pp. 359–364, Indiana University Press, Bloomington and Indianapolis, 1990.

Lykke-Anderson, J., Aagaard, C., Semionenkov, M., and Garrett, R.A., Archaeal Introns: Splicing, Intercellular Mobility and Evolution, *Trends Biochem. Sci.,* 22:326–331, 1997.

Maxwell, E.S., and Fournier, M.J., The Small Nuclear RNAs, *Ann. Rev. Biochem.,* 35:897–934, 1995.

Portugal, F.H., and Cohen, J.S., *A Century of DNA: A History of the Discovery of the Structure and Function of the Genetic Substance,* MIT Press, Cambridge, Mass., 1977.

Varmus, H., Retroviruses, *Science,* 240:1427–1435, 1988.

Watson, J.D., *The Double Helix,* Atheneum, New York, 1968.

Weintraub, H.M., Antisense RNA and DNA, *Sci. Amer.,* 262(1): 40–46, 1990.

QUESTIONS

1. Clearly define the following terms:
   a. genetics
   b. replication
   c. transcription
   d. sugar-phosphate backbone
   e. bacteriophage
   f. Chargaff's rule
   g. palindrome
   h. Hoogsteen base pairing
   i. linking number
   j. Alu family

2. Under physiological conditions, DNA ordinarily forms B-DNA. However, RNA hairpins and DNA-RNA hybrids adopt the structure of A-DNA. Considering the structural differences between DNA and RNA, explain this phenomenon.

3. What structural features of DNA cause the major groove and the minor groove to form?

4. There are several structural forms of DNA. List and describe each form.

5. Describe the higher-order structure of DNA referred to as supercoiling.

6. List three biological properties facilitated by DNA supercoiling.

7. Explain in general terms how polyamines aid in achieving the highly compressed structure of DNA.

8. List three differences between eukaryotic and prokaryotic DNA.

9. Describe the structure of a nucleosome.

10. Describe the structural differences between RNA and DNA.

11. In contrast to the double helix of DNA, RNA exists as a single strand. What effects does this have on the structure of RNA?

12. What are the three most common forms of RNA? What roles do they play in cell function?

13. Z-DNA derives its name from the zig-zagged conformation of phosphate groups. What features of the DNA molecule allow this distinctive structure to form?

14. Jerome Vinograd found that circular DNA from a polyoma virus separates into two distinct bands when it is centrifuged. One band consists of supercoiled DNA and the other relaxed DNA. Explain how you would identify each band.

15. 5-Bromouracil is an analogue of thymine that normally pairs with adenine. However, 5-bromouracil frequently pairs with guanine. Explain.

16. There is one base pair for every 0.34 nm of DNA and the total contour length of all the DNA in a single human cell is 2 m. Calculate the number of base pairs in a single cell. Assuming that there are 10^{14} cells in the human body, calculate the total length of DNA. How does this estimate compare to the distance from the earth to the sun (1.5×10^8 km)?

17. A DNA sample contains 21% adenine. What is its complete percentage base composition?

18. A cyclic unstressed double-stranded DNA chain with 500 bp has 50 helical turns. An enzyme alters the DNA chain so that it has 4 negative supercoils. What are the linking numbers for the DNA molecule before and after the alteration?

19. The flow of genetic information is from DNA to RNA to protein. In certain viruses, the flow of information is from RNA to DNA. Does it appear possible for that information flow to begin with proteins? Explain.

20. HIV is a retrovirus. Suggest reasons why the development of a vaccine to prevent HIV infection is such a difficult undertaking.

21. You wish to isolate mitochondrial DNA without contamination with nuclear DNA. Describe how you would accomplish this task. (Refer to Appendix B, Supplement 16.)

22. The melting temperature of a DNA molecule increases as the G-C content increases. Explain. (Refer to Appendix B, Supplement 16.)

GENETIC INFORMATION

Chapter Seventeen

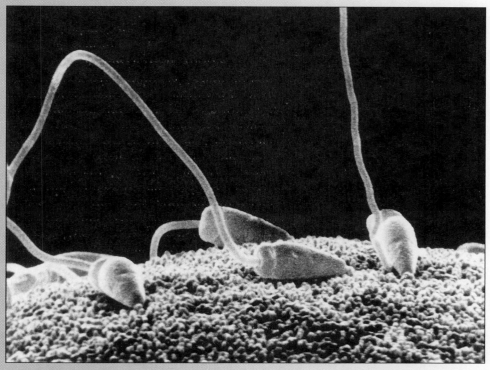

The penetration of a sperm cell into an egg cell triggers a programmed pattern of gene expression. The development of a fertilized egg into a new individual organism is encoded in the DNA within the egg and sperm.

OBJECTIVES

After you have studied this chapter, you should be able to answer these questions:

1. What is semiconservative replication?

2. What is a replisome and what role does it play in DNA synthesis?

3. What are the major enzymatic activities involved in DNA replication in E. coli?

4. What are Okazaki fragments?

5. How does DNA replication differ in prokaryotes and eukaryotes?

6. How is DNA damage repaired?

7. What are the mechanisms of excision repair and light-induced repair?

8. What is the difference between general recombination and site-specific recombination?

9. What are the major differences between prokaryotic and eukaryotic transcription?

10. What are ribozymes?

11. How do E. coli cells regulate lactose metabolism?

12. By what mechanisms are eukaryotic genes regulated?

Within the past several years, the investigation of genetic inheritance that began with Mendel has reached an explosive stage. Information concerning how living cells store, use, and inherit genetic information now constantly increases. Biochemists and geneticists are painstakingly revealing the principles of nucleic acid function. The practical consequences of work done in the 1980s have already been considerable. In addition to numerous medical applications (e.g., diagnostic tests and therapies), recombinant DNA technology has allowed investigators access to the inner workings of living organisms that has never before been possible. Not only are long-standing questions about living processes more accessible to investigation, but previously unimaginable medical strategies now seem almost commonplace. For example, the replacement of defective genes with their normal counterparts is now being investigated in clinical trials. Recent experience suggests it is impossible to predict the outcome of nucleic acid research. Whatever happens, it will certainly be exciting.

The ability of living organisms to function in the midst of a chaotic environment ultimately depends on the timely flow of information. The information requirements of the simplest of single-cell organisms, the bacteria, are impressive enough. Recall, for example, a few of the tasks that such organisms must perform to maintain the living state. They must acquire and metabolize nutrients, dispose of waste products, and evade predators while maintaining and repairing their component parts. In addition, when suitable environmental conditions exist, these organisms must invest relatively large resources in reproductive processes. In bacteria, information about their external and internal environments is acquired largely through receptors. For example, they detect a certain nutrient in their external environment when a few molecules bind to specific receptors (referred to as chemoreceptors) located in or near their plasma membranes. Motile cells (e.g., bacteria that possess flagella) can move toward a nutrient source because of information processing. The chemoreceptors, as well as several other components, operationally link the binding of nutrient to the purposeful movement of flagella so that the organism can efficiently move toward a food source. The structure of all the molecules in this mechanism (from the chemoreceptor to those in the flagella) are specified by the base sequence in the organism's genome. The organism's continued functioning, which requires consuming nutrients, depends ultimately on the coordinated expression of genetic information.

In comparison, the coordination of function in even the simplest of multicellular organisms requires a significant increase in information storage and management capacity. In addition to providing the information required for the moment-by-moment existence of individual cells, the genomes of multicellular organisms must provide the instructions for both the differentiation of each cell type and the mechanisms by which all their functions are coordinated. Consider, for example, that germinating plant seeds give rise to such different tissues as leaves, the vascular bundles, and roots. The mechanisms by which each tissue differentiates are now believed to involve differential gene expression, the unfolding of a precise pattern inherited from the previous generation. The plant embryo consists of two principal groups of cells, one that generates a root and another that evolves into a shoot (which later develop into the plant's stems and leaves). The differentiation of these cells is the result of genes being turned off and on according to a specific plan or pattern. This pattern is governed by each cell's chromosomes and signals received from other cells. Although the cell types are vastly different, a similar process appears to occur during the development of an animal from a fertilized egg.

Because of their inherent complexity and despite decades of investigation, developmental processes in multicellular organisms are still not completely understood.

The mechanisms by which the genomes of organisms direct developmental processes are being actively investigated with recombinant DNA technology. Recombinant DNA technology, which splices DNA molecules from different sources together, has revolutionized biochemical research. Two important techniques used in recombinant DNA technology (i.e., molecular cloning and the polymerase chain reaction (PCR)), are described in Appendix B, Supplement 17.

Each organism's genome is a library of detailed instructions (or blueprints) called genes. Genes are the master documents from which expendable copies, that is, various RNA molecules, are transcribed. The transcription of so-called **structural genes** forms the mRNA molecules that precisely specify the primary sequence of polypeptides. As mentioned, the other types of RNA formed by transcription are the means of processing and translating genetic information.

The instructions encoded in the primary structure of DNA also facilitate the transmission of genetic information from parent to offspring. DNA replication, which synthesizes identical copies, occurs in both germ cells (the cells that become gametes, the haploid cells from which the next generation is produced) and somatic cells (differentiated cells, other than the germ cells, that form a multicellular body). Because of their vital role in directing all living processes, the structural integrity of DNA sequences must be maintained. However, as described, DNA is constantly exposed to myriad physical and chemical processes that damage its structure. Fortunately, most living organisms possess several DNA repair mechanisms that usually eliminate virtually all damaged sequences.

Finally, DNA provides yet another mechanism for living organisms to respond to changes in their environment. Genetic variations that arise from mutations as well as **recombination** (the reassortment of DNA sequences) are the raw material of evolution. Through natural selection these variations may confer survival advantages on individual organisms in changing environments.

This chapter provides an overview of the mechanisms that living organisms use to synthesize the nucleic acids DNA and RNA and direct cellular processes. The chapter begins with a discussion of several aspects of DNA replication, repair, and recombination. This is followed by descriptions of the synthesis and processing of RNA. The chapter ends with a section devoted to gene expression, the mechanisms cells use to produce gene products in an orderly and timely manner.

17.1 GENETIC INFORMATION: REPLICATION, REPAIR AND RECOMBINATION

Because of the strategic importance of DNA, all living organisms must possess the following features: (1) rapid and accurate DNA synthesis and (2) genetic stability provided by effective DNA repair mechanisms. Paradoxically, the long-term survival of species also depends on genetic variations that allow them to adapt to changing environments. In most species these variations arise predominantly from genetic recombination, although mutation also plays a role. In the following sections, the mechanisms that prokaryotes and eukaryotes use to achieve these goals are discussed. Prokaryotic genetic information processes are more completely understood than those of eukaryotes. Because of their minimal growth requirements, short generation times, and relatively simple genetic composition, prokaryotes (especially *Escherichia coli*) are excellent subjects for investigations of genetic mechanisms. In contrast, multicellular eukaryotes possess several properties than hinder genetic investigations. The most formidable are long generation times (often months or years) and extraordinary difficulties in identifying gene products (e.g., enzymes or structural components). (A common tactic in genetic research is to induce mutations, then observe changes in or the absence of a specific gene product.) Unfortunately, because of their considerable complexity, in higher organisms this method has identified very few gene products. Recombinant DNA techniques are being used to circumvent this obstacle.

DNA Replication

DNA replication must occur before every cell division. The mechanism by which DNA copies are produced is similar in all living organisms. After the two strands separate, each subsequently serves as a template for the synthesis of a complementary

FIGURE 17.1

Semiconservative DNA Replication.

As the double helix unwinds at the replication fork, each old strand serves as a template for the synthesis of a new strand.

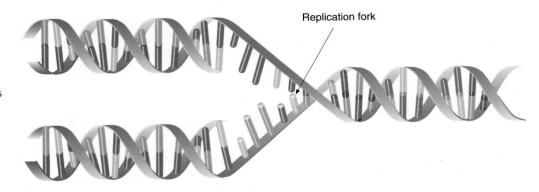

Replication fork

FIGURE 17.2

The Messelson-Stahl Experiment.

CsCl centrifugation of *E. coli* DNA can discriminate between DNA grown in ^{15}N media (1) and in ^{14}N media (2). (In (3), $^{15}N,^{15}N$ (heavy) DNA and $^{14}N,^{14}N$ (light) DNA separate in the CsCl gradient.) When *E. coli* cells enriched in ^{15}N are grown in ^{14}N for one generation, all cellular DNA is of intermediate density ($^{14}N,^{15}N$) (4). After two generations in ^{14}N media, the cells contain equal amounts of light DNA and intermediate DNA (5). After three generations, the amount of light DNA has increased, while the amount of intermediate DNA has remained constant (6).

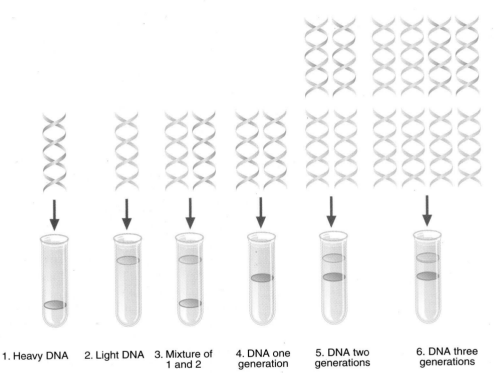

1. Heavy DNA   2. Light DNA   3. Mixture of 1 and 2   4. DNA one generation   5. DNA two generations   6. DNA three generations

strand (Figure 17.1). (In other words, each of the two new DNA molecules contains one old strand and one new strand.) This process, referred to as **semiconservative replication,** was first demonstrated in an elegant experiment (Figure 17.2) reported in 1958 by Matthew Messelson and Franklin Stahl. In this classic work, Messelson and Stahl took advantage of the different densities of DNA labeled with the heavy nitrogen isotope ^{15}N (the most abundant nitrogen isotope is ^{14}N). After *E. coli* cells were grown for 14 generations in growth media whose nitrogen source consisted only of $^{15}NH_4Cl$, the ^{15}N-containing cells were transferred to growth media containing the ^{14}N isotope. At the end of both one and two cell divisions, samples were removed. The DNA in each of these samples was isolated and analyzed by CsCl density gradient centrifugation. (Refer to Appendix B, Supplement 2 for a description of density gradient centrifugation.) Because pure ^{15}N-DNA and ^{14}N-DNA produce characteristic bands in centrifuged CsCl tubes, this analytical method discriminates between DNA molecules containing large amounts of the two nitrogen isotopes. When the DNA isolated from ^{15}N-containing cells grown in ^{14}N medium for precisely one generation was centrifuged, only one band was observed. Because this band occurred halfway between where ^{15}N-DNA and ^{14}N-DNA bands would normally appear, it seemed reasonable to assume that the new DNA was a hybrid molecule, that is, it contained one ^{15}N strand and one ^{14}N strand. (Any other means of replication would create more than one band.) After two cell divisions, extracted DNA was resolved into two discrete bands containing equal amounts of ^{14}N,^{14}N-DNA and ^{14}N,^{15}N-DNA, a result that also supported the semiconservative model of DNA synthesis.

In the years since the Messelson and Stahl experiment, many of the details of DNA replication have been discovered. As mentioned, much of this work was accomplished by using *E. coli* and other prokaryotes.

QUESTION

17.1

Two alternative models of DNA replication were considered before the Messelson-Stahl experiment. The *conservative model* proposed that the old (parental) strands remain in one of the two cells resulting from cell division, while the two new strands are dispersed into the other cell. In the *dispersive model,* double-stranded DNA breaks up into fragments during replication. After replication is complete, the old and new fragments are randomly reconstituted into two separate DNA molecules. If either of these models had proven to be correct, what CsCl banding pattern would Messelson and Stahl have observed?

DNA Synthesis in Prokaryotes   DNA replication in *E. coli* has proven to be a complex process that consists of several basic steps. Each step requires certain enzyme activities.

1. **DNA unwinding.** As their name implies, the helicases are ATP-requiring enzymes that catalyze the unwinding of duplex DNA. The principal helicase in *E. coli* is Dna B protein, a product of the dnaB gene.

2. **Primer synthesis.** The formation of short RNA segments called **primers,** required for the initiation of DNA replication, is catalyzed by primase, an RNA polymerase. Primase is a 60 kD polypeptide product of the dnaG gene. A multienzyme complex containing primase and several auxillary proteins is called the **primosome.**

3. **DNA synthesis.** The synthesis of a complementary DNA strand by forming phosphodiester linkages between nucleotides base-paired to a template strand is catalyzed by large multienzyme complexes referred to as the DNA polymerases. DNA polymerase III (pol III) is the major DNA-synthesizing enzyme. The pol III holoenzyme comprises at least ten subunits. The core polymerase is formed from three subunits: α, ε and Θ. The subunit τ allows two core enzyme complexes to form a dimer. The β protein (also called the sliding clamp protein) is composed of two subunits. It forms a ring around the template DNA strand. The γ-complex (or brace protein), composed of five subunits, recognizes single DNA strands with primer and transfers the β-protein to the core polymerase. The tether created by the β-protein allows the holoenzyme to remain associated with the DNA template as replication proceeds. The DNA replicating machine, called the **replisome,** comprises two copies of the pol III holoenzyme, the primosome, and DNA unwinding proteins. DNA polymerase I (pol I), also called the Kornberg enzyme, after its discoverer Arthur Kornberg (Nobel Prize in Physiology or Medicine, 1959), is a DNA repair enzyme. Pol I also plays a role in the timely removal of RNA primer during replication. The function of DNA polymerase II (pol II) is not understood, although it appears to be similar to pol I. In addition to a $5' \rightarrow 3'$ polymerizing activity, all three enzymes possess a $3' \rightarrow 5'$ exonuclease activity (e.g., the ϵ subunit in pol III). (An exonuclease is an enzyme that removes nucleotides from an end of a polynucleotide strand.) Pol I also possesses a $5' \rightarrow 3'$ exonuclease activity.

4. **Joining DNA fragments.** Discontinuous DNA synthesis (described below) requires an enzyme, referred to as a ligase, that joins the newly synthesized segments.

5. **Supercoiling control.** DNA *topoisomerases* prevent tangling of DNA strands, which can block further unwinding of the double helix. Tangling is a very real

possibility, since the double helix unwinds rapidly (as many as 50 revolutions per second during bacterial DNA replication). Topoisomerases are enzymes that alter the linking number of closed duplex DNA molecules. The terms *topoisomerase* and *topoisomers* (circular DNA molecules that differ only in their linking numbers) are derived from *topology* a branch of mathematics that investigates the features of geometric structures that do not change with bending or stretching. As described, topoisomerases alter supercoiling, a property of DNA. When appropriately controlled, supercoiling can facilitate the unzipping of duplex molecules. Type I topoisomerases produce transient single-strand breaks in DNA; type II topoisomerases produce transient double-strand breaks. In prokaryotes, a type II topoisomerase, called DNA gyrase, helps separate strands by pumping negative supercoils into the circular DNA molecule. (In contrast, eukaryotic type II topoisomerases catalyze only the removal of superhelical tension.)

The replication of the circular *E. coli* chromosome (Figure 17.3) begins at a precise initiation site referred to as *oriC* and proceeds in two directions. After a type I topoisomerase "nicks" the DNA (a *nick* is a single-strand break that allows for a temporary local relaxation of the molecule), replication proceeds outward in both directions. As the two sites of active DNA synthesis (referred to as **replication forks**) move farther away from each other, a "replication eye" forms. Because an *E. coli* chromosome contains one initiation site, it is considered a single replication unit. A replication unit, or **replicon,** is a DNA molecule (or DNA segment) that contains an initiation site and appropriate regulatory sequences.

When DNA replication was first observed experimentally (using electron microscopy and autoradiography), investigators were confronted with a paradox. The bidirectional synthesis of DNA as it appeared in their research seemed to indicate that continuous synthesis occurs in the $5' \rightarrow 3'$ direction on one strand and in the $3' \rightarrow 5'$ direction on the other strand. (Recall that DNA double helix has an antiparallel configuration.) However, all the enzymes that catalyze DNA synthesis do so in the $5' \rightarrow 3'$ direction only. It was later determined that only one strand, referred to as the *leading strand,* is continuously synthesized in the $5' \rightarrow 3'$ direction. The other strand, referred to as the *lagging strand,* is also synthesized in the $5' \rightarrow 3'$ direction but in small pieces (Figure 17.4). (Reiji Okazaki and his colleagues provided the experimental evidence for discontinuous DNA synthesis.) Subsequently, these pieces (now called **Okazaki fragments**) are covalently linked together by DNA ligase. (In prokaryotes such as *E. coli*, Okazaki fragments possess from 1000 to 2000 nucleotides.)

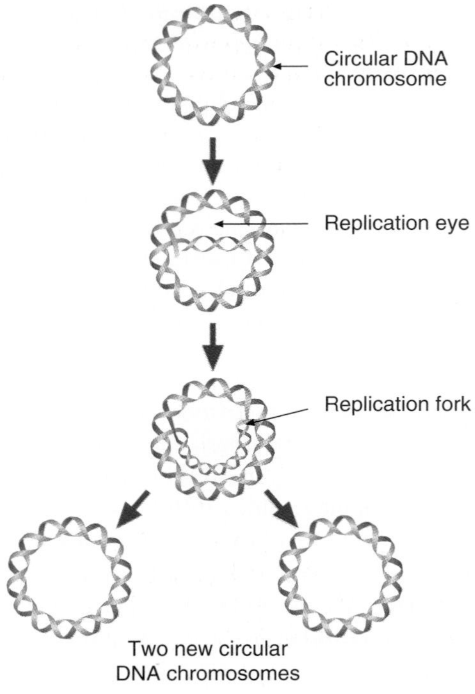

The initiation of replication is a complex process involving several enzymes, as well as other proteins. Replication begins when *Dna A protein* (52 kD) binds to four 9-bp sites within the oriC sequence. When additional DnaA monomers (between 20 and 40 copies) bind to oriC, a process that requires ATP and a histone-like protein (HU), a nucleosome-like structure forms. As the DnaA-DNA complex forms, localizing "melting" of the DNA duplex in three nearby 13-bp sequences causes a small segment of the double helix to open up. *DnaB* (a 300 kD helicase composed of six subunits), com-

FIGURE 17.3

Replication of Prokaryotic DNA.

As DNA replication of a circular chromosome proceeds, two replication forks can be observed by using autoradiography. The structure that forms is called a replication eye.

Circular DNA chromosome

Replication eye

Replication fork

Two new circular DNA chromosomes

DNA Replication at a Replication Fork.

The 5′ → 3′ synthesis of the leading strand is continuous. The lagging strand is also synthesized in the 5′ → 3′ direction but in small segments (now referred to as Okazaki fragments).

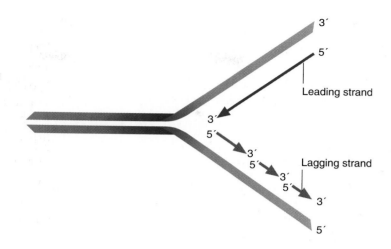

FIGURE 17.5

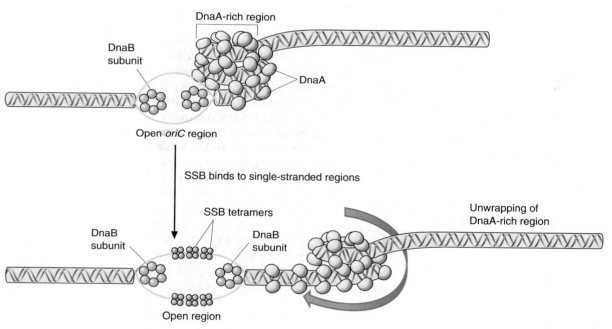

Replication Fork Formation.

After DnaA and DnaB binding, the DnaB helicase separates the duplex DNA strands at two replication forks. The binding of SSB to newly formed ssDNA prevents reassociation of the single strands. As the replication forks advance, the DnaA-DNA region unwraps. Subsequently DnaA is displaced.

plexed with *DnaC* (29 kD), then enters the open oriC region. Then DnaC is released. The replication fork moves forward as DnaB, assisted by topoisomerases, unwinds the helix (Figure 17.5). As DNA unwinding proceeds, DnaA is displaced. The single strands are kept apart by the binding of numerous copies of single-stranded DNA binding protein (SSB). SSB, a tetramer, may also protect vulnerable ssDNA segments from attack by nucleases.

A model of DNA synthesis at a replication fork is illustrated in Figure 17.6. For pol III to initiate DNA synthesis, an RNA primer must be synthesized. On the leading strand, where DNA synthesis is continuous, primer formation occurs only once per replication fork. In contrast, the discontinuous synthesis on the lagging strand requires primer synthesis for each of the Okazaki fragments. The primosome travels along the lagging strand and stops and reverses direction at intervals to synthesize a

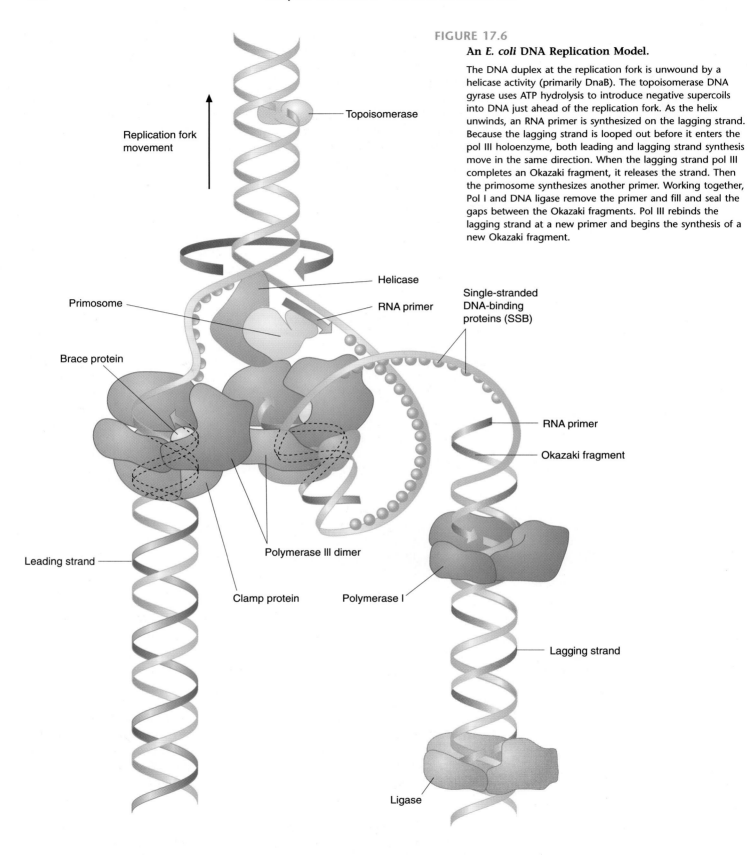

Replication fork movement

Topoisomerase

Helicase

Primosome

RNA primer

Single-stranded DNA-binding proteins (SSB)

Brace protein

RNA primer

Okazaki fragment

Leading strand

Polymerase III dimer

Clamp protein

Polymerase I

Lagging strand

Ligase

FIGURE 17.6

An *E. coli* DNA Replication Model.

The DNA duplex at the replication fork is unwound by a helicase activity (primarily DnaB). The topoisomerase DNA gyrase uses ATP hydrolysis to introduce negative supercoils into DNA just ahead of the replication fork. As the helix unwinds, an RNA primer is synthesized on the lagging strand. Because the lagging strand is looped out before it enters the pol III holoenzyme, both leading and lagging strand synthesis move in the same direction. When the lagging strand pol III completes an Okazaki fragment, it releases the strand. Then the primosome synthesizes another primer. Working together, Pol I and DNA ligase remove the primer and fill and seal the gaps between the Okazaki fragments. Pol III rebinds the lagging strand at a new primer and begins the synthesis of a new Okazaki fragment.

short RNA primer. Subsequently, pol III synthesizes DNA beginning at the 3′ end of the primer. As lagging strand synthesis continues, the RNA primers are removed by Rnase H and replaced by DNA segments synthesized by pol I. DNA ligase joins the Okazaki fragments.

As illustrated in Figure 17.6, the synthesis of both the leading and lagging strands are coupled. The tandem operation of two pol III complexes requires that one strand (the lagging strand) be looped around the replisome. When the lagging strand pol III complex completes an Okazaki fragment, it releases the duplex DNA. Once it does so, the primosome moves in and synthesizes another RNA primer.

Despite the complexity of DNA replication in *E. coli,* as well as its rate (as high as 1000 base pairs per second per replication fork), this process is amazingly accurate (approximately one error per 10^9 to 10^{10} base pairs per generation). This low error rate is largely a consequence of the precise nature of the copying process itself (i.e., complementary base pairing). However, both pol III and pol I also proofread newly synthesized DNA. Most mispaired nucleotides are removed (by the 3′ → 5′ exonuclease activities of pol III and pol I) and then replaced. Several postreplication repair mechanisms (discussed below) also contribute to the low error rate in DNA replication.

Replication ends when the replication forks meet on the other side of the circular chromosome at the termination site, the *ter* (τ) region. The ter region is composed of a pair of 20-bp inverted repeat ter sequences separated by a 20-bp segment. Each ter sequence prevents further progression of one of the replication forks when a 36 kD *ter binding protein* (TBP) is bound. How the two daughter DNA molecules separate is not understood, although a type II topoisomerase is believed to be involved.

DNA Synthesis in Eukaryotes   Although the principles of DNA replication in prokaryotes and eukaryotes have a great deal in common (e.g., semiconservative replication and bidirectional replicons), they also have significant differences. Not surprisingly, these differences appear to be related to the size and complexity of eukaryotic genomes.

1. **Timing of replication.** In contrast to rapidly growing bacterial cells, in which replication occurs throughout most of the cell division cycle, eukaryotic replication is limited to a specific period referred to as the S phase (Figure 17.7). It is now known that eukaryotic cells produce certain proteins (Section 17.3) that regulate phase transitions within the cell cycle.

2. **Replication rate.** DNA replication is significantly slower in eukaryotes than in prokaryotes. The eukaryotic rate is approximately 50 nucleotides per second per replication fork. (Recall that the rate in prokaryotes is about ten times higher.) This discrepancy is presumably due, in part, to the complex structure of chromatin.

3. **Replicons.** Despite the relative slowness of eukaryotic DNA synthesis, the replication process is relatively brief, considering the large sizes of eukaryotic genomes. For example, on the basis of the replication rate mentioned above, the replication of an average eukaryotic chromosome (approximately 150 million base pairs) should take over a month to complete. Instead, this process usually requires several hours. Eukaryotes use multiple replicons to compress the replication of their large genomes into short periods (Figure 17.8).

4. **Okazaki fragments.** From 100 to 200 nucleotides long, the Okazaki fragments of eukaryotes are significantly shorter than those in prokaryotes.

Although many eukaryotic replication enzymes are generally similar to their prokaryotic counterparts, they do possess their own distinctive properties. For example, eukaryotes have five DNA polymerases designated α, β, δ, ε and γ. DNA polymerase α, composed of four subunits, synthesizes lagging strands. Two of its associated subunits have primase activity, while the largest subunit is a polymerase. The

FIGURE 17.7

FIGURE 17.7

The Eukaryotic Cell Cycle.

Interphase (the period between mitotic divisions) is divided into several phases. DNA replication occurs during the synthesis or S phase. The G₁ (first gap) phase is the time between mitosis and the beginning of the S phase. During the G₂ phase, protein synthesis increases as the cell readies itself for mitosis (M phase). After mitosis, many cells enter a resting phase (G₀).

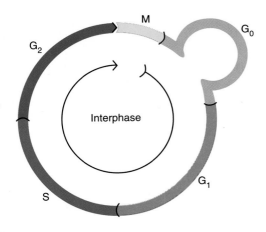

FIGURE 17.8

Multiple Replicon Model of Eukaryotic Chromosomal DNA Replication.

A short segment of a eukaryotic chromosome during replication.

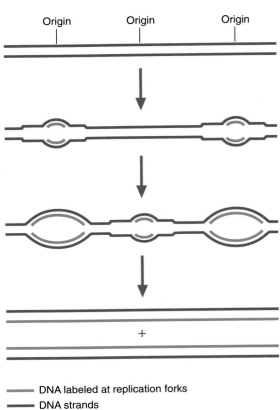

—— DNA labeled at replication forks
━━ DNA strands

function of the fourth subunit is unresolved. Much less is known about the other eukaryotic DNA polymerases except for the following. DNA polymerase δ, which synthesizes leading strands, binds to *PCNA* (proliferating cell nuclear antigen). PCNA is a sliding clamp protein similar in function to β-protein in *E. coli*. DNA polymerase δ also possesses a $3' \rightarrow 5'$ exonuclease activity. DNA polymerase β is believed to be involved in DNA repair. DNA polymease ε, which possesses a $3' \rightarrow 5'$ exonuclease activity, plays an as yet poorly resolved role in both DNA replication and repair. DNA polymerase γ catalyzes the replication of the mitochondrial genome. The DNA polymerase activity of chloroplasts remains poorly characterized.

The eukaryotic replication process is complicated by both the large size of eukaryotic genomes and the complex structure of chromatin. Attempts to reconstitute eukaryotic replication in the laboratory have been unsuccessful. For example, the enzyme that catalyzes the removal of RNA primer has not yet been identified.

DNA Repair

The natural (or background) rate of spontaneous mutation is remarkably constant among species at about 0.1 to 1 mutation per gene per million gametes in every generation. This estimate appears low, but most mutations are deleterious. Because of the complexity of living processes, most genetic changes can be expected to reduce the viability of the organism. For example, humans (an extraordinarily complex species) lose approximately 50% of conceptions because of genetic factors.

Mutations take many forms, from point mutations (single base changes) to gross chromosomal abnormalities (see Section 16.1). As described, they are caused by factors that include the chemical properties of the bases themselves and various chemical processes (e.g., depurinations and oxidative stress), as well as the effects of xenobiotics (foreign molecules). Not surprisingly, cells possess a wide range of DNA repair mechanisms.

The importance of maintaining the structural integrity of DNA is reflected in the variety of repair mechanisms employed by living cells. Prominent examples of these mechanisms include excision repair, photoreactivation repair, and recombinational repair.

In **excision repair,** mutations are excised by a series of enzymes that remove incorrect bases and replace them with the correct ones. Figure 17.9 illustrates the excision of a thymine dimer in bacteria. The process begins with the detection of a distorted DNA segment by a repair endonuclease, referred to as an excision nuclease or excinuclease. The excinuclease cuts the damaged DNA and removes a single-stranded

FIGURE 17.9

Excision Repair of a Thymine Dimer in *E. coli.*

Uvr A, a damage recognition protein, detects helical distortion caused by DNA adducts such as thymine dimers (a). It then associates with Uvr B to form the A_2B complex. After binding to the damaged segments, A_2B forces DNA to bend. UvrA then dissociates (b). The binding of the nuclease Uvr C to Uvr B (c) and the action of the helicase Uvr D (d) results in the excision of a 12 nucleotide DNA strand (12-mer). After UvrB is released (e) the excision gap is repaired by pol I (f).

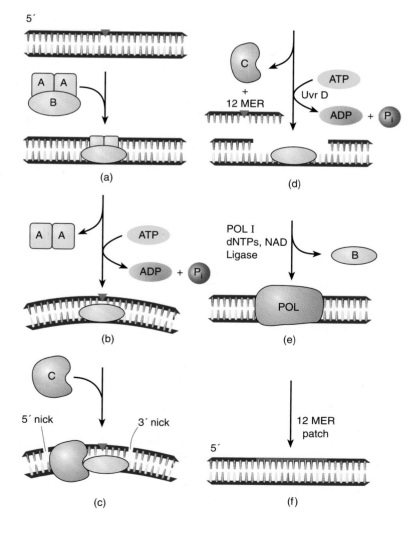

sequence about 12 or 13 nucleotides long. (In eukaryotes, 27 to 29 nucleotide sequences are removed.) In *E. coli* the excinuclease is composed of three proteins: Uvr A, Uvr B and Uvr C. Uvr A identifies the damaged site and associates with Uvr B to form a complex (A_2B). After A_2B bends the DNA, Uvr A dissociates from Uvr B-DNA. Uvr C then binds to Uvr B, cutting the damaged DNA strand 4 or 5 nucleotides to the 3′ side of the thymine dimer. Then Uvr C cuts the strand 8 nucleotides to the 5′ side. Uvr D, a helicase, releases Uvr C and the thymine dimer-containing oligonucleotide. The excision gap is repaired by pol I and DNA ligase. Although excision repair in humans is poorly understood, it is probably more complicated than the prokaryotic process because it involves at least 7 polypeptides. Several of these molecules are missing in xeroderma pigmentosum patients. (In xeroderma pigmentosum, skin lesions and skin cancer are caused by exposure to UV radiation.)

In **photoreactivation repair,** or **light-induced repair,** pyrimidine dimers are restored to their original monomeric structures (Figure 17.10). In the presence of visible light, photoreacting enzyme (PR enzyme) cleaves the dimer, leaving the phosphodiester bonds intact. Light energy captured by the enzyme's flavin and pterin chromophores breaks the cyclobutane ring.

Recombinational repair can eliminate certain types of damaged DNA sequences that are not eliminated before replication. For some structural damage (e.g., pyrimidine dimers), replication is interrupted because the replication complex detaches, moves beyond the damage, and reinitiates synthesis. One of the daughter DNA molecules has a gap opposite the pyrimidine dimer. This gap can be repaired by exchanging the corresponding segment of the homologous DNA molecule. After the recombination process, the newly opened gap in the homologous molecule can easily be repaired by DNA polymerase and DNA ligase, since an intact template is present. The remaining damaged segment (the pyrimidine dimer) can then be eliminated by other repair mechanisms. Recombinational repair closely resembles genetic recombination, which is discussed next.

DNA Recombination

Recombination, often referred to as genetic recombination, can be defined as the rearrangement of DNA sequences by exchanging segments from different molecules. The process of recombination, which produces new combinations of genes and gene fragments, is primarily responsible for the diversity among living organisms. More important, the large number of variations made possible by recombination can allow species opportunities to adapt to changing environments. In other words, genetic recombination is a principal source of the variations that make evolution possible.

There are two forms of recombination: general and site-specific. **General recombination,** which occurs between homologous DNA molecules, is most commonly observed during meiosis. (Recall that meiosis is the form of eukaryotic cell division in

KEY *Concepts*

DNA is constantly exposed to chemical and physical processes that alter its structure. Each organism's survival depends on its capacity to repair this structural damage.

FIGURE 17.10

Photoreactivation Repair of Thymine Dimers.

Light provides the energy for converting the dimer to two thymine monomers. No nucleotides are removed in this repair mechanism. Many species, including humans, do not possess photoreacting enzyme (PR) enzyme activity.

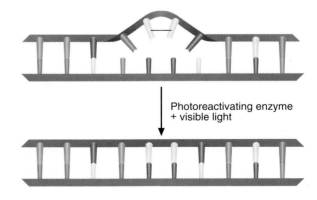

Photoreactivating enzyme + visible light

which haploid gametes are produced.) A similar process has been observed in some bacteria. In **site-specific recombination,** the exchange of sequences from different molecules requires only short regions of DNA homology. These regions are flanked by extensive nonhomologous sequences. Site-specific recombinations, which depend more on protein-DNA interactions than on sequence homology, occur throughout nature. For example, this mechanism is used by a bacteriophage to integrate its genome into the *E. coli* chromosome. In eukaryotes, site-specific recombination is responsible for a wide variety of developmentally controlled gene rearrangements. Gene rearrangements may be at least partially responsible for cell differentiation in complex multicellular organisms. One of the most interesting examples of gene rearrangement is the generation of antibody diversity in mammals. A variation of site-specific recombination, sometimes referred to as **transposition,** moves certain sequences, called **transposable elements,** from one chromosome or chromosomal region to another.

General Recombination   General recombination requires the precise pairing of homologous DNA molecules. Figure 17.11 illustrates the currently accepted model for general recombination. This model, proposed by Robin Holliday in 1964 and based on his genetic investigation in fungi, involves the following essential steps:

1. Two homologous DNA molecules become paired.
2. Two of the DNA strands, one in each molecule, are cleaved.
3. The two strand segments crossover, thus forming a Holliday intermediate.
4. DNA ligase seals the cut ends.
5. A second series of DNA strand cuts occurs.
6. DNA polymerase fills any gaps, and DNA ligase seals the cut strands.

Because the Holliday intermediate can migrate, more base pairs may be in the recombination event.

During meiosis, homologous recombination involves the intimate association of two pairs of chromatids. (Chromatids are identical daughter chromosomes that are still attached to each other by a centromere. The chromatids in each pair are called sister chromatids.) During this process, sometimes referred to as synapsis, an exchange of DNA sequences can occur between homologous segments of nonsister chromatids.

In bacteria, general recombination appears to be involved in several forms of intermicrobial DNA transfer.

1. **Transformation. In transformation,** naked DNA fragments enter a bacterial cell through a small opening in the cell wall and are introduced into the bacterial genome. (Recall Fred Griffith's experiment.)

2. **Transduction. Transduction** occurs when bacteriophage inadvertently carry bacterial DNA to a recipient cell. After a suitable recombination, the cell uses the transduced DNA.

3. **Conjugation.** Certain bacterial species are known to engage in **conjugation,** an unconventional sexual mating that involves a donor cell and a recipient cell. The donor cell possesses a specialized plasmid that allows it to synthesize a sex pilus, a filamentous appendage that functions in a DNA exchange process. (Recall that plasmids are small extrachromosomal circular DNA molecules that replicate independently of the cell's chromosome.) After the pilus attaches to the surface of the recipient cell, a fragment of the donor's genetic material is transferred. The transferred DNA segment can be integrated into the recipient's chromosome by recombination or it may exist outside it in plasmid form.

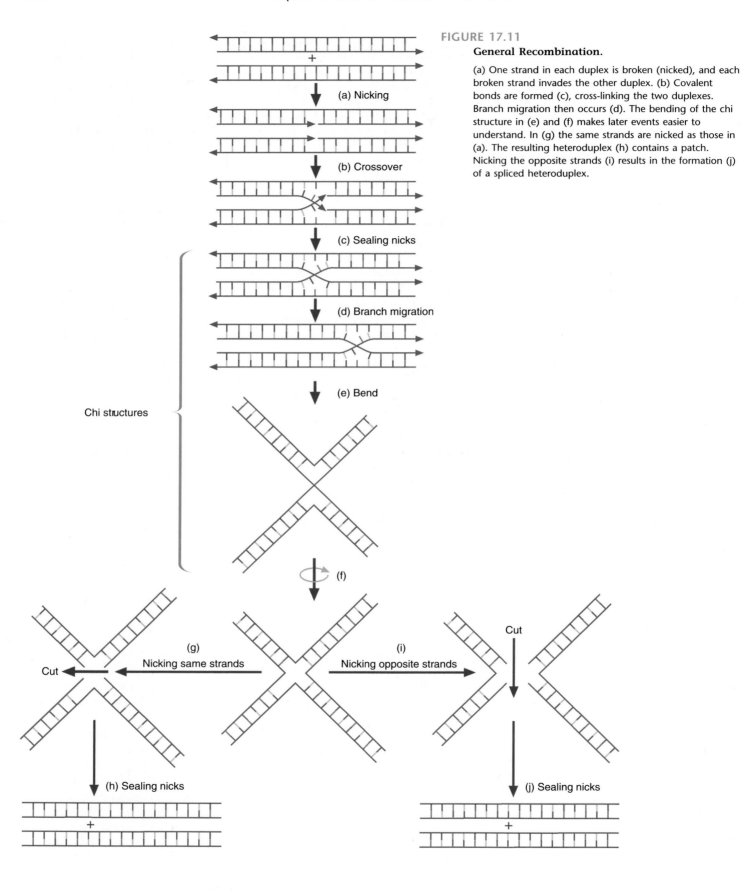

FIGURE 17.11

General Recombination.

(a) One strand in each duplex is broken (nicked), and each broken strand invades the other duplex. (b) Covalent bonds are formed (c), cross-linking the two duplexes. Branch migration then occurs (d). The bending of the chi structure in (e) and (f) makes later events easier to understand. In (g) the same strands are nicked as those in (a). The resulting heteroduplex (h) contains a patch. Nicking the opposite strands (i) results in the formation (j) of a spliced heteroduplex.

17.2

Bacterial conjugation has medical consequences. For example, certain plasmids contain genes that code for toxins. The causative agent of a new deadly form of food poisoning is a strain of *E. coli* (*E. coli* 0157) that synthesizes a toxin that causes massive bloody diarrhea and kidney failure. This toxin is now believed to have originated in Shigella, another bacterium that causes dysentery. Similarly, the growing problem of antibiotic resistance is the result, in part, of the spread of antibiotic-resistance genes among bacterial populations. Antibiotic resistance develops because antibiotics are over used in medical practice and in livestock feeds. Suggest a mechanism by which this extensive use promotes antibiotic resistance. (Hint: The high-level use of antibiotics acts as a selection pressure.)

Site-Specific Recombination As mentioned, site-specific recombination depends more on protein-DNA interactions than on DNA-DNA sequence homology. For example, the integration of bacteriophage λ into the *E. coli* chromosome (Figure 17.12) requires only short recognition sequences. A site-specific viral enzyme, called integrase, is largely responsible for promoting this recombinational event.

Transposition Barbara McClintock, a geneticist working with corn (maize), reported in the 1940s that certain genome segments can move from one place to another. Because chromosomes were believed to consist of genes in a fixed and unvarying order, it was not until 1967, when transposable elements (or **transposons**)

FIGURE 17.12

Insertion of the Bacteriophage λ Genome into the *E. coli* Chromosome.

(a) The λ DNA circularizes as the single-stranded cos sequences anneal. (b) Insertion occurs through site-specific recombination between short homologous phage and bacterial sequences.

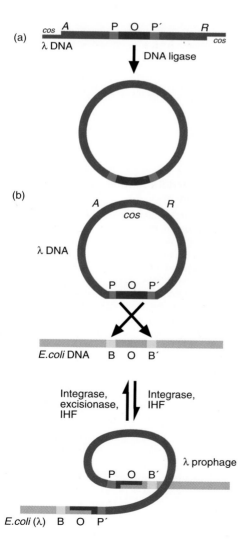

were discovered in *E. coli,* that scientists began to comprehend that genomes are not as stable as they had once thought. (In recognition of her monumental contribution to genetics, Dr. McClintock received the 1983 Nobel Prize in Physiology or Medicine.) Transposons (also referred to as "jumping genes") have now been observed in a wide variety of organisms in addition to bacteria, for example, various fungi, plants, and animals.

Several bacterial transposons, referred to as insertion elements (IS elements), consist only of a gene that codes for a transposition enzyme (i.e., transposase), flanked by short DNA segments called inverted repeats (Figure 17.13). (Inverted repeats are short palindromes.) More complicated bacterial transposable elements, called composite transposons, contain additional genes, several of which may code for antibiotic resistance. Because transposons can "jump" between bacterial chromosomes, plasmids, and viral genomes, transpositions are now believed to play an important role in the spread of antibiotic resistance among bacteria.

Two transposition mechanisms have been observed.

1. **Replicative transposition.** During transposition a replicated copy is inserted into the new location (target site), leaving the original transposon at its original site (Figure 17.13). In replicative transposition an intermediate referred to as a cointegrate forms. An additional enzyme, called resolvase, catalyzes a site-specific recombination that allows the resolution of the cointegrate into two separate molecules.

2. **Nonreplicative transposition.** The transposable element is spliced out of its original site (donor site) and is inserted into the target site. Then the donor site must be repaired. If it cannot be repaired by the cell's DNA repair system, the consequences may be lethal.

Whether transposition is replicative or nonreplicative, short duplications of target site DNA segments are generated by the staggered cleavage catalyzed by transposase (Figure 17.14).

FIGURE 17.13

Bacterial Insertion Elements.

(a) An insertion sequence. (b) A composite transposon. (c) Insertion of a transposon (Tn3) into bacterial DNA. The insertion process involves the duplication of the target site. (Also refer to Figure 17.14.)

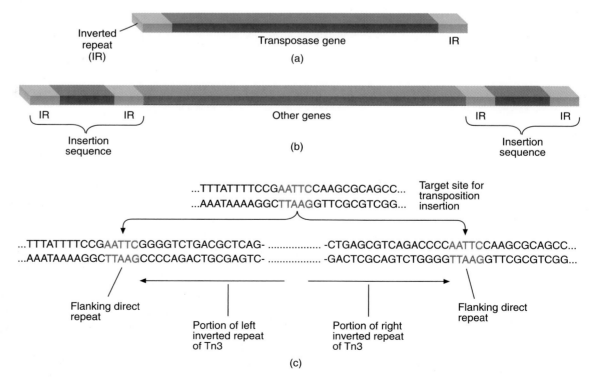

FIGURE 17.14

Formation of Duplicated Target Site Sequences During Transposition.

(a) Host DNA is cut (see arrows) in a staggered fashion (b). (c) The transposon (blue) is covalently attached at both ends to a strand of host DNA. (d) After the gaps are filled in by a DNA polymerase activity, there are nine base pair repeats of host DNA (red) flanking the transposon.

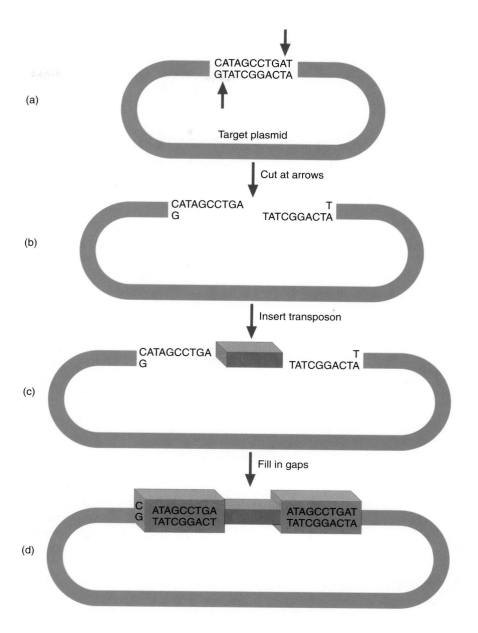

Some transposons found in eukaryotes resemble those found in bacteria. For example, the Ac element, the maize transposon that was first described by McClintock, is composed of a transposase gene flanked by short inverted repeats. (McClintock referred to the Ac transposon as a "controlling element" because it appeared to control the synthesis of the pigment anthocyanin in corn kernels.) Many other eukaryotic transposons, however, have somewhat different structures than those observed in bacteria. Instead of inverted repeats, many eukaryotic transposons, such as the Ty transposon in yeast, possess long terminal repeats (LTR), sometimes referred to as delta repeats. More important, the transposition mechanisms of Ty and many other eukaryotic transposons involve an RNA intermediate and bear a remarkable resemblance to the replicative phase of the retroviral life cycle. Ty contains genes for reverse transcriptase and RNase H, as well as several other genes required for its reverse transcription and integration into target DNA. Because these so-called retrotransposons lack the genes required to synthesize an envelope, they can move only within the genome, that is, they cannot exit as do the retroviruses.

The movement of transposons within a genome makes several genetic changes. The insertion and excision of transposons often cause duplications, inversions, and deletions. Depending on the changes and their location, the effects of transposons can be viewed either as disruptive and damaging or as providing opportunities for genetic diversity. Some effects of transposition are observed as changes in gene expression, a topic that is discussed in Section 17.3.

17.3

One of the fascinating aspects of complex organisms such as mammals is the existence of gene families (groups of genes that code for the synthesis of a series of closely related proteins). For example, several different types of collagen are required for the proper structure and function of connective tissues. Similarly, there are several types of globin gene. It is currently believed that gene families originate from a rare event in which a DNA sequence is duplicated. Some gene duplications provide a selective advantage by providing larger quantities of important gene products. In others, the two duplicate genes evolve independently. One copy continues to serve the same function, while the other eventually evolves to serve another function. Can you speculate about how gene duplications occur? Once a gene has been duplicated, what mechanisms introduce variations?

17.2 TRANSCRIPTION

As with all aspects of nucleic acid function, the synthesis of RNA molecules is a very complex process involving a variety of enzymes and associated proteins. Recall that RNA molecules are transcribed from the cell's genes. As RNA synthesis proceeds, the incorporation of ribonucleotides is catalyzed by RNA polymerase, sometimes referred to as DNA-directed RNA polymerase. The reaction catalyzed by all RNA polymerases is

$$\text{NTP} + (\text{NMP})_n \rightarrow (\text{NMP})_{n+1} + \text{PP}_i$$

Because the nontemplate or plus (+) strand has the same base sequence as the RNA transcription product (except for the substitution of U for T), it is also called the **coding strand** (Figure 17.15). By convention, the direction of the gene, a segment of double-stranded DNA, is the same as the direction of the coding strand. Because the template DNA strand, also called the minus (−) strand, and the newly made RNA molecule are antiparallel, the polymerization proceeds from the 3′ end to the 5′ end of the gene. As noted, transcription generates several types of RNA of which rRNA, tRNA, and mRNA are directly involved in protein synthesis (Chapter 18).

Transcription in Prokaryotes

The RNA polymerase in *E. coli* catalyzes the synthesis of all RNA classes. With a molecular weight of about 450 kD, RNA polymerase is a relatively large complex (Figure 17.16). It is composed of five types of polypeptide: α, β, β', ω, and σ (sigma). The core enzyme (α_2, β, β', and ω) catalyzes RNA synthesis. The transient binding of the σ factor to the core enzyme allows it to bind both the correct template strand and the proper site to initiate transcription. A variety of σ factors have been identi-

DNA
(+)
5′— TTTGGACAACGTCCAGCGATC —3′ Nontemplate strand

3′— AAACCTGTTGCAGGTCGCTAG —5′ Template strand
(−)

RNA
5′—UUUGGACAACGUCCAGCGAUC—3′

FIGURE 17.15

DNA Coding Strand.

One of the two complementary DNA strands, referred to as the template (−) strand, is transcribed. The RNA transcript is identical in sequence to the nontemplate (+) or coding strand, except for the substitution of U for T.

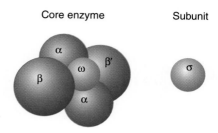

FIGURE 17.16

E. coli RNA Polymerase.

The *E. coli* RNA polymerase consists of two α subunits and one each of β, β′, and ω subunits. The transient binding of a σ subunit allows binding of the core enzyme to appropriate DNA sequences.

fied. For example, in *E. coli*, σ^{70} is involved in the transcription of most genes, whereas σ^{32} and σ^{28} promote the transcription of heat shock genes and the flagellin gene, respectively. (As its name suggests, flagellin is a protein component of bacterial flagella.) The superscript indicates the protein's molecular weight in kilodaltons.

The transcription of an *E. coli* gene is outlined in Figure 17.17. The process consists of three stages: initiation, elongation, and termination. Each is discussed briefly.

The initiation of transcription involves binding RNA polmerase to a specific DNA sequence called a **promoter.** Although their promoters are variable in size (from 20 to 200 bp), two short sequences at positions about 10 and 35 bp away from the transcription initiation site are remarkably similar among various bacterial species. (In these sequences, called **consensus sequences,** specific nucleotides are usually found at each site. The sequences shown in the figure are named in relation to the transcription starting point, the −35 region and the −10 region. The −10 region is also called the *Pribnow box,* after its discoverer.) RNA polymerase slides along the DNA until it reaches a promoter sequence. Once the enzyme binds to the promoter region, a short DNA segment near the Pribnow box unwinds. Transcription begins with binding the first nucleoside triphosphate (usually ATP or GTP) to the RNA polymerase complex. A nucleophilic attack by the 3′-OH group of the first nucleoside triphosphate on the α-phosphate of a second nucleoside triphosphate (also positioned by base pairing in an adjacent site) causes the first phosphodiester bond to form. (Because the phosphate groups of the first molecule are not involved in this reaction, the 5′ end of prokaryotic transcripts possesses a triphosphate group.) If the transcribed sequence reaches a length of about ten nucleotides, the conformation of the RNA polymerase complex changes; for example, the σ factor is released, and the initiation phase ends. As soon as an RNA polymerase has initiated transcription and moved beyond the promoter site, another RNA polymerase can move in, bind to the site, and start another round of RNA synthesis.

Once the σ factor detaches and the affinity of the RNA polymerase complex for the promoter site decreases, the elongation phase begins. The core RNA polymerase

FIGURE 17.17

A Typical E. coli Transcription Unit.

If RNA polymerase can bind to the promoter, DNA transcription begins at +1, downstream from the promoter in bacteria, translation of RNA begins immediately.

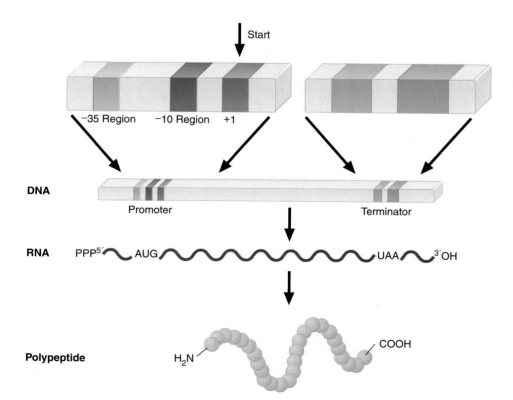

FIGURE 17.18

Transcription Initiation in *E. coli.*

(a) A transcription bubble forms as a short DNA segment unwinds. An RNA-DNA hybrid forms as transcription progresses. The bubble moves to keep up with transcription as DNA unwinds before it and rewinds behind it. (b) Transcription induces coiling. Positive supercoils form ahead of the bubble, while negative supercoils form behind it.

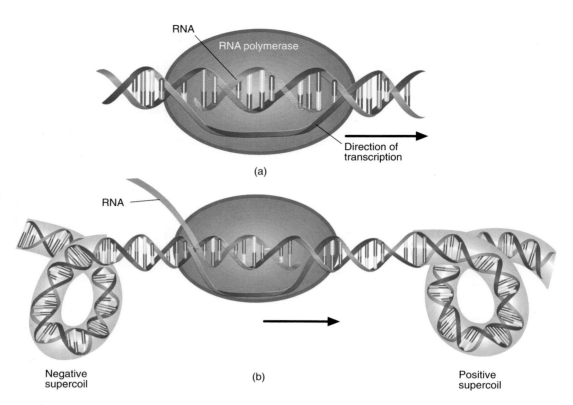

converts to an active transcription complex as it binds several accessory proteins. As RNA synthesis proceeds in the $5' \rightarrow 3'$ direction (Figure 17.18), the DNA unwinds ahead of the *transcription bubble* (the transiently unwound DNA segment in which an RNA-DNA hybrid has formed). The unwinding action of RNA polymerase creates positive supercoils ahead of the transcription bubble and negative supercoils behind the bubble. (As the bubble moves down the gene, it is said to move "downstream." Any structure or activity in the other direction is said to be "upstream." Also recall that negative supertwists aid to unwind the double helix.) The incorporation of ribonucleotides continues until a termination signal is reached.

Termination sequences contain palindromes. The RNA transcript of the DNA palindrome forms a stable hairpin turn. Apparently, this structure causes the RNA polymerase to slow or stop and partially disrupts the RNA-DNA hybrid structure. In some termination sequences, referred to as ρ (rho)-independent termination sites, several (about six) uridine residues follow the hairpin structure. Because U-A base pair interactions are weak, the short U-A sequence promotes the dissociation of the newly synthesized RNA from the DNA strand. In ρ-dependent termination the ρ factor (an enzyme that catalyzes the ATP-dependent unwinding of RNA-DNA helices) promotes the dissociation of the RNA polymerase complex from the RNA-DNA hybrid. ρ binds to RNA and not to RNA polymerase.

In prokaryotes, mRNA is used immediately in protein synthesis. In fact, protein synthesis begins while transcription is ongoing. However, mature rRNA and tRNA molecules are produced from larger transcripts by posttranscriptional processing. The RNA processing reactions for *E. coli* rRNA are outlined in Figure 17.19. The *E. coli* genome contains several sets of the rRNA genes 16S, 23S, and 5S. (Each set of genes is called an **operon.**) In the primary processing step, the polycistronic 30S transcript is methylated and then cleaved by several RNases into a number of smaller segments. Further cleavage by different RNases produces mature rRNAs. A few tRNAs are also produced. The other tRNAs are produced from primary transcripts in a series of processing reactions in which they are trimmed down by several RNases. In the last step of tRNA processing, a large number of bases are altered by several modification reactions (e.g., deamination, methylation, and reduction).

KEY *Concepts*

During transcription, an RNA molecule is synthesized from a DNA template. In prokaryotes this process involves a single RNA polymerase activity. Transcription is initiated when the RNA polymerase complex binds to a promoter sequence.

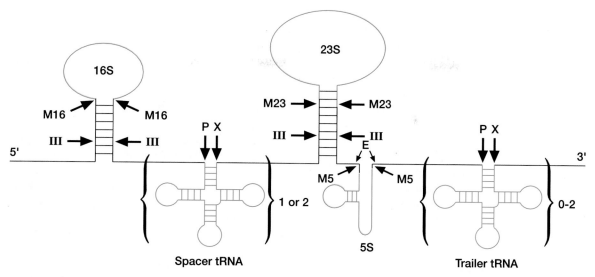

FIGURE 17.19

Ribosomal RNA Processing in _E. coli._

Each rRNA operon encodes a primary transcript that contains one copy each of 16S, 23S and 5S rRNAs. Each transcript also encodes one or two spacer tRNAs and as many as two trailer tRNAs. Posttranscriptional processing involves numerous cleavage reactions catalyzed by various RNases and splicing reactions. RNase P is a ribozyme.

Transcription in Eukaryotes

KEY _Concepts_

Transcription in eukaryotes is significantly more complex than its counterpart in prokaryotes. In addition to requiring three RNA polymerases, the eukaryotic process requires the combined binding of numerous transcription factors before RNA polymerase can initiate transcription.

Although DNA transcription in prokaryotes and in eukaryotes resemble each other, there are significant differences, apparently because of the greater structural complexity of eukaryotes. For example, most of the chromatin in eukaryotic cells is at least partially condensed at any time. Yet to be transcribed, DNA must be sufficiently exposed and accessible for RNA polymerase activity. Similarly, proper cell function depends on the timely transport of a wide variety of transcription products across the nuclear membrane into the cytoplasm. Eukaryotes appear to have solved these and other complex problems with equally complex solutions. For example, eukaryotic transcription is regulated by a vast number of transcription factors that must be precisely assembled before transcription can begin. Some of these factors influence transcription when bound to DNA sequences that are spatially removed from the promoter region they influence. Transport problems appear to have been solved in part by certain processing reactions (briefly discussed below) that allow each transcription product to be exported through a nuclear pore. (Nuclear pore complexes are complicated multisubunit structures. It is currently believed that transport through the pore complex occurs when RNA and protein molecules bind to specific receptors. Transcriptionally active nuclei may possess larger numbers of nuclear pores.)

Principally because of the complexity of eukaryotic genomes, eukaryotic DNA transcription is not understood as completely as the prokaryotic process. However, the eukaryotic process is known to possess the following unique features:

1. **RNA polymerase activity.** Eukaryotes possess three nuclear RNA polymerases, each of which differs in the type of RNA synthesized, subunit structure, and relative amounts. RNA polymerase I, which is localized within the nucleolus, transcribes the large rRNAs. The precursors of mRNA and most snRNAs are transcribed by RNA polymerase II, while RNA polymerase III is responsible for transcribing the precursors of the tRNAs and 5S rRNA. Each polymerase possesses two large subunits and several (six to ten) smaller subunits. For example, the two large subunits of RNA polymerase II, the enzyme that transcribes the majority of eukaryotic genes, have molecular weights of 215 and 139 kD. The number of smaller subunits varies among species, for example, plants possess eight, while vertebrates have six. Some of the smaller subunits are also present in the other two RNA polymerases. In contrast to the prokaryotic RNA polymerase,

the eukaryotic enzymes cannot initiate transcription themselves. Various transcription factors must be bound at the promotor before transcription can begin.

2. **Promoters.** The promoter sequences in eukaryotic DNA are larger, more complicated, and more variable than those of prokaryotes. Many promoters for RNA polymerase II contain consensus sequences, referred to as the TATA box, which occur about 25–30 bp upstream from the transcription initiation site. As in Figure 17.20, the binding of the transcription factor TFIID to the TATA box is the first step in the assembly of the RNA polymerase II transcription complex. The frequency of transcription initiation is often affected by binding certain transcription factors to upstream elements such as the *CAAT box* and the *GC box.* The activity of many promoters is affected by *enhancers,* regulatory sequences that may occur thousands of base pairs upstream or downstream of the gene they affect. (In yeast these sequences are called upstream activator sequences, or UAS.) The effects of enhancers can be complex. For example, a single gene may be controlled by the combined activities of several enhancers. Hormone response elements (Section 15.4) often act as enhancers.

3. **Processing.** Posttranscriptional processing occurs in both prokaryotes and eukaryotes. The most notable differences between the two types of organisms lies in the processing of mRNA. In contrast to prokaryotic mRNA, which usually requires little or no processing, eukaryotic mRNA are the products of extensive editing. Shortly after the transcription of the primary transcript begins, a modification of

FIGURE 17.20

Transcription Initiation in Eukaryotes.

Before RNA polymerase II can begin transcribing a gene, transcription factors must assemble into a complex. The process begins when TFIID binds to a TATA sequence. TFIID, which consists of *TBP* (a TATA binding protein) and several associated proteins, binds to and unwinds the DNA duplex in the TATA sequence. Then TFIIB binds and, later, TFIIE, TFIIH, and TFIIJ. TFIIF binds directly to RNA polymerase II. Because of phosphorylation reactions catalyzed by TFIIH, the RNA polymerase II becomes active and begins transcription.

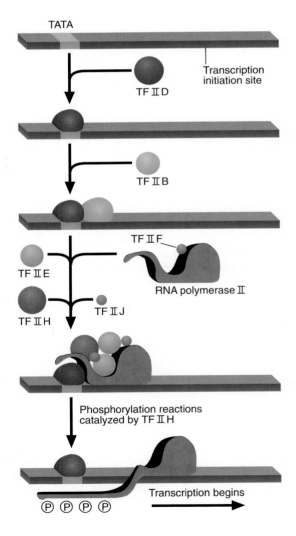

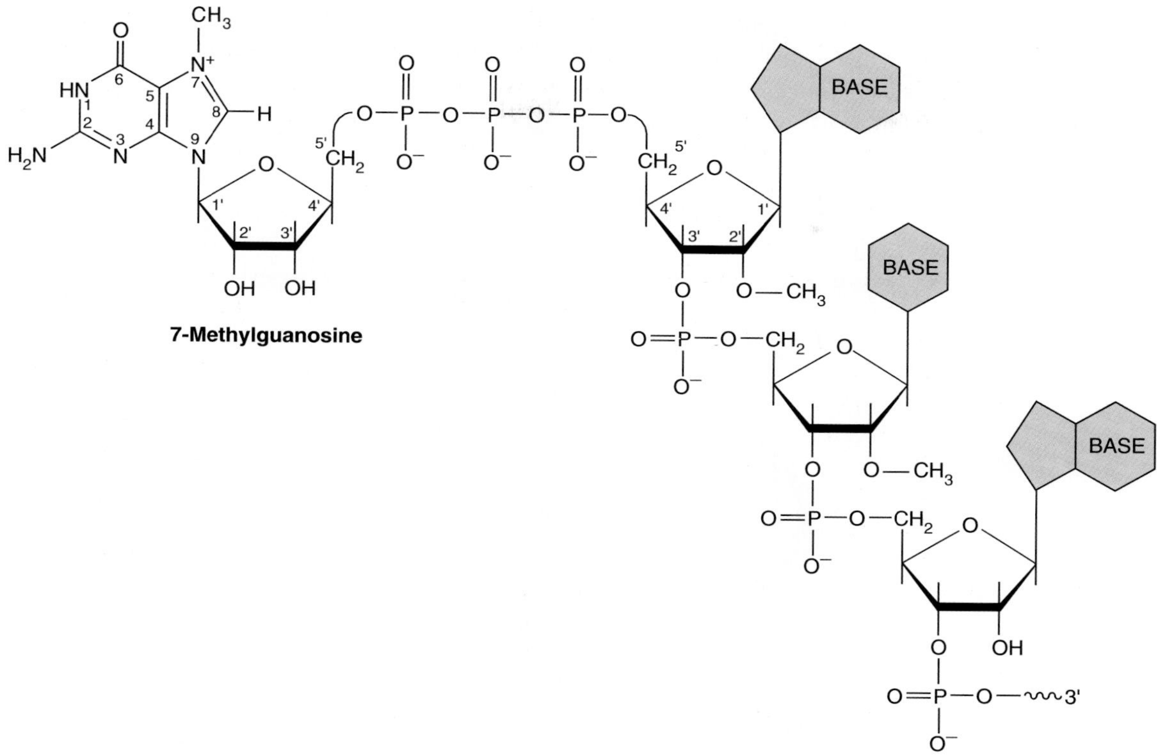

7-Methylguanosine

FIGURE 17.21

The Methylated Cap of Eukaryotic mRNA.

The cap structure consists of a 7-methyl guanosine attached to the 5' end of an RNA molecule through a unique 5' → 5' linkage. The 2'-OH of the first two nucleotides of the transcript are methylated.

the 5' end called *capping* occurs. The cap structure (Figure 17.21), which consists of 7-methylguanosine linked to the mRNA through a triphosphate linkage, protects the 5' end from exonucleases and promotes mRNA translation by ribosomes. For unknown reasons, RNA polymerase II transcribes well past the functional end of the primary transcript. After transcription has terminated, the transcript is cleaved at a specific site near the sequence AAUAAA. Immediately afterwards, from 100 to 250 adenylate residues are added by poly(A)polymerase to the 3' end. This *poly A tail* is believed to have several functions, which include protection of the mRNA from nucleases and promoting the export of mRNA into the cytoplasm and their translation by ribosomes. (Some mRNAs, such as histone mRNAs, do not contain poly A tails.) The most dramatic and complex processing reactions are those that remove introns. During this process (illustrated in Figure 17.22), which is referred to as **splicing,** each intron is excised in an unusual configuration called a *lariat.* Splicing takes place within a **spliceosome,** a multicomponent structure (40-60 S) containing several snRNAs, as well as several proteins. An unusual and unanticipated example of splicing, discovered in 1982 by Tom Cech, is the self-splicing by pre-rRNA molecules in the protozoan *Tetrahymena.* These catalytic RNA molecules, now called **ribozymes,** have also been found in several other organisms.

17.3 GENE EXPRESSION

Ultimately, the internal order most essential to living organisms requires the precise and timely regulation of gene expression. It is, after all, the capacity to switch genes on and off that enables cells to respond efficiently to a changing environment. In multicellular organisms, complex programmed patterns of gene expression are responsible for cell differentiation as well as intercellular cooperation.

RNA Splicing.

mRNA splicing begins with the nucleophilic attack of the 2'-OH of a specific adenosine on a phosphate in the 5' splice site. A lariat is formed by a 2', 5' phosphodiester bond. In the next step, 3'-OH of Exon 1 (acting as a nucleophile) attacks a phosphate adjacent to the lariat. This reaction releases the intron and ligates the two exons.

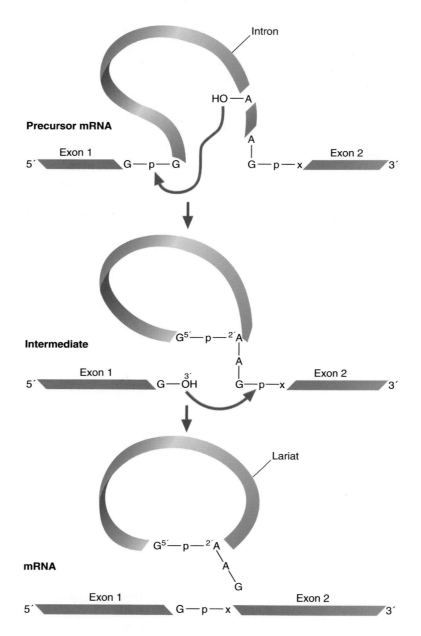

The regulation of genes, as measured by their transcription rates, is the result of a complex hierarchy of control elements that coordinate the cell's metabolic activities. Some genes, referred to as **constitutive** or housekeeping **genes,** are routinely transcribed because they code for gene products (e.g., glucose-metabolizing enzymes, ribosomal proteins, and histones) required for cell function. In addition, in the differentiated cells of multicellular organisms, certain specialized proteins are produced that cannot be detected elsewhere (e.g., hemoglobin in red blood cells). Genes, which are expressed only under certain circumstances, are referred to as *inducible.* For example, the enzymes that are required for lactose metabolism in *E. coli* are synthesized only when lactose is actually present and glucose, the bacterium's preferred energy source, is absent.

Most of the mechanisms used by living cells to regulate gene expression involve DNA-protein interactions. At first glance, the seemingly repetitious and regular structure of B-DNA appears to make it an unlikely partner for the sophisticated binding with myriad different proteins that must occur in gene regulation. As noted in Chapter 16, however, DNA is somewhat deformable, and certain sequences can be curved

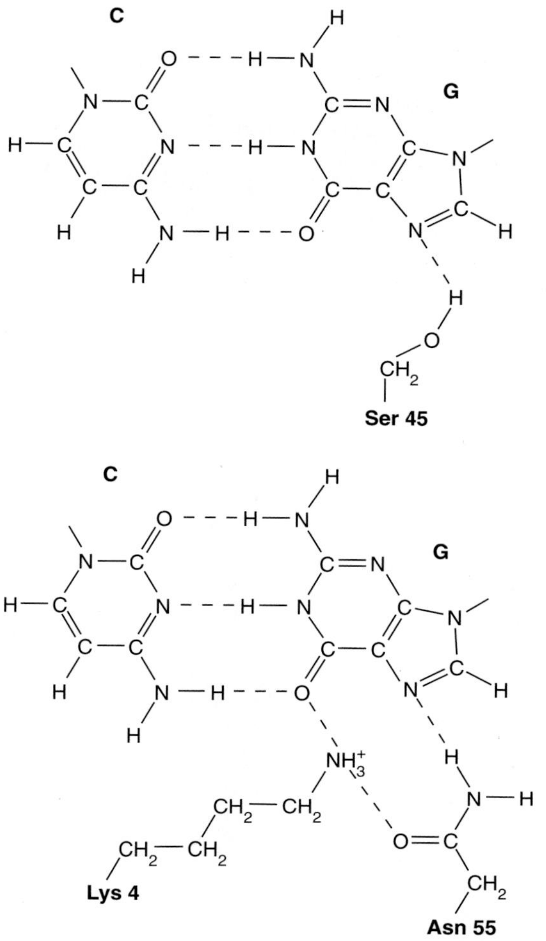

FIGURE 17.23

Amino Acid–Nucleotide Base Interactions During Protein-DNA Binding.

These examples are taken from structure studies of the binding of λ repressors to DNA. (λ is a virus that infects *E. coli*.)

or bent. In addition, it is now recognized that the edges of the base pairs within the major groove (and to a lesser extent the minor groove) of the double helix can participate in sequence-specific binding to proteins. Numerous contacts (often about 20 or so) involving hydrophobic interactions, hydrogen bonds, and ionic bonds between amino acids and nucleotide bases result in highly specific DNA-protein binding. Several examples of amino acid–nucleotide base interactions are illustrated in Figure 17.23.

The three-dimensional structures of a number of DNA regulatory proteins that have been determined have surprisingly similar features. In addition to usually possessing twofold axes of symmetry, many of these molecules can be separated into families (Figure 17.24) on the basis of the following structures: (1) helix-turn-helix, (2) helix-loop-helix, (3) leucine zipper, and (4) zinc finger. DNA-binding proteins, many of which are transcription factors, often form dimers. For example, a variety of transcription factors with leucine zipper motifs form dimers as their leucine-containing α-helices interdigitate (see Figure 5.18c.) Because each protein possesses its own unique binding specificity, and these and many other transcription factors can combine to form homodimers (two identical monomers) and heterodimers (two different monomers), a large number of unique gene regulatory agents are formed.

Considering the obvious complexity of function observed in living organisms, it is not surprising that the regulation of gene expression has proven to be both remarkably complex and difficult to investigate. For many of the reasons stated, knowledge about prokaryotic gene expression is significantly more advanced than that of eukaryotes. Prokaryotic gene expression was originally investigated, in part, as a model for the study of the more complicated gene function of mammals. Although it is now

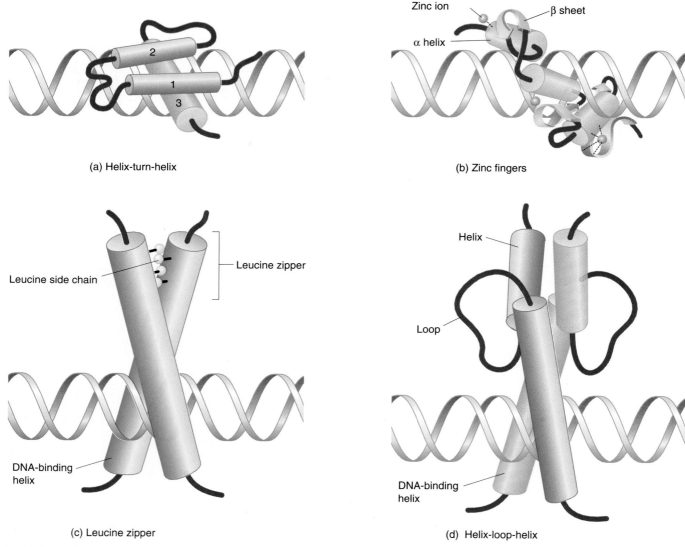

FIGURE 17.24

DNA-Protein Interactions

Gene regulatory proteins contain certain structural motifs: (a) helix-turn-helix, (b) zinc fingers, (c) leucine zipper, and (d) helix-loop-helix.

recognized that the two genome types are vastly different in many respects, the prokaryotic work has provided many valuable insights into the mechanisms of gene expression. In general, prokaryotic gene expression involves the interaction of specific proteins (sometimes referred to as regulators) with DNA in the immediate vicinity of a transcription start site. Such interactions may have either a positive effect (i.e., transcription is initiated or increased) or a negative effect (i.e., transcription is blocked). In an interesting variation, the inhibition of a negative regulator (called a *repressor*) activates affected genes. (The inhibition of a repressor gene is referred to as derepression.) Eukaryotic gene expression uses these mechanisms as well as several others, including gene rearrangement and amplification and various complex transcriptional, RNA processing, and translational controls. In addition, the spatial separation of transcription and translation inherent in eukaryotic cells provides another opportunity for regulation: RNA transport control. Finally, eukaryotes (as well as prokaryotes) also regulate cell function through the modulation of proteins through covalent modifications.

In this section, several examples of control of gene expression are described. The discussion of prokaryotic gene expression focuses on the lac operon. The *lac operon* of *E. coli,* originally investigated by Francois Jacob and Jacques Monod in the 1950s, remains one of the best-understood models of gene regulation. Despite a daunting ignorance about eukaryotic gene expression, a significant number of the pieces in this marvelous puzzle have been revealed. The section ends with a brief discussion of recent discoveries concerning growth factor–triggered gene expression.

Gene Expression in Prokaryotes

As described, the highly regulated metabolism of prokaryotes such as *E. coli* allows these organisms to respond rapidly to a changing environment to promote growth and survival. The timely synthesis of enzymes and other gene products only when needed prevents wasting energy and nutritional resources. At the genetic level, the control of inducible genes is often effected by collections of structural and regulatory genes called operons. Investigations of operons, especially the lac operon, have provided substantial insight into how gene expression can be altered by environmental conditions. Similarly, as described, investigations of viral infections of prokaryotes have furnished relatively unobstructed views of certain genetic mechanisms. The infection of *E. coli* by bacteriophage λ has been especially instructive.

The Lac Operon   The lac operon (Figure 17.25) consists of a control element and structural genes that code for the enzymes of lactose metabolism. The control element contains the promoter site, which overlaps the operator site. (In prokaryotes the *operator* is a DNA sequence involved in the regulation of adjacent genes that binds to a repressor protein.) The promoter site also contains the CAP site, described below. The structural genes Z, Y, and A specify the primary structure of β-galactosidase, lactose permease, and thiogalactoside transacetylase, respectively. β-Galactosidase catalyzes the hydrolysis of lactose, which yields the monosaccharides galactose and glucose, whereas lactose permease promotes lactose transport into the cell. Because lactose metabolism proceeds normally without thiogalactoside transacetylase, its role is unclear. A repressor gene i, directly adjacent to the lac operon, codes for the lac repressor protein, a tetramer that binds to the operator site with high affinity. (There are about ten copies of lac repressor protein per cell.) The binding of the lac repressor to the operator prevents the functional binding of RNA polymerase to the promoter (Figure 17.26).

Without its inducer (allolactose, a β-1,6-isomer of lactose) the lac operon remains repressed because the lac repressor binds to the operator. When lactose becomes available, a few molecules are converted to allolactose by β-galactosidase. Allolactose then binds to the repressor, changing its conformation and promoting dissociation from the operator. Once the inactive repressor diffuses away from the operator, the transcription of the structural genes begins. The lac operon remains active until the lactose supply is consumed. Then the repressor reverts to its active form and rebinds to the operator.

FIGURE 17.25
The Lac Operon in *E. coli.*

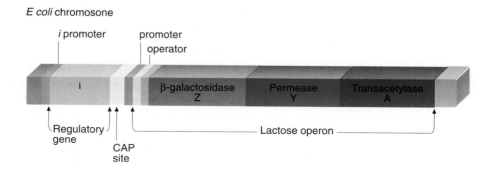

**Function of the Lac
Operon.**

(a) The repressor gene i encodes
repressor that binds to the
operator when lactose (the
inducer) is not present. (b)
When lactose is present, its
isomer allolactose binds to
repressor protein, thereby
inactivating it. (Not shown: the
effect of glucose on the lac
operon. Refer to the text for a
discussion of this topic.)

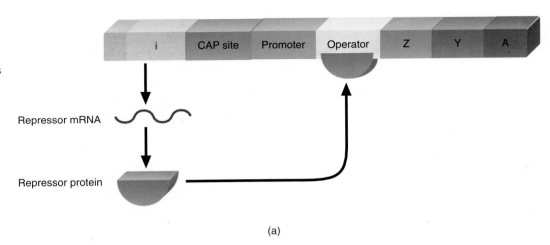

(a)

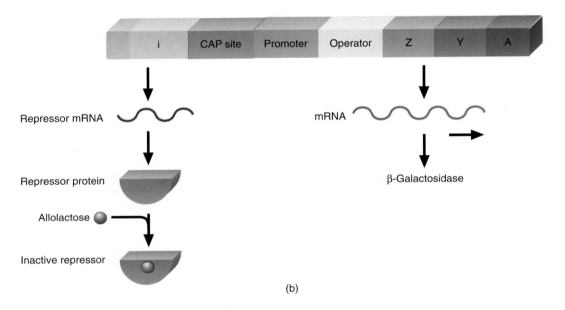

(b)

Glucose is the preferred carbon and energy source for *E. coli*. If both glucose and
lactose are available, the glucose is metabolized first. Syntheses of the lac operon enzymes are induced only after the glucose has been consumed. (This makes sense because glucose is more commonly available and has a central role in cellular metabolism. Why expend the energy to synthesize the enzymes required for the metabolism
of other sugars if glucose is also available?) The delay in activating the lac operon is
mediated by a catabolite gene activator protein (CAP). CAP is an allosteric homo-
dimer that binds to the chromosome at a site directly upstream of the lac promoter
when glucose is absent. CAP is an indicator of glucose concentration because it binds
to cAMP. The cell's cAMP concentration is inversely related to glucose concentration because glucose transport depresses adenylate cyclase activity. The binding of
cAMP to CAP, which occurs only when glucose is absent and cAMP levels are high,
causes a conformational change that allows the protein to bind to the lac promotor.
CAP binding promotes transcription by increasing the affinity of RNA polymerase
for the lac promoter. In other words, CAP exerts a positive or activating control on
lactose metabolism.

KEY *Concepts*

Constitutive genes are routinely
transcribed, whereas inducible
genes are transcribed only under
appropriate circumstances. In
prokaryotes, inducible genes and
their regulatory sequences are
grouped into operons.

Gene Expression in Eukaryotes

As noted, eukaryotic genomes are vastly larger and more complex than those of prokaryotes. Presumably, these differences can be accounted for, at least in part, by the obstacles that confront each type of organism. Eukaryotes typically lead more complicated lives than prokaryotes. This is especially true of multicellular eukaryotes. For example, among the higher animals and plants, numerous differentiated cells in each individual organism are derived from the genome in a fertilized egg. By its very nature, development requires an orderly and sequential expression of a vast number of genes. In addition, as described, the sustained life of multicellular organisms requires intercellular coordination, which involves changes in gene expression. In recent years, progress in the investigation of eukaryotic gene expression has been made largely because of molecular cloning and other recombinant DNA techniques.

QUESTION 17.4

Recently, several cases of infection caused by a rare, virulent strain of group A streptococcus have been reported. In approximately 25–50% of these cases (reported in Great Britain and the United States), infection resulted in necrotizing fasciitis, a rapidly spreading destruction of flesh, often accompanied by hypotension (low blood pressure), organ failure, and toxic shock. If antibiotic treatment is not initiated within three days of exposure to the bacterium, gangrene and death may result. Similar cases were reported in the 1920s. However, these earlier cases had a significantly lower fatality rate, although antibiotics were not then available. (Physicians reported treating affected areas by washing with acidic solutions.)

Group A streptococci are converted into the pathogenic form by becoming infected themselves with a certain virus. This virus's genome contains a gene that codes for a tissue-destroying toxin. Can you describe in general terms how a viral infection might cause a permanent change in the pathogenicity of a group A streptococcus bacterium? Considering the apparent difference in virulence between the bacterium in the 1920s and the present, is there any method for determining whether the same strain of group A streptococcus is responsible for both sets of cases? Preserved specimens of infected tissue specimens from these early cases are available. (Hint: Refer to Appendix B, Supplement 17.)

Current evidence indicates that eukaryotic gene expression is regulated by the following mechanisms:

1. **Gene rearrangements.** The differentiation of certain cells involves gene rearrangements, e.g., the rearrangements of antibody genes in B lymphocytes. Transposition (p. 503) is also believed to affect gene regulation.

2. **Gene amplification.** During certain stages in development, the requirement for specific gene products may be so great that the genes that code for their synthesis are selectively amplified. Amplification occurs via repeated rounds of replication within the amplified region. For example, the rRNA genes in various animals (most notably amphibians, insects, and fish) are amplified within immature egg cells (called oocytes). rRNA amplification is apparently required because of the enormous requirement for protein synthesis during the early developmental stages of fertilized eggs.

3. **Transcriptional control.** A significant amount of gene regulation occurs through selective transcription. There appear to be two major influences on eukaryotic transcription initiation: chromatin structure and gene regulatory proteins. During interphase of the cell cycle, chromatin is observed in two forms. Heterochromatin is so highly condensed that it is transcriptionally inactive. A small portion of each cell's heterochromatin occurs in all of an individual organism's cells. Other portions of heterochromatin differ in a tissue-specific pattern. Euchromatin, a less condensed form of chromatin, has varying levels of transcriptional activity. Transcriptionally active euchromatin is the least condensed. Inactive euchromatin is somewhat more condensed (but less so than that observed in heterochromatin). The mechanism by which chromatin reversibly condenses is unknown. However,

histone covalent modifications (e.g., acetylation, phosphorylation, and methylation) are believed to be involved. As described, a wide variety of gene regulatory proteins affect transcription by either activating or repressing genes. Several proposed mechanisms for protein-mediated gene repression are illustrated in Figure 17.27. Transcription factors may also be involved in changes in transcription start sites.

4. **RNA processing.** Cells often use alternative RNA processing to control gene expression. For example, alternative splicing results in the different forms of α-tropomyosin, a structural protein produced in tissues. The selection of alternative sites for polyadenylation also affects mRNA function. For example, such a change is involved in the switch, during the early phase of B lymphocyte differentiation, from producing membrane-bound antibody to that of secreted antibody. As noted, poly A tails have several roles in mRNA function (p. 511). In general, mRNAs with longer poly A tails are more stable, thereby increasing their opportunities for translation.

5. **RNA transport.** As mentioned, eukaryotes regulate molecular traffic into and out of the nucleus. Nuclear export signals, for example, capping and association with specific proteins, are believed to control the transport of processed RNA molecules through nuclear pore complexes.

6. **Translational control.** Eukaryotic cells can respond to various stimuli (e.g., heat shock, viral infections, and cell cycle phase changes) by selectively altering pro-

FIGURE 17.27

Proposed Mechanisms for Eukaryotic Gene Repression.

(a) Transcription factor proteins compete for binding to the same regulatory sequence. (b) Both activating and repressing proteins bind to DNA but at different sites. The repressor blocks transcription by binding to and masking activating sites on the activator. (c) The repressor factor binds to a transcription factor bound to DNA, thus preventing a transcription complex from assembling.

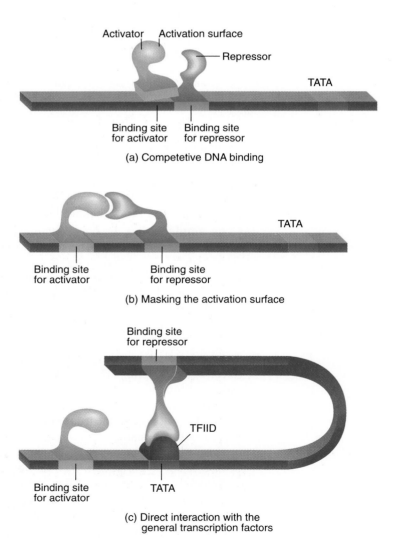

(a) Competetive DNA binding

(b) Masking the activation surface

(c) Direct interaction with the general transcription factors

tein synthesis. The covalent modification of several translation factors has been observed to alter the overall protein synthesis rate and/or enhance the translation of specific mRNAs. For example, the phosphorylation of the protein eIF-2 affects the rate of hemoglobin synthesis in rabbit reticulocytes (immature red blood cells).

In contrast to single-cell organisms in which cell growth and cell division are governed largely by nutrient availability, the proliferation of cells in multicellular organisms is carefully regulated. This regulation (a complex process that is understood only in general terms) is known to involve both positive and negative controls. Positive control is exerted largely by binding growth factors to specialized cell receptors. The initiation of cell division typically requires binding a variety of such factors. Negative controls appear to involve several checkpoints at different stages of the cell cycle. Recall that the cell cycle is divided into four phases: M, G_1, S, and G_2 (Figure 17.7). Investigations of mutant cell types reveal that checkpoints occur within G_1 (in yeast cells it is referred to as START), G_2, and M phases. The cell is held in apparent stasis at these checkpoints to both ensure that required cellular processes have occurred (e.g., sufficient cell growth in G_2 or alignment of the chromosomes in M) and allow regulation by external signals (i.e., hormones and growth factors). The cell cycle control system is believed to involve interactions between two groups of proteins: the *cyclin-dependent protein kinases* (Cdks), a group of enzymes that activate key proteins, and the *cyclins,* a group of regulatory proteins that bind to Cdks. Cell proliferation is also inhibited by *tumor suppressor genes.* Well-known examples of these genes include Rb (so named because of its role in retinoblastoma, a childhood eye cancer) and the p53 gene, which regulates cell cycle progression. When a certain amount of DNA damage occurs, cell cycle progression is arrested. This inhibition provides an opportunity for DNA repair. If the damage is unrepairable, the p53 gene product initiates a complex destructive mechanism called programmed cell death or **apoptosis.**

The positive effects exerted by growth factors are now believed to include gene expression that specifically overcomes the inhibitions at the cell cycle checkpoints, especially the G_1 checkpoint. The binding of growth factors to their cell surface receptors initiates a cascade of reactions that induces two classes of genes.

1. **Early response genes.** These genes are rapidly activated, usually within 15 minutes. Among the best-characterized early response genes are the jun, fos, and myc protooncogenes. (**Protooncogenes** are normal genes that, if mutated, can promote carcinogenesis. Refer to Special Interest Box 17.1.) Each of the jun and fos protooncogene families codes for a series of transcription factors containing leucine zipper domains. Both jun and fos proteins form dimers that can bind DNA. Among the best-characterized of these is a jun-fos heterodimer, referred to as AP-1, which forms through a leucine zipper interaction. Although myc gene expression is known to be critically important in normal cell function (i.e., the inappropriate expression of myc is found in several types of cancer), the biochemical function of this gene family remains unresolved. However, myc gene products probably act as transcription factors.

2. **Delayed response genes.** These genes are induced by the activities of the transcription factors and other proteins produced or activated during the early response phase. Among the products of the delayed response genes are the Cdks, the cyclins, and other components required for cell division.

Work in a large number of laboratories has made progress in resolving some of the early events in growth factor-triggered cell division. A summary of selected aspects of this process is outlined in Figure 17.28.

As mentioned, many growth factors (e.g., PDGF) bind to receptors that possess a cytoplasmic domain with tyrosine kinase activity. When such receptors are activated, various intracellular proteins are phosphorylated. One of the consequences of the phosphorylation cascade is the activation of ras, a family of plasma membrane–associated guanine nucleotide–binding proteins. When ras is active, it binds GTP. (Ras activation

CARCINOGENESIS

Cancer is a group of diseases in which genetically damaged cells proliferate autonomously. Such cells cannot respond to normal regulatory mechanisms that ensure the intercellular cooperation required in multicellular organisms. Consequently, they continue to proliferate, thereby robbing nearby normal cells of nutrients and eventually crowding surrounding healthy tissue. Depending on the damage they have sustained, abnormal cells may form either benign or malignant tumors. Benign tumors, which are slow-growing and limited to a specific location, are not considered cancerous and rarely cause death. In contrast, *malignant* tumors are often fatal because they can undergo metastasis. (In metastasis, cancer cells migrate through blood or lymph vessels to distant locations throughout the body.) Wherever new malignant tumors arise, they interfere with normal functions. When life-sustaining processes fail patients die.

Cancers are classified by the tissues affected. The vast majority of cancerous tumors are carcinomas (tumors derived from epithelial tissue cells such as skin, various glands, breasts, and most internal organs). In the leukemias, cancers of the bone marrow, excessive leukocytes are produced. Similarly, the lymphocytes produced in the lymph nodes and spleen proliferate uncontrollably in the lymphomas. Tumors arising in connective tissue are called sarcomas. Despite the differences among this diverse class of diseases, they also have several common characteristics, among which are the following:

1. **Cell culture properties.** When grown in culture, most tumor cells lack contact inhibition, that is, they grow to high density in highly disorganized masses. (Normal cells grow only in a single layer of cells.) In contrast to normal cells, cancer cell growth and division tend to be growth factor–independent, and these cells often do not require attachment to a solid surface. The hallmark of cancer cells, however, is their immortality. Normal cells undergo cell division only a finite number of times, whereas cancer cells can proliferate indefinitely.

2. **Origin.** Each tumor originates from a single damaged cell. In other words, a tumor is a clone derived from a cell in which heritable changes have occurred. The genetic damage consists of mutations (e.g., point mutations, deletions, and inversions) and chromosomal rearrangements or losses. Such changes result in the loss or altered function of molecules involved in cell growth or proliferation. Tumors typically develop over a long time and involve several independent types of genetic damage. (The risk of many types of cancer increases with age.)

The transformation process in which an apparently normal cell is converted or "transformed" into a malignant cell consists of three stages: initiation, promotion, and progression.

During the *initiation* phase of carcinogenesis, a permanent change in a cell's genome provides it with a growth advantage over its neighbors. Most initiating mutations affect protooncogenes or tumor suppressor genes. Protooncogenes code for a variety of growth factors, growth factor receptors, enzymes, or

TABLE 1

Selected Oncogenes*

Oncogene	Function
sis	Platelet-derived growth factor
erbB	Epidermal growth factor receptor
src	Tyrosine-specific protein kinase
raf	Serine/threonine-specific protein kinase
ras	GTP-binding protein
jun	Transcription factor
fos	Transcription factor
myc	DNA-binding protein (transcription factor?)

*Abnormal versions of protooncogenes that mediate cancerous transformations.

transcription factors that promote cell growth and/or cell division. Mutated versions of protooncogenes that promote abnormal cell proliferation are called **oncogenes** (Table 1). Because tumor suppressor genes suppress carcinogenesis, their loss also facilitates tumor development. Recall that Rb and p53 are tumor supressors. Other examples are FCC and DCC, which are both associated with susceptibility to colon cancer. The functions of most tumor suppressor genes are unknown. However, it now appears that p53 codes for a 21-kD protein that normally inhibits Cdk enzymes. Recent evidence indicates that other damaged or deleted tumor suppressor genes may code for enzymes involved in DNA repair mechanisms. The damage that alters function of protooncogenes and tumor suppressor genes is caused by the following:

1. **Carcinogenic chemicals.** Most cancer-causing chemicals are mutagenic, that is, they alter DNA structure. Some carcinogens (e.g., nitrogen mustard) are highly reactive electrophiles that attack electron-rich groups in DNA (as well as RNA and protein). Other carcinogens (e.g., benzo[a]pyrene) are actually procarcinogens, which are converted to active carcinogens by one or more enzyme-catalyzed reactions. (Refer to Chapter 19.)

2. **Radiation.** Some radiation (UV, X-rays, and γ-rays) is carcinogenic. As noted, the damage inflicted on DNA includes single- and double-strand breaks, pyrimidine dimer formation, and the loss of both purine and pyrimidine bases. Radiation exposure also causes ROS to form. ROS may be responsible for most of radiation's carcinogenic effects.

3. **Viruses.** Viruses appear to contribute to the transformation process in several ways. Some introduce oncogenes into a host cell chromosome as they insert their genome. (Viral oncogenes are now recognized as sequences that are similar to normal cellular genes that have been picked up accidentally from a previous host cell. To distinguish viral oncogenes and their cellular counterparts, they are referred to as v-onc and c-onc, respectively.) Viruses can also affect the expression of cellular protooncogenes through inser-

tional mutagenesis, a random process in which viral genome insertion inactivates a regulatory site or alters the protooncogene's coding sequence. Most virus-associated cancers have been detected in animals. Only a few human cancers have been proven to be associated with viral infection.

Tumor development can also be promoted by chemicals that do not alter DNA structure. So-called **tumor promoters** contribute to carcinogenesis by two principal methods. By activating components of intracellular signaling pathways, some molecules (e.g., the phorbol esters) provide the cell a growth advantage over its neighbors. (Recall that phorbol esters activate PKC because they mimic the actions of DAG.) The effects of many other tumor promoters are unknown but may involve transient effects such as increasing cellular Ca^{2+} levels or increasing synthesis of the enzymes that convert procarcinogens into carcinogens. Unlike initiating agents, the effects of tumor promoters are reversible. They produce permanent damage only with prolonged exposure after an affected cell has undergone an initiating mutation.

Following initiation and promotion, cells go through a process referred to as progression. During *progression,* genetically vulnerable precancerous cells, which already possess significant growth advantages over normal cells, are further damaged. Eventually, the continued exposure to carcinogens and promoters makes further random mutations inevitable. If these mutations affect cellular proliferative or differentiating capacity, then an affected cell may become sufficiently malignant to produce a tumor. A proposed sequence of the events in the development of colorectal cancer is outlined in Figure 17A.

An Ounce of Prevention . . .

Because of the enormous cost and limited success of cancer therapy, it has become increasingly recognized that cancer prevention is cost-effective. Recent research indicates that the majority of cancer cases are preventable. For example, over one-third of cancer mortality is directly caused by tobacco use, while another one-third of cancer deaths have been linked to inadequate diets. Tobacco smoke, which contains thousands of chemicals, many of which are either carcinogens or tumor promoters, is responsible for most cases of lung cancer and contributes to cancers of the pancreas, bladder, and kidneys, among others. Diets that are high in fat and low in fiber content have been associated with increased incidence of cancers of the large bowel, breast, pancreas, and prostate. Other dietary risk factors include low consumption of fresh vegetables and fruit.

In addition to providing sufficient antioxidant vitamins, many vegetables (and fruits to a lesser extent) contain numerous nonnutritive components that actively inhibit carcinogenesis. Some carcinogenesis inhibitors (e.g., organosulfides), referred to as *blocking agents,* prevent carcinogens from reacting with DNA or inhibit the activity of tumor promoters. Other inhibitors, referred to as *suppressing agents* (e.g., inositol hexaphosphate), prevent the further development of neoplastic processes that are already in progress. Many nonnutritive food components (e.g., tannins and protease inhibitors) possess both blocking and suppressing effects. In general, these molecules very effectively protect against cancer because many of them inhibit the arachidonic acid cascade and oxidative damage. Apparently low-fat, high-fiber diets that are rich in raw or fresh leafy green, cruciferous, and allium vegetables (e.g., spinach, broccoli, and onions), as well as fresh fruits, are a prudent choice for individuals seeking to reduce their risk of cancer.

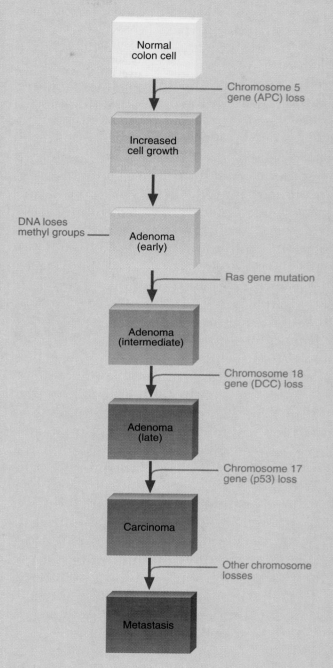

FIGURE 17A

The Development of Colorectal Cancer.

Colorectal cancer develops over a long period of time. Because somatic cells are diploid, the loss of the tumor suppressor genes APC, DCC, and p53 usually requires two mutations. Recent research suggests that mutations in genes involved in DNA repair processes may occur during the early stages of colon cancer. (An adenoma is a precancerous epithelial tumor.)

is mediated by guanine nucleotide-releasing proteins, or GNRPs, which cause ras to release GDP and bind a GTP. Ras becomes inactive when GTP is hydrolyzed, a reaction catalyzed by GTPase activating proteins, or GAPs.) Ras activation is also promoted by a class of tyrosine kinases, referred to as src, which become associated through noncovalent bonding to a class of cell surface receptors, referred to as tyrosine kinase–associated receptors, each of which lacks an intracellular phosphorylating domain of its own. Examples of such receptors include those for growth hormone and a variety of local mediators of blood cell proliferation (e.g., interleukin 2, a powerful proliferative signal for T cells). Finally, ras has a complex relationship with protein kinase C (PKC). (Recall that when activated by DAG, PKC phosphorylates several cellular proteins. DAG results from phosphatidyl inositol cleavage.) The relationship between ras and PKC appears to be complex because they seem to have overlapping functions.

Both ras and PKC initiate phosphorylation cascades. In the example in Figure 17.28, ras binding activates raf-1, a serine/threonine kinase. Once it is active, raf-1 phosphorylates MAP kinase kinase (MAP stands for "mitogen-activated protein"; a mitogen is any molecule that stimulates cell division). MAP kinase kinase, in turn, phosphorylates both a tyrosine and a tryptophan of MAP kinase. (This unusual reaction appears to en-

KEY *Concepts*

The following mechanisms appear to be involved in the regulation and expression of eukaryotic genes: gene rearrangements, gene amplification, transcriptional control, RNA processing, RNA transport, and translational control.

FIGURE 17.28

Eukaryotic Gene Expression Triggered by Growth Factor Binding.

Eukaryotic cells require stimulation by several growth factors to begin cell division. In addition, growth factors often act through several different mechanisms at the same time. For example, PDGF induces gene expression via DAG synthesis (PKC activation) and via ras activation. The red arrows indicate an activating process.

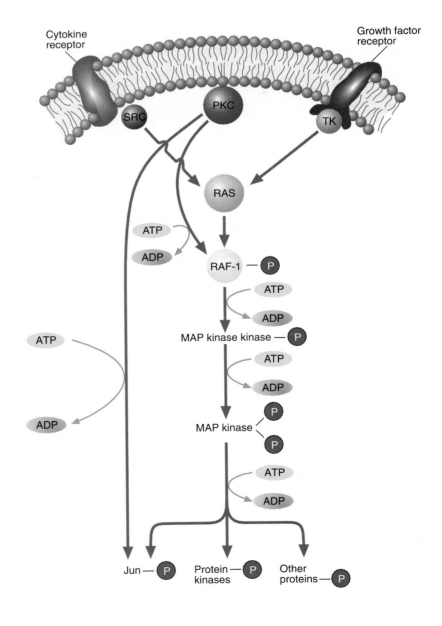

sure that MAP kinase is activated only by MAP kinase kinase.) Active MAP kinase then phosphorylates a variety of cellular proteins. Among these are jun (shown in the figure) and elk-1. (Phosphorylated elk-1 promotes the transcription of the fos gene.) Phosphorylated jun and fos protein then combine to form the transcription factor AP-1. AP-1 subsequently promotes the transcription of several delayed response genes.

QUESTION 17.5

The mechanism by which light influences plant gene expression is referred to as *photomorphogenesis*. Because of serious technical problems with plant cell culture, relatively little is known about plant gene expression. However, certain DNA sequences, referred to as *light-responsive elements* (LRE), have been identified. On the basis of the gene expression patterns observed in animals, can you suggest (in general terms) a mechanism whereby light induces gene expression? (Hint: Recall that phytochrome is an important component of light-induced gene expression.)

KEY WORDS

apoptosis, *519*
coding strand, *506*
conjugation, *501*
consensus sequence, *507*
constitutive gene, *512*
excision repair, *499*
general recombination, *500*
light-induced repair, *500*
Okazaki fragment, *494*

oncogene, *520*
operon, *508*
photoreactivation repair, *500*
primer, *493*
primosome, *493*
promoter, *507*
protooncogene, *519*
recombination, *491*
recombinational repair, *500*

replication fork, *494*
replicon, *494*
replisome, *493*
ribozyme, *511*
semiconservative replication, *492*
site-specific recombination, *501*
spliceosome, *511*
splicing, *511*

structural gene, *491*
transduction, *501*
transformation, *501*
transposable element, *501*
transposition, *501*
transposon, *503*
tumor promoter, *521*

SUMMARY

DNA structure and function are so important for living organisms that they must possess efficient mechanisms for the rapid and accurate synthesis of DNA. DNA synthesis, referred to as replication, occurs by a semiconservative mechanism, that is, each of the two parental strands serves as a template to synthesize a new strand.

There are several types of DNA repair mechanisms. These include excision repair, photoreactivation, and recombinational repair.

Genetic recombination, a process in which DNA sequences are exchanged between different DNA molecules, occurs in two forms. In general recombination, the exchange occurs between sequences in homologous chromosomes. In site-specific recombination, the exchange of sequences requires only short homologous sequences.

DNA-protein interactions are principally responsible for the exchange of largely nonhomologous sequences.

The synthesis of RNA, sometimes referred to as DNA transcription, requires a variety of proteins and associated proteins. Transcription initiation involves binding an RNA polymerase to a specific DNA sequence called a promoter. Regulation of transcription differs significantly between prokaryotes and eukaryotes.

The control of transcription, referred to as gene expression, is still poorly understood. However, because of intensive research, many of the details in several examples of gene expression in both prokaryotes and eukaryotes are now known.

SUGGESTED READINGS

Bernstein, C., and Bernstein, H., *Aging, Sex and DNA Repair*, Academic Press, New York, 1991.

Cech, T.R., RNA as an Enzyme, *Sci. Amer.*, 255(5):64–75, 1986.

Cohen, S. N., and Shapiro, J.A., Transposable Genetic Elements, *Sci. Amer.*, 242:40–49, 1980.

Grunstein, M., Histones as Regulators of Genes, *Sci. Amer.*, 267(4): 68–74, 1992.

Kornberg, A., and Baker, T.A., *DNA Replication*, 2nd ed., W.H. Freeman, New York, 1992.

McKnight, S.L., Molecular Zippers in Gene Regulation, *Sci. Amer.*, 264(4):54–64, 1991.

Mitchell, P.J., and Tjian, R., Transcriptional Regulation in Mammalian Cells by Sequence-Specific DNA Binding Proteins, *Science*, 245:371–378, 1989.

Moses, P.B., and Chua, N.-H., Light Switches for Plant Genes, *Sci. Amer.*, 258(4):88–93, 1988.

Ptashne, M., How Eukaryotic Transcriptional Activators Work, *Nature*, 335:683–689, 1988.

Vos, J.-M. H., *DNA Repair Mechanisms: Impact on Human Diseases and Cancer*, R.G. Landes, Austin, 1995.

Watson, J.D., Gilman, M., Witkowski, J., and Zoller, M., *Recombinant DNA*, 2nd ed., W.H. Freeman, New York, 1992.

QUESTIONS

1. Clearly define the following terms:
   a. chemoreceptors
   b. structural genes
   c. recombination
   d. semiconservative replication
   e. replisome
   f. oriC
   g. transcription
   h. protooncogene
   i. splicosome
   j. oncogene
2. How does negative supercoiling promote the initiation of replication?
3. Explain how the Messelson-Stahl experiment supports the semi-conservative model of DNA replication.
4. In the Messelson-Stahl experiment, why was a nitrogen isotope chosen rather than a carbon isotope?
5. List and describe the steps in prokaryotic DNA replication. How does this process appear to differ from eukaryotic DNA replication?
6. Indicate the stage of DNA replication when each of the following enzymes is active:
   a. helicase
   b. primase
   c. DNA polymerases
   d. ligase
   e. topoisomerase
   f. DNA gyrase
7. The bidirectional synthesis of DNA implies that one strand is synthesized in the $5' \rightarrow 3'$ direction and the other in the $3' \rightarrow 5'$ direction. However, all known enzymes that synthesize DNA do so in the $5' \rightarrow 3'$ direction. How did Reiji Okazaki explain this paradox?
8. DNA is polymerized in the $5' \rightarrow 3'$ direction. Demonstrate with the incorporation of three nucleotides into a single strand of DNA how the $5' \rightarrow 3'$ directionality is derived.
9. In eukaryotes the DNA replication rate is 50 nucleotides per second. How long does the replication of a chromosome of 150 million base pairs take? If eukaryotic chromosomes were replicated like those of prokaryotes, the replication of a genome would take months. Actually, eukaryotic replication takes only several hours. How do eukaryotes achieve this high rate?
10. Mutations are caused by chemical and physical phenomena. Indicate the type of mutation that each of the following reactions or molecules might cause:
    a. ROS
    b. caffeine
    c. a small alkylating agent
    d. a large alkylating agent
    e. nitrous acid
    f. intercalating agents
11. How can viruses cause mutations?
12. There are three principal mechanisms of DNA repair. What are they and how do they operate?

13. Describe two forms of genetic recombination. What functions do they fulfill?
14. Although genetic variation is required for species to adapt to changes in their environment, most genetic changes are detrimental. Explain why genetic mutations are rarely beneficial.
15. General recombination occurs in bacteria, where it is involved in several types of intermicrobial DNA transfer. What are these types of transfer and by what mechanisms do they occur?
16. There appears to be insufficient genetic material to direct all the activities of several types of eukaryotic cell. Explain how genetic recombination helps to solve this problem.
17. What are the two types of transposition mechanisms and by what mechanisms do they occur?
18. Mustard gas is an extremely toxic substance that in large amounts severely damages lung tissue when it is inhaled. In small amounts, mustard gas is a mutagen and carcinogen. Considering that mustard gas is a binfunctional alkylating agent, explain how it inhibits DNA replication.
19. Within cells, cytosine slowly converts to uracil. To what type of mutation would this lead in DNA molecules? Why is this not a problem in RNA?
20. Adjacent pyrimidine bases in DNA form dimers with high efficiency after exposure to UV light. If these dimers are not repaired, skin cancers can result. Melanin is a natural sun-screen produced by melanocytes, a type of skin cell, when the skin is exposed to sunlight. Individuals who spend long periods over many years developing a tan eventually acquire thick and highly wrinkled skin. Such individuals are also at high risk for skin cancer. Can you explain, in general terms, why these phenomena are related?
21. A correlation has been found among species between life span and the efficiency of DNA repair systems. Suggest a reason why this is so.
22. Phorbol esters have been observed to induce the transcription of AP-1-influenced genes. Explain how this process could occur. What are the consequences of AP-1 transcription? What role would intermittent exposure to phorbol esters have on an individual's health?
23. Because of overuse of antibiotics and/or weakened governmental surveillance of infectious disease, several diseases that had been thought to be no longer a threat to human health (e.g., pneumonia and tuberculosis) are rapidly becoming unmanageable. In several instances, so-called "superbugs" (microorganisms that are resistant to almost all known antibiotics) have been detected. How did this circumstance arise? What will happen in the future if this process continues?
24. Retinoblastoma is a rare cancer in which tumors develop in the retina of the eye. The tumors arise because of the loss of the Rb gene, which codes for a tumor suppressor. Hereditary retinoblastoma usually occurs during childhood. Such individuals inherit only one functional copy of Rb. Explain why nonhereditary retinoblastoma usually occurs later in life.

PROTEIN SYNTHESIS

Chapter Eighteen

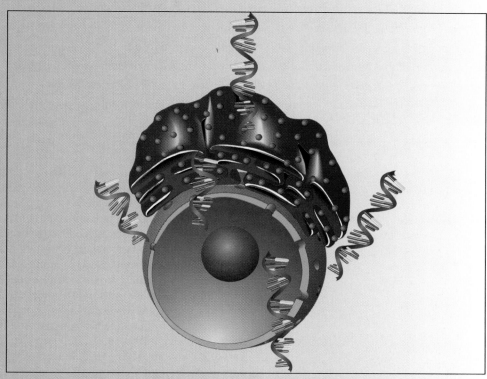

In eukaryotes the nucleic acids used in protein synthesis are produced within the nucleus. Some polypeptides are inserted into the RER membrane during or after synthesis.

OBJECTIVES

After you have studied this chapter, you should be able to answer these questions:

1. What is the RNA world hypothesis?
2. What is the genetic code and what role does it play in gene expression?
3. What is the wobble hypothesis?
4. Why does the functioning of the aminoacyl-tRNA synthetases constitute a second genetic code?
5. What are the principal phases of translation?
6. What role does polypeptide targeting play in cellular metabolism?
7. What are the principal differences between prokaryotic and eukaryotic translation processes?
8. What types of prominent posttranslational modifications are found in prokaryotes and eukaryotes? What functions do these chemical alterations serve?
9. What is the significance of the signal hypothesis?
10. How do prokaryotes and eukaryotes appear to control translation?
11. What functions do molecular chaperones serve?

Proteins are perhaps the most dynamic and varied class of biomolecules. As described, in addition to providing structural components, proteins are largely responsible for promoting many of the most dynamic aspects of living processes. The roles that proteins play in living organisms are astounding. In addition to the unbelievably diverse catalytic proteins, protein receptors mediate the actions of untold numbers of signal molecules, many of which are also proteins. The uniqueness of each cell type is due almost entirely to the proteins it produces. It is not surprising, therefore, that the expression of most genes alters patterns of protein synthesis. Because of their strategic importance in the cellular economy, protein synthesis is a regulated process. Although control is also of major importance at the transcriptional level, control of the translation of genetic messages allows for additional opportunities for regulation. This is especially true in multicellular eukaryotes, whose complex lifestyles require amazingly diverse regulatory mechanisms.

Protein synthesis is an extraordinarily complex process in which genetic information encoded in the nucleic acids is translated into the 20 amino acid "alphabet" of polypeptides. In addition to translation (the mechanism by which a nucleotide base sequence directs the polymerization of amino acids), protein synthesis can also be considered to include the processes of posttranslational modification and targeting. Posttranslational modification consists of chemical alterations cells use to prepare polypeptides for their functional roles. Several modifications assist in targeting, which directs newly synthesized molecules to a specific intracellular or extracellular location.

In all, at least 100 different molecules are involved in protein synthesis. Among the most important of these are the components of the ribosome, a sophisticated ribonucleoprotein machine that rapidly and precisely decodes genetic messages. Since their discovery in the 1950s, detailed structural investigations of prokaryotic and eukaryotic ribosomes have revealed significant details. For example, ribosome structure appears to be highly conserved among species. Although there are notable differences among the ribosomes of various species, similarities in the three-dimensional structures of rRNA and ribosomal proteins are even more striking. As noted, despite differences in rRNA base sequences, the secondary structure of these molecules is amazingly similar.

Recent research has revealed an even more interesting facet of ribosomal structure and function. Because of the well-documented catalytic properties of protein, it was presumed from the beginning of ribosomal research that the principal aspect of ribosomal function, that is, peptide bond formation, is catalyzed by a ribosomal protein component. rRNA was believed to serve as a structural framework for the translation process. However, it is becoming increasingly apparent that rRNA, far from being an inert scaffold for protein synthesis, plays a central and very active role. For example, it has recently been discovered that peptidyl transferase, the enzymatic activity that catalyzes the formation of peptide bonds, resides in the 23S rRNA component of bacterial ribosomes. Removing protein from bacterial ribosomes while leaving remaining rRNA molecules relatively intact does not substantially affect their capacity to catalyze protein synthesis. In addition, various evidence has linked specific rRNA molecules to roles in tRNA and mRNA binding, ribosomal subunit association, proofreading, and some regulatory aspects of translation (e.g., binding translation factors).

Investigations into the roles of ribosomal proteins have also provided some surprises. For example, several ribosomal proteins possess DNA binding domains (e.g., zinc finger and helix-turn-helix motifs). Additionally, ribosomal proteins may possess functions other than protein synthesis. Certain eukaryotic ribosomal proteins have DNA repair activities. These findings, as well as several others, support the **RNA world hypothesis,** a concept devised several decades ago to explain how life evolved.

One of the principal issues of biochemical evolution (the study of the molecular origins of life) has been whether DNA or protein appeared first. As discussed, the information encoded in DNA is required to direct the synthesis of proteins. However, certain proteins are required for DNA replication. According to the RNA world hypothesis, neither DNA nor protein came first. Instead, the first living cells that evolved contained a molecule that possessed both informational and catalytic properties, namely, RNA. In this paradigm, RNA molecules could replicate themselves to some degree and could catalyze primitive reactions. Later, as living cells became more complex, DNA was acquired to serve as a more stable information storage molecule. As interesting as these ideas are, however, there is as yet insufficient data to support them. Many investigators now believe that modern cells (especially their ribosomal components) may provide clues to the origins of life, a process that began over three billion years ago.

The focus of this chapter is modern protein synthesis. The chapter begins with a discussion of the genetic code, the mechanism by which nucleic acid base sequences specify the amino acid sequences of polypeptides. This is followed by discussions of protein synthesis as it occurs in both prokaryotes and eukaryotes. The chapter ends with a description of the mechanisms that convert polypeptides into their biologically active conformations.

KEY *Concepts*

In modern organisms, proteins are synthesized by ribosomes, which translate an RNA base sequence into an amino acid sequence.

18.1 THE GENETIC CODE

It became apparent during early investigations of protein synthesis that translation is fundamentally different from the transcription process that precedes it. During transcription the language of DNA sequences is converted to the closely related dialect of RNA sequences. During protein synthesis, however, a nucleic acid base sequence is converted to a clearly different language (i.e., an amino acid sequence), hence the term "translation." Because mRNA and amino acid molecules have little natural affinity for each other, researchers (e.g., Francis Crick) knew that a series of adaptor molecules must mediate the translation process. This role was eventually assigned to tRNA molecules (Figure 16.17).

Before adaptor molecules could be identified, however, a more important problem had to be solved: deciphering the genetic code. The **genetic code** can be described as a coding dictionary that specifies a meaning for each base sequence. Once the importance of the genetic code was recognized, investigators speculated about its dimensions. Because only four different bases (G, C, A, and U) occur in mRNA and 20 amino acids must be specified, it appeared that a combination of bases coded for each amino acid. A sequence of two bases would specify only a total of 16 amino acids (i.e., $4^2 = 16$). However, a three-base sequence provides more than sufficient base combinations for translation (i.e., $4^3 = 64$).

The first major breakthrough in assigning mRNA triplet base sequences (later referred to as **codons**) came in 1961, when Marshall Nirenberg and Heinrich Matthaei performed a series of experiments using an artificial test system containing an extract of *Escherichia coli* fortified with nucleotides, amino acids, ATP, and GTP. They showed that poly U (a synthetic polynucleotide whose base components consist only of uracil) directed the synthesis of polyphenylalanine. Assuming that codons consist of a three-base sequence, Nirenberg and Matthaei surmised that UUU codes for the amino acid phenylalanine. Subsequently, they repeated their experiment using poly A and poly C. Because polylysine and polyproline products resulted from these tests, the codons AAA and CCC were assigned to lysine and proline, respectively.

Most of the remaining codon assignments were determined by using synthetic polynucleotides with repeating sequences. Such molecules were constructed by enzymatically amplifying short chemically synthesized sequences. The resulting polypeptides, which contained repeating peptide segments, were then analyzed. The information obtained from this technique, devised by Har Gobind Khorana, was later supplemented with a strategy used by Nirenberg. This latter technique measured the capacity of specific trinucleotides to promote tRNA binding to ribosomes.

TABLE 18.1

The Genetic Code

		Second Position								
		U		**C**		**A**		**G**		
First position (5′ end)	**U**	UUU UUC	Phe	UCU UCC	Ser	UAU UAC	Tyr	UGU UGC	Cys	U C
		UUA UUG	Leu	UCA UCG		UAA UAG	STOP	UGA UGG	STOP Trp	A G
	C	CUU CUC	Leu	CCU CCC	Pro	CAU CAC	His	CGU CGC	Arg	U C
		CUA CUG		CCA CCG		CAA CAG	Gln	CGA CGG		A G
	A	AUU AUC	Ile	ACU ACC	Thr	AAU AAC	Asn	AGU AGC	Ser	U C
		AUA AUG	Met	ACA ACG		AAA AAG	Lys	AGA AGG	Arg	A G
	G	GUU GUC	Val	GCU GCC	Ala	GAU GAC	Asp	GGU GGC	Gly	U C
		GUA GUG		GCA GCG		GAA GAG	Glu	GGA GGG		A G

(Third position (3′ end) shown in rightmost column)

The codon assignments for the 64 possible trinucleotide sequences are presented in Table 18.1. Of these, 61 code for amino acids. The remaining three codons (UAA, UAG, and UGA) are *stop* (polypeptide chain terminating) signals. AUG, the codon for methionine, also serves as a *start* signal (sometimes referred to as the *initiating codon*). The genetic code is now believed to possess the following properties:

1. **Degenerate.** Any coding system in which several signals have the same meaning is said to be degenerate. The genetic code is partially degenerate because most amino acids are coded for by several codons. For example, leucine is coded for by six different codons (UAA, UUG, CUU, CUC, CUA, and CUG). In fact, methionine (AUG) and tryptophan (UGG) are the only amino acids that are coded for by a single codon.

2. **Specific.** Each codon is a signal for a specific amino acid. The majority of codons that code for the same amino acid possess similar sequences. For example, in each of the four serine codons (UCU, UCC, UCA, and UCG) the first and second bases are identical. Consequently, a point mutation in the third base of a serine codon would not be deleterious.

3. **Nonoverlapping and without punctuation.** The mRNA coding sequence is "read" by a ribosome starting from the initiating codon (AUG) as a continuous sequence taken three bases at a time until a stop codon is reached. A set of contiguous triplet codons in an mRNA is called a **reading frame.** The term **open reading frame** describes a series of triplet base sequences in mRNA that do not contain a stop codon.

4. **Universal.** With a few minor exceptions (described below) the genetic code is universal. In other words, examinations of the translation process in the species that have been investigated have revealed that the coding signals for amino acids are always the same.

KEY *Concepts*

The genetic code is a mechanism by which ribosomes translate nucleotide base sequences into the primary sequence of polypeptides.

18.1

As described, DNA damage can cause deletion or insertion of base pairs. If a nucleotide base sequence of a coding region changes by any number of bases other than three base pairs, or multiples of three, a frame shift mutation occurs. Depending on the location of the sequence change, such mutations can have serious effects. The following synthetic mRNA sequence codes for the beginning of a polypeptide:

5'-AUGUCUCCUACUGCUGACGAGGGAAGGAGGUGGCUUAUCAUGUUU-3'

First, determine the amino acid sequence of the polypeptide. Then determine the type of mutations that have occurred in the following altered mRNA segments. What effect do these mutations have on the polypeptide products?

a. 5'-AUGUCUCCUACUUGCUGACGAGGGAAGGAGGUGGCUUAUCAUGUUU-3'
b. 5'-AUGUCUCCUACUGCUGACGAGGGAGGAGGUGGCUUAUCAUGUUU-3'
c. 5'-AUGUCUCCUACUGCUGACGAGGGAAGGAGGUGGCCCUUAUCAUGUUU-3'
d. 5'-AUGUCUCCUACUGCUGACGGAAGGAGGUGGCUUAUCAUGUUU-3'

Codon-Anticodon Interactions

tRNA molecules are the "adapters" that are required for the translation of the genetic message. Recall that each type of tRNA binds a specific amino acid (at the 3' terminus) and possesses a three-base sequence called the **anticodon.** The base pairing between the anticodon of the tRNA and an mRNA codon is responsible for the actual translation of the genetic information of structural genes. Although codon-anticodon pairings are antiparallel, both sequences are given in the 5' → 3' direction. For example, the codon UGC binds to the anticodon GCA (Figure 18.1).

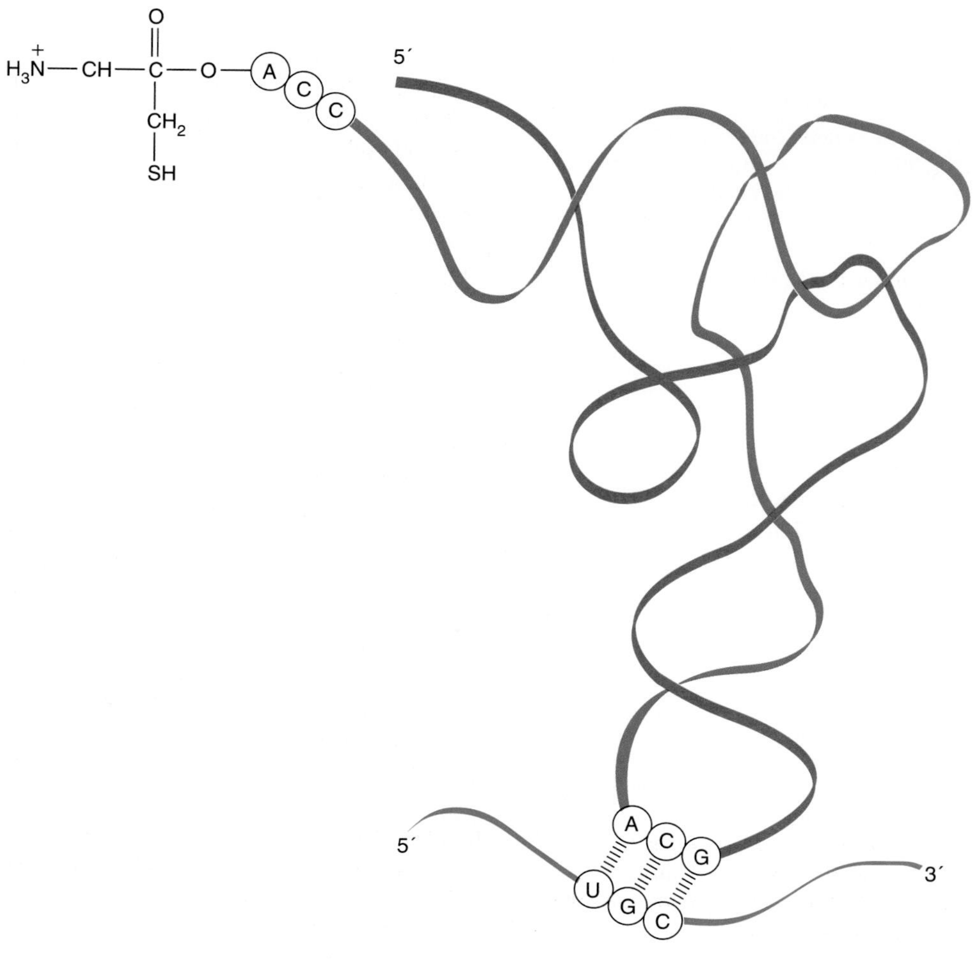

FIGURE 18.1

Codon-Anticodon Base Pairing.

The pairing of the codon UGC with the anticodon GCA ensures that the amino acid cysteine will be incorporated into a growing polypeptide chain.

Once the genetic code was broken, researchers anticipated the identification of 61 types of tRNAs in living cells. Instead, they discovered that cells often operate with substantially fewer tRNAs than expected. Most cells possess about 50 tRNAs, although lower numbers have been observed. Further investigation of tRNAs revealed that the anticodon in some molecules contains uncommon nucleotides, such as inosinate (I), which typically occur at the third anticodon position. (In eukaryotes, A in the third anticodon position is deaminated to form I.) As tRNAs were investigated, it became increasingly clear that some molecules recognize several codons. In 1966, after reviewing the evidence, Crick proposed a rational explanation, the **wobble hypothesis.**

The wobble hypothesis, which allows for multiple codon-anticodon interactions by individual tRNAs, is based principally on the following observations:

1. The first two base pairings in a codon-anticodon interaction confer most of the specificity required during translation. Recall that most redundant codons specifying a certain amino acid possess identical nucleotides in the first two positions. These interactions are standard (i.e., Watson-Crick) base pairings.

2. The interactions between the third codon and anticodon nucleotides are less stringent. In fact, nontraditional base pairs (i.e., non-Watson-Crick) often occur. For example, tRNAs containing G in the 5′ (or "wobble") position of the anticodon can pair with two different codons (i.e., G can interact with either C or U). The same is true for U, which can interact with A or G. When I is in the wobble position of an anticodon, a tRNA can base pair with three different codons, since I can interact with U or A or C.

A careful examination of the genetic code and the "wobble rules" indicates that a minimum of 31 tRNAs are required to translate all 61 codons. An additional tRNA for initiating protein synthesis brings the total to 32 tRNAs.

Chloroplast and mitochondrial genomes encode fewer tRNAs than nuclear genomes. Chloroplasts possess about 30 tRNAs, whereas mitochondria have only 24. It has been suggested that mitochondria can function with a reduced complement of tRNAs because of the smaller size of their genomes. Chloroplasts and mitochondria are also unique in another respect. Many of the known variations in the genetic code appear in these organelles. The most dramatic and best-documented of these variations are in animal mitochondria. In the mitochondria of humans and other vertebrates, for example, AUA and UGA code for methionine and tryptophan instead of isoleucine and a stop signal, respectively. Similarly, two codons (AGA and AGG), which ordinarily code for arginine, code instead for stop signals. The significance of these alterations is unclear at present.

KEY *Concepts*

The genetic code is translated through base pairing interactions between mRNA codons and tRNA anticodons. The wobble hypothesis explains why cells usually have fewer tRNAs than expected.

The Aminoacyl-tRNA Synthetase Reaction: The Second Genetic Code

Although the accuracy of translation (approximately one error per 10^4 amino acids incorporated) is lower than those of DNA replication and transcription, it is remarkably higher than one would expect of such a complex process. The principal reasons for the accuracy with which amino acids are incorporated into polypeptides include codon-anticodon base pairing and the mechanism by which amino acids are attached to their cognate tRNAs. The attachment of amino acids to tRNAs, considered the first step in protein synthesis, is catalyzed by a group of enzymes called the aminoacyl-tRNA synthetases. The precision with which these enzymes esterify each specific amino acid to the correct tRNA is now believed to be so important for accurate translation that their functioning has been referred to collectively as the **second genetic code.**

In most organisms there is at least one aminoacyl-tRNA synthetase for each of the 20 amino acids. (Each enzyme links its specific amino acid to any appropriate tRNA. This is important, since in most cells many amino acids have several cognate tRNAs each.) The process that links an amino acid to the 3′ terminus of the correct tRNA

FIGURE 18.2

Formation of Aminoacyl-tRNA.

Each aminoacyl-tRNA synthetase catalyzes two sequential reactions in which an amino acid is linked to the 3′ terminal ribose residue of the tRNA molecule.

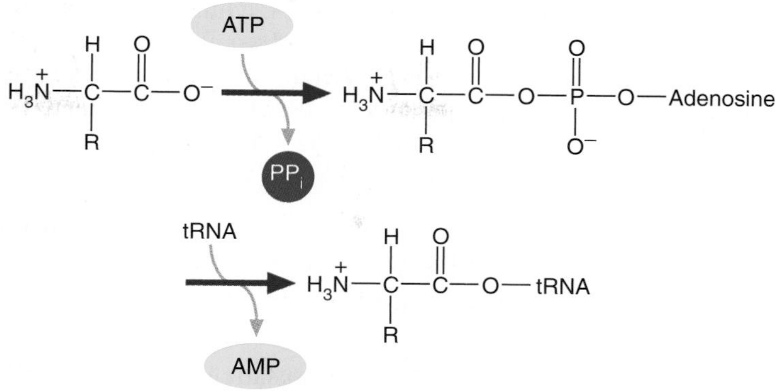

consists of two sequential reactions (Figure 18.2), both of which occur within the active site of the synthetase:

1. **Activation.** The synthetase first catalyzes the formation of aminoacyl-AMP. This reaction, which activates the amino acid by forming a high-energy mixed anhydride bond, is driven to completion through the hydrolysis of its other product, pyrophosphate. (An **anhydride** is a molecule containing two carbonyl groups linked through an oxygen atom. The term **mixed anhydride** describes an anhydride formed from two different acids, for example, a carboxylic acid and phosphoric acid.)

2. **tRNA linkage.** A specific tRNA, also bound in the active site of the synthetase, becomes attached to the aminoacyl group through an ester linkage. (Depending on the synthetase, the ester linkage may be through the 2′-OH or 3′-OH of the ribose moiety of the tRNA's 3′-terminal nucleotide. Subsequently, the aminoacyl group can migrate between the 2′-OH and 3′-OH groups. Only the 3′-aminoacyl esters are used during translation.) Although the aminoacyl ester linkage to the tRNA is lower in energy than the mixed anhydride of aminoacyl AMP, it still possesses sufficient energy to form peptide bonds.

The sum of the reactions catalyzed by the aminoacyl-tRNA synthetases is as follows:

$$\text{Amino acid} + \text{ATP} + \text{tRNA} \rightarrow \text{aminoacyl-tRNA} + \text{AMP} + \text{PP}_i$$

The product PP_i is immediately hydrolyzed with a large loss of free energy. Consequently, tRNA charging is an irreversible process. Because AMP is a product of this reaction, the metabolic price for the linkage of each amino acid to its tRNA is the equivalent of two molecules of ATP.

The aminoacyl-tRNA synthetases are a diverse group of enzymes that vary in molecular weight, primary sequence, and number of subunits. Despite this diversity, each enzyme efficiently produces a specific aminoacyl-tRNA product relatively accurately. As was mentioned, the specificity with which each of the synthetases binds the correct amino acid and its cognate tRNA is crucial for the fidelity of the translation process. Some amino acids can easily be differentiated by their size (e.g., tryptophan versus glycine) or the presence of positive or negative charges in their side chains (e.g., lysine and aspartate). Other amino acids, however, are more difficult to discriminate because their structures are similar. For example, isoleucine and valine differ only by a methylene group. Despite this difficulty, isoleucyl-tRNA[ile] synthetase usually synthesizes the correct product. However, this enzyme occasionally also produces valyl-tRNA[ile]. Isoleucyl-tRNA[ile] synthetase, as well as several other synthetases, can correct such a mistake because it possesses a separate *proofreading* site. Because of its size, this site binds valyl-tRNA[ile] and excludes the larger isoleucyl-tRNA[ile]. After its binding in the proofreading site, the mixed anhydride bond of valyl-tRNA[ile] is hydrolyzed.

Aminoacyl-tRNA synthetases must also recognize and bind the correct tRNA molecules. For some enzymes (e.g., glutaminyl-tRNA synthetase), anticodon structure is an important feature of the recognition process. However, several enzymes appear to recognize other tRNA structural elements, e.g., the acceptor stem, in addition to or instead of the anticodon.

18.2 PROTEIN SYNTHESIS

An overview of protein synthesis is illustrated in Figure 18.3. Despite its complexity and the variations among species, the translation of a genetic message into the primary sequence of a polypeptide can be divided into three phases: initiation, elongation, and termination.

1. **Initiation.** Translation begins with **initiation,** when the small ribosomal subunit binds an mRNA. The anticodon of a specific tRNA, referred to as an *initiator tRNA,* then base pairs with the initiation codon AUG. Initiation ends as the large ribosomal subunit combines with the small subunit. There are two sites on the complete ribosome for codon-anticodon interactions: the P (peptidyl) site and the A (acyl) site. In both prokaryotes and eukaryotes, mRNAs are read simultaneously by numerous ribosomes. An mRNA with several ribosomes bound to it is referred to as a **polysome.** In actively growing prokaryotes, for example, the ribosomes attached to an mRNA molecule may be separated from each other by as few as 80 nucleotides.

2. **Elongation.** During the **elongation** phase the polypeptide is actually synthesized according to the specifications of the genetic message. (As the message is read in the $5' \rightarrow 3'$ direction, polypeptide synthesis proceeds from the N-terminal to the C-terminal.) Elongation begins as a second aminoacyl-tRNA becomes bound to the ribosome in the A site because of codon-anticodon base pairing. Peptide bond formation is then catalyzed by peptidyl transferase. During this reaction (referred to as *transpeptidation*) the α-amino group of the A site amino acid (acting as a nucleophile) attacks the carbonyl group of the P site amino acid (Figure 18.4). Because of peptide bond formation, both amino acids are now attached to the A site tRNA. The now uncharged P site tRNA is released from the ribosome. (There is some evidence that a discharged tRNA lingers briefly in another site within the ribosome referred to as the E, or exit, site.) The next step in elongation involves **translocation,** whereby the ribosome is moved along the mRNA. As the mRNA moves, the next codon enters the A site, and the tRNA bearing the growing peptide chain moves into the P site. This series of steps, referred to as the *elongation cycle,* is repeated until a stop codon enters the A site.

3. **Termination.** During **termination** the polypeptide chain is released from the ribosome. Translation terminates because a stop codon cannot bind an aminoacyl-tRNA. Instead, a protein releasing factor binds to the A site. Subsequently, peptidyl transferase (acting as an esterase) hydrolyzes the bond connecting the now-completed polypeptide chain and the tRNA in the P site. Translation ends as the ribosome releases the mRNA and dissociates into the large and small subunits.

In addition to the ribosomal subunits, mRNA and aminoacyl-tRNAs, translation requires an energy source (GTP) and a wide variety of protein factors. These factors perform several roles. Some have catalytic functions; others stabilize specific structures that form during translation. Translation factors are classified according to the phase of the translation process that they affect, that is, initiation, elongation, or termination. The major differences between prokaryotic and eukaryotic translation appear to be due largely to the identity and functioning of these protein factors.

Regardless of the species, immediately after translation, some polypeptides fold into their final form without further modifications. Frequently, however, newly synthesized polypeptides are modified. These alterations, referred to as **posttranslational modifications,** can be considered to be the fourth phase of translation. They include

KEY *Concepts*

Translation consists of three phases: initiation, elongation, and termination. After their synthesis, many proteins are chemically modified and targeted to specific cellular or extracellular locations.

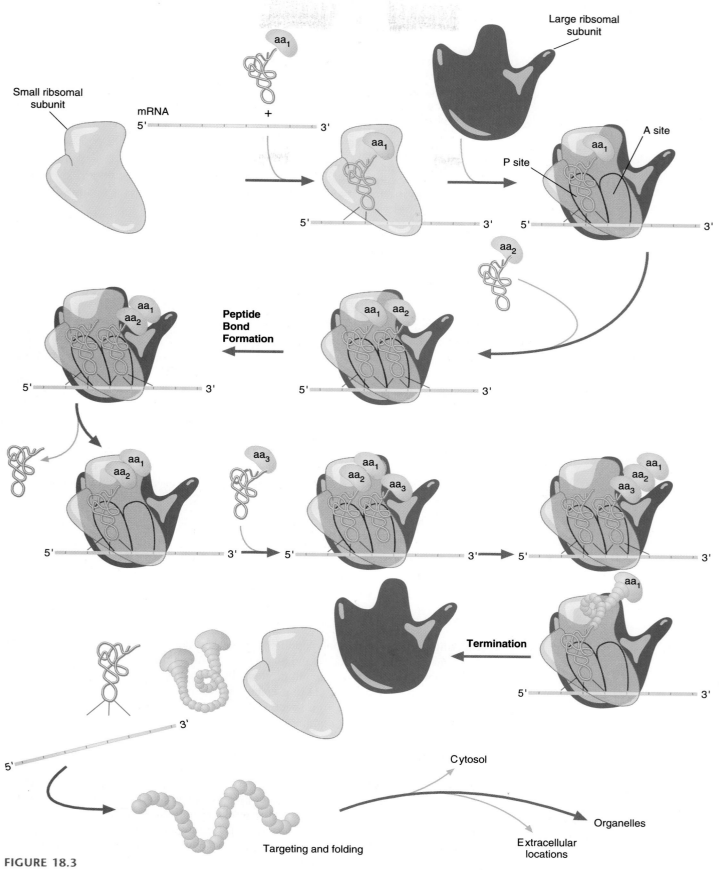

FIGURE 18.3

Protein Synthesis.

No matter what the organism, translation consists of three phases: initiation, elongation, and termination. The elongation reactions, which include peptide bond formation and translocation, are repeated many times until a stop codon is reached. Posttranslational reactions and targeting processes vary according to cell type.

FIGURE 18.4

Peptide Bond Formation.

During the first elongation cycle peptide bonds form because of the nucleophilic attack of the A site amino acid's amino group on the carboxyl carbon of the methionine residue in the P site. Because a peptide bond has formed, both amino acids are now attached to the A site tRNA.

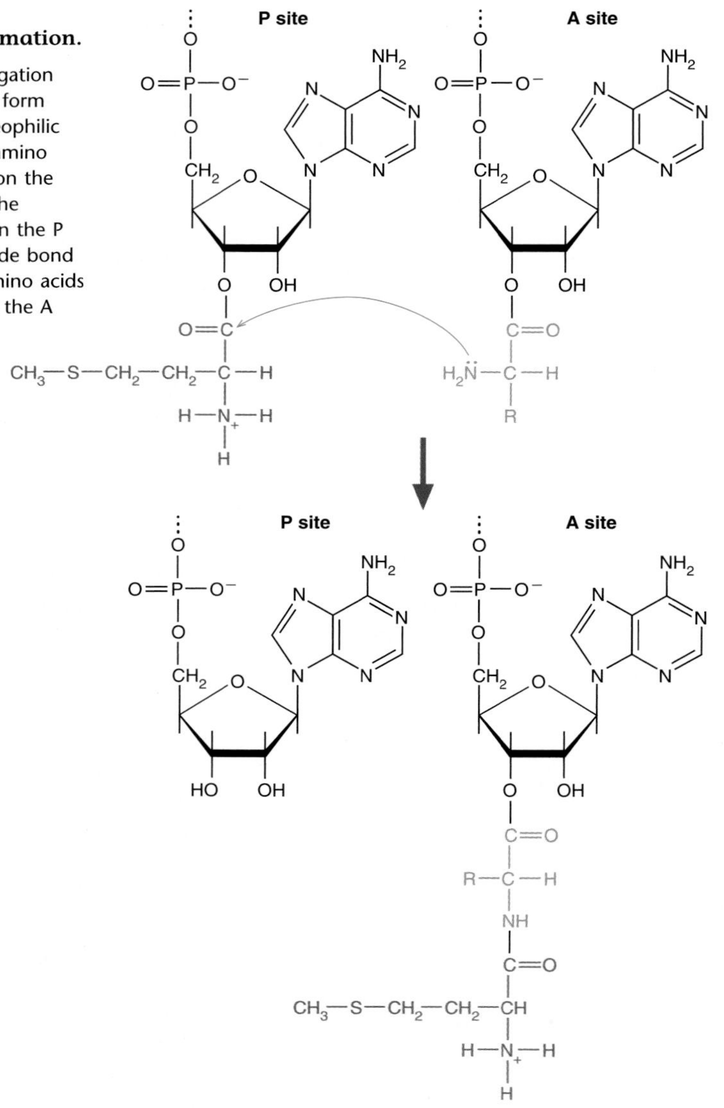

removing portions of the polypeptide by proteases, adding groups to the side chains of certain amino acid residues, and inserting cofactors. Often, individual polypeptides then combine to form multisubunit proteins. Posttranslational modifications appear to serve two general purposes: (1) to prepare a polypeptide for its specific function and (2) to direct a polypeptide to a specific location, a process referred to as **targeting.** Targeting is an especially complex process in eukaryotes because proteins must be directed to many destinations. In addition to cytoplasm and the plasma membrane (the principal destinations in prokaryotes), eukaryotic proteins may be sent to a variety of organelles (e.g., mitochondria, chloroplasts, lysosomes, peroxisomes).

Although there are many similarities between prokaryotic and eukaryotic protein synthesis, there are also notable differences. In fact, these differences are the basis for the therapeutic and research uses of several antibiotics (Table 18.2). Consequently, the details of prokaryotic and eukaryotic processes are discussed separately. Each discussion is followed by a brief description of mechanisms that control translation.

Prokaryotic Protein Synthesis

Translation is relatively rapid in prokaryotes. For example, an *E. coli* ribosome can incorporate as many as 20 amino acids per second. (The eukaryotic rate, at about 50 residues per minute, is significantly slower.) Recall that prokaryotic ribosomes are composed of a 50S large subunit and a 30S small subunit.

TABLE 18.2

Selected Antibiotic Inhibitors of Protein Synthesis

Antibiotic	Action
Chloramphenicol	Inhibits prokaryotic peptidyl transferase
Cycloheximide	Inhibits eukaryotic peptidyl transferase
Erythromycin	Inhibits prokaryotic peptide chain elongation
Streptomycin	Binding to 30S subunit causes mRNA misreading
Tetracycline	Binding to 30S subunit interferes with aminoacyl-tRNA binding

Initiation As described, translation begins with forming an initiation complex (Figure 18.5). In prokaryotes this process requires three initiation factors (IFs). *IF-3* has previously bound to the 30S subunit, preventing it from binding prematurely to the 50S subunit. As an mRNA binds to the 30S subunit, it is guided into a precise location (so that the initiation codon AUG is correctly positioned) by a purine-rich sequence referred to as the **Shine-Dalgarno sequence.** The Shine-Dalgarno sequence (named for its discoverers, John Shine and Lynn Dalgarno) occurs a short distance upstream from AUG. It binds to a complementary sequence contained in the 16S rRNA component of the 30S subunit. Base pairing between the Shine-Dalgarno sequence and the 30S subunit provides a mechanism for distinguishing a start codon from a codon specifying methionine. Each gene on a polycistronic mRNA possesses its own Shine-Dalgarno sequence and an initiation codon. The translation of each gene appears to occur independently, that is, translation of the first gene in a polycistronic message may or may not be followed by the translation of subsequent genes.

In the next step in initiation, *IF-2* (a GTP-binding protein with a bound GTP) binds to the 30S subunit, where it promotes the binding of the initiating tRNA to the initiation codon of the mRNA. The initiating tRNA in prokaryotes is N-formylmethionine-tRNA (fmet-tRNA^{fmet}). (After a special initiator tRNA is charged with methionine, the amino acid residue is formylated in an N^{10}-THF-requiring reaction. The enzyme that catalyzes this reaction does not bind met-tRNA^{met}.)

The initiation phase ends as the GTP molecule bound to IF-2 is hydrolyzed to GDP and P_i. GTP hydrolysis presumably causes a conformational change that binds the 50S subunit to the 30S subunit. Simultaneously, IF-2 and IF-3 are released. The role of *IF-1* remains unresolved.

Elongation Elongation, consists of three steps: (1) positioning an aminoacyl-tRNA in the A site, (2) peptide bond formation, and (3) translocation. As noted, these steps are referred to collectively as an elongation cycle.

The prokaryotic elongation process begins when an aminoacyl-tRNA, specified by the next codon, binds to the A site. Before it can be positioned in the A site, the aminoacyl-tRNA must first bind EF-Tu-GTP. The elongation factor *EF-Tu* is a GTP-binding protein (See Special Interest Box 18.1) involved in positioning aminoacyl-tRNA molecules in the A site. After the aminoacyl-tRNA is positioned, the GTP bound to EF-Tu is hydrolyzed to GDP and P_i. GTP hydrolysis releases EF-Tu from the ribosome. Then a second elongation factor, referred to as *EF-Ts,* promotes EF-Tu regeneration by displacing its GDP moiety. EF-Ts is then itself displaced by an incoming GTP molecule (Figure 18.6).

After positioning the second aminoacyl-tRNA in the A site, the formation of a peptide bond is catalyzed by peptidyl transferase. (Recall that the peptidyl transferase activity is now believed to reside in the 23S rRNA component of the 50S subunit.) The energy required to drive this reaction is provided by the high-energy ester bond linking the P site amino acid to its tRNA. (During the first elongation cycle, this amino acid is formylmethionine.) As described, the now uncharged tRNA occupying the P site leaves the ribosome.

FIGURE 18.5
FIGURE 18.5
**Formation of the
Prokaryotic Initiation
Complex.**

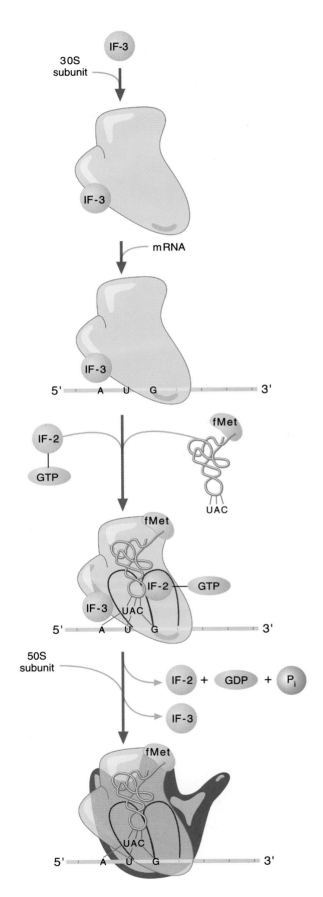

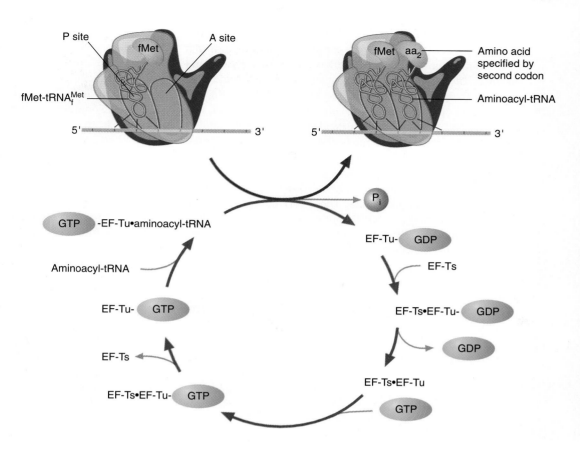

FIGURE 18.6

The EF-Tu-EF-Ts Cycle in *E. coli*.

Before EF-Tu can bind an aminoacyl-tRNA, its GDP moiety must be replaced by GTP. The binding of EF-Ts to EF-Tu (GDP) displaces GDP. EF-Ts is then itself displaced by an incoming GTP.

For translation to continue, the mRNA must move, or translocate, so that a new codon-anticodon interaction can occur. Translocation requires binding another GTP-binding protein, referred to as *EF-G*. GTP hydrolysis provides the energy required for the ribosomal conformational change that is apparently involved in moving the peptidyl-tRNA (the tRNA bearing the growing peptide chain) from the A site to the P site. The unoccupied A site then binds an appropriate aminoacyl-tRNA to the new A site codon. After EF-G is released, the ribosome is ready for the next elongation cycle. Elongation continues until a stop codon enters the A site.

Termination   The termination phase begins when a termination codon (UAA, UAG, or UGA) enters the A site. Three **releasing factors** (RF-1, RF-2, and RF-3) are involved in termination. The codons UAA and UAG are recognized by RF-1, whereas UAA and UGA are recognized by RF-2. (The role of RF-3 is unclear. It may promote RF-1 and RF-2 binding.) This recognition process, which involves GTP hydrolysis, alters ribosome function. The peptidyl transferase, which is transiently transformed into an esterase, hydrolyzes the bond linking the completed polypeptide chain and the P site tRNA. Following the polypeptide's release from the ribosome, the mRNA and tRNA also dissociate. The termination phase ends when the ribosome dissociates into its constituent subunits.

Posttranslational Modifications   As each nascent polypeptide emerges from the ribosome, it begins to fold into its final three-dimensional shape. (See p. 552) As mentioned, most of these molecules also undergo a series of modifying reactions that prepare them for their functional role. Most of the information concerning posttranslational modifications has been obtained through research on eukaryotes. However, prokaryotic polypeptides are known to undergo several types of covalent alterations.

1. **Proteolytic processing.** Several cleavage reactions may occur. These include removing the formylmethionine residue and signal peptide sequences. (**Signal**

EF-TU: A MOTOR PROTEIN

Multisubunit complexes such as the ribosome function as biological machines, that is, they are composed of interacting parts that perform work. Mechanical work is the product of force and distance. The design of each biological machine ensures that precisely the correct amount of applied force creates the appropriate amount and direction of movement to complete a specific task. This force is usually provided by nucleotide-binding protein components. These molecules, referred to as NTPases, function as *motor proteins* (mechanochemical transducers), because the hydrolysis of nucleotides such as ATP and GTP change their conformations to promote ordered conformational changes in adjacent molecules. These NTP-hydrolysis driven conformational changes, which principally occur in localized structural units called switches, alter the affinity of the NTPase for other molecules. EF-Tu is a well-researched example of a GTP-binding motor protein.

As described, EF-Tu is the protein factor that positions aminoacyl-tRNA molecules in the A site of prokaryotic ribosomes. EF-Tu possesses three domains. Domain 1 contains a GTP binding site and two switch regions. Domain 2 is connected to domain 1 through a pliable peptide segment. In its active GTP-binding form (EF-Tu-GTP), the elongation factor possesses a binding site for an aa-tRNA. After aa-tRNA binding, the entire structure is referred to as the ternary complex. All three domains of EF-Tu are involved in t-RNA binding. For example, the TψC stem of t-RNA molecules (see Figure 16.17) interacts with several amino acid residues in domain 3. The binding of aa-tRNA projects the anticodon away from the ternary complex so it is free to interact with mRNA codons.

During protein synthesis, the interaction of EF-Tu-GDP (the inactive form) with EF-Ts releases GDP. The subsequent binding of GTP in the domain 1 nucleotide binding site changes the conformation in the two switch regions. These changes bring domains 1 and 2 close together, forming a binding cleft. Once an aa-tRNA has been bound in the cleft, the ternary complex enters the ribosome where the aa-tRNA anticodon binds reversibly to an mRNA codon in the A site. When a ternary complex contains a cognate aa-tRNA, a conformation change in the ribosome triggers a conformation change in the EF-Tu nucleotide binding site. The subsequent hydrolysis of GTP causes domains 1 and 2 to move apart, thus allowing the release of the aa-tRNA. This feature of EF-Tu function has been described as a timing mechanism. The timer is activated when the ternary complex binds to the ribosome. The relatively slow rate of GTP hydrolysis provides sufficient time for the dissociation of incorrect codon-anticodon pairings. In contrast, the binding of a cognate aa-tRNA is so tight that there is sufficient time for GTP to undergo hydrolysis. Thus, the functioning of the ternary complex provides another mechanism for proofreading during translation in addition to that described for the aa-tRNA synthetases (p. 530).

peptides, or leader peptides, are short peptide sequences, typically near the amino terminal, that determine a polypeptide's destination. In bacteria, for example, a signal peptide is required to insert a polypeptide into the plasma membrane.)

2. **Glycosylation.** Although most glycoconjugates in prokaryotes are lipids, a few glycoproteins have been reported, for example, the cell surface glycoprotein of *Halobacterium.* The prokaryotic glycosylation mechanism, as well as its functional significance, is unknown.

3. **Methylation.** A group of enzymes, referred to as the protein methyltransferases, use S-adenosylmethionine to methylate certain proteins. For example, one type of methyltransferase found in *E. coli* and related bacteria methylates glutamate residues in membrane-bound chemoreceptors. The methyltransferase and a methylesterase are components in a methylation/demethylation process, which plays a role in a signal transduction mechanism involved in chemotaxis. (Recall that the capacity of a living cell to respond to certain environmental cues by moving toward or away from specific molecules is referred to as chemotaxis.)

4. **Phosphorylation.** In recent years, protein phosphorylation/dephosphorylation catalyzed by protein kinases and phosphatases have been revealed to be widespread among prokaryotes. Many of the purposes of these reactions remain unclear. However, roles for transient phosphorylation have been identified in chemotaxis and nitrogen metabolism regulation.

 Translational Control Mechanisms   Protein synthesis is an exceptionally expensive process. Costing four high-energy phosphate bonds per peptide bond (i.e., two bonds expended during tRNA charging and one each during A site-tRNA binding and translocation) it is perhaps not surprising that enormous quantities of energy are involved. For example, approximately 90% of *E. coli* energy production used in the synthesis of macromolecules may be devoted to the manufacture of proteins. Although the speed and accuracy of translation require a high energy input, the cost would be even higher without metabolic control mechanisms. These mechanisms allow prokaryotic cells to compete with each other for limited nutritional resources.

In prokaryotes such as *E. coli,* most of the control of protein synthesis occurs at the level of transcription. (Refer to Section 17.3 for a discussion of the principles of prokaryotic transcriptional control.) This circumstance makes sense for several reasons. First, transcription and translation are directly coupled; that is, translation is initiated shortly after transcription begins (Figure 18.7). Second, the lifetime of prokaryotic mRNA is usually relatively short. With half-lives of between 1 and 3 minutes, the types of mRNA produced in a cell can be quickly altered as environmental conditions change. Most mRNA molecules in *E. coli* are degraded by two exonucleases, referred to as RNase II and polynucleotide phosphorylase.

Despite the preeminence of transcriptional control mechanisms, the rates of prokaryotic mRNA translation also vary. A large portion of this variation is attributed to differences in Shine-Dalgarno sequences. Because Shine-Dalgarno sequences help select the initiation codon, sequence variations may affect the rate of translating genetic messages. For example, the gene products of the lac operon (β-galactosidase, galactose permease, and galactoside transacetylase) are not produced in equal quantities. Thiogalactoside transacetylase is produced at approximately one-fifth the rate of β-galactosidase. (Recall that the function of thiogalactoside transacetylase remains unknown and that lactose fermentation proceeds normally in mutant cells that are unable to produce this gene product.)

An interesting example of negative translational regulation in prokaryotes is provided by ribosomal protein synthesis. The approximately 55 proteins in prokaryotic ribosomes are coded for by genes located in 20 operons. Efficient bacterial growth

KEY *Concepts*

Prokaryotic protein synthesis is a rapid process involving several protein factors. Although most prokaryotic gene expression appears to be regulated at the transcriptional level, several types of translational regulation have been detected.

FIGURE 18.7

Transcription and Translation in *E. coli.*

(a) An electron micrograph of *E. coli* transcription and translation. (b) Diagram of (a). Note polyribosomes.

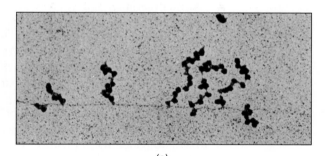

(a)

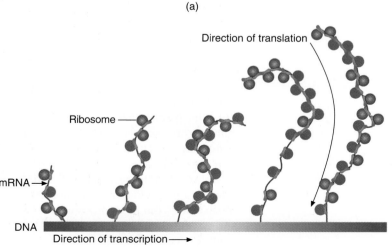

Direction of translation

Ribosome

mRNA

DNA

Direction of transcription →

(b)

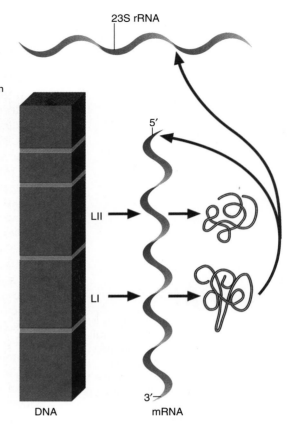

FIGURE 18.8

A Negative Translational Control.

The P_{L11} operon of *E. coli* is controlled by the level of one of its gene products. The ribosomal protein L1 can bind to 23S rRNA or its own mRNA. When excessive amounts of L1 accumulate, then L1 binds to the 5′ end of its own mRNA, thereby inhibiting the translation of the operon.

requires that their synthesis be coordinately regulated among the operons as well as with rRNA synthesis. For example, in the P_{L11} operon, which contains the genes for the ribosomal proteins L1 and L11, excessive amounts of L1 (i.e., more L1 molecules than can bind available 23S rRNA) inhibit P_{L11} mRNA translation (Figure 18.8). Apparently, L1 can bind to either 23S rRNA or P_{L11} mRNA. In the absence of 23S rRNA, L1 inhibits the translation of its own operon by binding to the 5′ end of P_{L11} mRNA.

Eukaryotic Protein Synthesis

Although the earliest work in protein synthesis (e.g., the discovery of aminoacyl-tRNA synthetases and the tRNAs) were made by using mammalian cells, translation investigators directed their attention to bacteria in the 1960s. This change occurred for a variety of reasons, including the relative ease of culturing bacterial cells and the perception that bacterial gene expression is simpler and more accessible than that in the more complex eukaryotes. Only in the 1970s, when the principles of prokaryotic translation were understood, did the eukaryotic process again become a focus of attention. Not surprisingly, the large, complex genomes of eukaryotic cells (especially those in multicellular organisms) are now known to require sophisticated regulation of translation (Section 17.3). A large number of protein factors assist in translation. In addition, the posttranslational modifications of eukaryotic polypeptides appear significantly more complex than those observed in prokaryotes. Considering the structural complexity of eukaryotes, it is inevitable that polypeptide targeting mechanisms are also quite intricate.

In this section the features that distinguish the three phases of eukaryotic translation from its prokaryotic counterparts are described. This is followed by a discussion of several of the most prominent forms of eukaryotic posttranslational modifications and targeting mechanisms. The section ends with discussions of translational control mechanisms and protein folding.

Initiation   Most of the major differences between the prokaryotic and eukaryotic versions of protein synthesis occur during the initiation phase. Among the reasons for the additional complexity of eukaryotic initiation are the following:

1. **mRNA secondary structure.** Recall that eukaryotic mRNA is extensively processed. Various processing reactions, which allow the transport of mRNA molecules out of the nucleus and prepare them for their functional roles, form extensive secondary structure. In addition, eukaryotic mRNAs associate with several types of protein. (An mRNA that is complexed with these proteins is sometimes referred to as a ribonucleoprotein particle.) Before an mRNA can be translated, its secondary structure must be altered, and certain proteins must be removed.

2. **mRNA scanning.** In contrast to prokaryotic mRNA, eukaryotic molecules lack Shine-Dalgarno sequences, which allow for the identification of the initiating AUG sequence. Instead, eukaryotic ribosomes "scan" each mRNA. This scanning is a complex (and poorly understood) process in which ribosomes bind to the capped 5′ end of the molecule and migrate in a 5′ → 3′ direction searching for a translation start site.

Eukaryotes use a more complex spectrum of initiating factors than prokaryotes. There are at least nine eukaryotic initiating factors (eIFs), several of which possess numerous subunits. The functional roles of most of these factors are still under investigation.

Eukaryotic initiation (Figure 18.9) begins when the small 40S ribosomal subunit binds to a complex composed of *eIF-2* (a GTP-binding protein), GTP, and an initiating species of methionyl-tRNA^{met} (met-tRNA$_i$). (eIF-2-GTP, which mediates the binding of the initiating tRNA to the 40S subunit, is regenerated from inactive eIF-2-GDP by *eIF-2B*, a guanine nucleotide–releasing protein. After GDP is released from eIF-2, GTP binding occurs.) The small (40S) subunit is prevented from binding to the large (60S) subunit during this phase of initiation because it is associated with *eIF-3*, a multisubunit protein. (This binding is also prevented by the association of eIF-6 with the 60S subunit.) The complex consisting of the small subunit, eIF-2-GTP, eIF-3, and methionyl-tRNA^{met} is referred to as a *40S preinitiation complex*. Subsequently, mRNA binds to the 40S complex to form a 40S *initiation complex*. This is an ATP-requiring process that involves several additional initiation factors (e.g., eIF-4A, eIF-4B, eIF-1, eIF-4F). *eIF-4F* binds to the cap structure at the 5′ end of the mRNA, whereas the binding of *eIF-4A* (an ATPase) and *eIF-4B* (a helicase) are believed to reduce its secondary structure. Identifying eukaryotic initiation factors has been confusing. For example, some factors have been revealed to be subunits of larger factors. *eIF-4E*, also referred to as cap-binding protein or *CBP I*, is one of several subunits of eIF-4F. eIF-4F is often referred to as *CBP II*.

Once the 40S initiation complex is formed, it scans the mRNA for a suitable initiation codon, which is usually an AUG near the 5′ end. The 40S complex then binds the 60S subunit (now dissociated from eIF-6) to form an *80S initiation complex*. The formation of the 80S complex involves the hydrolysis of the GTP associated with eIF-2, a process that requires *eIF-5*. The initiation phase ends as the initiation factors eIF-2, eIF-3, eIF-4A, eIF-4B, eIF-4F, and eIF-1 are released from the initiation complex.

Elongation   Figure 18.10 illustrates the eukaryotic elongation cycle as it is currently understood. Several elongation factors (eEFs) are required during this phase of translation. *eEF-1α* is a 50-kD polypeptide that mediates the binding of aminoacyl-tRNAs to the A site. After a complex is formed between eEF-1α, GTP, and the entering aminoacyl-tRNA, codon-anticodon interactions are initiated. If correct pairing occurs, eEF-1α hydrolyzes its bound GTP and subsequently exits the ribosome, leaving its aminoacyl-tRNA behind. If correct pairing does not occur, the complex leaves the A

FIGURE 18.9
Formation of the
Eukaryotic
Initiation
Complex.

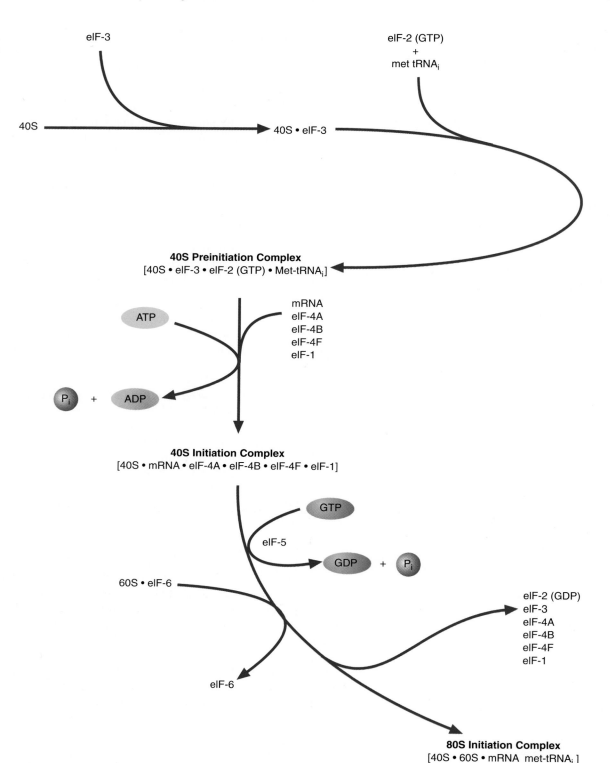

FIGURE 18.9
Formation of the Eukaryotic Initiation Complex.

site, thereby preventing incorrect amino acid residues from being incorporated. This process has been referred to as **kinetic proofreading.** In various fungi (e.g., yeast), another elongation factor, referred to as *eEF-3,* is also required in combination with eEF-1α for A site aminoacyl-tRNA binding.

During the next elongation step (i.e., peptide bond formation) the peptidyl transferase activity of the large ribosomal subunit catalyzes the nucleophilic attack of the

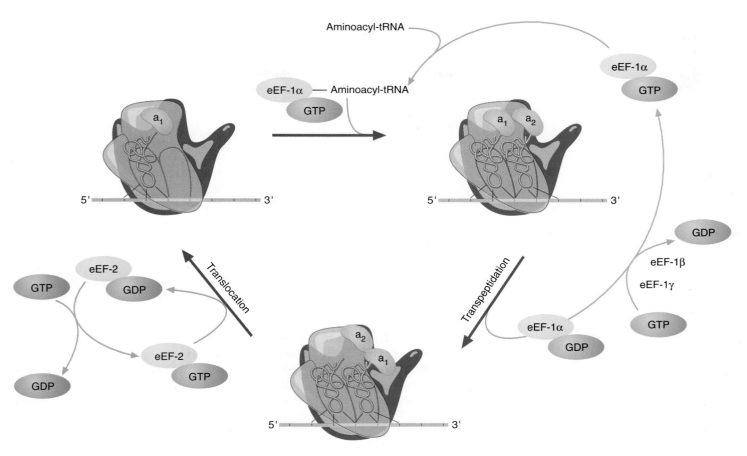

FIGURE 18.10

The Elongation Cycle in Eukaryotic Translation.

Elongation comprises three phases: (1) binding of an aminoacyl tRNA to the A site, (2) transpeptidation, and (3) translocation.

A site α-amino group on the carboxyl carbon of the P site amino acid residue. Apparently, eEF-1α dissociates from the ribosome immediately before transpeptidation. eEF-1β and eEF-1γ mediate the regeneration of eEF-1α by promoting an exchange of GDP for GTP. (Recall that a similar process involving EF-Tu and EF-Ts occurs in bacteria such as *E. coli.*)

Translocation in eukaryotes requires a 100-kD polypeptide referred to as *eEF-2,* which is also a GTP-binding protein. eEF-2-GTP binds to the ribosome at some as yet undetermined site during translocation. GTP is then hydrolyzed to GDP, and eEF-2-GDP is released. As noted, GTP hydrolysis provides the energy needed to physically move the ribosome along the mRNA. At the end of translocation a new codon is exposed in the A site.

Termination In eukaryotic cells two releasing factors, *eRF-1* and *eRF-3* (a GTP-binding protein), mediate the termination process. When GTP binds to eRF-3, its GTPase activity is activated. eRF-1 and eRF-3-GTP form a complex that binds in the A site when UAG, UGA, or UAA enter. Then GTP hydrolysis promotes the dissociation of the releasing factors from the ribosome. This step is soon followed by the release of mRNA and the separation of the functional ribosome into its subunits. As described, the release of the newly synthesized polypeptide is catalyzed by peptidyl transferase.

18.2   eEF-2 possesses a unique modification of a specific histidine residue called diphthamide. Mutant eukaryotic cells that cannot transform this histidine into diphthamide do not appear to be adversely affected. However, the ADP-ribosylation of the diphthamide residue of eEF-2 by toxins produced by *Corynebacterium diphtheriae* and *Pseudomonas aeruginosa* renders the factor inoperative.

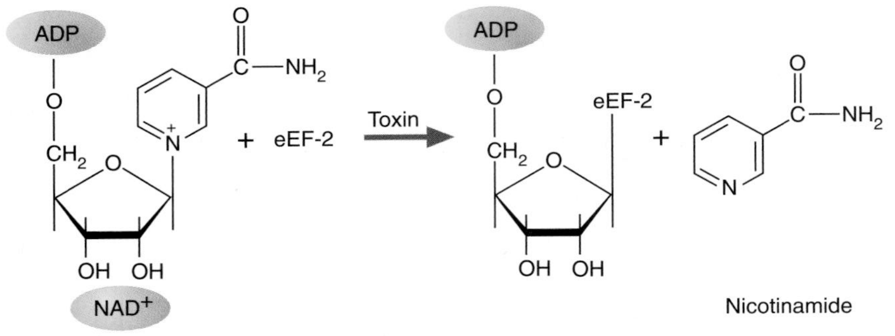

Cells die because they cannot synthesize proteins. The mechanism by which eEF-2 function is affected by ADP-ribosylation is unknown. Can you suggest any possibilities?

Posttranslational Modifications in Eukaryotes   Most nascent polypeptides undergo one or more types of covalent modifications. These alterations, which may occur either during ongoing polypeptide synthesis or afterwards, consist of reactions that modify the side chains of specific amino acid residues or break specific bonds. In general, posttranslational modifications prepare each molecule for its functional role and/or for folding into its native (i.e., biologically active) conformation. Examples of prominent posttranslational changes include the following:

1. **Proteolytic cleavage.** Typical examples of proteolytic cleavage include removing the N-terminal methionine residue, signal sequences (discussed below), and the conversion of inactive precursors to their active counterparts. Recall, for example, that certain enzymes, referred to as proenzymes or zymogens, are transformed into their active forms by cleaving specific peptide bonds. Inactive polypeptide precursors are called **proproteins.** The proteolytic processing of insulin (Figure 18.11) provides a well-researched example of the conversion of a nonenzyme protein into its active form. (The insulin precursor containing a signal peptide is referred to as pre-proinsulin. The inactive insulin precursor produced by removing the signal peptide is referred to as proinsulin.)

2. **Glycosylation.** Although a wide variety of eukaryotic proteins are glycosylated, the functional purpose of the carbohydrate moieties is not always obvious (see Special Interest Box 7.3). In general, secreted proteins contain complex oligosaccharide species, while ER membrane proteins possess high mannose species. The synthesis of the core N-linked oligosaccharide (common to all N-linked forms) is illustrated in Figure 18.12. The core oligosaccharide is assembled in association with phosphorylated dolichol. (Dolichol is a polyisoprenoid found within all cell membranes. Phosphorylated dolichol is found predominantly in ER membrane.)

3. **Hydroxylation.** Hydroxylation of the amino acids proline and lysine is required for the structural integrity of the connective tissue proteins collagen (Section 5.3) and elastin. Additionally, 4-hydroxyproline is also found in acetylcholinesterase (the enzyme that degrades the neurotransmitter acetylcholine) and complement (a complex series of serum proteins involved in the immune response). Three mixed-function oxygenases (prolyl-4-hydroxylase, prolyl-3-hydroxylase, and lysyl hydroxylase) located in the RER are responsible for hydroxylating certain proline and lysine residues. Substrate requirements are highly specific. For example, prolyl-4-hydroxylase hydroxylates only proline residues in the Y position

FIGURE 18.11

Proteolytic Processing of Insulin.

After the removal of the signal peptide, a peptide segment referred to as the C chain is removed by a specific proteolytic enzyme. Two disulfide bonds are also formed during insulin's posttranslational processing.

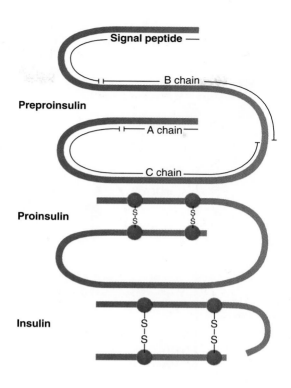

of peptides containing Gly-X-Y sequences, while prolyl-3-hydroxylase requires Gly-Pro-4-Hyp sequences (Hyp stands for hydroxyproline; X and Y represent other amino acids). Hydroxylation of lysine occurs only when the sequence Gly-X-Lys is present. (Polypeptide hydroxylation by prolyl-3-hydroxylase and lysyl hydroxylase occurs only before helical structure forms.) The synthesis of 4-Hyp is illustrated in Figure 18.13. Ascorbic acid (vitamin C) is required to hydroxylate proline and lysine residues in collagen. When dietary intake is inadequate, scurvy (Special Interest Box 7.1) results. The symptoms of scurvy (e.g., blood vessel fragility and poor wound healing) are effects of weak collagen fiber structure.

4. **Phosphorylation.** Examples of the roles of protein phosphorylation in metabolic control and signal transduction have already been discussed. Protein phosphorylation may also play a critical (and interrelated) role in protein-protein interactions. For example, the autophosphorylation of tyrosine residues in PDGF receptors precedes the binding of cytoplasmic target proteins.

5. **Lipophilic modifications.** The covalent attachment of lipid moieties to proteins improves membrane binding capacity and/or certain protein-protein interactions. Among the most common lipophilic modifications are acylation (the attachment of fatty acids) and prenylation (Section 9.1). Although the fatty acid myristate (14:0) is relatively rare in eukaryotic cells, myristoylation is one of the most common forms of acylation. N-myristoylation (the covalent attachment of myristate by an amide bond to a polypeptide's amino terminal glycine residue) has been shown to increase the affinity of the α subunit of certain G proteins for membrane-bound β and γ subunits.

6. **Methylation.** Protein methylation serves several purposes in eukaryotes. The methylation of altered aspartate residues by a specific type of methyltransferase promotes either the repair or the degradation of damaged proteins. Other methyltransferases catalyze reactions that alter the cellular roles of certain proteins. For example, methylated lysine residues have been found in such disparate proteins as ribulose-2,3-bisphosphate carboxylase, calmodulin, histones, certain ribosomal proteins, and cytochrome c. Other amino acid residues that may be methylated

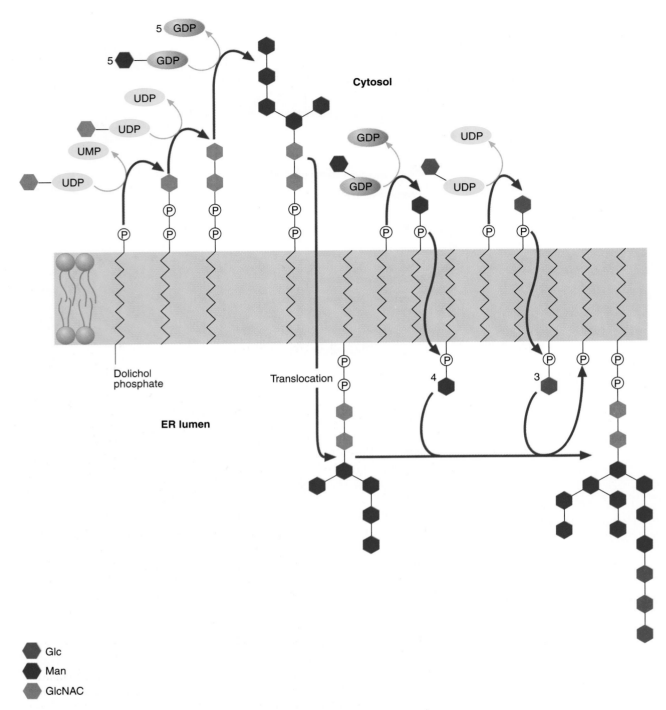

Glc
Man
GlcNAC

FIGURE 18.12

Synthesis of Dolichol-linked Oligosaccharide.

In the first step, GlcNAc-1-P is transferred from UDP-GlcNAc to dolichol phosphate (Dol-P). The next GlcNAc and the following five mannose residues are then transferred from nucleotide-activated forms. After the entire structure flips to the lumenal side of the membrane, each of the remaining sugars (four mannoses and three glucoses) is transferred first to Dol-P and then to the growing oligosaccharide. Then N-glycosylation of protein takes place in the ER in a one-step reaction catalyzed by a membrane-bound enzyme called glycosyl transferase.

Hydroxylation of Proline.

Ascorbic acid and ferrous iron are cofactors of prolyl-4-hydroxylase, the enzyme that catalyzes the hydroxylation of the C-4 position of certain prolyl residues in nascent polypeptides. Ascorbic acid, acting as a reducing agent, prevents the oxidation of the iron atom cofactor of the enzyme.

Prolyl residue + O_2 + α-Ketoglutarate → (Prolyl hydroxylase, Ascorbate Fe^{2+}) → 4-Hydroxyprolyl residue + CO_2 + Succinate

include histidine (e.g., histones, rhodopsin, and eEF-2) and arginine (e.g., heat shock proteins and ribosomal proteins).

7. **Disulfide bond formation.** Disulfide bonds are generally found only in secretory proteins (e.g., insulin) and certain membrane proteins. (Recall that "disulfide bridges" are strong bonds that confer considerable structural stability on the molecules that contain them.) As described (Section 5.3), cytoplasmic proteins generally do not possess disulfide bonds because of the presence of various reducing agents in cytoplasm (e.g., glutathione and thioredoxin). Because the ER has a nonreducing environment, disulfide bonds form spontaneously in the RER as the nascent polypeptide emerges into the lumen. Although some proteins have disulfide bridges that form sequentially as the polypeptide enters the lumen (i.e., the first cysteine pairs with the second, the third residue pairs with the fourth, etc.), this is not true for many other molecules. Proper disulfide bond formation for these latter proteins is now presumed to be facilitated by **disulfide exchange.** During this process, disulfide bonds rapidly migrate from one position to another until the correct structure is achieved. An ER enzymatic activity, referred to as protein disulfide isomerase, is now believed to catalyze this process.

Targeting Despite the vast complexities of eukaryotic cell structure and function, each newly synthesized polypeptide is normally directed to its proper destination. Considering that translation takes place in the cytoplasm (except for certain molecules that are produced within mitochondria and plastids) and that a wide variety of polypeptides must be directed to their proper locations, it is not surprising that the mechanisms by which cellular proteins are targeted are complex. Although this process is not yet completely understood, there appear to be two principal mechanisms by which polypeptides are directed to their correct locations: transcript localization and signal peptides. Each is briefly discussed below.

It is generally recognized that cells often have asymmetric protein distributions within the cytoplasm. For example, mature *Drosophila* eggs contain a gradient of bicoid, a protein that plays a critical role in gene regulation during development. A high concentration of bicoid in the anterior portion of the egg is required for the normal development of anterior body parts (i.e., head segments), whereas the low bicoid concentration in the posterior portion of the egg cytoplasm promotes the development of posterior body parts. (If posterior cytoplasm is removed from one egg and substituted for anterior cytoplasm in a second egg, two sets of posterior body parts appear in the larva that develops from the recipient egg.) It is now believed that cytoplasmic protein gradients are created by **transcript localization,** that is, the binding of specific mRNA to receptors in certain cytoplasmic locations. It is known that bicoid mRNA is transported from nearby nurse cells into the developing oocyte (an immature egg cell). Once in the oocyte, bicoid mRNA binds via its 3′ end to certain components of the anterior cytoskeleton. After the mature egg is fertilized, translation of bicoid mRNA, coupled with protein diffusion, gives rise to the concentration gradient.

Polypeptides destined for secretion or for use in the plasma membrane or any of the membranous organelles must be specifically targeted to their proper location. Several types of these proteins possess sorting signals referred to as signal peptides. Each signal peptide sequence helps insert the polypeptide that contains it into an appropriate membrane. Signal peptides generally consist of a positively charged region followed by a central hydrophobic region and a more polar region. Although many signal peptides occur at the amino terminal, they may also occur elsewhere along the polypeptide.

The **signal hypothesis** was proposed by Gunter Blobel in 1975 to explain the translocation of polypeptides across RER membrane. Subsequent investigations revealed significant information concerning the insertion of polypeptides through the RER membrane. For this reason the discussion primarily focuses on this organelle. This is followed by a brief description of polypeptide uptake by other organelles.

As soon as a portion of a nascent polypeptide (containing the signal peptide) emerges from a ribosome, a **signal recognition particle** (SRP) (a large complex consisting of six proteins and a small RNA molecule) binds to the ribosome. As a consequence of this binding, translation is temporarily arrested. The SRP then mediates binding of the ribosome to the RER via **docking protein,** a heterodimer also referred to as SRP receptor protein. Once binding to the RER has occurred, translation restarts, and the growing polypeptide inserts into the membrane. (The simultaneous translocation of a polypeptide during ongoing protein synthesis is sometimes referred to as **cotranslational transfer.)** As translation begins, the SRP is released. An integral membrane protein complex, referred to as a **translocon,** is believed to mediate polypeptide translocation. The translocon consists of a hydrophilic transmembrane pore and several proteins that facilitate polypeptide translocation and processing. It is presumed that GTP hydrolysis provides the energy to push polypeptides across the RER membrane, since both the SRP and SRP receptor bind GTP. In **posttranslational translocation,** previously synthesized polypeptides are pulled across the RER membrane by an ATP-binding peripheral translocon-associated protein (hsp 70).

The fate of a targeted polypeptide depends on the location of the signal peptide and any other signal sequences (discussed below). For soluble secretory proteins, transmembrane transfer is usually followed by removing of an N-terminal signal peptide by signal peptidase (Figure 18.14), a process that releases them into the ER lumen. Such molecules usually undergo further posttranslational processing. The initial phase of the translocation of transmembrane proteins is similar to that of secretory proteins. For these molecules, the amino terminal signal peptide serves as a *start signal* that remains bound in the membrane as the remaining polypeptide sequence is threaded through the membrane. So-called "single-pass" transmembrane proteins possess a *stop transfer signal* (or stop signal), which prevents further transfer across the membrane (Figure 18.14b). Membrane proteins with multiple membrane spanning segments possess a series of alternating start and stop signals (Figure 18.14c).

Most proteins that are translocated into the RER are directed to other destinations. After they undergo initial posttranslational modifications, both soluble and membrane-bound proteins are transferred to the Golgi complex via transport vesicles that bud off from the ER and fuse with Golgi membrane (Figure 18.15). (Proteins that must be retained within the ER possess retention signals. In most vertebrate cells this signal consists of the carboxyterminal tetrapeptide Lys-Asp-Glu-Leu, often referred to with the one-letter abbreviations KDEL.) Within the Golgi complex, proteins undergo further modifications. For example, N-linked oligosaccharides are processed further, and O-linked glycosylation of certain serine and threonine residues occurs. Lysosomal proteins are targeted to the lysosomes by adding a mannose-6-phosphate residue. It is still unclear what signals direct secretory proteins to the cell surface (via exocytosis) or promote the delivery of plasma membrane proteins to their destination, al-

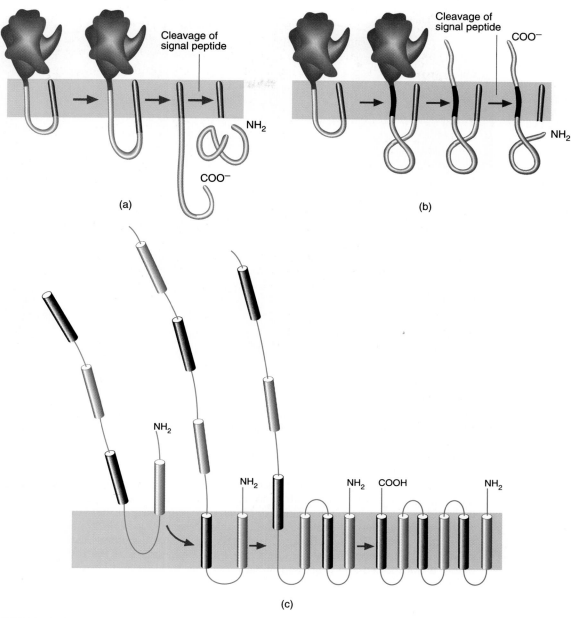

FIGURE 18.14

Cotranslational Transfer of Polypeptides Through the ER Membrane.

(a) Transfer of a secretory protein. (b) Transfer of a transmembrane protein. (c) Transfer of a multiply-folded membrane protein. For the sake of clarity the transfer apparatus has been omitted from the diagrams. In addition, the ribosome has been omitted from (c). The shaded segment is a signal peptide. The black segment is a stop transfer signal.

though a "default mechanism" has been proposed. (In default mechanisms, the absence of a signal results in a specific sequence of events.)

As noted, although mitochondria and chloroplasts produce several of their own proteins, a variety of other proteins are produced on cytoplasmic ribosomes and subsequently imported. Again, specific signal sequences are required. The import of polypeptides into these organelles is complicated by the presence of several membranes. (Recall that both mitochondria and chloroplasts contain several compartments created by internal membranes.) Consequently, polypeptide transfer in these organelles

FIGURE 18.15

The ER, Golgi, and Plasma Membrane.

Transport vesicles transfer new membrane components (protein and lipids) and secretory products from the ER to the Golgi complex, from one Golgi cisterna to another, and from the trans-Golgi network to other organelles (e.g., lysosomes) or to the plasma membrane.

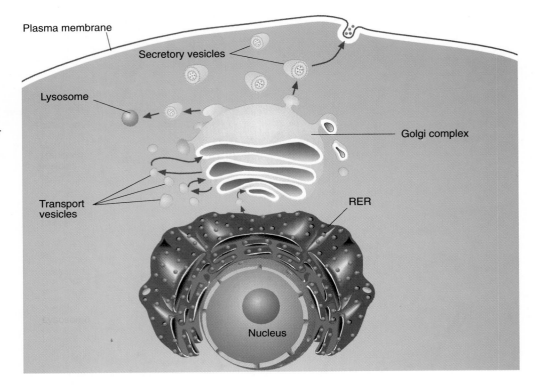

often involves several signal sequences. An example of this import mechanism (the targeting of cytochrome c_1 to the inner membrane space of mitochondria) is illustrated in Figure 18.16.

QUESTION	18.3	The mechanism involved in the posttranslational transport of proteins into chloroplasts has so far received only limited attention. However, the import of plastocyanin into the thylakoid lumen has been determined to require two import signals near the N-terminal of the newly synthesized protein. Assuming that chloroplast protein import resembles the import process for mitochondria, suggest a reasonable hypothesis to explain how plastocyanin (a lumen protein associated with the inner surface of the thylaloid membrane) is transported and processed. What enzymatic activities and transport structures do you expect are involved in this process?

Translation Control Mechanisms   Eukaryotic translation control mechanisms are proving to be exceptionally complex, substantially more so than those observed in prokaryotes. In eukaryotes these mechanisms appear to occur on a continuum, from *global* controls (i.e., the translation of a wide variety of mRNAs are altered) to *specific* controls (i.e., the translation of a specific mRNA or small group of mRNAs are altered). Although most aspects of eukaryotic translational control are currently unresolved, the following features are believed to be important:

1. **mRNA export.** The spatial separation of transcription and translation afforded by the nuclear membrane appears to provide eukaryotes with significant opportunities for gene expression regulation. Only a small portion of RNA produced within the nucleus ever enters the cytoplasm. Most discarded RNA is probably nonfunctional introns. Although there is insufficient evidence to indicate that the export of processed mRNAs is selectively blocked, this circumstance remains a distinct possibility. Export through the nuclear pore complex is known to be a carefully controlled, energy-driven process whose minimum requirements include the presence of a 5'-cap and a 3' poly A tail.

FIGURE 18.16

FIGURE 18.16

Posttranslational Transport of Cytochrome c_1 into a Mitochondrion.

After its synthesis in cytoplasm, cytochrome c_1 must be translocated into the mitochondrial inner membrane space. (Recall that cytochrome c_1 is a component of complex III of the ETC.) The targeting of cytochrome c_1 requires two sequences. The first targets the polypeptide to the matrix. After this sequence is removed by a protease, the second sequence targets the molecule to the inner membrane space. The second targeting sequence is then also removed. After folding and binding a heme, the molecule associates with complex III in the inner membrane.

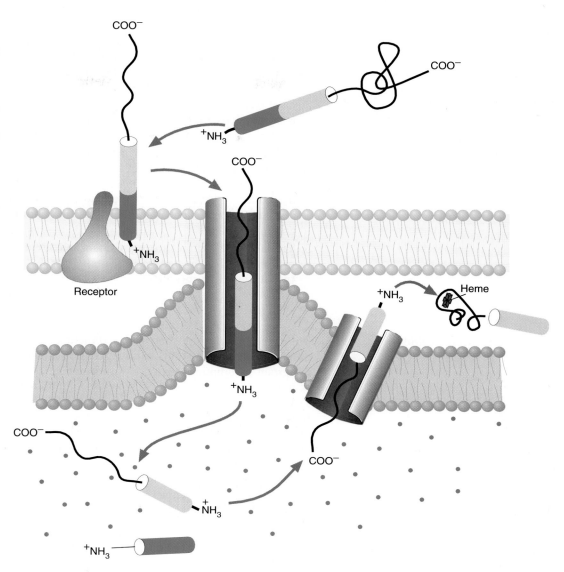

KEY *Concepts*

Eukaryotic protein synthesis is slower and more complex than its prokaryotic counterpart. In addition to requiring a larger number of translation factors and a more complex initiation mechanism, the eukaryotic process also involves vastly more complicated posttranslational processing and targeting mechanisms. Eukaryotes use a wide spectrum of translational control mechanisms.

2. **mRNA stability.** In general, the translation rate of any mRNA species is related to its abundance, which is in turn dependent on both its rates of synthesis and degradation. mRNA half-lives range from about 20 minutes to over 24 hours. Several features of mRNA structure are known to affect its stability, that is, its capacity to avoid degradation by various nucleases. These include the presence of several destabilizing sequences whose secondary structures are substrates for nucleases as well as certain stabilizing sequences. (mRNAs often appear to contain multiple destabilizing sequences.) The binding of specific proteins to certain sequences can also affect mRNA stability. Finally, reversible adenylation and deadenylation of the 3′ end of mRNA strongly influence both its stability and its translational activity. After processing within the nucleus, most mRNAs are transported into the cytoplasm possessing poly A tails containing between 100 and 200 nucleotides. As time passes, many poly A tails progressively shorten to no fewer than 30 residues when the entire mRNA is degraded. In certain circumstances the poly A tail of some mRNAs is selectively elongated or shortened. For example, mRNAs in mature oocytes are "masked" by removal of most of their poly A tail nucleotides. After fertilization these mRNAs are reactivated by adding adenine nucleotides.

3. **Negative translational control.** The translation of certain specific mRNAs is known to be blocked by binding repressor proteins to sequences near their 5′ ends. A well-researched example is provided by ferritin synthesis control. Ferritin, an iron storage protein that is found predominantly in hepatocytes, is synthesized in response to high iron concentrations. Ferritin mRNA contains an iron response element (IRE) that binds an iron-binding repressor protein. When cellular iron concentrations are high, the large number of iron atoms binding to the repressor protein cause it to dissociate from the IRE. Then ferritin mRNA is translated.

4. **Initiation factor phosphorylation.** The phosphorylation of eIF-2 in response to certain circumstances (e.g., heat shock, viral infections, and growth factor deprivation) has been observed to decrease protein synthesis generally. However, the translation of certain mRNA increases. For example, hsp synthesis increases in response to heat shock and other stressful conditions. The specific mechanisms are unknown.

5. **Translational frameshifting.** Certain mRNAs appear to contain structural information that, if activated, results in a +1 or −1 change in reading frame. This **translational frameshifting,** which has been most often observed in cells infected by retroviruses, allows more than one polypeptide to be synthesized from a single mRNA.

The Folding Problem

The direct relationship between a protein's primary sequence and its final three-dimensional conformation, and by extension its biological activity, is among the most important assumptions of modern biochemistry. One of the principal underpinnings of this paradigm is a series of experiments reported by Christian Anfinsen in the late 1950s. Anfinsen (Nobel Prize in Chemistry, 1972), working with bovine pancreatic RNase, demonstrated that under favorable conditions a denatured protein could refold into its native and biologically active state (Figure 18.17). This discovery suggested that the three-dimensional structure of any protein could be predicted if the physical and chemical properties of the amino acids as well as the forces that drive the folding process (e.g., bond rotations, free energy considerations, and the behavior of amino acids in aqueous environments) were understood. Unfortunately, after several decades of painstaking research with the most sophisticated tools available (e.g., X-ray crys-

FIGURE 18.17

The Anfinsen Experiment.

Ribonuclease denatured by 8 M urea and a mercaptan (RSH; a reagent that reduces disulfides to sulfhydryl groups) can be renatured by removing the urea and RSH and air oxidizing the reduced disulfides.

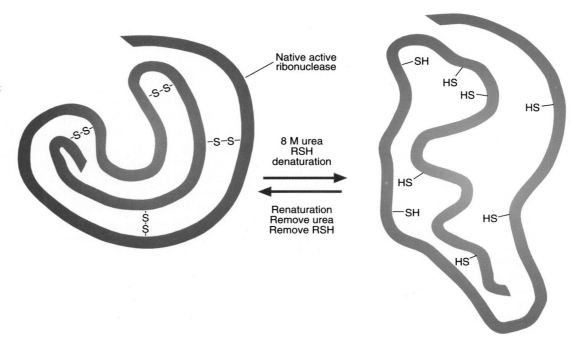

tallography and NMR in combination with site-directed mutagenesis and computer-based mathematical modeling), progress has been limited. (**Site-directed mutagenesis** is a recombinant DNA technique in which specific sequence changes can be introduced into cloned genes.) Briefly, such work has revealed that protein folding is a stepwise process in which secondary structure formation (i.e., α-helix and β-pleated sheet) is an early feature. Hydrophobic interactions appear to be an important force in folding. In addition, amino acid substitutions experimentally introduced into certain proteins reveal that changes in surface amino acids rarely affect the protein's structure. In contrast, substitutions of amino acids within the hydrophobic core often lead to serious structural changes in conformation.

The limitations of the traditional protein-folding model (i.e., interactions between amino acid side chains alone force the molecule to fold into its final shape) are highlighted by the following considerations:

1. **Time constraints.** The time to synthesize proteins routinely ranges from a few seconds to no more than a few minutes. According to one prominent attempt to explain folding, a newly made polypeptide tries out all possible conformations until the most stable one is achieved. Calculations of the time required for every bond in a small protein molecule to rotate until the final biologically active form is achieved indicate that an astronomical number of years would be required. Even when only a smaller number of possible bond rotations are considered, the time required for folding is still measured in years. Therefore most researchers have concluded that protein folding is not a random process based solely on primary sequence.

2. **Complexity.** The calculations required in the mathematical models of protein folding based on physical data (e.g., bond angles and degrees of rotation) are overwhelmingly complex. Therefore it appears unlikely at this time that they alone can resolve the principles of what appears in living organisms to be an astonishingly fast and elegant process.

Over the past several years, it has become increasingly clear that protein folding and targeting in living cells are aided by a group of molecules now referred to as the **molecular chaperones.** These molecules, most of which appear to be hsps (heat shock proteins), apparently occur in all organisms. Several classes of molecular chaperones have been found in organisms ranging from bacteria to the higher animals and plants. In addition, they are also found in several eukaryotic organelles, such as mitochondria, chloroplasts, and ER. There is a high degree of sequence homology among the molecular chaperones of all species so far investigated. The properties of several of these important molecules are described next.

Molecular Chaperones   Molecular chaperones apparently assist unfolded proteins in two ways. First, during a finite time between synthesis and folding, proteins must be protected from inappropriate protein-protein interactions. Some proteins must remain unfolded until they are inserted in an organellar membrane, for example, certain mitochondrial and chloroplast proteins. Second, proteins must fold rapidly and precisely into their correct conformations. Some must be assembled into multisubunit complexes. Investigations of protein folding in a variety of organisms reveal that two major molecular chaperone classes are involved in protein folding.

1. **Hsp70s.** The **hsp70s** are a family of molecular chaperones that bind to and stabilize proteins during the early stages of folding. Numerous hsp70 monomers bind to short hydrophobic segments in unfolded polypeptides, thereby preventing aggregation. Each type of hsp70 possesses two binding sites, one for an unfolded protein segment and another for ATP. Release of a polypeptide from a hsp70 involves ATP hydrolysis. Mitochondrial and ER-localized hsp70s are required for transmembrane translocation of some polypeptides.

2. **Hsp60s.** Once an unfolded polypeptide has been released by hsp70, it is passed on to a member of a family of molecular chaperones referred to as the **hsp60s**

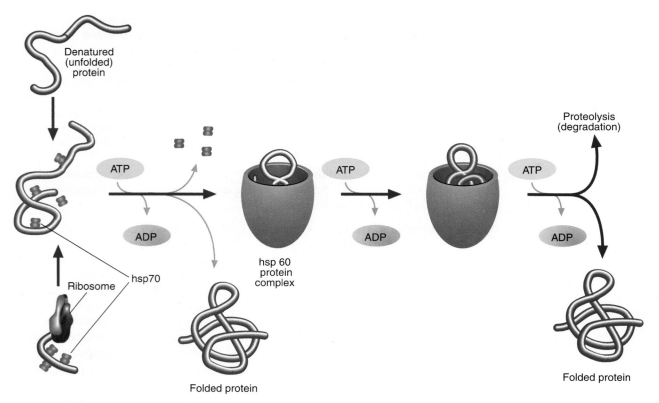

FIGURE 18.18

The Molecular Chaperones.

Molecular chaperones bind transiently to both nascent proteins and unfolded proteins (i.e., denatured by stressful conditions). The members of the hsp70 family stabilize nascent proteins and reactivate some denatured proteins. Many proteins also require hsp60 proteins to achieve their final conformations. If a protein cannot be salvaged, the molecular chaperones help destroy it.

(also called the **chaparonins** or *Cpn 60s*), which mediate protein folding. The hsp60s form a large structure composed of two stacked 7-subunit rings that unfolded proteins enter. In an ATP-requiring process, hsp60 then facilitates the transformation of an unfolded molecule into a properly folded one. Many of the details of this process remain unresolved.

In addition to promoting the folding of nascent protein, molecular chaperones direct the refolding of protein partially unfolded as a consequence of stressful conditions. If refolding is not possible, molecular chaperones promote protein degradation. A diagrammatic view of protein folding is presented in Figure 18.18.

KEY WORDS

anhydride, *531*
anticodon, *529*
chaparonin, *554*
codon, *527*
cotranslational transfer, *548*
disulfide exchange, *547*
docking protein, *548*
elongation, *532*
genetic code, *527*
hsp60, *553*

hsp70, *553*
initiation, *532*
kinetic proofreading, *542*
mixed anhydride, *531*
molecular chaperone, *553*
open reading frame, *528*
polysome, *532*
posttranslational modification, *532*

posttranslational translocation, *548*
proprotein, *544*
reading frame, *528*
releasing factor, *537*
RNA world hypothesis, *526*
second genetic code, *530*
Shine-Dalgarno sequence, *535*
signal hypothesis, *548*
signal recognition particle, *548*

signal peptide, *537*
site-directed mutagenesis, *553*
targeting, *534*
termination, *532*
transcript localization, *547*
translational frameshifting, *552*
translocation, *532*
translocon, *548*
wobble hypothesis, *530*

SUMMARY

Protein synthesis is a complex process in which information encoded in nucleic acids is translated into the primary sequence of proteins. During the translation phase of protein synthesis, the incorporation of each amino acid is specified by one or more triplet nucleotide base sequences, referred to as codons. The genetic code consists of 64 codons: 61 codons that specify the amino acids and three stop codons. Translation also involves the tRNAs, a set of molecules that act as carriers of the amino acids. The base-pairing interactions between codons and the anticodon base sequence of tRNAs result in the accurate translation of genetic messages. Translation consists of three phases: initiation, elongation, and termination. Each phase requires several types of protein factors. Although prokaryotic and eukaryotic translational mechanisms bear a striking resemblance to each other, they differ in several respects. One of the most notable differences is the identity and function of the translation factors.

Protein synthesis also involves a set of posttranslational modifications that prepare the molecule for its functional role, assist in folding, or target it to a specific destination. These covalent alterations include various proteolytic processing, the addition of groups to certain amino acid side chains, and the insertion of cofactors.

Prokaryotes and eukaryotes differ in their usage of translational control mechanisms. In addition to variations in Shine-Dalgarno sequences, prokaryotes also use negative translational control, that is, the repression of the translation of a polycistronic mRNA by one of its products. In contrast, a wide variety of eukaryotic translational controls have been observed. These mechanisms range from global controls in which the translation rate of a large number of mRNAs is altered to specific controls in which the translation of a specific mRNA or small group of mRNAs is altered.

One of the most important aspects of protein synthesis is the folding of polypeptides into their biologically active conformations. Despite decades of investigation into the physical and chemical properties of polypeptide chains, the mechanism by which a primary sequence dictates the molecule's final conformation is unresolved. It has become increasingly clear that most proteins require molecular chaperones to fold into their final three-dimensional conformations.

SUGGESTED READINGS

Arnez, J.G., and Moras, D., Structural and Functional Considerations of the Aminoacylation Reaction, *Trends. Biochem. Sci.*, 22(6): 211–216, 1997.

Craig, E.A., Chaperones: Helpers Along the Pathways to Protein Folding, *Science,* 260:1902–1903, 1993.

Ellis, R.J., and Hemningsen, S.M., Molecular Chaperones: Proteins Essential for the Biogenesis of Some Macromolecular Structures, *Trends Biochem. Sci.,* 14:339–342, 1989.

Gilbert, W., The RNA World, *Nature,* 319:618, 1986.

Noller, H.F., Hoffarth, V., and Zimniak, L., Unusual Resistance of Peptidyl Transferase to Protein Extraction Procedures, *Science,* 256:1416–1419, 1992.

Rothman, J.E., and Wieland, F.T., Protein Sorting by Transport Vesicles, *Science,* 272:227–234, 1996.

Saks, M.E., Sampson, J.R., and Abelson, J.N., The Transfer RNA Identity Problem: A Search for Rules, *Science,* 263:191–197, 1994.

Schatz, G., and Dobberstein, B., Common Principles of Protein Translocation across Membranes, *Science,* 271:1519–1526, 1996.

Stansfield, I., Jones, K.M., and Tuite, M.F., The End in Site: Terminating Translation in Eukaryotes, *Trends Biochem. Sci.,* 20(12):489–491, 1995.

Weissman, J.S., All Roads Lead to Rome? The Multiple Pathways of Protein Folding, *Chem. Biol.,* 2:255–260, 1995.

Welch, W.J., How Cells Respond to Stress, *Sci. Amer.,* 268(5): 56–64, 1993.

QUESTIONS

1. The three-dimensional structures of ribosomal RNA and ribosomal protein are remarkably similar among species. Suggest reasons for these similarities.

2. Describe the RNA world hypothesis. What are its principal assumptions?

3. List and describe four properties of the genetic code.

4. What two observations prompted the wobble hypothesis?

5. Explain the significance of the following statement: The functioning of the aminoacyl-tRNA synthetases is referred to as the second genetic code.

6. Describe the two sequential reactions that occur in the active site of aminoacyl-tRNA synthetases.

7. Although aminoacyl-tRNA synthetases make few errors, occasionally an error does occur. How can these errors be detected and corrected?

8. What are the three phases of protein synthesis? Describe the principal events in each phase. What specific roles do translation factors play in both prokaryotic and eukaryotic translation processes?

9. Determine the codon sequence for the peptide sequence glycylserylcysteinylarginylalanine. How many possibilities are there?

10. Indicate the phase of protein synthesis during which each of the following processes occurs:
    a. A ribosomal subunit binds to a messenger RNA.
    b. The polypeptide is actually synthesized.
    c. The ribosome moves along the codon sequence.
    d. The ribosome dissociates into its subunits.

11. Estimate the minimum number of ATP and GTP molecules required to polymerize 200 amino acdis.

12. Discuss the role of GTP in the functioning of translation factors.

13. What are the major differences between eukaryotic and prokaryotic translation?

14. Posttranslational modifications serve several purposes. Discuss and give examples.

15. Describe how the base pairing between the Shine-Dalgarno sequence and the 30S subunit provides a mechanism for distinguishing a start codon from a methionine codon. What is the eukaryotic version of this mechanism?

16. What are the three steps in the elongation cycle?

17. What are the major differences between eukaryotic and prokaryotic translation control mechanisms?

18. Describe how kinetic proofreading takes place.

19. Clearly define the following terms:
    a. targeting
    b. scanning
    c. codon
    d. reading frame
    e. molecular chaperones
    f. disulfide exchange
    g. proofreading site
    h. signal peptide
    i. glycosylation
    j. negative translational regulation

20. Given an amino acid sequence for a polypeptide, can the base sequence for the mRNA that codes for it be predicted?

21. Describe the structure and function of the signal recognition particle.

22. Describe the function of the translocon in cotranslational transfer.

23. Describe how eukaryotic mRNA structure can affect translational control.

24. Because of the structural similarity between isoleucine and valine, the aminoacyl-tRNA synthetases that link them to their respective tRNAs possess proofreading sites. Examine the structures of the other α-amino acids and determine other sets of amino acids whose structural similarities might also require proofreading.

25. What advantages are there for synthesizing an inactive protein that must subsequently be activated by posttranslational modifications?

26. In general terms, describe the intracellular processing of a typical glycoprotein that is destined for secretion from a cell.

27. Describe the problems associated with determining a polypeptide's final three-dimensional shape using its primary structure as a guide.

28. Describe the roles of the most prominent molecular chaperones in protein folding.

BIOTRANSFORMATION

Chapter Nineteen

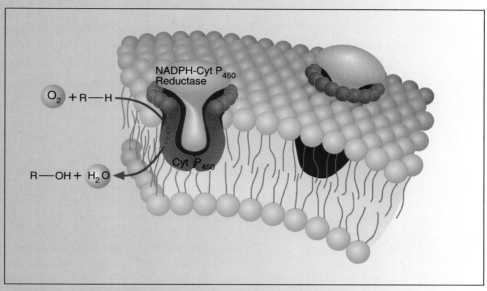

Cytochrome P_{450} and Cytochrome P_{450} Reductase are components of an electron transport system used to oxidize both endogenous and exogenous molecules.

OBJECTIVES

After you have studied this chapter, you should be able to answer these questions:

1. *What molecules are considered primary metabolites? secondary metabolites?*

2. *What is biotransformation and what purpose does it serve?*

3. *What types of reactions occur during phase I of biotransformation?*

4. *What types of reactions characterize phase II of biotransformation?*

5. *How do mammals and plants differ in their phase III processing of xenobiotics?*

6. *What are the differences between the reactions catalyzed by cytochrome P_{450} and flavin-containing monooxygenases?*

7. *What are the consequences of drug tolerance and cross-tolerance?*

8. *How is ethanol detoxified? Why is this drug a serious threat to health?*

9. *What roles do plants play in processing environmental toxins?*

Perhaps one of the most remarkable features of living organisms is their adaptability. Throughout its immense and incomprehensibly rich history, life on earth has constantly been challenged by an ever changing and frequently hostile environment. Only species capable of adapting to or even profiting from changes (e.g., in weather patterns, food supply, or predation) on this turbulent planet survived. Among the survival tactics living organisms use is the synthesis of molecules that either enhance reproductive potential or protect against hostile conditions. An even more interesting adaptive feature of living processes is their capacity to convert noxious foreign molecules to less dangerous products. In this chapter the principal enzymes that modern organisms use to perform this feat are described. This story, however, contains a cautionary note. There are limits to the damage that ecosystems can sustain. Rapid and overwhelming toxic environmental changes can outstrip the capacity of living processes to adapt.

Life on earth has never been easy. In the beginning, noxious substances spewed from ancient volcanoes, and intense, unfiltered ultraviolet light threatened newly emerging life forms. Later, primordial forest fires initiated by electrical storms and increasing atmospheric oxygen created even more poisons. As living organisms became more numerous, another source of toxins arose, one created by the chemical warfare between species. In an unending evolutionary process driven almost entirely by competition for limited resources, species have produced an astonishing array of offensive and defensive molecular weapons to improve their chances of survival. Some of these molecules protect from predators. For example, many plants produce odious or toxic molecules that discourage herbivorous organisms from eating their leaves. Similarly, many of today's antibiotics were originally used by the producing organisms to inhibit the growth of nearby microbial competitors or predators. Living organisms also produce molecules that protect them from abiotic factors. For example, most plants possess a variety of molecules (e.g., the carotenoids) that protect them from intense light and radiation, as well as numerous antioxidants (e.g., α-tocopherol and ascorbic acid).

Perhaps the most intriguing aspect of interspecies chemical warfare is that it is constantly escalating. There is apparently no strategy that cannot eventually be overcome or coopted by some other organism. Through a variety of mechanisms (e.g., gene duplications followed by random mutations, transposition, or intraspecies gene transfers) a species under selection pressure may develop a means of thwarting the defenses of another organism. For example, several herbaceous plants produce toxic molecules called the pyrrolizidine alkaloids, which deter herbivores. Danaiad butterflies can consume the leaves of these plants because the butterflies possess enzymes that convert the alkaloids into nontoxic derivatives that act as pheromones (sex attractants). Similarly, herbivores that can convert antifeedant molecules to harmless metabolites have an adaptive advantage over species that cannot.

When selection pressure is high, adaptive mutations spread quickly through a population. Any organism that cannot protect itself against serious environmental threats may be significantly less likely to reproduce and/or may die prematurely. When an entire species cannot adapt rapidly enough to changing environmental circumstances, it is threatened with extinction.

In the modern world, living organisms have been exposed to new threats. Since the beginning of the Industrial Revolution (ca. A.D. 1760), hundreds of thousands of new compounds produced in an enormous number of industrial processes have been distributed throughout most ecosystems. As before, the species that can adapt to these new threats survive. Those that cannot do not survive. It is now estimated that toxic chemicals and other environmental threats (e.g., human overpopulation and deforestation) now cause the extinction of between several dozen to hundreds of species every year.

The chemical weapons that allow living organisms to withstand chemical and physical threats are sometimes referred to as secondary metabolites. **Secondary metabolites,** many of which are produced only by small groups of species (or even single species) are so named because they are derived from primary metabolites. (**Primary metabolites,** such as sugars, amino acids, fatty acids, and the intermediates of the energy-generating pathways, are molecules that are absolutely required for maintenance of the living state.) The precise functions of many secondary metabolites remain obscure. However, the majority of those whose functions are known serve in a variety of protective roles (e.g., against predators or abiotic factors such as intense light or oxidative and osmotic stresses). In addition, some secondary metabolites enhance reproductive potential by increasing the likelihood that mates find each other (e.g., pheromones) or ensuring adequate food supplies by eliminating competing species (e.g., antibiotics and toxins). Examples of secondary metabolites are found among the alkaloids, the terpenes, glycosides, and numerous fatty acid and amino acid derivatives.

Living organisms are also highly dependent on a vast array of enzymes for protection from **xenobiotics** (foreign molecules). Among bacterial populations, most of these enzymes are highly specific, both in the type of reaction catalyzed and in the species possessing them. Often such enzymes allow a bacterium to convert a toxic molecule into a useful (or at least harmless) metabolite. A new environmental technology, referred to as **bioremediation,** uses certain bacterial species to clean up groundwater and soils contaminated with organic chemicals. Many of these organisms can use certain organic contaminants as carbon sources.

In **biotransformation,** a series of enzyme-catalyzed processes in which toxic substances are converted into less toxic metabolites, the enzymes generally possess broad specificities. Biotransforming mechanisms have been most thoroughly investigated in mammals and plants. In mammals, biotransformation is used principally to convert toxic molecules, which are usually hydrophobic, into water-soluble derivatives so that they may be more easily excreted. In plants, biotransformed derivatives are usually safely stored in the vacuole or cell wall. The enzymes that catalyze the biotransformation of xenobiotics are similar to several of the enzymes that dispose of hydrophobic endogenous molecules.

In this chapter the principal biotransformation mechanisms of mammals and plants are described. The discussion begins with descriptions of the classes of biotransformation reactions in mammals. This is followed by descriptions of the processing of selected exogenous and endogenous molecules. The chapter ends with a brief discussion of plant xenobiotic biotransformation.

19.1 BIOTRANSFORMATION REACTIONS

Although biotransformation reactions occur in several locations within the cell (e.g., the cytoplasm and mitochondria), most occur within the ER. Cell types also differ in their biotransforming potential. In general, cells located near the major points of xenobiotic entry into the body (e.g., liver, lung, and intestine) possess greater concentrations of biotransforming enzymes than others.

Biotransformation processes have been differentiated into two major types. During **phase I,** reactions involving oxidoreductases and hydrolases convert hydrophobic substances into more polar molecules. **Phase II** consists of reactions in which metabolites containing appropriate functional groups are conjugated with substances such as glucuronate, glutamate, sulfate, or glutathione. In general, conjugation dramatically improves solubility, which then promotes rapid excretion. (In animals excretion of biotransformed molecules is sometimes referred to as phase III.) Although many substances undergo these phases sequentially, a significant number do not. For example, some molecules are excreted as phase I metabolites, while others undergo only phase II reactions. Adding to the complexity of this process is that, as a result of variations in enzyme concentrations, availability of cosubstrates, and the order in which the reactions occur, certain substances may be converted into more than one end product. However, despite these and other complications, basic biotransformation patterns have emerged. Several well-researched examples of phase I and phase II reactions are described below. In the following discussions, the term **detoxication** describes the process by which a toxic molecule is converted to a more soluble (and usually less toxic) product. The more familiar term **detoxification** implies the correction of a state of toxicity, that is, the chemical reactions that produce sobriety in an inebriated person.

Phase I Reactions

Phase I reactions usually convert substrates to more polar forms by introducing or unmasking a functional group (e.g., —OH, —NH$_2$, or —SH). Many phase I enzymes are located in the ER membrane, but others such as the dehydrogenases (e.g., alcohol dehydrogenases and peroxidases) occur in the cytoplasm, while still others (e.g., monoamine oxidase) are localized in mitochondria. The predominant enzymes of microsomal oxidative metabolism are the monooxygenases, sometimes referred to as mixed function oxidases. They are so named because in a typical reaction, one molecule of oxygen is consumed (reduced) per substrate molecule, one oxygen atom appearing in the product and the other in a molecule of water. Monooxygenases can carry out an immense variety of chemical reactions. Some of these reactions form highly unstable (and therefore toxic) intermediates.

There are two major types of microsomal monooxygenases, both of which require NADPH as an external reductant: the cytochrome P$_{450}$ system and flavin-containing monooxygenases. The **cytochrome P$_{450}$ system,** which consists of two enzymes (NADPH–cytochrome P$_{450}$ reductase and cytochrome P$_{450}$) is involved in the oxidative metabolism of many endogenous substances (e.g., steroids and bile acids) as well as the detoxication of a wide variety of xenobiotics. **Flavin-containing monooxygenases** catalyze redox reactions involving a diverse group of xenobiotics.

Cytochrome P$_{450}$ Electron Transport Systems

In cytochrome P$_{450}$ electron transport systems, found in microsomal and inner mitochondrial membranes, two electrons are transferred one at a time from NADPH to a cytochrome P$_{450}$ protein by NADPH–cytochrome P$_{450}$ reductase. The latter enzyme is a flavoprotein that contains both FAD and FMN in a ratio of 1:1 per mole of enzyme. In addition to its role in the cytochrome P$_{450}$ system, the reductase is also believed to be involved in the function of heme oxygenase (Section 19.2).

The hemoproteins referred to as cytochrome P$_{450}$ are so named because of the complexes they form with carbon monoxide. In the presence of the gas, light is strongly absorbed at a wavelength of 450 nm. Well over 100 cytochrome P$_{450}$ genes have been identified so far. Each gene codes for a protein with a unique specificity range. Cytochrome P$_{450}$ proteins found in the liver have broad and overlapping specificities. For example, molecules as diverse as alkanes, aromatics, ethers, and sulfides are routinely oxidized. In contrast, cytochrome P$_{450}$ proteins in the adrenal glands, ovaries, and testes that add hydroxyl groups to steroid molecules have narrow specificities. Despite this diversity, all cytochrome P$_{450}$ isozymes contain one mole of heme. In addition, the cytochrome P$_{450}$s are similar in their physical properties and catalytic mechanisms.

Despite an enormous variety of substrates, all of the oxidative reactions catalyzed by cytochrome P$_{450}$ may be viewed as hydroxylation reactions (i.e., an OH group appears in each reaction) (Figure 19.1). The general reaction is as follows:

$$R—H + O_2 + NADPH + H^+ \rightarrow ROH + H_2O + NADP^+$$

where R—H is the substrate.

The oxygenation reaction is initiated when the substrate binds to oxidized cytochrome P$_{450}$ (Fe^{3+}). This binding promotes a reduction of the enzyme substrate complex by an electron transferred from NADPH via cytochrome P$_{450}$ reductase

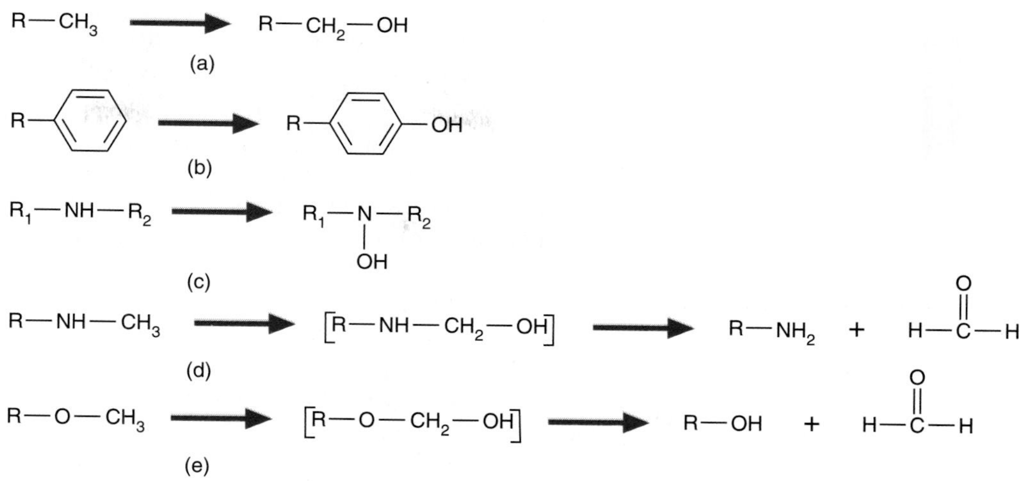

FIGURE 19.1

Diverse Substrates Oxidized by Cytochrome P$_{450}$ Isozymes.

Among the reactions catalyzed by cytochrome P$_{450}$ are (a) aliphatic oxidation, (b) aromatic hydroxylation, (c) N-hydroxylation, (d) N-dealkylation, and (e) O-dealkylation.

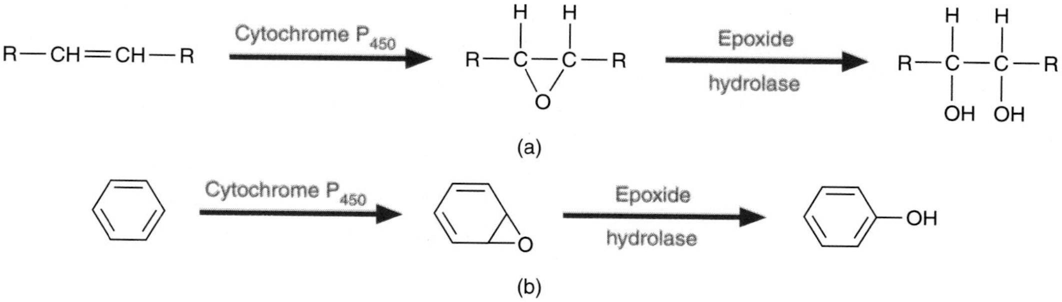

FIGURE 19.2

Conversion of Substrates to Alcohols Via an Epoxide Intermediate.

(a) Aliphatic substrates, such as hydrocarbons. (b) Aromatic substrates, such as benzene.

(Fe^{2+}—substrate). After reduction, cytochrome P$_{450}$ can bind O$_2$. Then the electron from heme iron is transferred to the bound O$_2$, thus forming a transient Fe^{3+}—O$_2^-$— substrate species. (If the bound substrate is easily oxidized, it can be converted into a peroxy radical. See below.) A second electron transferred from the flavoprotein results in the generation of a Fe^{3+}—O$_2^=$—substrate complex. This brief association ends when the oxygen-oxygen bond is broken. One atom is released in a water molecule, while the other remains bound to heme. After a hydrogen atom or electron is abstracted from the substrate, the oxygen species (now a powerful oxidant) is transferred to the substrate. The cycle ends by releasing the product from the active site. Depending on the nature of the substrate, the product is either an epoxide or an alcohol.

Epoxides are highly reactive and have been shown to bind irreversibly to DNA, RNA, and proteins. Considerable effort has therefore been expended in investigating the formation and subsequent biotransformation of these intermediates, which have been implicated in carcinogenesis. Many epoxides are hydrolyzed to diol products (molecules containing two adjacent —OH groups) by epoxide hydrolase, a microsomal enzyme (Figure 19.2). In most cases the diols that are formed are less reactive and less toxic than the parent epoxide. However, with some polycyclic hydrocarbons (e.g., benzo[a]pyrene) the diols that are formed are precursors for carcinogenic metabolites (Section 19.2).

Hydrogen peroxide and organic hydroperoxides products of cytochrome P$_{450}$–catalyzed reactions are dangerous oxidants. As described (Section 11.4), cell membranes are especially vulnerable because of peroxide attack on the unsaturated fatty acids of phospholipids. The principal protection against peroxidation is glutathione peroxidase (refer to Figure 11.27).

19.1 The cytotoxicity of xenobiotics is often a consequence of biotransformation reactions. Some of the most dangerous products of these reactions are free radicals. For example, the hepatotoxic effects of the solvent carbon tetrachloride are due in large part to its conversion to a free radical (Figure 19.3). Among these effects are the inhibition of protein synthesis, the binding of carbon tetrachloride metabolites to lipids, and lipid peroxidation. After referring to Figure 11.25, determine the product of a carbon tetrachloride–derived free radical reacting with oleic acid. It has been observed that carbon tetrachloride eventually destroys cytochrome P_{450} activity. Can you suggest how and why this damage occurs?

FIGURE 19.3

The Bioactivation of Carbon Tetrachloride (CCl_4).

L = lipid molecule.

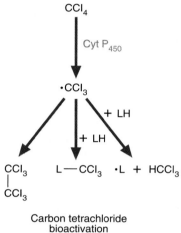

Carbon tetrachloride bioactivation (L = Lipid)

The Flavin-Containing Monooxygenases   Flavin-containing monooxygenases catalyze an NADPH- and oxygen-requiring oxidation of substances (primarily xenobiotics) bearing functional groups containing nitrogen, sulfur, or phosphorus. Among the most extensively studied examples is the enzyme isolated from hog liver, where it accounts for about 1% of cellular protein. The hog enzyme, a tetramer consisting of identical subunits (63 kD), is an integral membrane protein localized primarily in microsomal membrane. Each purified polypeptide contains one mole of FAD.

The catalytic cycle of flavin-containing monooxygenases begins with binding NADPH and transferring two electrons to the flavin moiety. Oxygen then adds to the reduced flavoprotein to produce an unusually stable peroxyflavin (Figure 19.4). Subsequent substrate binding is followed by a direct oxygenation. One atom of oxygen is transferred to the substrate, while the second is released in a water molecule. In the last step, $NADP^+$ is released from the enzyme.

Flavin-containing monooxygenases differ from all other mammalian monooxygenases in that the substrate is not required for oxygen reduction. The substrate need not fit precisely in the catalytic site. Consequently, substrates of widely different structures can be easily accommodated, and virtually any compound can be oxygenated

KEY *Concepts*

Phase I reactions are used to inactivate biologically active biomolecules and/or form more polar derivatives. During phase II, molecules with appropriate functional groups are conjugated to second substrates. Among the most frequently used second (or donor) substrates in these reactions are glucuronic acid, glutathione, sulfate, and certain amino acids.

FIGURE 19.4

The Enzyme Peroxyflavin Complex.

This complex is formed by a flavin-containing monooxygenase and molecular oxygen.

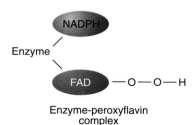

Enzyme-peroxyflavin complex

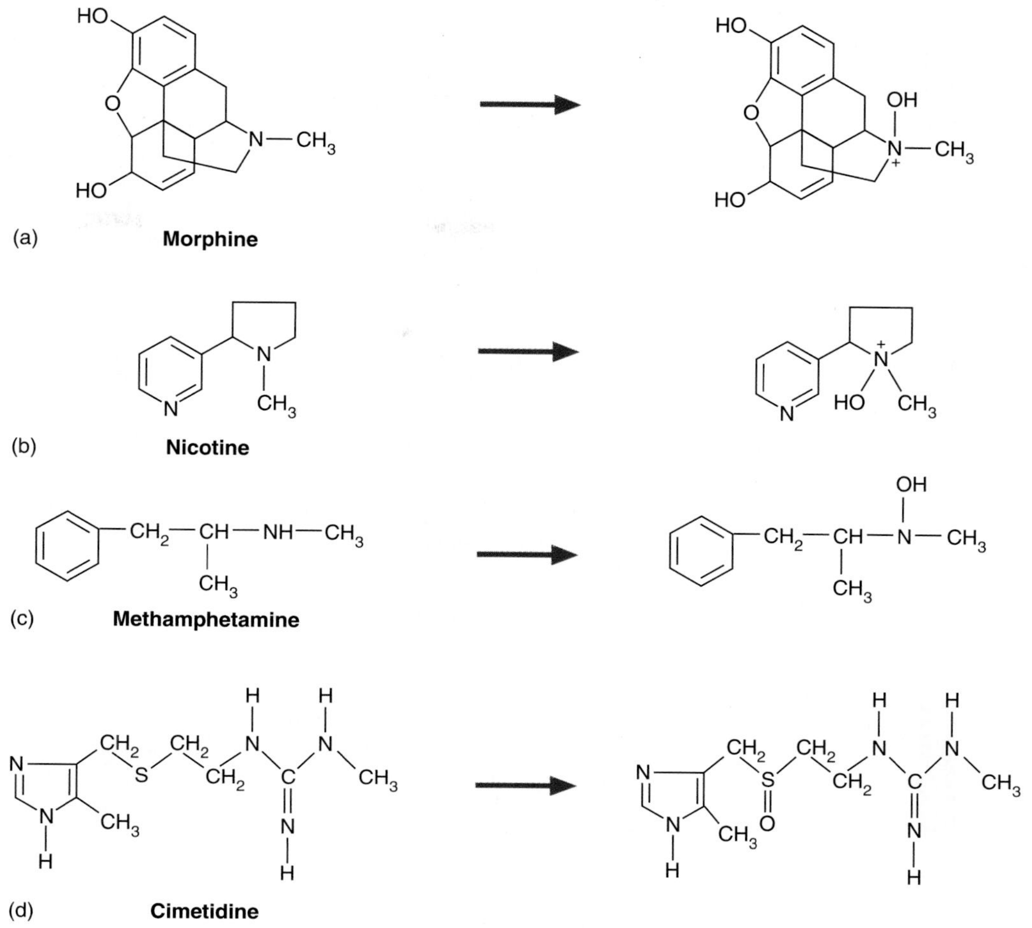

FIGURE 19.5

Selected Substrates and Initial Oxidation Products of Flavin-Containing Monooxygenases.

(a) Morphine (an alkaloid and an opiod narcotic), (b) nicotine (an insecticidal alkaloid), (c) methamphetamine (a synthetic central nervous system stimulant), and (d) cimetidine (an antiulcer drug).

(a)   **Morphine**

(b)   **Nicotine**

(c)   **Methamphetamine**

(d)   **Cimetidine**

by this mechanism. Examples of xenobiotic substrates are illustrated in Figure 19.5. The flavin-containing monooxygenases usually discriminate between xenobiotics and cellular nucleophiles. How they do so is unknown.

Phase II Reactions

The function of conjugation reactions is to inactivate biologically active substances and/or form more polar (and therefore more easily excretable) derivatives. During this process, lipophilic metabolites, bearing functional groups that can act as acceptors, undergo enzyme-catalyzed reactions along with second (or donor) substrates. Among the most frequently used donor substrates are glucuronic acid, glutathione (Section 13.3), sulfate, and amino acids.

Conjugation with glucuronic acid is one of the major mechanisms involved in the mammalian conversion of lipophilic substances to polar, water-soluble molecules. Once formed, glucuronides are usually rapidly removed from the body in either urine or bile. The enzyme activity responsible for the reaction (UDP-glucuronosyltransferase) is embedded in the cytoplasmic face of both rough and smooth ER. Four classes of glucuronides have been characterized: O-, N-, S-, and C-glucuronides. (Refer to Table 19.1.) Endogenous substances that undergo conjugation with glucuronic acid include certain steroid molecules, such as aldosterone (an important electrolyte-regulating hormone) and bilirubin (the degradation product of heme). However, considerably more attention has been paid to the glucuronidation of xenobiotics. This interest has arisen not only because of its role in the detoxication of noxious compounds, but also because certain glucuronides have been implicated in carcinogenesis. For example, N-glucuronides of aromatic amines have been implicated in urinary bladder tumor formation.

TABLE 19.1

Glucuronide Classes

Glucuronide Type	Substrate	Example
O-glucuronide	$R\!-\!CH_2\!-\!\overset{*}{O}H$	Phenacetin
	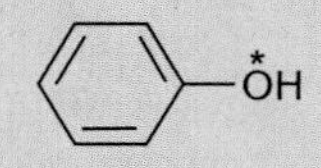 ⬡$-\overset{*}{O}H$	Phenol
	$R\!-\!\overset{\overset{\displaystyle O}{\|\|}}{C}\!-\!\overset{*}{O}H$	Salicylates
N-glucuronide	$R\!-\!CH_2\!-\!\overset{*}{N}H_2$	Aniline
	$R\!-\!\overset{\overset{\displaystyle O}{\|\|}}{C}\!-\!\overset{*}{N}H_2$	Meprobamate (sedative)
S-glucuronide	$R\!-\!CH_2\!-\!\overset{*}{S}H$	Methimazole (antithyroid drug)
C-glucuronide	$R\!-\!CH_2\!-\!\overset{*}{C}H\!=\!CH\!-\!R_1$	Phenylbutazone (antiinflamatory)

* = attachment site for glucuronyl group.

QUESTION

19.2 The conjugation of many toxic xenobiotics detoxicates them and rapidly eliminates them from the body. However, glutathione conjugation of some xenobiotics (especially halogenated hydrocarbons) forms more reactive species. For example, the glutathione derivative of 1,2-dibromoethane is eventually converted to the episulfonium ion, an alkylating agent shown below, which has been observed to form DNA adducts.

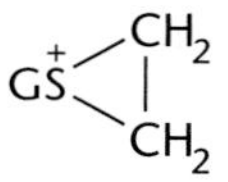

$$G\overset{+}{S}\!\!<\!\!\overset{\displaystyle CH_2}{\underset{\displaystyle CH_2}{\displaystyle |}}$$

Liver and kidney damage and tumors have been associated with the conjugation of GSH with certain xenobiotics. In general terms, suggest how and why such damage occurs.

19.2 BIOTRANSFORMATION PROCESSES

As medical research has progressed, it has become increasingly obvious that biotransformation processes have important implications for human health. The putative role of various xenobiotics in carcinogenesis, for example, has had an impact on research efforts. Similarly, efforts by the pharmaceutical industry to develop effective and safe medications have resulted in a major economic investment in medical research. In addition, the processing and disposition of physiologically important biomolecules have received considerable attention from medical researchers investigating the etiology of disease processes. In this section, several well-researched examples of the processing of exogenous and endogenous molecules will be discussed.

Biotransformation of Selected Exogenous Molecules

Xenobiotics ("strangers to life") are generally considered to fall into three categories: foods, drugs, and poisons. However, almost anything taken into the body has poisonous aspects. The Swiss physician Paracelsus (1493–1541) recognized this principle when he wrote in 1538, "What is not a poison? All things are poisons, and nothing is without toxicity. Only the dose permits anything not to be poisonous."

Considering the enormous exposure that animals have to the nonnutritive natural products found in food as well as to the myriad human-made xenobiotics, it is surprising that physiological catastrophes do not occur more often than they do. It has been postulated that the detoxication systems of the animal body have evolved over the last billion or so years because of the selection pressure provided principally by exposure to plant metabolites and decayed plant products. For example, in addition to nutrient molecules (e.g., starch, sugars, protein, fat, minerals, and vitamins), most vegetables and fruits contain small amounts of dozens or even hundreds of nonnutritive organic molecules. Examples of nonnutritive molecules include various aldehydes, ketones, alcohols, esters, tannins, steroids, phenols, and terpenes, among others. Amazingly, animals have devised mechanisms for using or disposing of most of these molecules.

As noted, a significant amount of the research into the detoxication mechanisms of xenobiotics has been concerned with pharmaceuticals. For this reason, the following discussion focuses on the biotransformation of drugs. Although biotransformation usually reduces the toxicity of xenobiotics, certain molecules are inadvertently made more toxic. Benzo[a]pyrene, a constituent of coal tar, tobacco smoke, and charcoal-broiled food, is now considered to be a classic example of xenobiotic bioactivation.

Phenacetin   Phenacetin is a neutral, lipid-soluble analgesic (pain reliever) used to treat mild to moderate pain. The molecule is converted by cytochrome P$_{450}$ into the more water-soluble para-acetamidophenol (acetaminophen). Typically, this latter molecule is then conjugated with glucuronic acid to form para-acetamidophenyl-β-D-glucuronide, an extremely water-soluble substance, which is then rapidly excreted (Figure 19.6).

Phenacetin processing reveals a feature of biotransformation processes commonly observed with numerous xenobiotics. Acetaminophen is a significantly more effective pain reliever than phenacetin, the molecule from which it is derived. In addition, when it is taken in excess, phenacetin is toxic, presumably because of its by-product acetaldehyde. For this reason, acetaminophen is now dispensed in place of the parent compound.

Phenobarbitol   Phenobarbitol, a derivative of barbituric acid, is often prescribed as a sedative or antiepileptic drug. Approximately 10–25% of the drug is eliminated unaltered by the kidneys. The remainder is inactivated by microsomal enzymes, which convert it to a para-hydroxyphenyl derivative, most molecules of which are excreted as sulfate esters (Figure 19.7.)

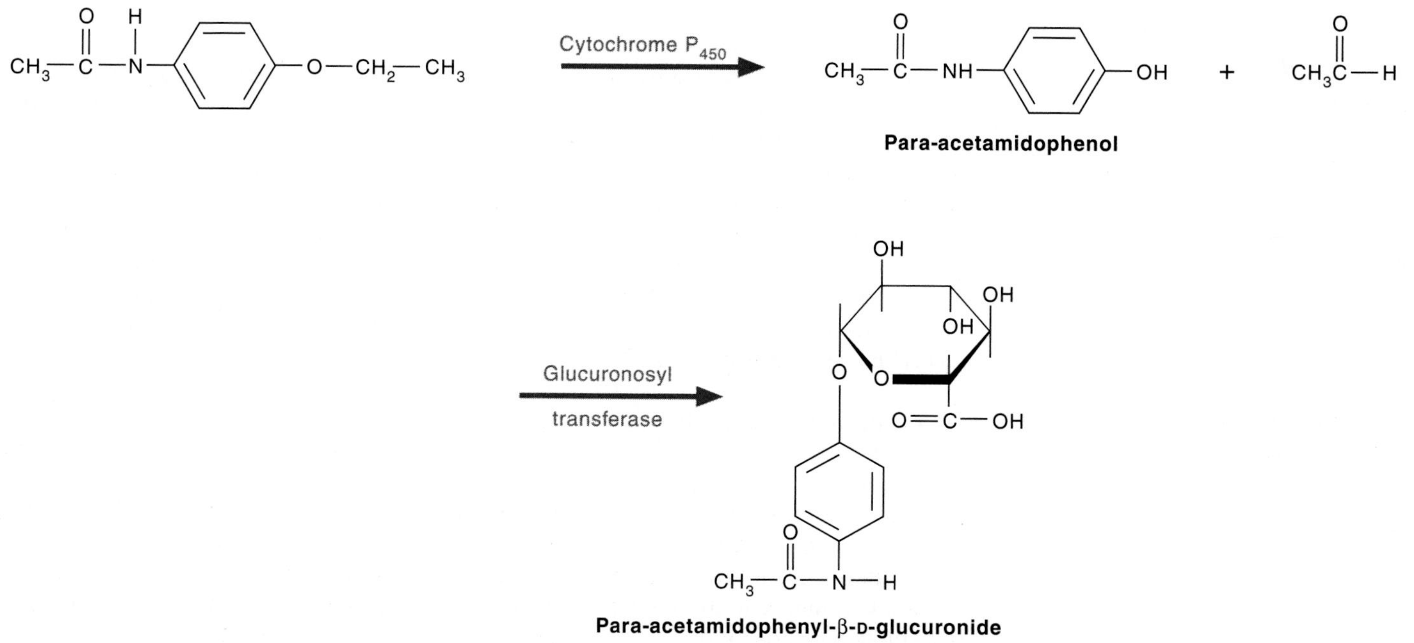

KEY *Concepts*

Animals have adapted enzymes used primarily to dispose of toxic or worthless nonnutritive plant molecules for use with modern drugs and environmental contaminants. These enzymes sometimes convert certain xenobiotics into more toxic derivatives.

Para-acetamidophenol

Para-acetamidophenyl-β-D-glucuronide

FIGURE 19.6

Biotransformation of Phenacetin.

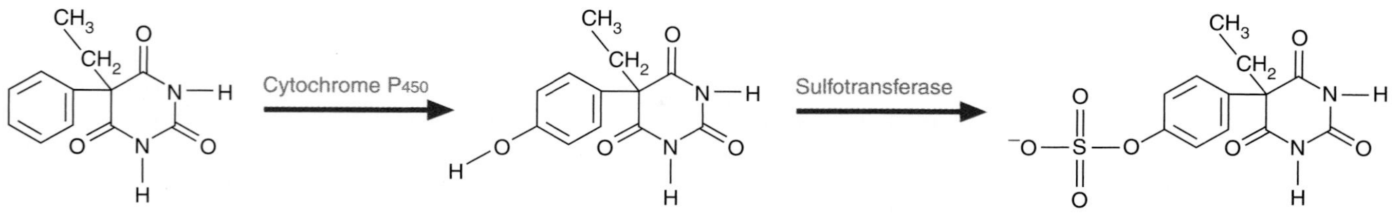

FIGURE 19.7

Biotransformation of Phenobarbitol.

Benzo[a]pyrene

Cytochrome P$_{450}$

Epoxide hydrolase

HO OH

Benzo[a]pyrene-7,8-dihydrodiol

Cytochrome P$_{450}$

HO OH

DHEP-BP

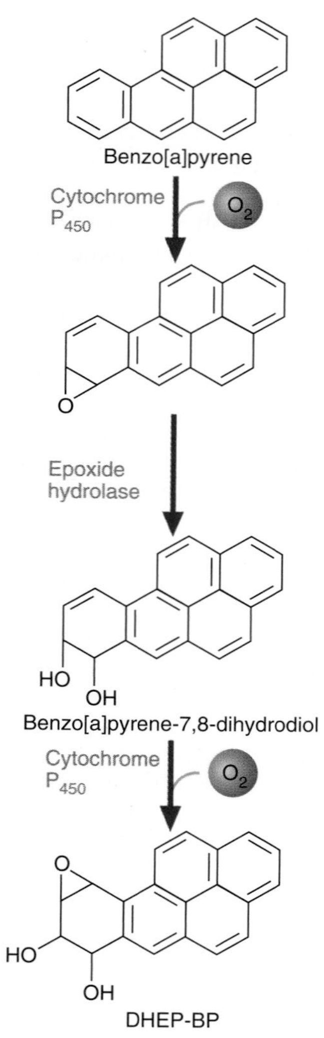

FIGURE 19.8

Proposed Mechanism for the Synthesis of DHEP-BP.

Phenobarbitol consumption leads in a short time to a state of **tolerance,** in which the patient's responsiveness to the drug is reduced. Larger and larger doses are required to achieve the same level of symptomatic relief. This phenomenon, which is not unique to phenobarbitol, is accompanied by a proliferation of SER membrane, as well as increases in total protein and microsomal protein (especially cytochrome P$_{450}$ and related enzymes) per gram of liver. Additionally, changes in membrane lipid composition have been observed (e.g., cholesterol/phospholipid ratios and phospholipid fatty acid composition). Apparently, increased phenobarbitol dosage is required because the drug is being processed in the liver more quickly than before. The mechanism by which phenobarbitol induces the synthesis of additional cytochrome P$_{450}$ is not understood.

Cross-tolerance with other drugs may also occur. Any drug that requires the phenobarbitol-induced enzyme for its detoxication will also require dosage increases to maintain its clinical effectiveness. A classic example is the anticoagulant warfarin. (Warfarin is an inhibitor of blood clotting. Its structure is similar to that of vitamin K$_2$, a required cofactor in prothrombin synthesis. Prothrombin is one of several proteins involved in the blood-clotting process.) When phenobarbitol and warfarin are administered together, the warfarin soon becomes ineffective. If phenobarbitol is withdrawn after warfarin dosage has been increased, the patient quickly becomes susceptible to internal hemorrhaging. In this circumstance a warfarin overdosage has occurred because without the inducing effects of phenobarbitol, the induced enzymes quickly revert to their previous levels. Consequently, warfarin accumulates in the body and significantly inhibits blood clotting.

Benzo[a]pyrene   The aromatic hydrocarbon benzo[a]pyrene (BP) is now recognized as a mutagenic carcinogen. Once in the body, BP is converted (depending on its mode of entry) by tissues (e.g., lung, oral mucosa, intestine, and liver) to a large number of metabolites by microsomal enzymes. These include phenols, dihydrodiols, quinones, and epoxides. The biotransformation process generates numerous electrophilic intermediates that can covalently bond to cellular macromolecules, especially DNA. It is now generally accepted that the binding of one such metabolite, 7,8-dihydroxy-9,10-epoxy-7,8,9,10-tetrahydrobenzo[a]pyrene (DHEP-BP) to DNA, and not BP itself, is an important factor in the induction of certain tumors. A proposed synthesis of DHEP-BP is illustrated in Figure 19.8.

Cocaine   Taken in small amounts, cocaine is a central nervous system (CNS) stimulant that induces a sense of euphoria and well-being. Higher doses lead to respiratory depression, cardiac arrest, and/or convulsions. Because the drug is addictive, users move inexorably toward taking larger and larger doses. An alkaloid obtained from the coca tree *Erythroxylon coca,* cocaine exerts its CNS effects by blocking the reuptake of neurotransmitters such as norepinephrine, dopamine, and serotonin at nerve terminals. In addition to mood-altering effects, this process also results in increased heart rate, dilated pupils, and constricted blood vessels.

Cocaine is biotransformed by two different pathways. Most of the molecules (approximately 90%) are hydrolyzed to biologically inactive products by esterases, which

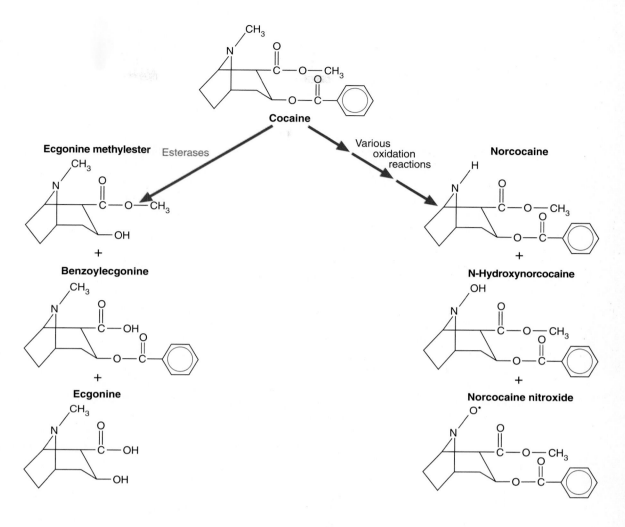

FIGURE 19.9

Hydrolytic and
Oxidative
Metabolism of
Cocaine.

are found predominantly in blood plasma. The remaining cocaine molecules are oxidized by microsomal enzymes, primarily those of the liver (Figure 19.9). The products of this latter pathway (especially norcocaine nitroxide, a free radical) appear to be responsible for the severe liver damage that often accompanies cocaine addiction. Recent studies have indicated that both cytochrome P_{450} and flavin-containing monooxygenase are involved in this conversion.

QUESTION

19.3

Cannabis sativa, the hemp plant, contains a psychoactive substance known as *tetrahydrocanabinol* (THC) (Figure 19.10). Smoking marijuana, which consists of the dried leaves and flowering tops of hemp plants, induces euphoria and a feeling of relaxation. However, even small doses cause short-term memory losses and impairment of motor skills. Chronic users progressively lose interest in jobs and social relationships and risk hallucinations and paranoid thoughts. On the basis of the knowledge you have acquired in this chapter, suggest likely end products of tetrahydrocanabinol biotransformation.

FIGURE 19.10

Tetrahydrocannabinol.

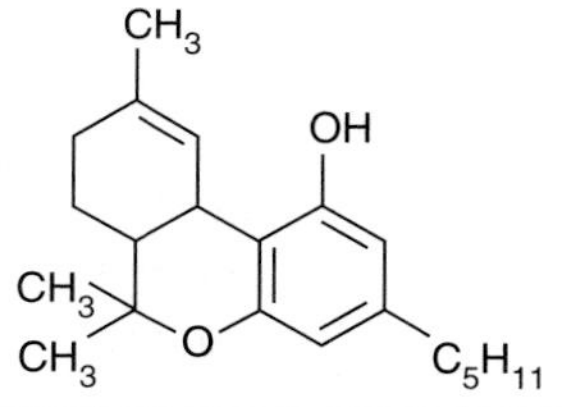

ALCOHOL AND THE LIVER: A CAUTIONARY TALE

Poets and connoisseurs have been extolling the virtues of alcoholic beverages for thousands of years. Indeed, the social use of wines, beers, and other "spirits" has long been associated with conviviality and gracious living. In addition, when consumed in moderate amounts, certain beverages confer health benefits. For example, resveratrol, a component of red wine, is now believed to reduce the risk of coronary heart disease. (Resveratrol is also found in red grapes.) However, it is now generally recognized that alcohol (ethanol) is a drug, the abuse of which has very serious personal and social consequences. When alcohol has been abused over a long period, *alcoholism* can result. Alcoholism is a complex addiction whose causes are still unresolved. Its costs are truly staggering. For example, annual alcohol-related health care costs and productivity losses in the United States are measured in the hundreds of billions of dollars. In addition to its long-term effects on the central nervous system (e.g., impaired judgment and dysfunctional behavior), excessive alcohol consumption is associated with damage to numerous other organs (e.g., the cardiovascular and gastrointestinal systems, among others). Chronic alcohol abuse increases the risk of certain types of cancer (e.g., oral, esophageal, and pancreatic cancers). The most tragic effect of alcohol abuse is *fetal alcohol syndrome*, a set of developmental abnormalities caused by maternal drinking during pregnancy. The manifestations of fetal alcohol syndrome range from mild mental retardation to catastrophic abnormalities such as microcephaly (an abnormally small head), heart valve defects, and profound mental retardation.

Because ethanol is a small water-soluble molecule, it is rapidly absorbed from the gastrointestinal tract. Although some ethanol is metabolized in the stomach or eliminated unchanged through the kidneys and lungs, most molecules are degraded in the liver. Therefore it is not surprising that alcohol abuse and liver disease are so closely associated. Interestingly, the first documented association between alcohol abuse and "scirrhus" (also referred to as "gin-drinker's liver") occurred in the 1790s. The more familiar term **cirrhosis** was introduced in the early nineteenth century. Currently, cirrhosis (a term most often used in reference to the liver) refers to a degenerative condition in which living cells are replaced by fibrous connective tissue. Despite two centuries of medical interest in cirrhosis, the mechanisms by which ethanol damages the liver have remained unknown until recently. Until approximately 20 years ago it was believed that alcoholic liver disease was principally caused by malnutrition. Although malnutrition is a contributing factor (see below), the specific hepatotoxic effects of ethanol are now relatively well understood. These effects are described after a brief discussion of ethanol detoxification in the hepatocyte.

Ethanol Detoxification

The major mechanism by which ethanol is detoxified involves the cytoplasmic enzymatic activity of alcohol dehydrogenase (ADH). Recall that ADH catalyzes the oxidation of ethanol to form acetaldehyde (Section 6.4). (ADH also catalyzes the oxidation of other alcohols, such as methanol. The normal function of ADH may be the oxidation of the small amounts of alcohols produced by microbial fermentation processes within the intestine.)

$$CH_3-CH_2-OH \; + \; \boxed{NAD^+} \; \xrightarrow{ADH} \; CH_3\overset{\displaystyle O}{\overset{\|}{C}}-H \; + \; \boxed{NADH} \; + \; \boxed{H^+}$$

Soon after its production, acetaldehyde is converted to acetate by aldehyde dehydrogenase, the most important isozyme of

which is located within the mitochondrial matrix.

$$CH_3-\overset{\displaystyle O}{\overset{\|}{C}}-H \; + \; \boxed{NAD^+} \; + \; \boxed{H_2O} \; \xrightarrow{\text{Aldehyde}\atop\text{dehydrogenase}} \; CH_3-\overset{\displaystyle O}{\overset{\|}{C}}-O^- \; + \; \boxed{NADH} \; + \; \boxed{H^+} \; + \; \boxed{H_2O}$$

Most of the acetate produced in this reaction is metabolized further in other tissues, such as cardiac and skeletal muscles.

When the concentration of ethanol in hepatocytes becomes sufficiently high, it is also detoxified by the microsomal ethanol-

oxidizing system (MEOS). MEOS consists of an ethanol-inducible cytochrome P_{450} isozyme and NADPH cytochrome P_{450} reductase.

$$CH_3CH_2-OH + \boxed{NADPH} + \boxed{H^+} + \boxed{O_2} \longrightarrow CH_3-\overset{\displaystyle O}{\overset{\|}{C}}-H + \boxed{NADP^+} + 2 \boxed{H_2O}$$

MEOS also catalyzes the oxidation of other xenobiotics (e.g., CCl_4 and acetaminophen) in addition to other alcohols. Induction of MEOS therefore alters detoxication rates for these substances. In addition, the ethanol-inducible cytochrome P_{450} apparently possesses a unique capacity to convert certain xenobiotics (e.g., CCl_4) to toxic metabolites. This phenomenon may explain the increased risk associated with alcohol consumption combined with exposure to tobacco-derived products or industrial solvents.

A third and significantly less important mechanism for ethanol detoxification involves catalase, an enzyme found predominantly within peroxisomes. It is unclear whether this peroxide-dependent process is physiologically significant.

$$CH_3CH_2-OH \;+\; H_2O_2 \xrightarrow{\text{Catalase}} CH_3-\overset{\displaystyle O}{\overset{\|}{C}}-H \;+\; H_2O$$

Hepatotoxic Effects of Ethanol

Ethanol's toxic effects are principally mediated by acetaldehyde, free radicals, and altered cellular redox conditions. The mechanisms by which each damages hepatocytes are briefly described below.

Acetaldehyde is an extremely toxic molecule because of its propensity to covalently bond to proteins and to promote lipid peroxidation. Because protein adduct formation is nonspecific, a wide variety of enzymes and structural proteins can be affected. In addition, the secretion of proteins from liver cells can be retarded by exposure to acetaldehyde. Acetaldehyde promotes lipid peroxidation principally by depleting cells of GSH.

As described above, excessive consumption of ethanol induces MEOS. Increased concentration of ethanol-inducible cytochrome P_{450} generates excessive ROS. Another possible source of tissue-damaging ROS is the increased activity of xanthine oxidase (Section 14.3), from excessive increased purine degradation.

The excessive production of NADH created by increased activity of ADH and aldehyde dehydrogenase has several effects. The most notable is depressing the citric acid cycle activity by transferring reducing equivalents into mitochondria by shuttle mechanisms. Because liver cells have no control mechanism for regulating ethanol metabolism, ethanol is the preferred energy substrate.

Because of the liver's important role in metabolism, damage caused by alcohol is widespread. A few of the most obvious alterations in vital hepatocyte functions caused by ethanol-induced toxicity include the following:

1. **Hepatic steatosis.** Hepatic steatosis (or "fatty liver"), the abnormal accumulation of fat within the liver, is one of the earliest manifestations of alcohol abuse. The promotion of fat synthesis is principally the result of excessive NADH production. Fat also accumulates because of decreased fat oxidation and decreased lipoprotein secretion, both of which are caused by multiple degenerative changes in cell structure. As fat accumulates, hepatocytes are less and less able to perform their metabolic roles.

2. **Hyperlactacidemia.** Hyperlactacidemia (high blood lactate concentration) results from decreased conversion of pyruvate to acetyl-CoA, a consequence of depressed citric acid cycle activity. (Recall that excess pyruvate is converted to lactate.) When lactate production becomes excessive, it is released into blood. High blood levels of lactate contribute to reducing the kidney's capacity to excrete uric acid, thereby promoting hyperuricemia (Special Interest Box 14.4).

3. **Malnutrition.** The relationship between excessive alcohol consumption and nutrition is complex. Because of ethanol's high caloric value (29 kJ/g), alcoholic beverages often displace other foods. Many modern alcoholic beverages possess few if any other nutrients. (Ancient beer and wine may have been more nutritious than their modern counterparts because they were probably not filtered.) In addition, chronic excessive alcohol consumption so damages the gastrointestinal tract that the digestion and absorption of food is compromised. Finally, ethanol metabolism promotes the depletion of the body's stores of vitamins (e.g., vitamins A, B, C, and E) and minerals (e.g., selenium).

Biotransformation of Selected Endogenous Molecules

Among the most carefully investigated examples of endogenous substances that undergo phase I and phase II type reactions are those in steroid metabolism. Additionally, considerable effort has been devoted to elucidating the details of the conversion of heme to bilirubin. Representative examples of steroid transformations are briefly outlined below. This is followed by a detailed description of bilirubin metabolism.

Steroid Biotransformations The metabolism of steroids is extremely complicated. A variety of cells and cellular organelles figure prominently in producing and processing these powerful biomolecules. Several biotransforming enzymes play key roles in both the synthesis and the inactivation of most steroid molecules. For example, the cytochrome P_{450} enzyme 21-hydroxylase (see Fig. 10.28) found in the adrenal cortex catalyzes the initial reaction in the conversion of progesterone into aldosterone, a potent mineralocorticoid. In addition a variety of glucocorticoids, androgens, and estrogens are produced in a series of reactions initiated by the 17-α-hydroxylation of progesterone. The enzyme that catalyzes this reaction, also a cytochrome P_{450} protein, is found in the ER of all steroid-producing cells. In the adrenal glands, cortisol is a principal product, while testosterone is primarily produced in the testes. The ER of ovaries and placenta contain a complex of three enzymes, referred to collectively as the aromatase system. The aromatase system converts testosterone to 17-β-estradiol, a potent estrogen (Figure 19.11). The first of these enzymes, 19-hydroxylase, is also a cytochrome P_{450} protein.

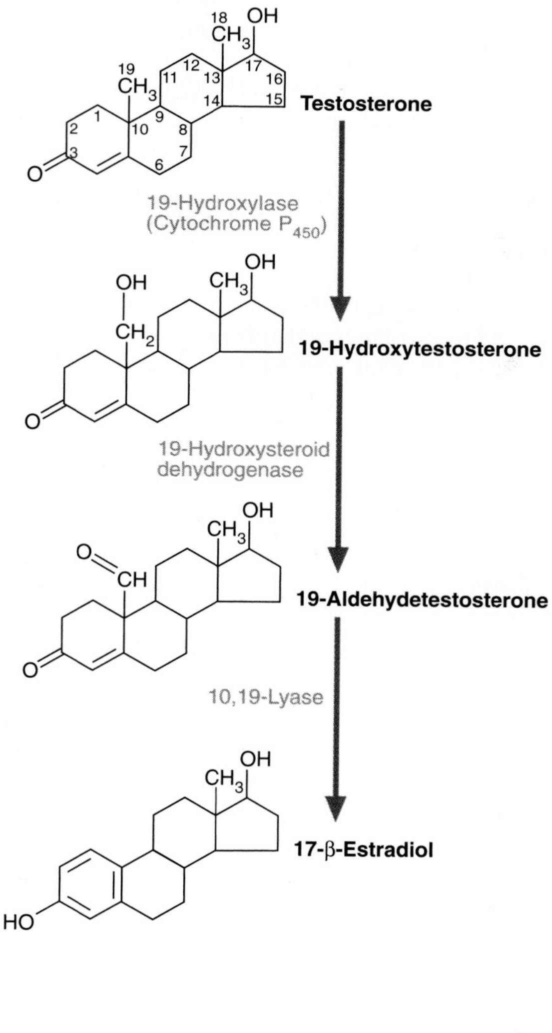

FIGURE 19.11

The Aromatase System.

Three enzyme activities in the ER of cells within ovaries and placenta convert testosterone into 17-β-estradiol. 19-Hydroxylase is a cytochrome P_{450} isozyme.

KEY *Concepts*

Biotransforming enzymes, especially a variety of cytochrome P_{450} isozymes and conjugating enzymes, are used in many metabolic processes.

Biotransforming enzymes also play an important role in the metabolism of the steroid molecule vitamin D. Vitamin D is a metabolite synthesized in the skin from 7-dehydrocholesterol in a reaction catalyzed by light (vitamin D_3, cholecalciferol) (Figure 19.12) or obtained from dietary sources (vitamin D_2, ergocalciferol). Whatever its source, vitamin D is absolutely essential to maintain calcium homeostasis in animals. The major biologically active form of vitamin D_3 is 1,25-dihydroxyvitamin D_3 [1,25$(OH)_2D_3$], which is produced in a series of reactions in kidney and liver. (Vitamin D_2 metabolism is similar.) After its synthesis in skin, vitamin D_3 is transported to the liver bound to *vitamin D–binding protein.* Once inside a liver cell, vitamin D undergoes a 25-hydroxylation reaction. Although enzymes capable of catalyzing this reaction are found in both microsomes and mitochondria, the microsomal 25-hydroxylase (a cytochrome P_{450} protein) is primarily responsible for the physiological processing of the vitamin. 25-Hydroxyvitamin D_3 is subsequently transported by *vitamin D transport protein* to the kidney, where it undergoes a further hydroxylation, catalyzed by a mitochondrial 1-α-hydroxylase (also a cytochrome P_{450} protein) to form the active hormone 1,25-dihydroxyvitamin D_3.

Because steroid ring systems are apparently too complex to be degraded to smaller molecules such as acetate, several mechanisms have evolved to protect cells from the accumulation of these toxic hydrophobic molecules. Conjugation reactions (principally involving glucuronic acid and sulfate), alone or in combination with hydroxylation and reduction reactions, alter the solubility characteristics of numerous steroid molecules. Because steroid metabolism is complex and the detoxication reactions are apparently random, an enormous number of steroid derivatives are excreted. For example, the following derivatives of estriol (an end product of estrogen metabolism) have been detected in the urine of pregnant women: estriol-3-glucuronide, estriol-16-glucuronide, estriol-3-sulfate, and estriol-3-sulfate-16-glucuronide, as well as estriol itself.

Heme Biotransformation Bilirubin is the product of a series of reactions that degrade the heme groups of various hemoproteins. This conversion, which occurs

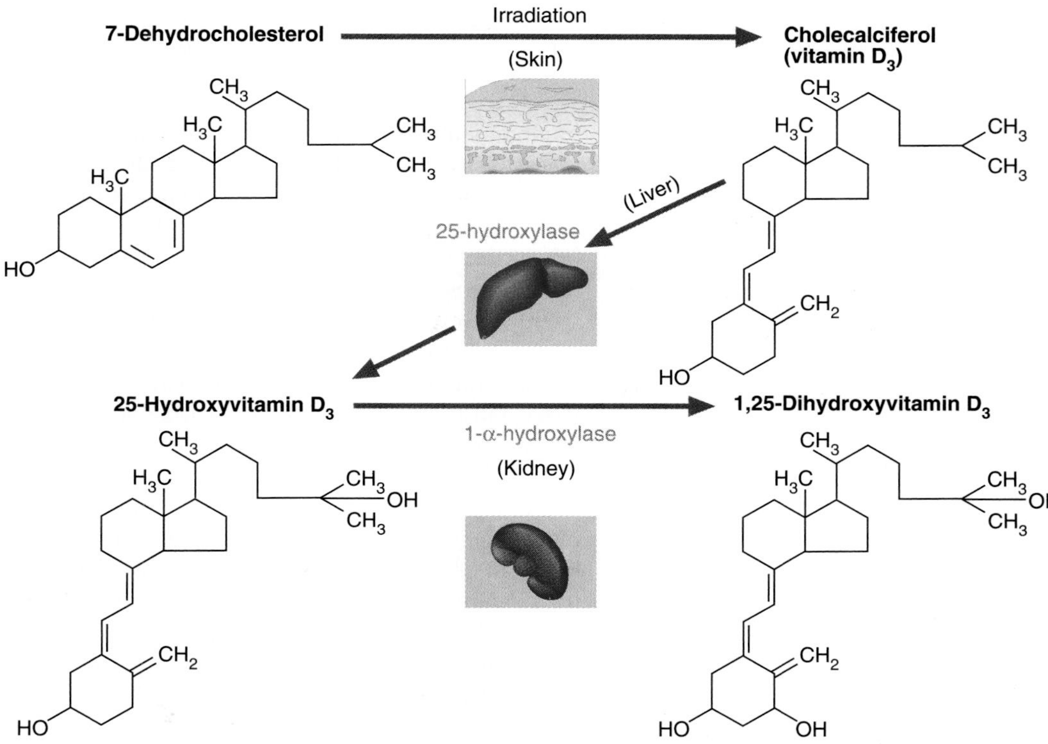

FIGURE 19.12
Biotransformation of Vitamin D_3.

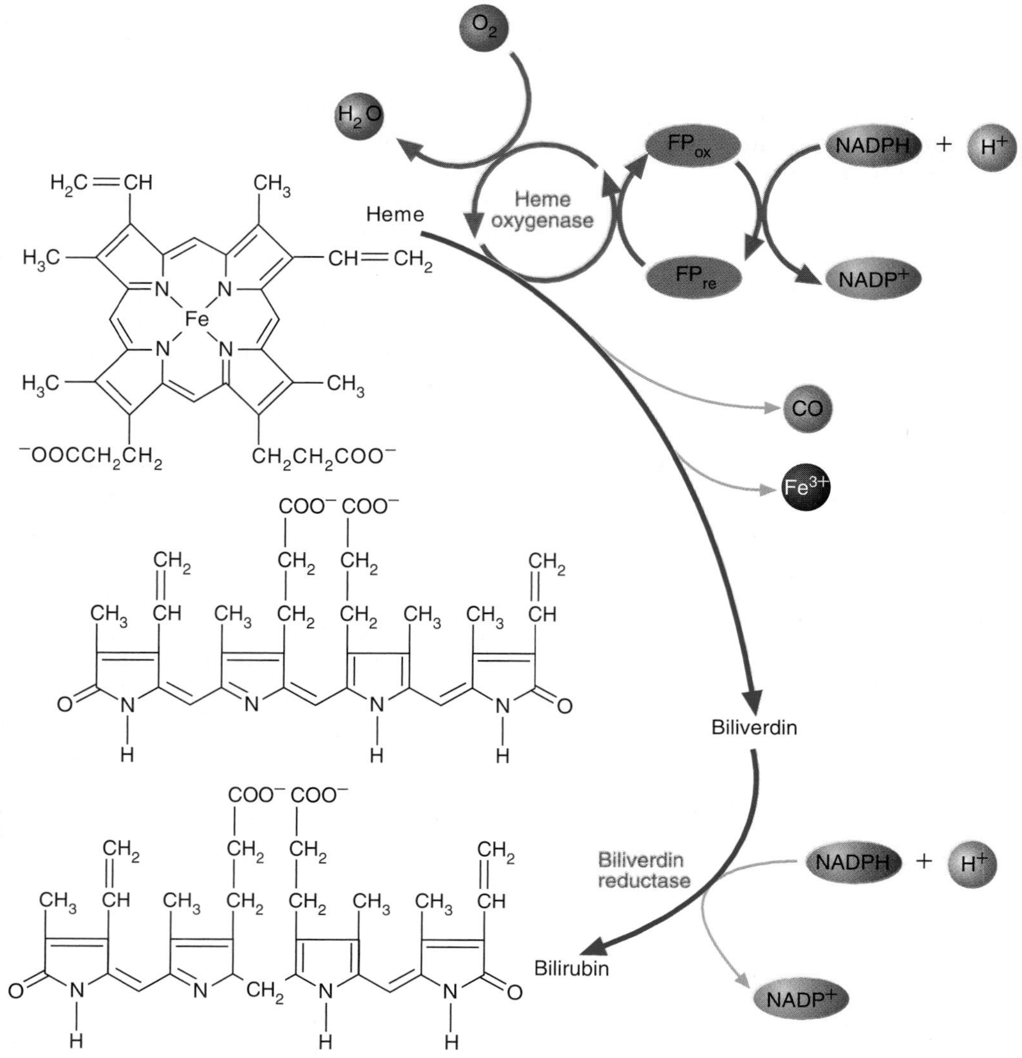

FIGURE 19.13

Bilirubin Synthesis.

Heme oxygenase, which catalyzes the conversion of free heme groups to biliverdin and CO, functions as part of a microsomal electron transport system similar to that of cytochrome P_{450} (FP = NADPH-cytochrome P_{450} reductase.) Heme oxygenase requires 3 O_2 and 5 NADPH. Biliverdin reductase can use NADPH or NADH as a reductant.

predominantly in the reticuloendothelial cells of liver, spleen, and bone marrow, takes place in two phases. During the first phase, heme is oxidized by heme oxygenase, a microsomal enzyme that is a component of an electron transport system similar to that of cytochrome P_{450}. The products of this reaction are the dark green pigment biliverdin and carbon monoxide (CO) (Figure 19.13). During the second phase, biliverdin is converted into bilirubin in a reaction catalyzed by the cytoplasmic enzyme biliverdin reductase. Meanwhile, the CO diffuses out of the cell, is then transported by blood to the lungs, and released there.

The product bilirubin is a very toxic compound. For example, it is known to inhibit RNA and protein synthesis and carbohydrate metabolism in the brain. Mitochondria appear to be especially sensitive to its effects. Bilirubin is also a metabolically expensive molecule to produce. For example, bilirubin is virtually insoluble in water, because of intramolecular hydrogen bonding. Therefore, sophisticated transport mechanisms and conjugation reactions are required for excretion (Figure 19.14). Because bilirubin creates so many problems, considerable effort has been devoted to elucidating its purpose. (Many species, such as amphibians, reptiles, and birds, excrete the water-soluble precursor biliverdin.) Because it reacts with peroxyradicals, bilirubin may act as an antioxidant. During bilirubin transport in blood the radical scavenging pigment is distributed throughout the circulatory system. (The association of bilirubin with the plasma protein albumin protects cells from the molecule's toxic effects.)

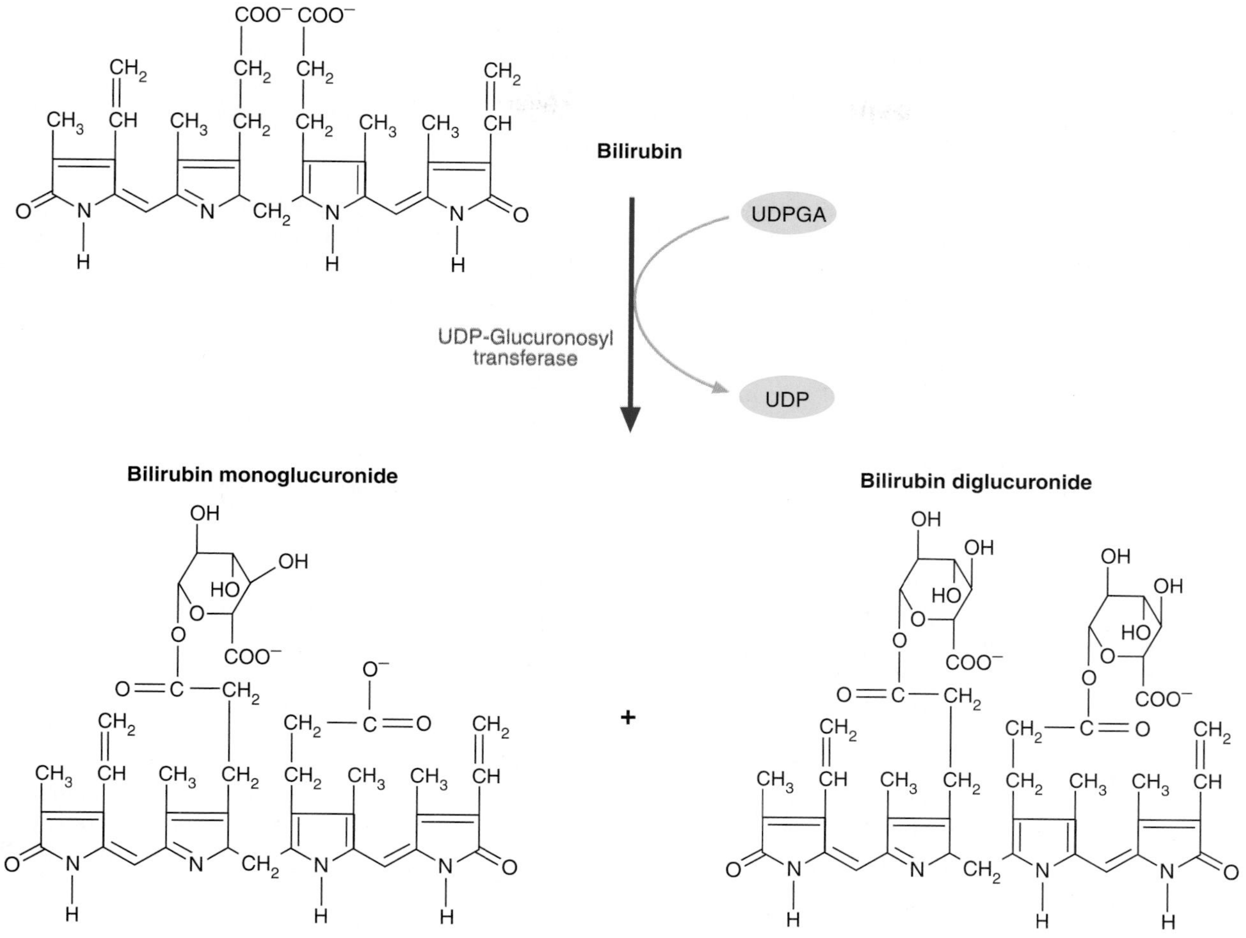

FIGURE 19.14

Bilirubin Conjugation.

Before bilirubin is excreted in bile, its propionyl carboxyl groups are esterified with glucuronic acid to form both monoglucuronides and diglucuronides. (UDPGA = UDP-glucuronic acid.) The diglucuronide is the major form produced in many animals. In a number of species, especially mammals, bilirubin conjugation is required for efficient secretion into bile.

19.3 PLANT BIOTRANSFORMATION PROCESSES

Because of the sheer size of their biomass (approximately 10^{14} kg dry weight), plants have a profound impact on the global environment. In addition to providing the foundation of the food chain, plants serve as a "sink" for environmental pollutants. It has been recognized for several years that plants possess a capacity to metabolize xenobiotics that bears a striking resemblance to animal biotransformation processes. For example, plant biotransformation is also composed of three phases. A variety of xenobiotics are oxidized by phase I reactions. A significant number of these molecules subsequently undergo phase II conjugation reactions. In contrast to animals, which largely depend on excretion as the ultimate method for detoxication, phase III in plants usually consists of storage in vacuoles or translocation into the cell wall.

Unfortunately, plant biotransformation mechanisms are limited. For example, few plants can protect themselves from the damaging effects of ozone (O_3) or sulfur dioxide (SO_2). The annual economic loss due to ozone-damaged crops alone is estimated to be at least several billion dollars. In addition, plant detoxication mechanisms rarely degrade xenobiotics completely. Plants protect themselves by transferring toxic molecules into inaccessible compartments. When a plant is consumed, xenobiotics are released from these compartments. Such molecules may or may not be successfully detoxicated by the consuming organisms (e.g., herbivorous insects and animals). Often, xenobiotics become concentrated in the herbivore's tissues. Consumption of these organisms by others often results in a further concentration of xenobiotics (**bioaccumulation**). The concentration of toxic substances within organisms at the top of a food chain may be several thousand times that found in plants. Plant biotransformation mechanisms, like those of animals, sometimes convert xenobiotics into mutagenic and/or carcinogenic metabolites.

Phase I Reactions

Currently, relatively little is known about xenobiotic oxidation reactions in plants. However, most phase I reactions appear to be catalyzed by peroxidases and (to a lesser extent) by cytochrome P_{450} proteins. Phase I enzymes appear either identical or similar to the enzymes that oxidize endogenous molecules.

Peroxidases   The peroxidases, which catalyze one-electron oxidation reactions, play an important role in xenobiotic oxidation because of their broad substrate specificities. The general reaction catalyzed by the peroxidases is as follows.

$$RH_2 + H_2O_2 \longrightarrow RH\cdot + 2\,H_2O$$

where RH_2 is the substrate.

The free radical products of this reaction can form dimers or be incorporated into polymers (e.g., lignin). Not surprisingly, peroxidase activity is now suspected to play an important role in the conversion of certain xenobiotics (e.g., polycyclic aromatic hydrocarbons and N-nitrosoamines) into more toxic metabolites.

Cytochrome P_{450}   In contrast to their counterparts in mammalian liver, plant cytochrome P_{450} proteins have narrow substrate specificities. In addition, their amounts in plant microsomes are significantly lower than those observed in animals. In some plant species, cytochrome P_{450} proteins can be induced by phenobarbitol. Plants often possess several types of NADPH–cytochrome P_{450} reductase.

Phase II Reactions

A variety of enzymes that catalyze xenobiotic conjugation reactions have been observed in plants.

1. **O- and N-glucosyl transferases.** A significant number of xenobiotics (especially those with chlorinated substituents) are converted to β-D-glucosides.
2. **O- and N-malonyl transferases.** The malonyl transferases also prefer chlorinated xenobiotics. These enzymes also catalyze the malonation of β-D-glucosides. Adding an O-malonyl group to a glucoside may allow some processed xenobiotics to transfer into the vacuole.
3. **Glutathione transferases.** Plants can form a variety of glutathione derivatives of xenobiotics. For example, species such as maize, which can form a glutathione derivative of the herbicide atrazine (a photosynthetic inhibitor), resist its toxic effects. Glutathione conjugates are sometimes also converted to N-malonate derivatives.

KEY *Concepts*

Plants have an enormous capacity for processing xenobiotics. Like mammals, plants oxidize hydrophobic molecules and conjugate the products to a variety of donor substrates. Instead of excreting xenobiotics, plants often store them in inaccessible compartments. Principal examples of plant conjugation reactions include O- and N-glycosylation, O- and N-malonation, and the formation of glutathione derivatives.

QUESTION 19.4

When the fungus *Gibberella fujikuroi* infects rice plants, the rate of stem elongation is rapidly accelerated. As described previously (Special Interest Box 15.3), this fungus produces the plant hormone gibberellin. Apparently, rapid gibberellin-induced cell elongation allows the fungus to invade a plant more easily. The fungus may produce gibberellin because requisite genes have been transferred from plants. Can you suggest a strategy for testing this hypotheses? What are the problems and possibilities when plants develop mechanisms to defend themselves from gibberellin-producing microbial pathogens?

KEY WORDS

bioaccumulation, *574*	cytochrome P_{450} system, *560*	phase I, *559*	secondary metabolite, *559*
bioremediation, *559*	detoxication, *560*	phase II, *559*	tolerance, *566*
biotransformation, *559*	detoxification, *560*	primary metabolite, *559*	xenobiotic *559*
cirrhosis, *568*	flavin-containing monooxyge-nase, *560*		

SUMMARY

Over millions of years, living organisms have evolved a wide variety of mechanisms to promote their survival in a very dangerous world. Among these are the synthesis of an enormous number of secondary metabolites that protect against various biotic and abiotic threats or promote reproductive success. Living organisms have also developed enzymatic mechanisms of inactivating and/or disposing of xenobiotics. These mechanisms, collectively referred to as biotransformation, typically convert hydrophobic molecules into more polar and usually less toxic derivatives. In mammals, most biotransformed xenobiotics are usually excreted. Plants store biotransformed xenobiotics in inaccessible compartments such as the vacuole and the cell wall.

Biotransformation consists of two principal phases. In phase I, molecules are oxidized. In mammals the principal enzymes involved in phase I are cytochrome P_{450} isozymes and flavin-containing monooxygenase, which both use NADPH as an external reductant. In plants, most phase I reactions appear to be catalyzed by the peroxidases, although a few reactions have been observed to be catalyzed by cytochrome P_{450} isozymes.

During phase II, molecules with appropriate functional groups are converted into more polar and usually less toxic derivatives by conjugation reactions. In mammals the most common conjugation reactions involve glucuronic acid, glutathione, sulfate and amino acids. In plants, O- and N-glucosylation, O- and N-malonation, and glutathione derivatives are commonly observed.

Many of the enzyme activities involved in biotransformation are also involved in processing endogenous metabolites, especially in steroid metabolism and degrading heme to bilirubin. It has been suggested that as living organisms evolved, they gradually coopted the enzymes used in metabolism to detoxicate xenobiotics.

SUGGESTED READINGS

Chadwick, D.J., and Marsh, J. (Eds.), *Bioactive Compounds from Plants,* John Wiley and Sons, New York, 1990.

Chadwick, D.J., and Whelan, J. (Eds.), *Secondary Metabolites: Their Function and Evolution,* John Wiley and Sons, New York, 1992.

Coleman, J.O.D., Blake-Kolff, M.M.A., and Davies, T.G.E., Detoxification of Xenobiotics by Plants: Chemical Modification and Vacuolar Compartmentation, *Trends Plant Sci.,* 2(4):144–151, 1997.

Guengerich, F.P., Cytochrome P450 Enzymes, *Amer. Sci.,* 81(5): 440–447, 1993.

Guengerich, F.P., Influence of Nutrients and other Dietary Materials on Cytochrome P-450 Enzymes, *Am. J. Clin. Nutri.,* 61(3 Suppl): 651S–658S, 1995.

Jakoby, W.B., Detoxication: Conjugation and Hydrolysis, in Arias, I.M., Jakoby, W.B., Popper, H., Schachter, D., and Shafritz, D.A. (Eds.), *The Liver: Biology and Pathobiology,* 2nd ed., pp. 375–388, Raven Press, New York, 1988.

Lieber, C.S., Herman Award Lecture, 1993: A Personal Perspective on Alcohol, Nutrition and the Liver, *Amer. J. Clin. Nutr.,* 58: 430–432, 1993.

Lieber, C.S., Ethanol Metabolism, Cirrhosis, and Alcoholism, *Clin. Chim. Acta,* 257(1):59–84, 1997.

Pickett, C.B., and Lu, A.Y.H., Glutathione-S-Transferases: Gene Structure, Regulation and Biological Function, *Ann. Rev. Biochem.,* 58:743–764, 1989.

Sandermann, H., Plant Metabolism of Xenobiotics, *Trends Biochem. Sci.,* 17:82–84, 1992.

Ziegler, D.M., Detoxication: Oxidation and Reduction, in Arias, I.M., Jakoby, W.B., Popper, H., Schachter, D., and Shafritz, D.A. (Eds.), *The Liver: Biology and Pathobiology,* 2nd ed., pp. 363–374, Raven Press, New York, 1988.

QUESTIONS

1. Clearly define the following terms:
   a. bioaccumulation
   b. detoxication
   c. detoxification
   d. phase I reactions
   e. phase II reactions
   f. secondary metabolite
   g. bioremediation
   h. biotransformation

2. The struggle for survival has promoted living organisms to produce an astonishing array of offensive and defensive molecular weapons. List three methods by which organisms use these weapons against predators.

3. Describe several processes by which the chemical arsenal of plants and animals may have arisen.

4. During periods of extreme selection pressure, adaptive modifications spread quickly through the population. Describe how this phenomenon might occur.

5. Describe three functions of secondary metabolites.

6. For each of the following classes of compounds, give several examples of a secondary metabolite:
   a. alkaloids
   b. terpenes
   c. glycosides

7. In both plants and animals, biotransformation aims to protect against toxic compounds. How do these groups of organisms protect themselves?

8. What are the two stages of biotransformation? In general, what processes occur at each stage?

9. What are the principal enzyme systems in microsomal oxidative metabolism?

10. List and describe the components of the cytochrome P_{450} electron transport system. What is the role of each component?

11. Of what type are all of the reactions catalyzed by the cytochrome P_{450} system? Give examples.

12. Describe in detail the mechanism for the cytochrome P_{450}-mediated conversion of benzene to phenol.

13. Which three functional groups are oxidized by the flavin-containing monooxygenases?

14. How do flavin-containing monooxygenases differ from all other mammalian monooxygenases?

15. What is the function of conjugation reactions in the biotransformation process?

16. List three commonly used donor substrates used in conjugation reactions.

17. Into what three categories do xenobiotics fall? Give several examples for each category.

18. In addition to its mood-altering effects, cocaine use increases heart rate, dilates pupils, and constricts blood vessels. Explain why these symptoms occur. How is cocaine disposed of in the body? Why is liver damage so common among long-term cocaine users?

19. Parathion is a toxic insecticide. Although malathion is also a powerful insecticide, it is much less toxic than parathion. Examine the structures of each compound and indicate why their toxicities are so unequal.

20. Tetrachloroethylene was originally used to extract seed oils from grains such as corn. The residual cornmeal, containing traces of tetrachloroethylene, was then fed to cattle. The cattle developed liver disease and died. Studies indicated that the toxicity of tetrachloroethylene was due to a bioactivated derivative. From your knowledge of bioactivation processes, suggest a compound that might have damaged the liver.

Tetrachloroethylene

21. What is the difference between a xenobiotic and a nutrient?

22. Rifampicin, a drug used to treat tuberculosis, is a potent cytochrome P_{450} inducer. What do you think would be the effect of taking rifampicin simultaneously with an oral contraceptive that is also metabolized by a cytochrome P_{450} system?

23. Phenobarbital and alcohol are two potent drugs that should never be consumed simultaneously. When they are consumed together, alcohol and phenobarbital exert synergistic effects (i.e., their combined biological effects are significantly greater than if they were consumed separately). Both drugs are metabolized by the same cytochrome P_{450} system. Do you think that this cometabolism would account for the observed lethal effects of this combination of drugs? Phenobarbital is a sedative and alcohol is a depressant drug.

24. Describe the hepatotoxic effects of ethanol. Why is this organ so susceptible to excessive alcohol consumption? Why are smokers at higher risk for developing cancer if they also consume alcohol?

25. N-glucuronides have been implicated in bladder tumors. Suggest a reason why this organ is especially susceptible to such a problem.

26. What is cross-tolerance and how can it develop?

Parathion

Malathion

Solutions

A p p e n d i x   A

C h a p t e r   1

End-of-Chapter Questions

1. The principles that are central to the understanding of living organisms are
   a. cells are the structural units of life
   b. living processes require regulated chemical reactions
   c. all organisms use the same fundamental reactions
   d. all organisms use similar molecules
   e. nucleic acids contain the basic information for growth and reproduction

3. The functional group(s) in each molecule are
   a. aldehyde   b. carboxylic acid and amino
   c. sulfhydryl   d. ester
   e. alkene   f. amide
   g. ketone   h. alcohol

5. Define each of the following terms:
   a. biochemistry—the study of the molecular basis of life
   b. oxidation—loss of electrons
   c. reduction—gain of electrons
   d. active transport—energy required to move substances against a concentration gradient
   e. leaving group—a molecular group displaced during a reaction
   f. elimination—loss of an atom or group
   g. isomerization—a shift of atoms or groups within a molecule
   h. nucleophilic substitution—displacement of an atom or group by an electron-rich species
   i. reducing agent—atom or group oxidized during an oxidation/reduction reaction
   j. oxidizing agent—atom or group reduced during an oxidation/reduction reaction

7. DNA contains the cell's genetic information; RNA is involved in the expression of that information, i.e., in the synthesis of proteins.

9. Plants dispose of waste products mainly by degradation or storage in vacuoles and cell walls.

11. Saturated compounds contain only carbon-carbon single bonds, whereas unsaturated compounds contain carbon-carbon double or triple bonds.

13. The most important advantages of multicellular organisms over unicellular organisms are
    a. multicellular organisms have a stable internal environment.
    b. cells can be specialized (division of labor).
    c. there is a more efficient use of resources.
    d. sophisticated functions can be accomplished by such organisms.

15. Define the following terms:
    a. metabolism—the sum of the chemical reactions carried out in a living cell
    b. nucleophile—an atom or group with an unshared pair of electrons that is involved in a displacement reaction
    c. cytoskeleton—the protein framework of a cell
    d. electrophile—an electron-deficient species
    e. energy—the ability to do work

17. Prokaryotes are easily obtained single-celled organisms that are much less complex than the eukaryotes. The assumption is made that the basic life processes are the same. Prokaryotes are easier to study and handle than are multicellular organisms.

19. Give an example of each of the following reaction processes:
    a. nucleophilic substitution—the reaction of glucose with ATP to produce glucose-6-phosphate and ADP
    b. elimination—the dehydration of 2-phosphoglycerate to form phosphoenolpyruvate
    c. oxidation/reduction—the conversion of ethyl alcohol to acetaldehyde
    d. hydrolysis—the cleavage of a phosphate bond in ATP by water

21. Important ions found in living organisms are Na^+, K^+, and Cl^-.

23. While biochemical reactions and organic chemical reactions must conform the same physical laws, the control exhibited by a biochemical process far exceeds the capabilities of the organic chemist. In addition, biochemical reactions are integrated in an extraordinarily complex manner that makes possible the phenomena of life.

25. In addition to being an important energy source, carbohydrates are important structural molecules for the cell and have a role in intracellular and intercellular communication.

27. Nucleotides participate in energy forming and generating reactions. Much of the energy available to drive biochemical reactions is stored in ATP.

29. Animal cells produce waste products such as carbon dioxide, ammonia, urea, and water.

C h a p t e r   2

In-Chapter Questions

2.1 Volume of a prokaryotic cell
$$\pi r^2 h = 3.14 \times (0.5~\mu m)^2 \times 2~\mu m = 1.57~\mu m^3$$
$$= 2~\mu m^3$$
Volume of a eukaryotic cell
$$4/3~\pi r^3 = (4 \times 3.14 \times (10~\mu m)^3)/3 = 4200~\mu m^3$$
Approximately 2100 prokaryotic cells could fit in one liver cell.

2.2 Lipid molecules would accumulate in the cells. The cell function is eventually compromised, and the cells die.

2.3 The cyanobacteria obtain a stable environment and constant supply of nutrients. The eukaryotic organism is assured a constant supply of food.

2.4 Refer to Figures 2.2 and 2.4.

End-of-Chapter Questions

1. The cell is the basic structural unit of life.

3. The function of the bacterial cell components is as follows:
   a. mesosome—unknown
   b. nucleoid—contains DNA

c. plasmid—site of extrachromosomal DNA
d. cell wall—protection and support
e. pili—attachment to other cells
f. flagella—locomotion

Refer to Figure 2.2 for a view of a bacterial cell structure.

5. Indicate which of the following structures are present in eukaryotes or prokaryotes.
   a. nucleus—eukaryotes
   b. plasma membrane—eukaryotes and prokaryotes
   c. endoplasmic reticulum—eukaryotes
   d. mesosome—prokaryotes
   e. mitochondria—eukaryotes
   f. nucleolus—eukaryotes

7. Lysosomes digest food molecules, components of dead cells, and extraneous extracellular material.

9. The evidence that supports the endosymbiotic hypothesis are
   a. Symbiosis has been observed between eukaryotes and prokaryotes.
   b. Mitochondria and chloroplasts are about the same size as prokaryotes.
   c. The ability of these organelles to synthesize DNA and proteins is similar to prokaryotes.
   d. Prokaryotes, mitochondria, and chloroplasts reproduce by binary fission.
   e. The ribosomes of chloroplasts and mitochondria are similar in size and function to those of prokaryotes.
   f. Traces of RNA found in other eukaryotic cell structures suggest that these also arose by symbiotic fusion.

11. The thick mucoid coat would help to prevent antibodies from binding to cell structures used by the immune system for recognition, thereby interfering with the immune response.

13. Two essential functions of the nucleus are metabolic regulation and storage of the genetic material.

15. In peroxisomes, molecular oxygen is used to oxidize organic molecules. The hydrogen peroxide produced must be destroyed before it can damage the cell. Mitochondria use molecular oxygen to liberate energy from food molecules.

17. Various plasma membrane proteins are involved in the transport of materials across the plasma membrane. Many also function as enzymes and cell receptors.

19. The Golgi apparatus sorts and packages protein and other cell products.

21. The immobilization maintained by the cytoskeleton promotes an ordered state.

Chapter 3

In-Chapter Questions

3.1 Electrostatic interactions, hydrogen bonds, hydrophobic interactions

3.2 Ammonia "ice" would be expected to be less dense than liquid ammonia. See Figure 3.9 for the analogous structure of water.

3.3 The equilibrium would shift to the right to replace lost bicarbonate, and the acid concentration would increase. The resulting condition is called acidosis.

3.4 The equation for osmotic pressure M is:
$$\pi = nMRT \quad \text{where } \pi = 2.06 \times 10^{-3} \text{ atm}$$
$$n = 1$$

$$R = 0.0821 \text{ L atm/mole K}$$
$$T = 298 \text{ K}$$

Substitute these values into the equation and solve for M.

$$2.06 \times 10^{-3} \text{ atm} =$$
$$(1)M(0.0821 \text{ L atm/mole K})(298 \text{ K})$$

$$M = 2.06 \times 10^{-3} \text{ atm}/(0.0821 \text{ L atm/mole K})(298 \text{ K})$$
$$= 2.06 \times 10^{-3} \text{ atm}/24.4658 \text{ L atm/mole}$$
$$= 0.000084199 \text{ mole/L}$$

$$1.5 \text{ g/L} = 0.000084199 \text{ mole/L}$$

$$1 \text{ mole} = 1.5 \text{ g/.00008419} =$$

$$17816.84286 = 18,000 \text{ g/mole}$$

End-of-Chapter Questions

1. **c** and **d** are acid/conjugate base pairs.

3. The effective buffer range is between pH 7 and 8.

5. Only **d** can form hydrogen bonds with like molecules; **a, b, c,** and **d** all can form hydrogen bonds with water.

7. The equation for freezing point depression is
$$\Delta T = -K_f MN$$

where

ΔT = temperature change
$-K_f$ = freezing point depression constant
M = molality of the solution
N = number of particles

For 2.8 M sodium nitrate,

$$\Delta T = (-1.86°C \text{ kg/mole})(2.8 \text{ mole/kg})(2)$$
$$= -10.4°C$$

for 3 M glucose

$$\Delta T = (-1.86°C \text{ kg/mole})(3 \text{ mole/kg})(1)$$
$$= -5.58°C = -6°C$$

9. The interactions between the following molecules and ions are
   a. water and ammonia—hydrogen bonds
   b. lactate and ammonium ion—electrostatic interactions
   c. benzene and octane—van der Waals forces
   d. carbon tetrachloride and chloroform—van der Waals forces
   e. chloroform and diethyl ether—van der Waals forces

11. The sugar solution has a high concentration of solute. This pulls water out of any organisms that are present, thus killing them and preserving the fruit.

13. Define the following terms:
   a. hydrogen bond—an attraction between hydrogen bound to an electronegative atom and unshared electrons on nearby electronegative atoms
   b. pH—the negative log of the hydrogen ion concentration
   c. buffer—a weak acid/conjugate base mixture that resists changes in pH
   d. osmotic pressure—the applied pressure that will just stop the movement of water across a semipermeable membrane
   e. colligative properties—the properties of a system that are based on the number of particles
   f. isotonic—two solutions of equal solute concentration
   g. amphipathic—molecules containing both polar and nonpolar goups
   h. hydrophobic interactions—interactions that tend to exclude water
   i. dipole—a molecule with unequal charge distributions

j. induced dipole—an unbalanced charge distribution that results from the presence of a nearby charge.

15. Water forms a great many hydrogen bonds. In order to raise its temperature, some of these bonds must be broken. Ammonia has a similar ability to form hydrogen bonds and would also be expected to have a high heat capacity. Methane cannot form hydrogen bonds and would therefore be expected to have a low heat capacity.

17. **b**, **c**, and **e** are all weak acids because they are not completely ionized in water.

19. Buffering capacity is a function of the amount of buffer (weak acid/conjugate base) that is present. In order to increase the buffering capacity, the weak acid and its conjugate base should be added in the same ratio as they were present in the original buffer.

21. Colligative properties depend on the number of particles present and not on the particular group; **a** and **f** are dependent on the species present and not on the number of particles.

23. The regular crystal lattice of the ice crystal is more open than the tightly hydrogen-bound liquid water. If ice were more dense than water, ice formed in lakes and oceans would sink to the bottom and eventually only a narrow layer at the surface would remain liquid. Then aquatic life could not survive.

25. In the presence of bicarbonate any acid that ionizes produces carbon dioxide, which is exhaled. The pH of the blood then remains virtually unchanged.

27. The pH scale is derived using the ionization constant of water. To establish a pH scale for another solvent, the ionization constant of that solvent would have to be used, and the pH scale would be different from the pH scale for water.

29. No. The carbonic acid and carbonate react to produce bicarbonate. It is possible to have buffer systems of carbonic acid and bicarbonate and a system of bicarbonate and carbonate.

31. The contribution from the ionization of water must be considered. The hydrogen ion concentration is 10^{-8} M from acid and 10^{-7} M from water for a total acid concentration of 1.1×10^{-7} M acid. The pH is therefore equal to $-\log 1.1 \times 10^{-7}$, i.e., 6.96.

33. The extreme electronegativity of the oxygen polarizes the O—H bond of water and makes the hydrogen electron-deficient. Because the unshared pairs of electrons on the oxygen are available for bonding, an electrostatic interaction develops.

35. Carbon dioxide is present in the blood in sufficient quantities to make it effective as a buffer; phosphate is not. In cells the phosphate concentration is much higher, and it can therefore act as an effective buffer.

C h a p t e r   4

In-Chapter Questions

4.1 $\Delta G' = \Delta G^{\circ\prime} + 2.303\ RT \log\ [ADP][P_i]/[ATP]$

$R = 1.987$ cal/mole · K

$T = K$

[ADP] = 0.00135 M   [P_i] = 0.00465 M
[ATP] = 0.004 M $T = 273.15 + 37 = 310$ K

$\Delta G^{\circ\prime} = -7.3$ kcal/mole

$\Delta G' = -7.3 + 2.303(1.987)(310)\log(.00135)(.00465)/(.004)$

$= -7.3 + 1.418(\log.00157)$

$= -7.3 - 4 = -11.3$ kcal/mole

4.2 The values for the oxidation state of the functional group carbon (indicated in bold) are

a. CH_3CH_2OH $0 - 1 - 1 + 1 = -1$
b. $CH_3CH{=}O$ $0 - 1 + 2 = +1$
c. $CH_3C(OH){=}O$ $0 + 1 + 2 = +3$

4.3 Amount of ATP required to walk a mile
= (100 kcal/mile)/7.3 kcal/mole
= 13.7 moles/mile $\times$ 507 g/mole = 6945.2 g/mile = 6950 g/mile

Amount of glucose required to produce 100 kcal through ATP
= (100 kcal)/(0.4)(686 kcal/mole)
= 100 kcal/274.4 kcal/mole
= 0.36 mole
= 0.36 mole $\times$ 180 g/mole = 65.6 g of glucose

End-of-Chapter Questions

1. a. thermodynamics—the study of the heat and energy transformations in a chemical reaction
   b. endergonic—chemical reactions that absorb energy
   c. enthalpy—a measure of the heat evolved during a reaction
   d. free energy—a measure of the tendency of a reaction to occur
   e. high-energy bond—a bond that liberates large amounts of free energy when broken
   f. reduction—a gain of electrons
   g. redox potential—a measure of the tendency of a group to lose an electron
   h. phosphate group transfer potential—the tendency of a phosphate bond to undergo hydrolysis

3. For a reaction to proceed to completion, the total overall ΔG° must be negative. This is true of answers **a, b, d** and **e.** Answer **d** is very close to equilibrium.

5. ATP + glutamate + $NH_3 \longrightarrow$ ADP + P_i + glutamine
   ATP + $H_2O \longrightarrow$ ADP + P_i $\Delta G^{\circ\prime} = -7.3$ kcal/mole
   Glutamine + $H_2O \longrightarrow$ glutamate + NH_3
   $\hspace{6cm}\Delta G^{\circ\prime} = -3.4$ kcal/mole

   Reverse the second equation and add the $\Delta G^{\circ\prime}$ values.
   ATP + $H_2O \longrightarrow$ ADP + P_i $\Delta G^{\circ\prime} = -7.3$ kcal/mole
   Glutamate + $NH_3 \longrightarrow$ glutamine + H_2O
   $\hspace{6cm}\Delta G^{\circ\prime} = +3.4$ kcal/mole

   ATP + glutamate + $NH_3 \longrightarrow$ ADP + P_i + glutamine
   $\hspace{6cm}\Delta G^{\circ\prime} = -3.9$ kcal/mole

7. **d:** Reaction rates cannot be determined from energy values.

9. **b, d, e,** and **f** are true statements concerning free energy change.

11. The energy liberated by the hydrolysis of 12.5 moles of ATP is 12.5 mole $(-7.3$ kcal/mole) = 91.3 kcal. The energy required to produce 12.5 moles of ATP is 273 kcal. The efficiency is $[(91.3)/273] \times 100 = 33.4\%$.

13. $\Delta G^{\circ} = -\log$ (ADP)(G-6-P)/(ATP)(glucose)
    -4 kcal/mole $= -\log$ (1)(G-6-P)/(1)(glucose)
    $-4 = -\log$ (G-6-P)/(glucose)
    $4 = \log$ (G-6-P)/(glucose)
    (G-6-P)/(glucose) = 10,000:1

15. a. false
    b. true
    c. true
    d. false
    e. false
    f. true
    g. false

17. Even though only a few molecules are involved, the laws of thermodynamics are obeyed.

19. ΔG is the most useful criteria of spontaneity because it takes into consideration both entropy and enthalpy factors.

21. It is necessary to know the following information to determine $\Delta G°'$: temperature, concentration of reactants and products, and $\Delta G°$.

Chapter 5

In-Chapter Questions

5.1 The structure for penicillamine-cysteine disulfide is

```
        COOH            COOH
         |               |
   H—C—NH₂      H—C—NH₂
         |               |
  CH₃—C—S ———— S—C—H
         |               |
        CH₃              H
```

5.2 The trait is recessive, and two copies of the gene are required for full expression of the disease. Primaquine produces a strong oxidizing agent (hydrogen peroxide). In the absence of reducing agents such as NADPH, the peroxide does extensive damage to the cell. No, a high peroxide level in blood cells is damaging to the malarial parasite and is selected for in regions where malaria occurs.

5.3 Collagen is a protein found in connective tissue, tendons, and the cornea of the eye. Failure of collagen to form properly would weaken these tissues. Symptoms include cataracts and easily torn ligaments and blood vessels.

5.4 BPG stabilizes deoxyhemoglobin. In its absence oxyhemoglobin would therefore form more easily. Fetal hemoglobin binds BPG poorly and therefore has a greater affinity for oxygen.

End-of-Chapter Questions

1. A protein is a polymer containing more than 50 amino acid residues comprising one or more polypeptide chains. A peptide is a polymer containing less than 50 amino acid residues. A polypeptide is a polymer containing greater than 50 amino acids.

3. The structure and net charge of arginine at various pH levels is as follows:

pH	Structure	Net Charge
1	$H_2N—C—N—CH_2—CH_2—CH_2—C—COOH$ (with ‖NH_2 and $+NH_2$, $+NH_3$, H)	+2
4	$H_2N—C—N—CH_2CH_2CH_2—CHCOO^-$ (with ‖$+NH_2$, $+NH_3$, H)	+1
7	$H_2N—C—NH—CH_2CH_2CH_2\,CH—COO^-$ (with ‖$+NH_2$, $+NH_3$)	+1
10	$H_2N—C—NH—CH_2CH_2CH_2—CHCOO^-$ (with ‖$+NH_2$, NH_2)	0
12	$H_2N—C—NHCH_2CH_2CH_2—CH—COO^-$ (with ‖$+NH_2$, NH_2)	0

5. a. The correct name for the peptide is cysteinylglycyltyrosine.

b. The three-letter abbreviation structure is H_2N—Cys—Gly—Tyr—COOH.

7. Six functions of protein in the body are catalysis, structure, movement, defense, regulation, and transport.

9. a. The α carbon is the carbon next to the carboxyl group in an amino acid.
   b. The isoelectric point is the pH at which an amino acid is electrically neutral.
   c. A peptide bond is an amide bond between two amino acids.
   d. A hydrophobic amino acid is one with a nonpolar side group.

11. a. polyproline: left-handed helix
    b. polyglycine: β pleated sheet
    c. Ala-Val-Ala-Val-Ala-Val: α helix
    d. His-Gln-Phe-Thr-Trp: β pleated sheet

13. a. heat—hydrogen bonding (secondary and tertiary structure)
    b. strong acid—hydrogen bonding and salt bridges (secondary and tertiary structure)
    c. saturated salt solution—salt bridges (tertiary structure)
    d. organic solvents (alcohol)—hydrophobic interactions (tertiary structure)

15. Amino acids with basic side groups such as lysine, arginine, or tyrosine would contribute to a high pI value.

17. The remainder of the protein acts to help hold the active site in the correct conformation and to shield it from extraneous compounds. In addition, other parts of the protein may be active in transport, recognition, or binding the protein within the cell.

19. The amide bond is more highly resonance-stabilized than is the ester bond. This stabilization strengthens the peptide bond.

21. The peptide bonds on the surface of the chymotrypsin are not readily cleaved by the enzyme.

23. Refer to Appendix B, Supplement 5.1 for protein purification procedures.

25. In sequencing a polypeptide, the next residue in a sequence is determined by the increase in height of the peak on an amino acid analyzer. If there is already a large amount of that amino acid present in the solution, it would be difficult if not impossible to detect any change in peak height. Small fragments have only a few amino acids, and this problem does not occur.

27. The structure of β-endorphin is

Tyr-Gly-Gly-Phe-Met-Thr-Ser-Glu-Lys-Ser-Gln-Thr-Pro-Leu-Val-Thr-Leu-Phe-Lys-Asn-Ala-Ile-Val-Lys-Asn-Ala-His-Lys-Lys-Gly-Gln

29. The following fragments are produced when bradykinin is treated with the indicated reagents:
    a. Carboxypeptidase
       Arg and Arg-Pro-Pro-Gly-Phe-Ser-Pro-Phe
    b. Chymotrypsin
       Arg-Pro-Pro-Gly-Phe, Ser-Pro-Phe, and Arg
    c. Trypsin
       Arg and Pro-Pro-Gly-Phe-Ser-Pro-Phe-Arg
    d. DNFB/DNP-Arg and the following amino acid residues 3 Pro, 1 Gly, 2 Phe, 1 Ser, and 1 Arg

Chapter 6

In-Chapter Questions

6.1 Dialysis removes the formaldehyde, formic acid, and methanol that build up in the bloodstream. The bicarbonate neutralizes the acid produced and helps offset the resultant acidosis. The ethanol competitively binds with the alcohol dehydrogenase. This slows the dehydrogenation of the methanol and allows time for the kidneys to excrete it.

6.2 a. Menkes' Syndrome—injections of copper salts into the blood would avoid intestinal malabsorption and provide the copper necessary to form adequate levels of ceruloplasmin and offset the symptoms of the disease.

   b. Wilson's disease—zinc induces the synthesis of metalothionein, which has a high affinity for copper. Some organ damage can be averted because metalothionein sequesters copper and prevents this toxic metal from binding to and inactivating susceptible proteins and enzymes. Penicillamine forms a complex with copper in the blood. This complex is transported to the kidneys where it is excreted.

6.3 The patient that failed to show improvement probably had a higher level of acetylating enzymes. The patients' dosage should be based on their biotransforming capability and not on their body weight.

End-of-Chapter Questions

1. a. activation energy—the minimum energy required to bring about a reaction

   b. catalyst—a substance that speeds up a reaction without being consumed

   c. active site—the part of an enzyme directly responsible for catalysis

   d. coenzyme—a small molecule needed to enable the enzyme to function

   e. velocity of a chemical reaction—the change in concentration of a reactant with time

   f. half-life—the time needed to consume half the reactant molecules

   g. turnover rate—the number of moles of substrate converted per second per mole of enzyme

   h. katal—one mole of substrate converted to product per second

   i. noncompetitive inhibitor—the inhibitor binds to the enzyme but not at the active site

   j. repression—prevention of polypeptide synthesis

3. Cells regulate enzymatic reactions by using genetic control (certain key enzymes are synthesized in response to changing metabolic needs), covalent modification (certain enzymes are regulated by the reversible interconversion between their active and inactive forms, a process involving covalent changes in structure), allosteric regulation (binding effector molecules to pacemaker enzymes alters catalytic activity), and compartmentation (preventing wasteful "futile cycles" by physical separation of opposing biochemical processes within cells).

5. From the data, the reaction is first order in pyruvate and ADP, zero order for Pi. The overall order of the reaction is second order.

7.
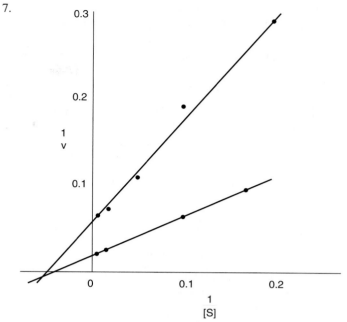

Horizontal intercept $= -1/K_m$
Vertical intercept $= 1/V_{max}$
Slope $= K_m/V_{max}$
The inhibition is noncompetitive.

9.
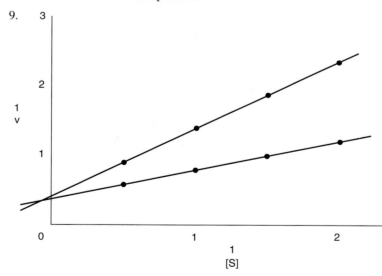

The enzyme's activity decreases. The inhibition is competitive.

11. Three reasons that the regulation of biochemical processes are important are maintenance of an ordered state, conservation of energy, and response to environmental cues.

13. In the concerted model, the substrate and activators bind to the relaxed conformation. This binding shifts the equilibrium to the R conformation. In the sequential model, binding the activator molecule changes the conformation of the enzyme to a shape more favorable to binding substrate. When oxygen binds to hemoglobin, the first oxygen molecule binds slowly. It, however, introduces a conformational change that makes the sequential binding of the second, third, and fourth oxygen molecules much easier.

15. Transition metal ions are useful as enzyme cofactors because they have high concentrations of positive charge, can act as Lewis acids, and can bind to two or more ligands at the same time.

17. The activation energy for the reaction of ethyl alcohol with oxygen is quite high and consequently the reaction is slow.

19. At the start of a reaction, the concentrations of the reactants and products are known precisely and, since equilibrium has not yet been established, presumably only the forward reaction is taking place.

21. The substrate binds to the active site and stabilizes the enzyme in that conformation.

Chapter 7

In-Chapter Questions

7.1 The sugars belong to the following classes
   a. aldotetrose   b. ketopentose   c. ketohexose

7.5 The larger insoluble glycogen molecule does not contribute to the osmotic pressure of the cell. A similar amount of free glucose would burst the cell.

End-of-Chapter Questions

3. In D family sugars the OH on the chiral carbon farthest from the carbonyl is on the right in a Fischer projection. So both (+) glucose and (−) fructose are D sugars despite their rotation of plane-polarized light in opposite directions.

5. Heteropolysaccharides are made up of more than one type of monosaccharide residue but homopolysaccharides have only one. Examples of a homopolysaccharide and a heteropolysaccharide are starch and hyaluronic acid, respectively.

7. a. nonreducing   b. nonreducing   c. reducing
   d. nonreducing   e. reducing

9.
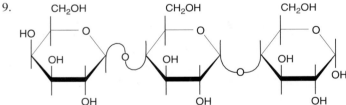

11. The thick proteoglycan coat acts to protect bacteria by preventing the binding of antibodies to their surface antigens.

13. Ribulose has four possible stereoisomers (i.e., 2^2) while sedoheptulose has sixteen (i.e., 2^4).

15. a. Glycogen stores glucose.
   b. Glycosaminoglycans are components of proteoglycans.
   c. Glycoconjugates may serve as membrane receptors.
   d. Protoglycans provide strength, support, and elasticity to tissue.
   e. Hormones such as FSH and enzymes such as RNAse are glycoproteins.
   f. Polysaccharides play important roles in the storage of carbohydrate (starch and glycogen) and the structure of plants (cellulose).

17. a. The carboxyl and sulfate groups are charged and help to bind water.
   b. Hydrogen bonding plays an important role in these interactions.

19. Mannuronic acid has four chiral centers and therefore sixteen possible stereoisomers.

21. The hydrogen bonding between chains produces a three-dimensional network containing many small holes and channels. Small molecules can pass through, but larger molecules are excluded because they cannot fit into the channels.

23. A reducing sugar reduces Cu(II) in Benedict's reagent. This reduction takes place because the hemiacetal portion of a sugar can form an aldehyde functional group.

25. Numerous carbohydrate groups protect glycoproteins from denaturation.

27. a. D-erythrose and D-threose are epimers.
   b. D-glucose and D-mannose are epimers.
   c. D-ribose and L-ribose are enantiomers.
   d. D-allose and D-galactose are diastereoisomers.
   e. D-glyceraldehyde and dihydroxyacetone are an aldose-ketose pair.

Chapter 8

In-Chapter Questions

8.1 The enzyme deficiencies prevent the breakdown of glycogen. Because the synthetic enzymes are active, some glycogen continues to be produced and causes liver enlargement. Because of the liver's strategic role in maintaining blood glucose, defective debranching enzyme causes hypoglycemia (low blood sugar).

8.2 The large excess of NADH drives the conversion of pyruvate to lactate.

8.3 Chromium is acting as a cofactor.

8.4 In the absence of oxygen, energy is produced only through glycolysis, an anaerobic process. Glycolysis produces less energy per glucose molecule than aerobic respiration, hence more glucose must be metabolized to meet the energy needs of the cell.

8.5 At three strategic points, glycolytic and gluconeogenic reactions are catalyzed by different enzymes. For example, phosphofructokinase and fructose-1,6-diphosphatase catalyze opposing reactions. If both reactions occur simultaneously (i.e., in a futile cycle) to a significant extent, ATP hydrolysis in the reaction catalyzed by phosphofructokinase releases large amounts of heat. If the heat is not quickly dissipated, an affected individual could die of hyperthermia.

8.6 In gluconeogenesis pyruvate is converted to oxaloacetate. NADH and H^+ are required to reduce glycerate 1,3-bisphosphate to glyceraldehyde-3-phosphate. NAD^+ is the oxidized form of NADH also produced in this reaction. ATP is needed to provide the energy to carboxylate pyruvate to oxaloacetate and phosphorylate glyceraldehyde-3-phosphate to glycerate-1,3-bisphosphate. Both of these reactions also produce ADP and P_i. GDP converts oxaloacetate to phosphoenolpyruvate. This reaction is also the source of GDP and P_i. Water is involved in the hydrolysis reactions of ATP to ADP and P_i, the conversion of phosphoenolpyruvate to 2-phosphoglycerate, and the hydrolysis of glucose-6-phosphate to glucose. Six protons are formed when four molecules of ATP and two molecules of GTP are hydrolyzed.

8.7 Without glucose-6-phosphatase activity, the individual cannot release glucose into the blood. Blood glucose levels must be maintained by frequent consumption of carbohydrate. Excess glucose-6-phosphate is converted to pyruvate, which is then reduced by NADH to form lactate.

End-of-Chapter Questions

1. The two major processes of metabolism are anabolism and catabolism. In anabolic pathways, complex molecules such as

polysaccharides and proteins are synthesized. In catabolic pathways, larger biomolecules are degraded to smaller molecules. Examples of catabolic processes include the conversion of glucose and fatty acids to carbon dioxide and water.

3. Acetyl-CoA is the most important product of stage 2 metabolism. The oxidation of acetyl CoA in stage 3 to carbon dioxide and water results in the synthesis of ATP.

5. a. The term *amphibolic* describes a biochemical pathway that functions in both anabolism and catabolism.
   b. Homeostasis is a complex physiological mechanism in which a constant internal environment is maintained.
   c. A target cell possesses receptors for specific chemical messengers such as hormones and growth factors.
   d. A second messenger is a molecule that mediates the intracellular action of a hormone.
   e. A limit dextrin is the product when glycogen is degraded to its major branch points.

7. Phosphorylation of glucose upon entering the cell is important because it prevents glucose from leaking out of the cell and promotes binding to enzymes.

9. In such an individual, following a carbohydrate meal blood glucose levels would be higher than normal. Recall that the kinetic properties of hexokinase D allow the liver to remove excess glucose from blood. Skeletal muscle would accumulate some additional glycogen, but most excess glucose would be used to synthesize triacylglycerol in adipocytes, a process that is promoted by insulin. A significant amount of liver glycogen is synthesized from glucose produced by gluconeogenesis.

11. Epinephrine promotes the conversion of glycogen to glucose by activating adenylate cyclase. Adenylate cyclase then initiates a reaction cascade that activates glycogen phosphorylase.

13. In liver, fructose is metabolized more rapidly than glucose because its metabolism bypasses two regulatory steps in the glycolytic pathway: the conversion of glucose to glucose-6-phosphate and fructose-6-phosphate to fructose-1,6-bisphosphate. Recall that fructose-1-phosphate is split into glyceraldehyde and DHAP, both of which are subsequently converted to glyceraldehyde-3-phosphate.

15. Two common oxidizing agents in anaerobic metabolism are NAD^+ and $NADP^+$.

17. Gluconeogenesis occurs mainly in the liver. It is activated by processes such as fasting and exercise that deplete blood glucose.

19. a. Lactate stimulates gluconeogenesis.
    b. ATP stimulates gluconeogenesis.
    c. Pyruvate stimulates gluconeogenesis.
    d. Glycerol stimulates gluconeogenesis.
    e. AMP inhibits gluconeogenesis.
    f. Acetyl-CoA stimulates gluconeogenesis.

C h a p t e r   9

In-Chapter Questions

9.1 The product of complete hydrogenation would be hard and therefore not useful as a margarine.

9.2 When soap and grease are mixed, the hydrophobic hydrocarbon tails of the soap insert (or dissolve) into the oil droplet. The oil droplet becomes coated with soap molecules. The hydrophilic portion of the soap molecules allows the soap-oil complex to be dispersed in water.

9.3 The phospholipid of the surfactant, which posses a polar head group and two hydrophobic acyl groups, disrupts some of the intermolecular hydrogen bonds of the water thereby decreasing the surface tension.

9.4 Carvone and camphor are monoterpenes; abscisic acid is a sesquiterpene.

9.5 The transport mechanisms discussed in the chapter fit into the following categories:

sodium channel: uniporter
glucose permease: symporter
Na^+-K^+-ATPase: antiporter

End-of-Chapter Questions

1. a. Lipids are naturally occurring substances that dissolve in hydrocarbon solvents.
   b. Hormones active within the synthesizing cell are called autocrine regulators.
   c. Amphipathic compounds contain both hydrophobic and hydrophilic groups.
   d. A terpene containing three isoprene units is called a sesquiterpene.
   e. The lipid bilayer is the basic structural feature of biological membrane.
   f. Prenylation is the process of attachment of isoprene units to a protein after it has been synthesized.
   g. Fluidity is a measure of the resistance of membrane components to movement.
   h. Chylomicrons are large lipoprotein complexes of extremely low density.
   i. A voltage-gated channel is opened in the membrane by changes in membrane voltage.

3. The lipid classes to which each structure belongs are as follows:
   a. triacylglycerol   b. steroid
   c. wax ester   d. unsaturated fatty acid
   e. phosphatidyl choline   f. sphingolipid

5. The ordered water molecules surrounding each phospholipid molecule are released from the polar heads. Order is lost and entropy increases.

7. The ceramide component of glycolipids is hydrophobic and acts in an analogous manner to the hydrophobic groups in the phospholipids. The hydrophilic sugar portion of the glycolipid can form hydrogen bonds with the water and is analogous to the charged portion of the phospholipid.

9. Cholesterol would stiffen the cell membrane. However, the presence of a cell wall would make incorporation of exogenous cholesterol into the underlying cell membrane improbable.

11. Many transmembrane and peripheral proteins are attached to the cytoskeleton and therefore are not free to move in the phospholipid bilayer.

13. For a phospholipid to move from one side of the bilayer to the other, the polar head must move through the hydrophobic portion of the phospholipid membrane. This process requires a significant amount of energy and is therefore relatively slow.

15. Most of the cholesterol in plaque results from the ingestion of LDL by the foam cells that line the arteries. High blood plasma LDL therefore promotes atherosclerosis. Because the coronary arteries are narrow, they are especially prone to occlusion by atherosclerotic plaque.

17. Eicosanoids are derived from arachidonic acid. Medical conditions in which it is advantageous to suppress the synthesis of eicosanoids are anaphalaxis, allergies, pain, the inflammation caused by injury, and fever.

19. The indicated compounds are classified as follows:
    a. monoterpene   b. monoterpene
    c. sesquiterpene   d. polyterpene
    e. diterpene   f. triterpene

21. HDL scavenges free cholesterol and transports it to the liver where it is converted to bile acids and excreted. This reduces total cholesterol in the serum and helps prevent plaque formation.

23. The sodium gradient created by the Na^+-K^+-ATPase in the plasma membrane of kidney tubule cells transports the glucose. This is an example of secondary active transport.

25. The fluidity of the membrane allows for flexible movement. Any breaks that do occur expose the hydrophobic core of the membrane to an aqueous environment. Hydrophobic interactions spontaneously move the broken ends together and, in combination with certain other components of cell membrane resealing mechanisms (e.g., cytoskeleton and calcium ions), the membrane reseals.

Chapter 10

In-Chapter Questions

10.1 If there is a connection between female sex hormone levels and VLDL secretion, injection of estrogen into a male rat should have the following effects: There should be a timely and measurable increase in VLDL secretion. This process requires a concomitant increase in the synthesis of the components of VLDL, i.e., apoproteins, triacylglycerols, phospholipid, and cholesterol. FABP synthesis should increase in response to the increased intracellular fatty acid concentrations.

10.2 In cells without observable intact peroxisomes, the absence of ether-type lipids or the buildup of long chain fatty acids suggests that the organelle is not present. In addition, histochemical tests can be employed to diagnose Zellweger syndrome. For example, radioactive isotope-labeled antibodies to peroxisomal marker enzymes can be used to determine if peroxisomal function is present in cells.

10.3 Because steroids inhibit the release of arachidonic acid, their use shuts down the synthesis of most if not all eicosanoid molecules, hence their reputation as potent anti-inflammatory agents. Aspirin inactivates cyclooxygenase and prevents the conversion of arachidonic acid to PGG_2, the precursor of prostaglandins and thromboxanes. Aspirin is not as effective an anti-inflammatory agent as the steroids because it shuts down only a portion of eiconanoid synthetic pathways.

10.4 Following the hydrolysis of sucrose, both monosaccharide products enter the bloodstream and travel to the liver, where fructose is converted to fructose-1-phosphate. Recall that the conversion of fructose-1-phosphate to glyceraldehyde-3-phosphate bypasses two regulatory steps. Consequently, more glycerolphosphate and acetyl-CoA (the substrates for triacylglycerol synthesis) are produced. High blood glucose concentrations that result from this consumption of excessive amounts of sucrose trigger the release of larger than normal amounts of insulin. One of the functions of insulin is to promote fat synthesis.

10.5 The structure of cortisol differs from that of cortisone in that the C-11 hydroxyl group is replaced by a carbonyl group. Cor-

tisone can be administered to Addison's disease patients because it is converted to cortisol by 11-β-hydroxysteroid dehydrogenase.

10.6 The higher activity of HMG-CoA reductase in obese patients in combination with a high calorie diet increases the synthesis of cholesterol.

End-of-Chapter Questions

1. a. Biosynthesis from new materials is termed *de novo*.
    b. Oil bodies are structures in plants that store triacylglycerol.
    c. β-Oxidation is the main pathway for the degradation of fatty acids in which two-carbon fragments in the form of acetyl-CoA are removed from the carboxyl end of fatty acids.
    d. The rate at which molecules are degraded and replaced is called turnover.
    e. Thiolytic cleavage involves the cleavage by thiolase between the α and β carbons of a β ketoacid during β-oxidation to produce a molecule of acetyl-CoA and a new acyl-CoA.
    f. Autoantibodies are defense proteins that bind to surface antigens of the patient's own cells as if they were foreign.
    g. Acetyl-CoA molecules are condensed to produce acetone, β-hydroxybutyrate, and acetoacetate (i.e., the ketone bodies) in a process called ketogenesis.

3. Peroxisomes have β-oxidation enzymes that are specific for long chain fatty acids, whereas mitochondria possess enzymes that are specific for short and moderate chain length fatty acids. In addition, the first reaction in the peroxisomal pathway is catalyzed by a different enzyme than the mitochondrial pathway. The $FADH_2$ produced in the first peroxisomal reaction donates its electrons to O_2 directly, (forming H_2O_2) instead of UQ as in mitochondria. The processes are similar in that acetyl-CoA is derived from the oxidation of fatty acids.

5. Three differences between fatty acid synthesis and β-oxidation are the following: (1) The two pathways take place in different cell compartments, i.e., synthesis in the cytoplasm and β-oxidation within the mitochondria. (2) The intermediates of fatty acid synthesis and β-oxidation are linked through thioester linkages to ACP and CoASH respectively. (3) The electron carrier for fatty acid synthesis is NADPH while those for β-oxidation are NADH and $FADH_2$.

7. In the short term, hormones alter the activity of preexisting regulatory enzyme molecules. For example, the binding of glucagon inhibits acetyl-CoA carboxylase. Long-term effects of hormones usually involve changes in the pattern of enzyme synthesis in target cells. For example, insulin promotes the synthesis of the enzymes involved in lipogenesis (e.g., acetyl-CoA carboxylase and fatty acid synthase).

9. The head group of a phospholipid is hydrophilic. Movement across the membrane requires desolvation of the water molecules surrounding it and insertion of this charged head into a hydrophobic region. Both of these steps require energy and are therefore unlikely.

11. Because of the presence of a methyl substituent on the β-carbon, the fatty acid first undergoes one cycle of α-oxidation. The resulting molecule, now shorter by one carbon atom, then undergo one cycle of β-oxidation. The products of this latter process are two moles of propionyl-CoA.

13. In the oxidation of butyric acid by the β-oxidation pathway, one mole of $FADH_2$, one mole of NADH, and two moles of acetyl-CoA are produced.

15. Enoyl-CoA isomerase converts the naturally occurring *cis* double bond at Δ^3 to a *trans* double bond at Δ^2, the correct position for the next round of β-oxidation.

17. Normally, phospholipids do not spontaneously translocate across the membrane. Flippase, however, promotes the translocation of phospholipids containing choline across the membrane. This type of phospholipid is found in unusually high concentration on the lumenal side of the ER membrane.

19. a. Hydrophobic interactions are probable between the enzyme and the lipid in the micelle.
    b. For phospholipases to be drawn into the micelle, they must have a hydrophobic surface.

21.

The indicated bond is cleaved by glucocerebrosidase.

Chapter 11

In-Chapter Questions

11.1 Propionyl-CoA can be reversibly converted to succinyl-CoA, an intermediate in the citric acid cycle. Oxaloacetate, a downstream intermediate of this cycle can be converted to PEP. PEP is then converted to glucose via gluconeogenesis. Adipic acid undergoes one round of β-oxidation to yield acetyl-CoA and succinyl-CoA. As just described, succinyl-CoA is sequentially converted to oxaloacetate, PEP, and then to glucose.

11.2 Fluoroacetate is converted to fluoroacetyl-CoA. This substance then reacts with oxaloacetate to produce fluorocitrate. Fluorocitrate is toxic because it inhibits aconitase, the enzyme that normally converts citrate to isocitrate, hence the buildup of citrate. In the plant, the fluoroacetate is stored in vacuoles away from the mitochondria.

11.3 DNP is a lipophilic molecule that binds reversibly with protons. It dissipates the proton gradient in mitochondria by transferring protons across the inner membrane. The uncoupling of electron transport from oxidative phosphorylation causes the energy from food to be dissipated as heat. DNP causes liver failure because of insufficient ATP synthesis in a metabolically demanding organ.

11.4 Disregarding proton leakage and assuming that the glycerol phosphate shuttle is in operation, 38 ATP would be produced from the aerobic oxidation of a glucose molecule. If the malate shuttle is in operation, only 36 ATP would be produced.

11.5 The larger selenium atom holds its electrons less tightly than sulfur. Selenium is more easily oxidized and therefore acts as a better scavenger for oxygen than does sulfur.

11.6 The SH groups reduce hydrogen peroxide or trap hydroxyl radicals to form water. An example of a nonsulfhydryl group-containing molecule that should be capable of this activity is vitamin C.

11.7 Low levels of G-6-PD in combination with a high level of oxidized GSH cause high oxidative stress. Without antioxidant protection, red cell membranes become fragile, a condition that eventually causes hemolytic anemia.

11.8 The phenolic groups of both molecules are responsible for their antioxidant activity.

End-of-Chapter Questions

1. a. Aerotolerant anaerobes are organisms that grow without oxygen but are protected against its toxic effects.
   b. Anaplerotic reactions replenish substrates used in biosynthetic reactions.
   c. Glyoxysomes are plant organelles that possess glyoxylate cycle enzymes.
   d. The chemical coupling hypothesis postulates that a high energy intermediate generated by electron transport is used to drive ATP formation.
   e. The chemiosmotic coupling theory proposes that a proton gradient created by the mitochondrial electron transport system drives ATP synthesis.
   f. An ionophore is a hydrophobic molecule that inserts into membrane and dissipates osmotic gradients.
   g. Respiratory control there refers to the regulation of aerobic respiration by ADP.
   h. The term *ischemia* indicates an inadequate blood flow.
   i. *Aerobic respiration* is the process in which oxygen generates energy from food molecules.

3. a. Obligate anaerobes avoid oxygen by living in reduced environments.
   b. Aerotolerant anaerobes generate energy from fermentation. They use structural barriers or oxygen scavenging molecules to protect themselves from oxygen's toxic effects.
   c. Depending on the availability of oxygen, facultative anaerobes generate energy by fermentation or aerobic respiration.
   d. Obligate aerobes require oxygen for energy production.

5. The citric acid cycle is an important component of aerobic respiration. The NADH and $FADH_2$ produced during oxidation-reduction reactions of the cycle donate electrons to the mitochondrial ETC. Citric acid cycle intermediates are also used as biosynthetic precursors.

7. The glyoxylate cycle is a modified version of the citric acid cycle that allows certain organisms (e.g., plants or some microorganisms) to grow by utilizing two-carbon molecules such as acetyl-CoA, acetate, or ethanol. The glyoxylate cycle allows for the net synthesis of larger molecules from two-carbon molecules because the two decarboxylation reactions of the citric acid cycle are bypassed.

9. Three functions of citric acid in the cytoplasm are serving as a precursor of acetyl-CoA used in the synthesis of cholesterol and fatty acids, serving as a precursor of oxaloacetate used in the synthesis of certain amino acids, and serving as an inhibitor of PFK-1, a major regulatory enzyme in glycolysis.

11. Processes believed to be driven by mitochondrial electron transport are ATP synthesis, the pumping of calcium ions into the mitochondrial matrix, and the generation of heat by brown fat.

13. According to the chemiosmotic theory, an intact inner mitochondrial membrane is required to maintain the proton gradient required for ATP synthesis.

15. The translocation of three protons is required to drive ATP synthesis. The fourth proton drives the transport of ADP and P_i.

17. Oxygen is widely used as an energy source because it makes possible greater energy yields from food molecules and it is readily available.

19. ROS damage cells by inactivating enzymes, depolymerizing polysaccharides, breaking DNA, and destroying membrane.

21. A genetic defect with survival value is G-6-PD deficiency. The lower NADPH and GSH concentrations in the red blood cells of G-6-PD-deficient individuals inhibit the growth of *Plasmodium* (the malarial parasite), a major killer of humans.

23. Aspartic acid and alanine can be produced by transamination reactions using oxaloacetate and pyruvate, respectively, both citric acid cycle intermediates.

25. The oxidation of one mole of ethanol to acetyl-CoA produces two moles of NADH. The conversion of acetate to carbon dioxide and water through the citric acid cycle produces 3 NADH, 1 $FADH_2$, and 1 GTP. Assuming that the aspartate-malate shuttle is in operation, each cytoplasmic NADH yields 2.25 ATP for a total of 4.5 ATP. Each mitochondrial NADH yields 2.5 ATP for a total of 7.5 ATP. Each molecule of $FADH_2$ yields 1.5 ATP for a total of 1.5 ATP. Each GTP yields 0.75 ATP for a total of 0.75 ATP. The total ATP produced by the oxidation of ethanol is therefore 14.25 ATP.

Chapter 12

In-Chapter Questions

12.1 Excessive light promotes the formation of ROS, which damage proteins such as D1. β-Carotene is an antioxidant that prevents some of this damage.

12.2 The process of phosphorylation changes the conformation and the binding characteristics of the complex and thereby alters its capacity to bind to PSII.

12.3 Of the herbicides discussed, paraquat and DCMU are most hazardous to humans. Paraquat generates free radicals that can attack cell components. DCMU poisons the electron transport complex.

12.4 Reactions of the Calvin cycle strongly resemble the pentose phosphate pathway.

End-of-Chapter Questions

1. a. A photosystem is a molecular complex that absorbs light and converts it into chemical energy.
   b. A reaction center is a pigment-protein component of a photosystem that mediates the conversion of light energy into chemical energy.
   c. Light reactions are a mechanism whereby the electrons energized by light are subsequently used in ATP and NADPH synthesis.
   d. In dark (or light independent) reactions, ATP and NADPH are used in carbohydrate synthesis.
   e. The chloroplast is a chlorophyll-containing organelle in which photosynthesis occurs.

f. In photorespiration, light promotes the consumption of oxygen and the wasteful release of carbon dioxide.

3. The three primary photosynthetic pigments are (1) the chlorophylls, which absorb blue and green wavelengths of light; (2) the carotenoids, which serve as antenna pigments and protect from ROS; and (3) the xanthophylls, which also serve as antenna pigments.

5. Excited molecules can return to the ground state by several means. (1) Fluorescence (light is absorbed at one wavelength and emitted at a longer wavelength). (2) Resonance energy transfer (energy is transferred to neighboring molecules through interaction with adjacent molecular orbitals). (3) Oxidation-reduction (an excited electron returns to its ground state by reducing another molecule). (4) Radiationless decay (an excited molecule returns to the ground state and loses its excess energy as heat). Of these processes, oxidation-reduction and resonance energy transfer are important in photosynthesis. Oxidation-reduction is important in the release of oxygen and synthesis of ATP. Resonance energy transfer passes light energy to chlorophyll from accessory pigments.

7. During the light reactions of photosynthesis, light driven electron transport results in ATP and NADPH synthesis.

9. The oxygen evolving system is referred to as a clock because it involves five oxidation-reduction states that must be completed in order.

11. Chloroplast structure allows the transfer of excited electrons from one pigment to another by resonance energy transfer until they are trapped as chemical energy or reemitted. ATP synthesis can occur because a proton gradient is established across the thylakoid membrane.

13. Oxidative phosphorylation and photophosphorylation use many of the same molecules in their reactions and both are linked to an electron transport system. However, chloroplasts use light energy to drive redox reactions while mitochondria use the energy of chemical bonds to drive redox reactions. In contrast to mitochondrial inner membrane, thylakoid inner membrane is permeable to magnesium and chloride ions. Therefore, the electrochemical gradient across the thylakoid membrane consists mainly of a proton gradient.

15. The electrons energized by light energy are normally transferred to carbon dioxide eventually. In the absence of carbon dioxide, the energy of the excited electrons of chlorophyll is released as fluorescence.

17. High oxygen concentrations promote photorespiration.

19. Chloroplasts possess DNA similar to that of modern cyanobacteria as well as protein synthesizing machinery. In addition, they multiply by binary fission as do bacteria.

21. Under conditions of high temperature, the carbon dioxide compensation point of C3 plants rises because the oxygenase activity of Rubisco increases more rapidly than the carboxylase activity.

23. The herbicides kill marine photosynthesizing organisms, thereby inhibiting world-wide O_2 production.

25. The optimal wavelengths for photosynthesis appear to be 400–500 nm and 600–700 nm.

27. The oxygen-evolving complex of PSII exists in five transient oxidation states (S_0 through S_4), collectively referred to as a clock. Because oxygen evolution occurs only when the S_4 state has been reached, several light bursts are required.

C h a p t e r   1 3

In-Chapter Questions

13.1 Because of its close structural similarity to folic acid, methotrexate is a competitive inhibitor of the enzyme dihydrofoliate reductase. (Recall that this enzyme converts folic acid to its biologically active form, THF.) Rapidly dividing cells require large amounts of folic acid. Methotrexate prevents the synthesis of THF, the one-carbon carrier required in nucleotide and amino acid synthesis. It is therefore toxic to rapidly dividing cells, especially those of certain tumors.

13.2 The synthetic pathway for melatonin.

13.3 The reaction types involved in the synthesis of IAA from tryptophan are a deamination (removal of the α-amino group) and an oxidative decarboxylation.

Serotonin → N-Acetyltransferase → **5-Hydroxy-N-acetyltryptamine** → O-Methyltransferase / SAM → **Melatonin**

13.4 The disruption of function that accompanies brain damage has numerous consequences. Among these are reduced ATP synthesis and the unbalanced intracellular ion concentrations that result from oxygen deprivation. Under these conditions, when an oxygen supply is reestablished, ROS production increases the extent of brain cell damage. Among the ROS formed are superoxide, hydroxyl radical, singlet oxygen, and the peroxy-nitrite anion. The normal cellular defenses against ROS, such as NADPH, GSH, antioxidant vitamins such as vitamins C and E, and antioxidant enzyme systems, are quickly depleted because of disrupted synthesis and/or transport. When cells swell, they release the excitatory neurotransmitter glutamate, which stimulates some nearby cells to synthesize NO. Excessive and unregulated NO synthesis is extraordinarily dangerous to brain tissues under these circumstances, in part because the presence of elevated ROS concentrations causes the toxic peroxynitrite anion to be formed.

End-of-Chapter Questions

1. a. Essential amino acids are molecules that cannot be synthesized by an animal and must be obtained from its diet.
   b. Nitrogen balance is a physiological condition in which the body's nitrogen intake is equal to the nitrogen loss.
   c. In *de novo* pathways, molecules such as amino acids or nucleotides are synthesized from simple precursors.
   d. Biogenic amines are amino acids or amino acid derivatives that act as neurotransmitters.
   e. An excitotoxin is a molecule that can sufficiently overstimulate a cell, causing its death.
   f. A retrograde neurotransmitter is a molecule that is released from a postsynaptic cell and diffuses back to and promotes an action in a presynaptic cell.

3. Because the nitrogen molecule is very stable, forming nitrogen compounds requires a great deal of energy. In addition, only a few organisms can reduce nitrogen to form ammonium ions in a process known as nitrogen fixation.

5. Glutamate is synthesized from α-ketoglutarate by two means: (1) transamination, catalyzed by the aminotransferases (pyridoxal phosphate is a required coenzyme) and (2) direct amination, catalyzed by glutamate dehydrogenase. NADPH provides the reducing power for this reaction.

7. The positively charged nitrogen of the pyridinium ring strongly attracts electrons during PLP catalyzed reactions. Since no such structure exists in the benzene ring, benzene is less effective in attracting electrons.

9. Neurotransmitters are either excitatory or inhibitory. Excitatory neurotransmitters (e.g., glutamate and acetylcholine) promote the depolarization of a postsynaptic cell. Inhibitory neurotransmitters (e.g., glycine) inhibit action potentials in postsynaptic cells, i.e., they make the membrane potential more negative.

11. Glutamate is an excitatory neurotransmitter with stimulating effects on neurons that regulate bodily functions such as blood pressure and body temperature. Individuals who display symptoms after consuming monosodium glutamate apparently possess efficient mechanisms for transporting glutamate across the blood-brain barrier.

13. The two most prominent one-carbon carriers are folic acid and S-adenosylmethionine. THF, the biologically active derivative of folic acid, plays critically important roles in the synthesis of certain amino acids and nucleotides. SAM is a methyl donor in numerous reactions. Examples of methylated products include phosphatidylcholine, epinephrine, and carnitine.

15. Glutathione reduces free radicals (primarily the hydroxyl radical) to water. Other strong antioxidants, such as vitamin C, also convert the hydroxyl free radical to water.

17. No. The two fused rings of the pyrimidine ring system result in steric interaction with the pentose. Purine rings contain only one ring and steric interactions with the pentose are minimal.

19. The reactions involved in the synthesis of the purine ring are outlined in Figure 13.24 on p 399.

21. Radiolabel both the carbon and nitrogen of aspartate and glutamate. If the amino acid is used in the ring assembly, then both carbon and nitrogen should bear the label. If nitrogen exchange takes place, only the carbon atom will be labeled. In addition, isolating each intermediate allows the origin of each atom to be traced.

23. If lysine were transaminated, the ϵ amino group of the new keto acid (derived from lysine) would cyclize to form an intramolecular Schiff base. Consequently, the α-keto acid required to produce lysine by transamination cannot exist in appreciable quantities. Therefore, lysine cannot be produced by this reaction.

25. Arginine is normally synthesized by the urea cycle. In small children, the urea cycle is not fully functional. Consequently, arginine must be obtained from external sources.

27. When glycine is absent, the other substrate of the enzyme phosphoribosylglycinamide synthase, 5-phospho-β-D-ribosylamine accumulates.

29. Refer to Figure 13.27, p. 403.

31. In the transamination reaction involving alanine (an α-amino acid) and α-ketoglutarate (an α-ketoacid) the products are pyruvate (an α-keto acid) and glutamate (an α-amino acid). Refer to Figure 13.3 for the role pyridoxal phosphate plays in transamination reactions.

33. Breaking the bond perpendicular to the pyridinium ring generates a P orbital that can then interact with the π system of the pyridinium ring. The resulting delocalization of the charge stabilizes the carbanion.

Chapter 14

In-Chapter Questions

14.1 Certain intestinal bacteria can release ammonia from urea molecules that diffuse across the membrane into the intestinal lumen. Treatment with antibiotics kills these organisms, thereby reducing blood ammonia concentration.

14.2

$$^-O_2S-CH_2CH(NH_3^+)COO^- \xrightarrow{\text{decarboxylation}}$$

Cysteine sulfinate

$$-O_2S-CH_2CH_2NH_3^+ + CO_2$$

$$^-O_2S-CH_2CH_2NH_3^+ \xrightarrow{\text{oxidation}}$$

$$^-O_3S-CH_2CH_2NH_3^+$$

Taurine

14.3 The following amino acids are both ketogenic and glycogenic: phenylalanine, isoleucine, lysine, tryptophan, and tyrosine.

14.4 Acetylcholine is normally degraded rapidly by cholinesterase. Drugs that block the action of cholinesterase prevent this hydrolysis. Consequently, acetylcholine molecules remain in the synaptic cleft for an extended time. There they can rapidly and reversibly bind and rebind to a reduced number of functional acetylcholine receptors. This process promotes the depolarization of the muscle cells.

14.5

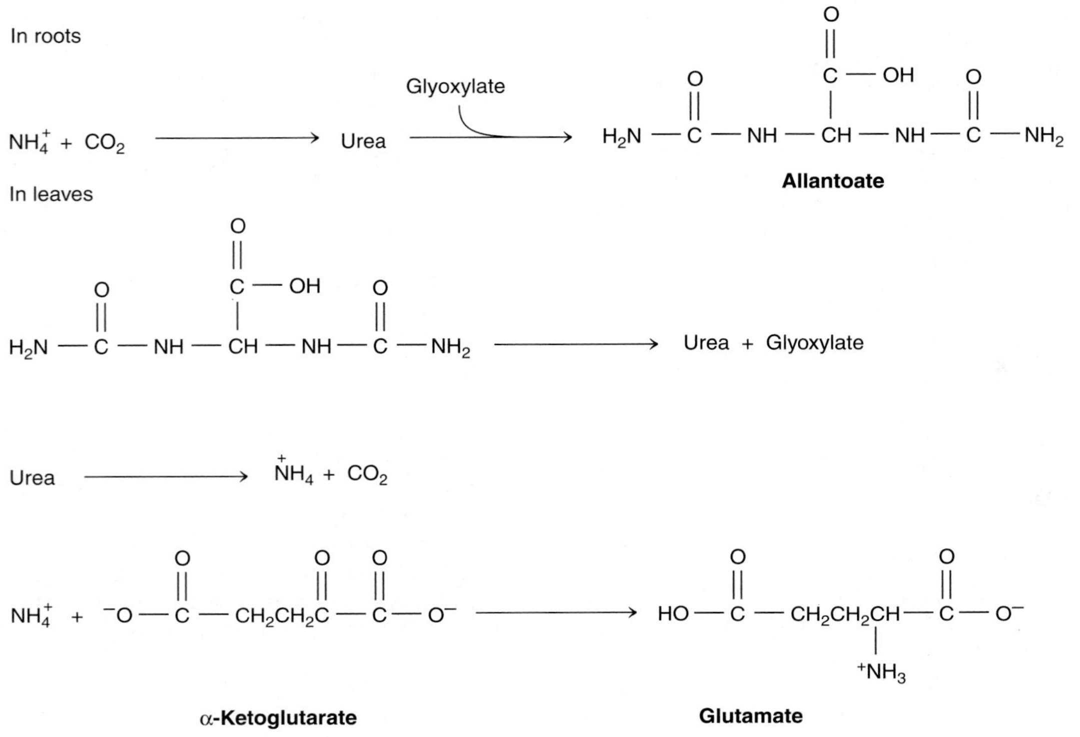

14.6 After undergoing transamination reactions, the carbon skeletons of β-alanine and β-aminoisobutyrate react with coenzyme A to form malonyl CoA and methylmalonyl CoA, respectively. Methylmalonyl CoA is converted to succinyl CoA by methylmalonyl CoA mutase.

End-of-Chapter Questions

1. a. In denitrification, nitrate is converted to atmospheric nitrogen.
   b. Ammonotelic animals excrete ammonia directly, i.e., without conversion to less toxic molecules.
   c. The term *protein turnover* describes a process in which the proteins of living organisms are constantly synthesized and degraded.
   d. Ubiquination is a eukaryotic mechanism whereby proteins destined to be degraded are first covalently linked to the small protein ubiquitin.

3. The process of protein turnover promotes metabolic flexibility, protects a cell from the accumulation of abnormal proteins, and is a key feature of organismal developmental processes.

5. The metabolic products of amino acid degradation are acetyl-CoA, acetoacetyl-CoA, pyruvate, α-ketoglutarate, succinyl-CoA, fumarate, and oxaloacetate.

7. Refer to the following pages for a description of the degradation of each amino acid.
   a. lysine   p. 419
   b. glutamate   p. 421
   c. glycine   p. 418
   d. aspartate   p. 422
   e. tyrosine   p. 419
   f. alanine   p. 418

9. Purine rings are degraded to uric acid. A significant percentage of uric acid is excreted in the urine. (Refer to Figure 14.12.)

11. In the muscle, pyruvate undergoes a transamination reaction and is converted to alanine. Alanine is then transferred to the liver, where it is reconverted to pyruvate. The amino group is transferred to α-ketoglutarate thus forming glutamate, which is subsequently oxidatively deaminated to form α-ketoglutarate and ammonia.

13. The NADH formed during the conversion of fumarate to aspartate results in the synthesis of approximately 2.5 ATP. Therefore, the net ATP requirement for urea synthesis is approximately 1.5 moles of ATP per mole of urea.

15. As the concentration of glutamate (as well as its deamination product ammonia) rises, the enzyme N-acetylglutamate synthase catalyzes the synthesis of N-acetylglutamate, an activator of carbamoyl phosphate synthetase I. The latter enzyme catalyzes the first committed step in urea synthesis.

17. Tetrahyrobiopterin is a cofactor in the oxidation of phenylalanine to form tyrosine. The sustained absence of this cofactor would result in a buildup of phenylalanine and the appearance of the symptoms of PKU.

19. These amino acids are intermediates of the urea cycle. Therefore, their addition stimulates the formation of urea.

21. Because of the structural similarities to purine, caffeine is converted to a variety of derivatives by xanthine oxidase (e.g., 1-methyluric acid and 7-methylxanthine).

23. The branched-chain amino amino acids (leucine, isoleucine, and valine) are metabolized in tissues, such as muscle, where they are principally used to synthesize nonessential amino acids.

25. In addition to being less toxic molecules than ammonia, urea and uric acid (the nitrogenous waste products of terrestrial animals) require significantly less water for their excretion. Recall that ammonotelic organisms excrete ammonia directly into surrounding water.

Chapter 15

In-Chapter Questions

15.1 High levels of stress combined with lack of sleep or distorted wake/sleep cycles (e.g., shift work) depress levels of thymic hormone secretion. Lower secretion rates of thymic hormone decrease the effectiveness of the immune system. A number of other hormones affect immune system functions. For example, melantonin inhibits the secretion of CRH and GnRF. The subsequent decreased release of ACTH, LH, and FSH causes decreased synthesis of the glucocorticoids and the sex hormones. Several sex hormones have been observed to modify the immune system. For example, estrogens stimulate the female immune system.

15.2 The high blood glucose levels in untreated diabetics result in the loss of increasingly large amounts of glucose in urine, a condition that causes dehydration. In the absence of useable glucose, the body rapidly degrades fats and proteins to generate energy. Hence Aretaeus' observation that in this disease excessive weight loss and excessive urination are related.

15.3 The inhibition of GTP hydrolysis causes the subunit of G_S protein to continue activating adenylate cyclase. In intestinal cells, this enzyme activity opens chloride channels, causing loss of large amounts of chloride ions and water. The massive diarrhea caused by this process quickly leads to serious dehydration and electrolyte loss.

15.4 Nitroglycerin molecules are hydrolysed in blood to yield NO. In turn NO relaxes smooth muscle cells in the walls of blood vessels. It is believed that NO activates guanylate cyclase, which subsequently promotes the intracellular sequestration of calcium ions that allows muscle cells to relax.

15.5 Both DAG and phorbol esters promote the activity of protein kinase C, which promotes cell growth and division. Phorbol esters provide initiated cells with a sustained growth advantage over normal cells. This condition is an early stage in carcinogenesis.

End-of-Chapter Questions

1. a. Ketoacidosis is a condition in which large amounts of ketone bodies occur in blood.
   b. In hyperlipoproteinemia, blood levels of lipoprotein are high.
   c. Neurophysins are proteins packaged with vasopressin and oxytocin in secretory granules that transport these hormones down axons from the hypothalamus to the posterior pituitary.
   d. Growth factors are a series of polypeptides and proteins that regulate the growth, differentiation, and proliferation of various cells.
   e. Chylomicron remnants are chylomicrons from which triacylglycerol molecules have been removed.

f. In down regulation, hormone receptor molecules are internalized by endocytosis.

g. G proteins are multisubunit proteins that bind GTP. They mediate transmembrane signaling.

3. a. Corticotropin stimulates steroid synthesis in the adrenal cortex.

b. Insulin promotes general anabolic effects, including glucose uptake by some cells and lipogenesis.

c. Glucagon promotes glycogenolysis and lipolysis.

d. Oxytocin stimulates uterine muscle contraction.

e. LH stimulates the development of cells in the ovaries and testis and the synthesis of sex hormones.

f. GnRF stimulates LH and FSH secretion.

g. Somatostatin inhibits GH and TSH secretion, as well as the secretion of gastrin and glucagon.

h. Vasopressin (or antidiuretic hormone) is involved in the regulation of blood osmolarity.

i. FSH promotes ovulation and estrogen synthesis in ovaries and sperm development in testis.

5. Hormones are molecules that function in intercellular communication, i.e., they regulate physiological responses throughout the body.

7. The second messenger is an effector molecule synthesized when a hormone (the first messenger) binds. It stimulates the cell to respond to the original signal. Second messengers also allow the signal to be amplified.

9. Long-term fasting or low-calorie diets are interpreted by the brain as starvation. The brain responds by lowering the body's BMR. One effect of this mechanism is that skeletal muscle mass is reduced because it is such a metabolically demanding tissue.

11. Increased mobilization of fatty acids provides an alternate energy source for muscle, thereby sparing glucose for the brain. In addition, glucagon stimulates gluconeogenesis, a pathway that utilizes amino acids derived from muscle.

13. To ensure proper control of metabolism, powerful hormones are synthesized in small quantities. Hormones also elicit responses in only specific target cells. They are metabolized quickly to ensure the precision of metabolic regulation.

15. The major recognized second messengers are (1) cAMP (a nucleotide synthesized from ATP by adenylate cyclase), (2) cGMP (a nucleotide synthesized from GTP by guanylate cyclase), (3) the components of the phosphoinositol cycle (DAG_2 and IP_3), which are cleaved from PIP_2 by phospholipase C, and (4) calcium ions, which are released from cellular stores in response to several stimuli.

17. Phorbol esters, found in croton oil, activate protein kinase C, an action that stimulates cell growth and division. However, unlike DAG, the molecule they mimic, phorbol esters continue to activate proteins kinase C for a prolonged time. This provides the affected cell with an advantage over unstimulated cells. Phorbol esters may transform a cell previously exposed to a carcinogenic initiating event into a cancerous cell whose unrestrained proliferation creates a tumor.

19. The steroid molecule is covalently bound to the matrix in a chromatographic column. The extract suspected of containing the steroid binding protein is then passed through the column. Any proteins remaining on the column are eluted by changing the salt concentration of the eluting buffer. After their isolation and purification, such proteins can be examined specifically for binding activity to the steroid.

21. In uncontrolled diabetes, large amounts of glucose are excreted in the urine. Excessive urine flow caused by the large amounts of water that are excreted along with the glucose dehydrates the body. Dehydration then usually triggers the thirst response.

23. The hormones trigger increased protein synthesis in skeletal muscle among other anabolic changes.

25. The functions of the kidney include (1) the excretion of water soluble waste products, (2) the reabsorption of electrolytes, amino acids, and sugars from the urinary filtrate, (3) regulation of pH, and (4) the regulation of the body's water content.

27. Metabolism of these amino acids generates ammonia, used to buffer blood pH.

Chapter 16

In-Chapter Questions

16.1 Other possible culprits are the superoxide free radical and singlet oxygen. The brain is susceptible to oxidative damage because of its proportionally large consumption of oxygen and because many brain cells cannot be replaced if damaged.

16.2 The antisence DNA sequence is 3'-CGTAAGCTTAACGTCT-GAGGACGTTAAGCCGTTA-5'; the mRNA sequence is 3'-CGUAAGCUUAACGUCUGAGGACGUUAAGCCGUUA-5'. The antisence RNA sequence is 3'-GCAUUCGAAUUGCA-GACUCCUGCAAUUCGGCAAU-5'.

16.3 In the original central dogma, the flow of genetic information is in one direction only, i.e., from DNA to the RNA molecules, which then direct protein synthesis. The altered diagram indicates that the RNA genome of some viruses can replicate their RNA genomes (using a viral enzyme activity referred to as RNA directed-RNA polymerase) or undergo reverse transcription (i.e., synthesize DNA from an RNA sequence).

16.4 In an HIV ELISA assay, HIV antigens (e.g., gp41 or gp120) are attached to an inert support, such as a plastic lab dish. An aliquot of a patient's blood serum is added to the dish. After a short time the serum is removed and an antihuman antibody attached to an enzyme is added. When the enzyme's substrate is added to the dish, a color change indicates that the patient's blood contains antibodies to HIV. In the Western blot confirming test for HIV, the viral proteins are separated on polyacrylamide gel electrophoresis. The proteins are then transferred to nitrocellulose strips by a blotting procedure. Then the nitrocellulose strips are incubated with the patient's blood serum. Any HIV antibodies present in the serum bind to the viral proteins. The strips are then incubated with enzyme-linked antihuman antibodies. After the addition of the substrate, any color change is measured. The amount of the patient's HIV antibodies (if any) is then precisely determined.

End-of-Chapter Questions

1. a. Genetics is the study of inheritance.

b. Replication is the process by which DNA strands are copied.

c. Transcription is the synthesis of RNA from a DNA template.

d. In polynucleotides the "backbone" is created by the 3′,5′-phosphodiester bonds that join the 5′-hydroxyl group of deoxyribose residues to the 3′-hydroxyl group of the sugar unit of another nucleotide.

e. Bacteriophage are virus that attack bacteria.

f. According to Chargaff's rules, there is a 1 : 1 ratio of adenine to thymine and guanine to cytosine, regardless of the source of DNA.

g. A palindrome is a sequence that reads the same in both directions.

h. Hoogsteen base pairing is a form of nonconventional base pairing that allows H-DNA to form.

i. Linking number is the number of times that DNA strands cross each other.

j. The Alu family is a group of short DNA sequences that occur over 500,000 times in the human genome.

3. The major and minor grooves of DNA arise because the glycosidic bonds in the two hydrogen bonded strands are not exactly opposite to each other.

5. Supercoiling is a process in which DNA bends and twists, allowing DNA strands to be packaged in compact chromosomes.

7. The polyamines are positively charged at pH 7, which promotes binding to the negative charges of the DNA backbone. Polyamine binding overcomes the mutual repulsion of the adjacent DNA chains, packing the chains more closely.

9. Nucleosomes are the structural units of chromatin. Each nucleosome consists of a left-handed supercoiled DNA segment wound around eight histone molecules.

11. RNA can coil back on itself to form complex three-dimensional structures.

13. The transition from B-DNA to Z-DNA can occur when the nucleotide base sequence is composed of alternating purine and pyrimidines (e.g., CGCGCG). Because alternate nucleotides can assume different conformations (syn or anti), these DNA segments form a left-handed helix. The phosphate groups in the backbone of this DNA conformation zig-zag, hence the name Z-DNA.

15. The electron withdrawing effect of the bromine increases the liklihood of uracil enol formation. This enol mimics the hydrogen bonding pattern of cytosine. Therefore, this base can be paired with guanine.

17. According to Chargaff's rules, if a DNA sample contains 21% adenine then it also contains 21% thymine. If the A-T content is 42%, then the G-C content is 58%. Consequently the guanine and cytosine percentages are both 29% in the DNA sample.

19. Because nucleotide base sequences and amino acid sequences are such completely different "languages," a complex mechanism is required for the "translation" of one type of information into another. In the absence of any evidence to the contrary it does not appear likely that information expressed in proteins can be utilized to direct the synthesis of nucleic acids.

21. Nuclei and mitochondria are separately obtained from source tissue by using cell homogenization followed by density gradient centrifugation. The nucleic acids in each organellar fraction are then extracted with the aid of detergents, solvents, and proteases (to remove proteins). RNA is removed by treating each sample with RNase. The DNA from both types of organelles is then further purified by centrifugation.

Chapter 17

In-Chapter Questions

17.1 If either of the alternate models were correct, the CsCl banding patterns would be different from those observed in the Messelson-Stahl experiment. The expected results for the conservative model would be two bands, i.e., one heavy DNA band and one light DNA band, after the first cell division. Expected results after the first cell division for the dispersive model would be a large number of bands of varying density.

17.2 When antibiotics are used in large quantities, the bacterial cells that possess resistance genes (acquired through spontaneous mutations or through intermicrobial DNA transfer mechanisms such as conjugation, transduction, and transformation) survive and even flourish. Because of antibiotic use, which acts as a selection pressure, resistant organisms (once only a minor constituent of a microbial population) become the dominant cells in their ecological niche.

17.3 Most gene duplications are apparently a consequence of accidents during genetic recombination. Examples of possible causes of gene duplication are unequal crossing-over during synapsis and transposition. After a gene has been duplicated, random mutations as well as genetic recombination may introduce variations.

17.4 Bacteria can permanently acquire the capacity to produce a toxin when the viral toxin gene becomes incorporated into the bacterial chromosome or a self-replicating plasmid. Comparison of modern group A streptococcus with the organism that caused a similar disease in the 1920s requires the production via PCR of a series of DNA probes from the modern organism. These probes are then used in an *in situ* hybridization investigation of preserved specimens to determine any similarities or differences between the two organisms.

17.5 Because phytochrome has been demonstrated to mediate numerous light induced plant processes, it appears reasonable to assume that it does so in part by interacting with light response elements (LRE) in plant cell genomes. Presumably, phytochrome influences gene expression by binding, either alone or as part of a complex, to various LREs when its chromophore is activated by light.

End-of-Chapter Questions

1. a. Chemoreceptors are protein receptors on or near the external surface of a cell's plasma membrane that bind specific chemicals or nutrients, thus triggering chemotaxis.

b. Structural genes are DNA segments that code for polypeptides.

c. Recombination is the rearrangement of DNA sequences involving the exchange of segments from different molecules.

d. In semiconservative replication, each new DNA molecule possesses one new strand and one old strand.

e. A replisome is the protein complex that replicates DNA molecules.

f. Ori C is the replication initiation site on the *E. coli* chromosome.

g. Transcription is the process in which an RNA molecule with a base sequence complementary to the template strand of DNA is synthesized.

h. Protooncogenes are normal genes that code for molecules that promote cell proliferation.

i. Spliceosomes are large complexes composed of proteins and snRNA in which exons are spiced together during RNA processing.

j. Oncogenes are mutated protooncogenes that promote the formation of cancerous tumors.

3. The single band that occurred in centrifuged CsCl tubes halfway between where ^{15}N-DNA and ^{14}N-DNA would normally appear after one generation confirmed that each replicated DNA molecule had one old ^{15}N-DNA strand and one new ^{14}N-DNA strand.

5. Briefly, prokaryotic DNA replication consists of DNA unwinding, RNA primer formation, DNA synthesis catalyzed by DNA polymerase and the joining of Okazaki fragments by DNA ligase. Prokaryotic DNA replication differs from the eukaryotic process in that prokaryotic replication is faster, and in prokaryotes the Okazaki fragments are longer.

7. DNA synthesis of the lagging strands occurs in the $5' \longrightarrow 3'$ direction in a series of small pieces that are later joined together by DNA ligase.

9. DNA replication time is calculated as follows:

$$\frac{150,000,000 \text{ base pairs}}{50 \text{ bases/s}} = 3 \times 10^6 \text{ s} = 34.5 \text{ days}$$

Consequently, approximately one month would be required for DNA replication. Eukaryotic DNA synthesis is significantly faster than expected because each chromosome contains multiple replication units (replicons).

11. Viruses can cause mutations that affect the expression of protooncogenes by inserting their genomes into host cell regulatory sequences, thereby inactivating them.

13. Genetic recombination promotes species diversity. General recombination, a process in which segments of homologous DNA molecules are exchanged, is most commonly observed during meiosis. In site-specific recombination, protein-DNA interactions promote the recombination of nonhomologous DNA. Transposition is an example of site-specific recombination.

15. Intermicrobial DNA transfer mechanisms include transformation, transduction, and conjugation. In transformation, DNA fragments that enter the bacterial cell through cell wall openings and through a recombination event are inserted into the chromosome or a plasmid. In transduction, fragments of bacterial DNA are transferred to a recipient cell by a virus. Transduced DNA may insert into recipient cell DNA through recombination. In conjugation, a donor cell produces a sex pilus that allows DNA to transfer to a recipient cell.

17. In replicatve transposition, a replicated copy of a transposable element is inserted into a new chromosome location in a process that involves the formation of an intermediate called a cointegrate. In nonreplicative transposition, sequence replication does not occur, i.e., the transposable element is spliced out of its donor site and inserted into the target site. The donor site must be repaired.

19. In DNA, if cytosine is converted to uracil, which forms a base pair with adenine, an AT base pair is substituted for a GC base pair. Such a change in RNA is not as important because RNA molecules are short lived and disposable. In contrast, because DNA molecules are the cell's permanent repository of genetic information, any change in base sequence may affect an organism's viability.

21. Because DNA is constantly exposed to disruptive processes, its structural integrity is highly dependent on efficient repair mechanisms. The life span of an organism is dependent on the health of its constituent cells, which is in turn dependent on the timely and accurate expression of genetic information. Consequently, the capacity of the organisms in a species to maintain the integrity of DNA molecules is an important factor in determining life span.

23. Antibiotic resistance arises because the overuse of antibiotics acts as a selection pressure, i.e., they provide a growth advantage for disease-causing organisms that possess resistant genes. So-called superbugs are organisms that are resistant to several types of antibiotics because they possess plasmids containing several resistance genes. If the circumstances that cause antibiotic resistance continue probably antibiotics will eventually become ineffective against most infectious diseases.

Chapter 18

In-Chapter Questions

18.1 The amino acid sequence of the beginning of the polypeptide is Met-Ser-Pro-Thr-Ala-Asp-Glu-Gly-Arg-Arg-Trp-Leu-Ile-Met-Phe. The mutation types in the altered mRNA sequences are (a) insertion of one base, (b) deletion of one base, (c) insertion of two bases, (d) deletion of three bases. The consequences of these mutations are altered amino acid sequences of the polypeptides produced from mRNA. In (a), (b), and (c) a frame shift occurs. Therefore the amino acid sequences past the mutation are different. In (d) no frame shift occurs because three bases are deleted. In this case, the only difference between the normal polypeptide and the mutated version is the deletion of a single amino acid.

18.2 The formation of an ADP-ribosylated derivative of eEF-2 affects the three-dimensional structure of this protein factor. Presumably protein synthesis is arrested because the ability of eEF-2 to interact with or bind to one or more ribosomal components is altered.

18.3 After the synthesis of the plastocyanin precursor in cytoplasm, the first import signal mediates the transport of the protein into the chloroplast stroma. After this signal is removed by a protease, a second import signal mediates the transfer of the protein into the thylakoid lumen. Plastocyanin then binds a copper atom, folds into its final three-dimensional structure, and associates with the thylakoid membrane.

End-of-Chapter Questions

1. Despite considerable species differences in the amino acid and nucleotide sequences of ribosomal proteins and RNA, respectively, the overall three-dimensional structures of these molecules are remarkably similar. This similarity is presumably due to high selection pressure. In other words, ribosomal function is such an important factor in species viability that evolution has conserved their tertiary structure.

3. The genetic code is degenerate (several codons have the same meaning), specific (each codon specifies only one amino acid),

and universal (with a few exceptions each codon always specifies the same amino acid). In addition, the genetic code is nonoverlapping and without punctuation (i.e., mRNA is read as a continuous coding sequence).

5. Because the accuracy of protein synthesis depends directly on codon-anticodon interactions, the specificity with which t-RNAs are linked to amino acids is critically important. The process in which the amino acid-tRNA synthetases catalyze the covalent binding of each of the t-RNAs with its correct amino acid has, therefore, been referred to as the second genetic code.

7. When errors in amino acid-tRNA binding do occur, they are usually the result of similarities in amino acid structure. Several aminoacyl-tRNA synthetases possess a separate proofreading site that binds incorrect aminoacyl-tRNA products and hydrolyses them.

9. One possible codon sequence for the peptide sequence is GGUAGUUGUAGAGCU. The number of possible codons for the amino acids in this peptide sequence is as follows: glycine (4), serine (6), cysteine (2), arginine (6), and alanine (4). The total number of possible codon sequences for this peptide sequence is therefore 1152.

11. Five high-energy phosphate bonds are required to incorporate each amino acid into a polypeptide (i.e., 3 GTP and 2 ATP). The polymerization of 200 amino acids requires 600 GTP and 400 ATP.

13. The major differences between prokaryotic and eukaryotic translation are speed (the prokaryotic process is significantly faster), location (the eukaryotic process is not directly coupled to transcription as prokaryotic translation is), complexity (because of their complex life styles, eukaryotes possess complex mechanisms for regulatory protein synthesis, e.g., eukaryotic translation involves a significantly larger number of protein factors than prokaryotic translation), and posttranslational modifications (eukaryotic reactions appear to be considerably more complex and varied than those observed in prokaryotes.)

15. Each Shine-Dalgarno sequence in a prokaryotic mRNA occurs near a start codon (AUG). The Shine-Dalgarno sequence provides a mechanism for promoting the correct alignment of the start codon on the ribosome (as opposed to a methionine codon) because it binds to a nearby complementary sequence in the 16S rRNA component of the 30S ribosome. Eukaryotic ribosomes identify the initiating AUG codon by binding to the capped 5' end of the mRNA and scanning the molecules for a translation start site.

17. The major differences between prokaryotic and eukaryotic translation control mechanisms are related to the complexity of eukaryotic gene expression. Features that distinguish eukaryotic translation include mRNA export (spatial separation of transcription and translation), mRNA stability (the half-lives of mRNA can be modulated), negative translational control (the translation of certain mRNAs can be blocked by the binding of specific repressor proteins), initiation factor phosphorylation (mRNA translation rates are altered by certain circumstances when eIF-2 is phosphorylated), and translational frame-shifting (certain mRNAs can be frame-shifted so that a different polypeptide is synthesized).

19. a. Targeting is a series of mechanisms that directs newly synthesized polypeptides to their correct cellular locations.

b. Scanning is a mechanism eukaryotic ribosomes use to locate a translation start site on an mRNA.

c. A codon is an mRNA triplet base sequence that specifies the incorporation of a specific amino acid into a growing polypeptide chain during translation or acts as a start or stop signal.

d. A reading frame is a set of contiguous triplet codons.

e. Molecular chaperones are molecules that assist in the folding and targeting of proteins.

f. Disulfide exchange is a mechanism that facilitates the formation of disulfide bridges in newly synthesized proteins.

g. Certain aminoacyl-tRNA synthetases possess a second active site, the proofreading site, which binds a specific tRNA if it is covalently bonded to an incorrect amino acid. After this binding, the t-RNA-amino acid bond is hydrolyzed.

h. Signal peptides are short peptide sequences that determine a polypeptide's destination, e.g., direct its insertion into a membrane.

i. Glycosylation is a posttranslational mechanism whereby carbohydrate groups are covalently attached to polypeptides.

j. In negative translational regulation, the translation of a specific mRNA can be blocked if a specific protein binds to a sequence near its 5' end.

21. A signal recognition particle (SRP) is a large complex composed of protein and RNA that binds to a ribosome that has begun translating a polypeptide possessing a signal peptide component. Once the SRP has bound to the ribosome, translation is temporarily arrested. The SRP then mediates ribosomal binding to docking proteins on the surface of a membrane (e.g., RER membrane). Translation subsequently recommences and the growing polypeptide inserts into the membrane.

23. Several features of mRNA structure affect mRNA stability and, consequently, its translation rate. These include destabilizing sequences that facilitate the action of nucleases, the binding of proteins to certain sequences, and reversible adenylation of the 3' end of mRNA.

25. The synthesis of an inactive protein that is subsequently activated by posttranslational modification allows the cell a greater degree of regulation.

27. One of the most significant problems associated with predicting the three-dimensional structure of a polypeptide based solely on its primary structure is that the calculations based on the forces that drive the folding process (e.g., bond rotations, free energy considerations, and the behavior of the amino acids in aqueous environments) are extraordinarily complex.

Chapter 19

In-Chapter Questions

19.1 The product of the reaction is

$$CH_3(CH_2)_7CH{=}CH(CH_2)_7COOH + \cdot CCl_3 \longrightarrow$$

$$CH_3(CH_2)_6\dot{C}HCH{=}CH(CH_2)_7COOH + HCCl_3$$

The trichloromethyl radical is sufficiently stable that it can diffuse from the active site of cytochrome P_{450} and react with and destroy the structural integrity of other portions of the protein complex.

19.2   Tissue damage caused by DNA adduct formation by certain types of GSH conjugates occurs in liver because of this organ's primary role in detoxification of xenobiotics. The kidney is susceptible to damage by GSH conjugated xenobiotics because of its role in their excretion.

19.3   There are numerous end products of tetrahydrocannabinol (Δ^9-THC) biotransformation. These include hydroxyl derivatives (e.g., Δ^9, 11-dihydroxy THC), which may also be conjugated with glucuronate, sulfate, and other conjugating agents.

19.4   A comparison of plant and fungal gibberellin genes can be made using nucleic acid techniques such as PCR and DNA sequencing to determine if the plant gibberellin genes are related to the fungal genes. Because gibberellin is a required growth factor, a plant cannot use this substance as a trigger for detection of the fungus. To mount a successful defense a plant must first detect another uniquely fungal molecule. Then a mechanism for inhibiting fungal growth and synthesizing gibberellin would have to be developed.

End-of-Chapter Questions

1. a. Bioaccumulation is a process whereby toxic substances become increasingly concentrated in the organisms in a food chain, e.g., in the herbivores that consume plants that contain xenobiotics such as insecticides.
    b. Detoxication is a biochemical process whereby toxic molecules are converted to more soluble (and usually less toxic) products that are readily excreted.
    c. Detoxification is a detoxication process that corrects a state of toxicity, i.e., brings an inebriated person to sobriety.
    d. During phase I reactions, oxidoreductases and hydrolases convert hydrophobic substances into more polar molecules.
    e. During phase II reactions, metabolites containing appropriate functional groups (e.g., —OH) are conjugated with substances such as glucuronate, glutamate, sulfate, or glutathione.
    f. Secondary metabolites are molecules derived from primary metabolites (e.g., sugars, amino acids, and fatty acids) that allow organisms to withstand chemical or physical threats or promote reproductive success.
    g. Bioremediation is a series of techniques that use bacterial species to decontaminate soil and ground water that contains organic chemicals.
    h. Biotransformation is a series of enzyme-catalyzed processes in which living organisms convert toxic substances into less toxic metabolites.

3. When organisms are exposed to toxic conditions, an individual who possesses the capacity (resulting from any combination of gene duplications, random mutations, transposition, or inter-species gene transfers) to produce a protective substance has a survival advantage.

5. Examples of functions served by secondary metabolites include protecting against predators or osmotic stress and enhancing reproductive potential.

7. In plants, biotransformation processes promote storage of partially or completely detoxicated molecules in inaccessible compartments such as cell walls or vacuoles. In contrast, animal biotransformation processes usually enhance the solubility of xenobiotics so their derivatives can be excreted.

9. The principal enzymes of microsomal oxidative metabolism are the cytochrome P_{450} isozymes and the flavin-containing monooxygenases.

11. All of the reactions catalyzed by the cytochrome P_{450} isozymes are hydroxylation reactions since an OH group appears in each reaction. Examples include the conversion of aliphatic hydrocarbons to alcohols ($R—CH_3 \longrightarrow R—CH_2OH$) and secondary amines to hydroxylated derivatives (e.g., $R_1—NH—R_2 \longrightarrow R_1—N(OH)—R_2$).

13. Substances bearing functional groups containing nitrogen, sulfur, or phosphorus are oxidized by the flavin-containing monooxygenases.

15. Conjugation reactions convert lipophilic substances to polar, water-soluble derivatives.

17. Xenobiotics (foreign molecules taken into the body) may be foods, drugs, or poisons. Examples include proteins and starch (food molecules), phenobarbital and aspirin (drugs), and insecticides and benzo[a]pyrene (poisons).

19. Malathion is less toxic to humans than it is to insects because humans can more easily hydrolyze malathion's ester functional groups.

21. Usually the term *xenobiotic* describes a nonnutritive foreign molecule. In contrast, nutrients are foreign molecules such as plant starch that provide the body with energy and biosynthetic precursor molecules.

23. The chronic overconsumption of alcohol results in the induction of the same cytochrome P_{450} isozyme as that induced by phenobarbital. When an alcoholic is not drinking alcohol, phenobarbital is then metabolized quickly so higher doses are required to achieve sedation. When both drugs are consumed together, ethanol competitively inhibits phenobarbital metabolism, thereby potentiating the latter drug's depressant effect on the central nervous system.

25. The kidney is susceptible to the damaging effects of toxic metabolites because it provides the principal mechanism for their excretion. Depending on the individual's exposure to toxic xenobiotics, the exposure of the kidney and the urinary bladder to biotransformed derivatives may be exceptionally high.

Techniques in Biochemistry Supplement

Appendix B

These are exciting times for biochemists! During the past fifty years, there has been a continuously accelerating revolution in our understanding of the functioning of living organisms. Much of this knowledge has been possible because of technological innovations. For example, the development of the electron microscope as a biological instrument by Keith Porter and his colleagues in the 1940s led to the discovery of organelles such as mitochondria and lysosomes. Other examples include X-ray crystallography (protein and nucleic acid structure determinations) and radioisotopes (metabolic pathway investigations). In the 1990s biochemists are seeking increasingly more rapid methods for determining DNA base sequences. These and other biochemical techniques exploit the physical and chemical properties of biomolecules.

Scientific research, however, is not a collection of techniques. At the heart of science is the passion and curiosity of the scientist who seeks to understand the natural world. A scientist tests his or her perception of a natural process, sometimes referred to as a paradigm, by designing and performing experiments. Success in scientific investigations depends on three principle factors:

1. The design of experiments that ask clear and well-thought out questions about the living system under investigation.
2. The effective use of currently available technologies.
3. The capacity of the scientist to interpret experimental data, and (if necessary) modify or discard paradigms if they are not supported by this data.

These features of the scientific method are interactive. The availability of a new technology allows scientists to ask new questions. As the research progresses and the limits of the technology are reached, newer and more sophisticated questions (made possible by the older methods) drive the development of newer and more sophisticated technologies.

Because of the rich store of knowledge acquired through biochemical research, producing a textbook that provides a concise and accurate description of this diverse subject is a daunting task. Unfortunately, however close authors come to fulfilling these goals, any textbook will fail by one important measure—it cannot adequately convey the excitement, frustration and intellectual challenge of doing science. This is an important issue for several reasons. First, knowledge of scientifically acquired information is incomplete without an understanding of how data is actually acquired and the limits of data reliability. Second, although creativity is the hallmark of successful scientific careers, many students as well as the public often view scientists as stereotypical number-crunching nerds. Indeed, the scientists are as dependent on intuition and imagination as are the most renowned artists. Consider, for example, the contributions to our understanding of living organisms made by the following scientists: Louis Pasteur (optical isomers and the concept that fermentation is caused by living yeast cells), Linus Pauling (protein structure, enzyme catalysis, bonding theory; recipient of the 1954 Nobel Prize in Chemistry and the 1963 Nobel Peace Prize), Frederick Sanger (sequencing of proteins and nucleic acids; recipient of the Nobel Prize in Chemistry in 1956 and 1980), and Rita Levi-Montalcini (discov-

ery of nerve growth factor; recipient of the 1987 Nobel Prize in Physiology and Medicine). Both scientists and artists seek to discern truth. Scientists differ from artists in one respect: they must submit their conceptions of reality (objective reality is the ultimate measure of scientific work) to skeptical colleagues who must be convinced by verifiable experimental results.

The technologies described in this appendix have been chosen because of their seminal importance in modern biochemistry. Because of the intimate relationship between technology and biochemical knowledge, students will find that an understanding of these methods will improve their comprehension of the subject.

Suggested Readings

Bronowski, J., *Science and Human Values,* Harper and Row, New York, 1965.

Burke, J., *The Day the Universe Changed,* Little, Brown, Boston, 1985.

Fischer, E.P. and Lipson, C., *Thinking About Science: Max Delbruck and the Origins of Molecular Biology,* W.W. Norton, New York, 1988.

Hoagland, M., *Toward the Habit of Truth: A Life in the Sciences,* W.W. Norton, New York, 1990.

Keller, E.F. *A Feeling for the Organism: The Life and Work of Barbara McClintock,* W.H. Freeman, 1983.

Kuhn, T.S., *The Structure of Scientific Revolutions,* University of Chicago Press, Chicago, 1970.

Medawar, P.B., *Advice to a Young Scientist,* Harper and Row, New York, 1979.

Supplement 2.1

Cell Technology

Of the hundreds of techniques that have yielded useful information concerning cell structure and function, several have had an enormous impact on biochemistry. These include cell fractionation, electron microscopy, and autoradiography.

Cell Fractionation

Cell fractionation techniques allow the study of cell organelles in a relatively intact form outside of cells. For example, functioning mitochondria can be used to study cellular respiration. In these techniques, cells are gently disrupted and separated into several organelle-containing fractions. Cells may be disrupted by several methods, but homogenization is the most commonly used. In this process, a cell suspension is placed in a glass tube. A specially designed pestle is then used to break the cell open. The homogenate is separated into several fractions during a procedure called **differential centrifugation.** An instrument called the ultracentrifuge generates enormous centrifugal forces that separate cell components on the basis of size, surface area, and relative density. (Forces as large as 500,000 times the force of gravity, or 500,000 g, can be generated in unbreakable test tubes placed in the rotor of an ultracentrifuge.) Initially, the homogenate is spun in the ultracentrifuge at low speed (700–1000 g) for 10–20 minutes. The heavier particles, such as the nuclei, form a sediment, or pellet. Lighter particles, such as mitochondria and lysosomes, remain suspended in the supernatant, the liquid above the pellet. The supernatant is then transferred to another centrifuge tube and spun at a higher speed (10,000 g) for 20 minutes. The resulting pellet contains mitochondria, lysosomes, and peroxisomes. The supernatant, which contains **microsomes** (small closed vesicles formed from ER during homogenization), is transferred to another tube and spun at 105,000 g for 120 minutes. Microsomes are deposited in the pellet, and the supernatant contains ribosomes, various cellular membranes, and granules such as glycogen, a carbohydrate polymer.

Often, the organelle fractions obtained with this technique are not sufficiently pure for research purposes. One method that is often employed to further purify cell fractions is **density-gradient centrifugation.** In this procedure the fraction of interest is layered on top of a centrifuge tube containing a solution that consists of a dense substance such as sucrose. (In such a tube the concentration of the sucrose increases from the top to the bottom of the tube.) During centrifugation at high speed for several hours, particles move downward in the gradient until they reach a level that has a density equal to their own. Cell components are then collected by puncturing the plastic centrifuge tube and collecting drops from the bottom. The purity of the individual fractions can be assessed by visual inspection using the electron microscope. However, assays for **marker enzymes** (enzymes that are known to be present in especially high concentration in specific organelles) are more commonly used. For example, glucose-6-phosphatase, the enzyme responsible for converting glucose-6-phosphate to glucose in the liver, is a marker for liver microsomes. Likewise, DNA polymerase, which is involved in DNA synthesis, is a marker for nuclei.

Electron Microscopy

The electron microscope (EM) permits a view of cell ultrastructure not possible with the more common light microscope. Direct magnifications as high as 1,000,000X have been obtained with the EM. Electron micrographs may be enlarged photographically to 10,000,000X. The light microscope, in contrast, magnifies an image to about 1000X. This difference is due to the greater resolving power of the EM. The **limit of resolution,** defined as the minimum distance between two points that allows for their discrimination as two separate points, is $0.2\mu m$ using the light microscope. The limit of resolution for the EM is approximately 0.5 nm. The lower resolving power of the light microscope is related to the wavelength of visible light. In general, shorter wavelengths allow greater resolution. The EM uses a stream of electrons instead of light to illuminate specimens. Because this electron stream has a much shorter wavelength than visible light, more detailed images can be obtained.

There are two types of EM: the transmission electron microscope (TEM) and the scanning electron microscope (SEM). Like the light microscope, the TEM is used for viewing thin specimens. Since the image in the TEM depends on variations in the absorption of electrons by the specimen (rather than on variations in light absorption), heavy metals such as osmium or uranium are used to increase contrast among cell components. The SEM is used to obtain three-dimensional views of cellular structure. Unlike the TEM, which uses electrons that have passed through a specimen to form an image, the SEM uses electrons that are emitted from the specimen's surface. The specimen is coated with a thin layer of heavy metal and then scanned with a narrow stream of electrons. The electrons emitted from the specimen's surface, sometimes referred to as "secondary electrons," form an image on a television screen. Although only sur-

face features can be examined with the SEM, this form of microscopy provides very useful information about cell structure and function.

Autoradiography

Autoradiography is used to study the intracellular location and behavior of cellular components. It has been an invaluable tool in biochemistry. For example, it was used to determine the precise sites of DNA, RNA, and protein synthesis within eukaxyotic cells. In this procedure, living cells are briefly exposed to radioactively labeled precursor molecules. Tritium (^{3}H) is the most commonly used isotope. The tritiated nucleotide thymidine, for example, is used to study DNA synthesis, since thymidine is incorporated only into DNA molecules. After exposure to the radioactive precursor, the cells are processed for light or electron microscopy. The resulting slides are then dipped in photographic emulsion. After storage in the dark, the emulsion is developed by standard photographic techniques. The location of radioactively labeled molecules is indicated by the developed pattern of silver grains.

Supplement 5.1

Protein Technology

The technology used to determine protein structure has changed significantly since the 1950s. For example, the determination of insulin's structure required the efforts of many scientists over ten years. Currently, a well-funded research team can determine the primary structure of a newly discovered protein in less than a year. Much of this time is devoted to devising efficient methods for the protein's extraction and purification. Because of automated technology, the amino acid sequence determination may require only a few days work.

A substantial portion of an investigator's time is devoted to the extraction and purification of proteins because of several formidable problems. The most prominent of these are the following:

1. Cells contain thousands of different substances. With rare exceptions (e.g., hemoglobin in red blood cells) the protein of interest exists in extremely small amounts. Separating a specific protein from a cell extract in sufficient quantities for research purposes is often a challenge to the investigator's endurance and ingenuity.

2. Proteins are often unstable and may require special handling. For example, they may be especially sensitive to pH, temperature, or salt concentration, among other factors. Problems in handling a protein may become apparent only after considerable time and effort have been expended. For example, the investigation of nitrogenase, the enzyme that catalyzes the reduction of N_2 to form NH_3, was hindered for years until it was discovered that the enzyme's activity is destroyed by contact with O_2.

The techniques for the isolation, purification, and initial characterization of proteins, which are outlined below, exploit differences of charge, molecular weight, and binding affinities. Many of these techniques apply to the investigation of other biomolecules.

Isolating Techniques

The first step in any project is to develop an assay for the protein of interest. Because the protein is typically extracted from source material that contains hundreds of similar molecules, the assay must be specific. In addition, the assay must be convenient to perform, since it will be used frequently during the investigation. If the protein is an enzyme, the disappearance of the substrate (reactant) or the formation of product can be measured. (This is usually accomplished by using a spectrophotometer, a machine that measures differences in the absorption of a specific wavelength of light.) Nonenzymatic proteins are often detected by employing antibodies. (Antibodies are proteins produced by an animal's immune system in response to a foreign substance.) Often antibodies, which bind only to specific structures called **antigens,** are linked to radioactive or fluorescent "tags" to enhance their visability.

Extraction of a protein begins with cell disruption and homogenization. (See Supplement 2.1.) This process is often followed by differential centrifugation and, if the protein is a component of an organelle, by density gradient centrifugation.

Purification

After the protein-containing fraction has been obtained, several relatively crude methods may be used to enhance purification. **Salting out** is a technique in which high concentrations of salts such as ammonium sulfate $[(NH_4)_2SO_4]$ are used to precipitate proteins. Because each protein has a characteristic salting-out point, this technique removes many impurities. (Unwanted proteins that remain in solution are then discarded when the liquid is decanted.) When proteins are tightly bound to membrane, organic solvents or detergents often aid in their extraction. Dialysis is routinely used to remove low-molecular-weight impurities such as salts, solvents, and detergents.

As a protein sample becomes progressively more pure, more sophisticated methods are used to achieve further purification. The most commonly used techniques include chromatography and electrophoresis.

Chromatography

Originally devised to separate low-molecular-weight substances such as sugars and amino acids, chromatography has become an invaluable tool in protein purification. There is a wide variety of chromatographic techniques. They can be used to separate protein mixtures on the basis of molecular properties such as size, shape, and weight or certain binding affinities. Often, several techniques must be used sequentially to obtain a demonstrably pure protein.

In all chromatographic methods, the protein mixture is dissolved in a liquid known as the **mobile phase.** As the protein molecules pass across the **stationary phase** (a solid matrix), they separate from each other because of their different distributions between the two phases. The relative movement of each molecule results from its capacity to remain associated with the stationary phase while the mobile phase continues to flow.

Three chromatographic methods commonly used in protein purification are ion-exchange chromatography, gel-filtration chromatography, and affinity chromatography. **Ion-exchange chromatography** separates proteins on the basis of their charge. Anion-exchange resins, which consist of positively charged materials, bind reversibly with a protein's negatively charged groups. Similarly, cation-exchange resins bind positively charged groups. After proteins that do not bind to the resin are removed, the protein of interest is recovered by an appropriate change in the solvent pH and/or salt concentration. (A change in pH alters the protein's net charge.) In **gel-filtration chromatography,** a column packed with a gelatinous polymer separates molecules according to their size and shape. Molecules that are larger than the gel pores are excluded and therefore move through the column quickly. Molecules that are smaller than the gel pores diffuse in and out of the pores, so their movement through the column is retarded. The smaller their molecular weight,

the slower they move. Differences in these rates separate the protein mixture into bands, which are then collected separately.

Affinity chromatography uses the unique biological properties of proteins. That is, it uses a special noncovalent binding affinity between the protein and a special molecule (the ligand). The ligand is covalently bound to an insoluble matrix, which is placed in a column. After nonbinding protein molecules have passed through the column, the protein of interest is removed by altering the conditions that affect binding (i.e., pH or salt concentration).

Electrophoresis

Because proteins are electrically charged, they move in an electric field. In this process, called **electrophoresis,** molecules separate from each other because of differences in their net charge. For example, molecules with a positive net charge migrate toward the negatively charged electrode (cathode). Molecules with a net negative charge will move toward the positively charged electrode (anode). Molecules with no net charge will not move at all.

Electrophoresis, one of the most widely used techniques in biochemistry, is nearly always carried out by using gels such as polyacrylamide. The gel, functioning much as it does in gel-filtration chromatography, also acts to separate proteins on the basis of their molecular weight and shape. Consequently, gel electrophoresis is highly effective at separating complex mixtures of proteins or other molecules.

There are two basic types of **polyacrylamide gel electrophoresis** (PAGE): tubes and slabs. The smaller tube gels are used for small amounts of material or for a low number of samples. Slab gels are used for a large number of samples because several can be run simultaneously.

Bands resulting from a gel electrophoretic separation may be treated in several ways. During purification, specific bands may be excised from the gel after visualization with ultraviolet light. Each protein-containing slice is then eluted with buffer and prepared for the next step. Because of its high resolving power, gel electrophoresis is often used to assess the purity of protein samples. Staining gels with a dye such as Coomassie brilliant blue is a commonly used method for quickly assessing the success of a purification step.

A variation of gel electrophoresis called SDS gel electrophoresis, used primarily for characterizing proteins, is discussed below.

Initial Characterization of Proteins

Because protein function is determined by its physical structure, a vast array of analytical techniques have been devised. A few basic biophysical and biochemical techniques are described below.

Amino Acid Composition

The first step in characterizing any protein is the determination of the number of each type of amino acid residue present in the molecule. The process for obtaining this information, referred to as the amino acid composition, begins with the complete hydrolysis of all peptide bonds. Hydrolysis is typically accomplished with 6 N HCl for 10–100 hours. (Long reaction times are required because of difficulties in the hydrolysis of the aliphatic amino acids, Leu, Ile, and Val.) Hydrolysis is followed by analysis of the resulting amino acid mixture, referred to as the **hydrolysate.** Two automated methods are now commonly used to analyze protein hydrolysates: (1) ion-exchange chromatography and (2) high-pressure liquid chromatography (HPLC).

In the ion-exchange method, the hydrolysate is applied to a column that contains a cation exchange resin. Initially, an acidic buffer (pH 3) is allowed to flow through the column. Because most amino acids have net positive charges at pH 3, they displace the bound positive ions and bind with the negatively charged resin. Amino acids with net negative charges are the first to be eluted. As the pH and ionic strength of the buffer are increased, the other amino acids are released sequentially from the column. As the amino acids emerge from the column, they are analyzed quantitatively. Heating the amino acids with ninhydrin yields a purple product, known as Ruhemann's purple:

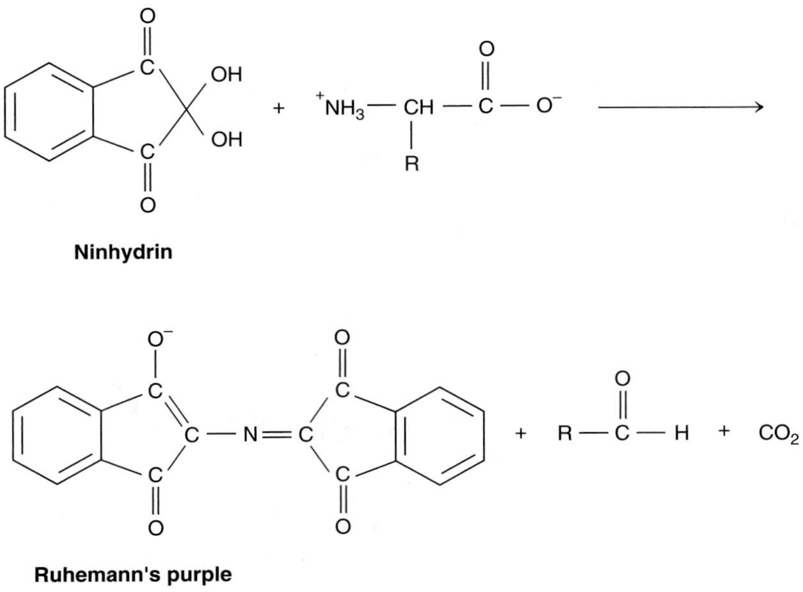

Ninhydrin

Ruhemann's purple

The reaction with the amino acid proline results in a yellow derivative. The amount of each amino acid is then determined by measuring the absorption of light by the ninhydrin derivatives.

In HPLC, after the hydrolysate is treated with compounds such as Edman's reagent (see below), the products are forced at high pressure through a stainless steel column packed with a stationary phase. Each amino acid derivative is identified according to its retention time on the column. Because the time required for amino acid analysis by HPLC (about 1 hour) is significantly shorter than that of other methods, it is becoming the method of choice.

Amino Acid Sequencing
Determining a protein's primary structure is similar to solving a complex puzzle. Several steps are involved in solving the amino acid sequence of any protein.

1. **Cleavage of all disulfide bonds.** Oxidation with performic acid is commonly used.

2. **Determination of the N-terminal and C-terminal amino acids.** Several methods are available to determine the N-terminal amino acid. In Sanger's method, the polypeptide chain is reacted with 1-fluoro-2,4-dinitrobenzene.

Sanger method

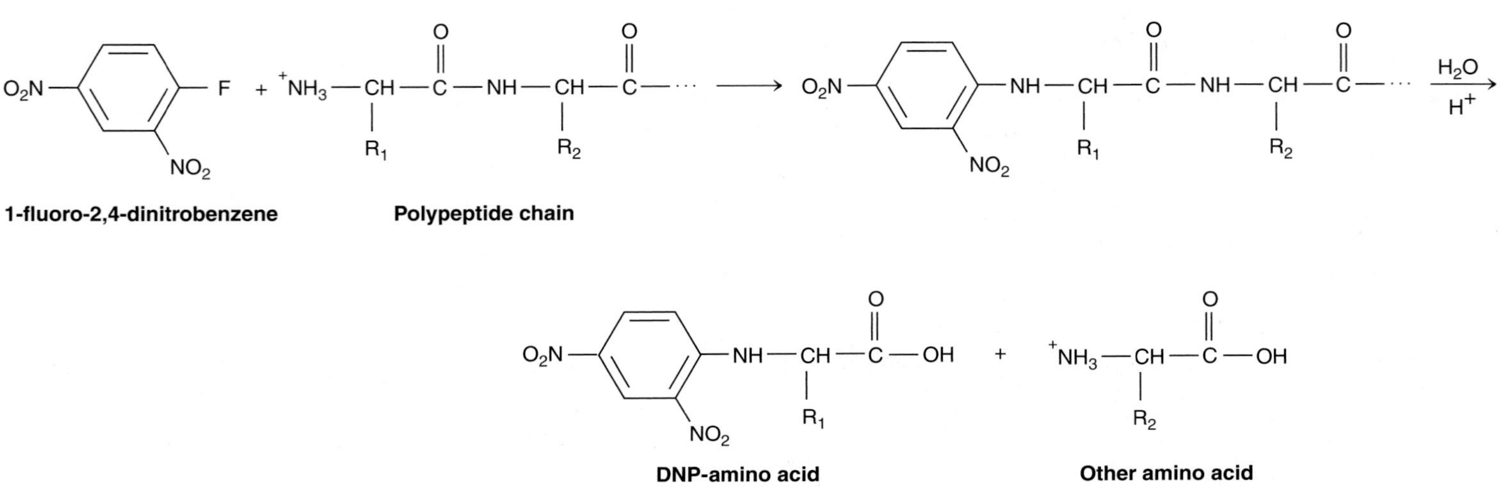

1-fluoro-2,4-dinitrobenzene   **Polypeptide chain**

DNP-amino acid   **Other amino acid**

Edman degradation

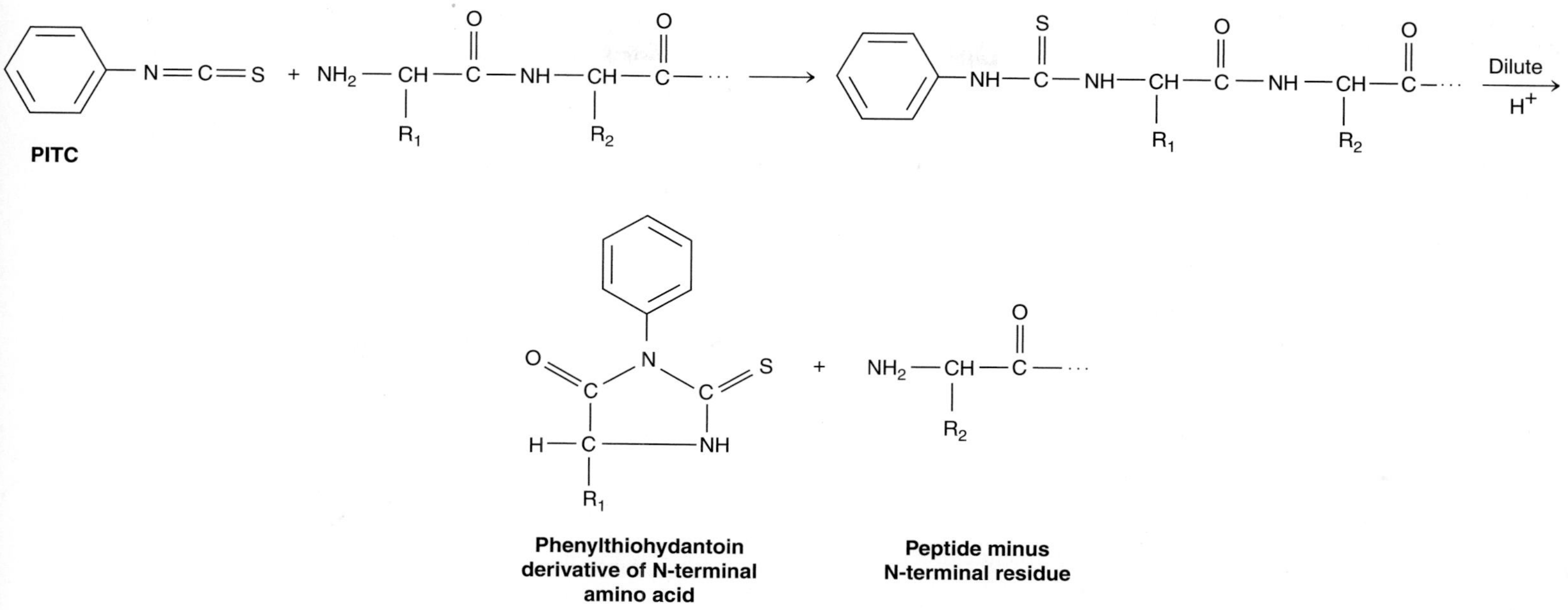

PITC

**Phenylthiohydantoin
derivative of N-terminal
amino acid**

**Peptide minus
N-terminal residue**

The dinitrophenyl (DNP) derivative of the N-terminal amino acid can be isolated and identified by ion-exchange chromatography after the polypeptide is hydrolyzed. A group of enzymes called the carboxypeptidases are used to identify the C-terminal residue. (Carboxypeptidases A and B, both secreted by the pancreas, hydrolyze peptides one residue at a time from the C-terminal end. Carboxypeptidase A preferentially cleaves peptide bonds when an aromatic amino acid is the C-terminal residue. Carboxypeptidase B prefers basic residues.) Because these enzymes sequentially cleave peptide bonds starting at the C-terminal residue, the first amino acid liberated is the C-terminal residue.

3. **Cleavage of the polypeptide into fragments.** Because of technical problems, long polypeptides cannot be directly sequenced. For this reason the polypeptide is broken into smaller peptides. The use of several reagents, each of which cuts the chain at different sites, creates overlapping sets of fragments. After the amino acid sequence of each fragment is determined, the investigator uses this information to work out the entire sequence of the polypeptide. There are several methods for fragmenting polypeptides. Of all the enzymes commonly used, the pancreatic enzyme trypsin is the most reliable. It cleaves peptide bonds on the carboxyside of either lysine or arginine residues. The peptide fragments, referred to as **tryptic peptides,** have lysine or arginine carboxyl terminal residues. Chymotrypsin, another pancreatic enzyme, is also often used. It breaks peptide bonds on the carboxyl side of phenylalanine, tyrosine, or tryptophan. Treating the polypeptide with the reagent cyanogen bromide also generates peptide fragments. Cyanogen bromide specifically cleaves peptide bonds on the carboxyl side of methionine residues.

4. **Determination of the sequences of the peptide fragments.** Each fragment is sequenced through repeated cycles of a procedure called the **Edman degradation.** In this method (see above), phenylisothiocyanate (PITC), often referred to as Edman's reagent, reacts with the N-terminal residue of each fragment.

Treatment of the product of this reaction with acid cleaves the N-terminal residue as a phenylthiohydantoin derivative. The derivative is then identified by comparing it with known standards, using electrophoresis or various chromatographic methods. (HPLC is most commonly used.) Because of the large number of steps involved in sequencing peptide fragments, Edman degradation is usually carried out by using a computer-programmed machine called a **sequenator.**

5. **Ordering the peptide fragments.** The amino acid sequence information derived from two or more sets of polypeptide fragments are next examined for overlapping segments. Such segments make it possible to piece together the overall sequence.

Several typical examples of primary sequence determination problems along with their solutions are given below.

Problem 1

Consider the following peptide:

Gly—Ile—Glu—Trp—Thr—Pro—Tyr—Gln—Phe—Arg—Lys

What amino acids and peptides are produced when the above peptide is treated with each of the following reagents?

a. Carboxypeptidase

b. Chymotrypsin

c. Trypsin

d. DNFB

Solution

a. Because carboxypeptidase cleaves at the carboxyl end of peptides, the products are

Gly—Ile—Glu—Trp—Thr—Pro—Tyr—Gln—Phe—Arg

and Lys

b. Because chymotrypsin cleaves peptide bonds in which aromatic amino acids (i.e., Phe, Tyr, and Trp) contribute a carboxyl group, the products are

Gly—Ile—Glu—Trp, Thr—Pro—Tyr,

Gln—Phe, and Arg—Lys

c. Trypsin cleaves at the carboxyl end of lysine and arginine. The products are

Gly—Ile—Glu—Try—Thr—Pro—Tyr—Gln—Phe—Arg

and Lys

d. DNFB tags the amino terminal amino acid. The product is

DNP—Gly—Ile—Glu—Trp—Thr—Pro—

Tyr—Gln—Phe—Arg—Lys

Hydrolysis then cleaves all the peptide bonds. DNP—Gly can then be identified by a chromatographic method.

Problem 2

From the following analytical results, deduce the structure of a peptide isolated from the Alantian orchid that contains 14 amino acids.

Complete hydrolysis produces the following amino acids: Gly (3), Leu (3), Glu (2), Pro, Met, Lys (2), Thr, Phe. Treatment with carboxypeptidase releases glycine. Treatment with DNFB releases DNP-glycine. Treatment with a nonspecific proteolytic enzyme produces the following fragments:

Gly—Leu—Glu, Gly—Pro—Met—Lys,

Lys—Glu, Thr—Phe—Leu—Leu—Gly,

Lys—Glu—Thr—Phe—Leu, Leu—Leu—Gly,

Glu—Thr—Phe, Glu—Gly—Pro, Pro—Met—Lys—Lys,

and Gly—Leu

Solution

The amino acid analysis provides information concerning the kind and number of amino acids in the peptide. The carboxypeptidase and DNFB results show that the carboxy and amino terminal amino acids are both glycine. Finally, by overlapping the fragments, the sequence of amino acids can be determined. Remember to start with a fragment that ends with the N-terminal residue, in this case, glycine.

Gly—Leu—Glu, Gly—Pro—Met—Lys, Lys—Glu,

Thr—Phe—Leu—Leu—Gly, Gly—Leu,

Glu—Gly—Pro, Pro—Met—Lys—Lys,

Lys—Glu—Thr—Phe—Leu, Leu—Leu—Gly

The overall structure then becomes

Gly—Leu—Glu—Gly—Pro—Met—Lys—Lys—

Glu—Thr—Phe—Leu—Leu—Gly

Review each piece of analytical data to ensure that it supports the final amino acid sequence.

Molecular Weight Determination

Several methods are available for determining the molecular weight of proteins. Gel-filtration column chromatography, SDS-PAGE, and ultracentrifugation are among the most commonly used.

Because gels act as the molecular sieve, there is a direct correlation between elution volume (V_e, the volume of solvent required to elute the protein from the column since it first contacted the gel) and molecular weight. When gel-filtration chromatography is used to determine molecular weight the gel column must be carefully calibrated. This is accomplished by careful measurement of several quantities. The total column volume V_t is equal to the sum of the volume occupied by the gel V_g and the void volume V_o, which is the volume occupied by the solvent molecules.

$$V_t = V_g + V_o$$

The molecular weight of the protein is determined by comparing its relative elution volume ($V_e - V_o/V_g$) to those of several standard molecules (Figure 1). V_e is the solvent volume required for the elution of a solute from the column.

A widely used variation of electrophoresis that determines molecular weight employs the powerful negatively charged detergent sodium dodecyl sulfate (SDS). In **SDS polyacrylamide gel electrophoresis** (SDS-PAGE) the detergent binds to the hydrophobic regions of protein molecules. As a result of binding SDS molecules, proteins denature and assume rodlike shapes. (This effect is also achieved by adding mercaptoethanol, which cleaves disulfide bridges.) Because most molecules bind SDS in a ratio roughly proportional to their molecular weights, during electrophoresis SDS-treated proteins migrate toward the anode (+ pole) only in relation to their molecular weight because of gel-filtration effects.

Estimates of molecular weight can also be obtained by using ultracentrifugation. Recall that ultracentrifugation separates components on the basis of size, surface area, and relative density. By using high centrifugal forces, the molecular masses of macromolecules such as proteins can be determined. By using an analytical ultracentrifuge, the rate at which such molecules sediment due to the influence of centrifugal force is optically measured.

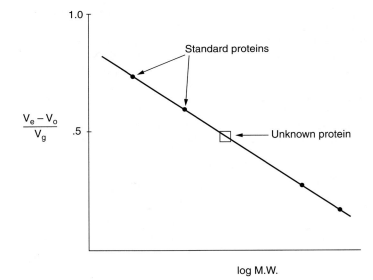

FIGURE 1

Molecular Weight Determination Using Cell Filtration Chromatography. The molecular weight of known standards is plotted against their elution values. This standard curve is then used to determine the molecular weight of the unknown protein.

X-Ray Crystallography

Much of the three-dimensional structural information about proteins was obtained by X-ray crystallography. Because the bond distances in proteins are approximately 15 nm, the electromagnetic radiation used to resolve protein structure must have a short wavelength. Visible light, whose wavelengths are 40,000–70,000 nm, clearly does not have sufficient resolving power for biomolecules. X-rays, however, have very short wavelengths (0.01–0.001 nm).

In X-ray crystallography, highly ordered crystalline specimens are exposed to an X-ray beam. As the X-rays hit the crystal, they are scattered by electrons in the crystal. The diffraction pattern that results is recorded on a photographic plate. The diffraction patterns are used to construct an electron density map. Because there is no objective lens to recombine the scattered X-rays, the image is reconstructed mathematically. Computer programs now perform these extremely complex and laborious computations.

Supplement 10.1

Membrane Methods

One of the most important principles revealed by membrane research is that the membranes of most living organisms have many structural similarities. These common features allow biochemists to apply (with caution) information gained from one membrane system to solving structural problems of other membranes. For example, the structural features of the red blood cell (rbc) membrane have proved to be valuable in studies of other membranes.

Membrane research requires reasonably pure specimens. Most membranes are obtained by a stepwise process that begins with cell homogenization, followed by several centrifugation steps. (Refer to Supplement 2.1 for details.) There are several reasons why rbc membrane is a popular choice in membrane research. First, because the rbc plasma membrane is the only membrane in the cell, it cannot be contaminated with intracellular membranes (a common problem in membrane research). Second, rbc membrane preparation is relatively simple. Rbc membrane (or "ghosts") are prepared by exposing red blood cells to a hypotonic solution. After washing, ghosts can be resealed to form "sealed ghosts," or they can be converted to numerous small vesicles by disruption and resealing. (Both inside-out and right-side-out vesicles can be produced.) Finally, rbc are easily obtained in large quantities from blood banks.

Membrane Composition

Once they are isolated, membranes are analyzed biochemically for lipid and protein composition. After they are extracted from membranes with organic solvents (e.g., chloroform), lipids are separated into classes by column chromatography. Phospholipids, the major lipid membrane component, are often resolved and identified by thin-layer chromatography. As mentioned, extracting and purifying membrane proteins require detergents. Commonly used detergents are triton X-100 and SDS (sodium dodecyl sulfate). Because membrane proteins require a hydrophobic environment to maintain their structure and biological activity, they are also investigated in the presence of detergent. (In the absence of detergent, integral membrane proteins often aggregate and precipitate.) For example, SDS-PAGE (Supplement 5.1) is often used to resolve membrane proteins.

Membrane Morphology

The arrangement of proteins within biological membranes can be directly observed by using freeze-fracture electron microscopy. In this technique, a rapidly frozen membrane is struck with a microtome knife. (A microtome is an instrument used for cutting thin sections of biological specimens for microscopic study.) The membrane often splits along the inner surfaces of the two lipid layers. In preparation for viewing in the electron microscope, the delicate membrane's inner surfaces are shadowed with a thin layer of heavy metal (usually platinum). Numerous intramembranous particles are commonly observed in freeze-fractured membranes.

Although X-ray crystallography (Supplement 5.1) has had limited use in determining membrane morphology, it usually provides low-resolution structural detail. High-resolution information requires highly ordered crystalline samples. Most membranes, however, contain an assortment of different proteins. In a few instances (i.e., biological membranes that possess only one type of protein), X-ray crystallography has provided high-resolution structural information. The "purple membrane" of the bacterium *Halobacterium halobium* is an excellent example. In the presence of O_2 this halophilic (salt-loving) organism depends on aerobic metabolism for energy production. If O_2 concentration is low and light intensity is high, the organism produces crystalline membrane patches called purple membrane. The pigment bacteriorhodopsin, the sole protein component of purple membrane, acts as a light-driven proton pump. (The pH gradient that results from the pumping activity of bacteriorhodopsin is used to synthesize ATP.) Relatively pure specimens of purple membrane are easily obtained by decreasing the salt concentration in the medium. The organism's ordinary membrane disintegrates, leaving the purple membrane intact.

Initial protein structure studies indicate that bacteriorhodopsin is a 248-residue polypeptide. Its amino-terminal residue is on the membrane's outside surface, and its carboxyl residue projects into the cytoplasm. Careful analysis of bacteriorhodopsin's primary sequence reveals seven peptide segments with amino acid sequences typical of α-helices. Using electron microscopy and X-ray crystallography, researchers determined that bacteriorhodopsin possesses seven α-helices, which are roughly perpendicular to the membrane.

Membrane Fluidity

Membrane fluidity is one of the most important assumptions of the fluid mosaic model of membrane structure. One measure of membrane fluidity, the ability of membrane components to diffuse laterally, can be demonstrated when cells from two different species are fused to form a **heterokaryon** (Figure 2). (Certain viruses or chemicals are used to promote cell-cell fusion.) The plasma membrane proteins of each cell type can be tracked because they are labeled with different fluorescent markers. Initially, the proteins are confined to their own side of the heterokaryon membrane. As time passes, the two fluorescent markers intermix, indicating that proteins move freely in the lipid bilayer.

Another technique, referred to as **fluorescence recovery after photobleaching (FRAP),** is also used to observe lateral diffusion. Cell plasma membranes are uniformly labeled with a fluorescent marker. Using a laser beam, the fluorescence in a small area is destroyed (or "bleached"). Using video equipment, the lateral movement of membrane components into and out of the bleached area can be tracked as a function of time.

Membrane fluidity can also be measured with membrane probes. (A *probe* is an atom or molecule whose presence in a membrane can be detected.) Probes provide information about their molecular environment. Nuclear magnetic resonance (NMR) spectroscopy has been valuable with probes that measure the movement of acyl chains

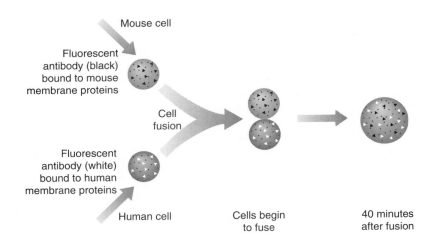

FIGURE 2

Fusion of Fluorescence-labeled Mouse and Human Cells. This experiment demonstrates that membranes are fluid and that proteins can move freely within the lipid bilayer. Metabolic inhibitors do not slow down protein movement, but lowering the temperature below 15°C does.

within membranes (as well as other aspects of membrane structure and function). NMR spectroscopy takes advantage of the different energy states of atomic nuclei. In the presence of electromagnetic radiation (in the microwave region) and a magnetic field, certain nuclei behave as spinning magnets. Atomic nuclei (e.g., ^1H, ^2H, ^{13}C, and ^{31}P) possess the property of spin. For example, ^1H has two nuclear spin states ($+\frac{1}{2}$ and $-\frac{1}{2}$), each of which has the same energy. In an applied magnetic field, the energy states are no longer equal. Transitions between the two states can be measured. (The energy difference between the two states is referred to as a *resonance frequency*.) In an ^1H-NMR spectrum the radio frequency at which a specific proton absorbs energy is determined by the intensity of the magnetic field. (An *NMR spectrum* is a graph of energy absorption versus resonance frequency.) This field is the sum of the applied magnetic field and the magnetic fields of nearby atoms. This last feature enables NMR to measure the environment of a particular proton.

NMR spectroscopy with ^2H or deuterium (D) has been especially useful because it has provided a detailed view of the interior of lipid bilayers. By chemically substituting D atoms for ^1H at specific positions along the hydrocarbon chains of phospholipid molecules, an investigator can measure the fluidity gradient within a membrane. D atoms in different environments experience different magnetic fields. If the membrane is relatively rigid, then the acyl chains are not free to rotate. Each D atom attached to a specific carbon atom is then locked into a different environment, and two separate signals are observed. If the membrane is totally fluid, the D atoms are completely free to move and can exchange environments. Then a single signal will be observed. The NMR signals move closer together as D is substituted for ^1H farther and farther away from phosphotidylcholine polar head groups. This observation has been interpreted in the following manner. Membranes possess a fluidity gradient. They are most rigid on their two surfaces and progressively more fluid toward their center.

Supplement 12.1

Photosynthetic Studies

Most of the technologies used in biochemical research have a variety of applications. This is certainly true of the following techniques used in photosynthetic research.

Spectroscopy

Spectroscopy measures the absorption of electromagnetic radiation by molecules. Instruments that measure this absorption, called spectrophotometers, can scan a wide range of frequencies. A graph of a sample's absorption of electromagnetic radiation is called an **absorption spectrum.**

In photosynthesis research, the relative absorbance of radiation by various plant components has been measured to determine their contribution to light harvesting. This work revealed that most light absorbance is accomplished by the chlorophylls and the carotenoids. The absorption spectra of several plant pigments are shown in Figure 3a. As expected, the chlorophylls absorb little light between 500 and 699 nm (green and yellow-green light). They do absorb strongly between 400 and 500 nm (violet and blue light) and between 600 and 700 nm (orange and red light).

If the effect of wavelength on the rate of photosynthesis is measured, an **action spectrum** is generated. Note in Figure 3b that the action spectrum of a typical leaf suggests that photosynthesis at specific wavelengths (e.g., 650 nm and 680 nm) uses light absorbed by chlorophylls b and a, respectively. Intact leaves absorb light more efficiently than pure pigments because in intact leaves nonabsorbed wavelengths are reflected from chloroplast to chloroplast. Every time an internal reflection occurs, a small percentage of the reflected wavelength is absorbed. Eventually, a significant percentage of the wavelengths that strike a leaf are absorbed.

In the 1950s, Robert Emerson used a more precise version of the action spectrum to investigate photosynthesis. When he measured the number of oxygen molecules produced per quantum of light absorbed over the visible spectrum, he observed that light with wavelengths longer than 690 nm are ineffective in promoting photosynthesis. However, if blue wavelengths are used in addition to the red ones, the photosynthetic rate (i.e., the rate of oxygen evolution) is significantly enhanced. This phenomenon, referred to as the **Emerson enhancement effect,** was later used to support the theory of two separate photosystems (PSI and PSII).

Another type of spectroscopy is known as **electron spin resonance spectroscopy (ESR).** In molecules that possess unpaired electrons, the energy of such electrons can be measured in a rapidly changing magnetic field. Because each electron generates its own magnetic field, it orients itself with or against an external field. (Elec-

(a)   (b)

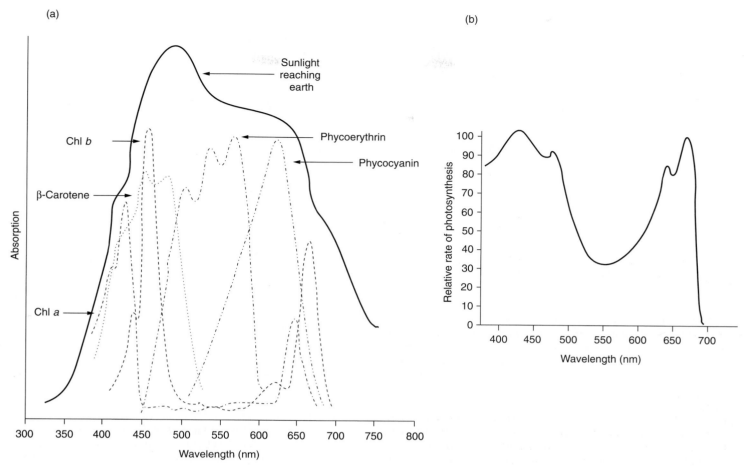

FIGURE 3

Light Absorbance Measurements in Photosynthesis Research. (a) Absorbance spectrum of visible light by photopigments. (b) Action spectrum.

trons are always affected by their molecular environments.) The ESR spectrum is a measure of the difference between these two energy levels. Although ESR is a valuable technique in many areas of biochemistry, it has been especially useful in photosynthesis research. For example, ESR played an important role in determining that the photon-absorbing component of photosynthetic reaction centers is a pair of chlorophyll molecules.

Photochemistry

Photochemistry is the study of chemical reactions that are initiated by light absorption. During photochemical reactions, chemical bonds may be cleaved when ions or radicals are formed. Excited molecules may also be isomerized or converted to oxidizing agents. Several techniques monitor photochemical events. These measure product formation or fluorescence or phosphorescence emission.

One of the more notable uses of photochemistry in photosynthetic research was a study that resulted in the discovery of the water-oxidizing clock. (Refer to Figure 12.11, p. 343.) Pierre Joliot and Bessel Kok studied PSII by measuring the evolution of O_2 when algae or chloroplasts were exposed to brief flashes of light after a period of darkness. (More recently, these experiments have been repeated using membranous vesicles into which PSII reaction centers were inserted.) In 1969, Joliot found that no O_2 is released on the first and second flashes. There is then a burst of O_2 evolution on the third flash. Subsequent O_2 evolution follows an oscillating pattern,

with a maximal amount being produced every fourth flash. In 1970, Kok suggested that the oxygen-evolving complex of PSII exists in five transient oxidation states, S_0 through S_4. After the first flash, P680 is converted to P680*. The clock, which provides the electrons that reconvert P680* to P680, releases O_2 when the S_4 state has been reached. It is now believed that the first burst of O_2 comes in the third flash because during a period of darkness the reaction center relaxes into S_1 rather than into S_0. Subsequently, of course, O_2 evolution peaks at every fourth flash. The dampening of the oscillations results from random inefficiencies in the absorption of light by the large number of photosystem complexes being measured.

X-Ray Crystallography

Although X-ray crystallography has been a valuable tool in determining molecular structure (Supplement 5.1), because investigators cannot crystallize hydrophobic biomolecules it has had limited use. Because many important photosynthetic components are found within membrane, this technique has not been useful in photosynthetic research. However, recently small amphipathic organic molecules have been used during the extraction and purification of membrane proteins, a process referred to as cocrystallization. Using this technique the structures of the reaction center in the rhodopseudomonads (a group of purple nonsulfur bacteria) have been determined. The structural information from X-ray crystallography

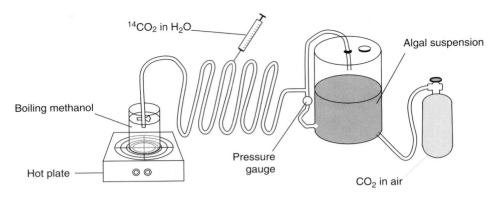

FIGURE 4

Calvin Apparatus for Investigations of CO_2 Fixation.

combined with the knowledge gained from spectroscopy has provided a coherent view of photosynthetic electron transport.

Radioactive Tracers

Because numerous reaction pathways occur simultaneously within living organisms, tracing specific biochemical pathways can be frustrating. However, if biomolecules can be "tagged" (labeled) with a tracer (a substance whose presence can be monitored), reaction pathways become easier to investigate. Radioactive isotopes have been very valuable in tracing the metabolic fate of labeled molecules.

Because the nucleus of a radioisotope is unstable, it decays to form a different nucleus. This process can be monitored by instruments that measure radiation emissions, such as Geiger counters and scintillation counters, or by autoradiography (Supplement 2.1).

One of the earliest radioactive tracers was ^{14}C, used by Melvin Calvin and his associates in the 1950s as they investigated carbon fixation in algae. To determine the pathway by which CO_2 is incorporated into carbohydrate, the Calvin team devised an ingenious apparatus (Figure 4). The labeling of reaction intermediates is limited to the first few stages of the carbon fixation pathway. Unlabeled CO_2 is bubbled into a transparent reservoir that contains a suspension of the alga *Chlorella*. After the reservoir is illuminated and photosynthesis is well underway, a stopcock is opened, and the algae are allowed to flow through a narrow glass tube into a beaker of boiling methanol. (Once algae enter boiling methanol, they are killed, and their metabolism is arrested.) Because $^{14}CO_2$ can be introduced at specific points along the tube, the exposure of the algae to ^{14}C can be precisely timed. Because photosynthesis continues as the algae flow in the tube, the organism's processing of the labeled carbon continues until the cells are killed in the methanol. The Calvin team analyzed the alcohol extract with paper chromatography and autoradiography. They determined the pathway by which carbon is assimilated in the algae by varying the exposure time. For example, the team found that, after a 5-second exposure to $^{14}CO_2$, most ^{14}C appears in glycerate-3-phosphate. After a 30-second exposure, most ^{14}C is found in hexose-phosphate.

Supplement 15.1

Hormone Methods

The detection and measurement of hormones is useful in both clinical and research laboratories. Before the discovery of modern methods, hormones could be detected only indirectly. For example, in the

old "rabbit test" for pregnancy, the patient's urine was injected into a female rabbit. The formation of corpora lutea within 24 hours in the animal's ovaries indicated that the patient was pregnant. (The corpus luteum is a structure formed from an ovarian follicle after ovulation.) The hormone that induces this transformation is now known to be human chorionic gonadotropin (HCG). (HCG has biological activity similar to that of luteinizing hormone (LH). It is produced only during pregnancy, by the placenta.) In addition to being awkward and time consuming, bioassays are usually insensitive and imprecise (e.g., the rabbit test is useful only after several weeks of pregnancy and is either positive or negative). In newer analytical techniques, vanishingly small amounts of specific hormones can be precisely measured. Two of the most commonly used methods are the radioimmunoassay (RIA) and the enzyme-linked immunosorbent assay (ELISA). Both techniques can detect any substance for which an antibody can be obtained.

Radioimmunoassay

In radioimmunoassays the concentration of a specific antigen is determined by measuring the competition of the unlabeled antigen with a known amount of the same antigen which is radiolabeled for binding to an antibody. (There must be too little antibody to bind all antigen molecules.) The determination of the amount of an antigen present in a biological sample (e.g., blood or urine) is based on the following principle. The unlabeled antigen diminishes the binding of labeled antigen by an amount that is proportional to the concentration of unlabeled antigen present in the sample. The unlabeled antigen concentration can be determined by using a standard curve. A *standard curve* is constructed by varying the known amounts of unlabeled antigen in the presence of a constant amount of antibody and radiolabeled antigen. The percentage of labeled antigen bound to antibody is plotted as a function of the known antigen concentrations (Figure 5). The antigen concentration in a research or clinical sample can then be determined from the graph.

ELISA

Enzyme-linked immunosorbent assays are similar to RIA in that they use antibodies and are about as sensitive. However, ELISA is safer and less expensive than RIA. For example, inexpensive pregnancy tests are now available that can measure HCG in urine accurately as early as 2 days after conception.

In general, ELISA involves the following steps:

1. An antibody specific for an antigen is attached to an inert surface, e.g., the bottom of a well in a polystyrene lab dish.

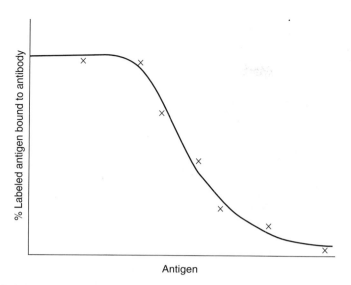

FIGURE 5

RIA Standard Curve. The linear portion of the curve is used to determine the antigen concentration in a sample.

2. A small sample of the biological specimen (e.g., blood or urine) is added to the dish. If the sample contains the appropriate antigen, then antigen-antibody binding occurs.

3. A second antibody is added that also specifically binds to the antigen. This antibody is covalently linked to an assayable enzyme.

4. The dish is rinsed to remove any unbound antibody or enzyme molecules.

5. Enzyme assays are used to determine the amount of antigen present. The enzyme converts a colored reagent to a colorless one, or vice versa. The change in color is proportional to the antigen concentration in the sample.

In an alternate method, a specific antigen is attached to the lab dish. Any antibodies capable of binding the antigen present in the biological specimen bind to the immobilized antigen. After unbound antibody is rinsed off, another antibody linked to an assayable enzyme that is specific for binding to the first antibody is added to the dish.

Supplement 16.1

Nucleic Acid Methods I

As noted regarding protein, the techniques used in the isolation, purification, and characterization of biomolecules take advantage of their physical and chemical properties. This is no less true for the nucleic acids. Most of the techniques used in nucleic acid research are based on differences in molecular weight or shape, base sequences or complementary base pairing. Techniques such as chromatography, electrophoresis and ultracentrifugation, which have been used successfully in protein research, have also been adapted to use with nucleic acids. In addition, other techniques have been developed that exploit the unique properties of nucleic acids. For example, under certain conditions DNA strands reversibly separate. One of several techniques that exploit this phenomenon, called **Southern blotting,** is often used to locate specific (and often rare) nucleic acid sequences. After brief descriptions of several techniques used to pu-

rify and characterize nucleic acids, two common methods for determining DNA sequences are outlined. More complex techniques are described in Supplement 17.1.

Once bacterial cells have been ruptured or eukaryotic nuclei have been isolated, their nucleic acids are extracted and deproteinized. This may be accomplished by several methods. Bacterial nucleic acid is often precipitated by treating cell preparations with alkali and lysozyme (an enzyme that degrades bacterial cell walls by breaking glycosidic bonds). Partially degraded protein is extracted by using certain solvent combinations (e.g., phenol and chloroform). Similarly, eukaryotic nuclei can be treated with detergents or solvents to release their nucleic acids. Depending on which type of nucleic acid is being isolated, specific enzymes are used to remove the other type. For example, RNase removes RNA from nucleic acid preparations leaving DNA intact. DNA is further purified by centrifugation. All nucleic acid samples must be handled carefully. First, nucleic acids are susceptible to the actions of a group of enzymes called nucleases. In addition to a variety of such enzymes that are released during cell extraction, nucleases can also be introduced from the environment, e.g., the experimenter's hands. Secondly, high molecular weight nucleic acids, primarily DNA, are sensitive to shearing stress. Purification procedures, therefore, must be gentle, applying as little mechanical stress as possible.

Techniques Adapted from Use with Other Biomolecules

Many of the techniques used in protein purification procedures have also been adapted for use with nucleic acids. For example, several types of chromatography (e.g., ion-exchange, gel filtration and affinity) have been used in several stages of nucleic acid purification as well as in the isolation of individual nucleic acid sequences. Because of its speed, HPLC has replaced many slower chromatographic separation techniques when small samples are involved.

A type of column chromatography, which uses a calcium phosphate gel called **hydroxyapatite,** has been especially useful in nucleic acid research. Since hydroxyapatite binds to double-stranded nucleic acid molecules more tenaciously than to single-stranded molecules, double-stranded DNA (dsDNA) can be effectively separated from single-stranded DNA (ssDNA), RNA, and protein contaminants by eluting the column with increasing concentrations of phosphate buffer. (The significance of separating dsDNA from ssDNA is discussed below.) The use of hydroxyapatite columns has recently been largely replaced by a form of affinity chromatography in which the column matrix molecules have been covalently bonded to avidin, a small protein that binds specifically to biotin. When a ssDNA binds to a biotinylated ssDNA, the resulting dsDNA binds to the column, while the rest of the sample passes through.

The movement of nucleic acid molecules in an electric field depends on both their molecular weight and their three-dimensional structure. However, because DNA molecules often have relatively high molecular weights, their capacity to penetrate some gel preparations (e.g., polyacrylamide) is limited. Although DNA sequences with less than 500 bp can be separated by polyacrylamide gels with especially large pore sizes, more porous gels must be used with larger DNA molecules. Agarose gels, which are composed of a crosslinked polysaccharide, are used to separate DNA molecules with lengths between 500 bp and approximately 150 kilobases (kb). Larger sequences are now isolated with a variation of agarose gel electrophoresis in which two electric fields (perpendicular to each other) are alternately turned on and off. DNA molecules reorient themselves each time the electric field alternates, resulting in a very efficient and precise separation of heterogenous groups of DNA molecules.

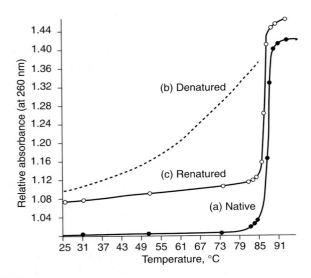

FIGURE 6

DNA Denaturation. (a) When native DNA is heated, its absorbance does not change until a specific temperature is reached. The "melting temperature" T_m of a DNA molecule varies with its base composition. (b) When the denatured DNA is cold, its absorbance falls, but along a different curve. It does not return to its original absorbance value. (c) Reannealed (renatured) DNA can be prepared by maintaining the temperature at 25° below the denaturing temperature for an extended period.

Density gradient centrifugation (Supplement 2.1) with cesium chloride (CsCl) has been widely used in nucleic acid research. At high speeds, a linear gradient of CsCl is established. Mixtures of DNA, RNA, and protein migrating through this gradient separate into discrete bands at positions where their densities are equal to the density of the CsCl. DNA molecules with high guanine and cytosine content are more dense than those with a higher proportion of adenine and thymine. This difference helps separate heterogeneous mixtures of DNA fragments.

Techniques that Exploit the Unique Structural Features of the Nucleic Acids

Several unique properties of the nucleic acids (e.g., absorption of UV light at specific wavelengths and their tendency to reversibly form double-stranded complexes) are exploited in nucleic acid research. Several applications of these properties are briefly discussed.

Because of their aromatic structures, the purine and pyrimidine bases absorb UV light. At pH 7 this absorption is especially strong at 260 nm. However, when the nitrogenous bases are incorporated into polynucleotide sequences, various noncovalent forces promote close interactions between them. This decreases their absorption of UV light. This **hypochromic effect** is an invaluable aid in studies involving nucleic acid. For example, absorption changes are routinely used to detect the disruption of the double-stranded structure of DNA or the hydrolytic cleavage of polynucleotide strands by enzymes.

The binding forces that hold the complementary strands of DNA together can be disrupted. This process, referred to as **denaturation,** is promoted by heat, low salt concentrations and extremes in pH. (Because it is easily controlled, heating is the most common denaturing method in nucleic acid investigations.) When a DNA solution is slowly heated, absorption at 260 nm remains constant until a threshold temperature is reached. Then the sample's absorbance increases (Figure 6). The absorbance change is caused by the un-

stacking of bases and the disruption of base pairing. The temperature at which one-half of a DNA sample is denatured, referred to as the melting temperature (T_m), varies among DNA molecules according to their base compositions. (Recall that there are three hydrogen bonds between G and C, while A and T only have two such bonds. More energy is required, therefore, to "melt" DNA molecules with high G and C content.) If the separated DNA strands are held at a temperature approximately 25°C below the T_m for an extended time, renaturation is possible. Renaturation, or reannealing, requires some time because the strands explore various configurations until they achieve the most stable one (i.e., paired complementary regions).

DNA melting is extraordinarily useful in **nucleic acid hybridization.** Single-stranded DNA from different sources associates (or "hybridizes") if there is a significant sequence homology (i.e., structural similarity). If a DNA sample is sheared into small uniform pieces, the rate of reannealing depends on the concentration of DNA strands and on the structural similarities between them. Reannealing rates have revealed valuable information about genome structure. For example, organisms vary in the number and types of unique sequences their genomes contain. (A unique DNA sequence occurs only once per haploid genome.) The relative number of unique and repeated sequences can be determined by constructing a $C_o t$ curve. ($C_o t$ is a measure of renaturation where C_o is the concentration of ssDNA in moles/liter and t is elapsed time in seconds.) $C_o t$ curves have demonstrated that the velocity of reannealing declines as genomes become larger and more complex. The frequency of unique and repeated DNA sequences determined by measuring reannealing rates for the mouse genome is shown in Figure 7.

Hybridization can also be used to locate and/or identify specific genes or other DNA sequences. For example, ssDNA from two different sources (e.g., tumor cells and normal cells) can be screened for sequence differences. If one set of ssDNA is biotinylated, then the double-stranded hybrids bind to an avidin column. If any unhybridized sequence is present, it passes through the column. Then it can be isolated and identified. In **Southern blotting** (Figure 8) radioactively labeled DNA or RNA probes (sequences with known identities) locate a complementary sequence in the midst of a DNA digest, which typically contains a large number of heterogeneous DNA fragments. (A DNA digest is obtained by treating a DNA sam-

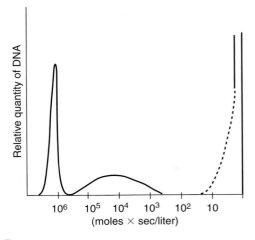

FIGURE 7

DNA Sequence Pattern of the Mouse Genome. The degree of repetitiveness in the segments of total mouse DNA is determined by measuring $C_o t$ values for several fractions of the genome. The dotted line indicates estimated values.

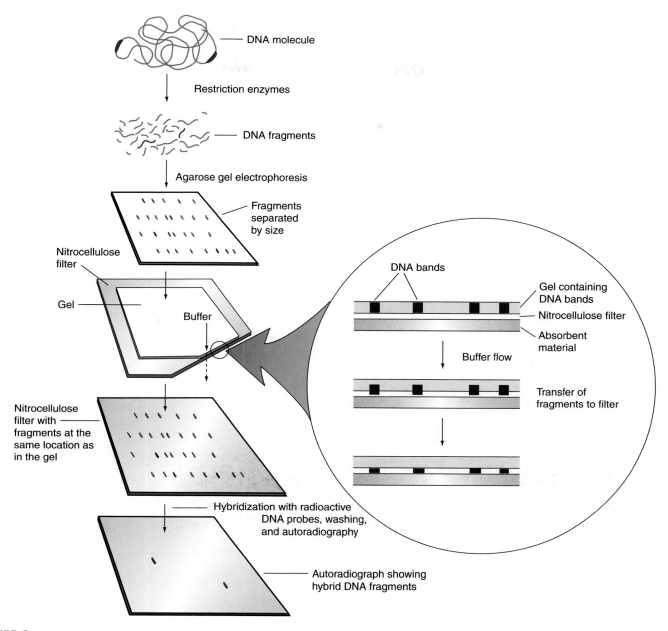

FIGURE 8

Southern Blotting. (1) DNA analysis begins with its digestion by a restriction enzyme. (2) DNA fragments are then separated by agarose gel electrophoresis. (3) The DNA fragments are transferred to nitrocellulose filter paper. (4) The ssDNA on the nitrocellulose filter paper is hybridized with radioactively labeled ssDNA probe. (5) Any hybridized DNA appears in an autoradiograph.

ple with restriction enzymes that cut at specific nucleotide sequences. Produced by bacterial cells, restriction enzymes protect bacteria against viral infection by cleaving viral DNA at specific sequences.) Once the DNA sample has been digested, the fragments are separated by agarose gel electrophoresis according to their sizes. After the gel is soaked in 0.5 M NaOH, a process that converts dsDNA to ssDNA, the DNA fragments are transferred to nitrocellulose filter paper by placing them on a wet sponge in a tray with a high salt buffer. (Nitrocellulose has the unique property of binding strongly to ssDNA.) Absorbent dry filter paper is placed in direct contact with nitrocellulose filter. As buffer is drawn through the gel and filter paper by capillary action, the DNA is transferred and becomes permanently bound

to the nitrocellulose filter. (The transfer of DNA to the filter is the "blotting," referred to in the name of this technique.) Subsequently, the nitrocellulose filter is exposed to the radioactively labeled probe, which binds to any ssDNA with a complementary sequence. For example, an mRNA which codes for β-globin binds specifically to the β-globin gene, even though β-globin mRNA lacks the introns present in the gene. Apparently, there is sufficient base pairing between the two single-stranded molecules that the gene can be located.

DNA Sequencing

The determination of DNA nucleotide sequences has provided valuable insights in such fields as biochemistry, medical science and evolutionary biology. The analysis of long DNA sequences begins with the formation of smaller fragments using one type of restriction enzyme. Each fragment is then sequenced independently by either the chemical cleavage method or the chain terminating method. As with protein primary structure determinations, these steps are repeated with a different set of polynucleotide fragments (generated by another type of restriction enzyme) which overlap the first set. Sequence information from both sets of experiments then orders the fragments into a complete sequence.

In the **chemical cleavage method** (Figure 9) developed by Allan Maxam and Walter Gilbert, the 5′ end of each strand of a DNA fragment is labeled with ^{32}P. This fragment is then cut into two pieces with another restriction enzyme. After they are separated by electrophoresis, each fragment is separately sequenced. (Each of these fragments contains only one radiolabeled atom.) The base sequence of each fragment is determined by subjecting separate aliquots to different chemical treatments, each of which causes strand cleavage. The products of these reactions are resolved according to their size by gel electrophoresis. The base sequence of the fragment is determined by observing the bands on an autoradiogram of the gel.

DNA sequencing by the **chain-terminating method** (Figure 10), developed by Frederick Sanger, also uses restriction enzymes to cleave large DNA segments into smaller fragments. Each fragment is separated into two strands, one of which is used as a template to produce a complementary copy. The sample is further divided into four test tubes. To each tube is added the substances required for DNA synthesis (e.g., the enzyme DNA polymerase and the four deoxyribonucleotide triphosphates). In addition, a ^{32}P-labeled primer (a short segment of a complementary DNA strand) is added to each tube. (By selecting the primer, the investigator can start DNA sequencing at specific sites.) Also present in each of the four tubes is a different 2′–3′ dideoxynucleotide derivative. (The dideoxy derivatives are synthetic nucleotide analogs in which the hydroxy groups on the 2′- and 3′-carbons have been replaced with hydrogens.) Dideoxynucleotides can be incorporated into a growing polynucleotide chain, but they cannot form a phosphodiester linkage with another nucleotide. Consequently when dideoxynucleotides are incorporated, they terminate the chain. Because small amounts of the dideoxynucleotides are used, they are randomly incorporated into growing polynucleotide strands. Each tube, therefore, contains a mixture of DNA fragments containing strands of different lengths. Each newly synthesized strand ends in a dideoxynucleotide residue. The reaction products in each tube are separated by gel electrophoresis and analyzed together by autoradiography. Each band in the autoradiogram corresponds to a polynucleotide that differs in length by one nucleotide from the one that precedes it. Note that the smallest polynucleotide appears on the bottom of the gel, because it moves more quickly than larger molecules.

Recently, an automated version of the Sanger method has become available. Instead of using radiolabeled nucleotides, it uses fluorescent tagged dideoxynucleotides. Because each dideoxy analog fluoresces a different color, the entire procedure is carried out in a single test tube. Afterwards the reaction products are loaded and run on a single electrophoresis gel. Then a computer analyzes the data.

Supplement 17.1

Nucleic Acids Methods II: Recombinant DNA Technology

Recombinant DNA technology is series of techniques that allow DNA molecules obtained from various sources to be cut and spliced

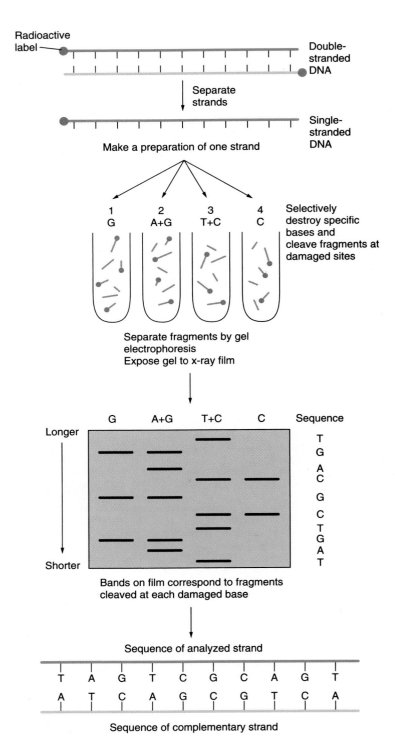

FIGURE 9

The Maxam-Gilbert Method. A restriction enzyme-generated dsDNA fragment is separated into single strands by gel electrophoresis under denaturing conditions (e.g., high temperature in the presence of a hydrogen bond disruptor such as formamide). Because of their different bases, the two ssDNA strands can be separated on the gel. Four aliquots of the ssDNA of interest are cleaved by chemical reactions that destroy one or two types of base. Each chemical treatment generates a series of labeled fragments that differ in length according to how far the destroyed base is from the ^{32}P-labeled end. The labeled fragments are then run side by side in acrylamide gels. After an autoradiogram is obtained, the band pattern is used to determine the base sequence of the DNA.

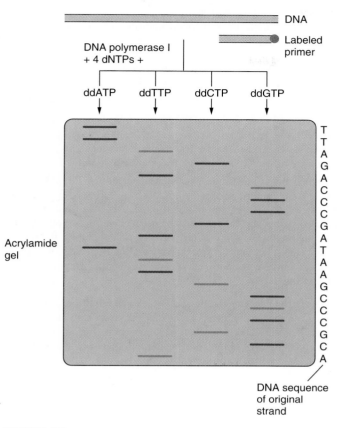

DNA

Labeled primer

DNA polymerase I + 4 dNTPs +

ddATP   ddTTP   ddCTP   ddGTP

Acrylamide gel

T
T
A
G
A
C
C
C
G
A
T
A
A
G
C
C
C
G
C
A

DNA sequence of original strand

FIGURE 10

The Sanger Chain Termination Method. A specific primer is chosen so that DNA synthesis begins at the point of interest. DNA synthesis continues until a radioactive dideoxynucleotide is incorporated and the chain terminates. Afterwards the products of the reactions are separated by gell electrophoresis and analyzed by autoradiography. The fragments migrate according to size. The sequence is determined by "reading" the gel.

together. These techniques have revolutionized biology and biochemistry because they have made investigations of genomes more accessible than ever before. For example, the large number of DNA copies required in DNA sequencing methods have been obtainable through molecular cloning and (more recently) the polymerase chain reaction. Commercial applications of recombinant DNA techniques have revolutionized medical practice. For example, human gene products such as insulin and growth hormone as well as certain vaccines and diagnostic tests are now produced in large quantities by bacterial cells into which recombinant genes have been inserted. Currently, several research groups are investigating the use of recombinant techniques in human gene therapy, a process in which (it is hoped) defective genes can be replaced by their normal counterparts.

Recombinant DNA techniques all depend on restriction enzymes discovered in bacteria in the 1970s. As noted, bacterial restriction enzymes cleave foreign DNA at specific sites. For example, the *E. coli* enzyme EcoRI cleaves DNA only at the sequence

5′ . . . . . GAATTC . . . . . 3′
3′ . . . . . CTTAAG . . . . . 5′

The bacterium protects its own DNA by methylating one of the adenines in the sequence. When a virus infects a cell that possesses

an active restriction enzyme and a corresponding methylase, its DNA is cleaved. Then bacterial DNase rapidly degrades the viral genome. The feature that makes restriction enzymes so useful to molecular biologists is the unique and precise cleavage specificity of each of many enzymes that have been isolated from bacteria. Some restriction enzymes make straight (or "blunt") cuts through both strands of the double helix. Others, however, make staggered cuts that leave so-called "sticky ends."

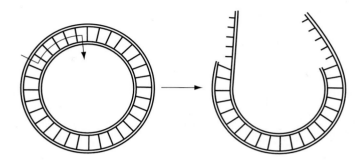

Sticky ends are useful because they allow a DNA fragment to join to others cleaved with the same enzyme.

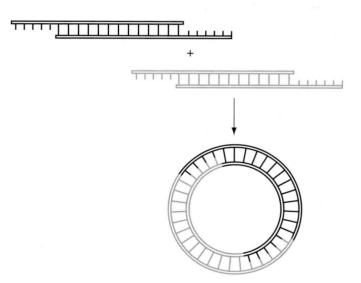

Molecular Cloning

In molecular cloning, a piece of DNA isolated from a donor cell (e.g., any animal or plant cell) is spliced into a **vector,** by which the gene of interest can be introduced into a host cell. The choice of vector depends on the size of donor DNA. For example, bacterial plasmids are often used to clone small pieces (15 kb) of DNA. Somewhat larger pieces (24 kb) are incorporated into bacteriophage λ vectors, while cosmid vectors are used for DNA fragments as large as 50 kb. Bacteriophage λ can be used as a vector because a substantial portion of its genome does not code for phage production and can therefore be removed. The removed viral DNA can then be replaced by foreign DNA. **Cosmids** are bacteriophage λ cloning vectors from which all viral DNA has been removed except **cos** sites, which are required for packaging DNA into phage heads. Consequently, after their introduction into a host cell, cosmids replicate like plasmids.

As noted, forming recombinant DNA requires a restriction enzyme, which cuts the vector DNA (e.g., a plasmid) open. After the sticky ends of the plasmid have annealed with those of the donor

DNA, a DNA ligase activity joins the two molecules covalently. Then the recombinant vector is inserted into bacterial cells. Each cell replicates it along with its own genome. Note that recombinant vectors must contain regulatory regions recognized by bacterial enzymes. Because antibiotic resistance genes are used in the vector and the growth media contains antibiotics, only recombinant bacteria reproduce in the culture. After a suitable time, many copies of the desired gene may be retrieved from the culture or a gene product can be harvested.

Polymerase Chain Reaction

Although cloning has been immensely useful in molecular biology, the polymerase chain reaction (PCR) is a more convenient method for obtaining large numbers of DNA copies. Using a heat-stable DNA polymerase from *Thermus aquaticus* (Taq polymerase), PCR can amplify any DNA sequence, provided that the flanking sequences are known. Flanking sequences must be known because, as described previously, PCR amplification requires primers. Priming sequences are produced by automated DNA synthesizing machines.

PCR begins by adding Taq polymerase, the primers, and the ingredients for DNA replication to a heated sample of the target DNA. (Recall that heating DNA separates its strands.) As the mixture cools, the primers attach to their complementary sequences on either side of the target sequence. Each strand then serves as a template for DNA replication. At the end of this process, referred to as a *cycle,* the copies of the target sequence have been doubled. The process can be repeated indefinitely, synthesizing an extraordinary number of copies. For example, by the end of thirty cycles a single DNA fragment has been amplified one billion times.

The Human Genome Project

The Human Genome Project is an intensive international effort to identify and sequence all the genes in the human genome. This information will be a resource for the investigation of hereditary diseases as well as normal gene structure and expression. The following techniques that have been used to analyze the 23 pairs of human chromosomes were developed to map the genomes of other organisms.

1. **Genomic libraries.** A cellular genome is extracted and fragmented with restriction enzymes. A so-called genomic library is obtained when each of the fragments is cloned. The nucleotide sequence of each cloned DNA fragment can then be determined.

2. **Chromosomal walking.** This technique allows for sequencing DNA fragments of several hundred kb. A series of overlapping fragments from the genome are cloned. Once the first fragment has been sequenced, it provides the sequence information required to construct a probe to identify the clone containing the neighboring DNA sequence. After the second clone is sequenced, the process is repeated until all of the overlapping fragments have been sequenced.

Because the human genome is complex and enormous, faster and more convenient techniques have been developed. For example, genomic libraries are now routinely constructed from individual human chromosomes obtained by using automated machines called fluorescence-activated cell sorters. Similarly, vast improvements in computers that store and retrieve sequence information have been made. Despite intense and innovative research, progress in sequencing the genome, as well as its cost-effectiveness, must be improved substantially for its goal to be met by the year 2005.

Glossary

acetal the family of organic compounds with the general formula $RCH(OR')_2$; formed from the reaction of two molecules of alcohol with an aldehyde.

acid a molecule that can donate hydrogen ions.

acidosis a condition in which the pH of the blood is below 7.35 for a prolonged time.

activation energy the threshold energy required to produce a chemical reaction.

active site the cleft in the surface of an enzyme where a substrate binds.

active transport the energy-requiring movement of molecules across a membrane against a concentration gradient.

acyl group the functional group found in derivatives of the carboxylic acids.

addition reaction a chemical reaction in which two molecules react to form a third molecule.

adduct the product of an addition reaction.

aerobic respiration the metabolic process in which oxygen is used to generate energy from food molecules.

alanine cycle a biochemical mechanism that transports ammonia from muscle to the liver and recycles carbon skeletons.

aldaric acid the product formed when the aldehyde and CH_2OH groups of a monosaccharide are oxidized.

alditol a sugar alcohol; the product when the aldehyde or ketone group of a monosaccharide is reduced.

aldol addition a reaction between two aldehyde molecules (or two ketone molecules) in which a bond is formed between the α-carbon of one and the carbonyl carbon of the other.

aldol cleavage the reverse of an aldol condensation.

aldol condensation an aldol addition involving the elimination of a water molecule.

aldonic acid the product when the aldehyde group of a monosaccharide is oxidized.

aldose a monosaccharide with an aldehyde functional group.

aliphatic hydrocarbon a nonaromatic hydrocarbon such as methane and cyclohexane.

alkaloid a class of naturally occurring molecules that have one or more nitrogen-containing rings; many of the alkaloids have medicinal and other physiological effects.

alkalosis a condition in which the blood pH is above 7.45 for a prolonged time.

alkylation the introduction of an alkyl group into a molecule.

alkyl group a simple hydrocarbon group formed when one hydrogen from the original hydrocarbon (e.g., methyl, CH_3—) is removed.

allosteric interaction a regulatory mechanism in which a small molecule, called an effector or modulator, noncovalently binds to a protein and alters its activity.

α-amino acid a molecule in which the amino group is attached to the carbon atom (the α-carbon) immediately adjacent to the carboxyl group.

amphibolic pathway a metabolic pathway that functions in both anabolism and catabolism.

amphipathic molecule a molecule containing both polar and nonpolar domains.

amphoteric molecule a molecule that can react as both an acid and a base.

amylopectin a type of plant starch; a branched polymer containing $\alpha(1,4)$ and $\alpha(1,6)$ glycosidic linkages.

amylose a type of plant starch; an unbranched chain of D-glucose residues linked with $\alpha(1,4)$ glycosidic linkages.

anabolism energy-requiring biosynthetic pathways.

anaerobic occurring in the absence of molecular oxygen.

analog a substance similar in structure to a naturally occurring molecule.

anaplerotic reaction a reaction that replenishes a substrate needed for a biochemical pathway.

anhydride the product of a condensation reaction between two carboxyl groups or two phosphate groups in which a molecule of water is eliminated.

anomer an isomer of a cyclic sugar that differs from another in the arrangement of groups around an asymmetric carbon.

antenna pigment a molecule that absorbs light energy and transfers it to a reaction center during photosynthesis.

anticodon a sequence of three ribonucleotides on a tRNA molecule that is complementary to a codon on the mRNA molecule; codon-anticodon binding results in the delivery of the correct amino acid to the site of protein synthesis.

antigen any substance able to stimulate the immune system; generally a protein or large carbohydrate.

antisense RNA an RNA molecule with a sequence complementary to that of an mRNA molecule.

apoenzyme the protein portion of an enzyme that requires a cofactor to function in catalysis.

apoprotein a protein without its prosthetic group.

apoptosis programmed cell death.

aromatic hydrocarbon a molecule that contains a benzene ring or has properties similar to those exhibited by benzene.

atherosclerosis deposition of excess plasma cholesterol and other lipids and proteins on the walls of arteries, decreasing artery diameter.

autocrine refers to hormonelike molecules that are active within the tissue or organ in which they are produced.

autoimmune disease a condition in which an immune response is directed against an animal's own tissues.

auxin a class of plant growth regulators.

base a molecule that can accept hydrogen ions.

base analog a molecule that resembles normal DNA nucleotides and can substitute for them during DNA replication, leading to mutations.

bioaccumulation the process that concentrates chemicals as they are processed through the food chain.

bioenergetics the study of energy transformations in living organisms.

biogenic amine an amino acid derivative that acts as a neurotransmitter (e.g., GABA and the catecholamines).

bioremediation the use of biological processes to decontaminate toxic waste sites.

biotransformation a series of enzyme-catalyzed processes in which toxic, and usually hydrophobic, molecules are converted into less toxic and more soluble metabolites.

branched chain amino acid one of a group of essential amino acids with branched carbon skeletons (leucine, isoleucine and valine).

buffer a solution that contains a weak acid or base and its salt and resists large pH changes when stronger acids or bases are added.

calorie a unit of energy equal to the quantity of heat necessary to raise the temperature of 1 g of water by 1 degree C; equivalent to 4.184 J.

Calvin cycle the major metabolic pathway by which CO_2 is incorporated into organic molecules.

carotenoid an isoprenoid molecule that either functions as a light-harvesting pigment or protects against ROS.

catabolism the degradation of fuel molecules and the production of energy for cellular functions.

cellular immunity immune system processes mediated by T cells, a type of lymphocyte.

chaparonin one of a family of molecular chaparones; also referred to as an hsp60.

Chargaff's rules in DNA, the equality of the concentrations of adenine and thymine, and of cytosine and guanine.

chemiosmotic coupling theory ATP synthesis is coupled to electron transport by an electrochemical proton gradient across a membrane.

chemolithotroph an organism that uses specific inorganic reactions to generate energy.

chiral molecule a molecule that has mirror-image forms.

chloroplast a chlorophyll-containing plastid.

chromatin the DNA-containing component of the eukaryotic nucleus; the DNA is almost always complexed with histones.

chromophore a molecular component that absorbs light of a specific frequency.

chromosome the physical structure, composed of DNA and some proteins, that contains the genes of an organism.

***cis* isomer** an isomer in which two substituents are on the same side of a double bond.

cistron a DNA sequence that contains the coding information for a polypeptide and the signals required for ribosome function.

citric acid cycle a biochemical pathway that degrades the acetyl group of acetyl CoA to CO_2 and H_2O as three molecules of NAD^+ and one molecule of FAD are reduced.

cloning a lab procedure that produces multiple copies of a gene.

C4 metabolism a photosynthetic pathway that produces a four-carbon molecule and avoids photorespiration in eukaryotic photosynthesizing organisms.

coding strand the DNA strand that has the same base sequence as the RNA transcript (with thymine instead of uracil).

codon a sequence of three nucleotides in mRNA that directs the incorporation of an amino acid during protein synthesis or acts as a start or stop signal.

coenzyme a small organic molecule required in the catalytic mechanisms of certain enzymes.

cofactor the nonprotein component of an enzyme (either an inorganic ion or a coenzyme) required for catalysis.

colligative property a property of solutions that depends on only the number of dissolved particles in solution.

conjugate acid the cation (or molecule) that results when a base reacts with a proton.

conjugate base the anion (or molecule) that results when an acid loses a proton.

conjugated protein a protein that functions only when it carries other chemical groups attached by covalent linkages or by weak interactions.

conjugate redox pair an electron donor and its electron acceptor form; for example, NADH and NAD^+.

conjugation reaction a reaction that may improve the water solubility of a molecule by converting it to a derivative that contains a water-soluble group.

consensus sequence the average of several similar sequences; for example, the consensus sequence of the -10 box of *E. coli* promoter is TATAAT.

constitutive gene a routinely transcribed gene that codes for gene products required for cell function.

cooperative binding a mechanism in which binding one ligand to a target molecule promotes the binding of other ligands.

Cori cycle a metabolic process in which lactate, produced in tissues such as muscle, is transferred to liver where it becomes a substrate in gluconeogenesis.

cotranslational transfer the insertion of a polypeptide across a membrane during ongoing protein synthesis.

Crassulacean acid metabolism a photosynthetic pathway that produces a four-carbon molecule (malate) and avoids photorespiration.

cruciform a crosslike structure in DNA molecules likely to form when a DNA sequence contains a palindrome.

cyclin one of a group of proteins that regulate the cell cycle.

cyclin-dependent protein kinase one of a group of enzymes that activate the cyclins.

cytokine a group of hormonelike polypeptides and proteins; also referred to as growth factors.

cytokinin a class of plant growth regulators.

cytoskeleton a set of protein filaments (microtubules, macrofilaments, and intermediate fibers) that maintain the cell's internal structure and allow organelles to move.

decarboxylation reaction in which a carboxylic acid loses CO_2.

denaturation the disruption of protein or nucleic acid structure caused by exposure to heat or chemicals leading to loss of biological function.

detoxication the conversion of a toxic molecule to a more soluble (and usually) less toxic product.

detoxification a process whereby a state of toxicity is reduced in severity; producing sobriety in an inebriated person.

dialysis a laboratory technique in which a semipermeable membrane is used to separate small solutes from larger solutes.

diastereomer a stereoisomer that is not an enantiomer (mirror image isomer).

dictyosome term often used for the Golgi complex in plants.

dipole a difference in charge between atoms in a molecule resulting from the unsymmetrical orientation of polar bonds.

disaccharide glycoside composed of two monosaccharide residues.

disulfide bridge a covalent bond formed between the sulfhydryl groups in two cysteine residues.

disulfide exchange an enzyme-catalyzed post-translational process in which correct disulfide bonds are formed, resulting in a biologically active protein.

DNA fingerprinting a laboratory technique used to compare DNA banding patterns from different individuals.

effector a molecule whose binding to a protein alters the protein's activity.

eicosanoid a hormonelike molecule that contains 20 carbons; most are derived from arachidonic acid; examples include prostaglandins, thromboxanes, and leukotrienes.

electron transport system a series of electron carrier molecules that bind reversibly to electrons at different energy levels.

electrophile an electron-deficient species such as a hydrogen ion (H^+).

electrostatic interaction noncovalent attraction between oppositely charged atoms or groups.

elimination reaction a chemical reaction in which a double bond is formed when atoms in a molecule are removed.

elongation the polypeptide chain growth phase during translation on ribosomes.

enantiomer a stereoisomer that is a mirror image of another.

endergonic reaction a reaction that does not spontaneously go to completion; the standard free energy change is positive and the equilibrium constant is less than 1.

endocrine hormone a hormone secreted into the bloodstream that acts on distant target cells.

endocytosis the process in which a cell takes up solutes or particles by enclosing them in vesicles pinched off from its plasma membrane.

endoplasmic reticulum a series of membranous channels and sacs that provides a compartment separate from the cytoplasm for numerous chemical reactions.

endothermic reaction a reaction that requires energy (as heat).

enediol the intermediate formed during the isomerization reactions of monosaccharides.

energy the capacity to do work.

enthalpy the heat content of a system; in a biological system it is essentially equivalent to the total energy of the system.

entropy a measure of the randomness or disorder of a system; a measure of that part of the total energy in a system that is unavailable for useful work.

enzyme induction a process in which a molecule stimulates increased synthesis of a specific enzyme.

enzyme kinetics the study of the rates of enzyme-catalyzed reactions.

epimer a molecule that differs from the configuration of another by one asymmetric carbon.

epimerization the reversible interconversion of epimers.

epoxide an ether in which the oxygen is incorporated into a three-membered ring.

essential amino acid an amino acid that cannot be synthesized by the body and must therefore be supplied by the diet.

essential fatty acid linoleic or linolenic acid, which must be supplied in the diet because they cannot be synthesized by the body.

eukaryotic cell a living cell that possesses a true nucleus.

excision repair a DNA repair mechanism that removes damaged nucleotides, then replaces them with normal ones.

exergonic reaction a reaction that spontaneously goes to completion as written; the standard free energy change is negative, and the equilibrium constant is greater than 1.

exocytosis the process in which an intracellular vesicle fuses with the plasma membrane, thereby releasing the vesicle contents into extracellular space.

exon the region in a split or interrupted gene that codes for RNA and ends up in the final product (e.g., mRNA).

exothermic reaction a reaction that releases heat.

facilitated diffusion diffusion across a membrane that is aided by a carrier.

fatty acid a long chain monocarboxylic acid that contains an even number of carbon atoms.

fermentation the anaerobic metabolism or degradation of sugars; an energy-yielding process in which organic molecules serve as both electron donors and acceptors.

fibrous protein a protein composed of polypeptides arranged in long sheets or fibers.

flavoprotein a conjugated protein in which the prosthetic group is either FMN or FAD.

fluid mosaic model the currently accepted model of cell membranes in which the membrane is a lipid bilayer with integral proteins buried in the lipid, and peripheral proteins more loosely attached to the membrane surface.

fluorescence a form of luminescence in which certain molecules can absorb light of one wavelength and emit light of another wavelength.

frame shift mutation deletion of one or more base pairs (but not multiples of three) from a DNA sequence.

free energy the energy in a system available to do useful work.

functional group a group of atoms that undergo characteristic reactions when attached to a carbon atom in an organic molecule or a biomolecule.

futile cycle a set of opposing reactions that can be arranged in a cycle, but usually do not occur simultaneously; functioning of such reactions in both directions is avoided by metabolic control mechanisms to prevent energy waste.

gene a DNA sequence that codes for a polypeptide, rRNA, or tRNA.

general recombination recombination involving exchange of a pair of homologous DNA sequences; it can occur at any location on a chromosome.

genetic code the set of nucleotide base triplets (codons) that code for the amino acids in proteins as well as start and stop signals.

genome the total genetic information possessed by an organism.

gibberellin one of a class of plant hormones that promote stem growth, flowering, and seed production.

globular protein a protein that adopts a rounded or globular shape.

glucocorticoid a steroid hormone produced in the adrenal cortex that affects carbohydrate, protein and lipid metabolism.

glucogenic amino acid a molecule whose carbon skeleton is a substrate in gluconeogenesis.

gluconeogenesis the synthesis of glucose from noncarbohydrate molecules.

glucosuria the presence of glucose in urine; a symptom of diabetes mellitus.

glycerol phosphate shuttle a metabolic process that uses glyceral-3-phosphate to transfer electrons from NADH in the cytosol to mitochondrial FAD.

glycoconjugate a molecule that possesses covalently bound carbohydrate components (e.g., glycoproteins and glycolipids).

glycogen a glucose storage molecule in vertebrates; a branched polymer containing $\alpha(1,4)$ and $\alpha(1,6)$ glycosidic linkages.

glycogenesis the biochemical pathway that adds glucose to growing glycogen polymers when blood glucose levels are high.

glycogenolysis the biochemical pathway that removes glucose molecules from glycogen polymers when blood glucose levels are low.

glycolysis the enzymatic pathway that converts a glucose molecule into two molecules of pyruvate; this anaerobic process generates energy in the form of two ATP molecules and two NADH molecules.

glycoprotein a conjugated protein in which carbohydrate molecules are the prosthetic group.

glycosaminoglycan a long unbranched heteropolysaccharide chain composed of disaccharide repeating units.

glycoside the acetal of a sugar.

glycosidic linkage an acetal linkage formed between two monosaccharides.

glyoxysome a type of peroxisome found in germinating seed in which lipid molecules are converted to carbohydrate.

Golgi apparatus (Golgi complex) a series of curved membranous sacs involved in packaging and distributing cell products to internal and external compartments.

G protein a protein that binds GTP, which activates the protein to perform a function; the hydrolysis of GTP to form GDP inactivates the G protein.

granum a folded portion of the thylakoid membrane.

growth factor an extracellular polypeptide that stimulates cells to grow and/or undergo cell division.

heat shock protein a protein synthesized in response to stress, e.g., high temperature.

hemiacetal one of the family of organic molecules with the general formula $RR'C(OR')(OH)$ formed by the reaction of one molecule of alcohol with an aldehyde or ketone.

hemoprotein a conjugated protein in which heme, an iron-containing organic group, is the prosthetic group.

Henderson-Hasselbach equation kinetic rate expression that defines the relationship between pH, pK_a, and the concentrations of the acid and base components of a buffer solution.

holoenzyme a complete enzyme consisting of the apoenzyme plus a cofactor.

holoprotein an apoprotein combined with its prosthetic group.

homologous protein a protein molecule whose amino acid sequences and functions are similar to those of another protein.

Hoogsteen base pairing nonconventional base pairing that stabilizes H-DNA.

hormone a molecule produced by specific cells that influences the function of distant target cells.

hormone response element a specific DNA sequence that binds hormone-receptor complexes; the binding of a hormone-receptor complex either enhances or diminishes the transcription of a specific gene.

humoral immunity the immunity that results from the presence of antibodies in blood and tissue fluid; also referred to as antibody-mediated immunity.

hydrocarbon a molecule that contains only carbon and hydrogen.

hydrogen bond the force of attraction between a hydrogen atom and a small, highly electronegative atom (e.g., O or N) on another molecule (or the same molecule).

hydrolase an enzyme that catalyzes reactions in which adding water cleaves bonds.

hydrolysis a chemical reaction that involves the reaction of a molecule with water; the process by which molecules are broken into their constituents by adding water.

hydrophobic interaction the association of nonpolar molecules when they are placed in water.

hyperammonemia a potentially fatal elevation of the concentration of ammonium ions in the blood.

hyperglycemia blood glucose levels that are higher than normal.

hyperosmolar possessing an osmotic pressure greater than that of normal blood plasma.

hyperosmolar hyperglycemic nonketosis severe dehydration in noninsulin dependent diabetics; caused by persistent high blood glucose levels.

hypertonic solution a concentrated solution with a high osmotic pressure.

hyperuricemia abnormally high level of uric acid in blood.

hypoglycemia blood glucose levels that are lower than normal.

hypotonic solution a dilute solution with a low osmotic pressure.

inducible gene a gene expressed only under certain conditions.

initiation the beginning phase of translation.

inner membrane the innermost membrane of mitochondria.

interferon one of a group of glycoproteins that have nonspecific antiviral activity (e.g., stimulation of cells to produce antiviral proteins) that inhibit the synthesis of viral RNA and proteins and regulate the growth and differentiation of immune system cells.

intermediate fiber a component of the cytoskeleton containing a heterogenous set of proteins.

intron a noncoding intervening sequence in a split or interrupted gene missing in the final RNA product.

ionophore a substance that transports cations across membranes.

isoelectric point the pH at which a protein has no net charge.

isomerase an enzyme that catalyzes the conversion of one isomer to another.

isomerization the reversible interconversion of isomers.

isoprenoid one of a class of biomolecules that contain repeating five-carbon structural units

known as isoprene units; examples include terpenes and steroids.

isothermic having a uniform temperature.

isotonic solution solutions with exactly the same particle concentration; having identical osmotic pressure.

isozyme one of two or more forms of the same enzyme activity with different amino acid sequences.

katal measure of the rate of enzyme activity; 1 katal (kat) is equal to the conversion of one mole of substrate to product per second.

ketoacidosis acidosis caused by an excessive accumulation of ketone bodies.

ketogenesis the synthesis of ketone bodies.

ketogenic amino acid a molecule whose carbon skeleton is a substrate for synthesizing fatty acids and ketone bodies.

ketone body acetone, acetoacetate, or β-hydroxybutyrate; produced in the liver from acetyl CoA.

ketosis accumulation of ketone bodies in blood and tissues.

kinetics the study of reaction rates.

Krebs bicycle a biochemical pathway in which the aspartate required in the urea cycle is generated from oxaloacetate, an intermediate in the citric acid cycle.

lactone a cyclic ester.

leaving group the group displaced during a nucleophilic substitution reaction.

Le Chatelier's principle a law that states that when a system in equilibrium is disturbed, the equilibrium shifts to oppose the disturbance.

ligand a molecule that binds to a specific site on a larger molecule.

ligase an enzyme that catalyzes the joining of two molecules.

light-independent reaction a photosynthetic reaction that can occur in the absence of light; also referred to as the Calvin cycle.

light-induced repair DNA repair in which the damaged sequences are repaired utilizing light energy; also referred to as photoreactivation repair.

light reaction a mechanism whereby electrons are energized and subsequently used in ATP and NADPH synthesis.

linking number the number of times one polynucleotide strand crosses over another polynucleotide strand.

lipid any of a group of biomolecules that are soluble in nonpolar solvents and insoluble in water.

lipogenesis the biosynthesis of body fat (triacylglycerol).

lipolysis the hydrolysis of fat molecules.

lipoprotein a conjugated protein in which lipid molecules are the prosthetic groups; a protein-lipid complex that transports water-insoluble lipids in blood.

lithotroph an organism that uses specific inorganic reactions to generate energy; also known as a chemolithotroph.

London dispersion force a temporary dipole-dipole interaction.

lyase an enzyme that catalyzes the cleavage of C–O, C–C, or C–N bonds, thereby producing a product containing a double bond.

lysogeny the integration of a viral genome into a host genome.

lysosome a saclike organelle capable of degrading most biomolecules.

lytic cycle a viral life cycle in which a virus destroys its host cell.

Maillard reaction nonenzymatic glycosylation of molecules possessing free amino groups (e.g., proteins).

malate-aspartate shuttle a metabolic process in which the electrons from NADH in the cytosol are transferred to mitochondrial NAD^+.

malate shuttle a metabolic process in which oxaloacetate is transferred by reversible conversion to malate from a mitochondrion to the cytoplasm.

membrane potential potential difference across the membrane of living cells; usually measured in millivolts.

messenger RNA an RNA species produced by transcription that specifies the amino acid sequence for a polypeptide.

metabolism the total of all chemical reactions in an organism.

metaloprotein a conjugated protein containing one or more metal ions.

micelle an aggregation of molecules having a nonpolar and a polar component, leaving the polar domains facing the surrounding water.

microfilament a component of the cytoskeleton composed of the protein actin.

microsome a membranous vesicle derived from fragments of endoplasmic reticulum obtained by differential centrifugation.

microtubule a component of the cytoskeleton composed of the protein tubulin.

mineralocorticoid a steroid hormone that regulates Na^+ and K^+ metabolism.

mitochondrion an organelle possessing two membranes in which aerobic respiration occurs.

mitogen a substance that stimulates cell division.

mixed anhydride an acid anhydride with two different R groups.

modulator a molecule whose binding to an allosteric site of an enzyme alters the enzyme's activity.

molecular chaperone a molecule that assists in protein folding; most are heat shock proteins.

molecular disease a disease caused by a mutated gene.

monosaccharide a polyhydroxy aldehyde or ketone with the formula $(CH_2O)_n$ where n is at least 3.

mutagen any chemical or physical agent that alters the nucleotide sequence of a gene.

mutation any change in the nucleotide sequence of a gene.

negative cooperativity a mechanism in which the binding of one ligand to a target molecule decreases the likelihood of subsequent ligand binding.

negative feedback a mechanism in which a biochemical pathway is regulated by binding a product molecule to a key enzyme in the pathway.

neurotransmitter a molecule released at a nerve terminal that binds to and influences the function of other nerve cells or muscle cells.

nitrogen fixation conversion of molecular nitrogen (N_2) into a reduced biologically useful form (NH_3) by nitrogen-fixing microorganisms.

nonessential amino acid an amino acid that can be synthesized by the body.

nonessential fatty acid a fatty acid that can be synthesized by the body.

nonpolar molecule a molecule that does not contain a dipole.

nuclear envelope the double membrane that separates the nucleus from the cytoplasm.

nuclear pore a channel through the nuclear envelope that allows molecules to pass between the cytoplasm and the nucleus.

nucleic acid a macromolecule composed of nucleotides; DNA and RNA are nucleic acids.

nucleolus a structure found in the nucleus when the nucleus is stained with certain dyes; it plays a major role in the synthesis of ribosomal RNA.

nucleophile an electron-rich atom or molecule.

nucleophilic substitution a reaction in which a nucleophile substitutes for an atom or molecular group.

nucleoside a biomolecule composed of a pentose sugar (ribose or deoxyribose) and a nitrogenous base.

nucleosome a repeating structural element in eukaryotic chromosomes, composed of a core of eight histone molecules with about 140 base pairs of DNA wrapped around the outside; an additional 60 base pairs connect adjacent nucleosomes.

nucleotide a biomolecule composed of a pentose sugar (ribose or deoxyribose), at least one phosphate group, and a nitrogenous base.

nucleus an organelle that contains the chromosomes.

oligomer a multisubunit protein in which some or all subunits are identical.

oligosaccharide an intermediate-sized carbohydrate composed of two to ten monosaccharides.

optical isomer a stereoisomer that possesses one or more chiral centers.

organelle a membrane-enclosed structure within a eukaryotic cell.

osmosis the diffusion of a solvent through a semipermeable membrane.

osmotic diuresis a process in which solute in the urinary filtrate causes excessive loss of water and electrolytes.

osmotic pressure the pressure forcing the solvent, water, to flow across a membrane.

outer membrane the porous external membrane of mitochondria.

oxidation an increase in oxidation number caused by the loss of electron(s).

oxidation-reduction (redox) reaction a reaction involving the transfer of one or more electrons from one reactant to another.

oxidative phosphorylation the synthesis of ATP coupled to electron transport.

oxidative stress excessive production of reactive oxygen species.

oxidized molecule a molecule from which one or more electrons have been removed.

oxidizing agent a substance that oxidizes (removes electrons from) another substance; the oxidizing agent is itself reduced in the process.

oxidoreductase an enzyme that catalyzes an oxidation-reduction reaction.

oxyanion a negatively charged oxygen atom.

pacemaker enzyme an enzyme that catalyzes the committed step in a biochemical pathway.

palindrome a sequence that provides the same information whether it is read forward or backward; DNA palindromes contain inverted repeat sequences.

passive transport transport across membrane that requires no direct energy input.

Pasteur effect the observation that glucose consumption is greater under anaerobic conditions than when O_2 is present.

pentose phosphate pathway a biochemical pathway that produces NADPH, ribose, and several other sugars.

peptide an amino acid polymer composed of fewer than 50 amino acid residues.

peptide bond an amide linkage in an amino acid polymer.

peroxisome an organelle that contains oxidative enzymes.

pH optimum the pH at which an enzyme catalyzes a reaction at maximum efficiency.

pH scale a measure of hydrogen ion concentration; pH is the negative log of the hydrogen ion concentration in moles per liter.

phosphate group transfer potential the tendency of a phosphorylated molecule to undergo hydrolysis.

phosphoglyceride a type of lipid molecule found predominately in membrane composed of glycerol linked to two fatty acids, phosphate and a polar group.

phosphoprotein a conjugated protein in which phosphate is the prosthetic group.

photoautotroph an organism that possesses a mechanism for transforming solar energy into other forms of energy.

photophosphorylation the synthesis of ATP coupled to electron transport driven by light energy.

photoreactivation repair a mechanism to repair thymine dimers using the energy of visible light.

photorespiration a light-dependent process occurring in plant cells actively engaged in photosynthesis that consumes oxygen and liberates carbon dioxide.

photosynthesis the trapping of light energy and its conversion to chemical energy, which then reduces carbon dioxide and incorporates it into organic molecules.

photosystem a photosynthetic mechanism composed of light-absorbing pigments.

plasma membrane the membrane that surrounds a cell, separating it from its external environment.

plasmid a circular, double-stranded DNA molecule that can exist and replicate independently of a bacterial chromosome; plasmids are stably inherited, but are not required for the host cell's growth and reproduction.

plastid an organelle found in plants, algae, and some protists in which carbohydrate is stores or synthesized.

point mutation a change in a single nucleotide in DNA.

polar molecule a molecule that has a permanent dipole resulting from an unsymmetrical electron distribution.

polymerase chain reaction a laboratory technique used to synthesize large quantities of specific nucleotide sequences from small amounts of DNA using a heat-stable DNA polymerase.

polypeptide an amino acid polymer with more than 50 amino acid residues.

polysaccharide a linear or branched polymer of monosaccharides linked by glycosidic bonds.

polyuria excessive urination; a symptom of diabetes insipidus and diabetes mellitus.

positive cooperativity a mechanism in which the binding of one ligand to a target molecule increases the likelihood of subsequent ligand binding.

postabsorptive the phase in the feeding-fasting cycle in which nutrient levels in blood are low.

postprandial the phase in the feeding-fasting cycle immediately after a meal; blood nutrient levels are relatively high.

posttranslational modification a set of reactions that alter the structure of newly synthesized polypeptides.

posttranslational transport the transfer of polypeptides across membrane after translation has been completed.

prenylation the covalent attachment of prenyl groups (e.g., fasnesyl and geranylgeranyl groups) to protein molecules.

primary metabolite a molecule such as a sugar, amino acid, fatty acid, or the intermediates of the energy-generating pathways, that is absolutely required for maintenance of the living state.

primary structure the amino acid sequence of a polypeptide.

primer a short RNA segment required to initiate DNA synthesis.

primosome a multienzyme complex involved in the synthesis of the RNA primers at various points along the DNA template strand during *E. coli* DNA replication.

proenzyme an inactive precursor of an enzyme.

prokaryotic cell a living cell that lacks a nucleus.

promoter the sequence of nucleotides immediately before a gene that is recognized by RNA polymerase and signals the start point and direction of transcription.

prosthetic group the nonprotein portion of a conjugated protein that is essential to the biological activity of the protein; often a complex organic molecule.

protein a macromolecule composed of one or more polypeptides.

proteoglycan a large molecule containing large numbers of glycosaminoglycan chains linked to a core protein molecule.

proteosome a multienzyme complex that degrades proteins linked to ubiquitin.

protomer a subunit of allosteric enzymes.

protonmotive force the force arising from a gradient of protons and a membrane potential.

protooncogene a normal gene that promotes carcinogenesis if mutated.

purine a nitrogenous base with a two-ring structure; a component of nucleotides.

pyrimidine a nitrogenous base with a single-ring structure; a component of nucleotides.

quantum theory a theory in physics that describes the behavior of particles (e.g., electrons) and their associated waves.

quaternary structure association of two or more folded polypeptides to form a functional protein.

racemization interconversion of enantiomers.

radical an atom or molecule with an unpaired electron.

reaction center the membrane-bound protein complex in a photosynthesizing cell that mediates the conversion of light energy into chemical energy.

reactive oxygen species a reactive derivative of molecular oxygen, including superoxide radical, hydrogen peroxide, the hydroxyl radical, and singlet oxygen.

reading frame a set of contiguous triplet codons in an mRNA molecule.

receptor a protein on the cell surface that binds to a specific extracellular nutrient molecule and facilitates its entry into the cell; other receptors bind chemical signals and direct the cell to respond appropriately.

recombination a process in which DNA molecules are broken and rejoined in new combinations.

redox potential a measure of the tendency of an electron donor in a redox pair to lose an electron.

reduced molecule a molecule that has gained one or more electrons.

reducing agent a substance that reduces the oxidation number of another reactant; the reducing agent is itself oxidized in the process.

reducing sugar a sugar that can be oxidized by weak oxidizing agents.

reduction the lowering of oxidation number by the gain of electron(s).

regulatory enzyme an enzyme that catalyzes a committed step in a biochemical pathway; also referred to as a pacemaker enzyme.

releasing factor a protein involved in the termination phase of translation.

replication the process in which an exact copy of parental DNA is synthesized using the polynucleotide strands of the parental DNA as templates.

replicon a unit of the genome that contains an origin for initiating of replication.

replisome the large complex of polypeptides, including the primosome, that replicates DNA in *E. coli.*

resonance energy transfer the transfer of energy from an excited molecule to a nearby molecule, thereby exciting the second molecule.

resonance hybrid a molecule with two or more alternative structures that differ only in the position of electrons.

respiratory burst an oxygen-consuming process in scavenger cells such as macrophages in which ROS are generated and used to kill foreign or damaged cells.

respiratory control the control of aerobic respiration by ADP concentration.

retrovirus one of a group of viruses with RNA genomes that carry the enzyme reverse transcriptase and form a DNA copy of their genome during their reproductive cycle.

ribosomal RNA the RNA present in ribosomes; ribosomes contain several types of single-stranded ribosomal RNA that contribute to ribosome structures and are also directly involved in protein synthesis.

ribosome a protein-RNA complex where protein is synthesized.

ribozyme self-splicing RNA found in several organisms.

rough ER a type of endoplasmic reticulum involved in protein synthesis.

salt bridge an electrostatic interaction in proteins between ionic groups of opposite charge.

saturated molecule a molecule that contains no carbon-carbon double or triple bonds.

secondary metabolite a molecule derived from a primary metabolite; many serve protective functions.

secondary structure folding of a polypeptide chain into local patterns such as α-helix and β-pleated sheet; secondary structure is maintained by hydrogen bonds between the amide hydrogen and the carbonyl oxygen of the peptide bond.

second messenger a molecule that mediates the action of some hormones.

semiconservative replication DNA synthesis in which each polynucleotide strand serves as a template for the synthesis of a new strand.

sense strand the DNA strand that RNA polymerase copies to produce mRNA, rRNA or tRNA.

serine protease one of a class of proteolytic enzymes that use the $—CH_2OH$ of a serine residue as a nucleophile to hydrolyze peptide bonds.

signal peptide a short sequence, typically near the amino terminal of a polypeptide, that determines its destination.

signal transduction mechanisms by which extracellular signals are received, amplified and converted to a cellular response.

site-directed mutagenesis a technique that introduces specific sequence changes into cloned genes.

site-specific recombination recombination of nonhomologous genetic material with a chromosome at a specific site.

small nuclear ribonuclear particle a complex of proteins and small nuclear RNA molecules that promotes RNA processing.

small nuclear RNA a small RNA molecule involved in removing introns from mRNA, rRNA, and tRNA.

smooth ER a type of endoplasmic reticulum involved in lipid synthesis and biotransformation.

solvation sphere a shell of water molecules that clusters around positive and negative ions.

somatomedin a polypeptide that mediates the growth-promoting action of growth hormone.

specific activity a measure of enzyme activity; the number of international units (I.U.) per milligram of protein (1 I.U. is the amount of enzyme that produces 1 μmole of product per minute).

spliceosome a multicomponent complex containing protein and RNA that is involved in the splicing phase of mRNA processing.

splicing the excision of introns during mRNA processing.

steady state a phase in an organism's life when the rate of anabolic processes is approximately equal to that of catabolic processes.

stereoisomer a molecule that has the same structural formula and bonding patterns as another but has a different arrangement of atoms in space.

stroma a dense, enzyme-filled substance that surrounds the thylakoid membrane within the chloroplast.

structural gene a gene that codes for the synthesis of a polypeptide or a polynucleotide with a nonregulatory function (e.g., mRNA, rRNA, or tRNA).

substrate the reactant in a chemical reaction that binds to an enzyme active site and is converted to a product.

substrate-level phosphorylation the synthesis of ATP from ADP by phosphorylation coupled with the exergonic breakdown of a high-energy organic substrate molecule.

subunit a polypeptide component of an oligomeric protein.

supersecondary structure a set of specific combinations of α-helix and β-pleated sheet structures in protein molecules.

symbiosis the living together or close association of two dissimilar organisms.

target cell a cell that responds to the binding of a hormone or growth factor.

targeting process that directs newly synthesized proteins to their correct destinations.

tautomer an isomer that differs from another in the location of a hydrogen atom and a double bond (e.g., keto-enol tautomers).

tautomerization chemical reaction by which two isomers are interconverted by the movement of an atom or molecular group.

termination phase in translation in which newly synthesized polypeptides are released from the ribosome.

tertiary structure the globular, three-dimensional structure of a polypeptide that results from folding the regions of secondary structure; folding results from interactions of the side chains or R groups of the amino acid residues.

thermodynamics the study of energy and its interconversion.

thiolytic cleavage cleavage of a carbon-sulfur bond.

thylakoid membrane an intricately folded internal membrane within the chloroplast.

tolerance the capacity acquired by an organism after the continual use of a drug to become less responsive to its effects.

transamination a reaction in which an amino group is transferred from one molecule to another.

transcription the process in which single-stranded RNA with a base sequence complementary to the template strand of DNA is synthesized.

transduction the transfer of genes between bacteria by bacteriophages.

transferase an enzyme that catalyzes the transfer of a functional group from one molecule to another.

transfer RNA a small RNA that binds to an amino acid and delivers it to the ribosome for incorporation into a polypeptide chain during translation.

trans isomer an isomer in which two substituents are on opposite sides of a double bond.

transition mutation a mutation that involves the substitution of a different purine base for the purine present at the site of the mutation or the substitution of a different pyrimidine for the normal pyrimidine.

transition state the unstable intermediate in catalysis in which the enzyme has altered the form of the substrate so that it now shares properties of both the substrate and the product.

translation protein synthesis; the process by which the genetic message carried by mRNA directs the synthesis of polypeptides with the aid of ribosomes and other cell constituents.

translocation movement of the ribosome along the mRNA during translation.

transposition the movement of a piece of DNA from one site in a genome to another.

transposons (transposable elements) a DNA segment that carries the genes required for transposition and moves about the chromosome; sometimes the name is reserved for transposable elements that also contain genes unrelated to transposition.

transsulfuration pathway a biochemical pathway that converts methionine to cysteine.

triacylglycerol an ester formed between glyceral and three fatty acids.

tumor promoter a molecule that provides cells a growth advantage over nearby cells.

turnover the rate at which all molecules in a structure are degraded and replaced with newly synthesized molecules.

turnover number the number of molecules of substrate converted to product each second per mole of enzymes.

ubiquination the covalent attachment of ubiquitin to proteins; it prepares proteins for degradation.

uncoupler a molecule that uncouples ATP synthesis from electron transport; it collapses a proton gradient by transporting protons across the membrane.

unsaturated molecule a molecule that contains one or more carbon-carbon double or triple bonds.

uronic acid the product formed when the terminal CH_2OH group of a monosaccharide is oxidized.

van der Waals forces a class of relatively weak, transient electrostatic interactions between permanent and/or induced dipoles.

vapor pressure the pressure exerted by a vapor in equilibrium with a liquid.

velocity the rate of a biochemical reaction; the change in the concentration of a reactant or product per unit time.

vitamin an organic molecule required by organisms in minute quantities; some vitamins synthesize coenzymes required for the function of cellular enzymes.

weak acid an organic acid that does not completely dissociate in water.

weak base an organic base that has a small but measurable capacity to combine with hydrogen ions.

work change in energy that produces a physical change.

writhing number the number of superhelical turns in a molecule's three-dimensional conformation.

xenobiotics foreign and potentially toxic molecules.

Z-scheme a mechanism whereby electrons flow between PSII and PSI during photosynthesis.

zymogen the inactive form of a proteolytic enzyme.

Credits

Index

Page numbers followed by f and t refer to figures and tables, respectively.

Names and Abbreviations of the Standard Amino Acids

Amino Acid	Three-Letter Abbreviations	One-Letter Abbreviations
Alanine	Ala	A
Arginine	Arg	R
Asparagine	Asn	N
Aspartic acid	Asp	D
Cysteine	Cys	C
Glutamic acid	Glu	E
Glutamine	Gln	Q
Glycine	Gly	G
Histidine	His	H
Isoleucine	Ile	I
Leucine	Leu	L
Lysine	Lys	K
Methionine	Met	M
Phenylalanine	Phe	F
Proline	Pro	P
Serine	Ser	S
Threonine	Thr	T
Tryptophan	Trp	W
Tyrosine	Tyr	Y
Valine	Val	V